The Mitchell Beazley Joy of Knowledge Library

History and Culture 1

Scientiam non dedit natura semina scientiae nobis dedit
"Nature has given us not knowledge itself, but the seeds thereof."
Seneca

The Joy of Knowledge Encyclopaedia is affectionately
dedicated to the memory of John Beazley 1932–1977,
Book Designer, Publisher and Co-Founder of the
publishing house of Mitchell Beazley Limited, by all
his many friends and colleagues in the company.

The Joy of Knowledge Library

General Editor: James Mitchell
With an overall preface by Lord Butler, Master of Trinity College,
University of Cambridge

The Mitchell Beazley Joy of Knowledge Library

History and Culture 1

Introduced by Christopher Hill, FBA, DLitt,

Master of Balliol College, University of Oxford

MITCHELL BEAZLEY

The Joy of Knowledge Library

Editorial Director	**Frank Wallis**
Creative Director	**Ed Day**
Project Director	**Harold Bull**

Volume editors
Science and The Universe — John Clark
Lawrence Clarke
The Natural World — Ruth Binney
The Physical Earth — Erik Abranson
Dougal Dixon
Man and Society — Max Monsarrat
History and Culture 1 & 2 — John Tusa
Roger Hearn
Time Chart — Jane Kenrick
Man and Machines — John Clark
The Modern World — John Clark
Fact Index — Stephen Elliott
Stanley Schindler
John Clark

Art Director	Rod Stribley
Production Editor	Helen Yeomans
Assistant to the Project Director	Graham Darlow
Associate Art Director	Anthony Cobb
Art Buyer	Ted McCausland
Co-editions Manager	Averil Macintyre
Printing Manager	Bob Towell
Information Consultant	Jeremy Weston
Sub-Editors	Don Binney
	Arthur Butterfield
	Peter Furtado
	Charyn Jones
	Jenny Mulherin
	Shiva Naipaul
	David Sharp
	John Smallwood
	Jack Tresidder
Proof-Readers	Jeff Groman
	Anthony Livesey
Researchers	Malcolm Hart
	Peter Kilkenny
	Ann Kramer
	Lloyd Lindo
	Heather Maisner
	Valerie Nicholson
	Elizabeth Peadon
	Jim Somerville
Senior Designer	Sally Smallwood
Designers	Rosamund Briggs
	Mike Brown
	Lynn Cawley
	Nigel Chapman
	Pauline Faulks
	Nicole Fothergill
	Juanita Grout
	Ingrid Jacob
	Carole Johnson
	Chrissie Lloyd
	Aean Pinheiro
	Andrew Sutterby
Senior Picture Researchers	Jenny Golden
	Kate Parish
Picture Researchers	Phyllida Holbeach
	Philippa Lewis
	Caroline Lucas
	Ann Usborne
Assistant to the Editorial Director	Judy Garlick
Assistant to the Section Editors	Sandra Creese
Editorial Assistants	Joyce Evison
	Miranda Grinling
Production Controllers	Jeremy Albutt
	Anthony Bonsels
	John Olive
	Barbara Smit
Production Assistants	Nick Rochez
	John Swan

The Joy of Knowledge Encyclopaedia
© Mitchell Beazley Encyclopaedias Limited 1976

The Joy of Knowledge History and Culture 1
© Mitchell Beazley Encyclopaedias Limited 1977

Artwork © Mitchell Beazley Publishers Limited
1970, 1971, 1972, 1973, 1974, 1975 and 1976
© Mitchell Beazley Encyclopaedias Limited 1976
© International Visual Resource 1972

ISBN 0 85533 109 7

Typesetting by Filmtype Services Limited, England
Photoprint Plates Ltd, Rayleigh, Essex, England

Printed in England by Balding + Mansell

Major contributors and advisers to The Joy of Knowledge Library

Fabian Acker CEng, MIEE, MIMarE; Professor Leslie Alcock; Professor H.C. Allen MC; Leonard Amey OBE; Neil Ardley BSc; Professor H.R.V. Arnstein DSc, PhD, FIBiol; Russell Ash BA(Dunelm), FRAI; Norman Ashford PhD, CEng, MICE, MASCE, MCIT; Professor Robert Ashton; B.W. Atkinson BSc, PhD; Anthony Atmore BA; Professor Philip S. Bagwell BSc(Econ), PhD; Peter Ball MA; Edwin Banks MIOP; Professor Michael Banton; Dulan Barber; Harry Barrett; Professor J.P. Barron MA, DPhil, FSA; Professor W.G. Beasley FBA; Alan Bender PhD, MSc, DIC, ARCS; Lionel Bender BSc; Israel Berkovitch PhD, FRIC, MIChemE; David Berry MA; M.L. Bierbrier PhD; A.T.E. Binsted FBBI (Dipl); David Black; Maurice E.F. Block BA, PhD(Cantab); Richard H. Bomback BSc (London), FRPS; Basil Booth BSc (Hons), PhD, FGS, FRGS; J. Harry Bowen MA(Cantab), PhD(London); Mary Briggs MPS, FLS; John Brodrick BSc(Econ); J.M. Bruce ISO, MA, FRHistS, MRAeS; Professor D.A. Bullough MA, FSA, FRHistS; Tony Buzan BA(Hons) UBC; Dr Alan R. Cane; Dr J.G. de Casparis; Dr Jeremy Catto MA; Denis Chamberlain; E.W. Chanter MA; Professor Colin Cherry D Sc(Eng), MIEE; A.H. Christie MA, FRAI, FRAS; Dr Anthony W. Clare MPhil(London), MB, BCh, MRCPI, MRCPsych; Professor Aidan Clarke MA, PhD, FTCD; Sonia Cole; John R. Collis MA, PhD; Professor Gordon Connell-Smith BA, PhD, FRHistS; Dr A.H. Cook FRS; Professor A.H. Cook FRS; J.A.L. Cooke MA, DPhil; R.W. Cooke BSc, CEng, MICE; B.K. Cooper; Penelope J. Corfield MA; Robin Cormack MA, PhD, FSA; Nona Coxhead; Patricia Crone BA, PhD; Geoffrey P. Crow BSc(Eng), MICE, MIMunE, MInstHE, DIPTE; J.G. Crowther; Professor R.B. Cundall FRIC; Noel Currer-Briggs MA, FSG; Christopher Cviic BA(Zagreb), BSc(Econ, London); Gordon Daniels BSc(Econ, London), DPhil(Oxon); George Darby BA; G.J. Darwin; Dr David Delvin; Robin Denselow BA; Professor Bernard L. Diamond; John Dickson; Paul Dinnage MA; M.L. Dockrill BSc(Econ), MA, PhD; Patricia Dodd BA; James Dowdall; Anne Dowson MA(Cantab); Peter M. Driver BSc, PhD, MIBiol; Rev Professor C.W. Dugmore DD; Herbert L. Edlin BSc, Dip in Forestry; Pamela Egan MA(Oxon); Major S.R. Elliot CD, BComm; Professor H.J. Eysenck PhD, DSc; Dr Peter Fenwick BA, MB, BChir, DPM, MRCPsych; Jim Flegg BSc, PhD, ARCS, MBOU; Andrew M. Fleming MA; Professor Antony Flew MA(Oxon), DLitt (Keele); Wyn K. Ford FRHistS; Paul Freeman DSc(London); G.S.P. Freeman-Grenville DPhil, FSA, FRAS, G.E. Fussell DLitt, FRHistS; Kenneth W. Gatland FRAS, FBIS; Norman Gelb BA; John Gilbert BA(Hons, London); Professor A.C. Gimson; John Glaves-Smith BA; David Glen; Professor S.J. Goldsack BSc, PhD, FINSTP, FBCS; Richard Gombrich MA, DPhil; A.F. Gomm; Professor A. Goodwin MA; William Gould BA(Wales); Professor J.R. Gray; Christopher Green PhD; Bill Gunston; Professor A. Rupert Hall LittD; Richard Halsey BA(Hons, UEA); Lynette K. Hamblin BSc; Norman Hammond; Peter Harbison MA, DPhil; Professor Thomas G. Harding PhD; Professor D.W. Harkness; Richard Harris; Dr Randall P. Harrison; Cyril Hart MA, PhD, FRICS, FIFor; Anthony P. Harvey; Nigel Hawkes BA(Oxon); F.P. Heath; Peter Hebblethwaite MA (Oxon), LicTheol; Frances Mary Heidensohn BA; Dr Alan Hill MC, FRCP; Robert Hillenbrand MA, DPhil; Catherine Hills PhD; Professor F.H. Hinsley; Dr Richard Hitchcock; Dorothy Hollingsworth OBE, BSc, FRIC, FIBiol,

FIFST, SRD; H.P. Hope BSc(Hons, Agric); Antony Hopkins CBE, FRCM, LRAM, FRSA; Brian Hook; Peter Howell BPhil, MA(Oxon); Brigadier K. Hunt; Peter Hurst BDS, FDS, LDS, RSCEd, MSc(London); Anthony Hyman MA, PhD; Professor R.S. Illingworth MD, FRCP, DPH, DCH; Oliver Impey MA, DPhil; D.E.G. Irvine PhD; L.M. Irvine BSc; E.W. Ives BA, PhD; Anne Jamieson cand mag(Copenhagen), MSc (London); Michael A. Janson BSc; G.H. Jenkins PhD; Professor P.A. Jewell BSc (Agric), MA, PhD. FIBiol; Hugh Johnson; Commander I.E. Johnston RN; I.P. Jolliffe BSc, MSc, PhD, ComplCE, FGS; Dr D.E.H. Jones ARCS, FCS; R.H. Jones PhD, BSc, CEng, MICE, FGS, MASCE, Hugh Kay; Dr Janet Kear; Sam Keen; D.R.C. Kempe BSc, DPhil, FGS; Alan Kendall MA (Cantab); Michael Kenward; John R. King BSc(Eng), DIC, CEng, MIProdE; D.G. King-Hele FRS; Professor J.F. Kirkaldy DSc; Malcolm Kitch; Michael Kitson MA; B.C. Lamb BSc, PhD; Nick Landon; Major J.C. Larminie QDG, Retd; Diana Leat BSc(Econ), PhD; Roger Lewin BSc, PhD, Harold K. Lipset; Norman Longmate MA(Oxon); John Lowry; Kenneth E. Lowther MA; Diana Lucas BA(Hons); Keith Lye BA, FRGS; Dr Peter Lyon; Dr Martin McCauley; Sean McConville BSc; D.F.M. McGregor BSc, PhD(Edin); Jean Macqueen PhD; William Baird MacQuitty MA(Hons), FRGS, FRPS; Professor Rev F.X. Martin OSA; Jonathan Martin MA; Rev Cannon E.L. Mascall DD; Christopher Maynard MSc, DTh; Professor A.J. Meadows; Dr T.B. Millar; John Miller MA, PhD; J.S.G. Miller MA, DPhil, BM, BCh; Alaric Millington BSc, DipEd, FIMA; Rosalind Mitchison MA, FRHistS; Peter L. Moldon; Patrick Moore OBE; Robin Mowat MA, DPhil; J. Michael Mullin BSc; Alistair Munroe BSc, ARCS; Professor Jacob Needleman; John Newman MA, FSA; Professor Donald M. Nicol MA PhD; Gerald Norris; Professor F.S. Northedge PhD; Caroline E. Oakman BA(Hons. Chinese); S. O'Connell MA(Cantab), MInstP; Dr Robert Orr; Michael Overman; Di Owen BSc; A.R.D. Pagden MA, FRHistS; Professor E.J. Pagel PhD; Liam de Paor MA; Carol Parker BA(Econ), MA (Internat. Aff.); Derek Parker; Julia Parker DFAstrolS; Dr Stanley Parker; Dr Colin Murray Parkes MD, FRC(Psych), DPM; Professor Geoffrey Parrinder MA, PhD, DD(London), DLitt(Lancaster); Moira Paterson; Walter C. Patterson MSc; Sir John H. Peel KCVO, MA, DM, FRCP, FRCS, FRCOG; D.J. Penn; Basil Peters MA. MInstP, FBIS; D.L. Phillips FRCR, MRCOG; B.T. Pickering PhD, DSc; John Picton; Susan Pinkus; Dr C.S. Pitcher MA, DM, FRCPath; Alfred Plaut FRCPsych; A.S. Playfair MRCS, LRCP, DObstRCOG; Dr Antony Polonsky; Joyce Pope BA; B.L. Potter NDA, MRAC, CertEd; Paulette Pratt; Antony Preston Frank J. Pycroft; Margaret Quass; Dr John Reckless; Trevor Reese BA, PhD, FRHistS; M.M. Reese MA (Oxon); Derek A. Reid BSc, PhD; Clyde Reynolds BSc; John Rivers; Peter Roberts; Colin A. Ronan MSc, FRAS; Professor Richard Rose BA(Johns Hopkins), DPhil (Oxon); Harold Rosenthal; T.G. Rosenthal MA(Cantab); Anne Ross MA, MA(Hons, Celtic Studies), PhD, (Archaeol and Celtic Studies, Edin); Georgina Russell MA; Dr Charles Rycroft BA (Cantab), MB(London), FRCPsych; Susan Saunders MSc(Econ); Robert Schell PhD; Anil Seal MA, PhD(Cantab); Michael Sedgwick MA(Oxon); Martin Seymour-Smith BA(Oxon), MA(Oxon); Professor John Shearman; Dr Martin Sherwood; A.C. Simpson BSc; Nigel Sitwell; Dr Alan Sked; Julie and Kenneth Slavin FRGS, FRAI; Professor T.C. Smout; Alec Xavier Snobel BSc(Econ); Terry Snow BA, ATCL; Rodney Steel; Charles S. Steinger MA, PhD; Geoffrey Stern BSc(Econ); Maryanne Stevens BA(Cantab), MA(London); John Stevenson DPhil, MA; J. Sidworthy MA; D. Michael Stoddart BSc, PhD; Bernard Stonehouse DPhil, MA, BSc, MInstBiol; Anthony Storr FRCP, FRCPsych;

Richard Storry; Charles Stuart-Jervis; Professor John Taylor; John W.R. Taylor FRHistS, MRAeS. FSLAET; R.B. Taylor BSc(Hons, Microbiol); J. David Thomas MA, PhD; D. Thompson BSc(Econ); Harvey Tilker PhD; Don Tills PhD, MPhil, MIBiol, FIMLS; Jon Tinker; M. Tregear MA; R.W. Trender; David Trump MA, PhD, FSA; M.F. Tuke PhD; Christopher Tunney MA; Laurence Urdang Associates (authentication and fact check); Sally Walters BSc; Christopher Wardle; Dr D. Washbrook; David Watkins; George Watkins MSc; J.W.N. Watkins; Anthony J. Watts; Dr Geoff Watts; Melvyn Westlake; Anthony White MA(Oxon), MAPhil(Columbia); Dr Ruth D. Whitehouse; P.J.S. Whitmore MBE, PhD; Professor G.R. Wilkinson; Rev H.A. Williams CR; Christopher Wilson BA; Professor David M. Wilson; John B. Wilson BSc, PhD, FGS, FLS; Philip Windsor BA, DPhil(Oxon), Roy Wolfe BSc(Econ), MSc; Donald Wood MA PhD, Dr David Woodings MA, MRCP, MRCPath; Bernard Yallop PhD, BSc, ARCS, FRAS · Professor John Yudkin MA, MD, PhD(Cantab), FRIC, FIBiol, FRCP.

The General Editor wishes particularly to thank the following for all their support:
Nicolas Bentley
Bill Borchard
Adrianne Bowles
Yves Boisseau
Irv Braun
Theo Bremer
the late Dr Jacob Bronowski
Sir Humphrey Browne
Barry and Helen Cayne
Peter Chubb
William Clark
Sanford and Dorothy Cobb
Alex and Jane Comfort
Jack and Sharlie Davison
Manfred Denneler
Stephen Elliott
Stephen Feldman
Orsola Fenghi
Professor Richard Gregory
Dr Leo van Grunsven
Jan van Gulden
Graham Hearn
the late Raimund von Hofmansthal
Dr Antonio Houaiss
the late Sir Julian Huxley
Alan Isaacs
Julie Lansdowne
Professor Peter Lasko
Andrew Leithead
Richard Levin
Oscar Lewenstein
The Rt Hon Selwyn Lloyd
Warren Lynch
Simon macLachlan
George Manina
Stuart Marks
Bruce Marshall
Francis Mildner
Bill and Christine Mitchell
Janice Mitchell
Patrick Moore
Mari Pijnenborg
the late Donna Dorita de Sa Putch
Tony Ruth
Dr Jonas Salk
Stanley Schindler
Guy Schoeller
Tony Schulte
Dr E. F. Schumacher
Christopher Scott
Anthony Storr
Hannu Tarmio
Ludovico Terzi
Ion Trewin
Egil Tveteras
Russ Voisin
Nat Wartels
Hiroshi Watanabe
Adrian Webster
Jeremy Westwood
Harry Williams
the dedicated staff of MB Encyclopaedias who created this *Library* and of MB Multimedia who made the IVR Artwork Bank.

History and Culture 1/Contents

Keystone

Lord Butler, Master of Trinity College,
Cambridge, knocks on the great door of
the college during his installation
ceremony on October 7, 1965

Preface

I do not think any other group of publishers could be credited with producing so comprehensive and modern an encyclopaedia as this. It is quite original in form and content. A fine team of writers has been enlisted to provide the contents. No library or place of reference would be complete without this modern encyclopaedia, which should also be a treasure in private hands.

The production of an encyclopaedia is often an example that a particular literary, scientific and philosophic civilization is thriving and groping towards further knowledge. This was certainly so when Diderot published his famous encyclopaedia in the eighteenth century. Since science and technology were then not so far developed, his is a very different production from this. It depended to a certain extent on contributions from Rousseau and Voltaire and its publication created a school of adherents known as the encyclopaedists.

In modern times excellent encyclopaedias have been produced, but I think there is none which has the wealth of illustrations which is such a feature of these volumes. I was particularly struck by the section on astronomy, where the illustrations are vivid and unusual. This is only one example of illustrations in the work being, I would almost say, staggering in their originality.

I think it is probable that many responsible schools will have sets, since the publishers have carefully related much of the contents of the encyclopaedia to school and college courses. Parents on occasion feel that it is necessary to supplement school teaching at home, and this encyclopaedia would be invaluable in replying to the queries of adolescents which parents often find awkward to answer. The "two-page-spread" system, where text and explanatory diagrams are integrated into attractive units which relate to one another, makes this encyclopaedia different from others and all the more easy to study.

The whole encyclopaedia will literally be a revelation in the sphere of human and humane knowledge.

Butler

Master of Trinity College,
Cambridge

The Structure of the Library

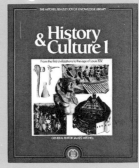

Science and The Universe

The growth of science
Mathematics
Atomic theory
Statics and dynamics
Heat, light and sound
Electricity
Chemistry
Techniques of astronomy
The Solar System
Stars and star maps
Galaxies
Man in space

The Physical Earth

Structure of the Earth
The Earth in perspective
Weather
Seas and oceans
Geology
Earth's resources
Agriculture
Cultivated plants
Flesh, fish and fowl

The Natural World

How life began
Plants
Animals
Insects
Fish
Amphibians and reptiles
Birds
Mammals
Prehistoric animals and
  plants
Animals and their habitats
Conservation

Man and Society

Evolution of man
How your body works
Illness and health
Mental health
Human development
Man and his gods
Communications
Politics
Law
Work and play
Economics

History and Culture

Volume 1 From the first
civilizations to the age of
Louis XIV

The art of prehistory
Classical Greece
India, China and Japan
Barbarian invasions
The crusades
Age of exploration
The Renaissance
The English revolution

History and Culture 1 is a book of popular history from the earliest records to the early eighteenth century. It is a self-contained book with its own index and its own internal system of cross-references to help you to build up a rounded picture of all but about 300 years of the history of the world.

It is one volume in Mitchell Beazley's intended ten-volume library of individual books we have entitled *The Joy of Knowledge Library*—a library which, when complete, will form a comprehensive encyclopaedia.

For a new generation brought up with television, words alone are no longer enough—and so we intend to make the *Library* a new sort of pictorial encyclopaedia for a visually oriented age, a new "family bible" of knowledge which will find acceptance in every home.

Seven other colour volumes in the *Library* are planned to be *Man and Society, The Physical Earth, The Natural World, History and Culture 2, Science and The Universe, Man and Machines*, and *The Modern World. The Modern World* will be arranged alphabetically: the other volumes will be organized by topic and will provide a comprehensive store of general knowledge rather than isolated facts.

The last two volumes in the *Library* will provide a different service. Split up for convenience into A-K and L-Z references, these volumes will be a fact index to the whole work. They will provide factual information of all kinds on peoples, places and things through approximately 25,000 mostly short entries listed in alphabetical order. The entries in the A-Z volumes also act as a comprehensive index to the other eight volumes, thus turning the whole *Library* into a rounded *Encyclopaedia*, which is not only a comprehensive guide to general knowledge in volumes 1–7 but which now also provides access to specific information as well in *The Modern World* and the fact index volumes.

Access to knowledge

Whether you are a systematic reader or an unrepentant browser, my aim as General Editor has been to assemble all the facts you really ought to know into a coherent and logical plan that makes it possible to build up a comprehensive general knowledge of the subject.

Depending on your needs or motives as a reader in search of knowledge, you can find things out from *History and Culture 1* in four or more ways: for example, you can simply browse pleasurably about in its pages haphazardly (and that's my way!) or you can browse in a more organized fashion if you use our "See Also" treasure hunt system of connections referring you from spread to spread. Or you can gather specific facts by using the index. Yet again, you can set yourself the solid task of finding out literally everything in the book in logical order by reading it from cover to cover: in this the Contents List (page 6) is there to guide you.

Our basic purpose in organizing the volumes in *The Joy of Knowledge Library* into two elements—the three volumes of A-Z factual information and the seven volumes of general knowledge—was functional. We devised it this way to make it easier to gather the two different sorts of information—simple facts and wider general knowledge, respectively—in appropriate ways.

The functions of an encyclopaedia

An encyclopaedia (the Greek word means "teaching in a circle" or, as we might say, the provision of a *rounded* picture of knowledge) has to perform these two distinct functions for two sorts of users, each seeking information of different sorts.

First, many readers want simple factual answers to straightforward questions, like "Who was Pompey?". They may be intrigued to learn that Pompey was a Roman general who lived between 106 BC and 58 BC. He formed the first triumvirate with Crassus and Julius Caesar, but then sided with the Senate against Caesar and was defeated by him at Pharsala in Greece. Such direct and simple facts are best supplied by a short entry and in the *Library* they will be found in the two A–Z *Fact Index* volumes.

But secondly, for the user looking for in-depth knowledge on a subject or on a series of subjects—such as "What are the hallmarks of Western art?"—short alphabetical entries alone are inevitably bitty and disjointed. What do you look up first—"art"? "Western art"? "modern art"? "the Renaissance"? "Gothic"? "aesthetics"? "Impressionism"?—and do you have to read all the entries or only some? You normally have to look up *lots* of entries in a purely alphabetical encyclopaedia to get a comprehensive answer to such wide-ranging questions. Yet comprehensive answers are what general knowledge is all about.

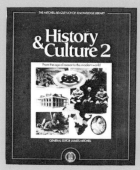

History and Culture

Volume 2 From the Age
of Reason to the
modern world

Neoclassicism
Colonizing Australasia
World War I
Ireland and independence
Twenties and the
  depression
World War II
Hollywood

Man and Machines

The growth of
  technology
Materials and techniques
Power
Machines
Transport
Weapons
Engineering
Communications
Industrial chemistry
Domestic engineering

The Modern World

Flags of the world
Nations of the world
Almanac
Atlas
Gazetteer

Fact Index A-K

The first of two volumes
containing 25,000 mostly
short factual entries
on people, places and
things in A-Z order. The
Fact Index also acts as
an index to the eight
colour volumes. In
this volume, everything
from Aachen to Kyzyl.

Fact Index L-Z

The second of the A-Z
volumes that turn the
Library into a complete
encyclopaedia. Like the
first, it acts as an
index to the eight
colour volumes. In this
volume, everything from
Ernest Laas to Zyrardow.

A long article or linked series of longer articles, organized
by related subjects, is clearly much more helpful to the
person wanting such comprehensive answers. That is why
we have adopted a logical, so-called *thematic* organization
of knowledge, with a clear system of connections relating
topics to one another, for teaching general knowledge in
History and Culture 1 and the six other general knowledge
volumes in the *Library*.

The spread system
The basic unit of all the general knowledge books is the
"spread"—a nickname for the two-page units that
comprise the working contents of all these books. The
spread is the heart of our approach to explaining things.

Every spread in *History and Culture 1* tells a story—
almost always a self-contained story—a story on the art of
prehistory, for example (pages 24 to 25) or Alexander the
Great (pages 80 to 81) or feudalism (pages 204 to 205) or the
politics of Europe from 1450 to 1600 (pages 240 to 241).
The spreads on these subjects all work to the same discipline,
which is to tell you all you need to know in two facing
pages of text and pictures. The discipline of having to get in
all the essential and relevant facts in this comparatively
short space actually makes for better results—text that has
to get to the point without any waffle, pictures and
diagrams that illustrate the essential points in a clear and
coherent fashion, captions that really work and explain the
point of the pictures.

The spread system is a strict discipline but once you get
used to it, I hope you'll ask yourself why you ever thought
general knowledge could be communicated in any other way.

The structure of the spread system will also, I hope
prove reassuring when you venture out from the things you
do know about into the unknown areas you don't know,
but want to find out about. There are many virtues in
being systematic. You will start to feel at home in all sorts
of unlikely areas of knowledge with the spread system to
guide you. The spreads are, in a sense, the building blocks
of knowledge. Like living cells which are the building
blocks of plants and animals, they are systematically
"programmed" to help you to learn more easily and to
remember better. Each spread has a main article of 850
words summarising the subject. The article is illustrated

by an average of ten pictures and diagrams, the captions
of which both complement *and* supplement the
information in the article (so please read the captions,
incidentally, or you may miss something!). Each spread,
too, has a "key" picture or diagram in the top right-hand
corner. The purpose of the key picture is twofold: it
summarises the story of the spread visually and it is
intended to act as a memory stimulator to help you to
recall all the integrated facts and pictures on a subject.

Finally, each spread has a box of connections headed
"See Also" and, sometimes, "Read First". These are
cross-reference suggestions to other connecting spreads.
The "Read Firsts" normally appear only on spreads with
particularly complicated subjects and indicate that you
might like to learn to swim a little in the elementary
principles of a subject before being dropped in the deep
end of its complexities.

The "See Alsos" are the treasure hunt features of *The
Joy of Knowledge* system and I hope you'll find them
helpful and, indeed, fun to use. They are also essential if
you want to build up a comprehensive general knowledge.
If the spreads are individual living cells, the "See Alsos"
are the secret code that tells you how to fit the cells
together into the body of general knowledge.

Level of readership
The level for which we have created *The Joy of Knowledge
Library* is intended to be a universal one. Some aspects of
knowledge are more complicated than others and so readers
will find that the level varies in different parts of the
Library and indeed in different parts of this volume,
History and Culture 1. This is quite deliberate: *The Joy of
Knowledge Library* is a library for all the family.

Some younger people should be able to enjoy and to
absorb most of the pages in this volume on Pepys's London,
for example, from as young as ten or eleven onwards—
but the level has been set primarily for adults and older
children who will need some basic knowledge to make
sense of the pages on the origins of Parliament or
thirteenth-century European learning, for example.

Whatever their level, the greatest and the bestselling
popular encyclopaedias of the past have always had one
thing in common—simplicity. The ability to make even

Main text Here you will find an 850-word summary of the subject.

Connections "Read Firsts" and "See Alsos" direct you to spreads that supply essential background information about the subject.

Illustrations Cutaway artwork, diagrams, brilliant paintings or photographs that convey essential detail, re-create the reality of art or highlight contemporary living.

Annotation Hard-working labels that identify elements in an illustration or act as keys to descriptions contained in the captions.

A typical spread Text and pictures are integrated in the presentation of comprehensive general knowledge on the subject.

Captions Detailed information that supplements and complements the main text and describes the scene or object in the illustration.

Key The illustration and caption that sum up the theme of the spread and act as a recall system.

complicated subjects clear, to distil, to extract the simple principles from behind the complicated formulae, the gift of getting to the heart of things: these are the elements that make popular encyclopaedias really useful to the people who read them. I hope we have followed these precepts throughout the *Library*: if so our level will be found to be truly universal.

Philosophy of the Library

The aim of *all* the books—general knowledge and *Fact Index* volumes—in the *Library* is to make knowledge more readily available to everyone, and to make it fun. This is not new in encyclopaedias. The great classics enlightened whole generations of readers with essential information, popularly presented and positively inspired. Equally, some works in the past seem to have been extensions of an educational system that believed that unless knowledge was painfully acquired it couldn't be good for you, would be inevitably superficial, and wouldn't stick. Many of us know in our own lives the boredom and disinterest generated by such an approach at school, and most of us have seen it too in certain types of adult books. Such an approach locks up knowledge instead of liberating it.

The great educators have been the men and women who have enthralled their listeners or readers by the self-evident passion they themselves have felt for their subjects. Their joy is natural and infectious. We remember what they say and cherish it for ever. The philosophy of *The Joy of Knowledge Library* is one that precisely mirrors that enthusiasm. We aim to seduce you with our pictures, absorb you with our text, entertain you with the multitude of facts we have marshalled for your pleasure—yes, *pleasure*. Why not pleasure?

There are three uses of knowledge: education (things you ought to know because they are important); pleasure (things which are intriguing or entertaining in themselves); application (things we can do with our knowledge for the world at large).

As far as education is concerned there are certain elementary facts we need to learn in our schooldays. The *Library*, with its vast store of information, is primarily designed to have an educational function—to inform, to be a constant companion and to guide everyone through

school, college and other forms of higher education.

But most facts, except to the student or specialist (and these books are not only for students and specialists, they are for everyone) aren't vital to know at all. You don't *need* to know them. But discovering them can be a source of endless pleasure and delight, nonetheless, like learning the pleasures of food or wine or love or travel. Who wouldn't give a king's ransom to know when man really became man and stopped being an ape? Who wouldn't have loved to have spent a day at the feet of Leonardo or to have met the historical Jesus or to have been there when Stephenson's *Rocket* first moved? The excitement of discovering new things is like meeting new people—it is one of the great pleasures of life.

There is always the chance, too, that some of the things you find out in these pages may inspire you with a lifelong passion to apply your knowledge in an area which really interests you. My friend Patrick Moore, the astronomer, who first suggested we publish this *Library* and wrote much of the astronomy section in the volume on *Science and The Universe*, once told me that he became an astronomer through the thrill he experienced on first reading an encyclopaedia of astronomy called *The Splendour of the Heavens*, published when he was a boy. Revelation is the reward of encyclopaedists. Our job, my job, is to remind you always that the joy of knowledge knows no boundaries and can work untold miracles.

In an age when we are increasingly creators (and less creatures) of our world, the people who *know*, who have a sense of proportion, a sense of balance, above all perhaps a sense of insight (the inner as well as the outer eye) in the application of their knowledge, are the most valuable people on earth. They, and they alone, will have the capacity to save this earth as a happy and a habitable planet for all its creatures. For the true joy of knowledge lies not only in its acquisition and its enjoyment, but in its wise and loving application in the service of the world.

Thus the Latin tag "Scientiam non dedit natura, semina scientiae nobis dedit" on the first page of this book. It translates as "Nature has given us not knowledge itself, but the seeds thereof."

It is, in the end, up to each of us to make the most of what we find in these pages.

The Structure of this Book

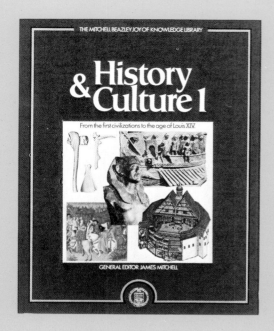

History and Culture 1 is a book about man's history from the time of the Sumerians, 6,000 years ago, to the age of Louis XIV, about 300 years ago. It covers all but a few centuries of recorded history—which are the subject of Volume 2—in such a way that the growth of world civilization is traced from its beginnings in Mesopotamia to the glories of the palace of the Sun King. Although the story centres on Europe, the simultaneous rise of Eastern civilizations is not neglected: there are spreads on China, India and Japan and also on the civilizations of the Americas.

Although this is a book of world history, with a balanced coverage of events, a disproportionately large amount of space is devoted to Britain and the Commonwealth. The deliberate editoral intention has been to make the coverage of our national heritage so dense that if the relevant spreads were separated from the chronology of world events into which they fall, they would form a complete text in themselves. In all, the coverage of British and Commonwealth history and culture amounts to a third of the two volumes.

Before itemizing the contents of *History and Culture 1* I'm going to assume that you—just like me when I began planning the book—are coming to history more as a "know-nothing" than as a "know-all". Incidentally, knowing nothing can be a great advantage as a reader—or as an editor, as I discovered in the early days of selecting topics for this book. If you admit to knowing nothing, but want to extend your knowledge, you ask awkward questions all the time. I spent much of my time as General Editor of this *Library* asking acknowledged experts awkward questions and refusing to be fobbed off with complicated answers I could not understand. As a result, *History and Culture 1*, like every other book in this *Library*, has been through the sieve of my personal ignorance in its attempt to inform simply and understandably.

If your only knowledge of history is a sketchy acquaintance with the history of your own country, I suggest that you start with Dr Christopher Hill's introduction on pages 16 to 19. He discusses the historian's job, what sources he should use and what questions he should ask, and concludes that the real nature of the task is to discover what kind of questions ordinary men and women were answering in the past when they did whatever it was they did—stormed the Bastille or fought the Battle of Waterloo or sailed with Drake. This discussion of the writing of history will enable you to read what follows with an informed eye and add to your enjoyment of the fascinating story of man's past.

Dr Hill also sounds a warning against the view of history presented by official sources, which up until the last few centuries make up the bulk of our surviving evidence on what happened in the past, and points out that it is only recently that we began to have adequate records of the lives of ordinary people. "If we are not merely to repeat historical mythology, the self-selected, self-justifying legends of past ruling classes, we are up against great difficulties." It would be presumptuous of me to pretend that The Joy of Knowledge volumes on *History and Culture* have surmounted those difficulties, but we have tried to be aware of them and as we approach the modern era, and our records of the lives of ordinary people improve, our writers have endeavoured to make use of those records to present a rounded picture.

Treatment of the subject
Both volumes on history and culture in *The Joy of Knowledge Library* tackle their subject chronologically. The story begins in the first volume with the earliest civilizations of which written records exist and ends in the second with a consideration of the current political situation in Europe. Europe figures largely in our treatment, principally because so many of the modern world's attitudes to life derive from European models, notably Greece and Rome. Taking Europe as the main element in the story also enables us to assess simultaneous events in other civilizations, such as the Chinese.

In both volumes there are frequent pauses in the chronological treatment so that social and economic progress, and the development of science, can be discussed. In a similar way, spreads on the history of the arts fall adjacent to those on the history of the civilizations that bore the most significant artistic fruit. To give a full picture of any given age, nation or trend, works of art have often been used to illustrate specific historical events and thus achieve greater impact.

History and Culture 1, like most volumes in *The Joy of Knowledge Library,* tackles its subject topically on a two-page spread basis. Although the spreads are self-contained, you may find some of them easier to understand if you read certain basic spreads first. Those spreads are illustrated here. They are "scene-setters" that will give you an understanding of the major civilizations of the past and of how historians approach their subjects. They will also demonstrate how the material in *History and Culture 1* is organized. The eight spreads are:

Time Chart

Finally, both volumes contain the relevant parts of an extensive time chart that begins in 4,000 BC. The chart covers milestones in politics, religion and philosophy, music, literature, art and architecture, and science and technology. A special section covers British national events. The chart may be used in three ways. First, readers who wish to get the flavour of a period may do so by reading the introductory paragraphs to each subject. These paragraphs, which are set in bold type, sum up the events covered on each spread of the time chart. Secondly, the chart may be used to follow the development of a particular discipline, such as philosophy, through the ages. Thirdly, the chart may be used to discover what progress was being made in other fields at the time a particular discovery was made or a particular event took place. Thus if you find a reference to the invention of the steam engine in the main body of the book, you may turn to the time chart and discover against what political, philosophical, artistic and scientific background the discovery was set. The periods of time covered by each spread of the chart shorten as the present day is approached so that proportionately more space is given to recent events.

Plan of the book

Because of our chronological treatment, neither volume of *History and Culture* lends itself to an easy division into topics. But it is possible to select certain cultures whose progress is charted regularly through the book and other major events—such as the Renaissance in this volume— which are treated in depth at the appropriate time in the story. In *History and Culture 1,* these are the subjects that receive such attention.

The beginnings of history

The first thirty spreads in *History and Culture 1* cover, broadly speaking, the events of the last 4,000 years before the birth of Christ. The story centres on the Middle East— on the Sumerians, the Babylonians, the Egyptians, the Minoans, the Hittites, the Phoenicians, the Assyrians, the Hebrews and the Persians—but it pauses regularly to discuss what was happening in Britain, India, China, Mexico, Peru and Africa. Many of the illustrations in these spreads are of the art of the various civilizations—not only does the art constitute an important part of the historical record (it can, with modern techniques, be dated quite precisely), it also gives us some of our best insights into the character of those civilizations.

Britain

Spreads on Stone Age Britain, and Bronze and Iron Age Britain, are the first of 110 spreads in the two volumes of *History and Culture* that are devoted solely or largely to the history, art and architecture of Britain, the British Empire and the Commonwealth. The first volume takes the story up to the age of Marlborough.

Greece

The influence of classical Greece on our present-day Western society is incalculable. We begin the story in 1,200 BC with the arrival of the Dorians. The next 400 years saw the development of a national consciousness and the adoption of the city state as the most important political, economic and social unit. Supreme among the city states by the 5th century BC was Athens, which under Cimon and Pericles became a maritime empire. Athens dominated mainland Greece until it was defeated by the Spartans in 404 BC. From the golden age of Athens came many of the most influential Western ideas in art, literature, philosophy and science—a machine with an elaborate system of internal gears, discovered in 1900 and dating from about 65 BC, has recently been interpreted as a Greek "computer".

Rome

At about the same time as Athens was making its intellectual contribution to prosperity, Republican Rome was creating a code of law that would have an equally far-reaching effect on our society. The most important aspect of the code was the classification of laws in order to protect the rights of citizens. At the instigation of the plebeians, the code was published in 450 BC. Many of the principles of Roman law are still in use today.

We trace the story of Rome from its beginnings to the final occupation of the empire by barbarians in the middle of the fifth century AD. There are two spreads on Roman art and one on literature.

India, China and Japan
The principal civilizations of the East are discussed in both volumes of *History and Culture*; in this volume the story begins with the history of India to 500 BC. The earliest evidence of a literate culture in India dates from about 2300 BC when the Indus civilization emerged from the prehistoric age. The hieroglyphic script that civilization used has yet to be deciphered. At about the same time, China laid down the roots of a culture that —largely because of the country's natural barriers—has remained unbroken to the present day. The Chinese were to prove to be the Greeks and Romans of the East: Japan, especially, owes its script and much of its philosophy, religion and art to Chinese models, although the adaptations that the Japanese made to those models eventually produced a unique culture. In this volume, fourteen spreads tell the story of the development of the Eastern civilizations.

Islam
The forces of Islam, at first represented by Arabs and Moors and later by the Turks, were to prove a factor in European history for twelve centuries; indeed, remnants of the Ottoman Empire remained intact in the Balkans until the end of World War I. The interaction between the forces of Christendom and the forces of Islam, part religious, part political, forms a recurrent theme through this volume.

The Americas
Unknown to Europe until the sixteenth century, three great civilizations flourished in Middle and South America—the Maya, Aztec and Inca. Europe was to plunder and destroy them. The brilliance of these civilizations—the Mayans developed mathematics and were the first to use a symbol for zero—is the subject of half a dozen spreads.

The Renaissance
The Renaissance—the rebirth of learning—brought Europe out of the Middle Ages and gave it some of its finest art. This was a period dominated by such men as Machiavelli, Michelangelo, Leonardo da Vinci, Rabelais, Erasmus, Thomas More, Shakespeare, Francis Bacon, Cervantes and Descartes. In this volume it is given uninterrupted treatment in a series of spreads that reproduce some of the major

works of art and lead naturally to the story of the Reformation—that arose from the Renaissance.

Science and technology
Although there are separate volumes in *The Joy of Knowledge Library* on science and technology, the history of their development properly finds its place in *History and Culture*. This volume brings the story up to 1700.

Economics and society
The way people lived in the past, and the development of economic systems, is part and parcel of history—indeed, as Christopher Hill points out in his introduction to this volume, social history is increasingly becoming a major theme of today's historians. In this volume you will find spreads on Greek and Roman society, on the rise of banking, on European economy in the Middle Ages, and on particular socio-political systems such as feudalism.

Religion
In the period covered by this volume, religion played an important role in men's lives. The birth and development of Buddhism, Christianity and Islam are traced, together with that of significant philosophies such as Confucianism.

History and Culture 1 ought, if we have done our job properly, both to inform you and enthral you. It takes the story of this world to the point where the Middle Ages give way to the modern era. There are lessons to be learned from that story and I am optimist enough to believe that the world has learned some of them.
Whether I am right, you must be the judge. But whatever view you take, I hope you will find the pages of this book as fascinating as I did.

History and culture

Dr Christopher Hill

Master of Balliol College, University of Oxford

The word "history" is ambiguous. On the one hand it means everything that has ever happened in the past; on the other it refers to what has been written about what is known of the past. History in the first sense is unknowable in its totality. A man living in a small village does not know what goes on behind its hedges: most of what has happened to most of humanity is irrecoverable. We could argue about how much this matters, but the fact is surely indisputable.

History is therefore an uncertain discipline. The survival of evidence is often quite haphazard. If we study the distant past, the evidence of archaeology seems random, though it may be true that artefacts like the pyramids and Stonehenge were built to survive and so tell us a good deal about the aspirations of those who planned their erection. But this does not help us to knowledge of those who actually built them, who may or may not have shared the ideological concerns of their social superiors. In later centuries, as we get written or printed evidence, we appear to know more about what life was like. But writing may be used to conceal or distort as well as to reveal; printing has throughout most of the history of most countries been subject to censorship. And again, what happens to survive may give a very lopsided view.

Historians of the present generation are increasingly conscious of their ignorance and of the possibility that much of written history is misleading and superficial. Until the last few centuries the vast bulk of our surviving evidence derives from ruling-class and governmental sources. No sociologist would feel that he could express useful views on what England was like after talking to a Treasury official and a professor of art history. Yet much of the evidence from which we write the history of the last 2,500 years is as flimsy as that. Consequently, all historians have to guard very carefully against falling for "the illusion of the epoch", against accepting at face value the assumptions of ruling classes and intellectuals of the past. The reason why we think of the Middle Ages in Western Europe as "an age of faith" may be only that those who knew how to write were almost exclusively ecclesiastics.

In the century of the common man we have become embarrassingly aware of how little we know about the lives of ordinary people until relatively recently. About women and children – three-quarters of the human race – we are even more ignorant. We can know a few members of the ruling class as individuals in classical Greece or Rome, in Chaucer's or Shakespeare's England. It is virtually impossible to achieve

History manipulated– the two men standing on the platform steps as Lenin speaks are Kamenev and Trotsky. After Stalin had denounced them both as counter-revolutionaries, "history" was revised and both were removed from the records.

Art as history –
with written records tending to concentrate on the activities of rulers, paintings such as Witte's "Adriana van Heusden and her daughter at the fish market" are often our only source of information about the lives of ordinary people.

such knowledge about the bottom 80 to 90 per cent of the population until we approach very modern times. So if we are not merely to repeat historical mythology, the self-selected, self-justifying legends of past ruling classes, we are up against great difficulties.

It is not a matter only of ignorance: there is probably a great deal of real distortion in many of our sources. Most of what we know about ancient and medieval heresies and witchcraft, for instance, comes from men so prejudiced against heretics and witches that they were prepared to torture and burn them to death. Most surviving accounts of slave revolts in antiquity, and of peasant revolts in the Middle Ages, come from men utterly devoid of sympathy with the rebels' cause. Nor is it only a matter of literary sources, in which the bias is often so obvious that it can be allowed for. Historians of an earlier generation, brought up on the admirable work of T. F. Tout and Lewis Namier, cherished the view that if we could only get behind literary sources into the archives, to government documents, then we should be on firm ground. There is of course a sense in which this is true. A tax is a tax: government archives can tell us how much was collected and when. But the more governmental archives are opened up, the more aware historians become of the truth in William Blake's dictum: "Nothing can be more contemptible than to suppose

public records to be true". Their apparent objectivity is frequently spurious. Contemporary civil servants, aware that their archives will soon be opened up to the historian, naturally take steps to remove any evidence that they do not wish posterity to see; anyone who has worked in the Civil Service can give examples of this.

But it is not only a matter of deliberate suppression. What is involved is also the unconscious assumptions and prejudices of the administrators, who are no more immune from national and class bias than the authors of "literary sources". Professor G. R. Elton's *Policy and Police: the Enforcement of the Reformation in the Age of Thomas Cromwell,* gives a mass of fascinating evidence about what ordinary Englishmen thought and said in the 1530s. But it is ordinary people seen by those trying to govern them. We have little reason to suppose that the lower classes had any more sympathy with the objectives of their governors than the latter had with theirs; the lower we go down the social scale, the less confidence central government had in its subjects. The events of the English Civil War of 1640 to 1660 showed how justified this lack of confidence was. But it should make us sceptical of the view from Whitehall.

Similarly, an historian has argued recently that since there is no evidence in ecclesiastical archives that the Pilgrim

Fathers were persecuted, then the *Mayflower* Pilgrims must have had some other reason for braving the terrible journey across the Atlantic to unknown shores in 1620. Historians who wrote about the Peterloo Massacre of 1819 solely from British Home Office sources managed to convince themselves that the victims and eye-witnesses of the massacre grossly exaggerated. We smile at the story of the French general who was discovered amid scenes of catastrophic retreat in March 1918 dictating a dispatch describing his successful advance. When questioned, he replied: "*Mais – c'est pour l'histoire*". But suppose his dispatch happened to be the only document that survived? His side won the war, after all; and throughout most history the defeated leave little evidence. "If we be overcome," Henry of Navarre wrote. "we shall die condemned heretics."

Whatever we think about the specific examples I have cited, it is clear that official documents are no more going to tell us the whole story than diaries, private letters and other unofficial sources. Edward I circulated to monasteries his version of his controversy with Scotland in order to get it into the chronicles. Official handouts about the deposition of Richard II deceived even so great an historian as Stubbs. Under Mary I the records were weeded out in order to elimate evidence of heresy. Almost any official document arising from any government department is engaged in making a case. It therefore omits some facts and arbitrarily emphasizes others. What it leaves out is no doubt well known to contemporary readers, less so to the historian.

The historian's job is to piece together the bits of evidence that happen to survive and to make what sense he can of them. He must approach all his sources with a great deal of scepticism. The fact that a document is official does not mean that it is impartial; if it has remained unpublished for 500 years this could be because it was not worth publishing. Superstitious belief in manuscripts, in archives, can be as misleading as naive acceptance of accounts of events written by participants in order to justify themselves in the eyes of posterity.

What I am saying is perhaps less alarming than it appears. The factual background of most written history is secure enough. No future discovery is likely to shake our belief that a Norman invading army defeated King Harold's English troops near Hastings in 1066, that Charles I of England was executed on 30 January 1649, or that Napoleon lost the Battle of Waterloo. But what the Normans were doing there, what ordinary people thought in 1066, 1649 or 1815 – these are matters on which we are still largely ignorant.

Some progress has been made. Application of anthropological and sociological techniques has enabled the *Annales* school of French historians, and historians like K. V. Thomas and E. P. Thompson in England, to cast light on hitherto dark and mysterious areas. We are becoming aware that until the coming of industrialization – and perhaps later – magical beliefs and practices dominated the lives of the population. Religious and other beliefs about which we hear a great deal in traditional sources may in fact have mattered very little to the mass of the people. Historians no longer speak glibly about "ages of faith", are no longer surprised that the bastardy rate was higher in Puritan England than in the Catholic France of the seventeenth century.

Too many historians who believed they were being "objective" were merely ignoring the distorting lenses though which they observed past history. It is easier for a present-day academic to enter into the mode of thought of a bureaucrat in ancient Egypt or nineteenth century England than it is for him to imagine how ancient Egyptian peasants or nineteenth-century Luddites felt. For this reason there was a short-lived reaction against some of the great trail-blazing writers who opened up the history of ordinary peope – R. H. Tawney, G. D. H. Cole, the Hammonds. The pioneers made mistakes, but these can be corrected. More important they wrote with imagination, and they were sceptical of official attitudes, and

this remains of permanent value.

Here we come up against the question of the historian's commitment. He is likely to write better history if he thinks it matters. He may be able to avoid reproducing the illusions of past epochs – mainly ruling-class illusions – if he asks questions which derive from his own society. (He may of course introduce the illusions of his own epoch, but since this is very different from most past epochs the danger, though real, can be avoided.) What is important is that the historian should be aware of what he is doing and should make clear to his readers what he is doing. This seems to me better than thinking you are being objective when you are merely blinkered. The historian must be sceptical of his sources *all* his sources. He must learn to live imaginatively in the society about which he writes, to participate in it.

History, it has been well said, offers a series of answers to which we do not know the questions. The historian's difficult job is to reconstruct the questions from the recorded answers. This is easier to do when the questions are obviously brash and new: the French Revolution was a question for British radicals, Darwin's *The Origin of Species* was a question for Victorian evangelicals. But historians still argue about the nature of the questions to which the French (or English, or American, or Russian or Chinese) Revolution was the answer. Most difficult to reconstruct are the questions that were taken for granted by those who answered them, or questions whose novelty is obscured by later events. If a man becomes or stays a Baptist or a Quaker today, for example, the questions include his conviction of the truth of the doctrine of the sect, and may include his desire to please, or to shock, his parents and friends. In the later seventeenth century we should have to add a question about readiness to endure persecution. But it is much more difficult to be certain what questions George Fox and John Bunyan were answering, even though they wrote so much about themselves. For in their case there was no clearly agreed body of doctrine to be taken or left; and since each of them was the first of his family to be converted to his faith, there was no question of pleasing his family or friends – though there may have been a question about displeasing them. It is hard to reconstruct the questions that faced men in the 1650s, to which they produced answers deceptively similar to the answers which eighteenth-century Quakers or Baptists gave. But the initial answer, that of Fox and Bunyan, presupposes a very special type of question: the later answers are to different, perhaps simpler questions.

Any serious history, it seems to me, deals with questions. The answers, the narrative, are known. The narrative can be rearranged; but the true originality of the historian lies in identifying questions that seem new to us because they approximate to the questions men and women were originally answering.

If I am right in defining the historian's job as discovering past questions to which we know the answers, this would help to explain why history has to be rewritten in every generation. New bits of experience in the present open our eyes to questions which men had to answer in the past. To deal only with the historiography of the English Revolution, nineteenth-century radicals rediscovered the Levellers, twentieth century socialists rediscovered the Diggers. The counter-culture of the later twentieth century helped us to see the significance of Ranters and other drop-out groups in the England of the 1640s and 1650s. His experience in the present helps the historian to sharpen and refine his account of the questions being answered in the past. If history has any use, it is in deepening our awareness of the process by which society sets questions that men and women, willy nilly, have to answer. It should help us to become more aware of the unconscious process of answering questions which is our daily life.

Today – by contrast with most of the past – the questions set are not wholly out of control. The more we comprehend

the question-setting process, the greater the hope of our being able to change the questions so as to get the right answers, instead of continuing to take the questions for granted as if they had been set by some eternal and omnipotent examiner. For we ourselves participate in setting the questions, even if only by passivity, by allowing the present questions to be set. We cannot influence the questions set to Fox and Bunyan, but Fox and Bunyan did: they took the whole examining system by the scruff of the neck and worked out new rules. They were not of course as successful as they would have wished: their successors could not keep the pace up; they made compromises themselves. Nevertheless, the questions never looked quite the same again once the Society of Friends existed and *Pilgrim's Progress* had been written.

Collingwood adjured historians to "think the thoughts of the past". A better, if clumsier, way of expressing it might be to identify the questions that were set to the men and women of past ages. This is less idealistic (in the philosophical sense), for the questions are set by society – or by the historical process, if you like. The historian is not primarily interested in the random thoughts of the past; he has to be selective. His concern should be to identify the major questions that men were in fact answering when – for instance – they executed Charles I, established the protectorate of Oliver Cromwell, and restored Charles II.

The sort of thoughts that Collingwood believed to be important were those that passed through Caesar's mind when he crossed the Rubicon. Crossing the Rubicon has however become rather a dangerous obsession for philosophers of history. What matters is not what was actually passing through Caesar's mind at that moment, for this is surely unknowable: he might have been wondering what had caused his indigestion. The important question is the one to which the act of crossing the Rubicon was an answer, the question posed by the political and social set-up of Roman society which Caesar and his armies were about to recast. It is not a personal question, to be answered as it were by saying "It must have been the cheese that gave me indigestion". The answer is the whole series of political actions, involving thousands of people as well as Caesar, of which crossing the Rubicon is a dramatic symbol. The question is vastly complex and only a historian who knows a great deal about Roman civilization in the first century BC can even approximate to formulating it correctly. It may be that Caesar never consciously asked himself the question to which crossing the Rubicon was the answer. But his actions answered it nevertheless. "Reasons and opinions concerning acts are not history", Blake observed, "acts themselves alone are history."

The historian's task, then, is to discover the questions that men and women of a past age were answering and to formulate them in the closest possible approximation to the way in which contemporaries would have formulated them if they had been conscious historians. It would have sounded pretty boring and pretentious if I had started with that sort of definition of history, but I hope it makes sense now. The good historian must above all be a questioner. He must question the assumptions of the past and of previous historians; he must question his own assumptions and prejudices; and he must force the past to yield up the questions which were being asked, the problems which were being set, as they were experienced by the people who lived in the period which he is studying. The broader his sympathies, the more he is likely to succeed in this imaginative task.

The tools of history

In his search for knowledge of the past, the historian uses various sources. The earliest historians relied largely on eye-witnesses, hearsay and word-of-mouth tradition. But England's first true historian, the Venerable Bede (673–735), also used written documents. Historians today extend their search even more widely. Government records for taxation and parish records tell about economic affairs. Art and poetry reveal cultural and spiritual life. Place-names show what diverse peoples have come to form a nation. Archaeology exposes many aspects of both rural and urban life that are described in no documents. And for the long prehistoric era leading up to the invention of writing, archaeology provides the only evidence.

Written records of King Arthur

The historian of early times often has to learn to use documents compiled long after the events they describe, or which copy, perhaps inaccurately, records that were contemporary. Or he may find legends and traditions wrapped round a kernel of fact, so that he has to peel off later fictions from a central truth.

A good example of the many tools used by historians to decipher fact from the most obscure evidence is the search for a historical basis of the semi-legendary King Arthur. In the period after the collapse of the Roman Empire in the fifth century AD, Anglo-Saxons from the far side of the North Sea and Scots from Ireland seized much of Britain from the native Britons and Picts, and in doing so created the basic racial mix of the British people. Our primary source for the recording of these events, the sixth-century British monk Gildas, describes a wholesale slaughter of the Britons by the pagan Saxons.

Gildas also records a great British victory over the invaders at Mount Badon [8], a battle that came to be associated with the name of Arthur. Most people think of Arthur as the idealized chivalrous king described by Thomas Malory in *Le Morte d'Arthur* (1470), or *The Idylls of the King*, by Alfred, Lord Tennyson (1809–92), or in even later versions of these romantic tales. It is often assumed that such a figure must be mythical.

But by his study of the sources the historian can remove successive layers of romance

to reach the historical truth. Arthur was first described as a great emperor in the twelfth century in the *History of the Kings of Britain* by Geoffrey of Monmouth (*c*. 1100–54). Three centuries earlier the British writer Nennius represented Arthur as the war-leader of the Britons against the Saxons [1]. In his own day, records kept in a Welsh monastery refer to the victory of Badon, (now thought to have been fought *c*. 490), and to Arthur's death along with Mordred's at the battle of Camlann.

This establishes Arthur as an historic personage – a great warrior, although not a king, who defended Britain against the Saxons. This is all the written documents can tell us. For further information we have to rely on the findings of the archaeologist.

The techniques of archaeology

Archaeology studies the physical remains of the past, and so gains insight into the material culture and living conditions of the characters of history. If documents prove that Arthur was a great war leader, then archaeology can show the kind of base that he may have

1 Medieval manuscripts were handwritten, and errors or extraneous ideas might occur during later copying. This 9th-century account of the 12 battles that Arthur is said to have fought is by the Welsh monk Nennius, and is thought to have been a summary of an early Welsh poem of which no other record exists. None of the battle sites can be located with certainty, and the description of the battle at Mt Badon "in which 960 fell in a single attack by Arthur" shows how the inflation of Arthur into a superhuman hero was under way by the time of Nennius. Arthur is called "leader of battles" for the British kings, a statement more likely to have been factual.

2 Cadbury Castle, Somerset, has been identified with Camelot, Arthur's court, since at least 1540. But in fact Camelot was simply the invention of 12th-century poets. Archaeological excavation, however, shows that Cadbury might instead have been the strong base or rallying point that Arthur needed to defend Britain against the Anglo-Saxons. The hill-fort had been first built *c*. 500 BC, but was refortified with a stone and timber rampart and gates that can be dated to AD 460–540, the years in which Arthur flourished. The fort was strategically situated to resist any westward drive of the Anglo-Saxons from Wessex towards the Bristol Channel.

3

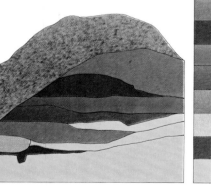

	Late Saxon town wall
	Arthurian rampart AD 460–550
	Roman period (site abandoned)
	Iron Age IV (ended AD 45–61)
	Iron Age III
	Iron Age II
	Iron Age I 500 BC
	Late Bronze Age

3 Stratification, the overlaying of earlier buildings by later ones, provides the basic clues by which archaeologists unravel the story of the past, particularly for a complex defensive system such as at Cadbury Camelot. Illustrated here is a schematic representation of the side of a trench excavated through the innermost of the four banks of the hillfort. At the base is the ground surface where the first Iron Age defence was built *c*. 500 BC. Below this was found pottery of the late Bronze Age, *c*. 1000 BC. The Iron Age Rampart I had a frame of wood that soon decayed, making it necessary to build Rampart II. This also collapsed and was followed by two other ramparts. The last, Rampart IV, was destroyed by the Romans in AD 45–61. After being abandoned for a long time, a new bank rampart was built. This was overlaid by a Late Saxon town wall about AD 1010, so it must be earlier than that, but later than the Iron Age. Contemporary pots suggest the date 460–550, the period of Arthur.

4 Tintagel Castle, Cornwall, on a superbly defensible headland, was built in its present form in the 12th century. It has been linked with Arthurian legend since the work of Geoffrey of Monmouth, but probably was used in Arthur's day as a Celtic monastery. Mediterranean style pots of that period have been found there.

4

fortified [2, 6], and the weapons, jewellery and equipment that his followers used.

A wide armoury of techniques is used to discover ancient sites, including walking the countryside, air-photography which gives a map-like view of sites [2, 4, 7], and geophysical prospecting to reveal ancient fireplaces, rubbish pits and other disturbances of the ground. Excavation is a precisely controlled process that seeks to recover the faintest traces of collapsed or decayed buildings [6].

For dating, the archaeologist traditionally relies on stratification [3], or layering, to reveal how later buildings lie on top of earlier ones; and typology [5] which compares developing types of the same class of object. The techniques of typology are familiar to anyone who can estimate the date and make of a motor car by its appearance.

Dating archaeological finds

In historical periods, written references to sites, inscriptions, and dated coins, all help to establish chronology. Stratification and typology can only establish the relative age of sites and objects; not how old they are in absolute years. For this several scientific techniques have been developed. Some of these depend on radioactive decay, measuring the extent of the decay of the radioactive carbon present in all living things.

Another practice, that of tree-ring dating or dendrochronology, studies the annual growth rings of trees. The width of the rings varies from year to year in a distinctive rhythm. Counting back from the present, it is possible to match early phases of the rhythm on living trees with that on timber beams in ancient buildings. The tree-ring count may be extended backwards for thousands of years.

These techniques are most useful in prehistory, but they can also be applied to historic times. A clear example of this is the dating of the Winchester Round Table [Key]. Typological study of the carpentry suggests a date in the mid-fourteenth century. Radiocarbon and tree-ring dates are consistent with this. Our historical knowledge of the growth of chivalrous ideals points to the same period. The table is part of the Arthurian legend; but it tells us nothing about the historic war leader of the Britons.

The Winchester Round Table can be shown, using historical tools, to have been made almost 1,000 years later than the period of the real King Arthur. History thus separates fact from fiction.

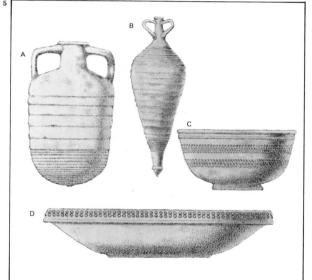

5 The study of pots is central to the archaeologist's interpretation of the past. These reconstructed two-handled jars, [A, B] found at Cadbury, each about 50 cm (20 in) tall, once contained Mediterranean wine for church services and princely feasts. The red dish [D] was also from the Mediterranean. Other pots have crosses inside the bowls and may have had a liturgical use. The grey bowl [C] came from Bordeaux, perhaps with wine in wooden casks. Such pots found on sites in Britain reveal trade links and are important for dating, because in Greece similar pots can be dated by coins found with them, and these dates can then be transferred to the British sites.

Observation platform

Earth bank
Limestone slabs
Lias limestone facing

6 Reconstructions of the buildings can be attempted from the archaeologist's study of the foundations, as in this gateway and rampart of the Arthurian period at Cadbury Castle. The remains consisted of the lower stones of the wall face; the pits in which the timber posts had stood; and dark stains in the ground where wood had rotted. If the tower has been correctly restored, its design may copy that of Roman military gate towers. But the rampart itself, with its use of timber and unmortared stone is very primitive. It reveals the sharp decline of technology in the post-Roman centuries.

7 Glastonbury, Somerset, has many ancient legends. In 1191 monks at Glastonbury Abbey claimed to have found the tombs of Arthur and his wife Guenevere. All trace of these tombs subsequently disappeared. The claim attracted great interest at the time, when the legend of Arthur was beginning to spread beyond purely Welsh legend. Although it may have been made to attract visitors to the monastery to contribute to a building programme, it is possible that the pair were indeed buried in the grounds of the Celtic monastery that originally occupied the site. Mediterranean wine jars of the 5th and 6th centuries have been found on Glastonbury Tor, (shown here), about 20km (12 miles) northwest of Cadbury.

8 Places linked with Arthur (historic or legendary), are widespread in Britain. Most of the names connected with Arthur or with King Mark and Tristan (Trusty) have no genuine historical significance. They show a popular habit of naming ancient ruins after long-dead heroes. But Mote of Mark and Cadbury Castle both have produced pottery of c. AD 500 and were fortified in Arthur's period – the 5th and 6th centuries. Killibury is probably Arthur's court of Celliwig, that was mentioned in early Welsh traditions. At least six sites are suggested, on place-name grounds, as possible locations for the historic battle of Mt Badon, Arthur's most important victory. The most probable is a hill near Bath.

Possible sites of battle of Mt Badon
Places associated with Arthurian legend
0   100km

Arthur's Seat
Trusty's Hill
Mote of Mark
Round Table
Camulodunum
Arthur's Stone
Caerwent Caerleon
Arthur's Stone
Liddington
Camulodunum
Glastonbury
Cadbury
Tintagel Callington
Killibury
Winchester

The development of archaeology

Man's interest in his own remote past effectively began with the Renaissance, although the urge to know about our ancestors is an ancient one, Nabonidus, for instance, last King of Babylon, excavated the foundation stone of a temple 3,200 years earlier than his reign to find out how old it was.

The knowledge of the Renaissance
During the Renaissance two important things happened. The texts of classical authors, such as Lucretius' *De rerum natura,* became widely disseminated by the printing press, so that their discussion of earlier ages of stone, bronze and iron and of the evolution of society from savagery through barbarism into civilization was firmly implanted in the educated minds of the age. Secondly, the age of exploration revealed the New World of America where a stone age technology was still in active use and where the civilization of the Aztec was just developing bronze.

The Renaissance tradition spread from Italy northwards to France and England, where it was reflected in Henry VIII's Palace of Nonsuch, and collecting Greek and Roman statues and vases became a fashionable occupation for the wealthy in Elizabethan and Jacobean England.

The middle classes had a classical education in the grammar schools, where they covered the range of Shakespeare's knowledge but lacked the financial resources to collect or go on the Grand Tour. Their attention therefore turned to the antiquities of their own region. In the work of scholars such as John Leland (c.1506-52) and William Camden (1551–1623) [1], the English antiquarian tradition was born in the sixteenth century, continuing to develop through the seventeenth with such notable figures as John Aubrey (1626–97), Thomas Browne (1605–82) and, later, William Stukeley (1687–1765).

Developments in Scandinavia
In Scandinavia the work of Johan Bure in Sweden and Ole Worm (1588–1644) in Denmark in the first half of the seventeenth century led to a parallel development, with more state involvement and a greater degree of protection being extended to antiquities.

A similar attitude in England followed the creation of the Royal Society in 1660.

The period of the Enlightenment saw speculation on the social origins of man, and from the stimulus of John Locke (1632–1704), the French and Scottish schools of thought evolved the notion of social typology (the study of types) and the development of society from the family through the band and the tribe to civilized urban groups with kings. This idea, itself a revival of classical thought, was later to prove crucial in the emergence of social anthropology under Edward Tylor (1832–1917) and Lewis Morgan (1818–81).

In the latter part of the eighteenth century evidence accumulated to show that the earth was very old, much older than the biblical date for its creation of 4004 BC. From the work of the geologist James Hutton (1726–97) in 1785 to the publication of the *Principles of Geology* by Charles Lyell (1797–1875) in 1833, a revolution in thought occurred in which the earth was recognized as immensely old and biblical chronology as wrong [Key]. In 1859 the

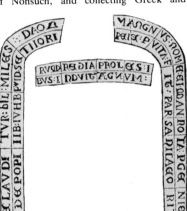

1 **The earliest illustration** of an archaeological monument to appear in a book was a typographical arrangement depicting in a stylized manner the inscription to the hermit Magnus, a prince from Scandinavia. This still stands outside the church of St John sub Castro at Lewes in Sussex. The picture appeared in the late 16th century in the 2nd edition of William Camden's famous *Britannia,* the first serious work in English on the subject of antiquities.

2 **Stonehenge,** in Wiltshire, is an outstanding megalithic monument that has long been one of the most notable and controversial sites in Britain. King James I sent Inigo Jones (1573–1652) to draw plans of it and Jones's account, describing it as a Roman temple, was published in the 1660s, setting off a violent argument in which it was ascribed by various scholars to the Danes, Saxons, Druids and ancient Britons. The first of these views shows the influence of contemporary Scandinavian scholars, particularly Ole Worm. In the late 17th century it was attributed to the Druids although it is much older than this.

3 **The great mound at Grave Creek, Miss.,** was excavated in the mid-19th century, more than 50 years after President Jefferson had undertaken a similar excavation in Virginia. A great stimulus to American archaeology was the continued presence of Stone Age Indians.

4 **Hissarlik in Turkey** was claimed by Heinrich Schliemann as the site of Homer's Troy. Schliemann was obsessed with uncovering the Homeric world and in the 1870s and 1880s excavated at Mycenae and Ithaca as well as Troy. At Hissarlik he uncovered a collection of Bronze Age jewellery which he claimed was "Priam's Treasure" and smuggled out of Turkey. He thought the second of the seven superimposed settlements he excavated was Homer's Troy, but it is now known to be too early.

Royal Society in London heard two of its most distinguished members accept the antiquity of man and in the same year Charles Darwin's *On the Origin of Species by Means of Natural Selection* raised new speculations as to where man had come from.

Discovery in the Near East

The middle of the nineteenth century was also the period when the great Near East civilizations were discovered. Mesopotamia saw the work of Austen Layard (1817–94) in the 1840s at Nineveh and the fierce rivalry of French and British archaeologists to loot the mounds of Assyrian and Babylonian sculpture. In the 1870s Heinrich Schliemann (1822–90) [4] dug at Troy and Mycenae and brought to the world the glories of Bronze Age Greece, a previously unknown civilization, the ancestor of which was uncovered at Knossos in Crete from 1900 onwards.

The increased length of man's history and the multitude of new discoveries of the prehistoric period were brought within a chronological scheme of successive Stone, Bronze and Iron Ages. This was first applied

to museum material by Christian Thomsen (1788–1865) in Copenhagen in 1816, it was subsequently proved stratigraphically (by the geological study of strata) by his successor Jens Worsaae (1821–85) and elaborated internally by the Swede Oscar Nontelius (1843–1921).

A concern for better excavation methods to acquire archaeological information was typified by the work of Augustus Pitt-Rivers (1827–1900), who between 1880 and 1900 set a standard still unrivalled for comprehensive recording. At the same time in Egypt Flinders Petrie (1853–1942) was trying to do the same under far worse conditions.

In a sense the development of prehistoric studies that began with Gordon Childe (1892–1957) in the 1920s continues. The ancestry of man is being pushed still further back in time under the impact of the work of Louis Leakey (1903–72) in Africa. The parallel development of archaeological thought in the Americas affects Old World ideas more strongly (and vice versa), and archaeology is becoming more of a unified discipline across the world.

This flint hand axe, (top and side views shown), came from Hoxne, Suffolk. John Frere sent it to the Society of Antiquaries in 1797, suggesting that it had a great age, "even beyond that of the known world". Frere's attitude reflects contemporary ideas in geology and about man's antiquity.

5 Howard Carter (1873–1939), right, directed the excavation of the tomb of Tutankhamen. In 1922 he located the entrance, cleared it and discovered "wonderful things" inside after a long and frustrating campaign. Tutankhamen was a boy king who died aged 18; the construction of later tombs above his own covered the entrance and preserved it from almost certain looting.

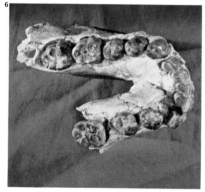

6 The jawbone of one of man's earliest ancestors, *Homo habilis*, was found at Olduvai Gorge in Tanzania by Louis and Mary Leakey. Their work there and the later extension of it to Lake Rudolf by their son Richard has taken the ancestry of man back (in less than 20 years) from under one million to nearly five million years. "Handy man" was found to be nearly 1.75 million years old.

7 Sir Mortimer Wheeler (1890–1976) worked on sites from Roman Wales to the mysterious Indus civilization in Pakistan. He was also founder of the Institute of Archaeology at London University and did much to make archaeology popular.

8 The Inca village of Machu Picchu, high in the Andes, was discovered in 1912 by Hiram Bingham (1875–1956) of Yale University. It was the first time that a late Inca settlement in such a good state of preservation had been found by archaeologists.

The art of prehistory

Man's aesthetic sense seems to have developed recognizable forms of expression about 30,000 years ago, as the modern form of man, *Homo sapiens sapiens*, annihilated, displaced or absorbed *Homo sapiens neanderthalensis*. The detectable differences between these two sub-species of human being are confined to a heavier bone structure in Neanderthal man, who was otherwise as upright and large-brained as modern man. But other differences in behaviour and adaptation to the environment can be guessed at from archaeological evidence; a predisposition to create art is one of them.

The "childhood of art"
With few exceptions, all of the world's earliest art is associated with the hunting economy of *Homo sapiens sapiens* [Key]. Neanderthal man had developed ritual, perhaps religion, and the practice of ceremonial burial of his dead, but had not developed a way of expressing himself pictorially in a lasting medium. Nevertheless the earliest art, which is found in caves in southwestern France and northeastern Spain, is of some

sophistication. When the art of this Upper Palaeolithic period (35,000–10,000 years ago) was first discovered, in 1875, it was received with incredulity. "It is the childhood of art, but not the art of a child", said one eminent French archaeologist, and until 1900 many regarded it with suspicion. Marcelino de Sautuola (died 1888), the discoverer of the great halls of Altamira in northern Spain, was charged with having forged the paintings and it was not until other paintings were discovered in the French sites of Pair-non-Pair and La Mouthe, concealed beneath Palaeolithic cultural deposits, that the sceptics were forced to recant.

The man who finally demonstrated the antiquity of Palaeolithic art was Abbé Henri Breuil (1877–1961). The evidence for Palaeolithic art is more or less confined to western Europe, although isolated sites such as Kapova in the Urals of Russia, Levanzo in Sicily and Beldibi and Belbasi in Turkey exist; and a lively tradition of rock art in northern and Saharan Africa at places such as Tassili N'Ajjer [5, 6] continues down into the Neolithic period (5,000 years ago).

Palaeolithic art is divided by archaeologists into two major categories – parietal, consisting of paintings, engravings and sculptures on the roofs, walls and floors of caves from which they cannot be removed, and mobiliary, consisting of small portable objects. To the archaeologist each type has its advantages and its frustrations; cave art, while often undated, is in its original position and context, while mobiliary art is often in well-dated archaeological layers but without any context that might shed light on its use. Mobiliary art consists of single isolated figures; cave art of many adjacent figures, sometimes related to each other.

Distribution of cave art
Cave art is confined to a small area of France and Spain: the Dordogne valley, the high Pyrenees and the Cantabrian mountains. More than 100 caves are known, including such famous sites as Lascaux [1], Altamira, Les Trois Frères, Montespan and Font de Gaume. In few of these, however, is the art firmly dated except by its style, and the art at Rouffignac has long been suspect.

1 **The famous animal paintings** in the cave of Lascaux are in a style similar to the parietal art found at many other sites in the Dordogne district of France. The cave was found accidentally in 1940 by a group of boys who were out with their dog and is one of the most recent of Palaeolithic cave art finds. Most of the animals in this portion of the Lascaux paintings are aurochs or European bison (*Bos primigenius*) which became extinct in Poland in the 17th century. They are drawn so that all four legs and both horns are visible.

2 **A wall of the cave of Pech-Merle** in France shows two spotted horses with more spots surrounding them and heads and necks in solid colour. The bodies are full, the limbs attenuated, a feature of much Palaeolithic art. The two horses overlap and were clearly painted at different times, but whether hours or centuries apart is not known. There are also hand prints made by blowing powdered pigment around a hand held on the wall.

3 **The small limestone figurine** known as the "Venus of Willendorf" from a Gravettian site in Austria is 20,000 to 25,000 years old. The large breasts, belly and buttocks are typical of this form of mobiliary art.

4 **Engraved figures** from the cave of Addaura in Sicily show humans in a variety of poses; overlapping suggests that they were not all done together. Engraving, with painting and low relief sculpture, is characteristic of the Upper Palaeolithic.

36,000-year-old cave picture tells explosive story

By John Lichfield
IN PARIS

Scientists believe that they have identified the oldest known images of erupting volcanoes, daubed in red and white pigments over other cave paintings in south-eastern France around 36,000 years ago.

The puzzling and apparently abstract images were first found in 1994 among startlingly precise paintings of lions, mammoths and other animals at a complex of caverns at Chauvet in the Ardèche.

A team of French scientists now believe that the surging, fountain-like images are the only example in Europe of prehistoric paintings of landscapes or natural phenomena. The oldest images of volcanoes previously identified were drawn 8,000 years ago at Catalhoyuk in central Turkey.

The findings of the French team – published at the weekend on the scientific website PLoS One – could transform conceptions about prehistoric art. The cave paintings at Chauvet are already among the oldest, most beautiful and most elaborate in the world.

Like all other known cave art in Europe, they depict animals and – in the case of Chauvet – human hands. If the volcano thesis is accepted, historians may have to revise their theories

about the meaning and purpose of cave paintings.

The claims are based on a new geological survey which dates volcanic eruptions in the nearby Bas-Vivarais area to between 30,000 and 40,000 years ago – coinciding with the

period when Chauvet was occupied by humans.

Carbon-dating of drawings both beneath and above the separate crimson and white "volcano" images suggests that they were drawn during this time.

The cave art depicts volcanoes that last erupted 30,000 to 40,000 years ago

{i} The nearest Vivarais volcano was 22 miles north-west of from the Chauvet caves, and was **depicted in the paintings.**

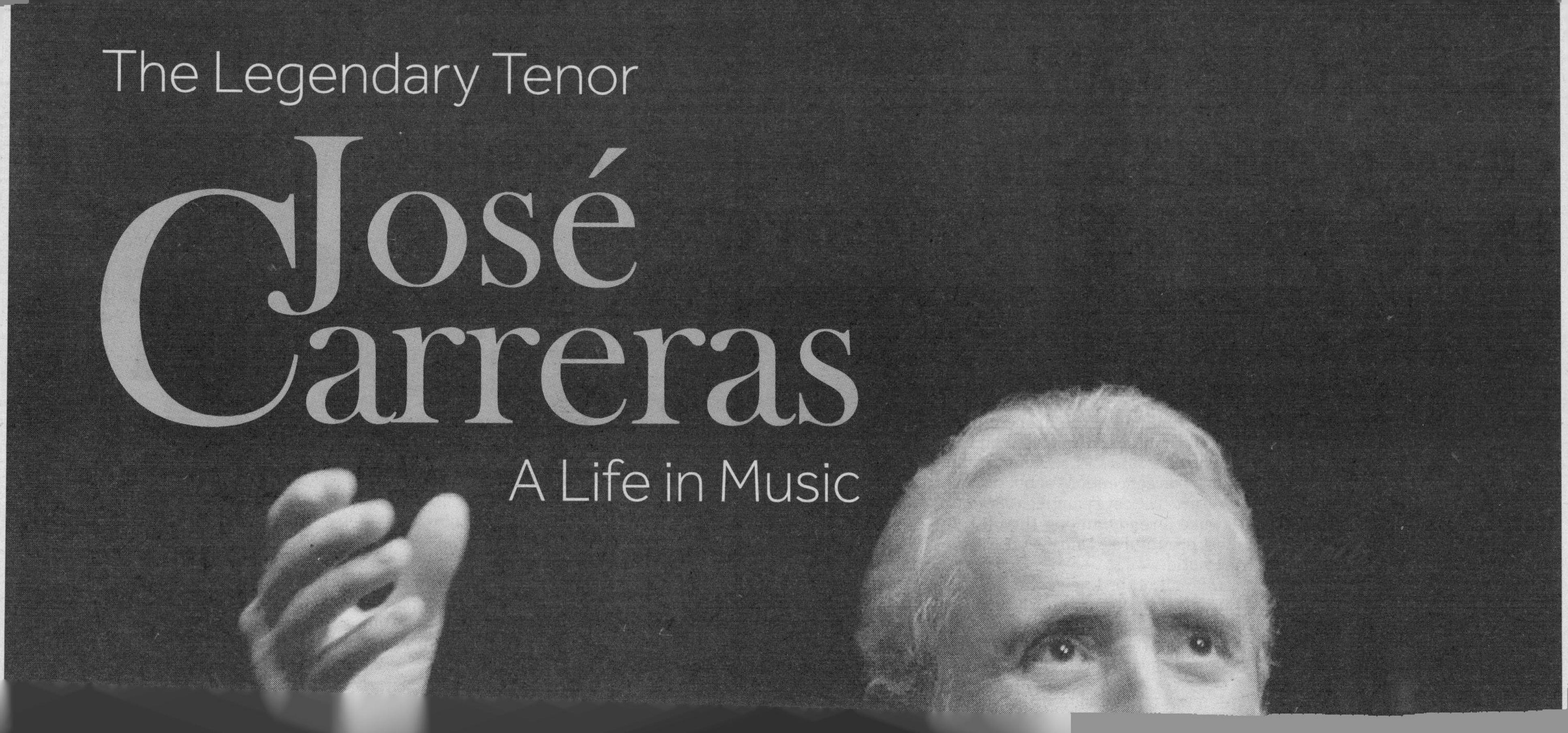

The Legendary Tenor

José Carreras

A Life in Music

On present evidence the earliest dated site is that of La Ferrassie, where painted blocks were found in Aurignacian culture levels dating from about 29,000 years ago. Other paintings and engravings and bas-relief sculpture were found in higher Aurignacian levels and thus the three major techniques of parietal art are all found from the earliest period onwards. Sites such as Pair-non-Pair may date from the late Aurignacian or the early Gravettian period about 25,000 years ago, while Gargas in the Pyrenees is definitely Gravettian, as also is the site at Laussel.

The brief Solutrean culture, confined to the same area as the cave art and noted for its superb flint work, has art at Le Roc de Sers. A reindeer-hunting final Palaeolithic phase in western Europe, the Magdalenian, gave birth to the astonishing polychrome paintings of Altamira. It has been suggested that most of the cave art sites are of this last period.

Mobiliary art is also found from the Aurignacian onwards, as an adjunct to the increasing use of bone and antler for tool-making. The best-known class of object is that of "Venus" figurines, stylized females with the buttocks and belly overemphasized and the head and legs reduced to stumps. The Willendorf Venus [3] is more individual than most. The Venuses date mainly from the Gravettian and like it are distributed through eastern Europe, even as far as Siberia. Their gross features may be the expression of some artistic scheme of proportions, rather than simply reflecting interest in woman's fertility.

Art or magic?

A number of explanations have been advanced for cave art, the simplest of which is that it is "art for art's sake". The theory of Breuil, accepted for many years, was that it was connected with obtaining magical power over the animal hunted by drawing it and also with the totemic association of man and animal species. More recent suggestions that the paintings may be sexually symbolic have not found wide acceptance. Some scholars are using comparisons with the more recent rock art of the Kalahari and the Australian Aborigines to cast further light on the mind of prehistoric man.

A group of warriors, hunters or dancers ornament a cave called Cingle de la Mola Remigia in the Gasulla Gorge of Castella province, Spain. The naked figures of running men are painted in a monochrome grey. Each man carries a longbow and a sheaf of arrows – standard hunting (and presumably also fighting) equipment during the late glacial period when this cave was decorated.

5 These striding men are painted on rock in the Tassili plateau of Algeria in the heart of the Sahara Desert. The figures themselves are shown in solid colour and attention is focused on the decoration covering their limbs and bodies, probably representing body-painting with natural pigments such as is still practised by the Nuba of Sudan. Body art may well have existed in the Palaeolithic period.

6 Tassili art persisted into the Neolithic period or later, after the camel had been domesticated, as is shown by this painted scene on the wall of a cave in the Tassili Plateau at the site of Ir Itinen. The figures in solid red depict a camel caravan and its drovers and the camels include both loaded adult animals and, at the lower left of the paintings, what appear to be younger camels.

7 These small amber carvings, including a stylized rendition of a bear, are from Siberia. Amber, a fossil resin found mainly on the shores of the Baltic Sea, was used by man from the Mesolithic period of 10,000 years ago onwards.

8 A shrine dating from 6000 or 7000 BC has been reconstructed at the Anatolian Neolithic site of Çatal Hüyük. This site has claims to being one of the world's earliest towns, with a population of several thousand. During excavations, many animal shrines dedicated to the herds on which the economy depended were found, decorated with elaborate paintings and modelled reliefs. The shrines and most of the houses seem to have been entered through the roof.

9 Rock engravings found in the Val Camonica, northern Italy, date from the Bronze Age. Some of these open-air carvings, which are pecked into smooth glaciated rock surfaces, show elaborate scenes, including chieftains in procession. This example (filled with coloured pigment to assist photographic visibility) depicts three reindeer and two schematic human figures. The significance of these scenes is unknown and the variety suggests many motives.

10 Scandinavian rock engravings from Bohuslan and Grevenvaenge in Denmark are similar in technique to the Val Camonica carvings, and of even later date. They depict a longship [A] with three human figures or perhaps a mast and steering oar as well; and two warriors in great horned helmets brandishing battleaxes [B]. The presence of a smooth rock surface presented an irresistible challenge to prehistoric graffitists throughout Europe and in Asia Minor.

Beginnings of agriculture

Until the Neolithic or New Stone Age the domestication of plants and animals was little practised. Agriculture developed at different times and rates in different places, and took various forms: plant-cultivation, pastoralism and mixed farming. Although the switch from total dependence on food gathering to the beginnings of food production was a gradual evolution, its long-term effects were nothing short of a revolution. Every food plant and animal of importance today was domesticated during the Neolithic.

The first farmers and herdsmen
Early in the eighth millennium BC cereals were cultivated and animals herded between latitudes 30° and 40° N, over an area stretching for a thousand miles from Anatolia to Iran [1]. This region offered a variety of wild plants and animals that could be domesticated. Early man found wheat and barley growing on the uplands, and goats [4] and sheep grazing on the slopes. Indeed, the wide variety of ecological zones and natural resources would have enabled hunters, fishermen and foodgatherers to live a semi-

sedentary life even after the end of the last Ice Age, about 8000 BC.

Initially the growth of agriculture and herding was slow and haphazard. Wild grain would have been collected and dropped round the settlement and hunters probably captured young animals and brought them home. In these early stages of farming it is not easy to distinguish between wild and domesticated flora and fauna. Animals evolved smaller forms and there were gradual changes, for example in the size and shape of horns. From the number of bones unearthed at Zawi Chemi Shanidar in northern Iraq, it seems that sheep may have been herded on the Iranian plateau as early as 8500 BC and goats not long after.

Evidence of cereal cultivation is harder to prove; cereals are preserved only in exceptional circumstances. Grain carbonized as a result of fire, or impressions left in clay ovens or storage pits, are often the only clues that remain. However, it is known that selection, promoted by some significant mutations, resulted in higher-yielding grain. Some of the evidence provided by Neolithic sites may be

purely circumstantial – the discovery of sickles need not necessarily imply cultivation. Experiments in Anatolia using stone-bladed reaping knives [5] showed that one family would have been able to collect enough wild wheat during three weeks' work to provide for them for a whole year.

Early Neolithic settlements
Most of the early Neolithic sites were located near springs and were occupied for thousands of years. Enormous mounds or "tells" accumulated from the remains of mud-brick houses and generations of rubbish provide archaeologists with a rich store of information. One of the most thoroughly investigated sites is Tell es Sultan at Jericho, which must have housed as many as 3,000 people during the "Pre-Pottery" Neolithic period around 7000 BC. Catal Hüyük in Anatolia, the largest known trading centre, was four times the size of Pre-Pottery Jericho by 6000 BC and may have been a sizeable settlement much earlier, although the earliest levels have not yet been excavated. It probably had a monopoly of nearby sources of obsidian,

CONNECTIONS

See also
24 The art of prehistory
30 The Sumerians 4000–2000 BC
42 Stone Age Britain

In other volumes
258 Man and Society
158 The Physical Earth
224 The Physical Earth

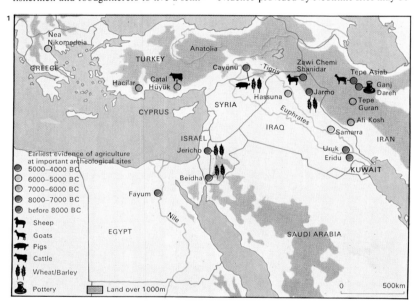

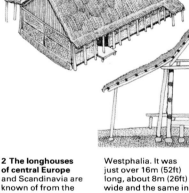

1 The main area of food production stretched from the Zagros mountains in Iran to the Taurus mountains of Anatolia and down the Jordan Rift. Earliest signs of herding are in Iran c. 8500 BC. Irrigation was carried out in the Tigris/Euphrates basin c. 5000 BC so that the alluvial land could be farmed.

3 Tassili N'Ajjer, an 800km-long (500 miles), eroded sandstone plateau in the Sahara, contains tens of thousands of rock paintings and engravings. The cattle and wild animals that flourished in the equable climate about 3000 BC are superbly depicted, but no other evidence exists to help identify the ancient inhabitants of this region.

2 The longhouses of central Europe and Scandinavia are known of from the first half of the 5th millennium BC. The one shown here is a reconstruction of an excavation at Deiringsen-Ruploh in Westphalia. It was just over 16m (52ft) long, about 8m (26ft) wide and the same in height. Long houses were normally rectangular, but the trapezoidal shape (as here) seems to have been a recognized variant. Many longhouses appear to have been divided in two, but whether this was to separate living quarters from storage is not known. The buildings were constructed on a timber frame.

4 The bezoar *(Capra hircus aegagrus)*, the wild ancestor of domesticated goats, still lives in the mountains of southwestern Asia. Goats and sheep were kept in herds before 8000 BC, the first animals to be domesticated.

5 Grain was first harvested with sickles made from stone insets in a bone or wooden shaft. This example comes from the Fayum, Egypt.

the chief material for tool-making: copper and marble were also obtained from the mountains and shells from the Mediterranean. Pottery, weaving and other arts and crafts reached a high standard. Religious beliefs centred on the worship of a fertility goddess, depicted in plaster reliefs, and a cult of the dead: wall paintings show vultures hovering over headless human corpses. However, the dominant theme in the shrines was the bull symbol of virility.

Reasonably secure in their settlements and supported by an agrarian economy Neolithic populations increased rapidly. With irrigation – which may have been introduced at Jericho – it was possible to expand into the lowlands of Mesopotamia. A settlement at Hassuna in the north dates from before 6000 BC, and a thousand years later the alluvial plains of the Tigris/Euphrates were being exploited at Eridu in the far south. Mesopotamia now became the heartland of Neolithic culture, laying the foundations for Sumerian civilization.

Meanwhile food production spread from Anatolia to Greece and on into Europe.

Land cleared by slashing and burning enabled agriculturalists to exploit the fertile soils of the Danube basin; and at the same time the megalith builders [12] were moving round the coasts of western Europe.

The staple crops of different cultures

The civilizations of western Asia and Europe were founded on wheat and barley, which are adapted to temperate climates and the subtropics, while millet, rice and maize are better suited to the tropics. Rice was cultivated in India earlier than in China, where millet was the main crop (and the pig the main domesticated animal) until about 2000 BC. In subsaharan Africa edible tuber cultivation may have preceded that of cereals (millet and sorghum), although there is no conclusive archaeological evidence. However, the evidence of numerous rock paintings shows that about 3000 BC pastoralists were able to find grazing all over the Sahara [3] in the period before the area became entirely desert. In Mexico primitive maize was cultivated by 5000 BC, but there were no fully sedentary populations in America before 1500 BC.

KEY

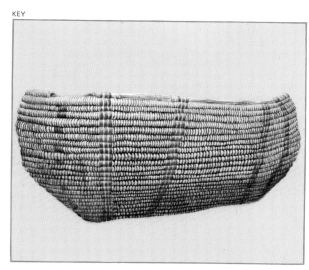

Wheat and barley were the staples of Neolithic economy in the Middle East. By about 4500 BC these cereals had spread from there to Egypt. This 40cm-long (16in) basket is made of coiled flax and may have been used for sowing. It was discovered in a grain storage pit in the Fayum.

6 Skara Brae, in the Orkney Islands, was occupied by herdsmen and fishermen around 1800 BC.

As there was no wood on the island houses and furniture were made of stone. One village includes

eight connected huts, each with a hearth, two slabs for beds, and a dresser and wall cupboards.

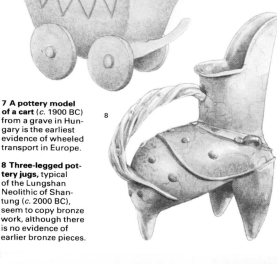

7 A pottery model of a cart (c. 1900 BC) from a grave in Hungary is the earliest evidence of wheeled transport in Europe.

8 Three-legged pottery jugs, typical of the Lungshan Neolithic of Shantung (c. 2000 BC), seem to copy bronze work, although there is no evidence of earlier bronze pieces.

9 This beaker from Siyalk, Iran, was made about 4000 BC, by which time the potter's wheel was in use. The earliest Neolithic settlements in western Asia have no pottery, the first appearing about 7000 BC. In Japan the Jómon culture pottery predates agriculture by several millennia.

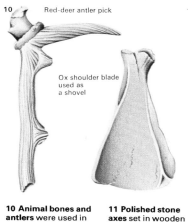

Red-deer antler pick

Ox shoulder blade used as a shovel

Neolithic flint axe

10 Animal bones and antlers were used in flint mining, which was an important Neolithic industry in western Europe. At Grime's Graves, Norfolk, shafts were sunk through the chalk and galleries tunnelled to reach the flint layers.

11 Polished stone axes set in wooden hafts were used for forest clearance and carpentry. The flints and other rocks used for these highly efficient tools were shaped in "axe factories". Their provenance tells us much about trade routes.

12 Mnajdra, one of the many fine megalithic temples in Malta, is trefoil-shaped and dates from the early Copper Age (c. 2800 BC). The first colonists of the island arrived from Sicily about 4000 BC. A thousand years later, collective burials in rock-cut chambers eventually gave rise to elaborate temples unique to Malta. Built of huge upright slabs surmounted by corbelled blocks, they are often decorated with carvings. There was a cult of the dead, with altars for animal sacrifices; corpulent female statues indicate the worship of a mother-goddess. The megalithic tradition extended from the eastern Mediterranean around the coasts of western Europe.

Early western Asia

The historical term "western Asia" comprises the modern states of Turkey, Syria, Lebanon, Israel, Jordan, Saudi Arabia, Iraq, Iran, and perhaps Afghanistan. In antiquity the centre of the stage was occupied successively by Sumerians, Babylonians, Assyrians, and Persians, with Hittites, Hebrews and Phoenicians also playing their part, and Elamites, Hurrians, Urartians and Aramaeans – to list only some of the best known – in the wings. Their civilizations were mostly lost to knowledge after their fall, and so remained through the Dark Ages; and their recovery was long delayed by restricted access to and exploration of their lands.

The development of writing
Rapid advances in knowledge came in the mid-nineteenth century, very much as a result of the decipherment of the Old Persian cuneiform script [Key], which through trilingual inscriptions (Old Persian, Babylonian and Elamite) provided the key to the immeasurably greater bulk of Assyrian, Babylonian and Sumerian texts in cuneiform writing of a more difficult kind.

Cuneiform developed from the pictographic script first used by the Sumerians in southern Mesopotamia late in the fourth millennium BC. Their language fell largely into disuse early in the second millennium, but the cuneiform script in which it had been written was used by their political and cultural heirs, the Babylonians, to express their own Semitic language, known as Akkadian, although it was entirely different from Sumerian. Other speakers of Akkadian, in particular the Assyrians, also used cuneiform. The same script, with relatively minor variations, was used elsewhere in western Asia to express other quite different languages, such as Hittite, Elamite, Hurrian and Urartian; and the same principle – varying groups of cuneiform (wedge-shaped) impressions – was used for the simpler Old Persian script of the Achaemenid Empire and for the alphabetic script of the Phoenicians.

Deciphering the inscriptions
The excavations and study of the cuneiform-inscribed clay tablets and of similar inscriptions on stone and other materials was stimu-

lated by the announcement, in 1872 of the decipherment of an Assyrian version of the biblical story of the Flood, found on a tablet at Nineveh in northern Iraq. Understanding other cuneiform inscriptions was made easier by the Semitic character of the Akkadian language in which many were written – it was related in varying degrees to such known languages as Hebrew, Aramaic and Arabic. The information provided by the inscribed material (nearly all in cuneiform, but with additions particularly from Old Testament Hebrew, Aramaic and hieroglyphic Hittite) has been supplemented by further surface discoveries and, particularly for the prehistoric (generally pre-writing) periods in western Asia, by archaeological excavation.

The geographical background
There are significant geographical variations in the territories covered by the ancient civilizations of western Asia. The coastlands of the Black Sea and the Mediterranean give way more or less steeply to the Anatolian plateau of central Turkey, mainly watered – around its central desert – by rivers flowing

1

1 The drama of ancient western Asia was centred on the riverine plains and adjacent hills and mountains between the Mediterranean and Caspian seas and the Persian Gulf. The area was dominated by the Tigris and Euphrates, the land between being Mesopotamia, but the term is used loosely for a wider area round the rivers. The silt they and their tributaries produced and the water diverted from them by irrigation helped the growth of the towns and later the capitals of Babylonia and Assyria, whose power and conflicts provided the framework for the history of their times until their decline in the seventh and sixth centuries BC.

2 A sun-dried clay figurine from a ninth-century BC palace at Nimrud in northern Iraq bears the inscription, impressed in the clay on the back in Assyrian cuneiform, "Come in, favourable demon; go out, evil demon". Such figurines lay in the brick-built foundation-boxes that are commonly found in the corners and beside the door jambs of important Assyrian buildings of the first millennium.

3 A main route from Trabzon and the eastern Black Sea coastal area of Turkey leads southwards up the 2,000m (6,560ft) Zigana Pass, which is open for most of the year. The route continues through high, mountainous country to Erzurum, Tabriz, and Teheran, thus serving as a north-western gateway to Armenia and Persia. It also provided a trade route between those lands and places accessible by ship from Trabzon. The Persian "Royal Road" between Susa in Persia and Sardis in the west used a more southerly route through Diyarbekir and Ankara, where the going was easier. The lands south of Trabzon are typical of much of eastern and northeastern Anatolia.

4 The wheel was probably invented in Mesopotamia during the fourth millennium. At that time it was constructed not of rings with spokes but of solid circular discs of planks clamped together, with a "tyre" of broad-headed nails driven into the outer rim. Spokes came later, but the earliest type of wheel still survives in remote areas; this one, for example, is found in a small village in eastern Turkey near Lake Van. Although many notable monuments had been built without the aid of the wheel, including the pyramids, its introduction greatly facilitated farming and transport and proved central to the development and spread of Mesopotamian civilization.

towards those two seas, with the Hittite capital in north central Anatolia close to the modern village of Bogazköy within the bend of the River Kızıl Irmak.

In the south, the Taurus range impedes the way up from the Cilician plain, while in the extreme east and northeast, high ranges (including Mount Ararat, 5,185m [17,011ft] high) restrict movement and cradle the sources of the Euphrates and Tigris. These two rivers, after flowing westwards to begin with, both turn southeast to form most of the Fertile Crescent, watering the plains of Mesopotamia where the capital cities – notably Babylon, Ashur and Nineveh – of their major empires were built [1]. To the west of Mesopotamia lies the desert, and still farther west are the Jordan and Orontes valleys and eventually the Mediterranean coastlands. East of Mesopotamia, valleys wind through the harsh mountains of the Zagros [3] to the dry Iranian plateau with the Elamite capital Susa and the Achaemenid Persian capitals Pasargadae and Persepolis.

Excavations have shown that the settlements of prehistoric man in some of these areas date back to as early as 9000 BC. Agriculture began sometime later. The cultivation of grain ("man's most precious artefact") assured a steady food supply, which led successively to expanding populations, release of labour for pursuits other than food-producing, specialization in craft, trade, government and religion, and so to the development of civilized societies from villages and towns to kingdoms and empires.

Wild cereal plants grew over much of western Asia, and their cultivation seems to have come first in the highland zones, notably in the Zagros mountains of eastern Iraq and western Iran and on the south Anatolian plateau. But as the grain-producing capacity of an inhabited area was bound to be a major factor in accelerating or retarding man's development, the fertility of the Tigris and Euphrates valleys, particularly with the addition of artificial irrigation from those rivers and from their main tributaries (the Habur, Upper and Lower Zab, and Diyala), gave the predominant prosperity in western Asia to the civilization of the Mesopotamian and southern plains.

Darius the Great of Persia (*r.* 521–486 BC) defeated nine princes to secure his throne. He recorded his achievement in detail on the face of the rock at Bisutun in western Iran, on the route from Mesopotamia to Teheran. The sculptures show two attendants behind Darius himself, his right foot planted on the pretender Gaumata, and nine rebels in captivity. The adjacent rock faces bear 11 short cuneiform inscriptions identifying Darius and his foes, with a long trilingual inscription that confirmed and amplified modern knowledge of the scripts, previously known only by similar, much shorter trilinguals.

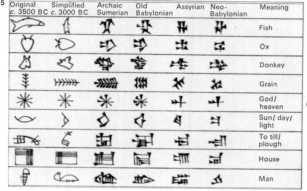

Original *c.* 3500 BC	Simplified *c.* 3000 BC	Archaic Sumerian	Old Babylonian	Assyrian	Neo-Babylonian	Meaning
						Fish
						Ox
						Donkey
						Grain
						God/heaven
						Sun/day/light
						To till/plough
						House
						Man

5 Sumerian writing probably evolved from the needs of public economy and administration. As the Sumerian city states developed, records were needed of goods moving in and out of the towns. Clay or gypsum tags were originally attached to objects and bore a seal-impression identifying the owner; line drawings of the objects followed. Drawings were gradually simplified to signs. Later the sign for a common word such as *ti* (arrow) was used for "ti" sounds generally. (*Ti* also meant "life".) The shift to phonetic representation led to the development of written symbols for entire languages.

6 Fragments of ivory boards each 33cm (13 in) by 15cm (6in) were found covered in sludge at the bottom of a well in a royal palace at Nimrud. The boards had a raised margin round a recessed portion that probably contained a mixture of beeswax and pigment as a base for a cuneiform text. This unrecessed board carries Sargon's name and the title of the "book" of omens taken from celestial observations.

7 Hollow "barrel cylinders" of clay were at times used by the Assyrians for recording texts in cuneiform. This one, about 17cm (6.7in) by 10cm (4in), refers to Esarhaddon and gives a general summary of his conquests and achievements. It also tells of a new palace he has built, with roof-beams of cedar from the Amanus mountains and "doors of sweet-smelling cypress wood". It dates from about 670 BC.

8 This monolithic basalt water-tank of the time of Sennacherib (*r.* 704–681 BC) was reassembled from small fragments, and measures about 3.2m (9.7ft) square by 1m (39in) high. Four corner-figures, almost in the round, represent the god Ea holding a water-dispensing bottle, as do the four figures facing outwards from the centre of each of the four sides. Two priests in fish-garments and holding ritual vessels turn to each of these figures. Two of the sides carry the inscription identifying the king. The detailed interpretation is uncertain, but the water element is clearly treated in an arcane sense; Ea, as god of the deep and of knowledge, may be intercessor between heaven and earth.

The Sumerians 4000–2000 BC

The partial excavation of the city of Eridu, about 19km (12 miles) southwest of "Ur of the Chaldees", the biblical home of Abraham, has produced the earliest settlements so far discovered in Babylonia, and has vindicated the Sumerians' tradition of its antiquity – they thought of it as the first of the five cities that existed before the Flood. Subsequent pre-dynastic periods of settlement, spanning more than the fourth millennium BC, are named, after the relevant sites, Ubaid, Uruk, and Jamdat Nasr.

The earliest dwellings at Eridu apparently consisted partly of reed huts and partly of mud-brick houses. The following pre-dynastic periods progressed successively through painted pottery, fishing-boats and hunting-slings, flint-headed hoes and hard-baked sickles, writing and the potter's wheel, plough and chariot, sculpture in the round, and vessels of silver, copper and lead.

Writing appeared in the Uruk period, a little before 3000 BC; whether the Sumerians brought it with them, or were already there (having possibly come from Khuzistan, at the foot of the Zagros), is not certain; but there seems no doubt that they invented the art.

The list of Sumerian kings includes a number of monarchs before the Flood, although their position in the archaeological framework and their history are unknown. Whether the Flood of Sumerian and Hebrew tradition is represented by the barren strata of silt found at Ur, Kish and Shuruppak is far from certain. One Ziusudra of Shuruppak, whom we are told survived the Flood, is famous from the *Epic of Gilgamesh* as the man who preserved the seed of living things; but it was at Kish that "kingship was restored to earth" after the flood.

The First Dynasty of Ur

Little else is known of these kings and leadership passed after a struggle to the famous Gilgamesh of Uruk [8], and from his successors there to the First Dynasty of Ur. The names of some verified monarchs belong to this period, as does probably the splendid Royal Cemetery. Of two vaults at the bottom of a deep shaft, one (partly plundered) presumably contained the king's remains, and the other those of Queen Shub-ad (or Pu-abi) on a bed, magnificently adorned and accompanied by female attendants.

Priceless treasures survived in these and the other major graves: golden harps, bull-headed lyres, a golden dagger with lattice sheath, gold and silver florally decorated combs, golden bowls, huge boat-shaped earrings, thousands of beads in gold, silver and cornelian, and much more; as well as the Royal Standard of Ur showing scenes of war and peace in shell and lapis lazuli.

Little reliable information survives about the history of the period following the First Dynasty of Ur, except at Lagash, where the royal line was inaugurated by Ur-Nanshe, whose power is attested by buildings, works of art (some archaic and crude), and inscriptions, including references to cargoes of timber arriving from the Persian Gulf, and giving the impression of inexperience in the use of writing. His grandson Eannatum rose to a supreme position in Sumer and defeated Mari, on the Euphrates, and Subur, perhaps in the north; the supremacy of Lagash was confirmed by his nephew Entemena.

The history of the rest of the Early

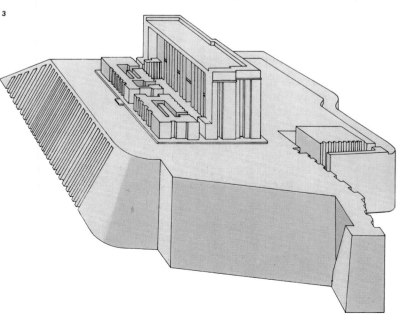

1 Mount Ararat, on which, according to the book of Genesis, Noah's Ark came to rest after the Flood, appears also in Sumerian tradition. It is probably more correctly vocalized as Urartu, the name of a kingdom that existed to the south of the mountain during the earlier part of the first millennium BC. The armies of this kingdom pressed long and hard on parts of the northern frontiers of Assyria.

2 Most of the Fertile Crescent was made up of the Syrian and lower reaches of the Euphrates and Tigris rivers. In the northern parts of this region there was some rainfall and natural fertility, and in the plains of Sumer and Akkad in the south-east, crops were encouraged by the silt of the rivers and by elaborate irrigation works. The need for these was early recognized by the area's growing civilizations.

3 Temple architecture dating from about the end of the fourth millennium is best represented by the White Temple at Uruk, the latest surviving shrine of an irregular mound that was probably an ancestor of the later, more regular ziggurats. The White Temple, approached by three ramps, was built of sun-dried bricks, whitewashed and buttressed, with an altar inside; the building's corners faced points of the compass. A ziggurat, a many-staged temple mount, may have been an attempt to bridge the gulf between man and the gods. It was strongly felt that man should offer residence to a deity, and the erection of a temple-tower may have bolstered belief in contact with such superhuman powers.

4 Gudea of Lagash, shown here, was the son-in-law of Ur-Baba, who brought to his city enough wealth to undertake extensive public works. He also patronized a school of sculptors who soon began to produce the finest masterpieces in hardstone. Gudea himself left inscriptions describing the religious observances and daily life of his time. He also enumerated the timbers and ornamental stones used in rebuilding the house of his god Ningirsu, on which he spent nearly all his wealth.

Dynastic period in Lagash and indeed in the whole of Sumer is largely ill-attested, but Urukagina (reigned *c.* 2378–*c.* 2371 BC) with a surprising political maturity instituted social reforms – some apparently intended to lighten burdens imposed on the population by governors and priests. He fell to Lugalzaggisi of Umma, who in turn fell, after a substantial and evidently successful reign, to the great Sargon of Akkad (reigned *c.* 2371–*c.* 2316 BC) [7].

Sargon the Great, King of Akkad

Sargon rose from obscurity to overthrow Lugalzaggisi, and to subdue the rest of Sumer, Syria, perhaps part of Asia Minor, and much, apparently, of the mountain area of southwestern Iran. Revolt followed, but his grandson Naram-Sin ruled gloriously for 37 years. Sargon's line fell in *c.* 2230 BC to the Gutian tribes from the north or northeast, whose sovereignty left little mark on history and few monuments (although Lagash emerged to a period of great prosperity about that time under Ur-Baba and Gudea) [4], and who were expelled by Utu-khegal of

Uruk; his deputy at Ur, Ur-Nammu [6], seems to have overthrown him, and so founded the dynasty of Ur.

The Third Dynasty of Ur

Ur-Nammu did not take the title "King of the Four Regions" – perhaps acknowledging his relatively limited authority – but assumed the new title "King of Sumer and Akkad". His 18-year reign was a time of considerable wealth and power, as shown by his many great building works, including the restoration of direct communication by water with the Gulf. His successor Shulgi (reigned *c.* 2095–*c.* 2048 BC) extended his territories in the northeast and east, dealing among others with the Gutians and the Hurrians; he also made his literary mark in his letters and royal hymns, and claims to have been a master performer on eight musical instruments. His second successor Shu-Sin (reigned *c.* 2038–*c.* 2030 BC) had to face the threat of western incursion; Ibbi-Sin (reigned *c.* 2029–*c.* 2006 BC), who claimed victory over these Amorites (under Ishbi-Erra of Mari), later saw his city fall to the Elamites.

The Mesopotamian harvest was won only after a long struggle against fierce heat and lack of rainfall. Fertility, celebrated on this seal, took on a central religious significance and the actions of nature were believed to be ruled by the gods.

5 The Royal Standard, dating from about 2600 BC, was found in one of the greatest tombs at Ur. It was apparently carried by an attendant wearing a peculiar bead headdress. It shows fully manned four-wheeled chariots, perhaps referring to victory [A], with domestic scenes on the reverse [B]. If it was a "standard" it was very small – 47cm by 20cm (18.5in by 7.5in) for a public display of royal wealth and success.

8 King Gilgamesh of Uruk had more legends told about him than any other hero of Babylonian history. The surviving Assyrian *Epic of Gilgamesh* was based on a much larger body of legend in Sumerian. This was so muddled that the king's career is quite unclear, although some of the stories may have been based on fact. He probably repaired a sanctuary at Nippur, and almost certainly built the city wall of Uruk.

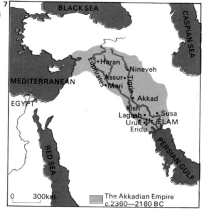

6 Ur-Nammu (*r. c.* 2113– *c.* 2096 BC), King of Ur, to whom this Sumerian seal was dedicated, may not have been a great warrior. But he did publish certain laws dealing, among other things, with sexual offences and wrongs committed in connection with the lands of others.

7 The world's first great empire, under Sargon of Akkad, extended so far that rebellion was almost inevitable, and this evidently occurred even before his death.

The Babylonians 2000–323 BC

The ferocious sacking and fall of Ur in 2006 BC allowed the Semitic-speaking Amorites under Ishbi-Erra to establish more effectively a dynasty that ruled at Isin for more than two centuries. A few years earlier another Semitic-speaking dynasty arose at Larsa, of slightly longer duration; and the two dynasties in parallel dominated Babylonia for a century until a third power was established – unopposed by them – consisting of more Semitic-speaking Amorites at Babylon early in the nineteenth century.

However, early in the eighteenth century BC, Larsa, under Rim-Sin "the true shepherd", overcame Isin and became the sole major contender with Babylon for the domination of the land [5].

Hammurabi and his laws

The first five kings of the new dynasty at Babylon were mainly preoccupied with defensive and religious building and by canal-clearing, with little extension of territory. It was left to Hammurabi to engage in victorious campaigns that left his empire stretching from Mari on the Euphrates in the northwest to Elam in the east, and by defeating Larsa to succeed to the traditional "kingship" of Sumer and Akkad.

Apart from his achievement of this relatively ephemeral empire, Hammurabi's fame rests mainly on his code of laws [8], written in Akkadian, a Semitic tongue that had by then – in parallel with political developments – become the principal language of Mesopotamia. Sumerian was retained mainly for religious use, although the civilization it expressed was absorbed by the Semites and continued to flourish.

No evidence has yet been found for the application of Hammurabi's laws in contemporary documents, nor was any appeal made to them; their standing and function are therefore unclear. But they may well have been an attempt to unify practice – notably in land tenure – among diversely regulated areas, with more uniform arrangements perhaps already prevailing in some matters not covered by them.

The reigns of Hammurabi's successors were long and undisturbed, and although in the later eighteenth century BC mention was made in Babylon of the alien Kassites (probably from the mountains in the northeast, and possibly Aryans), it was evidently an attack by the Hittite king Mursilis I in or soon after 1595 BC that brought the long-remembered destruction of Babylon and the downfall of Hammurabi's dynasty. But Mursilis can hardly have contemplated permanent conquest, and the void was filled by a Kassite dynasty later credited with a 576-year rule.

The dark ages of Babylon

Babylonia absorbed the Kassites, and during a dark age of more than 200 years little was heard of them. In the mid-fourteenth century BC the Kassite king married the daughter of the king of Assyria. But the alliance led to wars that resulted in the temporary conquest and occupation of Babylonia by the outstanding Assyrian soldier-king Tukulti-Ninurta I in 1235 BC. Essentially the Kassites retained Babylonia, but their dynasty fell to the Elamites in 1157 BC.

The Elamites lost political control of Babylonia before the end of the century; it passed to a second dynasty of Isin which,

CONNECTIONS

See also
30 The Sumerians
4000–2000 BC
52 The Hittites
1700–1200 BC
56 The Assyrians
1530–612 BC
28 Early western Asia

2 **Lilith**, with talons and feathered legs, was a Babylonian-Assyrian goddess who survived in Jewish lore into the Christian era. Traditionally a sinister bringer of death, in this clay relief she holds what may be a measuring rope to indicate the span of man's life. She is mentioned in an early fragment of the *Epic of Gilgamesh*, which also gives some independent evidence corroborating the biblical Flood. The profile used in narrative reliefs was less suited to representing the deity in actual rites; and reliefs over the altars of shrines show the goddess in a frontal view, perhaps to establish a relationship with the worshippers.

1 **The great ziggurat of the moon god** at Ur [A] was begun, according to King Nabonidus (r. 555–539 BC), by Ur-Nammu (r. 2113–2096 BC), but may well conceal the remains of an older tower from as far back as the predynastic period. Nabonidus says it was continued but left unfinished by Ur-Nammu's son Shulgi; Nabonidus himself made good the stairways with new treads a metre above the old and raised the level of the terrace. Different in many respects from the Mesopotamian ziggurats, and the largest known, – 100m (328ft) square – is the ziggurat at Dur-Untash in Elam [B], near Susa.

3 **This Babylonian tablet**, not yet fully understood, appears to be concerned with theoretical geometry. Most Babylonian mathematical texts are contemporary with the dynasty of Hammurabi (r. 1792–1750 BC); the rest are datable to the last three centuries BC. The earlier history of the Old Babylonian group is not known, beyond the evidence of innumerable economic-administrative texts from the earliest period of Mesopotamian writing, whose number system, based on 60, was retained by the Old Babylonians. But although the content of Old Babylonian mathematics reached a level which can be compared with that of the early Renaissance – it was elementary compared with that of the Greeks.

under Nebuchadrezzar I (reigned 1124–1103 BC), ended Elamite interference. The Isin Dynasty fell after little more than 100 years of political stability. The ensuing age of uncertainty and civil disturbance was relieved by the inauguration of the Eighth Dynasty of Babylon in 977 BC, and for a century Babylonia maintained close contact with the developing power of Assyria.

Shalmaneser III [Key] of Assyria (reigned 858–824 BC) was called upon to help quell a rebellion in Babylonia, at which time the powerful Chaldaean tribes of southern Babylonia were first making their appearance. Wars with Assyria and anarchy at home preceded the emergence of Tiglath-Pileser III (reigned 744–727 BC) as a strong king of Assyria who at length assumed the Babylonian crown. His successor Shalmaneser V (reigned 726–722 BC) ruled both countries for five years, but both Sargon II (reigned 721–705 BC) and Sennacherib (reigned 704–681 BC) found strong antagonists in the Chaldaeans.

Babylon's fortunes varied widely in the seventh century, until the rise of the unknown "son of a nobody" Nabopolassar (reigned 625–605 BC) inaugurated a great age of Babylonian civilization under the neo-Babylonian or Chaldaean Dynasty.

The rise of the neo-Babylonians

Babylon helped the Medes in the overthrow of Nineveh and the Assyrians in 612 BC. Her brilliant commander Nebuchadrezzar II (reigned 604–562 BC) destroyed Jerusalem and carried off its inhabitants, erected great monuments and buildings, which made Babylon one of the Seven Wonders of the world – this is the period of the "Hanging Gardens".

Nebuchadrezzar's son was murdered, and the decay of Babylonia accelerated under the pious antiquarian Nabonidus [1]; Babylon fell without a fight before the Achaemenid Persian king Cyrus the Great (c. 600–529 BC) in 539 BC. Xerxes (c. 519–465 BC) partly destroyed it in 482 BC; it might have been restored by Alexander the Great (356–323 BC) had he not died there, and thereafter, although its astronomical schools survived, Babylon passed into history.

The throne-base of Shalmaneser III found at Nimrud has an inscription on the horizontal surfaces. It includes a separate section referring to the king's campaigns of 851 and 850 BC in Babylonia, in which he helped the Babylonian king to defend his throne against a rebellion. This relief carving on the western vertical face shows Shalmaneser [right] and probably the Babylonian king [left], each with an attendant, under a canopy. They are shaking hands – a unique representation in Mesopotamian art of this modern gesture; whether it implies equality, Babylonian subservience or neither, is unknown.

4 Relief bricks from Babylon, some showing bulls and mythical creatures, once formed part of the Processional Way. Like most excavated Babylonian remains they are of the neo-Babylonian period.

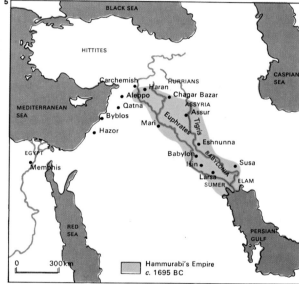

5 Babylon dominated western Asia, more or less, for thirteen and a half centuries. It lay on the lower course of the River Euphrates, an advantage much increased by the development of irrigation systems.

Hammurabi's Empire c. 1695 BC

6 Early clay tablets reveal the importance of sheep and goats – they also use the signs for merchant, cattle and donkey – in the economy of early Sumerian communities. The Akkadian period improved on the quality of the early tablets in the first-ever recording of a Semitic language – Old Akkadian. Their development into the tablets of Babylonia (those shown are Old Babylonian) and Assyria culminated in the calligraphy of the scribes of Assurbanipal's library at Nineveh.

7 Naram-Sin (c. 2254–c. 2218) of Akkad, whose Semitic line was an interlude in Sumerian history presaging the supremacy of the Semitic Babylonians, portrayed himself on this stele triumphing over the eastern Iraqi king of Lullubi.

8 The diorite stele of Hammurabi was carried off from Babylon by an Elamite invader, perhaps in the 12th century, and taken to Susa, where it was found in the winter of 1901–2. The text is topped by a bas-relief showing Hammurabi receiving the commission to write the laws from Shamash the sun god, god of justice. The Elamites apparently chiselled off parts of the text, but most of these survive on other copies of the code. It has a prologue and an epilogue in semi-poetic style.

Egypt: the Old and Middle Kingdoms

Successive prehistoric cultures designated by the names of Badarian, Naqada I and Naqada II have been identified from archaeological remains in Upper Egypt but remains from Lower Egypt are scanty and make any sound historical judgment rather difficult. It would appear that two distinct kingdoms evolved in Lower and Upper Egypt [1] and that the unification of the country was brought about by the victory of Upper Egypt over Lower Egypt c. 3100 BC. Even after this unification, however, the peoples of ancient Egypt continued to call their country the Two Lands.

The divine kings of Dynasty I
Traditionally the first ruler of Dynasty I and conqueror of Lower Egypt is known as Menes [2] and he is credited with the foundation of the national capital of Memphis, just south of the Nile Delta. The reigning king was regarded as the living embodiment of the falcon-god Horus and hence divine. During the first dynasty and the next farmers began to use the plough extensively, and irrigation was probably introduced. A national government evolved and writing was developed.

With Dynasty III began the period known as the Old Kingdom (c. 2686–2181 BC). The most prominent ruler of Dynasty III was Zoser, for whom the Step-Pyramid complex was built [Key]. Under Snofru, the founder of Dynasty IV, the first true pyramid was constructed and the technique of building pyramids was perfected under his successors Khufu, Khephren and Menkaure. The construction of the pyramids [4] entailed enormous expenditure and organization, and the weakening of the power of the crown at the end of Dynasty IV led to the abandonment of the more expensive techniques.

During Dynasty V the cult of the sun-god Re regained national pre-eminence and the rulers undertook the construction of solar temples for his worship. At the end of Dynasty V magical texts were inscribed on the walls of the burial chamber of the royal pyramids to ensure the safe passage of the ruler's spirit to the after-world. During Dynasties IV to VI periodic campaigns were undertaken against tribesmen in the Sinai peninsula and large-scale expeditions were dispatched to Nubia to extort or trade for

ivory, gold and other precious materials.

At the end of Dynasty VI the growth in power of the provincial governors, the nomarchs, led to the steady weakening of the control of the Memphis hierarchy. Rival dynasties appeared in Heracleopolis and Thebes and the country was plunged into civil war. The confusion was compounded by the infiltration of Asiatic tribesmen from Sinai into the fertile regions of the Delta.

A period of confusion ends
This chaotic and ill-documented era is known as the First Intermediate Period (c. 2181–c. 2050 BC). Montuhotep II of Dynasty XI, Prince of Thebes [7], overcame his rivals and reunited Egypt under his rule, although the many nomarchs retained considerable power. He expelled the Libyan and Bedouin raiders, inaugurating the Middle Kingdom in Egypt (c. 2050–1786 BC). Montuhotep II had a mortuary temple and tomb built for himself at Deir el-Bahari on the west bank of the Nile opposite Thebes. Montuhotep II's descendant Montuhotep IV was succeeded in unknown circumstances by his vizier

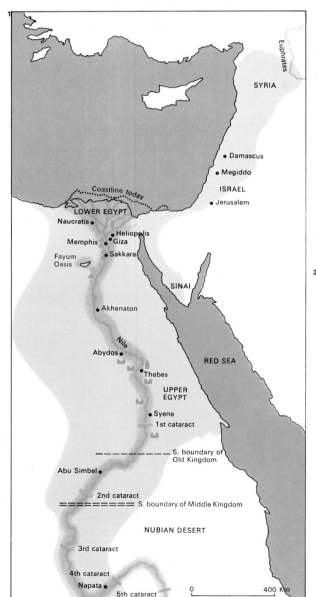

Pyramids
Obelisks
Temples
Maximum extent of Egyptian Empire (New Kingdom, c. 1450 BC)
Fertile region

1 Egypt divides naturally into two areas – Lower Egypt, which consists of the Nile Delta region, and the long, narrow strip of Upper Egypt, which is confined on both sides by desert. The annual flooding of the Nile brought water for land irrigation and also deposited a rich topsoil. Egypt's agricultural prosperity was based on the river's unfailing predictability.

2 Egypt's unification was brought about by war. Victories of the early rulers are commemorated on ceremonial palettes such as this one depicting Menes, the first ruler of all Egypt.

3 The monarch's public image in the Old and Middle Kingdoms was widely different, a change reflected in the royal statuary. The figure of King Menkaure of Dynasty IV [A] with his queen is an idealized portrait of the god-king. The aftermath of the First Intermediate Period saw the development of a more realistic and intense style, as in the statue of Senusret III [B].

Amenemhat I, who founded Dynasty XII in about 1991 BC.

Amenemhat I founded a new capital at Lisht south of Memphis because Thebes was too far south to serve as an efficient capital. To strengthen his hold on the throne he circulated a spurious prophecy concerning his rise to power and appointed his son Senusret I as co-ruler. Despite these precautions Amenemhat I was assassinated after 30 years of rule but his son managed to secure the throne. The powers of the provincial nobility that might rival the throne were suppressed under Senusret III. Amenemhat I had begun the conquest of Nubia and this expansion was completed under Senusret III, who fixed the southern border at Semna. Apart from one attested incursion into southern Palestine no steps appear to have been taken to exercise direct control in Palestine or Syria, but trade links were strongly maintained. A line of fortifications was built by Amenemhat I in Sinai to deter possible invasions. The rulers of Dynasty XII sponsored a vast land reclamation project in the Fayum area.

The Second Intermediate Period (c. 1786–1570 BC) was marked by a decline in the power of the central government. During Dynasty XIII Asiatic invaders broke through the Egyptian defences and infiltrated the country. These foreigners are commonly known as the Hyksos, although the term should properly be applied to the chiefs only. The Hyksos eventually secured control of most of the country, with the aid of new military weapons such as chariots, and were the founders of Dynasty XV (c. 1674–1570 BC), although it is most unlikely that they ever exercised direct control over Thebes and the south. More probably Thebes was forced to acknowledge the supremacy of the Hyksos ruler in his new capital of Avaris in the Delta.

The invaders are expelled

The Hyksos did not rule as foreigners but adopted Egyptian titles and Egyptian culture. As their power weakened, the princes of Thebes of Dynasty XVII were emboldened openly to reject Hyksos rule and, after several campaigns, Kamose and his successor, his brother Ahmose, succeeded in taking Avaris and expelling the invaders.

King Zoser's Step-Pyramid at Sakkara, Memphis, was the first major Egyptian building in stone. It was supposedly designed by his vizier Imhotep, who was later deified.

1 Subterranean chamber
2 Queen's chamber
3 King's chamber
4 Entrance
5 Corridor
6 Corridor
7 Grand chamber
8 Vault
9 Shafts

4 The Great Pyramid erected at Giza by Khufu, second ruler of Dynasty IV, provides a typical example of the art of the pyramid builder. It is built of limestone blocks and was originally faced with fine white limestone. It was approximately 146m (480ft) in height. After two changes in plan the final resting place of the body, called the King's Chamber, was constructed of granite and approached via the Grand Chamber. The sarcophagus is still in place.

5 Scribes, such as this one depicted in an Old Kingdom statue, were the key to the smooth functioning of the Egyptian administration. Papyrus was used for the recording of daily accounts and business, but very few pieces have survived from this period.

6 In the afterlife the same things were felt to be required as in this life. Everyday items, even down to models of retainers like these soldiers, were therefore placed in tombs.

7 King Montuhotep II is shown being embraced by Re, the sun-god, in a painted relief from the king's mortuary temple at Deir el-Bahari. The king also prepared tombs for Nefru, his sister and queen, and for several of his concubines. Later rulers of Dynasty XI built similar temples opposite Thebes on the Nile's west bank.

8 Egypt's lack of certain raw materials, notably timber, resulted in the growth from early dynastic times of a flourishing trade between Egyptian and Syrian ports. Egypt's influence in Syria and Palestine during the rich period of the Middle Kingdom is reflected in the Egyptian statuary and jewellery found in those regions.

India: prehistory to 500 BC

The earliest evidence of a literate culture in India dates from about 2300 BC when the Indus civilization emerged from the prehistoric age. This Indus or Harappan civilization (named after the town of Harappa) had its principal centres in the Indus valley, now mainly in Pakistan, but extended westwards to the present Iranian border, eastwards to beyond Delhi and southwards to the Gulf of Broach [1]. Its main cities were at Harappa in the Punjab and Mohenjo-daro in Sind, but there were also a number of smaller towns, including the port of Lothal.

The nature of settlements

The cities show advanced town planning and the remains testify to a high and diverse material culture [2, 3]. A considerable part of the now fairly arid Indus valley must have been brought under cultivation to yield the surplus crops with which to feed the city populations. Unless there has been a complete change of climate in this area since that time it is obvious that fields were irrigated.

Such a sophisticated civilization required a form of writing. Thousands of steatite (soapstone) seals [Key] have been discovered. In addition to representations of animals, men and gods these present brief inscriptions in as yet undeciphered hieroglyphic script. The seals, some of which have been found as far away as Syria, were probably merchants' seals attached to goods.

The end of the Indus

Even less is known about the end of the Indus civilization than about its origins. After flourishing for five or six centuries (c. 2300–1750 BC) without undergoing much change it completely disintegrated following a brief decline. Although natural calamities cannot completely be excluded, it now seems that the Indus cities were ravaged by invading nomadic horsemen in the eighteenth century BC. The latter are usually identified as Indo-Aryans, for whose presence there is, however, no reliable evidence until about four centuries after the end of the Indus civilization. It is therefore more likely that the Indus cities were conquered by tribesmen from the mountains who, in their turn, gave way to the Indo-Aryans.

By the thirteenth century BC, the Indo-Aryans – split into numerous tribes who fought each other no less fiercely than the earlier inhabitants – had occupied the Punjab. They subsequently spread into the Ganges valley and southwards into Gujerat and Maharashtra.

A vast collection of religious hymns written in archaic Sanskrit, the *Rigveda*, dates from this early phase (c. 1200–1000 BC). Apparently preserved by oral tradition, the hymns are addressed to many different deities whose help is implored in military and agricultural pursuits. The four Vedic texts, of which the *Rigveda* is the foremost, spawned expositions and commentaries of which the *Upanishads* are the most celebrated. The gods, such as Indra, are usually conceived of as anthropomorphic, but some features – for example speculation about the true nature of the sacrifice – which were to be become characteristic of Hinduism, are already distinguishable.

During the later Vedic period (c. 1000–550 BC) the Indo-Aryans, by then utilizing effective iron tools, spread over

CONNECTIONS

See also
62 India 500 BC–AD 300
120 Indian art to the Moguls
64 Buddha and Buddhism

In other volumes
220 Man and Society
260 Man and Society

1 **The brown areas on the map** mark the expansion of the Indus civilization, stretching southeast down the coast to beyond the Gulf of Broach and eastwards far beyond present Delhi. The westward expansion into Baluchistan is not shown. All over this vast area – Rupar and Lothal are about 1,600km (1,000 miles) apart – the Indus civilization was uniform, suggesting centralized control. Political control was facilitated by the nature of the land as this civilization flourished in relatively dry areas which, unlike the tropical rain forest, could be cultivated without iron tools.

2 **The ancient Indus city of Mohenjo-daro** was built according to a systematic plan with streets crossing at right-angles and houses opening onto the streets. Elaborate granaries have also been found.

Indus civilization 2300–1750 BC
Indus sites
Early historical sites c.1200–500 BC
Modern cities

Charsada
Taxila
Rupar
HIMALAYAS
Harappa
Kalibangan
Indrapat
Hastinapura
Ahichhhatra
Delhi
Indus
Ganges
Mohenjo-daro
Karachi
Lothal

0   600km

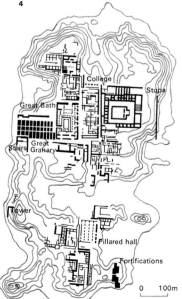

3 **The Great Bath** in the citadel of Mohenjo-daro was built of fine brickwork and presumably used for ceremonial functions. One of the most striking features of the Indus civilization was the importance attached to good water supplies. A well-built bathroom has been found in almost every house, which also possessed proper drainage. As a result each house has its own well, with a deep shaft which was kept in shape by means of terracotta hoops.

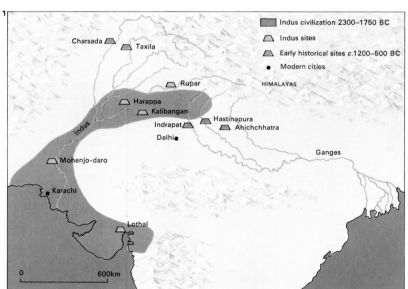

College
Stupa
Great Bath
Great Stairs
Great Granary
Tower
Pillared hall
Fortifications

0   100m

4 **The larger Indus cities** all consisted of a citadel mound and a lower city complex stretching eastwards from the citadel. The former was the site of most of the large public buildings such as, in Mohenjo-daro, the Great Bath, the granary and a building that on account of its shape (several courtyards, corridors, rooms and compartments) has tentatively been identified as a college. The lower city was the residential area. This ground plan of Mohenjo-daro shows the grid of the town in the lower part of the city and the houses, which vary greatly in size from those consisting of a single room to large residences boasting more than 20 rooms.

most of northern India including the Ganges valley, burning down the forest to cultivate the fertile land. The gradual progress of this expansion can be traced with the help of a distinctive type of pottery, Painted Grey ware. These people were not urbanized as the Harappans had been. During their expansion they mixed with earlier established forest tribes, introducing them to the horse, and their mode of life changed from semi-nomadic cattle breeders to settled farmers.

This change had important political and cultural implications. The tribal units gave way to kingdoms based not on kinship but on territory. The kingdoms were controlled by warrior classes (*kshatriyas*) headed by the king, and assisted by members of the powerful class of hereditary priests (*brahmins*). These two ruling classes controlled the free peasants, traders and craftsmen who constituted the third class (*vaisyas*), as well as the semi-servile labourers, hunters and fishermen who formed the lowest class of the *sudras*, partly descendants of forest tribes. Some of these, especially those whose way of life was considered unclean or repulsive by the Indo-Aryans, were assigned the status of untouchables. This marked the beginning of the complex caste system.

The rise of major cities

During this period major cities developed for the first time since the decline of the Indus cities. The most important were Hastinapura on the Ganges east of present-day Delhi and, in about 500 BC, Rajgir in southern Bihar, with its impressive walls. In the same period most of the basic concepts of Hinduism took shape: not only caste, but also the belief in the transmigration of the soul, in non-violence and in the holiness of the cow. These have all become lasting features of Indian civilization. While the Vedic sacrifice persisted and became more and more complicated, there was also a reaction among those who felt unsatisfied with formal religion and sought higher values in meditation. People of the ruling classes were encouraged to withdraw to a life of contemplation in the forest when their children no longer needed them. On this foundation Siddhartha Gautama (*c.* 563–*c.* 483 BC) instituted Buddhism.

Three Mohenjo-daro casts from seals supply valuable information about the ancient Indus civilization, revealing, for example, that cattle had already been domesticated. These seals represent a bull feeding from a manger, an elephant and a rhinoceros. The writing is hieroglyphic but scholars have not yet been able to decipher it.

5 The earliest true history of the Indo-Pakistani subcontinent began when, after the middle of the third millennium BC, a high civilization emerged in and about the Indus valley. Once established, the Indus valley civilization flourished for more than six centuries without undergoing any significant change. Its sudden end may have been due to Indo-Aryan invasions, but it seems more likely that these invasions took place when the Indus civilization had already disintegrated. The Indo-Aryans settled in villages in the Punjab and were divided into tribes, about which we know from the *Rigveda*. Between *c.* 1000 and 500 BC Indo-Aryan civilization gradually spread along the Ganges valley, as we can confirm from later Vedic literature, and early Hinduism took shape.

5

			EARLY VEDIC Ganges valley Pastoral economy	LATER VEDIC Ganges valley Urban economy

HARAPPAN CIVILIZATION
Indus valley
Urban economy

Aryan invasions

3000 BC	2500	2000		1500		1000	500 BC

7

7 Most seals from Mohenjo-daro show representations of animals, sometimes natural, sometimes composite or fantastic. This seal shows a three-headed, horned deity seated in an attitude that is reminiscent of the yoga *āsanas* of later times. The god is surrounded by a number of animals. This, among other features, recalls later representations of the god Shiva. The deity has therefore been identified as a proto-Shiva.

6 This limestone sculpture is one of the few surviving stone sculptures of the Indus civilization. It is 19cm (7in) high and is apparently a portrait statuette representing a bearded man with low, receding forehead, elongated eyes and thick lips. The sculpture reveals excellent taste and craftsmanship and gives some impression of the physical appearance of the ancient Indus peoples.

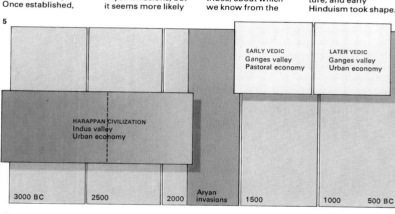

8

8 The modern inhabitants of the Mohenjo-daro area, like their predecessors, still use the river as their lifeline. The river was not only important for irrigation of the fields in the vicinity of its banks but it was also essential for communications. The striking uniformity of the Indus civilization would have been impossible without reasonable communications. Representations of boats have been found on stone and terracotta at Mohenjo-daro.

Minoan civilization 2500–1400 BC

Arthur Evans (1851–1941), the British archaeologist, excavating at Knossos on the island of Crete from 1900 to 1936, uncovered a richly decorated palace, an architectural masterpiece that indicated an early civilization more distinctive and sophisticated then any other European culture hitherto discovered [4]. Other similar palaces have since been revealed – by the Italians at Phaestos, by the French at Mallia, more recently by the Greeks at Zakro – and a balanced picture begins to emerge.

Beginnings of Minoan civilization

The Minoans (from the name of the legendary King Minos of Knossos), as Evans called the people who lived in Crete, appear mysteriously on the island at the start of the Bronze Age, about 3000 BC, perhaps as immigrants from Anatolia or, as recently argued, from Palestine. The contents of their circular tombs, most frequent on the plain of the Mesara, show that overseas contacts were maintained, contacts that gradually built up into a great trading network across the eastern Mediterranean [1].

By about 2000 BC, economic and social advance had spurred architecture to impressive achievements – the Old Palaces. Their details are much obscured by later additions and alterations, but enough survives at Knossos, at Mallia, and particularly under the west court at Phaestos, to show that building on a lavish scale was already being carried out. The implications for the social organization are of course even greater than for the technical abilities of the builders. The Minoans had achieved civilization.

One of the most obvious criteria of this is the use of writing. The earliest brief inscriptions in Crete are found on seals, where picture symbols appear to belong to a hieroglyphic script. Clay tablets found in the early palace at Phaestos show that by then a simpler syllabic writing had been devised for general use. It may well have been employed mainly for writing on some sort of paper or parchment, but such materials have failed to survive. No convincing translation of these texts, called Linear A, has yet been proposed, so its language remains unknown.

The exact nature of the island's political organization is difficult to discover. The extensive storage capacity of the palaces, for articles such as pottery as well as commodities like olive oil, suggests that they were economic as well as administrative centres, controlling territories within the island. Yet these territories had apparently nothing to fear from each other – there are no walled defences and few signs of weapons or soldiers. Knossos clearly held a leading position, perhaps by controlling overseas trade.

Height of Minoan culture

Demonstrated by finds of Minoan pottery from as far away as Egypt, by colonies on several Aegean islands, and by the permeation of the mainland culture, overseas trade goes far to explain the wealth of Crete at this period. But Minoan civilization cannot be explained as simply imported from abroad: it is far too individualistic for that.

After apparently natural disasters in about 1700 BC, the palaces were lavishly rebuilt and it is the ruins of these that can be seen today. Building was in limestone, within a timber-frame construction. This added a

CONNECTIONS

See also
40 The Greek mainland 2800–1100 BC

In other volumes
30 The Physical Earth

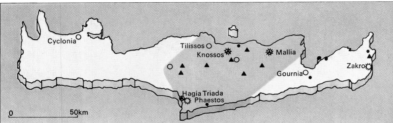

1
☐	Probably dominated by Knossos c.2000–c.1400
●	Town
○	Town with palace
☼	Great palace
▲	Cult centre

1 The long narrow island of Crete was the home of Minoan civilization, the first in Europe. With much of the interior mountainous, Mt Ida topping 2,400m (7,873ft), and the rugged south coast broken only by the plain of Mesara behind Phaestos, there seems little to explain civilization here. But the north coast has a gentler relief and is rich in olives and vines, and the surrounding seas encourage trade and contact over wide areas. The Minoans actively developed their agricultural wealth and foreign commerce.

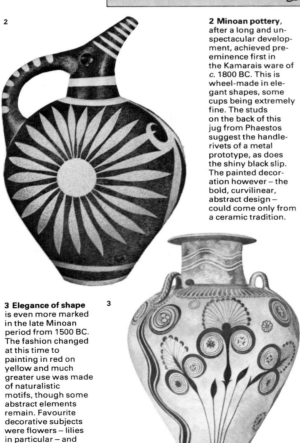

2 Minoan pottery, after a long and unspectacular development, achieved pre-eminence first in the Kamarais ware of c. 1800 BC. This is wheel-made in elegant shapes, some cups being extremely fine. The studs on the back of this jug from Phaestos suggest the handle-rivets of a metal prototype, as does the shiny black slip. The painted decoration however – the bold, curvilinear, abstract design – could come only from a ceramic tradition.

3 Elegance of shape is even more marked in the late Minoan period from 1500 BC. The fashion changed at this time to painting in red on yellow and much greater use was made of naturalistic motifs, though some abstract elements remain. Favourite decorative subjects were flowers – lilies in particular – and sea creatures, such as octopuses, nautili and shells among rocks and seaweed.

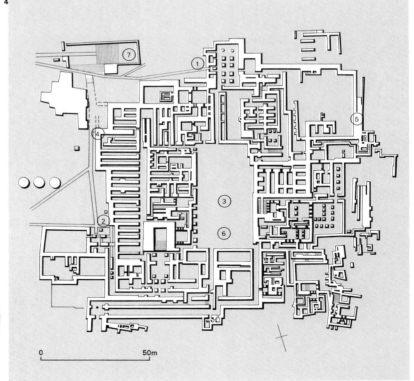

4 The Great Palace of Knossos had a long history. The plan shows it after its rebuilding in c. 1550 BC. The main entrances [1 and 2] lead by long corridors to the Central Court [3]. Beneath this, deposits have been found going back to 6000 BC. Storerooms for oil and other goods [4] and for pottery [5] hint at the wealth that poured into the building, which controlled at least all of central Crete. The administrative block fronted the court on the west and faced the great staircase to the domestic quarter on the east [6]. Outside the palace to the northwest, a processional way led to a theatre [7].

useful, if not always effective, resilience against earthquake shock, to which Crete is prone, but also added to the fire risk. Ranges of rooms opened on to the great central courts, with well-planned light wells to illuminate inner suites of chambers. Internal walls were plastered and often gaily decorated with the most elaborate and colourful frescoes [9]. Floors were frequently paved with alabaster slabs. But it is perhaps the drainage system, as advanced as any before the eighteenth century AD, which causes the most surprise. The overall impression is a convincing one of light, air and freedom – so different from that of the contemporary civilizations farther east and so much closer to modern ideals.

The art fully bears this out, whether it is a life-sized figure painted in the frescoes, decoration on the magnificent pottery [Key], or minute detail on the carved seal-stones [7, 8]. The colour and naturalism hold an immediate appeal to the modern world.

This great civilization's end is even more hotly debated than its origins, if only because there is so much more evidence on which to base the story. The most widely accepted version, but by no means the only one, would start from the cataclysmic eruption of the volcanic island of Thera, a little over 100km (63 miles) from the north coast of Crete, about 1450 BC. Much of Crete would have been plastered with poisonous ash, shattered by the shock waves of the explosion and pounded by monstrous tidal waves.

Decline and conquest
Of the major sites, only Knossos appears to have recovered from this destruction and here there are many signs of a profound change. The new rulers were warriors who used the Linear B script, now recognized as an adaptation of Linear A for writing an early version of the Greek language. This would all seem to point to a conquest of the island by mainlanders, Mycenaeans, who had seized the opportunity given them by the eruption of Thera to oust the Minoans from their control of the profitable sea-routes. Metropolitan Minoan was replaced by provincial Mycenaean and the palaces were lost to sight and memory.

This pottery figurine displays a courtly elegance typical of the Minoan civilization of Bronze Age Crete. It is 29.5cm (11.6in) high and was found in the Temple Repositories at Knossos, where it was buried c. 1500 BC. The tightly fitting bodice, open to expose the breasts, the embroidered apron and the long flounced skirt are shown frequently on seals and in frescoes. The lioness, if such it is, on her hat and the snakes in her hands are less usual and more sinister. She is probably the earth mother, whose worship is widely attested in shrines and pillar crypts on Minoan sites and in caves and hilltop sanctuaries throughout Crete.

5 This magnificent stone libation vase comes from the palace at Zakro, destroyed c. 1450 BC. It shows a sanctuary (centre, left) on a mountain peak, with wild goats on the roof. A bird flies past two pairs of horns used at consecration.

6 A gold pendant from Mallia shows the same love of nature in a very different technique. Two bees, wasps or hornets rest on a berry or honeycomb. It is 4.6cm (1.8in) across and is dated about 1550 BC.

7 On this seal impression, measuring 2.1cm (0.8in) across, two equestrian acrobats perform in a field full of flowers.

8 This seal, in reality only 1.5cm (0.6in) across, demonstrates the mastery of Minoan carvers. It is in blue chalcedony.

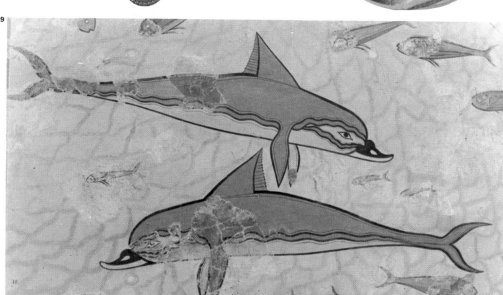

9 Fresco painting was another spectacular art of the Minoans, although it normally survives only in small fragments. This scene of dolphins and fish decorated the so-called Queen's Megaron in the domestic quarter of Knossos. Other frescoes also show birds, flowers, plants, animals and people.

The Greek mainland 2800–1100 BC

Homer's *Iliad* and *Odyssey*, written between about 800 and 700 BC, record the exploits of legendary Greek heroes around the time of the Trojan Wars. For centuries the fact that this heroic age ever existed about 1800–1100 BC was in doubt, until Heinrich Schliemann (1822–1890), a German archaeologist, recovered at Troy and Mycenae (from 1874) relics that supported the legends.

Bronze Age beginnings

This Greek Bronze Age, centred on the Greek mainland, began with the introduction of metal, important both for stimulating trade and for the acquisition of visible wealth. About this time, too, the grain economy of the plains of Thessaly and the north was replaced by one based on grain, olives and the vine, which flourished better in the south. By 2500 BC this economy supported a palace, the House of Tiles, at Lerna in the Plain of Argos. About 2200 BC however, Lerna and many contemporary settlements were destroyed by invaders from the northeast. These were probably the first inhabitants to speak a language recognizably Greek.

At this time the fast potter's wheel and the megaron, a hall with pillared porch, were introduced. Soon after, their civilization showed signs of influence from their contemporary, Minoan Crete [2].

In time, however, strong towns grew up, such as Mycenae. The first real evidence of the flowering of the mainland civilization appears in the shaft graves of Mycenae itself, dating largely from the sixteenth century BC. They contain real solid wealth, in gold, silver and bronze, in crystal, alabaster and clay. We can assume that these were the resting places of a princely or even royal family, ruling a rich and integrated society. Many of the objects [6, 8] reveal a contrast between polished and sophisticated craftsmanship, patently Cretan, and a very un-Cretan emphasis on weapons, armour and military scenes [7]. The owner of the stern gold mask [Key] was no soft courtier but a warring hero, an ancestor of those whom Homer portrayed.

These early Greeks also became seamen. A few of the objects from the shaft graves may be Egyptian work acquired through trade, rather than Cretan or Greek. Desirous

of more trade goods from abroad and supported by local agricultural wealth, the Greeks began overseas ventures of their own, to Troy, to Palestine and Egypt, and to the Lipari Islands beyond the "toe" of Italy.

Expansion of Mycenaean civilization

But their opportunity came when the power of the Minoans was destroyed, perhaps by the cataclysmic eruption of Thera *c.* 1450 BC. This freed the rich trade routes of the Mediterranean which the Cretans had hitherto controlled. The mainlanders, comparatively untouched by the disaster, made the most of the new situation. Their pottery, valued both for its own sake and for the perfumed oil exported in it, rose sharply in price in both the Levant and Egypt: it was prized, for example, in Akhenaton's newly built Egyptian capital Tell el Amarna, about 1350 BC. Cypriot copper was carried in their ships. A great westwards trade port grew up at Taranto in southern Italy, bringing more copper from Sardinia and the eastern Alps, and amber from the distant Baltic. Crete itself was occupied by the Mycenaeans, and

CONNECTIONS

See also
38 Minoan civilization 2500–1400 BC
68 The Greeks to the rise of Athens
84 Bronze and Iron Age Britain

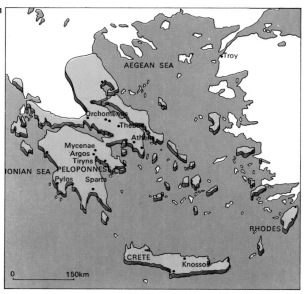

1 The Mycenaeans' homeland was the Peloponnese and adjacent parts of Greece, centred on the Argolid, and Mycenae itself. Here they built their distinctive version of the civilization already flourishing in Crete. As they took over the Minoans' sea trade, so their influence spread to the islands and coasts around the Aegean and beyond. This vast territory was never a unified state. More likely it was a collection of allied kingdoms.

2 The Mycenaeans were strongly influenced by the Minoans. This gold cup from Vaphio, near Sparta, if not made in Crete itself must at least have been the work of a mainland craftsman trained in the Minoan tradition.

3 The syllabic script came from Crete. Linear B script was extensively used for business documents, such as this stock list of herbs from Mycenae. The language used was an early form of Greek.

4 The fortified citadel of Tiryns, *c.* 1330 BC, typifies mainland architecture in the Bronze Age. A tortuous entrance passage [1] leads through the massive walls and an inner portico [2] leads to the first court. The administrative centre took up the whole of the inner court [3] and the megaron [4], with its great central hearth, opening on to it. The main structural fabric consisted of a wooden framework and columns, and sun-dried bricks.

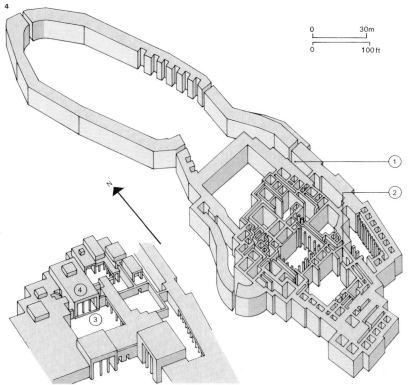

the Palace of Knossos was rebuilt as the seat of the new dynasty.

In the towns of Greece, craftsmen carried out their trades, producing fine metalwork, pottery and perishable goods. The towns themselves grew larger, stronger and better appointed. Walls were heightened and extended, with devices to ensure the safe supply of water. The palaces, still based on the traditional megaron plan, were now elaborately decorated with frescoes. The richly equipped tombs, now great corbelled tholoi (circular buildings first developed at this period) like the Treasury of Atreus [5], show the great architectural skill that had been attained by their builders.

To facilitate such a level of trade, craft and administration, writing was needed. The Minoans on Crete had developed a syllabic script of their own, still undeciphered. This is found inscribed on tablets and archaeologists have called it Linear A. The Mycenaeans adapted it, rather clumsily but adequately, for their Greek tongue (Linear B) [3]. In 1953 Michael Ventris, the English architect, deciphered Linear B and established beyond doubt that the Mycenaeans, Homer's Achaeans, were linguistically at least the true ancestors of the classical Greeks.

Mainland civilization declines

The closing stages of Bronze Age Greece are difficult to understand. By one account hardy frontiersmen from the northwest, the Dorians, overran the cities of the south and sacked them all, except Athens. By another, the Mycenaeans lost their expansionist drive towards 1200, engaged in civil war – the siege of Troy exemplifies this – and in effect destroyed themselves. At around this time, there was certainly great unrest over a wide area, and the bands which unsuccessfully attacked Egypt in 1225 BC and again in 1191 BC included Aegean peoples.

In mainland Greece, the succeeding age knew little of what had gone before. Shabby villages replaced the flourishing towns, simple pits the great tombs, and common pots the masterpieces in clay, silver and gold. The one advance in this dismal period, sometimes referred to as the "dark age", was the introduction of iron-working.

This gold mask shows a proud Mycenaean of *c.* 1550 BC. It was recovered from Shaft Grave V by Heinrich Schliemann in 1876.

5 Earlier shaft graves were succeeded at Mycenae by stone-built tholos tombs. The finest is the so-called Treasury of Atreus *c.* 1320 BC. A walled passage leads to a monumental door in the mound. Inside, a circular chamber is roofed by a corbelled vault. The bodies and grave goods were looted long ago, but other smaller tombs have also been found.

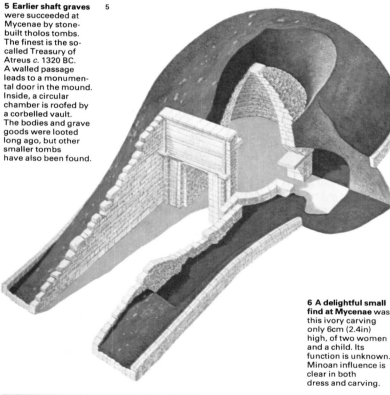

6 A delightful small find at Mycenae was this ivory carving only 6cm (2.4in) high, of two women and a child. Its function is unknown. Minoan influence is clear in both dress and carving.

7 These Mycenaean warriors may have a comic flavour to our eyes, but the discovery of fortifications, weapons and armour show that warfare was an important factor in life at the time.

8 Bronze daggers from shaft graves of *c.* 1550 BC reveal great artistic skill in their gold, silver and niello inlay. Some show sea creatures, one shows a hunting cat in Minoan style, but this one has a more robust mainland subject. A lion is attacking a deer while two more make their escape. The dagger is 23.5cm (9.25in) long and is one of the treasures discovered in Shaft Grave IV by Schliemann, the German archaeologist.

Stone Age Britain

Although traces of still earlier occupation are beginning to come to light, the first really clear evidence of human settlement in Britain belongs to a time about a quarter of a million years ago. At that time the country enjoyed a mild climate and favourable environmental conditions between two ice ages.

Swanscombe man and the ice ages
The population lived in small groups and occupied open-air sites along river valleys and lakesides. They were efficient hunters, catching large game such as now-extinct forms of elephant, and rhinoceros, as well as horse, wild ox, red deer and fallow deer. One of the most important sites of this period, at Swanscombe, just south of the River Thames in Kent, has yielded three fragments of the skull of a fossil man. Swanscombe man is usually classified as the earliest known example of the modern species *Homo sapiens*.

After the occupation of this period, which is sometimes known as the Great Interglacial, there are a number of sites that can be dated either to warmer phases within the next glaciation (*c.* 200,000–125,000 years ago) or to the last interglacial period (*c.* 125,000–100,000 years ago). These sites indicate that men continued to visit Britain, at least during the milder climatic phases. With the beginning of the last ice age, *c.* 100,000 years ago, modern understanding of human activity in Britain becomes much clearer. Probably on four separate occasions during warmer climatic phases, small groups of hunters moved into Britain across a land-bridge from the continent of Europe and settled in caves [3] and rock-shelters in England and Wales. Some of the best known sites of this date occur in Cheddar Gorge, Somerset, and there is another important group in Derbyshire. Occupation of Britain still remained very sparse and occasional.

Britain after the ice ages
About 14,000 years ago the climate began to become warmer once again and the ice sheets began their final retreat [1]. As conditions improved, the vegetation altered, and about 10,000 years ago the open tundra of the glacial periods began to be replaced by forests. At first these were forests of birch and pine, both cold-loving species; later hazel, oak, elm, lime and alder grew too, forming the so-called "Mixed Oak Forest". These changes in flora were accompanied by changes in fauna also: open-country animals such as the horse, reindeer, bison and mammoth disappeared and were replaced by woodland forms such as red deer, roe deer, elk, wild ox and wild boar.

The human population had to adapt to all these changes in climate, vegetation and animal life. Some groups, such as the one that settled at Star Carr in Yorkshire *c.* 7500 BC, adapted their way of life to forest conditions [5]. They used plant foods such as hazel nuts and learned to hunt the woodland animals that moved singly in the forest, unlike the herd animals that their ancestors had hunted on the tundra. They learned to use timber and they had the first true stone axes.

Other groups settled along the seashore to exploit coastal resources, such as fish and shellfish. During this period Britain became permanently an island. As the ice sheets melted, large quantities of water were released into the sea, and the sea-level (which had been much lower than at the

1 **Ice sheets** covered much of the British Isles at least four times during the Pleistocene period, also known as the Ice Age. During the last and best-known glacial period only half of Britain was covered by ice, but two ice ages earlier, much more of the country was affected – as far south as the valley of the Thames. Temperatures were much lower than today in the glacial periods, and arctic conditions prevailed even south of the ice itself. As a result, there was no human settlement at those times. But the ice ages were separated by warmer "interglacial periods", when temperatures were at least as warm as at present. Shorter mild phases, or "interstadials", occurred within the main ice ages themselves. In these warmer periods, men crossed the land-bridge linking Britain with the European continent and settled, only to retreat to their former homes when the cold returned.

Limits of
Devensian glaciation
c. 20,000 years ago
Wolstonian glaciation
c. 200,000 years ago
Anglian glaciation
c. 500,000 years ago

Star Carr
Creswell Crags
Cheddar Gorge   Swanscombe
Stonehenge   Woodhenge
Kents Cavern

0   100km

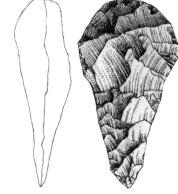

2 **Hand-axes made of flint** are among the earliest tools found in Britain. They could be used for cutting, scraping or hammering. Such axes occur throughout Europe, Africa and southern Asia.

3 **Kent's Cavern, Devon,** and similar caves elsewhere, provided shelter from the elements for the people who settled in Britain during the interstadial periods of the most recent ice age.

4 **A carving of a male figure,** scratched on an animal's rib bone, is probably about 12,000 years old, and was found in Pin Hole Cave, Derbyshire. The people who inhabited Britain during the last ice age did not decorate their caves with fine many-coloured paintings of animals as their contemporaries at Lascaux, France, or Altamira, Spain, did. Conditions of life at the extremities of Europe were probably too harsh to support such a high level of cultural activity.

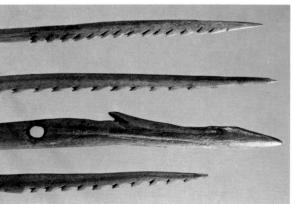

5 **Barbed points of antler** or other bone that might have been used for either hunting or fishing were numerous at Star Carr in eastern Yorkshire, one of the most fully explored Mesolithic sites in Britain. A lakeside platform of felled birch branches supported a small community there which lived by hunting the smaller animals such as deer in about 7500 BC. At that time, after the final retreat of the ice, open birch woodland began to give way to pine.

present during the glacial periods) rose. The last land-bridge connecting Britain to continental Europe was submerged c. 6000 BC. The Mesolithic people – as the hunting groups of the post-glacial era are known – probably used boats, both along the coasts and across stretches of water. Both Scotland and Ireland were settled for the first time during the Mesolithic period.

The introduction of agriculture

Some time before 4000 BC a new wave of settlers arrived in Britain from the European continent. These Neolithic immigrants introduced a whole new way of life based on farming, rather than on hunting and gathering. They cultivated wheat and barley and they bred domesticated cattle, swine, sheep and goats. They introduced new techniques, including the manufacture of pottery [6] and the production of ground and polished stone implements. These farmers were more numerous than the earlier hunters and they have left far clearer traces of their activities. Not many settlement sites of this period are known, but monumental tombs [Key] sur-

vive, as well as central meeting places or fairgrounds ("causewayed camps"), strange linear earthwork enclosures ("cursus monuments") and, rather later, the sanctuary sites ("henge monuments" [9]), of which the most famous are Avebury and Stonehenge. The sites from which they obtained the raw materials used for axes are also known, such as the mines at Grimes Graves in Norfolk and Great Langdale in Cumbria.

The first farmers in Britain probably had an egalitarian social organization without marked differences in rank or wealth. But it seems likely that by the later part of the Neolithic period, in the third millennium BC, a more hierarchical society had evolved. The construction of the five largest henge monuments probably involved about one million man-hours of work, while the two largest cursus monuments and the artificial mound of Silbury Hill [8] probably required more than ten million man-hours. The organization of manpower, materials and food supply required for these vast projects could have been undertaken only by a ranked society with a chief at its head.

Monumental tombs, such as this one at West Kennet, Wiltshire, were often built by the Neolithic inhabitants of Britain. Such tombs were used for many generations, and were closed after each burial and re- opened for the next. The bones of earlier burials were heaped together to make room for the new arrival.

6 The art of making pots was one of the skills introduced to Britain by the first farmers. The early pots were hand-made and fired at fairly low temperatures in open pit fires. It is therefore usually assumed that they were made domestically. But even such simple pots were sometimes traded over considerable distances.

7 This Neolithic wooden figure shows both male and female sexual characteristics and therefore probably represents a prehistoric deity. Evidence of the religious life of the early settlers is limited in Britain, because few cult figurines of the type common in Europe and Asia. have been found. This example comes from the Somerset levels.

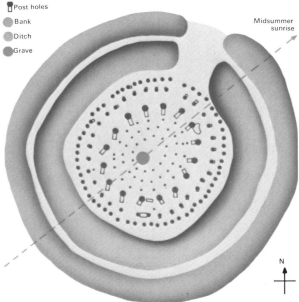

9 ▮ Post holes
● Bank
◖ Ditch
◗ Grave

Midsummer sunrise

8 Silbury Hill, Wiltshire, is close to another important Neolithic ritual site at Avebury. It is the largest man-made mound in Europe, and is conical in shape, 40m (131ft) high. It was built in three stages; the first consisted of a gravelly clay core capped by a turf stack within a ring of stakes and covered by mixed materials. This was immediately overlaid by the second stage, which consisted of a chalk mound and a surrounding ditch. In the third stage the ditch was filled in and the chalk mound ex- tended to cover a larger area. The whole process was continuous and took place about 2750 BC. Its purpose remains a mystery. Excavations have shown no sign of burials; it may have been a sanctuary with a temple on top.

9 Woodhenge is an example of a small henge (a circular Neolithic earthwork). It is 3 km (2 miles) north of Stonehenge, and is immediately south of the vast monument of Durrington Walls. Inside Woodhenge there are six concentric rings of holes that originally contained wooden posts. They are thought to be the remains of a wooden building with an open light in the centre. On the axis of the building, near the centre, is a grave of a child about three years old with a cleft skull, one of the few examples of ritual human sacrifice in Britain. The settings of the posts are oval in plan, and the long axis points in the direction of the midsummer sunrise, like the axis of Stonehenge. Woodhenge was built about 2300 BC.

N

China to 1000 BC

The Chinese have a unique place in history largely due to their natural barriers that protected their great land mass (the size of Europe) and their relative lack of contact with the outside world until the nineteenth century. Their ancient culture did not come to an end like the cultures of ancient Egypt and Greece, but has evolved unbroken from 2500 BC to the present day.

Earliest human remains

The earliest human remains so far discovered in China are those of the Peking and Lan-t'ien men, dating back well over half a million years. They were among the first men to make tools and evidence of their descendants has been found in caves in the hills near Chou-k'ou-tien, which contained scraping and cutting stones. These early people were followed by Upper Cavemen who lived about 50,000 years ago. They were able to make fire and lived by hunting, fishing and gathering fruits and edible roots. In the Neolithic age (c. 7000–1600 BC) people made needles, bows and arrows, and ground sharp edges on stones and shells. Antlers were fashioned into sickles and saws and an agricultural and pastoral society arose. The Neolithic period is also characterized by fine pottery [2, 3, 7].

The first farmers lived in beehive-shaped huts which were sunk into the ground for additional warmth and security and covered by thatched roofs supported on wooden posts. Agriculture was helped by the existence of a thick deposit of loess, a very fine soil that is thought to have been blown from the northwest. In some places in the Central Plain it is more than 60m (200ft) deep and, being exceptionally fertile, provides some of the finest agricultural land in the world. Like the Nile mud it is an excellent material for building rammed earth walls or for making bricks. The region's moderate climate supported wild horses, buffalo, deer, wild pigs, sheep and even rhinoceros.

The Central Plain lies between two great rivers – the Huang Ho [1] in the north and the Yangtze in the south. These rivers flow from west to east and carry the loess through the region. It is here that the Bronze Age culture of the Shang or Yin Dynasty was born (c. 1600–c. 1030 BC).

Fortunately the Chinese proved to be expert husbandmen and managed to maintain cereal crops on the same fields for thousands of years.

The writings of early historians

There is so far little evidence to support the earliest history of China as related by Chinese historians of the second century BC. They held that the country was first ruled by two groups of three and five emperors made up of legendary figures, who were followed by the Hsia, Shang and Chou dynasties. The existence of the Hsia Dynasty is still in doubt, but archaeological evidence has proved the existence both of the Shang Dynasty and of a central authority with a capital city, Great Shang near An-yang in north Honan. The Shang were mainly agricultural people but they are known for their mastery of bronze casting [Key]. Their bronze ritual vessels, bells and axes exhibit an unsurpassed quality of craftsmanship [5, 9].

Sacrifice played an important part in Shang culture, and people as well as animals were slaughtered to commemorate royal

1 The floodplains of the Huang Ho (seen here) and Yangtze rivers provided a natural center for the growth of Chinese civilization. The rich silt they carried and the annual irrigation encouraged settlements and the growth of agriculture. The Yangtze is the fourth longest river in the world. The Huang Ho or Yellow River derives its common name from its sludgy yellow-brown colour. It is subject to unpredictable floods caused by the melting snows; these have so often devastated the fertile plains that the river was known as "China's sorrow". It flows into the Yellow Sea today but has changed its course many times in its history.

2 This black pottery tripod vessel is known as a *li*. It represents the final and most developed phase of Neolithic pottery and comes from Lung-shan in the central Yellow River basin. The three legs are hollow and have the practical advantage of holding the pot upright on the embers and of offering a larger surface to the fire. Pots of this area are characterized by dark burnished surfaces.

3 This red pottery amphora was excavated from the Pan-p'o village site in Shensi province and is an excellent example of the red pottery ware of the Yang-shao culture. The pots were often painted with designs based on fish.

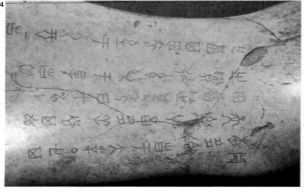

4 Oracle bones, usually made from the shoulder blades of oxen or the carapaces of tortoises, were used for divination. Questions in the form of pictures were scratched on the surface of the bone and a red-hot iron applied. The heat resulted in cracks radiating from the burn and these enabled the diviner to read the answer to the question. Pictograms were the earliest form of Chinese writing and the symbols are direct ancestors of the modern Chinese script.

5 The inscription and the human faces on this bronze ritual food vessel or *ting* (14th–11th centuries BC) allude to human sacrifice; the vessel may have been used in the rites. The sacrifice would have been an offering to an ancestor.

burials and to mark the erection of important buildings. One building excavated at An-yang disclosed guards outside the gates armed with halberds and others along the outside walls. Dogs were distributed along the walls and five chariots complete with charioteers and horses were buried in the central courtyard.

The royal tombs were even more exacting in their needs. Ramps led down to the pit [8] containing the coffin which lay in the centre of the deepest part, usually over a smaller grave containing the body of a dog. The pit contained the bodies of people, horses and dogs as well as chariots and all the furniture and household goods needed for the occupant in the next world. Some people had been beheaded and lay in groups of ten with their heads carefully laid in a separate place. Slaves were occasionally buried alive, and in one tomb 70 living people had apparently been buried with the dead.

The Shang Dynasty prospered. The new bronze tools made a variety of trade possible and exchange of foods became necessary. To facilitate trade, money was introduced in the form of cowrie shells, already one of the world's most popular currencies because of their valuable qualities of size and durability and because they were impossible to forge.

Calendars and cities

The Shang were a fairly sophisticated people and their astronomers produced an accurate calendar, based on the lunar month, which was corrected by the addition of seven extra lunar months over a period of 19 solar years.

Shang culture spread over central China and traces of it have been found in the Yangtze valley 640km (400 miles) to the south [6]. The Shang method of planning towns in squares can still be seen in Peking. The Shang regarded themselves as the centre of civilization and the name for China still remains Chung-kuo, the Middle Country. However, the rule of Shang was nearing its end. Frequent wars and the oppression of the people by the last ruler, Chou Hsin, finally drove the slaves into revolt. Chou Hsin perished in the flames that destroyed his palace and with this the Chou Dynasty began (*c.* 1030 BC).

During the Shang Dynasty a highly skilled bronze metallurgy was developed, much later than in the West although it was still the earliest bronzework in East Asia. The two main weapons used by the Shang people were the bow and the halberd. This bronze ritual halberd blade from the Shang Dynasty dates from the late 11th century BC.

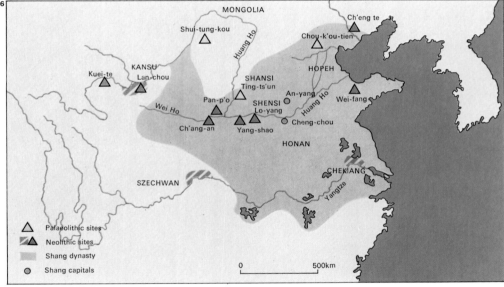

7 The Liang-chu Neolithic culture flourished within a relatively confined area in the region of Shanghai in the northern Chekiang province. It is characterized by a coarse brick-red or sometimes grey pottery, such as this pottery kettle *c.* 4000 BC, which frequently adopted curious shapes and forms of a type more usually associated with vessels made from bronze.

6 Neolithic China was centred on the fertile northern plains around the Huang Ho River. The first dynasty, the Shang, ruled first from Cheng-chou and later moved north to An-yang.

8 Two ramps lead down to the burial chamber (*c.* 1100 BC), in the centre of which the coffin is placed over a depression containing the body of a dog. Bodies of funeral sacrifice victims were carefully positioned on the ledge surrounding the coffin. Nothing was left to chance: furniture, clothes, household equipment, food and drink — everything that was needed on earth for the comfort of the deceased was buried with him.

9 Shang Dynasty bronzes were mostly discovered at An-yang, the Shang capital; their uses were ceremonial and funerary. Of the food vessels, *chiu* [A] cauldrons had flat bases; the *ting* [B] had three or four legs and handles. Wine beakers of the *chueh* type [C] had tripod bases, handles and spouts; the *ku* [D] was a deep cup with a flaring rim. Of the wine and water jugs, the *hsi-tsun* or *tsun* [E] had animal or bird forms; the *kuang* [F] had a lid in the back of the animal. The *hu* [G], *yu* [H] and *chih* [I] were wine or water jugs. Raised and depressed decoration with animal motifs, *t'ao-t'ieh*, was at first representative and later geometrically stylized.

Preclassic America to AD 300

The period between 2200 BC and AD 300 saw the rise of the first civilizations in both Mexico, where it is known as the Formative or Preclassic, and Peru. By 1500 BC agricultural villages in both Mesoamerica and South America were developing with craft specialization in such fields as ceramics; society was also moving towards stratification, expressed in the first public architecture. The first truly complex society, the Olmec culture, based in the tropical lowlands of the Mexican Gulf coast, had appeared by 1200 BC. The first great ceremonial centre was San Lorenzo Tenochtitlán, where the whole form of a natural ridge was altered by the construction of ridges and platforms.

Spread of Olmec culture
During excavations numerous pieces of monumental sculpture in volcanic stone were found, including giant heads deliberately defaced and buried [Key]. The stone itself came from the Tuxtla Mountains many kilometres away and the presence as well as the sophistication of the sculptures indicates an organized labour force, a body of

specialist sculptors and a powerful government whose patronage extended from sculptors to lapidaries and potters as well. A further range of sculpture and a large conical pyramid were found at another Olmec centre, La Venta.

Olmec trade spread far into the highlands of Mexico, Oaxaca being a source for various minerals, and the valley of the Río Balsas beyond Mexico City perhaps being the origin of the blue jade favoured by the Olmec; another possible source is even more distant, in Costa Rica. Rock carvings and cave paintings in Olmec style are known from western Mexico, and the carvings continue eastwards into El Salvador. Whether diffusion of this artistic stimulus was commercial, religious, diplomatic or military, it has a good claim to being the first pan-Mesoamerican style. There is some evidence that the Olmec possessed a system of numerical and perhaps other notation, but whether this can be described as "writing" in its true sense is still a matter for argument.

In Peru from about 1000 BC onwards a similar diffusion of the Chavín style occurred.

The style involves both birds of prey and feline-human compounds [1] with serpent attributes, and is full of arcane allusion. The style is named after the site of Chavín de Huántar, about 3,000m (9,750ft) up and just below and east of the crest of the Andes in central Peru, where a complex of massive stone buildings with subterranean galleries stand round formal courtyards. Some of the galleries contain sculptures, such as the *Lanzón* that stands in the deepest part of the main structure and portrays a mythical being.

Chavín influence on arts and crafts
The Chavín style is also found in pottery, and examples of it are known from as far north as Ecuador, while in textiles Chavín influence is found in the Paracas region [4] in southern Peru. Chavín influence continued until about 200 BC, and may have just continued in a provincial form in the sculptures of San Agustín in southern Colombia. The sites at San Agustín cover a wide area on the hills around the modern town and consist of megalithic chambered monuments and feline, fanged sculptures. They are dated

CONNECTIONS

See also
230 Mesoamerica 300–1521
232 Columbia and Peru 300–1534

In other volumes
260 Man and Society

1 **Human and feline characteristics** are combined in a sculpture of a monstrous head projecting from the wall of the Castillo, the main structure at Chavín de Huántar. The undressed but neat masonry contrasts with the sophistication of the carvings. The site lies in a high valley in the upper Amazon basin.

2 **Figures of *danzantes*** – dancers – at the Zapotec centre of Monte Albán, Oaxaca, Mexico, date from about 500 BC and resemble Olmec art of the same period and earlier. The abandoned poses, closed eyes and various details on the bodies suggest that the figures depict slain warriors.

3 **Two sculptured stone slabs** or stelae at Monte Albán bear short inscriptions in a form of hieroglyphic writing, including long bars and round discs that probably denote numbers, a bar equalling 5 and a disc 1. These stelae, dating from about 500 BC, are the earliest dated writing in the ancient Americas.

4 **Part of a richly decorated textile** from the desert cemeteries of the Paracas peninsula, on the coast of south-central Peru, shows a complex design that is repeated similarly on each of the many panels. These textiles have been preserved by the dry climate of the Peruvian coast, and provide an idea of what may have been lost elsewhere.

from AD 1–500, although more radiocarbon dates might vary this.

Elaborate gold work of the Chavín period is found in the Lambayeque valley and other regions of northern Peru. It initiated a tradition of working precious metals that continued until the Spanish conquest. In the Viru valley, south of Trujillo on the Peruvian coast, major centralized settlements developed during the early centuries AD, apparently in response to population pressure and competition for resources. Pucará in southern Peru was an important cultural centre, quite unrelated to Chavín.

From about 500 BC the rise of Zapotec culture [2] in the Valley of Oaxaca in southern Mexico accelerated. The first monumental inscriptions in hieroglyphic script from the New World [3] are found at the great hilltop centre of Monte Albán, together with large stone buildings and low-relief carvings of humans. Another structure, Mound J, could well have been an astronomical observatory. By the end of the Preclassic period Monte Albán already possessed the spacious formal layout, the huge open plazas and the massive platforms that mark its further development in the Classic.

The Valley of Mexico, after a period of Olmec influence, saw the rise of a number of independent political units round the margins of its broad, shallow lakes. Among these was Cuicuilco, which was destroyed about the time of Christ by an eruption [5].

Teotihuacán, a rival state in the northeast of the valley, grew in the next century into a centralized, planned urban settlement [6] which by AD 150 covered more than 20 sq km (7.7 sq miles), with a growing population. Teotihuacán trade extended a long way eastwards into the Maya area by about AD 400.

The beginning of the Maya culture

The Maya are the reason for the Classic period formally beginning at AD 300, because that is roughly when they started to erect monuments bearing inscriptions [8] in the Long Count, a complex calendar that enables Maya monuments [7] to be dated to within a day. It now seems that Maya culture itself acquired most of the attributes of civilization during the late Preclassic period.

This giant head of the Olmec period. found at San Lorenzo Tenochtitlán near the Gulf Coast of Mexico, dates from about 1200–900 BC. Several such heads were found at the site, and others are known from La Venta and a number of smaller sites including the early Laguna de los Cerros. They are thought to depict the Olmec rulers and are made of volcanic rock brought at least 100 km (60 miles) from the Tuxtla Mountains, giving strong evidence of the social control exerted. The Olmec culture is the first complex society in Mesoamerica to attain a level that can be described as "civilization" and it stimulated later developments.

5 The circular and stepped pyramid at Cuicuilco in the Valley of Mexico on the edge of present Mexico City was the main structure of a city that flourished at about the time of Christ. It was overwhelmed by an eruption of the volcano Xitle and buried under lava, but has recently been excavated and restored.

6 Teotihuacán, a great city in the northeastern basin of Mexico, began its dramatic rise about 100 BC and from AD 100–700 had a population approaching 200,000. The long Avenue of the Dead runs south from the Pyramid of the Moon to the citadel and market compound in the centre of the city.

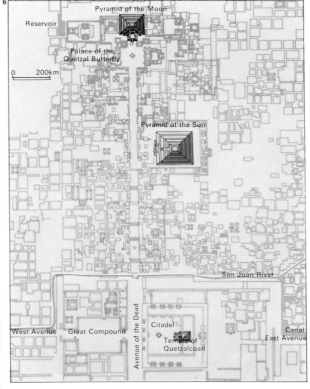

Pyramid of the Moon
Reservoir
Palace of the Quetzal Butterfly
0   200km
Pyramid of the Sun
San Juan River
West Avenue   Great Compound   Citadel   Canal East Avenue
Avenue of the Dead
Temple of Quetzalcoatl

7 Reconstruction of Pyramid E–VII–Sub shows a Preclassic temple of about AD 200–300 at the Maya site of Uaxactún, Guatemala, which archaeologists found preserved beneath a later structure.

8 Mayan writing has more than 800 hieroglyphs, few of which have been translated although the Mayan calendar is understood. This is part of an inscription on one of the stelae at Quiriguá, Guatemala.

Egypt 1570 BC to Alexander the Great

The New Kingdom in Egypt (c. 1570–1085 BC) dates from the victory of Theban forces over the Hyksos rulers of Egypt. New Kingdom Egypt was an empire extending from the northern Sudan to Syria and was one of the major powers of the ancient world. The god of Thebes, Amun, was elevated to the rank of principal deity of the realm and was identified with the sun-god Re in the form Amun-Re, king of the gods [1]. Much of the wealth that flowed into Egypt as a result of its conquests was directed towards the service of Amun and his priesthood.

The consolidation of Egyptian power
The early kings of Dynasty XVIII were primarily engaged in rendering Egypt safe from any further incursions by the Bedouin in the east or the Nubians in the south. After stabilizing the eastern frontier the Egyptian rulers undertook a series of campaigns to conquer the kingdom of Nubia in the south and to seize its gold mines. The lack of male heirs to the throne from the marriage of the ruler to his full sister led to the increasing importance of the royal heiresses. One such

was Hatshepsut (r.c. 1503– c. 1482 BC), wife and half-sister of Thutmose II, who bore her husband no sons. On his death the child of a concubine, Thutmose III (reigned c. 1504–1450 BC), was placed upon the throne and was presumably destined to marry his half-sister. However, his stepmother Hatshepsut later seized the throne for herself and ruled together with Thutmose III, who was allowed no real power.

On Hatshepsut's death, Thutmose III assumed effective control of the government and immediately embarked on a series of campaigns to subjugate the petty kingdoms in Palestine and most of Syria. He also completed the conquest of Nubia as far as Napata and his immediate successors continued his expansionist policies.

Under Amenhotep III (reigned 1417–1379 BC) the Egyptian court reached the height of its prestige, receiving tribute or trade goods from Syria, Mesopotamia, Anatolia, Crete and even Greece. His son Amenhotep IV or Akhenaton (reigned c. 1379–1362 BC) changed the Egyptian religion by his worship of the sun-god in the

form of Aton, the sun'[s] [...] capital to the new [...] (Amarna). Akhenaton [...] the opposition of the [...] gods and led to dome[stic] weakening of Egypt's [...] successors abandoned [...] was finally restored [...] Horemheb, the last [...] Upon Horemheb's de[ath] to his vizier Ram[eses I, 1320–1318 BC), who [...] (c. 1320–1200 BC).

Confrontation with the [...]
The aim of the rulers [...] restore Egypt's powe[r ...] and to confirm Egypt'[s ...] Palestine in the face [...] power of the Hittite E[mpire ...] to a major clash in Y[...] (reigned 1304–1237 B[C ...] Kadesh, where both t[he Hit]tites claimed victory. [...] concluded between t[he ...] Year 21 and sealed in [...]

1 The ancient cult of the sun-god Re had been eclipsed by the rise of the god Amun, until Akhenaton tried to suppress the worship of all gods except Re. The sun-god was represented in the form of the Aton or sun disc, adored here by his queen Nefertiti.

2 A

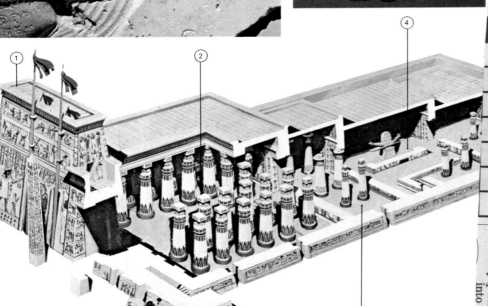

riage of Ramesses II with a Hittite princess. His successors were faced with increasing threats from Libyan tribesmen and piratical "Sea-Peoples", but these threats were contained by Ramesses III, (reigned *c.* 1198–1166 BC) of Dynasty XX [6].

Under Ramesses III's successors, who were also all named Ramesses, the power of the crown steadily declined in the face of increased Libyan incursions and the growth in the power of local governors, especially that of the high priest of Amun at Thebes. Egypt's foreign possessions in Palestine and Nubia were lost by the end of the dynasty and Egypt itself fell under the control of foreign rulers. Under Dynasty XXI, which ruled from the northern city of Tanis, Thebes was virtually independent under its high priests. The unity of Egypt was restored by the Libyan general Shoshenk I (reigned *c.* 935–914 BC), founder of Dynasty XXII. He installed his own son as high priest of Amun and attempted to restore Egypt's position as a great power by embarking on a major campaign in Palestine. Under his successors the unity of the country was broken by civil wars

and Egypt was partitioned into city states under independent dynasts, mostly of Libyan origin. An independent kingdom had emerged in Nubia in the south and the Nubian kings conquered Egypt in about 712 BC and founded Dynasty XXV.

Nubian rule was terminated by a series of Assyrian invasions that led to the nomination of the prince of Sais as puppet ruler of Egypt. With the help of Greek mercenaries, Psamtik I (reigned *c.* 664–610 BC) of Dynasty XXVI managed to impose his authority on the whole country and broke with Assyria.

Defeat by Babylonia and Persia
The rule of Dynasty XXVI (670–525 BC) marked a period of renewed prosperity but Egypt's hopes of restoring her position as a great power were defeated by the Babylonians at the Battle of Carchemish in 605 BC. Ultimately in 525 BC Egypt was absorbed by the Persians and although subsequent revolts re-established Egyptian independence briefly between 404 and 343 BC, the Persians reasserted their domination until the arrival of Alexander the Great in 332 BC.

Ramesses II ensured that his name would be remembered by his extensive building projects. He added the hypostyle hall to the temple of Amun at Karnak and made additions, including this colossus of himself, to the temple of Luxor. On the west bank at Thebes he built his mortuary temple – the Ramesseum. For his favourite queen, Nofretari, a tomb decorated with superb paintings was constructed in the Valley of the Queens. In Nubia he erected the great temple of Abu Simbel, as well as other temples elsewhere in Egypt. He also constructed the city of Pi-Ramesse, his northern capital, possibly using the labour of Hebrew slaves as mentioned in the Bible.

4 The brilliantly painted tombs of the New Kingdom at Thebes reflect the life led by Egyptians of all classes. It was believed that these scenes could be magically brought to life so that the dead man would not be bereft of his possessions and pleasures on entering the next world.

5 The economy of Egypt was agrarian, most of the people working on the land. In theory all land was held by the crown but in practice large estates were also held by the official classes and the temples. A limited number of peasant proprietors also owned tracts of agricultural land.

6 The invasion of the "Sea-Peoples" including the Philistines and maybe the ances-

tors of the Sicilians, was repulsed by Ramesses III at the end of the New Kingdom.

7 Most tombs in the New Kingdom include a *Book of the Dead* containing

spells intended to guarantee the safe passage of the deceased to the afterlife.

Africa: Kush and Axum

During the course of ancient Egyptian history, the armies of the pharaohs pushed the frontier of their empire ever farther south, along the axis and heart of their civilization, the Nile, towards tropical black Africa. By the time of the New Kingdom (*c.* 1500 BC) all the riverine lands as far as the Fourth Cataract – that is, in the middle of the great S bend of the Nile – had been conquered and to some extent settled by Egyptians. This country, later called Nubia, was known to the Egyptians as Kush and in time a typical late Egyptian civilization flourished there.

The emergence of the Kushite kingdom
By about 1000 BC the New Kingdom had fallen and Kush emerged as an independent state – independent not only politically from Egypt, but increasingly also culturally. In 200 years its rulers had grown so independent and powerful that in 725 BC they were able to march down the Nile and conquer the whole of Egypt, where they formed Dynasty XXV of the Pharaohs [2].

Kushite control of Egypt, however, was short-lived. Between 676 and 663 BC Assyrian armies invaded and devastated Egypt – first under Esarhaddon and later under Ashurbanipal. The Kushite pharaoh Taharqa retreated southwards to Kush. What had made the mighty Assyrian armies almost invincible over much of the Middle East was their possession of iron weapons, which were much superior to the bronze weapons of their foes. The leaders of Kush had learned a hard lesson and took with them the Assyrian knowledge of iron technology. This was to be the basis for the stability of Kush.

After the withdrawal from Egypt, the rulers of Kush expanded southwards, keeping to the valley of the Nile. The country on either side of the great river was more fertile than it became subsequently and could support large herds of cattle. By the sixth century BC the frontier of Kush had reached just to the south of present-day Khartoum, where the land was well wooded.

Dominance of the Meroë civilization
The power centre of the empire swung to the south, from the old capital of Napata (near the Fourth Cataract, where the surrounding land had become over-grazed) to Meroë, south of the Atbarah's confluence with the Nile. From this time, the empire is often referred to as Meroë, rather than Kush. Whereas some of the inhabitants of the northern part of Kush were black people, now nearly all the people in the country around Meroë were black and the empire became a black state.

Meroë had abundant iron-ore and wood with which to smelt it. The iron industry of the empire was on a large scale. Immense slag heaps still litter the landscape. The well-armed horsemen of the Meroë army were able to defend the settled lands from attacks of desert nomads. A flourishing trade was maintained with Ptolemaic Egypt, Arabia and even India, via the Red Sea.

By the beginning of the Christian era, however, the civilization of Kush/Meroë had begun to decline. This was the result of internal impoverishment, especially the drying up of once rich grazing and agricultural lands. Attacks by nomads became more difficult to contain, and the empire finally collapsed when invaded by a powerful army

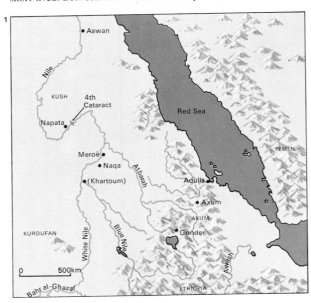

1 Civilizations of Kush/Meroë and Axum occupied the northeast corner of Africa: Kush/Meroë in the middle Nile valley south of Egypt and Axum on the high mountain escarpment of northern Ethiopia. Axum was the founding state of the Christian empire of Ethiopia.

2 Narwa was governor of Thebes during the reign of one of the Kushite line of pharaohs in the 8th and 7th centuries BC. The Kushites had moved north and ruled Egypt until they were defeated and driven back to their old lands by a new invader – the "iron armies" of the Assyrians.

3 The Lion God of Kush was engraved on a temple wall at Naqa (100 BC–AD 100), a centre south of Meroë. The carvings show an Indian influence.

4 An elephant and war captives form a temple frieze near Meroë, first century AD. Elephants were used for military and ceremonial purposes at Kush/Meroë.

from neighbouring Axum in AD 350.

The empire of Kush/Meroë had been in existence for over 1,000 years and its cultural achievements were rich and vital [3, 4]. They represented far more than an Africanized form of Egyptian culture. The inhabitants of the empire took their Egyptian heritage, borrowed more from the Hellenistic world and India, and fashioned something unique, including their own form of writing, a cursive script that has to date defied all efforts to decipher it [Key].

Axum, the rival empire in Ethiopia

Axum, the rival empire of Meroë, had its origin not on the African continent, but in Arabia. A number of small but prosperous states grew up in the Yemen early in the first millennium BC, one of which, Saba, was probably the Sheba of King Solomon's time. By about the seventh century BC Semitic-speaking people from the overpopulated Yemen spilt over the Red Sea to the Horn of Africa, settling as farmers on the north-eastern edge of the high plateau of Ethiopia. There they prospered and were able to domi-

nate the indigenous Kushistic-speaking peoples, many of whom gradually accepted the culture and the Semitic language of the newcomers. One group of these, called the Habashat, established a kingdom in the third century BC, centred on Axum [6, 7].

The Hellenistic Greeks were influential at the court of the kings of Axum and prepared the way for the reception of Christianity. The missionary responsible for the conversion of the kingdom in the fourth century AD was a Syrian called Frumentius (c. 300–c. 380). He was made bishop of Axum by the patriarch of Alexandria and within a few decades Christianity had been established there. The rulers of Christian Axum conquered parts of southern Arabia in the sixth century AD, before being driven out by Persian forces; from there on Axum went into gradual decline: the Muslim Arab conquest of Egypt disrupted her Red Sea trade and when the kingdom was devastated by nomads her power withered. Nevertheless, its political, religious and cultural traditions survived to be revived in the Middle Ages deep in the heartlands of Ethiopia.

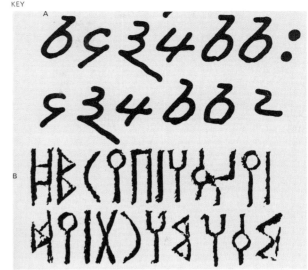

The mysterious **Meroitic script** [A] is still unde-ciphered. The Sabaean script [B] of Axum is the fore-runner of modern Amharic, the language of Ethiopia.

5 Christian art from vanished Christian kingdoms survived in a church at Faras, near Wadi Halfa, which has now been engulfed by the waters of the Aswan dam. Christian kingdoms succeeded Kush/Meroë after the empire collapsed in the fifth century AD shortly before Egypt became part of the realm of Islam.

6 Rich in products that drew Greek, Arabian and Indian merchants alike, Axum was a land of soaring crags, the highest (Rāsdajan) at 4,620m (15,158ft). One of the most difficult African lands to live in or invade, it was to enter European mythology as the home of the legendary Christian king, Prester John.

7 The tallest surviving stele, or obelisk, at Axum, is one of the many splendid monuments erected by the rulers of the kingdom of Axum as their wealth grew from trade passing through to Arabia and India.

The Hittites 1700–1200 BC

A hundred years ago knowledge of the Hittites was derived only from mentions in the Bible and in Mesopotamian records, but then, in the nineteenth century, discoveries began in Turkey of monuments and massive fortifications which had obviously survived from a flourishing and powerful civilization. This civilization was quickly identified as Hittite. What finally brought these peoples into vivid life was the discovery at Boğazköy, the site of the Hittite capital Hattusas, of more than 10,000 cuneiform tablets. The tablets were written in several languages, including Akkadian, Hattian, Hurrian, Luwian and Sumerian as well as Hittite, and were found to be the state archives.

Origins of Hittite civilization

Little is known about the earliest period of the Hittite civilization before contact with Mesopotamia. The Hittites came into Asia Minor before 2000 BC from Europe or southern Russia. In the middle of the seventeenth century BC King Hattusilas I seems to have united a number of small states and established his administrative centre at Hat-

tusas. Hittite power evidently grew rapidly, for in the early sixteenth century BC Mursilis I was able not only to occupy Syria but to march on Babylon, sacking the city and bringing to an end the dynasty of which Hammurabi was the greatest figure. The last king of the so-called Old Kingdom was Telepinus (reigned c. 1525–1500 BC), who is best known for his edict laying down rules of conduct for the nobles and the king.

After a period of which there are few records, the Hittite Empire began about 1460 BC. The first king of the new dynasty was apparently Tudhaliyas II, who is said to have captured Aleppo. Mitanni, a state founded by the Hurrians in northern Mesopotamia, seems to have kept further Hittite expansion in check.

The situation changed radically when Suppiluliumas seized the throne in about 1380. He fortified Hattusas and in a brilliant campaign conquered Mitanni and its Syrian satellites. Carchemish, which commanded an important crossing point on the Euphrates, remained independent, but Suppululiumas then led a second campaign, captured this

fortress and occupied all of Syria between the Euphrates and the ocean.

Suppululiumas died in about 1346 BC; his son Mursilis II proved equally capable, defeating the kingdom of Arzawa to the west and suppressing a revolt in Syria. The next king, Muwattalis, inherited a secure and prosperous empire. But Egypt was determined under Ramesses II (reigned 1304–1237 BC) to regain the possessions and influence she had lost in Syria. The two armies met at Kadesh [7] on the River Orontes. Egypt claimed an overwhelming victory, but the Hittites seem, at least, to have held their own and maintained their ascendancy in Syria. Relations between the two powers improved as Assyria became more of a threat, and in about 1284 BC they signed a treaty of friendship and non-aggression.

The end of the Hittite Empire

The records of the Hittite Empire end abruptly in the late thirteenth century BC, when Indo-European invaders known as the "Sea-Peoples" poured into the area from Europe. The Hittites fled southwards as the

1 The impressive sanctuary of Yazili-kaya is open to the air and is cut from the rock outside the walls of Hattusas, the Hittite capital. Spectacular reliefs are carved on the sides of two

chambers, which are approached through an elaborate gateway. They depict processions of gods, goddesses and kings, many of the deities being shown with their cult animals, weapons and symbols.

Some of the gods are the patrons of Hittite cities and all are named in hieroglyphic script. The principal scene shows the weather god Teshub standing on deified mountains; facing him are his

consort and his son and beneath him are the gods of the Hurrian pantheon. During the later years of the empire, Hittite religion came under strong Hurrian influence, as the carvings here illustrate.

2 The rock carving at Ivriz in the Taurus Mountains exemplifies the Aramaean style in Hittite sculpture, as is shown in particular by the god's cap and by the ringlets and profiles of both figures. But while the king wears an Assyrian cloak of a kind familiar from the royal statue from Malatya, the god's clothing has little that cannot be paralleled in purely Hittite sculptures, including the clear curved seam above the knees. The combination of relief and script facilitated the grouping (in 1880) of similar monuments reported from elsewhere in Asia Minor, notably at Boğazköy and Yazili-kaya, as Hittite. The inscription names the king as Warpala-was (about 730 BC), so dating the relief.

3 Among the chief deities in the Anatolian pantheon were the weather god, Teshub [A] and the sun-goddess of Arinna [B]. Teshub is shown in a ninth-century statue in his customary pose, wielding a flash of lightning; for unlike Mesopotamia, Anatolia is a land of cloud and storms. To the sun-goddess, shown in a pendant of the empire period, the king appealed for help in times of war and danger; she was supreme patroness of the Hittite state.

4 Most surviving Hittite sculpture takes the form of monumental bas-reliefs in stone, often with inscriptions in hieroglyphic (as distinct from cuneiform) Hittite. The almost total absence of sculpture in the round is partly compensated for by a few beautiful miniature figures like this gold statuette (4.2cm [1.7in] high) of a man, perhaps a king, wearing a full tunic with short sleeves. It was found at Yozgat, near Boğazköy, and dates from about the fourteenth century BC.

Phrygians overran Asia Minor. But although the homeland was lost, the Hittite way of life survived as the refugees established city states in northern Syria, which had so long been under Hittite control. Most of the archaeological evidence comes from this neo-Hittite revival rather than from the early imperial period. The writing of these petty kingdoms is hieroglyphic Hittite, but the language was essentially Luwian, so that we may assume that the Hittites did not necessarily form the majority of the population.

Structure of Hittite society

The neo-Hittite city states, which were often isolated geographically, were unable to unite effectively against the growing power of Assyria. After exacting tribute over many years and putting down occasional rebellions, Assyria, under Tiglath-Pileser III (reigned 744–727 BC), decided to incorporate the Syrian kingdoms into its empire, and by the end of the eighth century BC the Hittite Empire was little more than a memory.

During the period up to the end of the thirteenth century BC the Hittite king,

although perhaps not quite such a dominant figure as in Mesopotamia, made the final decision on military, religious and judicial matters [4]. On his death he joined the gods, and indeed "he became a god" was a euphemism for the death of a king. The queen was also important, having a role to play in state affairs. Nepotism was institutionalized, most of the highest offices falling to the king's relations. There seems to have been an exclusive caste of privileged nobility and landowners. The common people worked as farmers, craftsmen and labourers, and those of the servant class, while having some legal rights, were little better than slaves.

The brutality that so disfigures the history of Assyria was quite lacking among the Hittites. A city captured in war suffered grave retribution in that it was generally destroyed and its inhabitants enslaved, but we know of no mass killings or systematic torture such as Assyria's enemies could expect. Similarly, punishment for crime was generally based on restitution to the injured party or his relatives, even in the case of murder.

A double-headed eagle is carved on the back of a sphinx at Alaja Hüyük in Turkey, a city of the empire period. It is clutching in its talons two hares with their faces turned outwards like the eagle's heads. A figure, now badly damaged but perhaps a goddess, stood on the eagle's back. The same double-headed eagle motif appears also as a base for two standing figures among the sculptures of Yazili-kaya. A double-headed bird appears elsewhere in the ancient world, for example on an early geometric-period ivory from Sparta, and on a shrine at Taxila, and survives on the banners of the Kandyan chiefs.

5 In the sphinx gate at Alaja Hüyük in central Turkey, the sphinxes – unusually for Hittite art – are sculpted partly in the round. Many reliefs were found in the ruins, depicting animals, musicians, jugglers, and a shepherd with his flock. Pre-Hittite tombs of the third millennium were also discovered there and contained silver and bronze animal figures, golden jugs, goblets, and ornaments.

6 The excavation of Karatepe, northeast of Adana, revealed a bilingual inscription, one text in hieroglyphic Hittite and the other in Phoenician, which confirmed and amplified the decoding of the Hittite.

7 Use of the light war chariot, shown in this drawing of an Egyptian relief, accounted in part for the Hittites' military successes, particularly the drawn battle at Kadesh (1299). Hittite chariots differed from those of Egypt and other nations of the period in that they carried three men – one driver and two fighters – instead of the usual one driver and one fighter. The main tactical aim of the Hittites in battle was to draw the enemy into the open where the chariots would be most effective. But their planning was not based solely on that manoeuvre. A clever ruse, for example, caught the Egyptians entirely by surprise at Kadesh, and only bad luck prevented an overwhelming victory.

8 This basalt carving of a lion from Malatya, dated probably about 1000 BC, is fairly typical of the sculpture of the neo-Hittite period, when the use of stone lions to flank entranceways was not uncommon.

The Phoenicians 1500-332 BC

The ancient land of Phoenicia covered the coastal strip of modern Israel, Lebanon and Egypt. Its prosperity depended on the waters of the Mediterranean. Phoenician trading ships sailed remarkable distances, opening up new markets and protecting existing routes by establishing trading stations and colonies. The history of Phoenicia is the history of its great cities and colonies, particularly Byblos, Tyre, Sidon, Beirut and Ugarit.

Foreign influence on Phoenicia

Phoenician cities, like the city states of Classical Greece, maintained their independence from each other. However, independence from their more powerful neighbours was less easy to sustain. In the early sixteenth century BC, Egypt exacted tribute from the cities and then brought them totally under its control. Egyptian influence [6] remained strong in Phoenicia, but the cultural and physical presence of Mesopotamia became increasingly dominant, as the early Phoenician cuneiform script and the evidence of many artefacts makes clear. After a period of Hittite control the Assyrian king Tiglath-Pileser

I (reigned c. 1115–c. 1077 BC) received tribute from the Phoenician cities. Phoenicia then regained its independence of action. A period of great prosperity followed and with it came a remarkable expansion of power throughout and even beyond the Mediterranean.

However, as Assyria's power approached its peak, the Phoenicians again became a tributary people, in thrall first to Ashurnasirpal II and his son Shalmaneser III in the ninth century, and then to Tiglath-Pileser III in the eighth century. In the following century the armies of Esarhaddon overran most of Phoenicia. Tyre held out but eventually fell to the Babylonians in the reign of Nebuchadrezzar II (604–562 BC). Later Phoenicia became part of the Persian Empire of Cyrus the Great. It remained an important sea power, but the powerful contingent that it contributed to the fleet that Xerxes (c. 519–465 BC) led against Greece shared heavily in the defeat at the Battle of Salamis in 480 BC. Alexander the Great (356–323 BC) incorporated Phoenicia into his empire after his victory at the battle of Issus in 333

BC and his capture of Tyre in 332 BC.

From the earliest days of commerce with Egypt and Mesopotamia, the Phoenician economy relied on trade, importing gold, ivory, livestock and corn, and exporting timber, metals, cloth, glass and ships. Its trading vessels often carried the goods of other peoples, taking the produce of the Asian hinterland to Egypt, Greece and Cyprus and later to North Africa, Spain and the Mediterranean islands. Flourishing Phoenician industries included dyeing, which was centred on Sidon and Tyre, metalworking [2], glass-making, pottery and carvings in ivory and bone.

Phoenicia as a colonial power

Phoenicia's colonizing era began in the twelfth century. It had already planted settlements in Cyprus but it is likely that contacts about this time with the Mycenaeans sparked the imagination and the commercial acumen of Phoenician merchants. In the Aegean, Rhodes and probably Crete had Phoenician settlements, but there is only literary evidence of a Phoenician presence on the Greek

1 The Temple of the Obelisks at Byblos (the most powerful Phoenician city in the Egyptian period) dates from the middle Bronze Age. These open-air sanctuaries were dotted with stelae or pillars erected in honour of the gods or to mark cremations or the burial places of important objects.

2 The Phoenicians were skilled at working gold, an art they learnt from the Mycenaeans and Egyptians. This gold ring dates from the sixth or fifth century BC and was found at Tharros in Sardinia. The scarab on the ring depicts Bes the Egyptian dwarf god.

3 This limestone coffin dates from the thirteenth century BC, but was reused by King Hiram of Byblos in the early tenth century. The king is shown seated on a throne and flanked by winged sphinxes; the drooping lotus in his hand indicates that he has died. Before him is a food-laden table, and a procession of servants is approaching him. Above this scene is a typically Egyptian lotus frieze. Below are four lions which are more reminiscent of Assyrian or Hittite reliefs. On the lid is one of the earliest examples of Phoenician script.

4 King Eshmunazar II of Sidon, who perhaps reigned in the sixth century BC, was buried in this black basalt coffin of wholly Egyptian style. A Phoenician inscription warns against disturbing the body, and tells how Eshmunazar extended Sidon's dominance south to Joppa.

5 The tree of life, fertility symbol of the Babylonians and Assyrians, and the lotus of Egypt are combined on this Phoenician ivory plaque from an Assyrian palace. It shows both the extent of Phoenician trade, and the resultant diverse influences on their culture.

mainland. No more concrete evidence of Phoenician colonization in Italy exists, though there were close trading ties. But in the central and western Mediterranean, the Phoenicians established a chain of colonies, which made them dominant in the area.

Phoenicia also founded several colonies on the North African coast [9] at an early period, including Utica, Hadrumetum (modern Sousse) and Leptis Magna. But the most distinguished of all the settlements, eventually surpassing its mother city, Tyre, in power, was Carthage [7], which dominated – although it did not rule – the Phoenician colonies in the west. The most important of these were Gades (modern Cadiz), Ebesus (Ibiza), which Carthage founded in the middle of the seventh century BC, and Carthago Nova (Cartagena), founded by Hannibal's son-in-law Hasdrubal in 228 between the first and second of the Punic Wars (which ended with the destruction of Carthage by Rome).

A policy of colonization is often accompanied by an impulse towards exploration and the Phoenicians appear to have been enthusiastic explorers. The Greek historian Herodotus reports a story that Phoenician ships circumnavigated Africa, returning in three years to Egypt. Herodotus himself may not have credited this feat, but the details he gives make it possible that the remarkable voyage did take place. According to a much later Roman geographical work, the Carthaginian Himilco sailed from Spain up the Gallic coast to Brittany, and there is a possibility, although no archaeological evidence, that he reached Cornwall in an attempt to gain a share of the tin trade.

The legacy of a rich literature
Much light has been shed on the mythology and beliefs of the area in pre-Iron Age times by the discovery and decipherment of the cuneiform texts of Ugarit. Although few inscriptions or documents have survived of the once rich Phoenician literature, these relics are important because the Phoenician alphabet was the basis of the Greek alphabet. The Phoenicians spoke a Semitic language which they wrote in an alphabet of 22 consonants with no vowels.

This Phoenician ivory carving may be Ishtar, the Mesopotamian fertility goddess. Her Phoenician equivalent Anat (Astarte) was a chief helper of Baal, the Canaanite god.

6 Egyptian influence is a feature of almost all the decorated metal bowls found on Phoenician sites. Many, like this bowl (seventh century BC) from Amathus in Cyprus, also have Assyrian characteristics. The outer frieze depicts a Phoenician city besieged by Assyrians and Greeks (as in the detail here). The inner frieze includes Egyptian deities and also Phoenicians with Egyptian amulets.

7 The precinct of Tanit at Salammbo was used throughout Carthage's history. Tanit was another name for Anat or Astarte, the chief female deity of the Phoenicians. Many stelae with symbols of Tanit and dedicatory inscriptions have been discovered in the precinct, but the most sensational finds were thousands of urns containing the ashes of cremated babies – clear evidence of infant sacrifice in Carthage.

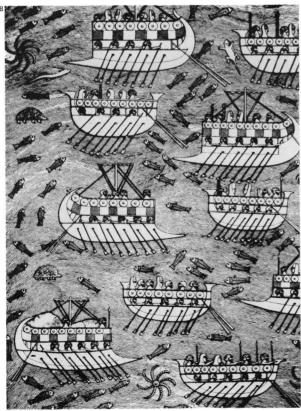

8 Luli, the king of Tyre and Sidon, fled to Cyprus in 701 BC. During his reign he formed alliances with Egypt and Judah, and resisted, as far as he could, continual Assyrian aggression.

Eventually, however, Sennacherib (r. 704–681 BC), King of Assyria, forced the departure depicted on this detail of a relief (which is now lost) from Nineveh. The relief shows two kinds of ship. The "long" ships, forerunners of the trireme, were used for war and exploration, and had a double bank of oars, with sails, a ram at the prow, and a high stern. The "round" ships, used for trading, also had two banks of oars but were sailless, with stern and prow of equal height. Both types of ship had steering oars on each side of the stern. The upper decks were hung with shields.

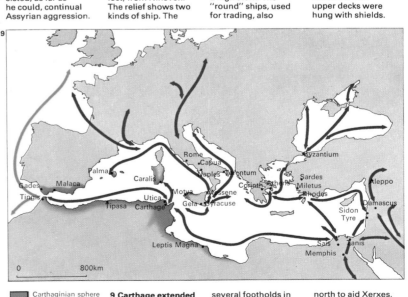

Carthaginian sphere of influence c. 323 BC
Mediterranean trade routes c. 375 BC
5th-century Phoenician exploration

9 Carthage extended Phoenician influence in Spain, Sardinia and North Africa. The Phoenicians had several footholds in western Sicily, and in 480 BC the Carthaginians made an abortive invasion of the north to aid Xerxes. At the end of the fourth century they waged an unsuccessful war against Syracuse.

55

The Assyrians 1530–612 BC

The Assyrians, a Semitic people in north-western Mesopotamia, enjoyed a material culture in the mid-third millennium comparable to that of the Sumerians and briefly emerged as a military force under Shamshi-Adad I early in the eighteenth century BC. But for the next four centuries the strongest kingdoms in Mesopotamia were those of the Kassites, who came from the Zagros Mountains to take over Babylonia in the south, and of the Hurrians, who established the state of Mitanni in northwestern Mesopotamia. In the 14th century BC Ashur-uballit led the revived Assyrians into Mitanni after that state had been shattered by Hittite invaders.

The empire established

Adadnirari I (reigned 1307–1275 BC) pushed Assyria's frontiers up to the Euphrates, and Shalmaneser I (reigned 1274–1245 BC) advanced northwards and finally crushed the Hurrians. The first of the really great Assyrian monarchs was Tiglath-Pileser I (reigned c. 1115– c. 1077 BC), who came to the throne after a period of instability in western Asia. By a succession of military campaigns, which combined brilliance and brutality, he extended Assyria's authority far into the north and northwest, overrunning Syria and exacting tribute even from the rich trading cities of Phoenicia.

Tiglath-Pileser succeeded, though with difficulty, in keeping back incursions by Aramaean peoples from the western desert. But after his death Aramaeans and Chaldaeans overran almost the whole of Mesopotamia, where a kind of Dark Age lasted for more than 150 years. When Adadnirari II came to the Assyrian throne in 912 BC his kingdom was a strip little more than 100 miles long and 50 miles wide. Yet by the middle of the seventh century BC Assyria was the largest and most powerful state in the civilized world.

Ashurnasirpal II (reigned 883–859 BC) [5] was mainly responsible for the restoration of Assyrian dominion. He built up the army into an irresistible fighting unit and led it to the shores of the Mediterranean. At his new capital at Calah (Nimrud) he built a vast palace, the doorway flanked by winged bulls. The glory of Assyria was paid for by the suffering of its foes, many inscriptions testifying to the tortures inflicted on soldiers and civilians alike. Ashurnasirpal's son Shalmaneser III (reigned 858–824 BC), his father's equal in brutality, continued to conduct annual campaigns [2], but with less success.

The surge of imperialism

A series of weak rulers and the growing power of the kingdom of Urartu to the north made Assyria's position a precarious one. The situation demanded a decisive and intelligent personality to meet the threat, and the Assyrians were fortunate to find such a one in Tiglath-Pileser III (reigned 745–727 BC). He reasserted Assyrian authority and established a uniform administration. He transformed conquered lands into provinces, each under its own governor and paying a fixed tribute to the central authority in addition to various taxes and duties. Although the governors had considerable authority, they were supervised by the central government, to which they sent regular reports through a remarkably effective system of posting stages. Tiglath-Pileser initiated a policy of

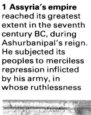

1 **Assyria's empire** reached its greatest extent in the seventh century BC, during Ashurbanipal's reign. He subjected its peoples to merciless repression inflicted by his army, in whose ruthlessness he gloried, and ruled through an efficient administrative system supervised by the central government. Assyrian hegemony collapsed, however, and was followed by a brief resurgence of Babylonian rule.

2 **Jehu, the King of Israel,** is shown bowing to Shalmaneser III (reigned 858–824 BC) on this panel from the gates of Balwat. The panels record many of Shalmaneser's campaigns – against Babylon, northern Mesopotamia, and Syria, which after several wars he failed to subdue entirely. But he was able to force Tyre and Sidon to pay tribute to him and in 841 BC Israel, too, was forced to become a tributary.

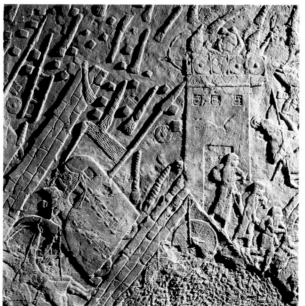

3 **Sennacherib's siege of Lachish,** a Judaean city, is portrayed on stone panels from Nineveh. The city's fall was followed by the submission of King Hezekiah of Judah. Many siege machines developed by the Assyrians are represented on stone panels such as the one shown here.

4 **A skin boat on the Tigris** carries perhaps building materials for Sennacherib's palace. Herodotus tells how circular hide boats floated down the Euphrates, carrying one or more donkeys in addition to the cargo. On arrival at Babylon, the boats were broken up and loaded on the donkeys for the return journey overland to Armenia.

mass deportations of defeated or rebellious peoples. By the end of his reign Urartu was no longer a threat and Babylonia, Palestine, Syria and Phoenicia were completely under Assyrian control.

Babylonia, however, was coming increasingly under the influence of the Chaldaean tribes, which dominated the surrounding country, and which remained bitterly hostile to the Assyrians. Sennacherib (reigned 704–681 BC) reacted with characteristic vigour, installing his son on the Babylonian throne. But in 694 BC the Babylonians revolted and invited the help of the king of Elam, who carried off the Assyrian prince to his own country. The brutal war that followed lasted five years and ended with the levelling of the holy city of Babylon by Sennacherib in 689 BC. He made his capital at Nineveh and initiated a vast programme of public works.

Sennacherib was murdered and his son Esarhaddon (reigned 681–669 BC) [Key] at once put in hand the rebuilding of Babylon. He was an able statesman who knew when to temper strength with mercy. Assyrian authority in the east remained supreme and the Mannaen buffer-state in the north was under Assyrian control. A revolt in Phoenicia was settled by deporting the inhabitants of Sidon and executing its king. But Esarhaddon's most spectacular success was against Egypt. The ruling dynasty there had been fomenting trouble in Phoenicia and Esarhaddon decided to subdue the country.

The fall of the empire

Ashurbanipal (reigned 668–c. 627 BC) quelled a rebellion in Egypt and again subdued that country, but as it was too large and too distant an area to occupy permanently [1] its administration was left to local princes. Eventually, Ashurbanipal was forced to withdraw from Egypt since he was involved in a large-scale rebellion headed by his brother, the King of Babylon, in alliance with Elam. Victorious in 648 BC, Ashurbanipal boasted of having the whole world at his feet. Although ruthless, he was also a man of learning. But Assyria could not survive the alliance of Nabopolassar, king in Babylon in 625 BC, with the Medes. Ashur soon fell and in 612 BC Nineveh was destroyed.

Esarhaddon, the great imperialist Assyrian ruler, is shown on a stele found at Zenjirli, which lies northeast of the Gulf of Iskenderun and northwest of Aleppo. The king towers over two suppliant prisoners, held by cords through their lips. The standing prisoner may be Ba'alu, King of Tyre, although in that case the stele represents Esarhaddon's wishes rather than the facts, for Ba'alu rejected his terms and the siege was probably concluded only under his successor. The kneeling, negroid, figure may represent either Tarku of Kush or his son Ushanakhuru, who was carried off with his family to Assyria; Tarku also retained control of his land.

5 Two figures, one of them (to judge from clothing and inscription) perhaps representing the ninth-century king Ashurnasirpal II of Assyria, stand on either side of a "tree of life". This motif, common in art throughout Western Asia, is a symbol of fertility conferred by the goddess Ishtar.

6 Ashurnasirpal's son Shalmaneser III built this dais for the throne in his palace at Fort Shalmaneser, Nimrud. Its vertical faces show relief carvings of tribute. The upper surface has shallow, round hollows to house the feet of throne and footstools, and is covered with inscriptions telling of his reign.

7 Painted decoration survived in parts of the palaces at Fort Shalmaneser, including the above figure dressed in a fish cloak with scale-covered legs and holding a pine cone in the fertilization gesture common in the reliefs of Ashurnasirpal.

8 Servants exercise hunting dogs in the royal park in this relief from the north palace of Ashurbanipal II at Nineveh. The Assyrians believed that the king was fulfilling a sacred duty in hunting wild animals, and lions were brought to the country so that the king could show his skill in lavish hunts. Ashurbanipal was a "mighty hunter before the lord"; and in his day the lions that infested thickets along the Euphrates destroyed not only flocks and herds but also human beings.

9 The valley of the River Zab, in western Iran, a tributary of the Tigris, is typical of the terrain covered by much of the Assyrian Empire. In much of the region mountains ensured that life and communications were centred on the river valleys.

The Hebrews 1200–322 BC

The name Hebrew is of uncertain origin and meaning. A similar name, Habiru, appears in documents of the fourteenth century BC describing races or classes of people, perhaps semi-nomads, who inhabited the northern fringes of the Arabian desert. "Hebrew" has been associated with Eber, the grandson of Shem (Genesis 24) and the forefather of all the Semitic peoples.

The twelve tribes of Israel
The early Hebrews were nomads and tradition tells of the migration of Abraham from Mesopotamia into Canaan, an area later called Palestine, on the borders of Egypt. Abraham's grandson Jacob, renamed Israel ("striver with God"), had 12 sons from whom descended the biblical 12 tribes of Israel. The Bible tells how Jacob's sons sold their brother Joseph into slavery in Egypt. When famine broke out in Canaan Joseph, who had found favour in Egypt, received his father and his brothers there and they prospered for many years. Much later, their descendants were enslaved, possibly during the reign of Seti I (reigned c. 1309–1291 BC);

their eventual emergence from captivity was organized by Moses.

Both Moses and the Exodus that he led have an historical stamp because to this day they remain in the consciousness of the Hebrew people [3]. The name Moses is Egyptian and legend tells of his youth at the royal court, so whether he was of completely Hebrew origin is debatable. Moses was told by God in a vision to deliver the children of Israel from the captivity of Egypt and lead them into the promised land of Canaan.

The date of this Exodus is uncertain, but it may have taken place in the fourteenth or thirteenth century BC. The captives escaped across the northern end of the Red Sea into the desert and went to Mount Sinai or Horeb. The Law, revealed to Moses on Sinai according to the biblical record, consisted of the Ten Commandments or Decalogue inscribed on stone and later kept in an ark or chest. The longer Law of Torah of "five books", the Pentateuch, is generally attributed to Moses.

In their present form much in these books suggests a settled agricultural community as

well as a nomadic existance. They include the death of Moses, and the later Temple and rituals were based on them. Moses was followed by Joshua and a series of judges and kings who led the invasion and gradual occupation of Canaan, the eventual goal. Saul was both prophet and king, successful in defeating some tribes such as the Amalekites, but was himself killed by the Philistines.

David, Israel's greatest king
David has always been regarded as an ideal Hebrew, second only to Moses. He was more successful than Saul in uniting the tribes and was an able administrator [1] as well as a poet – many of the Psalms are attributed to him. David captured Jerusalem from the Jebusites and made it his capital. The ark containing the Law, which had lodged in different places since the desert wanderings, was brought to Jerusalem, and when Solomon inherited this united kingdom he built the Temple in Jerusalem [2, 4, 5] for the ark and a larger palace for himself, financing both by taxation and using forced labour to construct them.

Discontent boiled over after Solomon's

CONNECTIONS

See also
52 The Hittites
1700–1200 BC
54 The Phoenicians
1500–332 BC
56 The Assyrians
1530–612 BC
60 The Persian Empire
of the Achaemenids

In other volumes
218 Man and Society

1 Ancient Israel
Phoenicia
Philistia
Assyrian Empire

1 Biblical tradition holds that the ancient Tribes of Israel escaped from slavery in Egypt in the 13th century BC under Moses. After conquering parts of Canaan in the 12th century, Israel reached its peak under David (c. 1000–960 BC), who took Jerusalem and subjugated the surrounding nations [A]. This empire divided after Solomon's death into two smaller and weaker nations, Judah in the south and Israel in the north [B], both declining in the face of Assyrian rule [C].

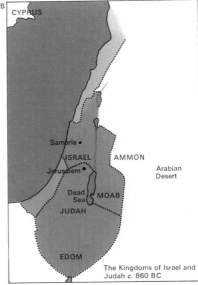

The Empire of David and Solomon c. 1000–930 BC

The Kingdoms of Israel and Judah c. 860 BC

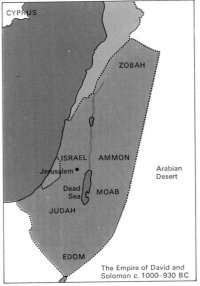

The Kingdom of Judah c. 700 BC

2 The Dome of the Rock, a Muslim shrine on the spot where Mohammed is believed to have risen to heaven, is on Mount Zion, the site of the Temple of Jerusalem. The Wailing Wall (the western wall of the second Temple) is a place of worship.

3 The Passover, the feast celebrating the deliverance of the Hebrews from captivity in Egypt, used to be introduced by a blast on a ram's horn. This woodcut is taken from a book of Jewish customs.

4 A seven-branched candlestick of the type used in the Temple of Jerusalem is shown here on the Arch of Titus in Rome, built after the destruction of Jerusalem by the Romans in AD 70.

death and the kingdom was irreparably divided. His weak son Rehoboam managed to hold only the southern country round Jerusalem, in a kingdom that came to be known as Judah (Judaea). His brother, Jeroboam, broke away with ten tribes to form a northern kingdom called Israel, with two rival shrines at Dan and Bethel. The following centuries saw the rivalries of Judah and Israel and the destruction of the northern kingdom by Sargon II of Assyria in 721 [7].

The small kingdom of Judah lingered on in semi-independence for more than a century, until finally it fell in 586 BC and most of its leading figures were taken in captivity to Babylon. The northern tribes of Israel had been scattered and lost (the ten lost tribes) and hence the return from exile after 539 BC was of the leaders of Judah.

The Hebrew prophets and monotheism

More important for world religion than these political events was the work of the Hebrew prophets. They were inspired men who were sometimes associated with, but often critical of, the official religion of their time. One of

the first, Samuel, was priest, prophet, seer and kingmaker. He chose Saul to be the first king of Israel, and also selected David as Saul's successor. Elijah and his servant and follower Elisha both denounced the prophets of Baal and as a result were hounded by the ruling monarchs of their time.

They were followed by a number of other prophets between the eighth and sixth centuries, whose messages were soon written down. Amos and Hosea preached at Bethel and in Israel, and Micah and Isaiah [6] in Judah. These men declared the unity of God and His demands of just behaviour from the people, a teaching that is termed "ethical monotheism". Through prophetic influence Deuteronomy, a "second law", was promulgated in Judah in 621 BC by King Josiah, who put down rival shrines and concentrated worship at Jerusalem. Jeremiah preached in Jerusalem before its fall in 586 BC and then went to Egypt, while Ezekiel went to Babylon. The latter denounced heathenism and planned the rebuilding of the Temple, whereas Jeremiah taught a more inward religion of a new covenant with God.

The Torah (an 18th-century copy is shown here) contains the five books of the Law given by God to Moses on Mount Sinai.

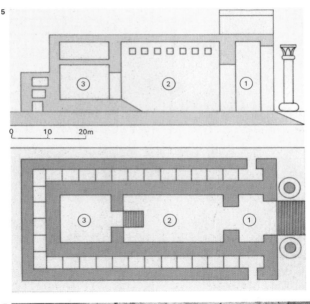

5 The first Temple of Jerusalem built by Solomon, of which nothing remains today, was a shrine for the ark, sacred vessels and offerings, with a courtyard for worshippers. It consisted essentially of a hall [1], shrine [2] and inner sanctum [3], or Holy of Holies, where only the high priest was admitted.

6 In Hebrew history a number of prophets arose who commented on society and rebuked the insincere practice of religion. Like Isaiah, depicted here by Michelangelo, the prophets taught belief in one God who was just and merciful and required similar qualities from His followers.

7 Evidence for the relations of the Hebrews with more powerful neighbours appears in a Mesopotamian relief showing the King of Israel, Jehu, paying tribute on his knees to Shalmaneser III, King of Assyria, about 840 BC. Israel was conquered by the Assyrians in 721 BC, and the southern kingdom of Judah eventually fell in 586 BC.

8 A Canaanite captive being led before Pharaoh was depicted in a temple of Ramesses III in Egypt. The Canaanites resembled the Hebrews in appearance, since both were Semites and had fine noses, long hair and beards. They were finally conquered by the Israelites about 1200 BC.

59

The Persian Empire of the Achaemenids

The Persian Empire, during the period of its height, between 550 BC and 480 BC, was the greatest in area and accomplishment the world had then known. Centred on an area comprising much of present-day Iran and Afghanistan it contained 40 million people and gave them common law, systems of coinage, postage and irrigation and a magnificent network of roads [1], as well as a liberal and unifying religion. Although the power of the Achaemenid dynasty was broken by Alexander the Great in 330 BC Persian influence revived under the Parthian and Sassanid empires and gave way to the Muslim Arab Empire only in the AD 600s.

Persia's early history

Persia is the natural bridge between Europe and Asia. Its history dates back to 6000 BC and the country contains some 250,000 archaeological sites [2], a thousand in the plain before Persepolis [5] alone. In about 1500 BC nomadic Aryans from the north arrived, giving the country the name Iran or "Land of the Aryans". In 549 BC their descendants, the Medes, were united with the

Persians in the south by Cyrus the Great who thus founded the Persian Empire, calling it the Achaemenid Empire after an ancestor.

Cyrus [Key] based his empire not merely on territorial conquest but also on international tolerance and understanding. The rights and religions of all the subject states were upheld and their laws and customs respected. After his victory in Babylon in 539 BC, which ended the Jewish captivity, he ordered the temple in Jerusalem to be rebuilt and more than 40,000 Jews left Babylonia and returned to Palestine. His army added the former realms of Assyria, Lydia and Asia Minor to the Persian Empire, making it the largest political organization of pre-Roman antiquity. The conquests of Cyrus had been carried as far as the Mediterranean in the west and the Hindu Kush in the east when he was slain in battle in 529 BC.

Cyrus was followed by his son Cambyses II (ruled 529–522 BC), who had none of his father's virtues but inherited his occasional vice of cruelty. Cambyses II began his reign by putting to death his brother Smerdis and then, lured by the wealth of Egypt, set out to

capture that country. Some 50,000 of his soldiers perished in the campaign and Cambyses unsuccessfully tried to put down the Egyptian religion. In a final outburst he killed his sister and wife Roxana, slew his son Prexaspes and buried 12 of his nobles alive. He died during his return journey to Persia.

The reign of Darius

Darius the Great (548–486 BC), who won a battle for the succession, had been the commander of the Ten Thousand Immortals, the elite of the Persian forces [7, 8]. His succession was marked by revolts among the conquered states, which he rapidly quelled. In Babylon 3,000 leading citizens were crucified. Realizing how vulnerable the vast empire was to any crisis he reduced military control in favour of wise administration and re-established his realms in a way that became a model of imperial organization. The result was a generation of order and prosperity. Having gained peace and stability at home Darius led his armies first across the Bosporus and the Danube to the Volga, then into the valley of the Indus. The Persian

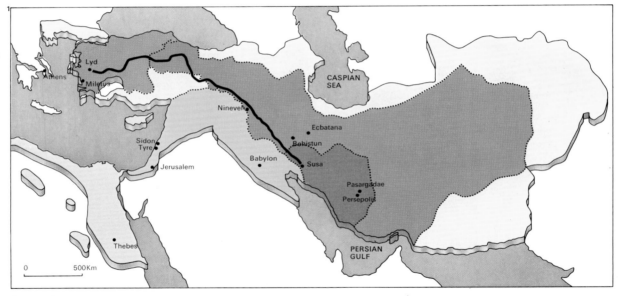

1 The Persian Empire in the Achaemenid period was administered through satrapies. To maintain contact with these provinces Darius created roads whose combined length was 2,700km (1,680 miles). At 111 staging posts fresh horses awaited the king's envoys, who could thus traverse the whole system in a week; it took merchant caravans 90 days.

- Kingdom of Persia
- Median Empire annexed 549 BC
- Lydian Empire annexed 546 BC
- Chaldean Empire annexed 538 BC
- Egyptian Empire annexed 525 BC
- Later conquests to 479 BC
- Royal Highway

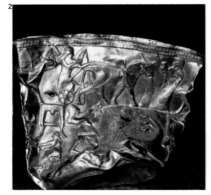

2 The crushed bowl of Hasanlu, exquisitely made in gold, shows a weather god in a chariot drawn by a bull, and a battle with a monster. It was found in 1958 during excavations of the citadel of Hasanlu at the northern end of the

Solduz valley and was clutched in the hands of a man's skeleton. He was probably trying to escape from a palace that collapsed in flames when the citadel was attacked in 800 BC. The bowl is now in the Teheran Museum Treasure Room.

3 Underground water tunnels, called *ghanats,* first introduced to Persia in Achaemenid times, carry mountain water across miles of desert safe from evaporation that would deplete surface canals. A one-man windlass is used to reach the tunnels through shafts sunk at intervals of some 10m (30ft). A digger must work alone in a tunnel only 40 by 60cm (2 by 3ft) in height and width, keeping the channel straight and accurately gauging the amount of fall needed to enable the water to flow steadily to its point of use. These unique water systems contributed as much to the progress of the Persian Empire as the wisdom of its rulers.

4 A coin of Xerxes I depicts him in an aggressive pose, but he is best known for leading the Persian forces to defeat at the hands of the Greeks at the Battle of Salamis. Xerxes inherited the empire from his father, Darius I, in 486

BC. In 484 he suppressed a usurper in Egypt in savage fashion and went on to quell a revolt in Babylonia with similar ruthlessness. After early successes in Greece he lost his fleet at Salamis; the Achaemenid decline dates from that point.

Empire had by that time achieved its greatest extent and influence.

Agriculture based on both grain and livestock was the mainstay of the country. Artificial irrigation was introduced by means of tunnels [3] many miles long.

The original religion of the country had been the worship of Mithras, identified with the sun, and of Anahita, goddess of water and fertility. This religion was later combined with the worship of a supreme being, Ahura Mazda [9], "the Wise Lord" of the sixth-century prophet Zoroaster, or Zarathustra. As creator and ruler of the world Ahura Mazda clothed himself with the firmament, the sun and moon were his eyes and all forms of nature were his: earth, fire [6], wind and water. To avoid polluting these natural elements Parsees (who still follow Zoroastrian beliefs in India) expose their dead on "towers of silence" to be devoured by vultures.

The invasion of Greece
Persia's monarchical form of government was supported by her people who believed that the sovereignty of individuals was best maintained by an individual sovereign, the "King of Kings". On the other hand the city state of Athens propounded the idea of democracy, except for slaves and non-citizens. Darius considered the Greek city states and their colonies a danger, and when Ionia revolted and received aid from Sparta and Athens, he crossed the Aegean but was defeated by an Athenian force at Marathon. In the midst of preparations for another attack upon Greece he died in 486 BC.

Xerxes (c. 519–465 BC), son of Darius, crossed the Hellespont with a vast army and defeated the Spartans at Thermopylae. But he was driven out of Europe in 479 BC after incurring the lasting hatred of the Greeks by burning the Acropolis at Athens. The Achaemenid Empire then declined until Alexander the Great (356–323 BC) from Macedon defeated the last of the dynasty, Darius III, at the Battle of Arbela (also called the Battle of Gaugamela) in 331 BC. He routed a huge Persian army and burnt Persepolis, possibly to avenge the destruction of the Acropolis. Thereafter, Persia formed part of the empire of Alexander.

The tomb of Cyrus the Great at Pasargadae commemorates an outstanding leader who united the Medes and Persians to form an empire that played an important intermediary role between the civilizations of East and West. Few kings have left such a reputation for tolerance to subject peoples.

5 The Palace of Persepolis was begun in 518 BC by Darius and was built mainly under Xerxes I in 486–485 BC. It owes much to its situation with its back to the mountain from which the great terrace was partly carved. The magnificent staircases leading to the terrace were wide enough for eight horsemen to ride abreast up the shallow steps. A procession of Immortals carved in stone decorates the sides of the staircases, followed by lines of courtiers – Medes and Persians – and subject peoples bearing tribute. Iron clamps filled with molten lead lock together some of the blocks of stone of which the terrace is built.

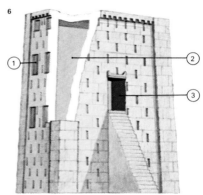

6 The so-called Fire Temple at Naqsh-i Rustam near Persepolis stands in front of a cliff in which the four tombs of Darius and his successors are carved. It is about 11 m (36ft) high with blind windows [1] of black limestone and a door [3] leading to an empty room [2]. Some authorities believe it to be a Zoroastrian temple for the sacred flame or for holding religious objects.

7 A   B

7 Persian warriors owed much of their success to their skill with bows. The arrows were carried in a quiver by the bowman [A] who wore leather shoes and cap and bore a short sword. A bodyguard of Darius the Great [B] wore long robes and carried a long spear with a cut-out shield. Such men, known as the Ten Thousand Immortals, were commanded by Darius during the campaign against Egypt and were the mainstay of his military achievements as emperor.

8 Depicted in colour on enamelled brickwork from the palace of Susa, one of the two capitals of the empire of Darius, is a soldier of the Ten Thousand Immortals holding a spear and carrying his bow and quiver. Darius rewarded his loyal bodyguard by having them portrayed on the walls of each palace he built.

9 Artaxerxes I (reigned 465–425 BC), a king of the Achaemenid dynasty, is enthroned in the Hall of a Hundred Columns at Persepolis beneath the winged Ahura Mazda, supreme god in the religion of Zoroaster, who was believed to direct the actions of the king as his viceroy, protecting the earth and its ruler.

India 550 BC – AD 300

The age of Buddha (c. 563–c. 483 BC) [3, 4] marked the beginning of a world religion and of important developments in the political and socio-economic fields. It was an age in which different religions (such as Jainism) emerged and a period of change in which established values were questioned.

Trade and political change
In the political field large and expansionist states developed, four of which dominated the scene by 500 BC. The most powerful of these was Magadha in southern Bihar with its capital originally at Rajgir, later at Pataliputra (Patna). The main economic asset was iron but power was also due to energetic rulers who gradually eliminated their rivals by force and diplomacy.

Cottage industries, such as textiles, pottery and metalcraft, flourished and were organized in guilds. Their produce, as well as agricultural surplus, was traded between various north Indian centres and with the Achaemenid (Persian) Empire in the west. Most trade was financed by bankers who supplied the means of transportation and took

the risks. Such activities favoured the rise of a prosperous class that included many who felt dissatisfied with the then rigid divisions of Hindu society. Such people often became enthusiastic patrons of Buddhism and other non-Hindu religions.

These developments were temporarily disturbed by the invasion in 327 BC of Alexander the Great (356–323 BC) who, after conquering the Persian Empire, set out to occupy its Indian provinces. But attracted by the legendary wealth of India, he advanced farther and managed to penetrate the Punjab. Alexander was, however, forced to retreat, soon to be followed by the governors whom he had appointed to rule the territories after his departure [5].

The glorious Mauryan age
The retreat of Alexander left behind a strong sense of Indian unity which found a leader in the young warrior Chandragupta (c. 321–c. 297 BC). First he liberated the western provinces and then he marched against Pataliputra, where he defeated the Nanda king of Magadha and so founded the Mau-

ryan Dynasty in 320 BC. In his bid for the throne, Chandragupta was assisted by his able and cunning minister Kautilya, who is regarded as the author of the most important Indian work on statecraft, the *Arthashastra*.

The Mauryan age (320–185 BC) was one of the most glorious periods of Indian history. Chandragupta controlled northern India from the Hindu Kush to Bengal and probably parts of southern India as well. The kingdom was largely centralized, partly because of a network of highways. In 305 BC Chandragupta concluded a treaty with the Greek Seleucus (c. 355–281 BC) who sent Megasthenes (c. 350–c. 290 BC) to the Indian court as an envoy to grant Chandragupta formal rights over Alexander's conquests in India.

The Mauryan Empire [1] reached its zenith under Chandragupta's grandson Ashoka (c. 274–c. 236 BC), the mightiest king of ancient India. At first Ashoka continued to follow traditional expansionist policies but, after a cruel campaign against Kalinga (the present state of Orissa), he renounced further conquests by force. Instead he substituted conquest by righteous-

CONNECTIONS

See also
118 India 300–1200
120 Indian art to the Moguls
80 Alexander the Great
64 Buddha and Buddhism

In other volumes
220 Man and Society

Ashoka's empire c. 250 BC
△ Rock and pillar edicts

1 The Mauryan Empire was the greatest of the states of ancient India. It comprised most of the subcontinent, except its southernmost portion, and most of Bengal and Sind. It also included significant parts of present Afghanistan, thus controlling the all-important overland communications with the Middle East. The size can be accurately established on the basis of the sites of the Ashokan rock and pillar edicts. Centralized administration of this vast empire was simplified by a network of highways connecting Pataliputra with the provincial centres.

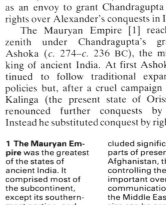

2 The rich sculpture of the eastern gateway of the great stupa at Sanchi shows both pious Buddhist stories and local deities belonging to folk religion but tolerated by Buddhism. The Buddhist stories concern either the important events in the life of Lord Buddha or stories relating to his earlier existences as a man or an animal. Popular religion is represented by various deities such as *yakshas* (tree spirits).

3 The Buddha is said to have converted people by performing miracles such as walking on the water, depicted here [A] in the carvings on the gateway at the Sanchi stupa. The presence of Lord Buddha, who is not represented in this early period, must be inferred. Another scene [B] shows Buddha's father paying homage to the tree under which Lord Buddha attained enlightenment. This time the presence of the Buddha has to be inferred from the tree.

4 The great stupa (dome-shaped shrine) at Sanchi was built in Ashoka's time, but the railings and the richly sculptured gateways were added during the following two centuries. Sanchi, situated at the centre of India where the main highways from east to west and from north to south crossed, was one of the principal Buddhist centres from at least the time of Ashoka to the 10th century AD; it also gave its name to a school of sculpture.

ness. At about that time Ashoka was converted to Buddhism and became one of its most fervent supporters.

To propagate his ideas Ashoka had edicts engraved on rocks and pillars [Key]. These edicts enjoined upon the population a common ideology based on the concept of loyalty towards one's elders and those in authority and also on justice and mercy for one's neighbours.

The Indo-Greeks and the Kushan Dynasty

The Mauryan Dynasty continued to rule over vast areas of India until 50 years after Ashoka's death but its authority soon declined. By about 185 BC the army commander Pushyamitra, in one of the earliest recorded military coups, overthrew the last Mauryan king and founded the Sunga Dynasty. The centre of the state was soon moved from Bihar to central India. Pushyamitra was a powerful ruler but his successors could not prevent incursions by the Bactrian Greeks, descendants of some of Alexander's generals. Some were repulsed by the Indians, while others founded short-lived kingdoms in the Punjab and elsewhere. Many of these were influenced by Indian culture and because of this they are usually called Indo-Greeks.

The Indo-Greeks were soon followed by invaders from central Asia and in AD 78 most of northern India was under the control of Kanishka (died *c.* AD 100) [6] of the Kushan Dynasty of Scythian origin (from central Asia). Like other Scythians, Kanishka was a pious Buddhist.

A century later, however, the Kushans had been expelled from India and a Hindu reaction followed. Sanskrit, the language of the sacred texts, developed into a medium of communication among the upper class, in administration and, above all, in literature. The great Sanskrit epics, the *Mahabharata* and the *Ramayana*, although incorporating much older tradition, were probably written down during this period, as were the *Laws of Manu*, the basic code of Hinduism. These three texts together incorporate the basic values of Hinduism and so laid the foundations of the classical age of India in the Gupta period (AD 320–550).

The Ashokan pillars are among the oldest and most splendid monuments of Indian art. They are built of sandstone with a special bright polish and rest on solid foundations below ground level. They are crowned with animal sculptures; this one at Lauriya Nandangarh in northern Bihar has a heraldic lion. Many pillars are also inscribed with Ashokan edicts proclaiming the emperor's authority. This pillar is inscribed with six edicts of the last phase of Ashoka's reign. Earlier scholars have emphasized foreign, especially Persian, influences, but recent research has established them in the Indian tradition.

5 A
Diodotus

B
Menander

5 After Alexander had left India in 325 BC, several of his governors founded small independent principalities in Bactria and northwestern India. Some of these Greeks were strongly influenced by the Indians. For example, the coin [A] of Diodotus (*fl.* 3rd century BC) compared with the coin of Menander [B] (*c.* 150 BC) shows Indian influence, particularly in the use of an Indian script.

6 Kanishka I, the greatest of the Kushan kings, ruled over vast areas of central Asia and also controlled most of northern India eastwards as far as Bihar. This modern illustration is based on a torso of Kanishka (as indicated by the inscription), to which the head has been added from one of Kanishka's coins. The Scythian mantle and boots convey a strong impression of power and authority.

7 According to Buddhist tradition, caves have been used by the community of Buddhist monks since early times, especially as shelters during the rainy season. Although they were gradually replaced by structural monasteries, the tradition was continued in parts of India where numerous caves were excavated between 100 BC and AD 900. The *chaitya* cave at Karli, Maharashtra (*c.* 50 BC) was not used as a dwelling but as a place of worship, as well as for monastic ceremonies. It consists of a long pillared hall with a small stupa, enshrining relics of Lord Buddha, at the end. The rich sculpture of the capitals of the pillars presents a striking contrast to the sober lines of the rest of the cave.

8 Another richly decorated part of the caves at Karli, Maharashtra, is the façade, which includes the space between the three entrances. In addition to traditional Buddhist scenes, there are usually representations of the pious donors of the caves and especially of the land which was donated for the requisites of the monks. The donors are usually couples, thus illustrating the high status of women in this period in western India.

Buddha and Buddhism

By the sixth century BC the semi-nomadic tribes of northern India had developed into settled agricultural communities ruled by oligarchies or royal dynasties. It was a time of social change and of new ideas. Chief among these, and evolved in opposition to the rituals and hardening caste system of Hinduism, were the philosophical and ethical teachings of Buddhism which were to develop into one of the greatest Oriental religions.

The life of Buddha

There are many differing accounts of the life of Buddha but the main outline seems clear. Siddhartha Gautama (c. 563–c. 483 BC), who was later to become the Buddha ("the Enlightened One"), was the son of Suddhodana, king of the Sakyas, and his queen Maya. His birthplace, Lumbini, is situated on the northern fringes of the Gangetic valley near Kapilavastu where he spent his early years. After an uneventful childhood the prince, struck by the problem of human suffering, decided to break with the past to seek the supreme truth in meditation. He left home secretly [1] and eventually, after years of seclusion, he attained "enlightenment" seated under the Bodhi tree at Bodhgaya near Gaya in southern Bihar [4]. This subsequently became one of the holiest places of Buddhism and saplings of the tree were taken to different Buddhist countries where they grew into new trees. Soon afterwards the Buddha delivered his first sermon in the Deer Park of Sarnath [5] near Varanasi "setting the wheel of the Law in motion" [Key].

Buddhist doctrine was a Middle Way, avoiding the extremes of mortification and indulgence. It accepted the basic concepts of Hinduism – rebirth and the law of karma, that a man's actions directly control his destiny – but concentrated on ethics as a means to salvation. For Buddha suffering was caused by desire. The abandonment of desire could be achieved by following the "noble eightfold path" of right living and actions. As a result nirvana, the state of bliss in which rebirth ended, would be attained. The ideal of nirvana [3] could best be attained by monastic discipline, but the order of monks (Sangha) depended on the entire community.

Buddha himself preached all over eastern India and received support from the rulers and the emerging merchant class. When at an advanced age he "entered nirvana", he left a monastic order but no written instructions.

Expansion of Buddhism

For two centuries Buddhism slowly expanded despite difficulties such as the animosity of the Hindu Brahmins who feared for their own privileges. However, when the Indian king Ashoka (reigned c. 269–232 BC) was converted to Buddhism his powerful patronage greatly favoured its expansion. The oldest extant stupas, distinctive monuments built to enshrine relics, belong to this period. Through Ashoka's influence Buddhism was introduced into Sri Lanka, where it has remained the established faith.

For the next few centuries Buddhism spread farther into India with centres located in central India (Bharhut, Sanchi), Maharashtra and Andhra Pradesh where Buddhist art and architecture flourished. In Maharashtra (Nasik, Karle and other places), many caves served as monasteries or halls for worship (the *chaitya* halls). Great stupas

CONNECTIONS

See also
118 India 500 BC–AD 300
122 China 1000 BC–AD 618
124 Confucius and Confucianism

In other volumes
220 Man and Society

1 The great departure from Kapilavastu is often represented in Buddhist art. Although the future Buddha left secretly at night, sculptors usually show the pomp of a royal procession.

2 In Buddhist philosophy, the Wheel of Existence consists of 12 spokes, each constituting a link in the ever-repeated cycle of life and death. The wheel will revolve as long as ignorance lasts.

12 Dukkha — Decay, old age suffering and death
11 Jati — Rebirth
10 Bhava — Subconscious process of becoming
9 Upadana — Clinging or attachment
8 Tanha — Desire or craving
7 Vedana — Sensation
6 Phassa — Contact of the senses
5 Satayatana — Development of the senses
4 Nama-rupa — Development of the psycho-physical organism
3 Vinnana — Consciousness
2 Sankhara — Longing to live
1 Avijja — Ignorance

Future   Past   Present

3 The fourth great event in Buddha's life is the nirvana, which is neither eternal life nor annihilation but an incomprehensible state of utter bliss. Here Buddha is shown in a symbolic representation of the state of nirvana. This huge sculpture at Gal Vihara, Polannaruva, Sri Lanka, dates from the twelfth century.

4 The Mohabodhi Temple, Bodhgaya, Bihar, marks the spot where Lord Buddha attained "enlightenment" seated under the Bodhi tree. This is perhaps the most hallowed spot in the Buddhist world. The present temple, which replaces an older foundation at the site, was built in the Gupta period (c. 320 – c. 550) but has been restored.

(Amaravati and Nagarjunakonda) also arose in the Andhra country.

Buddhism was always prone to divisions and councils organized to promote unity often had the opposite effect. Early in the Christian era there developed a fundamental division between the adherents of the Great Vehicle (Mahayana) and those of the Lesser Vehicle (Hinayana). The former had monks but emphasized the ideal of the pious layman continually assisting his fellow men by his wisdom and compassion. The historical Buddha and previous Buddhas were worshipped as deities and so were other beings who had taken the vow to become Buddhas (Bodhisattvas). Buddhas and Bodhisattvas received worship in the form of images, thus providing a strong incentive to Buddhist art, which soon included the image of Lord Buddha. Hinayana, however, kept closer to the other teachings.

The spread outside south Asia

From the beginning of the Christian era Buddhism spread outside south Asia [9]. Buddhism entered China in the first century

and subsequently spread to Korea and Japan. In Tibet it took root in the eighth century and developed into Lamaism.

In South-East Asia, too, Mahayana influence, mainly from Bengal, inspired great monuments such as Borobudur in central Java (ninth century) [7] and Ba-yon in Cambodia (twelfth century), but the great expansion of the Theravada school (the monastic version of Hinayana) gained momentum in the eleventh century when the Burmese king Aniruddha (reigned 1044–77) made Theravada the official religion. In the thirteenth century it became the official religion of Thailand and eventually spread to Laos and Cambodia.

Although it has not always guaranteed harmony between these countries, Buddhism has provided Sri Lanka and mainland South-East Asia with a firm ideology and a high standard of education. It has also contributed to great achievements in art and architecture. The astounding temple complex of Pagan (Burma), the splendid pagodas of Mandalay, Bangkok and Ayutha (Thailand), testify to the strength and inspiration of the faith.

In the life of Lord Buddha the first sermon is the third great event (after his birth and "enlightenment"). Buddhists call this event the "turning of the wheel of Law"; the wheel represents the cosmos, and the Law represents the Lord Buddha's philosophy, which offers an explanation of the mysteries of life. The presence of the wheel is suggested by the position of the hands. The sculpture of Sarnath dates from the Gupta period, (c. AD 500). The perfection of the physical shape reflects perfect knowledge; the elongated ears and almond-shaped eyes suggest profound concentration and the monastic robe reveals the perfect body.

5 The Wheel of Law and two deer on this seal suggest the Deer Park of Sarnath. The seal was recovered from the ruins of Nalanda, a great Buddhist monastery in Bihar. Founded in the fifth century, Nalanda became one of the great centres of Buddhist learning, attracting students even from Indonesia and China and enjoying the patronage of numerous kings. The Muslims largely destroyed it c. 1200.

6 One of the oldest-known Buddhist images in South-East Asia is this bronze Buddha from Vietnam, dating back to the fifth century. Its style was influenced by Buddhas from Amaravarti.

7 Borobudur, situated in the heart of Java, is often described as a stupa (dome-shaped shrine) but although a stupa crowns it, the rest of the structure predominates. It is a marvellous storehouse of sculpture with five lavishly decorated galleries exhibiting pious stories in 1,500 relief panels. It was built by the Sailendras, a Buddhist dynasty in Java from about 750 to 850.

8 The Mons of southern Thailand and the Thais of central Thailand became Buddhists by the third century. Despite the Khmer occupation, Buddhism continued to flourish. This Buddha head comes from Lopburi. Although conforming to Buddhist norms, it was modelled to correspond to the aesthetic ideals of the Thais.

9 The expansion of Buddhism outside India was a complicated process extending over many centuries. Here the broad outlines of its spread are shown, together with the percentage of Buddhists in various countries today. Buddhism was spread by monks who sometimes acted as advisers to kings.

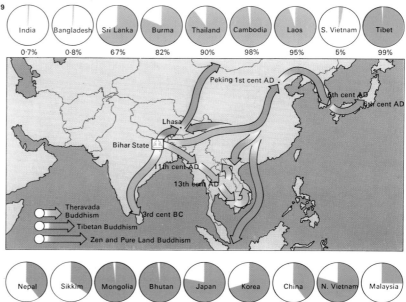

India	Bangladesh	Sri Lanka	Burma	Thailand	Cambodia	Laos	S. Vietnam	Tibet
0·7%	0·8%	67%	82%	90%	98%	95%	5%	99%

Peking 1st cent AD

Lhasa

Bihar State

11th cent AD

13th cent AD

Theravada Buddhism

Tibetan Buddhism

Zen and Pure Land Buddhism

3rd cent BC

Nepal	Sikkim	Mongolia	Bhutan	Japan	Korea	China	N. Vietnam	Malaysia
40%	35%	98%	98%	78%	70%	40%	80%	25%

Europe 1200–500 BC

By 1200 BC the Bronze Age was fully mature among the Urnfield peoples, predecessors of the Celts who were by 800–700 BC to become the dominant and most progressive element in European society for several hundred years. The Urnfield peoples lived north of the Alps and belong to the prehistoric phase of European evolution; their name was given to them by archaeologists because they introduced a new burial rite into Europe. Burial in barrows was replaced by cremation, the burnt bones then being placed in an urn and interred in large flat cemeteries or Urnfields.

Technological revolution
With some exceptions – Basques, Finns, Iberians – the Europeans were of Indo-European origin and thus had linguistic and cultural traditions in common. The Urnfield peoples, who were farmers living in villages or large homestead complexes, may have spoken some form of Celtic dialect as the distribution of their monuments corresponds with the oldest recognizable Celtic place-names. These Urnfield peoples are generally regarded as being "proto-Celtic" and there seems to be little difference between them and their immediate descendants, the Celts of the Hallstatt culture.

The Urnfield period saw not only improved agriculture in Europe but also great developments in metalworking. It was a time of expansion and of warfare, and archaeological evidence shows that the central Europeans possessed quite sophisticated bronze weapons. Towards the end of the period, about 800 BC, Europe experienced a major technological revolution initiated by the Celts. They introduced the use of iron into Europe north of the Alps, and with it all the superiority in weapons and edge tools which the use of iron made possible. The areas of initial development coincide closely with Urnfield settlements in the Danube and Rhine regions.

Apart from evidence of some contact with people from the Steppes, everything suggests a continuity of the indigenous population enriched and empowered by the discovery of iron. This first phase of Celtic culture is known to archaeologists as Hallstatt after a village in the Salzkammergut in Austria.

The Hallstatt site has provided dramatic and comprehensive evidence about the early Iron Age in Europe. The community there derived much of its wealth from large salt mines. Both the preserving and healing powers of salt were recognized at an early stage. But it was also a valuable trading commodity, being perhaps the chief export from the Hallstatt Celts to the Graeco-Etruscan world. Evidence from many sites points to the importance of this trade which passed between the two regions through Greek colonies in the northern Mediterranean. A leading colony was Massalia (Marseille) which was founded about 600 BC.

Workers and aristocrats
The equipment, dress, foods and eating habits of workers are all too often unknown in a society solely concerned with its aristocracy, the wealth of which is attested by the richness of the burials. Clothing and leather helmets have been preserved at Hallstatt. Their wooden bowls and other eating utensils have remained intact and numerous seeds

CONNECTIONS

See also
86 The Celts 500 BC–AD 450
84 Bronze and Iron Age Britain

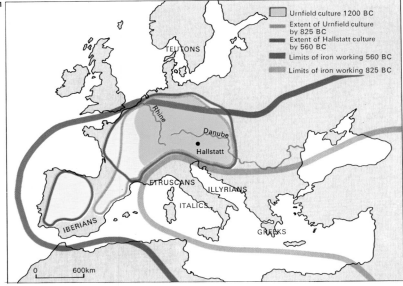

1 The Urnfield culture developed around the Danube and Rhine and by 825 BC had reached Spain. The Hallstatt culture, which began in the same place and superseded the Urnfield, by 560 BC covered most of Spain and Portugal. By then iron working, introduced from the east, had spread through most of Europe. The Hallstatt culture is named after a key archaeological site in Austria.

4 A gold armband from Dover of about 1000 BC anticipates Celtic work.

5 Irish goldsmiths made this gold-plated lead pendant about 800 BC.

2 In Irish tradition this stone with its spiral designs marked an entrance to the "otherworld". It dates from c. 2500 BC, foreshadowing Celtic designs of some 2,000 years later.

3 Hallstatt ceramic skill and restrained decorative elegance are shown in this painted pottery urn from Burrenhof, Germany, dating from about 620 BC.

6 Later Celtic symmetry is seen in a gold-plated Irish ring of about 1000 BC.

and fruit stones provide some indication of their typical diet.

Information about the aristocrats comes from a vast cemetery which points to great changes in burial customs and personal equipment coinciding with the development of a culture based on iron, a more abundant metal and cheaper for practical purposes than bronze. Chieftains were no longer cremated and placed in humble urns but were buried with much pomp.

Burial ceremony

The richest graves consisted of a wooden chamber, usually of oak and covered by a mound decorated by a commemorative stone figure [Key], sacred tree, or stone pillar. The deceased was placed under his status symbol, a four-wheeled wagon. Horse trappings were present, but there seems to be no evidence for the burial of the horses themselves; perhaps they were sacrificed to the gods and ritually eaten, or burnt. All the warrior's fine military equipment was placed in the tomb with him – his sword and spear, helmet, neck and arm ornaments and other things he had

valued. Pots no doubt full of ale or imported wine for the "other world" feast, and joints of meat were laid beside him to sustain him on his journey. According to the Old Irish tales, once the actual burial rite was completed, the ceremony would be followed by feasting and funerary games.

The Hallstatt period lasted until about 500 BC and laid the foundations for the full flowering of Celtic culture in Europe at that time. There was some intermarriage with the Teutonic peoples to the north – both Teuton and German are Celtic names – and the Celts were probably overlords, making up the aristocracy of these closely related peoples. The Hallstatt phase of the Celts manifested itself deep into Spain. Hallstatt art [3, 7, 8, 9] tended to be linear, influenced by Greek Geometric, the naturalism of Urnfield bird and animal art, and old indigenous motifs, magical and symbolic. It was to become an important element in the magnificent magico-religious art [11] of the La Tène period that followed, and in the important contribution of Irish art [4, 5, 6, 10] to the Western Church in early Christian times.

KEY

This stone statue of a Hallstatt warrior or divinized hero, dates from the end of the 6th century BC. The figure was found on a Hirshlanden burial mound, broken where it had fallen. The cult of graves and ancestors was strong among the Celts and this anthropomorphic figure links the prehistoric world with that of the historical Celtic world in all its flourishing vigour. In this sense, it epitomizes all that had gone before and anticipates the strength of Celtic culture in its later phase. The Hallstatt peoples had close trade links with the Etruscans, another important bronze-using culture.

7 Warfare, whether as necessity or as an aristocratic sport, dominated most aspects of Celtic culture. Emphasis was laid not only on efficient weapons but also on the appearance of the warriors.

Although not usually worn into battle, helmets were popular headgear. This bronze helmet from Italy probably dates from the Hallstatt period and is a particularly fine example of Celtic work.

8 A Greek bronze crater for mixing wine, weighing 208kg (460lb), was found in the burial of the priestess at Vix. It shows that as early as the 6th century BC the Celtic and Mediterranean worlds were in close contact. This may have been a gift of peace, friendship or trade.

9 The Celts were skilled metalworkers. They introduced iron into northern Europe by 800 BC but continued to use bronze with exquisite results. This detail of a bronze bowl handle from the 6th century gives some indication of the sophisticated craftsmanship of the Hallstatt culture.

10 This gold dress-fastener from County Galway (Castlekelly) in Ireland dates from about 700 BC and measures 28cm (11in) across. Such objects not only provide proof of Celtic pride

in personal ornamentation but also reveal the exquisite skill and harmony of Celtic craftsmen and their predecessors in their use of various metals as a medium of their art.

11 The severed head – the true godhead – with one, two or three faces was the supreme object of Celtic worship. This Janus head from a sanctuary in France about 250 BC looks back and also forward, possibly to the "otherworld".

The Greeks to the rise of Athens

Some time about 1200 BC, Greek civilization underwent a major change. Dorian tribes moved through the peninsula from the north and settled, mainly in the Peloponnese. The existing peoples were assimilated or confined into smaller areas, and Ionians and Aeolians in particular were pushed out across the sea to found cities in Asia Minor.

The development of institutions

Major social changes must have accompanied this upheaval, but little detail is known of the next 400 years; much of what we do know comes from the two epic poems by Homer, the *Iliad* and the *Odyssey*. It is, however, possible to identify several important developments that were of major significance for subsequent Greek civilization. The use of iron became widespread for tools and weapons [2]; a phonetic alphabet was developed and more or less standardized; and a feeling of national consciousness arose – a racial and intellectual identity bound up in the use of the word *Hellas* to describe the whole Greek world. At the same time a pantheon of gods that appear in Homer's poems was formed, religious ceremonies developed and the importance of shrines such as Delphi became generally accepted, and the city state was adopted as the most important political, economic and social unit.

The rugged geography of Greece played a major part in making the inhabitants of each valley regard themselves as a separate entity and Greece became a patchwork of small states each of which guarded its independence jealously [3]. They varied greatly in size; Athens succeeded in subduing Attica and no other states developed there, while Thebes failed to overcome Boeotia and at least 12 states existed there. Writing in the fourth century BC, Aristotle mentions the existence of at least 150 states.

The size of the city state was important for Greek political development because a firm belief developed that a satisfactory political unit must be small and totally independent. A Greek had a series of loyalties to *Hellas*, to his city state and to his tribe within the state – but loyalty to the city was the most powerful and the idea of a larger political unit was never developed. During this period the Greeks also evolved the belief that government must be based on a known constitutional and legal framework.

Agriculture, trade and politics

Economically, agriculture was all-important [Key] and as the cities developed, the link between them and the country remained strong. However, some states became increasingly involved in trade and industry [6] and a definite class of traders and craftsmen evolved. Politically, by the mid-seventh century BC most states had moved away from the monarchies described by Homer and were ruled by oligarchies of the richest and most powerful citizens. A privileged or noble class held all political, military and religious power and there was usually tension between them and the common people.

This tension was made worse by increasing population and a safety valve was found in colonization [4]. States organized expeditions to found new cities overseas, but once founded these new cities were completely independent of, although they

CONNECTIONS

See also
40 The Greek mainland 2800–1100 BC
70 Classical Creece
72 Classical Greek society
74 Greek literature and theatre
76 Greek art 1100–450 BC

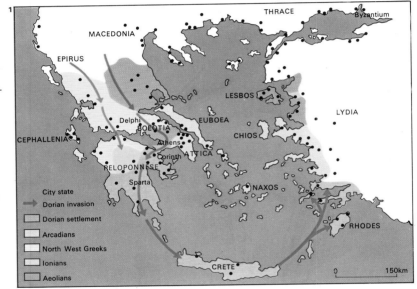

1 Little is known of the history of Greece between the arrival of the Dorians (c. 1200 BC) and the dawn of the Archaic period (c. 800 BC). Homer describes a society of small agricultural communities grouped round citadels and led by kings and aristocracies that still retained many of the features of Mycenaean civilization and had little trade or commerce. By the time of Hesiod (8th century BC), whose narrative poems give a description of the early Archaic period, trade and commerce were increasing, monarchy had generally been replaced by oligarchy and the city was becoming the focus of political life.

City state
Dorian invasion
Dorian settlement
Arcadians
North West Greeks
Ionians
Aeolians

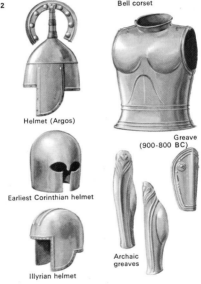

Bell corset

Helmet (Argos)

Greave (900–800 BC)

Earliest Corinthian helmet

Illyrian helmet

Archaic greaves

3 The landscape of Greece, with its limited fertile plains and harsh mountains, became a significant factor in the rise of small communities such as Corinth (shown here). Travel was difficult and economic and political life tended to be concentrated in confined areas. The Greeks were convinced that these small independent units were the most natural size for a satisfactory political life. This attitude, combined with constant diplomatic and military manoeuvring between states, meant that wider unity was never attempted, although the idea of a distinct Greek identity and a feeling of cultural superiority to outsiders did exist.

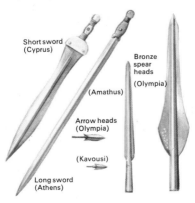

Short sword (Cyprus)

(Amathus)

Arrow heads (Olympia)

(Kavousi)

Long sword (Athens)

Bronze spear heads (Olympia)

2 Bronze arms and armour were used by the aristocratic warriors of Homer's poems, who fought in individual combat. But iron began to supersede bronze before 700 BC and tactics changed fundamentally, with heavily armed infantry (hoplites) fighting in a well-disciplined mass formation – the phalanx. Many hoplites came from outside the aristocracy (although they had to be wealthy enough to provide their own equipment) and this helped to weaken the old social order.

retained close links with, their mother cities. From about 750 BC migration established Greek cities throughout the Mediterranean.

Colonization did not solve the problem of class struggle and the archaic Greek world (Greece between 800 and 500 BC) saw the widespread appearance of tyrants, a word originally applied by the Greeks to men who seized power unconstitutionally and ruled – either well or badly – without legal backing.

New political systems
In Athens attempts were made by Solon (c. 640–c. 559 BC) to remove the economic problems that lay at the root of the class struggle and reform the legal framework of the state to protect the weak. Power was seized in 545 by Pisistratus (c. 600–527 BC) who increased the rights of the common people and brought the nobility under the rule of law. These developments formalized the democratic system that ruled Athens for the next two centuries.

The same pattern was followed in other states; tyrants rarely lasted long. Some states, such as Sparta with its peculiar constitution,

and Corinth with a strong oligarchy, avoided the tyrannical stage completely.

Until the middle of the sixth century BC the Greeks were largely unaffected by outsiders, apart from trading contacts [5]. The situation altered when the Lydians extended their power over the Greek states of Asia Minor and were succeeded in 546 by the Persians. Neither overlordship was oppressive, but in 512 the Greeks rose in revolt and succeeded in temporarily reasserting their independence. There was, however, little unity among them or support from the mainland, and in 499 the Persian emperor Darius was able to crush the revolt. He followed this up by reimposing his power in Thrace and Macedonia and then in 490 launching a major invasion of Greece.

Athens, as the largest state, was his immediate target and she called for help from the others, but before this could arrive her troops, who were significantly outnumbered, met and decisively defeated the Persians at Marathon. Darius' army withdrew, leaving the Greeks still disunited but convinced of their superiority over the "barbarians".

Greek agriculture, represented by the olive harvest shown on this 6th-century BC amphora, was always faced with the difficulty of providing sufficient food from the relatively small area of fertile land. The natural consequence of an expanding population faced with a limited food supply was emigration, first across the Ionian Sea and then throughout the Mediterranean. But overpopulation was only part of the reason for emigrating: the nature of aristocratic control of land resources meant that the distribution of agricultural land was by no means equitable; the resulting social tensions naturally encouraged the tendency to emigrate.

4 Colonization was used to send surplus and disaffected populations to found cities in new regions. The first phase began about 750 BC with expeditions to the west, where the Greeks found the Phoenicians already well established in many areas. They were able to settle in Sicily, southern Italy, France and Libya. About 650 BC the Greeks began to move into the Black Sea region until there were colonies round almost all its shores. By the 6th century BC, the colonies were sending enough food back to Greece to feed the expanding population and thus reduce emigration.

5 The Greeks traded with the Phoenicians, the Egyptians and the people of the Middle East as well as between colonies and mother cities. In states bordering the sea, such as Athens and Corinth, trade became an important source of wealth. Shipbuilding and navigation were therefore vital skills.

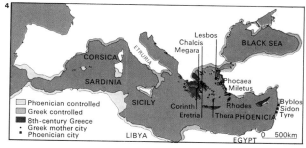

6 Trade and commerce were free to develop after the city states had been established, and the system of barter was replaced by more regulated methods. Various systems of weights and measures were developed (such as this one on an Attic black-figure amphora, showing men using a balance scale), but no single system became dominant. Precious metals were used for exchange either in the shape of weapons or as pieces valued by weight. By the end of the 7th century money had been invented. The issuing of coinage soon became the privilege of governments and not of individuals.

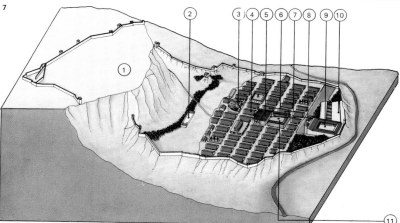

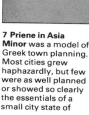

7 Priene in Asia Minor was a model of Greek town planning. Most cities grew haphazardly, but few were as well planned or showed so clearly the essentials of a small city state of only about 4,000 inhabitants. The citadel or Acropolis [1] stood at the top of the cliff with the main city below it. The principal features were the sanctuary of Demeter and other gods [2]; the huge theatre [3], possibly the oldest Hellenistic example, which could accommodate an audience of 5,000; housing blocks [4], each with four to six dwellings; buildings for the council and courts [5, 6]; the agora [11], which was the market-place and heart of the community, adjoining the main street [8]; the sanctuary of Zeus [7]; the gymnasium [9]; and the stadium [10].

Classical Greece

In 480 BC the Persians under Xerxes launched a second invasion of Greece. A large army advanced across the Hellespont and down the peninsula; it was briefly halted by the heroism of 300 Spartans under their king, Leonidas, who held the narrow pass at Thermopylae until outflanked. They fought to the last man. Athens was occupied but shortly afterwards its fleet, led by the statesman Themistocles (c. 528–c. 460 BC), annihilated the Persian army at Salamis [1]. Xerxes' army fell back northwards and in 479 BC it was defeated at Plataea by a largely Peloponnesian army under the Spartan Pausanias. If the Greeks had not defeated the Persians, the history of Europe might have been very different – a dominant Persian civilization would have left Europe with quite another set of values and institutions.

The golden age of Athens
The Persians withdrew but their threat remained and a defensive league of many of the Aegean islands and Greek states in Asia Minor was set up in 478 BC under Athenian leadership, with its headquarters at Delos [4].

A navy was established, maintained by contributions of money or ships.

Soon Athens, as the most powerful Greek state, came to dominate the Delian League and, under the guidance of her statesmen Cimon (died 449 BC) and Pericles (c. 490–429 BC), became a maritime empire in all but name. The democratic system was refined under Pericles, and the great buildings and cultural achievements that were to make Athens famous grew from trade.

Inevitably, the power of Athens aroused jealousy and fear, particularly in Sparta, which continued to dominate the Peloponnesian states, and Corinth, the other great trading state. There was sporadic warfare during the 450s BC and Athens built up a land empire in Megara, Boeotia and Achaea, but this was abandoned in 445 BC after concluding a truce with her rivals. Hostilities broke out again in 431 between the rival alliances and Athens, secure behind the Long Walls linking her with her major port of Piraeus [5], allowed the Spartans to invade Attica and instead concentrated on using her maritime power to wear down her enemies.

Pericles died of plague in 429, but his strategy was continued by Cleon (died 422 BC). Neither side could decisively defeat the other, and a peace was reached in 421. This lasted only two years before fighting broke out again; in 415 the war party in Athens led by Alcibiades (c. 450–404 BC) persuaded the people to launch a major expedition to invade Sicily and capture Syracuse. The venture was a ghastly failure, and in 413 the bulk of the Athenian army and navy were destroyed. Despite the disaster Athens fought on, although her enemies were now being financed by the Persians and several of her allies revolted. In 405 the remainder of the Athenian navy was surprised and destroyed at Aegospotami by Spartans under Lysander (died 395 BC), who then besieged the city. When it surrendered in 404, an oligarchy replaced democracy, and the Spartans took over the Athenian Empire.

Decline of Spartan rule
This Spartan supremacy was short-lived; spurred on by the exploits of Xenophon's Ten Thousand – Greek mercenaries who

CONNECTIONS

See also
68 The Greeks to the rise of Athens
72 Classical Greek society
76 Greek art 1100–450 BC
78 Greek art 450–31 BC
82 Greek science

In other volumes
156 Man and Society

Route of Persian army
Route of Greek army
Route of Persian navy
Island of Salamis

1 After the Persians overran Athens, the refugees fled with the Greek navy to Salamis. There, by a stratagem the Greeks trapped the Persians and destroyed an entire corps as well as 200 Persian ships for a loss of only 40 of their own.

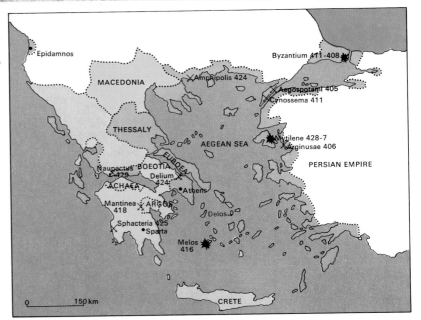

2 An Athenian trireme was the type of warship that defeated the Persians at Salamis and was the mainstay of the powerful navy built up under the direction of Themistocles at the beginning of the 5th century. The navy was significant in the democratic system, for it was largely manned by the poorer citizens and provided them not only with a livelihood but also with a source of pride and power. Since much of the wealth and power of Athens came from trade and her maritime empire, the common people could assert that they were the backbone of the state and should play a major part in its political life.

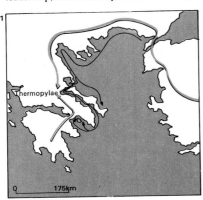

3 This Athenian four drachma coin shows the owl of Athena, who was the patron goddess of the city. Athens' naval and trading domination of the Aegean gave the city great wealth which was lavished on fine buildings and the arts, and allowed its citizens the leisure to participate in the democratic system or to make contributions to philosophy and literature. Even during its decline, Athens remained the acknowledged intellectual and artistic leader of Greece.

Sparta and allies (the Peloponnesian League) 431 BC
Athens and allies 431 BC
Neutral Greek states 431 BC
Revolts

4 After the defeat of the Persians, Athens used its navy to secure the Aegean against them and to liberate the Greek states in Asia Minor. Soon the defensive Delian League was turned into an aggressive empire with Athens intervening directly in the internal affairs of allied states. Sparta's fear of this expansion caused it to encourage revolts and oligarchic governments, and led in 431 to the start of the Peloponnesian war which engulfed most of Greece.

made an epic march across the Persian Empire – a crusade was launched to regain the freedom of the Greek states in Asia Minor, but the Persians encouraged the Spartans' allies to turn against them and the attempt had to be abandoned in 387, leaving them under Persian domination.

The next 50 years show the worst features of the Greek political system of the Classical age, with petty jealousies and continual military rivalry preventing the emergence of any wider unity. Athens recovered quickly, democracy was restored in 403 and by 377 she was again leading a naval confederacy against Sparta. However, it was Thebes that became the next dominant power when it destroyed the Spartan army at Leuctra in 371. After a decade she too began to decline, and Phocis, by capturing the treasure of Delphi, was able to hire mercenaries and set up a temporary mastery.

The Hellenistic age
Interstate rivalries continued and prevented the Greeks from realizing that a new threat was growing in the north. The Macedonians had hitherto been a loose confederation of tribes, but in 359 Philip II (382–336 BC) became king. A fine organizer, general and diplomat, he unified the tribes in his own kingdom and then went on to annex Thessaly in 352 and Thrace in 342.

The Athenians were the first of the Greeks to become aware of this new danger, for Philip's power threatened their lucrative trade routes to the Black Sea. However, years of relative peace had made the people of Athens apathetic and lacking in military zeal. A few orators such as Demosthenes (died 322 BC) tried to arouse them to the danger to their trade and to their city, and to exhort them to put an end to traditional rivalries so that the Greek states could unite against this new peril. Eventually a Greek league was formed, but in 338 Philip routed its armies at the battle of Chaeronea and occupied Thebes. Athens prepared to continue its resistance but there was little support elsewhere and at the Congress of Corinth a new league of Greek states was set up under Macedonian leadership. The independence of the free city states was at an end.

A Greek hoplite is shown killing his Persian enemy on this vase. The Greek Classical age is usually taken as beginning with Pausanias' defeat of the second Persian invasion in 479 BC and ending with the establishment of Macedonian power over all Greece in 320. Many of the great achievements of the Greeks flowed directly from the feeling of security and superiority that followed the Persian defeat, but with their devotion to the small city state and their obsession with political and military manoeuvring, the Greeks failed to develop a wider political unity which could have resisted the rising power of Macedon.

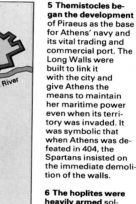

5 Themistocles began the development of Piraeus as the base for Athens' navy and its vital trading and commercial port. The Long Walls were built to link it with the city and give Athens the means to maintain her maritime power even when its territory was invaded. It was symbolic that when Athens was defeated in 404, the Spartans insisted on the immediate demolition of the walls.

6 The hoplites were heavily armed soldiers trained to fight in a highly disciplined phalanx (a solid formation). They were usually recruited from the merchant citizens who thereby won great political power.

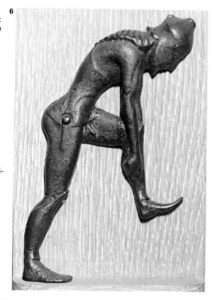

7 This wounded warrior is carved on the pediment of the temple of Aphaia in Aegina. Throughout the Classical age there was almost continual fighting between city states. This weakened the Greeks politically, but did not prevent a flowering of the arts, literature and philosophy. The achievements of Athens have tended to overshadow the fact that many other states, both on the Greek mainland and in Asia Minor, also produced great artists and writers.

8 The Temple of Athena at Delphi, whose oracle was presided over by Apollo, was part of the widely respected shrine in ancient Greece. Religion in the Classical age was an affair of sacrifices and ceremonies by the state or individual that contained little moral guidance or mystical experience. However, a belief in the same gods and legends did promote the feeling of a common culture which was added to by semi-religious festivals such as the Olympic games.

Classical Greek society

Classical Greece was the birthplace of many of the most influential Western ideas in art, literature, philosophy and science. Its other great contribution was in politics, for it was there that the ideals of democracy were first developed. In all these areas Athens was pre-eminent, a fact recognized even by her contemporaries. In two centuries she produced a succession of outstanding writers, artists, scientists and philosophers. Many who were not natives were attracted to the city, and there are few important figures in Greek cultural life who were not associated with Athens for at least part of their careers.

Athens and democracy

The city state of Athens, with an area of about 2,500 square kilometres (1,000 square miles), was the largest of the many city states into which Greece was divided. Her population at its peak was about 260,000, of which about 45,000 were male citizens and about 70,000 slaves. The rest were women, children and resident foreigners or *metics*. Corinth may have had a population of 90,000; Thebes, Corcyra and Acragas about

50,000 each; and the other states anything down to 5,000.

Politics in all these city states was a very intimate affair and this profoundly affected the Athenian concept of democracy [1]. It was based on direct participation, rather than representation, with every citizen having an equal opportunity to hold high office. In common with many other Greek states, Athens went through the transition from oligarchy (a small number of individuals holding power) to tyranny (a single all-powerful ruler) with struggles between rich and poor before the reforms of Cleisthenes in 508 BC established a democratic framework.

The essence of this democracy was the citizen body or *Demos*. Citizenship was a jealously guarded right, rarely given to foreigners and never to women [5] or slaves. All power was vested in the *Demos* which met in public assembly about every ten days [2]. There, any citizen could put forward proposals for laws or action which were discussed and voted on, and the civil and religious officials were chosen. Juries were selected from volunteers, and the business of the Assembly

was prepared by a Council of 500, the *Boulê*, elected by the ten tribes into which the citizens were divided.

Rights and duties

Every Athenian citizen had the right and duty to serve the state but, because there were more than 1,000 offices to be filled each year, the system could work only if there were enough men with both the time and inclination to devote their lives to public service. It is a remarkable fact that at no time was Athens short of able men to serve with little or no reward, and it was only during the fourth century BC that a small payment was introduced to help the poorest citizens to participate fully.

Athenians gained their wealth from land, trade and commerce. At the beginning of the fifth century Athens was a major exporter of pottery, oil and wine and an importer of fish, timber and wheat on which it was largely dependent. In Athens, in particular, a definite class of capitalists and an urban proletariat developed. Their leisure resulted from the widespread use of slaves to under-

1 The hub of Athenian democracy was the Pnyx where the Assembly of citizens gathered for its regular meetings. Public and social life was a gregarious and open-air affair with informal discussion, theatre and sports providing the most common interests.

2 Fifth-century Athenian democracy was based upon the power of the Assembly of citizens to vote on all major decisions. Public officials were responsible to the Assembly and were chosen by the ten citizen tribes for limited terms; only the ten military commanders or *strategoi* were elected and could serve for more than a year. Popular control over both the magistrates and the law could also be exercised through the courts where the large citizen juries had legislative as well as judicial powers and could try a law as unconstitutional.

3 A Negro slave follows a member of the leisured class. Athenian democracy, with its large number of official jobs being filled by unpaid or low-paid citizens, depended on a plentiful supply of men with the inclination and leisure to undertake them. Athenian civic responsibility and pride in the system meant that there was never a lack of volunteers and this was helped by the wealth from land and commerce which flowed into the city. The large-scale use of slaves freed citizens for public service. Athenian thinkers saw no contradiction between the individual rights and freedom on which their system was based and the slaves upon which it depended.

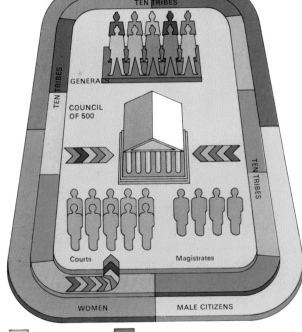

SLAVES — FOREIGNERS
TEN TRIBES
TEN TRIBES
GENERALS
COUNCIL OF 500
TEN TRIBES
Courts — Magistrates
WOMEN — MALE CITIZENS

☐ Assembly   ▨ Direct election
△ Nomination and lot

4 A heifer is led to sacrifice in a religious ceremony. Greek religion was supervised by the state – the correct prayers and sacrifices were carried out by elected priests or private individuals – but there was little of the moral certainty and interference in private affairs that characterized later religions. The gods were irrational and arbitrary and had to be placated, often by sacrifices; their conduct provided little guidance.

take many of the most basic jobs [3]. Despite this influence of the wealthy, it is also remarkable that Athenian democracy saw few of the direct confrontations between rich and poor of the kind that caused continual unrest in most other city states.

All citizens and *metics* were liable for military service, but usually only the more wealthy were called up because troops were expected to equip themselves. By the early seventh century BC the typical Greek soldier was a heavily – and expensively – armed hoplite infantryman [6]. At the height of the Peloponnesian war Athens put about 16,000 hoplites into the field; few other states could raise as many. In Athens the navy was especially important and had an intake of about 12,000 citizens a year.

Sparta: a military state
No other state reached such a fully developed system of democracy as Athens. Its main rival for much of the period was Sparta, whose political and social system is remembered as representing the political opposite of everything that democracy stands for. By 600 BC

Sparta had become a unique military state; Laconia and Messenia had been conquered and their populations either enslaved (*helots*) or deprived of political rights and forced to support the Spartans through taxation and food and manpower supplies.

To prevent revolt or secession, Sparta became a military camp. The citizen body was never large – probably never much more than 5,000 – but everyone was a professional soldier devoted from childhood to absolute discipline and the art of war. Two hereditary kings commanded the army in the field and were members of the ruling council of elders who were elected for life from citizens over 60. Five elected *ephors* had civil and judicial functions. No other Greek state approached Sparta in exclusiveness or xenophobia – even trade and commerce were largely disdained. A total refusal to admit new citizens led to a declining population and eventual defeat.

Sparta has remained the model of a closed and totally disciplined society, but it was Athens whose pursuit of individual freedom and democracy gave the modern world two of its most precious and lasting ideals.

Pericles (c. 490–429 BC) was the great Athenian statesman under whose leadership the city became the richest and most powerful Greek state. He was responsible for the building of the fine temples and monuments on the Acropolis.

5 Women in Classical Greece were the other great "slave" class; they had no political or legal rights and were excluded from all public affairs. Their place was in the home with the chil-dren. Their absence from much social life led to widespread development of homosexual rela-tionships and the institution of the *hetaira* – high-class courtesans outside conventional mores.

6 Spartan infantrymen (hoplites) were normally well and expensively equipped. The Spartan political system of a totally mobilized citizenry devoted to military service (boys were sent to barracks at the age of seven and not allowed other interests) was unique in Greece and gave Sparta far greater importance than its relatively small size justified. Most other states relied upon temporary conscription to fight the frequent wars that were endemic to Greek life. But a class of professional mercenaries did grow up during the 4th century, possibly as many as 50,000 in all, some fighting for the Persians.

7 Gods and goddesses with Athena holding a shield inscribed with an owl are shown on this Greek vase. The pantheon was featured in the epics of Homer and was the basis of the religion of Classical Greece. The gods were conceived of in human form with human emotions. They often interfered in human activities and could be invoked or calmed by prayer and sacrifice. There was little belief in an afterlife; religion was strictly temporal, devoted to a pleasant existence.

8 Athletic competitions and the cult of the well-trained and healthy body played an important part in everyday social life, the gymnasium or stadium being a popular meeting place where men could talk and hold political or philosophical discussions. The Greeks dated their history from the first Olympic games in 776 BC, a festival which, with contestants from all over the Greek world, provided an opportunity for Greeks to gather together as a nation.

73

Greek literature and theatre

Only a fraction of the literature of ancient Greece has survived but what we have is superb in its diversity, nobility and subtlety. Its influence on later European writers, both directly and through derived Latin literature, has been enormous.

The *Iliad* and the *Odyssey*

Homer [Key], one of the greatest names in world literature, still remains a shadow. We do not know whether his works are a compilation of lays, the writings of two or more minstrels or the masterpieces of one towering genius, although the last seems unlikely. What is indisputable is that the *Iliad* and the *Odyssey* are among the greatest of literary works. Composed probably in the eighth century BC, they describe semi-legendary characters 500 years earlier.

The *Iliad* portrays events towards the end of the Trojan War. The story has a substratum of fact, although basically it is a marvellous feat of poetic imagination. The shorter *Odyssey* [1] describes Odysseus' experiences on his return journey from the war. A little later than Homer, Hesiod wrote two epic poems with strong didactic elements, *Works and Days* and *Theogony*. The only other Greek epic poet of note was Apollonius Rhodius (born *c.* 295 BC).

The Greek poets in general preferred the greater freedom of metre and expression offered by the lyric form. Lyric poetry developed from songs sung to the lyre and reached its height during the seventh and sixth centuries BC, but only fragments have survived to demonstrate the passion, charm and vigour of the work of such poets as Terpander, Sappho and Alcaeus [5] of Lesbos, Alcman of Sparta, Anacreon of Ionia and Simonides of Ceos. Stesichorus, credited with the invention of the choral hymn celebrating the heroes of epic poetry, was a strong influence on Pindar (*c.* 522–*c.* 440 BC).

The Greek tragedians

Our debt to the Greeks is perhaps most marked in the field of drama. The 32 tragedies that have survived have moved audiences and inspired writers throughout the world for more than 2,000 years.

The three giants of Greek tragedy are Aeschylus (*c.* 525–456 BC), Sophocles (*c.* 496–*c.* 406 BC) and Euripides (*c.* 480–406 BC). Each has his own particular genius – Aeschylus his grandeur, Sophocles his ability to convey moral dilemmas and Euripides his insight into human nature. Greek tragedies were performed at festivals; three tragedies were given at a sitting plus one satyr play, a grotesque parody of ancient legends.

It is uncertain exactly how tragedy developed the sophisticated form that has come down to us. It seems likely that it grew out of choral songs performed at religious festivals. A semi-legendary poet named Thespis (hence "thespians", meaning actors) is reputed to have introduced a character who responded to the comments of the chorus. True drama involving conflict of character began when Aeschylus introduced a second actor. Sophocles used three actors and injected a greater degree of naturalism. Euripides offended many people by his cavalier treatment of the old subjects and the comparative realism of his themes. The motivation of his characters is convincing and his portraits of women memorable.

CONNECTIONS

See also
72 Classical Greek society
112 Roman literature

1 In the *Odyssey* Homer tells of the dangers and trials that beset the unfortunate Odysseus on his way home to Ithaca after fighting with the Greeks against Troy. Vase paintings show many incidents such as his encounter with the Sirens. He ordered his crew to stuff their ears with wax so that they would not be lured onto the rocks by the Sirens' song; but Odysseus was tied to the mast so that he could safely listen to them.

2 This terracotta statuette of the 4th century shows two actors playing the roles of drunken old men. As was customary in Greek comedy, the actors wore grotesque masks and weird costumes, in this case padded jerkins and tights. Like tragedy, comedy had three actors and a chorus and the plays were performed at festivals. And like tragedy, comedy probably came from choral songs, but of a ribald nature.

3 Seasons of Greek tragedy are presented annually at the magnificent theatre of Epidaurus, dating from the 4th century BC. The various parts of the theatre may be discerned. In front is the circular *orchestra* on which the chorus danced and sang. Behind it is the stage on which the actors stood. At the back of the stage is the *skene*, or permanent background, made of wood in early times and later of stone. The play shown is of Sophocles' *Electra*, the story of the revenge of Orestes, son of Agamemnon and brother of Electra, for his father's death. The actors are to the right, the chorus centre and left. Of Sophocles' work, seven complete tragedies survive with several hundred fragments of others. Immensely successful in 5th-century Greece, he remains admired for the skill with which he structured his plays, but even more for his characterization, especially that of his heroines.

Greek comedy is virtually synonymous with the 11 plays we have by Aristophanes (*c.* 448–*c.* 388 BC). The comedy is broad, the characters ludicrous, but the satire of the life, ideas and leading figures of the time is often sharp. He is not afraid to lampoon even the gods. Menander (*c.* 342–292 BC) wrote urbane comedies of manners which lacked the exuberance and bite of Aristophanes. Imitated and translated into Latin by Plautus and Terence, they have influenced Shakespeare and Molière.

The historians

The two greatest historians of ancient Greece had very different approaches to their craft. Herodotus (*c.* 485–425 BC) took as his overriding theme the struggle between Greece and Persia. He was an inveterate traveller and raconteur.

The first scientific historian was Thucydides (*c.* 460–*c.* 400 BC), who actually lived through the events that he described in his account of the Peloponnesian war. He obviously questioned participants and eyewitnesses on both sides, related cause to

effect and presented the facts in a concise, direct style. Among other Greek historians of note was Xenophon (*c.* 430–*c.* 354 BC), whose *Anabasis* describes the retreat of 10,000 Greeks across Asia Minor, and Polybius (*c.* 200–*c.* 120 BC), whose *Universal History* contains a penetrating account of the rise of Rome.

The two remaining branches of Greek literature are oratory and philosophy. Oratory was highly valued in Greece, although the written speeches we have are obviously partly literary creations. Of the leading orators, who included Lysias, Isocrates and Aeschines, Demosthenes (*c.* 383–322 BC) was by far the greatest. An active politician, he is best known for his eloquent speeches against Philip of Macedon and the speech in his own defence, "On the Crown". Of the philosophers, Aristotle's (384–322 BC) most polished literary work is lost, although the theories enunciated in his *Ethics*, *Politics* and *Poetics* have been extremely influential. The *Dialogues* of Plato (*c.* 427–347 BC) [7] on the other hand, show Greek prose at its best.

Homer is regarded as the father of Greek literature. His importance can be judged by his appearance on this 4th-century coin from Ios. The *Iliad* and *Odyssey* probably evolved from singers' tales but scholars do not know how the poems reached their present form, or if one great figure created them.

4 The chorus in a Greek tragedy is depicted on this vase as dancing to the metre of the verse and singing to the accompaniment of a flute. Particularly in its early period, Greek tragedy incorporated two styles. Dialogue between the actors, or the actors and chorus, was spoken in iambic rhythms. Choral odes, on the other hand, were based on many complex metres that reflected the emotions of the poetry.

5 A wine cooler of the 5th century shows the lyric poets Alcaeus and Sappho. Little remains of Sappho's nine books of verse except for one ode, but there is enough to show her as unequalled for the passion and tenderness of her love poems. Alcaeus, her contemporary and friend, is also known only by fragments, which express political thoughts and outpourings of love and hate.

6 No Grecian musical instruments have survived, but from vases and paintings it is known that the main instruments were the *aulos*, or flute [A], and two types of lyre, the *cithara* [B], and *lyra* [C]. *Auloi* were usually played in pairs, sometimes joined by a mouth band. The *cithara* was heavy and was used by paid musicians to accompany epic songs. The lighter *lyra* was used by amateurs.

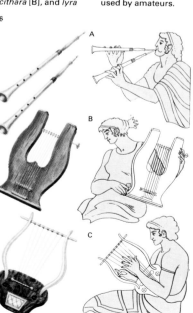

7 This Roman mosaic from a villa near Pompeii, shows the olive grove, dedicated to the Greek hero Academus, where Plato founded the Greek school of philosophy, called the Academy, in *c.* 387 BC. Plato, third from left, is shown here teaching at the Academy, which lasted until AD 529 when it was suppressed by the Romans. Plato was a master stylist, his *Dialogues* (ostensibly with Socrates) ranges from casual conversation to dramatic confrontation, and his work provides us with the prototype of the philosopher – questioning, wise and humble. His thought has been a major influence on Western philosophy up to the twentieth century. To Plato, reality perceived through the senses is a copy: the form or idea is the "real" reality. To achieve knowledge of ideas one must be capable of pure intellect or pure reason. As well as philosophy, Plato taught science and mathematics at the early Academy.

Greek art 1100–450 BC

Greek art and architecture underwent in seven centuries a gradual but profound change, from a primitive phase to a high level of refinement on the brink of the Classical period – the period that has exerted an enduring aesthetic and technical influence on the art of our own era. The objectivity and naturalness achieved in the Greek image of the human figure has remained the norm of later societies in the West. But it should be realized that because of the disappearance of textiles and woodwork and the paucity of early metalwork, the bare stone of ruined temples and the weathered surfaces of sculptures convey no impression of the original colour and variety of Greek art.

The origins of Greek art lie in the Mycenaean culture of the sixteenth century BC. Following its overthrow in the twelfth century there was a "dark age". The only examples of the material culture of this time are vases bearing simple linear devices in a style known as protogeometric [1]. The real birth of Greek art, however, was effectively a rebirth – the renaissance of the ninth and eighth centuries BC, by which time the

development of trade and the demands that a settled society made on its craftsmen produced a range of functional and aesthetically pleasing objects.

The zigzags and other motifs of the geometric art that flourished between c. 900–700 BC became progressively bolder and more elaborate and gradually figures of men and animals were included. Stylized figures were presented in a purely conceptual manner, with no attempt to depict perspective or movement. These early scenes of funeral processions and battles were a breakthrough; they were the first steps towards the narrative art that was to become the dominant feature of the work of Greek artists who drew on the heroic tales of the *Iliad* and *Odyssey* for inspiration.

The High Archaic period

The High Archaic period of 725–600 BC saw an Oriental influence, developing particularly out of the trade with the Phoenecians. Craftsmen from Cyprus and Syria worked in Crete and Attica and their innovations were assimilated by the native Greeks. Floral and

curving patterns were added to straight line motifs and distinctive styles emerged in the painting of Corinth and Sparta.

Athens, originally the acknowledged centre of geometric painting, engaged less in trade and colonization and for a time was eclipsed artistically. But as a direct result of the importation of ideas from Egypt, this period saw the first major works of Greek architecture. Based on the temple design of Egypt, wooden columns and other features were progressively replaced by their equivalents in stone and the "orders" of Greek architecture emerged, beginning with the Doric in the second half of the seventh century. Sculpture also developed under Eastern influence – originally employing moulds for the casting of small stereotyped images for shrines. By about 640, however, larger statues, derived from Egyptian models, were being manufactured. Plentiful domestic supplies of marble and ancillary materials, such as emery for smoothing, were exploited. During this time the tradition of the *kouros* figure [2] was established.

Although owing much in stance and

1 Protogeometric pottery, decorated with simple circles and wavy lines, probably originated in Athens before the end of the 11th century. As the potter's craft evolved, these motifs, often based on the patterns of basketworks and produced on a massive scale (up to 1.5m [5ft] high), became more complex with the introduction of zigzags, friezes and stylized figures of animals and men.

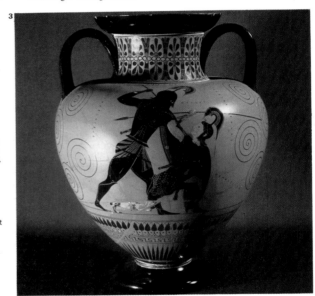

2 Kouroi are figures of naked youths that are derived from Egyptian examples. They are often used as grave markers and are typically elegant with clenched hands and one foot advanced.

3 A masterpiece of black-figure painting, this amphora was made and decorated by Exekias (*fl. c.* 550 BC–525 BC), the greatest painter of his time. It marks the rebirth after 600 BC of Athenian painting. Its detail captures the drama of the moment of death when, according to legend, the eyes of the Amazon queen, Penthesilea, met those of her slayer, Achilles, and she realized that he loved her.

4 From earliest times the Greek architect's chief concern was with temple building. Based on the traditional house plan, and originally constructed in wood, their designs evolved until in the 6th century BC stone superseded wood. Many of the elements from carpentry, such as the adze-like grooves in columns, were retained. Originating in mainland Greece, the first and longest enduring of the "orders" – the Doric – was established. Although modified by western Greek colonists, the temple of Hera I at Paestum, Italy, retains all the classic Doric attributes and is one of the best surviving examples of the style.

manufacturing technique to earlier Egyptian works, a distinctive Greek style emerged. The Greek preoccupation with athletic sports meant that figures of youths were characteristically naked and, through the requirements that this form imposed, sculptors were compelled to study anatomy in greater detail – an important factor in the development of Greek sculpture.

Corinth, a centre for pottery
By about 700 BC, Corinth had taken the lead as the great Greek pottery centre, and the black-figure technique pioneered and perfected there persisted for over 200 years. During the Archaic period, Greece was experiencing a new-found prosperity. Her colonial expansion was virtually complete and the development of stable cities provided the patronage that encouraged a wide range of artistic endeavours to flourish. Temple building increased and the Ionic order was introduced; temples became more elaborate and they were decorated with sculptured friezes of increasingly great artistry. An ever-growing understanding of human anatomy was applied both to a more naturalistic style of sculpture in the round and in relief and to the establishment in vase painting of a set of formal gestures and an iconography on which many unrivalled masterpieces of black-figure painting were based.

Athens and red-figure paintings
By 530 BC, red-figure painting had been invented [5] and Athens had become the chief centre of the style which, as a result of its superiority in allowing greater attention to detail, soon ousted the black-figure technique. The artistic feeling of the Early Classical period (c. 500–450 BC) was profoundly affected by devastating social, economic and especially psychological consequences of the Persian Wars (490–478 BC). The most striking examples of this are seen in the dramatic abandonment of the "Archaic smile" expression and its replacement by a more sober and objective rendering of faces. As the most sudden change in artistic outlook for many centuries it was to contribute greatly to the humanistic "detached calm" that characterizes the subsequent period.

One of the finest examples of the sculpture of the Attic period is the so-called "Mourning Athena". The blending of full and three-quarter profile views is boldly executed and the facial expression and relaxed and natural posture of the subject vividly convey a pensive mood appropriate to the purpose. It was once thought that the relief showed Athena mourning by a grave marker, but it is now held that it is a votive offering by a games victor and that the goddess stands beside a *terma*, the starting point of the race. The craftsmanship anticipates the perfection of the sculpture of the Classical age.

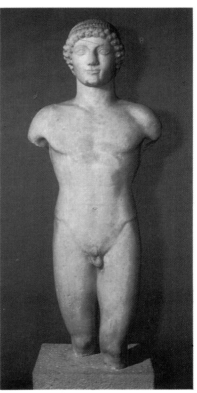

6 An outstanding example of freestanding sculpture from the late Archaic/early Classical period, the "Strangford Apollo" (after its former owner, Lord Strangford) marks the consummation of over a century in the evolution of *kouros* sculpture. The early stiffness and formality of style was replaced by an increasing awareness of anatomical accuracy in sculpture.

5 Red-figure vase-painting, in which the scenes are portrayed in sharp relief in red against an inky black background, began to supersede black-figure painting about 530 BC. This *crater* (a bowl for mixing wine and water) is a fine example of this stylistic evolution and depicts a scene from the Trojan Wars – Achilles and Hector in mortal combat. It marks a development that led to the mastery of draughtsmanship.

7 The "Ludovisi Throne" of the transitional late Archaic/early Classical period is remarkable for the fine quality of its moulding, particularly the effect of the clinging wet hair and clothes. Once in the collection of a Roman cardinal, it is of obscure origin and use but is perhaps part of an altar. The work of a western Greek artist, it shows the birth of Aphrodite from the sea, aided by two attendants.

8 The life-sized bronze statue of a charioteer, commissioned c. 470 BC as an offering in celebration of a victory by a Sicilian tyrant, Polyzalus of Gela, is of exaggerated proportions because it was designed to be viewed from below and partly concealed by the coachwork of the chariot in which it once stood. Although superficially architectural, the modelling and stance of the figure are subtle and brilliantly conceived, making it a classical masterpiece.

Greek art 450–31BC

After about 450 BC Greek art entered a phase regarded by succeeding generations as its great Classical period. It built on experience, took in new influences (especially from the East) and sought higher goals. Trends that had previously been evident, such as the concern with naturalism in sculpture, reached their peak in the work of such masters as Phidias, who is now regarded as the creator of Greek classical art.

Classical formality

Yet throughout the art of the High Classical period (c. 450–400 BC), and allied with its technical virtuosity, is restraint in emotion and a lack of extremes. The art is formal, with few representations of youth or old age and virtually no depictions of anger or pleasure. The concern of the Greek sculptor for an ideal, a perfection of form, led him to a style that was serene and restrained but impersonal and cold.

The High Classical period saw the resurgence of Athens as a great artistic centre, particularly under the ruler Pericles (c. 490–429 BC). After the destruction of the Persian wars, a rebuilding programme of unsurpassed vision produced some of the world's most remarkable buildings – the Parthenon [3] and Erectheum on the Athenian Acropolis especially stand out for their magnificence. The sculptures of the Parthenon offer a realism and grandeur of concept unrivalled in the history of art [Key]. Similarly splendid architectural sculpture, reliefs, votive statues and cult figures also date from this period – Phidias' massive statue of Zeus because the Temple at Olympia was universally regarded as one of the Seven Wonders of the Ancient World.

Although the heroic male figure remained the most typical, sculptors turned increasingly to representations of females, paying particular attention to draperies in, for example, Nike (the goddess of victory) figures [7]. This shift from the idealized portrayal of the nude male to the stylized draped female is an important element in the unformalizing process which was to alter Greek art. Among other arts, wall painting is known to have flourished in the High Classical period and indeed to have taken precedence over vase painting, but material evidence is lacking. Die-engraving, especially in Sicily, was developed as a fine art.

Portraiture develops

In the Late Classical period (c. 400–323 BC), increased prosperity and patronage led to a shift from the unemotional conventions of the earlier classical period. Encouraged first by the prestige-seeking rulers of the western Persian Empire, true portraiture developed as a rival to idealized representations of divinities. Similarly, monumental tomb sculpture was introduced as a celebration of status – it is no coincidence that from this period comes the superb sculptured tomb of King Mausolus of Caria at Halicarnassus, from which the word mausoleum is derived. This work, like that of the Alexander Sarcophagus [5], was produced by Greek craftsmen working on the fringes of the empire; their work and that of mainland Greek artists – Praxiteles [4], Scopas and Lysippus in particular – mark the transition from Classical idealism to Hellenistic realism. Lysippus, for example, Alexander the

Doric

Ionic

Corinthian

1 **The three main Greek "orders"** are the Doric, the earliest, whose capital has a deep abacus (the flat slab at the top) and widespreading echinus (the moulding beneath); the Ionic, with a thinner abacus and projecting spiral scrolls with rich carvings; and the Corinthian, deeper, more elaborate and, until its adoption by the Romans, less popular than the other two orders.

2 **"Poseidon of Artemisium"** - a representation of one of the most revered gods of the seafaring Athenians – presents an artistic bridge between the rather stylized portrait sculpture of the earlier ages of Greek art and the naturalism of the classical era. He is among the most splendidly modelled of all surviving bronze statues with his muscular body and fine head.

3 **The Parthenon** dominates the Acropolis and is one of the world's most impressive and important structures. It was erected by the Athenian ruler Pericles to replace the temple destroyed by the Persians in 480 BC. Under the overall direction of the sculptor Phidias, with Ictinus and Callicrates as architects, it took shape in the years after 447 and was completed by 438 BC, except for the sculptures, which took a further six years.

Designed as an expression of Athens' supremacy and civic pride, it was Greece's largest and most costly building and a fitting centre for the display of Phidias' ivory and gold statue of Athena. The Parthenon embodied many bold architectural innovations and the most cherished artistic conceptions of the day. Entirely of the Doric order, it has 8 frontal columns instead of the usual 6 and 17 in the peristyle. The entasis or bulging effect of the columns counters the optical illusion of concavity that would otherwise occur and their slight inward leaning, coupled with the outward tilt of the upper works, calculated to the minutest degree, ensure its harmonious proportions. The frieze and metopes, which were originally richly painted, were more refined, spectacular and skilfully executed than any predecessors and stand out among the world's greatest works of sculpture both in relief and in the round.

Great's official sculptor, produced works that, unlike some formal groups of earlier Classical times, could be viewed satisfactorily from virtually any angle. He also produced lifelike portraits of the patron. In response to the newfound wealth of the Greek Empire, both painting and jewellery flourished.

Realism in the Hellenistic period
The Hellenistic period (from the death of Alexander in 323 to the Battle of Actium in 31 BC) was the last phase in the history of Greek art. The development of court life stimulated a wide range of arts; Pergamon was established as a leading centre of architecture and sculpture, while distinctive local schools emerged in the Seleucid Empire and in Alexandria and Rhodes. Athens alone preserved many of the traditions of classical art and continued as the primary centre on the Greek mainland.

The aspirations of the rulers of the Alexandrian Empire led to a greater emphasis on portraiture, which often attempted to convey the quasi-divine status of its subjects. For the first time, representa-

tions of men predominated over those of gods and the original religious and mythological flavour of Greek art declined and was replaced by accurate – if still sometimes idealized – depictions of notable people. Both public and domestic architecture became increasingly grand while large-scale temple building declined. The more elaborate Ionic order was generally adopted in preference to the austere Doric, and the Corinthian order was used for the first time in exterior construction [1].

Works of sculpture such as the reliefs from the altar of Zeus, Pergamon, the Nike of Samothrace [7] and the Laocoön group depicted vigorous action and emotion [6]; femininity was portrayed in such works as the Venus of Milo; while physical suffering, anguish, old age and extreme youth were represented for the first time.

Towards the end of the Hellenistic period, a trend to the copying of earlier, classical works can be observed. Through this process and the Romans' large-scale absorption of its chief elements, the continuity of Greek art was assured.

The theme of the friezes from the Parthenon – the fight between the Lapiths (people from Thessaly) and centaurs at the wedding of the Lapith king Pirithaus – symbolizes the triumph of order and civilization over barbarism. This was a constantly recurring motif in High Classical art.

4 "Hermes carrying the infant Dionysus" is possibly the only surviving original work by the Athenian master sculptor Praxiteles, who was regarded as one of the greatest craftsmen of his age. This work displays a refinement of technique seldom encountered before the Hellenistic period. The idealized classical beauty and proportions of the figure of Hermes are enhanced by the impression of a graceful curving upward line as he playfully offers a bunch of grapes to the child-god Dionysus. This statue was discovered inside the Temple of Hera, at Olympia.

5 The "Alexander Sarcophagus" was commissioned by the last king of Sidon, who owed his position to Alexander the Great. It was both a tomb and a monument in praise of Alexander's achievements. It depicts scenes from his life, including a battle against the Persians and a spirited hunting scene. It was greatly influenced by the work of Lysippus, greatest of all fourth century sculptors, particularly in the modelling of the figures. It nevertheless conveys the vigour and dynamism with which Hellenistic art was inspired. It is also a fine example of Greek sculpture, showing traces of original paintwork.

7 The Nike (Winged Victory) of Samothrace – an example of the superb work of the Rhodes school of sculpture from the Hellenistic period – was probably made to commemorate a sea victory over Antiochus III of Syria and originally depicted Victory alighting on a ship's prow. The powerful, yet light movement of the figure is expressed by the swirling garment.

6 The famous statue depicting the death struggles of the Trojan prince Laocoön and his two sons was produced in Rhodes by Greek artists who specialized in compositions conveying violence and anguish, emotions which became characteristic of Hellenistic art. Like many Greek works of the time it was much admired by the Romans and was taken to Rome by the Emperor Titus in AD 69, where it was described by Pliny the Elder as "superior to anything produced in painting or sculpture". In 1506 the statue was rediscovered in the Baths of Titus and bought by Pope Julius II for the Vatican, where it was studied by such artists as Michelangelo, Bernini and Rubens.

Alexander the Great

Alexander, the son of Philip of Macedon and Olympia, princess of Epirus, was born at Pella in Macedon in 356 BC and died in Babylon in 323 BC. The pupil of the philosopher Aristotle between the ages of 13 and 16, he succeeded to the Macedonian throne in 336 BC on the assassination of his father, whose cavalry he had commanded two years earlier at the battle of Chaeronea. This battle had finally ended Athenian hopes of regaining the leadership of Greece, giving hegemony to the Macedonians who were, at best, peripheral Greeks. It also marked the victory of the soldier over the rhetorician, for it had been Demosthenes, the Athenian orator (c. 383–322 BC), who had most attacked the Macedonians and their king.

Securing the frontiers of Macedonia
During the next 13 years Alexander was to establish the greatest empire the ancient world had ever known [4], stretching from the Libyan frontier to the Punjab. His exploits gave rise to stories and legends in all the languages of Europe and many of those of Asia.

Upon his accession he marched south into Greece and, asserting Macedonian supremacy, had himself elected by the Greek League as a leader of an Asian expedition, one that had already been planned by his father. The oracle of Delphi hailed him as invincible. In 335 BC he campaigned towards the Danube, to secure Macedonia's northern frontier. On rumours of his death, a revolt broke out in Greece with the support of leading Athenians: Alexander marched south covering 386km (240 miles) in a fortnight. When the revolt continued he sacked Thebes, killing 6,000 people and enslaving the survivors, sparing only the temples and the house of Pindar (c. 522–c. 440 BC), one of the greatest of Greek poets. His base thus secured, he prepared for the campaign for which his father had raised him. He also needed the riches of Persia to pay his father's debts to the Macedonian army.

The victory over Darius
Alexander crossed the Hellespont in 334 BC, with 30,000 infantry, 5,000 cavalry and a corps of specialists. He paid a visit to Troy and at the River Granicus fought a battle that opened Asia Minor to his southward drive. He defeated Darius, the Persian king (who fled, leaving his family), at the Issus [2] in 333 BC and continued until his advance was held up at Tyre, which fell in July 332 BC.

After the sack of Tyre, Alexander went to Egypt, where he founded the city of Alexandria. On visiting the shrine of Amon at Siwah Oasis, he was greeted as pharaoh, son of Ammon, an event that gave rise to stories of his divine origin. In 331 BC, he marched eastwards to the Euphrates and fought another battle at Gaugamela, but Darius once again escaped; Alexander received the surrender of Babylon and Susa with their riches. In 330 BC he captured Persepolis; then he sent the Thessalians and Greeks home, apparently planning a Persian-Macedonian empire. While he was campaigning eastwards Darius was assassinated.

Alexander next marched into Afghanistan and Transoxiana, partly to pursue a rebellious general, and thence to Samarkand and Alexandria Eschate (present-day Leninabad). There were further revolts until

1 The battle of the Hydaspes (Jhelum) in the Punjab, shown here on this coin, was one of Alexander's most skilfully planned and executed victories, greatly extending his empire. His tactics were greatly influenced by the horse's natural fear of the elephant. The defeated Indian ruler Porus became an ally. During this battle Alexander's famous charger Bucephalus died and was given a full imperial funeral.

2 This mosaic from Pompeii, copied from a painting by Philoxenus, shows Alexander commanding his army against the Persians, under Darius, at the Battle of the Issus, 333 BC. Darius fled, leaving his queen and family together with a vast amount of wealth to the Macedonians.

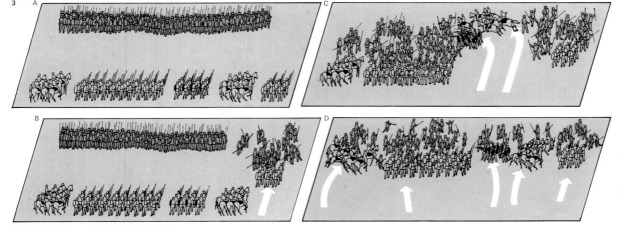

3 Alexander placed the phalanx at the centre of his battle order [A]. Fighting was initiated on the extreme right [B]. At the right moment Alexander would lead his companions, supported by the household infantry, in a charge that penetrated the gap in the enemy line [C]. As the enemy ranks broke, he wheeled his companions to take the flank to relieve the left and centre of his army. As the enemy retreated, he pressed home his advantage with his full force [D].

328 BC, the year of his marriage to Roxana, daughter of the king of Bactria, a marriage that seems to have been symbolic of East-West fusion. At the same time his absolutism increased. His murder of his commander and friend Clitus in a drunken brawl angered the Macedonians.

In 327 BC he led his men into India, one army marching through the Khyber Pass while the other, which he commanded, fought its way through Swat. The final battle was fought on the Hydaspes (Jhelum) [1]; after the defeat of Porus, an Indian prince, Alexander's soldiers refused to go farther.

Return from the East
On his return to Susa, Alexander found a state of corruption and oppression. He set about a ruthless campaign of punishment. This was followed by a scheme for settling Greeks and Macedonians in Asia and Asians in Europe as part of a plan for fusing the two regions. A more immediate project was the marriage of Alexander and Hephaestion (his closest friend and lover) to two of the daughters of Darius, while another 80 Macedonian officers married daughters of Persian nobles.

Asian soldiers had already been trained in Macedonian military methods and were now admitted to the army; others were recruited to the cavalry and Persian officers to the royal bodyguard. In 324 BC his 10,000 Macedonians, already disturbed by the new army policy, mutinied. Alexander ordered 13 of their leaders to be killed on the spot, appointed more Persians and Medes to Macedonian posts and transferred regimental names to what the Europeans considered barbarian regiments. The Macedonians then set off on the return march.

Plans were now made for a campaign into Arabia, but Alexander developed fever and on 13 June 323 BC he died, not yet 33 years old. His empire began to disintegrate almost at once as the various regional commanders assumed the titles of kings in their own right. Although Alexander was renowned primarily for his military conquests, his most enduring achievement was to extend the influence of the Greeks to a vast area of the ancient world, and to widen the basis of Hellenic culture through contact with the East.

Alexander is wearing the ram's horn headdress on this coin.

4 The extent of Alexander's campaign over 14 years explain the rapid collapse of his empire after his death. There was no consolidation and the vast distances precluded real integration.

5 A picture from the Flemish *Alexander Romance* (a 13th-century legend cycle) shows him preparing to fly in a basket attached to two griffins, which are to be lured upwards by a lump of meat on a stick. Another illustration from the same source shows him in a glass "submarine", an idea that also occurs in a Malay story depicting Alexander. Both pictures imply his dominion over land, sea and air.

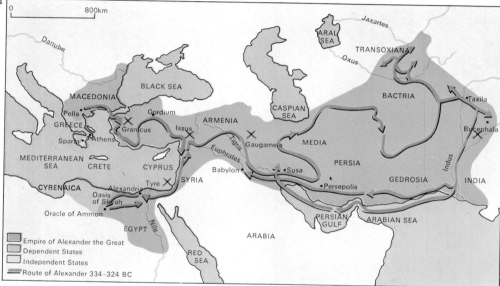

Empire of Alexander the Great
Dependent States
Independent States
Route of Alexander 334–324 BC

6 The Alexander legend is told in many languages, from Middle English to Malay, throughout the countries of Europe, the Middle East and India. He is depicted in many forms – Burgundian king, Armenian horseman, Persian prince. (One sultan in the 15th century had his own features depicted as Alexander's.) In the Muslim world he was held to be a pious follower of Islam who fought pagans and spread the faith. He was identified with the Two-horned One mentioned in the Koran, almost a thousand years after his death. In this Persian painting (1595–6) he is being asked to spare pagan idols.

Greek science

The earliest stirrings of Greek science are found in the eighth century BC in the Homeric poems with descriptions of the stars and an unusual concept of the universe as a sphere [1]. Other civilizations had been content with hemispherical skies but it was the Greek love of symmertical shapes that led them to the concept of a spherical universe.

Greek studies of the universe

The first Greek scientific men whose names are known – Thales, Anaximander and Anaximenes – came from the eastern seaboard of the Aegean and lived during the sixth century BC. Thales accepted the spherical universe and believed that water was the basic substance from which everything was formed. Anaximander, who thought some indefinable substance (not water) was the basic material, taught that the earth was cylindrical in shape and, like Anaximenes, he believed the heavenly bodies were holes in a dark sky through which shone a surrounding fiery zone. The most important sixth-century scientist was Pythagoras, best known for his proof of the relationship between the sides of

any right-angled triangle. He also investigated musical harmony, which led him to suggest that there was a divine relationship between numbers, music and the universe.

Significant developments in the fourth century BC followed the establishment by Plato (c. 427–347 BC) of an academy in Athens, where he laid great stress on the mathematical nature of the universe. His pupil Aristotle (384–322 BC), the greatest scientific philosopher of antiquity, set up his own academy – the Lyceum – also in Athens. He adopted the theory, first formulated by Empedocles in the fifth century BC, of the Four Elements – earth, air, fire and water – as the fundamental components of all matter. Astronomically, Aristotle discussed whether or not the earth moved in space but, on the basis of the available evidence, he tended to favour a fixed earth in the centre of the universe. Aristotle also discussed the nature of change and especially of motion, as well as teaching that there was a fundamental difference between celestial and terrestrial bodies. The former were eternal and changeless, and all change, he believed, must occur below the

sphere of the moon, the nearest body to the earth. He also rejected the idea, proposed in the fifth century BC by Democritus and Leucippus, that the universe is composed of separate and indestructible atoms.

Investigations in the pure sciences

Although it was Aristotle's views about the physical universe that exerted the most profound influence on science for the following 2,000 years, he was at his best in the biological field. He carefully described the compound stomach of ruminants such as the cow and the habits of bees and the diseases they suffered; he studied the placental dogfish that reproduces its young live rather as a mammal does and made a general study of sexual reproduction [2]. He also studied plants, although it was his friend and disciple Theophrastus (c. 372 – c. 286 BC), who was the founder of botanical science. Aristotle emphasized the important notion that there is a continuous order of being stretching from inanimate matter up to man.

Following the claim by Eudoxus (c. 408–c. 355 BC) that heavenly bodies moved

1

2

Aristotle's catfish
(*Parasilurus aristotelis*)

1 A Victorian illustration of the Homeric universe shows a disc-shaped earth with Greece as its centre, floating on water and surrounded by the sphere of the universe. The sun is rising in the east and the moon is shown high in the sky. Most Greek philosophers thought that the sun, moon and planets all orbited the earth: their innovatory idea was that the heavens were an all-embracing sphere, a more perfect shape than a dome or hemisphere. There was no evidence to support this view of the universe; it was adopted purely for aesthetic reasons. But the belief that the earth was at the centre of the universe remained until Copernicus' work in the 16th century.

2 Aristotle's biological observations include notes on bees and fish. He discovered that the male catfish takes upon himself to guard the eggs of the female until the young hatch. Since other catfish species behave differently, this was questioned until the 1850s, when Aristotle was proved correct by Louis Agassiz, who found that the North American male catfish (*Amieurus*) does the same.

3

4
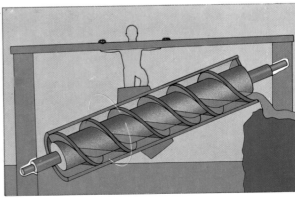

3 Archimedes in his bath (from a 16th-century engraving) is a reminder of his discovery that a body displaces fluid equivalent to its volume. This means that, regardless of shape, objects of equal density and weight displace the same quantity of fluid. Archimedes used the new principle to determine if a crown made for the king of Syracuse was of unalloyed gold. The crown displaced more water than the same weight of pure gold, proving that it contained other metal.

4 The Archimedean screw for raising water may have originated before Archimedes but was attributed to his mechanical skill. Driven by a handle or foot pedals, the spiralling screw rotates in a cylinder, drawing the water upwards.

in circular orbits round the earth, the third-century astronomer Aristarchus (310–230 BC) suggested that the sun, not the earth, was the centre of the universe and discussed the sizes of the sun and moon. But his theory of the universe proved unacceptable. A century later, Hipparchus (190–120 BC), the greatest observational astronomer of antiquity, discovered the precession of the equinoxes, compiled a catalogue of stars, calculated solar eclipses and had an advanced theory of the sun's motion. The third century BC is notable also for the mathematical physics of Archimedes (*c.* 287–212 BC) [3] and the establishment of the library and museum of Alexandria. There, for the next 500 years, research of an advanced kind was carried out; mathematics flourished with Euclid (*fl. c.* 300 BC) [Key] and Apollonius (*fl.* 250–220 BC), who was noted for his work on astronomy and conic sections. Eratosthenes (*c.* 276–194 BC) estimated the earth's circumference to within 400km (250 miles).

In the second century AD the astronomer Ptolemy (*c.* 90–168) produced his remarkable *Almagest*, a digest of Greek astronomy, and his *Geography*, using longitude and latitude as well as his own stereographic map projection [7]. The medical lore of Hippocrates of Cos in the fifth century BC was developed by Herophilus and Erasistratus in Alexandria two centuries later [6]. They dissected the human body, distinguishing veins from arteries, sinews from nerves, and generally laying the foundations of anatomy.

Applied sciences and the study of machines

The compound pulley was known in Aristotle's time and the screw for raising water was invented in Syracuse by Archimedes. Much Greek technology, however, was also developed at Alexandria where, in about 200 BC, Ctesibius designed water clocks [5] and a force-pump, and in the first century AD Hero not only developed the study of machines and pneumatic devices (although he failed to find practical uses for his inventions) but also improved surveying techniques. The discovery of a mechanically geared "computer" [8] for astronomical calculations suggests that Alexandrian technology had a wide influence.

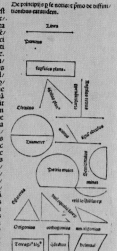

The great basic text on Greek geometry was Euclid's *Elements*. This page is from the first printed edition, 1482.

5 The Tower of Winds in Athens is surrounded at the top by sun-dials. It once held a large clepsydra (water clock), based on a design by Ctesibius. This had an elaborate siphoning system to maintain a constant water pressure so that the clock was reasonably accurate.

6 Greek medicine had a long and notable history; psychological medicine was practised in healing temples, as were herbal treatment and surgery. A miscellaneous collection of Greek surgical instruments shows, among other tools, forceps and scalpels of various designs.

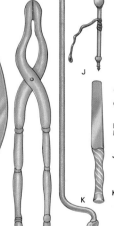

A Surgical knife with decorated handle and spatula.
B Small knife with spiral band of silver inlay.
C Bellied surgical knife made of bronze.
D Bistoury knife with ivory handle and steel blade mounted in bronze.
E Phlebotome used as a blood-letting knife.
F Lancet knife with steel point and guard.
G Fistula knife with curved blade cutting on one side and blunt end.
H Uvula forceps.
I Cautery used to burn away infection or stop bleeding.
J Drill driven by thong attached to the shaft and used to remove a weapon lodged deeply in the bone.
K Steel chisel used to divide distorted bone.

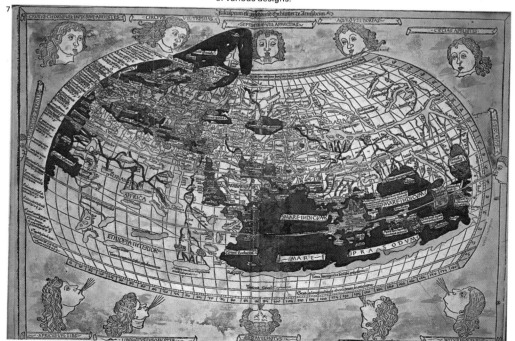

7 This map of the world, published in 1486, was based on one drawn by Ptolemy in Alexandria some 1,200 years earlier. That map drew on evidence from travellers, especially the Greek explorer Pythias of the 4th century BC.

8 A Greek astronomical "computer" was discovered in 1900 in a wreck dating from about 65 BC off the island of Antikythera and is shown in reconstruction here. It contained an extremely elaborate system of internal gears.

Bronze and Iron Age Britain

The end of the Neolithic period in Britain in the second half of the third millennium BC, was marked by major social and technological innovations. Collective burial under long barrows was replaced by a single burial under round ones; the first metal artefacts, of pure copper or a copper-arsenic alloy, are found in the burials; and a new type of pottery, known as the Bell Beaker [1], is found in burials and settlements throughout Britain. Traditionally these burials have been interpreted as marking the arrival of the so-called "Beaker people", metal prospectors from Europe. It is now thought, however, that the changes demonstrate internal developments in society rather than the arrival of a new people. The Beaker and copper dagger, already in use in Europe, were prestige objects for display by the increasingly important chiefs.

The Wessex culture

The Beaker period is sometimes described as the Copper Age, because of the predominance of copper tools at that time. In the full Early Bronze Age that began in the first half of the second millennium BC, the most notable feature was the emergence of the so-called Wessex culture in southern England. This was characterized by single burials under round barrows of specialized forms [2]. The rich grave goods that accompanied the burials included objects of gold [3], amber shale and faience as well as the more usual bronze, flint, bone and pottery.

The chieftains of that time, now perhaps united under a "paramount chief" holding sway over a larger area, may have been responsible for the last main period of construction at Stonehenge with its massive sarsen stones (it may possibly have been built earlier, in Beaker times). There is continuing controversy whether the Wessex culture was introduced to Britain by intrusive overlords or whether it arose locally, and also whether there was contact between the Wessex culture and Mycenaean Greece [3].

The end of the Early Bronze Age, c. 1400 BC, saw a hiatus in sepulchral and religious development and a notable geographical shift in the centres of wealth. Wessex and southwest Wales ceased to be as rich as before and new centres of wealth arose in north Wales, the Thames Valley and the Fens. At the same time the old sky-oriented religion that had continued throughout the Early Bronze Age (as continuing activity at Stonehenge demonstrates) seems to have died out. In its place there is evidence, from the Middle Bronze Age onwards, of a new water-dominated religion; fine metalwork was put in rivers, lakes and bogs, presumably as offerings to water gods.

The later Bronze Age

Few rich burials have been found from the Middle and Late Bronze Ages; the normal rite was now cremation, often with an urn but without grave goods, under a barrow or in a flat cemetery. However, many hoards of metalwork date from that time. These demonstrate an increasing quantity of bronze in circulation and the growing mastery of metalworking techniques achieved by the bronzesmiths. They also indicate the development of new forms of tools and weapons: palstaves, rapiers and sickles, for instance, appear in the Middle Bronze Age, whereas in the Late Bronze Age hoards, after

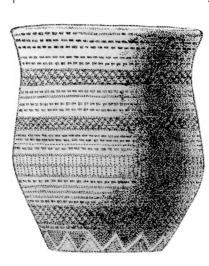

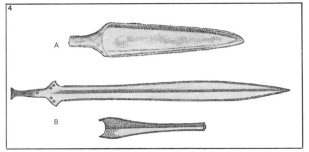

1 **Bell Beakers** were a distinctive type of pottery introduced into England in the late 3rd millennium BC and buried as prestige objects with chieftains. These pots were drinking vessels made in a fine hard ware with a burnished surface, varying in colour from red to dark brown or black. Their ornament was impressed in horizontal zones. Bell Beaker pottery has been found widely distributed in central and western Europe, and is especially common in parts of Spain and Portugal, southern France and Brittany, Czechoslovakia, Germany and the Low Countries. This distribution may reflect the movement of people prospecting for and trading in metal ores.

3 **This gold cup, found at Rillaton**, Cornwall, is the best-known of the many goods of rich and rare materials that have been found in the round barrows of southern Britain, left in the graves of the wealthy aristocrats of the first half of the 2nd millennium BC. Amber and faience (a blue or turquoise vitreous paste made from a mixture of sand and clay) have also been found. This cup has been compared with vessels in the contemporary Shaft Graves at Mycenae, as have other gold and faience objects from the barrows of the Wessex culture. But these objects could have been made in Britain without the benefit of Mycenaean example. There is no other evidence for strong Mycenaean influence in Britain.

2 **Burial in barrows was common** in the late Stone and Bronze Ages. The Neolithic people practised collective burial in monumental tombs with wooden or stone chambers under long or round mounds. Single burial under a round mound began at the end of the Neolithic period. This rite continued through the Bronze Age and cemeteries of these barrows, such as these at Windmill Hill, Wiltshire, are common in many parts of Britain. The practice of single burial, accompanied by grave goods, including some made of exotic and rare materials, suggests a changing emphasis in society, with stress now on the importance of the individual.

4 **More efficient tools and weapons** were developed during the Bronze Age. In the Early Bronze Age the chief personal weapon was the dagger [A] which lengthened into the full-length sword used for thrusting in the Middle Bronze Age. It was replaced by a slashing sword with a strong hilt and leaf-shaped blade (shown with a chape from its scabbard) only in the Late Bronze Age [B].

c. 1000 BC, there are socketed axes, true swords [4] and beaten metal objects, including shields and vessels. From the Middle Bronze Age onwards settlements are known; these include embanked enclosures and after c. 1200 BC the defended hilltop settlements known as hill-forts [7].

The first use of iron

The introduction of iron technology was not a sudden event. Some iron objects were occasionally used as early as c. 700 BC, but iron did not come into general use for another 200 years. The main factor in the development of the Iron Age was the continuing culture of the Bronze Age. The principal types of settlements – farmsteads, villages and hill-forts – all developed from Bronze Age antecedents, as did the round post-built house, which is one of the main Iron Age dwelling forms. However, there was certainly trade with Europe and, later in the Iron Age, there were also two limited movements of people from continental Europe. The first was a migration from France that brought the practice of chariot

burials [6] into Yorkshire perhaps as early as the late fifth century BC. The second was a movement from northeast France into southeast England in the first century BC. This can be identified with a historically known people, the Belgae, who were described by Julius Caesar (c. 100–44 BC).

Insular Iron Age communities were gradually developing an urban way of life during the second half of the first millennium BC, but the Belgae introduced a more organized urbanism already developed in Europe. They lived in large defended settlements known as *oppida*, which can be regarded as true towns; they were organized on a tribal basis and society was highly stratified, with a wealthy aristocracy at the top and slaves at the bottom, separated by a "middle class" of craftsmen and by the bulk of the peasantry who were engaged in primary production. They used the potter's wheel and they minted coins, both skills previously unknown in Britain. They traded abundantly with their Romanized relatives in Gaul and were partly Romanized by the time of the Claudian invasion of AD 43.

Decorated bronze mirrors are among the finest products of British craftsmen in the Early Iron Age and they must have been highly prized by their aristocratic women owners, with whom they were sometimes buried. This example, dating from the early 1st century AD, was found at Desborough, Northamptonshire. Its reflecting surface was made of polished bronze and its back was decorated with intricate patterns. The engraved area was probably set out with a pair of compasses; the outline was then engraved and the filling of parts of the pattern completed by chasing. Wear on the inside of the ring on the handle suggests that the mirror was suspended when it was not being used.

5 Dartmoor was an attractive area for prehistoric settlement and because it has been little exploited since early times it has many sites available for archaeological exploration. Stone was readily available there for use as a building material, and so the footings of walls of huts and of enclosures are still visible, as in this example at Grimspound. Few sites have yielded clear dating evidence but most were probably built in the Bronze Age. The huts are circular and vary in diameter from 3m (10ft) to 12m (39ft). Only the footings were of stone; the rest of the structure was timber, thatch and turf. The buildings were conical in shape.

6 Chariot burials have been found of some of the warrior aristocrats of the Iron Age in Yorkshire. In this example from Danes' graves, two bodies were buried in a crouched position on the floor of a rectangular grave pit under a round barrow. With them were the remains of a dismantled chariot and several items of harness fittings. In other burials the chariot was buried intact and on one occasion two horses had been buried with the vehicle. This special type of burial does not occur elsewhere in Britain but is well known in continental Europe. It may have been introduced by settlers from France, perhaps from Burgundy, where there are close parallels with the Yorkshire burials.

7 Hill-forts, or fortified settlements on hilltops, were first used in the later part of the Bronze Age, but most of them belong to the Iron Age. This example, Badbury Rings, Dorset, had a long development and may have lasted into the Roman period. The area enclosed is c. 7 hectares (17 acres), and is surrounded by two substantial banks and ditches, as well as a third, much slighter one outside them. One of the entrances turns inwards and the other has additional outworks. Hill-forts that have been excavated have revealed continuous intensive occupation, organized internal planning and some public buildings and provision for craftsmen. These hill-forts can be considered the first true urban settlements in Britain.

8 This large chalk-cut horse is situated just northwest of the hill-fort of Uffington Castle, Oxfordshire. It measures 110m (360 ft) in length and is a maximum of 40m (132 ft) in height and has been cut down to the natural chalk in broad terraces. Such hill figures are impossible to date accurately, but on stylistic grounds this horse is usually thought to have been cut in the late pre-Roman Iron Age, and to have been the first hill figure in Britain. But some scholars give it a much later date. Its dramatically curved and attenuated form resembles horses portrayed by Iron Age craftsmen in other media, such as those found on coins or on wooden buckets decorated in sheet bronze, known from Aylesford and Marlborough. It may have been the symbol of a local tribe.

The Celts 500 BC-AD 450

From the beginning of the second phase of Celtic culture, known to archaeologists as La Tène, our knowledge becomes much more detailed. The archaeological record is filled out by written accounts that give a fuller, more detailed identity to peoples previously distinguished by the remaining artefacts of their material culture alone. For the first 500 years of this period of European history, the story is basically that of the astonishing growth and development of the Celtic world and its gradual decline with the rise of Rome and the establishment of the Roman Empire.

The development of Celtic culture
The term La Tène is applied to this period of Celtic culture that extends from about 500 BC and is the period of their greatest attainments. Like Hallstatt for the first phase of c. 700–c. 500 BC, the name La Tène is taken from the name of a major archaeological site. In the nineteenth century a large variety of religious offerings [3] were found at La Tène on the shores of Lake Neuchâtel in Switzerland, and these findings, like those discovered at Hallstatt, testify to the changes in the

equipment and way of life experienced by the Celts. As with Celts of the Hallstatt culture, their altered life-style was developed more or less simultaneously, and with great rapidity, over much of Europe.

The Celts were still highly aristocratic in their social organization; kingship was common among the tribes and below the king were the warrior aristocracy and freemen farmers. The Celts had a highly evolved religion [5, 6] with the powerful priesthood – the Druids – which itself formed a major class. In addition there were slaves.

The Celts had taken to using a light two-wheeled war chariot pulled by two small ponies, and the artistic skill of their craftsmen in metalwork, always remarkable, had advanced to new heights. At the same time, a revolution had taken place in their art style. The old Urnfield and Hallstatt patterns, the fine animal art of the Scythians, Greek foliage motifs, and elements from styles formed much farther east had been merged into a brilliantly original art – subtle and elusive, and full of magical significance. From 500 BC this new aspect of Celtic culture

spread rapidly and by 300 BC it was dominant from the Baltic to the Mediterranean and from the Black Sea to the Atlantic.

Celtic society and warfare
The early Celts did not make written records, they recorded their history orally. However, unlike the earlier Hallstatt culture, that of La Tène is nevertheless well documented; much of our knowledge about the daily life of the Celts has been provided by contemporary Greek and Roman writers.

To them the Celts were tall, muscular and fair-skinned, qualities most attributable to the warrior class and reaching their ideal in the *gaestatae* ("spear-bearers") who were a highly specialized class of Celtic warriors.

Warfare was an essential part of the Celtic life. Armed with highly efficient iron weapons the Celts swept through central Europe in the fourth and third centuries, overcoming their Etruscan neighbours in about 400 BC, sacking Rome in 390 and plundering the shrine at Delphi in 279.

But the Celts were not only warriors. They achieved a high level of material

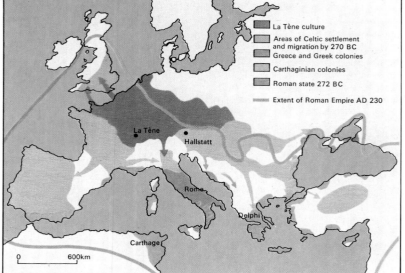

1 La Tène, a key archaeological site in Switzerland, is the name given to the period of Celtic culture in Europe that lasted from approximately 500 BC until the time of Roman expansion during the 2nd and 1st centuries BC. As the map shows, by about 270 BC the Celts had migrated into France and the Iberian Peninsula, parts of the British Isles and to some extent eastwards into central Europe. From this time, however, they gradually lost ground to Rome. By the mid-1st century BC the Roman dominance was assured.

La Tène culture

Areas of Celtic settlement and migration by 270 BC

Greece and Greek colonies

Carthaginian colonies

Roman state 272 BC

Extent of Roman Empire AD 230

2 This bronze wine flagon is one of a pair from Basse-Yutz in Lorraine. Inlaid with coral, it demonstrates elements that contributed to the subtlety and charm of La Tène art. The wolf-like animals forming the handles and on the lid link these pieces with the Bronze Age cult of water birds.

3 This silver-covered iron votive torque (bracelet) weighs about 6kg (13 lb). Both ends have sacred ox heads, each with its own twisted necklace. Too heavy to be worn, it was probably hung on a stone or wooden divine image. It comes from Trichtingen, Württemberg, in Germany.

4 Warfare was the main influence on Celtic architecture. Fortified dwellings built for defence were common. One such was the crannog [A]. Found mainly in Ireland, crannogs were artificial islands built of timber, clay, peat and brushwood. Dwellings were constructed on top of the island. This crannog, from the La Tène period, is in County Antrim. The most famous Celtic fortifications were the hill-forts (*oppida*) [B]. This hill-fort, seen from the air, is in Somerset.

civilization using their mastery of iron-working to open up new land and develop agriculture. Their economy was based on mixed farming – the cultivation of grains and vegetables, and cattle raising – and on trade. These two aspects – expansionist warfare and domestic settlement – can best be seen in their hill forts (*oppida*), the remains of which are scattered throughout Europe. Originally built as hill-top forts for defence only they varied considerably in size, some developing into major towns. One such was the *oppidum* at Bibracte near Autun in France which covered more than 130 hectares (330 acres) and had houses, streets and possibly shops.

The rise of the Roman Empire

It was the rise of Rome – a little town on the bank of the Tiber – into a great, organized civilization which, by 225 BC and after bitter fighting, defeated the Celts at Telamon in Italy, and took the first major step towards subduing them forever. Even after actual subjugation, the Celtic traditions and language continued to survive but in forms modified to agree with the needs of Roman

institutions. Only Ireland and much of Scotland escaped Roman domination in the British Isles. It was there that the old Celtic traditions and way of life survived and were written down by the scribes of a Celtic Church, of the fifth century AD and deeply sympathetic to the heritage of its people.

The Roman Empire lasted into the fifth century. But from the third century it was endangered by increasingly powerful "barbarians". It was these barbarians from the north, east of the Rhine and from the vast steppelands of the eastern continent who, during the fifth century AD, eventually wrecked the Roman Empire and laid the foundations of feudal Europe. The people of the Roman Empire had become used to living in towns and obeying a single, absolute government. The barbarians knew no such control, living in tribes without towns and led by chieftains who gained their power through the dictates of tribal custom and tradition. These pagan peoples struck deeply at the growing Christian Church in Europe, a Church to which Celtic missionaries from Ireland were to contribute profoundly.

A bronze figure of a dying Gaul was recovered from the hill-fort at Alesia, Alise-Ste-Reine, where Vercingetorix, the young Celtic leader, king and commander of all the Gallic confederates, made his last great stand against Julius Caesar in 52 BC. For a time he managed to repulse Caesar and cause him heavy losses, but after a final bitter battle he was defeated. In spite of the dignified and noble way he surrendered in order to save his fellow countrymen, Caesar held him prisoner in Rome for six years and then ordered his belated execution.

5 The god Cernunnos, "The Horned One", figures as the Lord of the Wild Beasts on one of the inner plates of the great votive silver cauldron found at Gundestrup, Jutland, Denmark. He wears the antlers of a stag.

6 The cock [A] was believed to avert evil with its crow and thus in Britain it was considered unlucky to eat it. The peculiar three horns on the bull [B] conveyed an idea of its supernatural qualities as the number three held religious significance.

7 Belief in the "evil eye" goes far back into Celtic minds. The eye on this stone figure of a boar god dates from about the 1st century BC.

7

6 A

B

8

8 An early example of La Tène art style is shown here. The bronze plaque from a double burial at Waldalgesheim (*c.* 325 BC) shows a mask-like face, a torque round the neck and, on the head, traces of a leaf-crown which is a token of divinity. The hands are raised in the Celtic attitude of prayer. This bronze is one of a pair.

9

9 Bronze decorated mirrors are among the most splendid pieces of Celtic art. This example, from a woman's grave at Birdlip, Glos, is pre-Christian.

The Etruscans

Far less is known about the Etruscans than about their contemporaries, the Greeks and Romans. The reason lies mainly in the total loss of their literature, which leaves them from our point of view inarticulate. But whereas their writings consist almost solely of brief funerary inscriptions, their other material remains offer much greater scope for archaeological research.

Origins of the Etruscans

The long-standing controversy over this people's origins seems at last to be dying down. Ancient records began it: Herodotus declared that the Etruscans sailed from Lydia in Anatolia to colonize a new territory in the west; and Dionysius of Halicarnassus argued that they were natives of Italy. Today both views are regarded favourably, and both are at least partly right.

Certainly the argument for continuity between the Villanovan settlements in northern Italy (c. 1100–700 BC) and cemeteries of the ninth century BC and the Etruscan cities and cemeteries of the early seventh century BC on the same sites is stronger than ever. Most of the population and much of the culture can be traced beyond this, back into the Bronze Age. Conversely, at several stages new ideas and cultural traits from abroad could well have arrived with bands of new immigrants.

The result was a great flowering of culture in the seventh and sixth centuries BC, which built on local foundations, themselves supported by the wealth of the metal resources of Etruria (present-day Tuscany) [1], but strongly influenced by Greece and many other east Mediterranean lands. Indeed this variety provides a strong argument against Herodotus' story of a homeland in Lydia.

Etruscan civilization as a whole is puzzling. The general impression is very Greek. Only on closer inspection of the details does it become apparent that, however Greek in inspiration, everything has been transmuted and absorbed. This, and the grafting on of other Asiatic ideas, like divination and tumulus tomb building, and the development of purely local traits like the high-quality black *bucchero* pottery, enable historians to think of Etruscan civilization as something in its own right, not merely a provincial version of the civilization of Greece. Yet politically the two peoples appear to have been bitter enemies, Etruscan history having as its constant theme the attempt to bar the Greeks from the Tyrrhenian Sea. By one sea battle, the Greeks were expelled from Corsica in 535 BC, at the height of the Etruscan political domination.

The Phoenicians played a part in this story too, as intermediaries between Etruria and the Orient. Here political relations were more cordial, Etruria and Carthage being often in alliance against their common enemy, the Greeks. At Pyrgi, a port of Cerveteri, a dedication was found to Uni (Juno) – the goddess Astarte – engraved on gold plates in parallel texts, one in Phoenician, the other one in Etruscan.

Clues to language

The tablets found at Pyrgi proved of great help in advancing our knowledge of the Etruscan language, the most important unifying link between the Etruscan peoples. Since they had adopted the Greek alphabet

1 The Etruscan home territory lay between the Arno and the lower Po. It was there that the distinctive civilization arose and reached its highest development. At its widest, c. 535 BC, Etruscan power extended to the Alpine foothills and south to the Gulf of Salerno. But this was not an empire; instead it was a loose federation of independent city states with no common front.

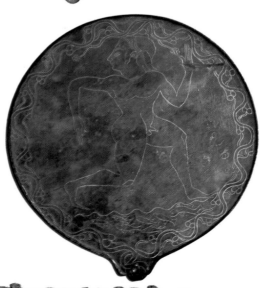

[Map: ALPS / Po / Melpum / Spina / Felsina / Arno / Volterra / ETRURIA / Populonia / Vetulonia / Perugia / Vulci / Chiusi / Tarquinia / Veii / Caere / Rome / APENNINES / TYRRHENIAN SEA / 0 — 300km]

Etruscan power at its greatest extent c. 540 BC
Etruscan sphere of influence

4 Jewellery reached a high level of development with the Etruscan civilization. The Etruscans were particularly skilled goldsmiths and a number of superb pieces decorated with fine granulation have survived. This fibula or safety pin dates from between 700–600 BC. It is an exquisite piece, demonstrating the quality and originality achieved by the Etruscans.

2 Bronze cists (caskets) held women's personal possessions, jewellery and toiletries. They were engraved with mythical scenes and were fitted with cast feet, handles and ornaments. This fine example from Palestrina (c. 300 BC) shows an engraving of Bellerophon holding Pegasus.

3 Mirrors of polished bronze assisted Etruscan ladies at their toiletries. The reverse side of mirrors such as this (c. 540–530 BC) was often engraved with mythical scenes – in this example Orion is crossing the sea – in an incised drawing. Mirrors were objects commonly buried with the dead.

5 The Apollo of Veii, together with figures of Mercury, Hercules and Latona, strode along the roof ridge of the Portonaccio temple at Veii. The sculptor was Vulca, who modelled Apollo in terracotta about 500 BC. Apollo has a frightening aspect, not in the least softened by his smile, and is different from the Apollos which the Greeks have left. The Etruscans clearly owed much to the Greeks, but they translated any borrowings into their own idiom.

with little change, there has never been any great problem in deciphering their writings. Translating them has proved much more difficult because all attempts at identifying the language with a known one have failed. It seems to have been the only survivor of those tongues spoken before Indo-European was introduced into the area. Little by little, however, the scholars' competence in translation increased, beginning with such tomb inscriptions as "X son of Y, aged Z years". But the few longer texts are obscure ritual documents, and yield so far little more than the general drift. The longest of such texts so far discovered was written on linen.

Etruscan culture

United by language, the Etruscans were less uniform in culture. There are clear differences (as in tomb architecture) between the Etruscan cities, mirroring their political independence. All, however, give a picture of an energetic, happy people with a bold and attractive art. Although much of the evidence comes from tombs [6, 7, 8], it is by no means a sombre picture. Their temples and cities too were architecturally exciting, and their civil engineering of a high order. A ship canal at Cosa, the channelling of streams below ground to prevent soil erosion, and their practice of tunnelling rivers below roads instead of building bridges over them, demonstrates these skills.

What survives from Etruscan civilization is to be found in the romantic ruined sites and cemeteries, and vast museum collections looted from these. But their place in history is assured by their contribution to Roman civilization. The end of the Etruscans probably came in the fourth century BC, because they had failed to present a common front and so fell individually to the Romans. However, the Romans' origins lay beneath the shadow of their rich and powerful neighbours and enemies. Much of the Greek civilization they adopted came to them not only through Etruscan hands but in Etruscanized form. What was non-Greek in Roman life was often pure Etruscan; for example, the realism of their portraiture and their religious practices and divination. Many Roman families had, and were proud of, Etruscan origins.

This married couple is sculpted in life-size on the lid of a terracotta sarcophagus from Cerveteri (Caere).

6 The Tomb of the Reliefs at Cerveteri is carved in the solid rock to represent a room in a house, with raftered ceiling supported on pillars, proto-Ionic capitals and bed niches in the walls. A series of funerary beds, each with two pillows, some even with a footstool and slippers awaiting their owners, is then added in stucco. Above the niches are displayed pieces of armour and weapons in relief – shields, helmets, leg armour, swords. The central pillars are decorated with a wide variety of tools and implements used in day-to-day life. This tomb is the only one of its kind.

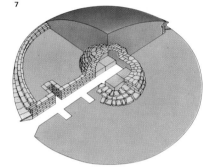

7 Types of tombs differed from city to city. At Populonia, a corbel-vaulted chamber of stone was covered by a hemispheric barrow within a stone retaining wall. The Tomb of the Chariots, shown here, contained two chariots fitted with bronze and iron decoration on ivory, and a bronze trumpet as well as gold jewellery and fragments of weapons and armour. The tomb dates from the 7th century BC. Many tumulus tombs were also erected at Cerveteri.

8 The Tarquinian tomb type was a rectangular chamber cut in the rock and approached by a sloping ramp. The chambers were then plastered and painted with vivid scenes of the highest beauty and significance. On the end wall of the Tomb of Hunting and Fishing (c. 520 BC), the funeral feast occupies the gable, with a frieze of mourning wreaths below. Fish leap all around the men in their boat, and a slinger ashore attempts to bring down one of the many gaily coloured birds that cover the ceiling. The funeral banquet was a common motif, coupled in other tombs with frescoes that pictured dancing and athletic contests in honour of the dead.

Early Rome: the kingdom and the republic

The origins of the city that founded the mighty Roman Empire are obscure. The traditional account, recorded by Livy, is that Romulus and Remus [Key] founded a city on the Palatine Hill [2] on 21 April 753 BC. At least the date agrees with the archaeological evidence, which shows that shepherds settled on the Palatine about the middle of the eighth century. They joined with other communities in the area in the early sixth century to establish a city around a site that later became the Roman Forum but which had for a period been used as a cemetery [1].

Of the six kings said to have followed Romulus the first three were almost certainly legendary, but the fourth, Tarquinius Priscus, was Etruscan. His reign marks the beginning of a period of Etruscan control.

Expansion under the republic
Etruscan rule and the monarchy ended simultaneously with the expulsion of King Tarquinius Superbus, traditionally in 509 BC. Two elected consuls and a senate, composed entirely of wealthy aristocrats known as patricians, controlled Rome's affairs.

During the next 200 years the policy in Italy was directed towards expansion, conquest and consolidation.

Soon after the foundation of the republic Rome played a leading part in the formation of the Latin League, an alliance of the cities of Latium, the western region of central Italy. In 390 BC the Celtic Senones and other Gallic peoples overran northern Italy and captured all Rome except the Capitol. They left on payment of a ransom and Rome profited from the bitter lesson by building the Servian Wall, which made it the most strongly defended city in Italy.

From this secure springboard Rome engaged confidently in a number of wars, which resulted in its obtaining undisputed mastery over the whole of Italy from the Po valley southwards. In 340 BC the cities of the Latin League rose against Rome but found themselves no match for their powerful partner. Rome imposed separate terms on each vanquished city, awarding some Roman citizenship, some part citizenship and punishing others, but in each case stipulating that the city should trade only with Rome. The

next struggle, against the Samnites in the south, was much harder. A Roman army was forced to surrender at the Battle of the Caudine Forks, but by 290 BC the war was won. Victory over the Greek cities then gave Rome control over the whole peninsula, a position it consolidated through alliances and the establishment of citizen colonies [4] but above all by the threat of its army.

By this time the plebeians had seen many of their grievances rectified. They had their own council, the *concilium plebis*, and their own officers, the tribunes. In 445 BC they received the right to marry into the patrician class and in 366 BC the first plebeian consul was elected. From 287 BC measures passed by the *concilium plebis* had the force of law. For the moment social conflict was muted although these changes had done little to alleviate the poverty of most of the plebeians.

The Punic Wars
Rome's advance into the south of Italy had brought it face to face with the Carthaginians (*Poeni*), who were ensconced in western Sicily and appeared to have designs on the

1 The Seven Hills of Rome are flat-topped spurs rising from a low, formerly marshy, plateau. The Palatine, Quirinal and Esquiline were the first to be settled, the Capitoline and Aventine the last. A unified city began to emerge when villages on the Palatine, Esquiline and Caelian came together in the early 7th century BC. The first wall round the city was built according to tradition in the 6th, which was apparently a period of great building activity under the Etruscan kings. The Cloaca Maxima, a sophisticated drainage system, and the temple of Jupiter Capitolinus both date from this early period.

1 1 Circus Maximus
   2 Circus Flaminius
   3 Senate House
   4 Record Office Temples
   5 Juno Moneto
   6 Jupiter Capitolinus
   7 Saturn
   8 Castor and Pollux
   9 Vesta
  10 Jupiter the Victor Markets, Bridges
  11 Aemilian Bridge
  12 Fabrician Bridge
  13 Fish Market
  14 Vegetable Market
  15 Cattle Market

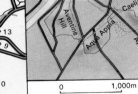

A Portico and Theatre of Pompey
B Shipyards

2 A village of huts on the Palatine, one of which is shown here in reconstruction, was discovered after World War II. It dates from the mid-8th century BC and so seems to support the legend of the foundation of Rome by Romulus and Remus. Crude in construction, these huts were the homes of the farmers and shepherds of Latin origin who are the earliest known inhabitants of the city. There is now archaeological evidence that Latium, the area surrounding Rome, was colonized in the 12th century BC by Late Bronze Age people who came from the east by sea, thus reinforcing the legend handed down to us by Livy that Rome's environs were settled by refugee Trojans under the leadership of Aeneas.

3 Aeneas, who escaped from Troy after the Trojan War, was said to have founded Rome. This legend was enshrined in the *Aeneid* by Virgil (70–19 BC). This fourth-century manuscript vividly illustrates the dangers of his voyage to Latium.

4 The formation of a new Roman colony was formally recognized when the founder guided a bronze plough, drawn by a bull and a cow, round its boundaries. Colonies were useful as a means of garrisoning vulnerable areas and also for moving surplus population and giving them employment. A few colonies were created as early as the 5th century, but the number increased as Rome's dominions spread because citizen colonies could be founded only in Roman territory.

eastern part. In the First Punic War (264–241 BC) Rome, after several near disasters, captured Sicily in 260 BC and Corsica and Sardinia the following year [7]. The ostensible cause of the Second Punic War (218–201 BC) was an attack by the Carthaginian general Hannibal on Saguntum, a city on the eastern coast of Spain allied to Rome. Hannibal with 40,000 men and a train of elephants made a remarkable march through Gaul and over the Alps [6]. He inflicted severe defeats on Roman armies at Trebbia, Lake Trasimene and Cannae, but then the delaying tactics of Fabius and the loyalty to Rome of most of its Italian allies began to have their effect. A crucial blow was the defeat at the Metaurus River in 207 BC of Hannibal's brother Hasdrubal, who was bringing reinforcements from Spain.

Rome opened up a second front, the young and brilliant Scipio (later given the title of Africanus Major) capturing Carthago Nova and then driving the Carthaginians out of Spain completely. In 204 BC Scipio led an invasion force from Sicily into Africa. Hannibal was forced to retire from Italy but the war ended with his defeat at Zama. The Third Punic War (149–146 BC) began when Carthage attacked Rome's ally Massinissa, king of Numidia (modern Algeria). After a desperate siege Carthage was captured in 146 BC and razed to the ground.

Influence of the Gracchi

During the second century BC Rome controlled almost the whole of the Mediterranean area, but storm clouds gathered at home. The patricians and rich plebeians kept a stranglehold on government and, perhaps of more immediate concern, on land. A champion of the poor arose in the person of Tiberius Gracchus, a tribune of the plebeians, who in 133 BC introduced a land bill intended to reduce drastically the large estates held by a few rich men. Rioters incited by apprehensive senators murdered Tiberius, but ten years later his brother Gaius tried to reduce the price of corn and generally to break the power of the senate. But proposals to extend the citizenship to all Rome's Latin allies were too radical brought about his political ruin and death.

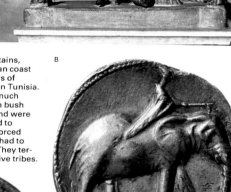

5 The three surviving columns of the Temple of Castor and Pollux in the Roman Forum were once part of a colonnade that ran round a shrine to the divine twins, who were also known as the Dioscuri. The tradition was that the temple was built following a battle between the Romans and the Latins at Lake Regillus in 499 or 496 BC. The twins fought on the Roman side and carried the news of the Romans' victory back to the city. The temple may, in fact, have been built much earlier, in the misty period of the kings. It lies within the boundary of the early Palatine villages. It was restored in the 2nd century BC and in the 1st century AD.

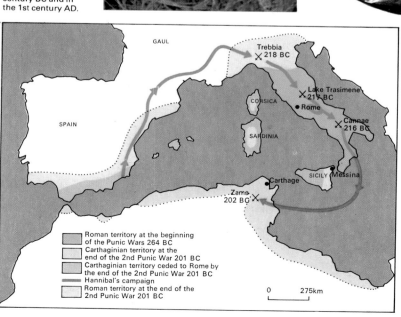

6 This Spanish coin shows Hannibal [A] on one side and an elephant [B] on the other. The elephants in the force that Hannibal led over the Alps were African forest elephants, which at that time were to be found round the Atlas Mountains, the Moroccan coast and the oasis of Ghadames in Tunisia. They were much smaller than bush elephants and were better suited to the kind of forced march they had to undertake. They terrified primitive tribes.

7 To make an effective challenge to Carthage's domination of the western Mediterranean, Rome had to become a naval power. A large fleet was constructed and equipped with boarding devices to allow for hand-to-hand fighting, at which the Romans excelled. As a result, after initial reverses they inflicted naval defeat on Carthage in the First Punic War.

8 This Roman soldier of the 6th century BC has a breastplate, helmet, sword and spears. Up to about 400 BC soldiers received no pay and only the rich could afford to do military service. A professional army was set up about 100 BC.

Map legend:
- Roman territory at the beginning of the Punic Wars 264 BC
- Carthaginian territory at the end of the 2nd Punic War 201 BC
- Carthaginian territory ceded to Rome by the end of the 2nd Punic War 201 BC
- Hannibal's campaign
- Roman territory at the end of the 2nd Punic War 201 BC

GAUL
Trebbia ✕ 218 BC
Lake Trasimene ✕ 217 BC
CORSICA
Rome
Cannae ✕ 216 BC
SPAIN
SARDINIA
Carthage
SICILY Messina
Zama 202 BC ✕

0   275km

Rome: the organization of the republic

Republican Rome (509–27 BC) is often regarded as a democratic state. Certainly later Roman writers looked back nostalgically on what seemed a golden age of political and social order and agreed morality. There was a theoretical balance between the powers of the magistrates, the Senate and the plebeians. But in fact most of the early republican period was an age of bitter conflict between the nobles of the Senate, anxious to keep their privileges, and the plebeians, equally anxious to have a share of the benefits. It was not until the third century that the plebeians were able to make laws on their own account. Their admission to all the magistracies only created an alliance of the nobility and rich ambitious plebeians, forming an oligarchy as exclusive as the former patrician nobility had been in earlier years.

The political structure of the republic

The Senate and the *concilium plebis* (people's assembly) formed the legislative branch of the republic and the magistrates [1] the executive. The 300 senators [5] were chosen from ex-magistrates by the censors (those who watched over public morals). The Senate was basically an advisory body with overriding financial control. It initiated legislation and in practice made the vital decisions on war and peace and foreign policy in general. It also assigned provinces to senior magistrates at the end of their year of office. The *concilium plebis*, which represented the plebeians, was able to initiate legislation, its resolutions having the force of law after 287 BC, but in general it merely accepted or rejected proposed legislation.

The chief executives during the Roman Republic were the two consuls. They were elected (from candidates proposed by the Senate) by the *comitia centuriata*, an assembly in which wealthy citizens had disproportionate voting strength. The consuls had immense power, presiding over the Senate and acting as supreme commanders in war. Beneath them were the praetors, whose numbers varied from two to eight and whose main duties concerned the administration of the law. Four quaestors were in charge of the state finances. The two censors supervised state contracts and public morals and, during the late republic, checked and if necessary expelled members of the Senate.

Streets, temples, public works, the grain supply and the public games were the responsibility of individual *aediles*. Twelve *lictors* preceded consuls and two preceded praetors, clearing the way and carrying bundles of rods known as *fasces*. The ten tribunes were entrusted with the defence of the plebeians, having the power of veto against actions of magistrates and against laws. In the second century BC the tribunes became entitled to sit in the Senate.

The power of the consuls

At the end of their year of office the *imperium* (power) of the consuls and praetors was transferred from Rome to one of the provinces. The Senate decided which provinces should be consular and which praetorian; to avoid corruption this decision was taken before the election of the magistrates. The actual province each magistrate received was decided by lot. *Imperium* did not extend beyond the province's boundaries, as was shown when Julius Caesar, by crossing the

CONNECTIONS

See also
90 Early Rome: the kingdom and the republic
94 Roman life
96 From the civil wars to Caesar's empire

In other volumes
276 Man and Society

1 **Most of the civil officers of Rome** [in the green band] had legal functions as well as administrative. The generals had military responsibilities only. The *pontifex maximus* was head of the state religion.

2 **The basilica at Pompeii**, measuring 56×21m (185×70ft), was built in the first century BC and was the centre of economic life. Most Roman cities had one of these rectangular roofed halls where business transactions were concluded. They generally stood near the Forum. Some basilicas included arcades and galleries and their design influenced later Christian basilicas.

Civil

Quaestors

Tribunes

Aediles

Censors

Praetors

Consuls

Generals

Pontifex maximus

Flamines (priests)

Military

Religious

3 **The demobilization of Roman soldiers** is portrayed on this relief (*c*. 1st century BC). The scribe on the left takes down details from a discharged soldier who holds his certificate, while others wait their turn.

4 **The *tabularium*** or record office [centre left] was built in 78 BC to the plans of Sulla (138–78 BC). It housed the state archives and was probably an annexe of the Treasury. The upper part is a palace (*c*. 1500).

Rubicon in 49 BC, in effect declared war on the Senate. Governors might keep their provinces for only one year or by *prorogatio* might have their command extended.

The most important religious figure was the chief priest, the *pontifex maximus*. He was the elected head of the college of priests, which also included the vestal virgins [7] (who kept the sacred fire burning in the Temple of Vesta) and the *flamines*, each of whom was responsible for the cult of one of the gods. Outside the college were the augurs who from certain signs, such as the behaviour of birds, decided whether a certain course of action was advisable.

The Romans worshipped many gods, most of them originally Greek. The chief god was Jupiter, to whom a temple was founded on the Capitol at the time of King Tarquinius Superbus (reigned 534–510 BC). Religious festivals were held throughout the year [6], among the most important being the Lupercalia, held on the Palatine on February 15, and the Saturnalia on December 17.

The security of the republic and its steady expansion into a great empire depended on the discipline and courage of its army. The record of victories over powers as formidable as Carthage, Macedonia and Syria was remarkable considering that it was still a citizen militia summoned by the consuls to meet specific emergencies. Only citizens with certain property qualifications were eligible, although an equal number of allies also served. Clearly the quality of commanders and centurions must have been high. Not until the reforms of Marius (157–86 BC) was a professional volunteer army formed.

Taxation and tax-collecting

Most Roman taxation was indirect and included customs duties and a number of special taxes, such as that on the freeing of slaves. The *tributum*, a direct tax, was levied mainly in time of war. The censors decided which taxes were applicable to the various classes, while the quaestors administered their collection. But more and more, tax-collecting in the provinces was farmed out to *publicani*. They were closely associated with the wealthy *equites*, the second social class, who waxed fat on the proceeds.

The magistrates of Rome had great power and influence. Civil lawsuits were heard first by a magistrate, such as the one shown here, before going to a judge for settlement. In the late republican period criminal cases were tried before special courts, where the penalties were generally exile, loss of citizenship or hard labour. The most important achievement of the Roman legal system was the classification of its code in order to clarify citizens' rights. The laws were published in 450 BC at the instigation of the plebeians. The body of Roman law, many of the principles of which are still very much in use, has had a vast influence in the West.

5 The Senate was the principal advisory body of Rome. This 19th-century painting by Maccari shows Appius Claudius persuading the Senate to reject peace proposals from Pyrrhus of Epirus in 280 BC.

6 In the Temple of Vespasian at Pompeii there are sacrificial scenes on an altar dedicated to the Roman imperial cult. Ritual sacrifices of animals such as pigs, sheep and bulls were an ancient element in the complex of Roman cults. Their purpose was to encourage divine beneficence.

7 The six Vestal Virgins lived in the House of the Vestals in the Forum. They began their service as children and could retire after 30 years. If they allowed the sacred fire to go out they were whipped by the *pontifex maximus*, under whose authority they came.

Roman life

In Rome, as in all societies, the sort of life the people led depended very much on the social class to which they belonged. In general the class system was based on wealth rather than birth, although often the two criteria merged. The highest class was composed of the members of the Senate, who for most of the republican period were *nobiles*, or nobles. The second class, the *equites*, or knights, derived most of their considerable wealth from business activities such as banking. Members of the upper classes wielded authority not only over their own families but as patrons over a number of semi-dependants known as clients. The third class of full citizens, the plebeians, after a long struggle won complete political equality. Once they were permitted to hold the magistracies they could move into the highest social class, because former magistrates were automatically *ex officio* members of the Senate.

The Roman slaves
At the base of the social pyramid were the slaves. The settled order of Roman life depended on the toil of slaves, who might well make up half the population of a town. A slave was under the absolute control of his master during the period of the republic.

During the imperial period the treatment of slaves improved somewhat, some protection being given to them against savage masters. More important, slaves were able to look forward more confidently to eventual *manumission*, or freeing. Many freedmen became extremely wealthy and influential and their sons became full Roman citizens. The life of slaves varied greatly, particularly between town and country. In the country they were likely to work long hours carrying out arduous tasks on *latifundia*, or estates, and on farms or in mills, while in the towns they might be comparatively well treated.

Careers and education
The son of a rich upper-class family might well have political ambitions. But he would generally start by training as a lawyer and become either an advocate in the courts or a legal consultant. Other professions were not thought really respectable. For much of the republic most doctors, architects and dentists were slaves or freedmen. Writers could expect to make little money from their work unless they could rely on some wealthy patron to encourage them, as Maecenas encouraged Virgil. But most Romans were far from rich or influential. Although manual labour was thought unworthy of a citizen, they struggled to make a living as bakers, shopkeepers and craftsmen of all kinds, generally employing one or two slaves to assist them. Some young men might enter the army which, after the reforms of Marius the consul at the end of the second century BC, was a professional force manned by voluntary recruits who received a small landholding on retirement.

Most children of Roman citizens received some formal education although only boys could expect to go beyond the primary stage. Under the empire some education was provided free for poor students but generally parent paid a small fee. Between the ages of 7 and 12, children received a somewhat rough-and-ready grounding in reading, writing and arithmetic from a *litterator*. They could then, if they chose, move on to a *gram-*

CONNECTIONS

See also
102 Life in Roman Britain
92 Rome: the organization of the republic
96 From the civil wars to Caesar's empire
98 Rome: the expansion of the empire
114 Rome: soldier emperors to Constantine
90 Early Rome: the kingdom and the republic

1 **The ruins of Pompeii** were preserved beneath ashes after the eruption, described by Pliny the Younger, of Vesuvius in AD 79. Excavations revealed a unique record of Roman daily life.

2 **This poulterer's shop** was located in Ostia, once the port of Rome. Cicero regarded shopkeeping as near the bottom of the list of employments suitable for people of taste – but better than dancing.

3 **Every upper-class Roman** visited baths – like the Stabian baths of Pompeii – daily with his oil flask, towels and other toiletries. He progressed from the tepid to the hot bath, from the sweat room to the cold bath. Refreshments and massage were available. Many baths had separate sections for women.

4 **All baking** was apparently done at home until the 2nd century BC. The exhausting work of turning the grinding mills was for slaves helped by donkeys. A bakery like this one was uncovered at Pompeii. The oven in it contained a large number of loaves, still intact and weighing about 900g (2lb) each.

maticus, under whom they studied Greek and Latin literature and received an introduction to geometry and advanced arithmetic. The third stage in schooling was study under a *rhetor*, or orator, to learn the principles of effective public speaking, an ability that was important for success in politics and in the law courts. Finally the more ambitious and wealthy students might progress to the equivalent of university level. They studied either law, under an established lawyer or at a law school, or Greek oratory and philosophy, which involved attendance at one of the great centres of Greek learning, the most popular of which were at Athens and Rhodes.

Spectator and participatory sports

Sports among the Romans were in two categories, spectator sports and participatory sports. In the first group were the public spectacles staged in the amphitheatres, the "bread and circuses" of which the satirist Juvenal contemptuously worte. These included musical performances, readings of verse and theatrical performances, which were sometimes staged with extraordinary lavishness. What really drew the crowds were gladiatorial contests to the death between men or between men and beasts, and also the thrilling and dangerous chariot races, which might be run between individual charioteers or between teams of chariots.

The Romans did not place the same emphasis on physical exercise as the Greeks, but many young men of the upper classes were enthusiastic riders and hunters. Many also enjoyed boxing and wrestling and there were several ball games, such as *harpastum* and *trigon*, the details of which are rather obscure. Children played many of the games that children have played through the ages – hoops, tops, pitch and toss, marbles, hide-and-seek and leap-frog. In general the adult Roman preferred gambling games above all other leisure activities, except perhaps drinking (there were about 120 taverns in Pompeii). Throwing dice, *tesserae*, and knucklestones, *tali*, were the favourite methods of gaming. Dice, which like modern dice carried numbers from one to six, were also used to determine the player's moves in various board games.

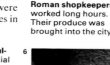

Roman shopkeepers worked long hours. Their produce was brought into the city at night to avoid the traffic congestion of the day. This relief of a greengrocer's shop in Ostia includes illustrations of various vegetables.

5 The life of Herculaneum, a residential town between Naples and Pompeii, was revealed only after arduous excavations; the eurption of Vesuvius in AD 79 covered it with a thick layer of mud. However, the mud preserved much of the original town. Here a mosaic shows a man and woman served by a slave.

6 The house of Menander at Pompeii was more a country villa than a town house. The main reception room in the centre of Roman villas was the *atrium*, which was luxuriously furnished with tiles, marbles and fresco paintings. In the middle of the floor was the *impluvium*, a pool into which rainwater fell.

7 A number of houses at Herculaneum have kept their second storeys. The house shown here, the Cása del Graticcio, still contains its original beds in two small rooms on the upper floor, together with other furnishings. The house was clearly divided into many small apartments; large blocks of flats, often shoddily built, were a common feature of towns and cities in imperial times.

8 This sandstone relief from Gaul shows a school scene: a teacher is seated between his two pupils who are opening their scrolls for the lesson. Discipline was severe and corporal punishment was frequent. Sons of wealthy families were escorted to and from school by a household slave or freedman known as a *paedagogus*, who was sometimes also a tutor. The children wore warm cloaks and thick shoes to protect them against the rigours of the northern climate.

9 Many Roman houses had a *lararium*, a private chapel to *lares*, spirits of the hearth to whom offerings of food and flowers were made. Closely related was the worship of the *penates*, the guardians of the store cupboard.

From the civil wars to Caesar's empire

The period between about 100 BC and 42 BC, leading up to the fall of the Roman Republic and its replacement by the Roman Empire, was one of disorder and disunity in which ambitious men used ruthless methods in their efforts to secure or maintain dominance. Julius Caesar (100–44 BC) was the outstanding figure but Marius, Sulla and Pompey all contributed to the end of the old form of government.

Political conflicts

In 108 BC Marius, champion of the popular party against undue senatorial power, was elected to the first of his seven consulships. An outstanding general, he proved less skilful as a politician and his influence began to wane. In 90 BC a new popular leader, Drusus, revived advocacy of an extension of citizenship to all Italians. His murder precipitated the Social War in which Rome was hard pressed by a rising of its allies, finally triumphing only by conceding most of the allies' demands.

The popular cause was left in some disarray. When most of the Eastern Empire rebelled under the lead of Mithridates (*c.* 133–63 BC), king of Pontus [3], command of the campaign of suppression was given to Sulla (138–78 BC), an aristocrat [5]. The popular party had the command transferred to Marius but Sulla forthwith marched on Rome, forcing Marius to flee. He pushed through measures against his opponents, then left for the East. Marius returned and began a slaughter of political enemies.

Sulla returned to Italy from the East in 83 BC and with the help of two powerful commanders, Pompey (106–48 BC) and Crassus (*c.* 115–53 BC), fought his way to Rome. There he began a reign of terror, massacring his opponents and making himself dictator. He drastically reduced the powers of the tribunes and the consuls and increased those of the Senate, but after he resigned in 79 BC the inability of the Senate to use its powers effectively was revealed. Corruption grew, particularly in the provinces. In Spain, Pompey fought a long campaign against Sertorius, a supporter of Marius. Meanwhile Spartacus (died 71 BC) led a slave uprising, terrorizing southern Italy for two years.

Pompey and Crassus joined forces to gain control of Rome, securing their illegal election as consuls in 70 BC. They soon swept away Sulla's legislation favouring the Senate and pinned their colours to the populist mast. Pompey demonstrated his military and organizational ability by speedily clearing the eastern Mediterranean of the growing menace of piracy. In 66 BC he was given virtually a free hand in the East.

The First Triumvirate

On his return Pompey received a cold reception from the Senate and he joined with Crassus and Julius Caesar [Key], who had returned from governing Spain, in forming the First Triumvirate. They forced through legislation by appealing to the Assembly over the heads of the Senate. Pompey received approval of his Eastern settlement and Caesar, after his consulship in 59 BC, was granted command in Gaul for five years.

In 55 BC Pompey and Crassus were consuls while Caesar received a five-year extension of his command. Crassus disappeared from the scene when an ill-conceived attack

1

1 Jugurtha, grandson of Masinissa, seized the throne of Numidia in 112 BC, killing several Romans in the process. Rome sent an army against him but the war dragged on in spite of some Roman successes. Marius, who became consul in 108 BC, led an army enlisted from the poor of Rome to Africa and defeated Jugurtha. After being exhibited as part of Marius' triumph, Jugurtha was strangled.

2 A relief from the Temple of Fortuna Primigenia at Praeneste shows a warship of the 1st century BC with soldiers prepared for hand-to-hand fighting.

3 Mithridates VI, King of Pontus, controlled the Crimea and much of southern Russia. In 88 BC he began a struggle against Rome that was to last for 25 years. He occupied Asia Minor and invaded Greece before making peace with Sulla. A second war in 83 BC was soon ended but a third Mithridatic war lasted from 74 to 66 BC, when Pompey made Pontus and Bithynia a Roman province. Mithridates killed himself when his son Pharnaces led an uprising against him.

4 Pompey the Great (Gnaeus Pompeius Magnus) lacked the military genius of Caesar and was less skilful politically. He was however a brilliant administrator and ruthless general who achieved power less by creating opportunities for himself than by waiting for situations to arise in which he would be called on to lead.

5 Lucius Cornelius Sulla made his military reputation in the Social War. Elected consul in 88 BC, he received the command against Mithridates. Later he used the office of dictator to massacre his opponents and force through pro-Senate legislation. He reformed the criminal law setting up *quaestiones* (new courts for particular crimes).

on Parthia ended with his death at the disastrous Battle of Carrhae. Pompey received Spain as his province but preferred to stay in Rome, intriguing with his supporters against Caesar, whose successful campaigns in Gaul [6, 8] were proving an embarrassment. Caesar, having conquered the whole of Gaul and made two exploratory invasions of Britain, was now ready to return to Rome backed by his devoted legions.

After fruitless attempts to reach a compromise or reconciliation, the Senate in 49 BC ordered Caesar to disband his army. Caesar at once crossed the Rubicon, the river dividing his province from Italy, and marched on Rome, so plunging Italy into civil war. Pompey hastily left for Greece, hoping to mobilize the resources of the East. Caesar soon mastered Rome and went on to crush forces favourable to Pompey in Spain, to defeat Pompey himself at Pharsalus in Thessaly in 48 BC, and finally to pursue him to Egypt where Pompey was murdered. Delaying his campaign, Caesar began his celebrated affair with the Egyptian queen, Cleopatra VII (69–30 BC) [9]. But within

two years he controlled Roman Africa. Finally he sealed his authority by quelling a revolt by Pompey's sons in Spain.

End of the republic

The main basis of Caesar's authority was the office of dictator, which he held twice before receiving it for ten years in 47 BC. He packed the Senate with his supporters, nominated some of the magistrates and although denying any ambition for kingship, finally accepted the dictatorship for life. Brutus (85–42 BC) and Cassius (died 42 BC) organized his assassination at a Senate meeting on 15 March 44 BC. They hoped to save the republic but it had been fatally weakened: the only question was who would be Rome's first emperor. Mark Antony (c. 83–30 BC), Lepidus (died 13 BC) and Octavian (63 BC–AD 14), Caesar's heir, were the three contestants. In 43 BC they formed the Second Triumvirate, issuing proscriptions for the deaths of many. Caesar's death was avenged when Octavian and Antony defeated Brutus and Cassius at Philippi in 42 BC. Octavian was soon to become sole ruler.

Julius Caesar, born in 100 BC, was the son of a leading patrician family. His aunt was the wife of the popular leader Marius and his own wife, Cornelia, was a daughter of Cinna, Marius' successor. These connections displeased the conservative Sulla and Caesar left Rome. On Sulla's death he returned and made a reputation as a barrister and political orator. Moving up the political hierarchy, he became quaestor in Spain. After Cornelia died, he married Sulla's wealthy granddaughter Pompeia and set out to become a popular party leader. His election as praetor for 62 BC provided a springboard for his swift rise to absolute power in Rome.

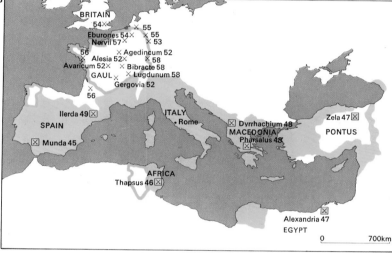

6 Successful campaigns were waged by Julius Caesar between 58 and 51 BC against the Helvetii, Belgae, Veneti and the Aquitani. He conquered the whole of Gaul and made it a new province, Transalpine Gaul. He twice landed in Britain, near Walmer or Deal in 55 and near Sandwich in 54 BC. The second expedition was on quite a large scale and Caesar penetrated northward beyond St Albans.

7 Standard equipment of a legionary in the later republican period was a *gladium* (short sword) and a *scutum* (shield).

Roman dominions in 63 BC
Conquests of Julius Caesar
× Campaigns in Gaul and Spain
⊠ Civil war campaigns

8 The triumphal arch at Orange, France, is the third largest Roman arch extant. It commemorates Caesar's victories over the Gauls with reliefs of prisoners-of-war and captured armour. The capture of the port of Massilia is commemorated by such designs as anchors, prows and ropes.

9 Cleopatra, Queen of Egypt, was the mistress of Caesar and Antony. She bore Caesar a son and followed him to Rome. After Caesar's death, she returned to Egypt, marrying Mark Antony in 37 BC.

10 A Roman siege tower was divided into several storeys and was up to 55m (180ft) tall according to the height of the wall to be attacked. It was hauled along a prepared causeway by ropes and capstans. Archers fired on the defenders from the upper storeys while on the bottom floor a battering-ram pounded the base of the wall. A boarding bridge could be let down from the top.

Rome: the expansion of the empire

Octavian became the undisputed master of the Roman Republic and Empire following his victory over Mark Antony at the Battle of Actium in 31 BC. There is no doubt that he intended to establish a personal dynasty, but he was too clever a politician to ignore the strength of republican feeling in Rome. When he returned from the East in 29 BC he ostensibly restored the republic and set out to establish absolute power within it.

Augustus: "first among equals"

The Senate voted Octavian the honorary titles of "Princeps" (first citizen) and "Augustus", by which he was known thereafter [Key]. Additionally, he received consular status and the power of a tribune with the right to summon the Senate, introduce business, veto decisions, nominate candidates for elections and issue edicts. This was how Augustus was able to influence the government of Rome and Italy and put in hand massive development programmes [2].

The government of the empire was divided between the older, settled provinces, governed by proconsuls elected by the Senate, and the newer, military provinces that were ruled through legates appointed by Augustus as a proconsul with special powers over all others [1]. The army was reorganized under the emperor's direct control into a force of 28 legions of professionals recruited for 20–25 years. It was drawn from Roman citizens plus an equal number of auxiliaries enlisted from provincial territories. This system, of emperors ruling with the Senate, was to endure for more than 200 years.

The most important immediate task was to restore order to the empire and secure its frontiers. Augustus's first expeditions were to Gaul and Spain, each of which was reorganized into three provinces. The River Danube line was secured with a series of military provinces garrisoned by large legionary forces [3], and attempts were made from 12 BC to push forward across the Rhine to the Elbe. But the annihilation of three legions under Varus by the Germans in AD 9 forced Augustus to accept the Rhine as his boundary. In the east, peace was made with the Parthian Empire and a buffer state was established in Armenia. Internally, brigandage and piracy were stamped out and taxation and the administration of Roman law put on a uniform basis.

Augustus died in AD 14 and under his successor, Tiberius (reigned 14–37), his policies of establishing order were continued so well that the incompetence of Caligula (reigned 37–41) caused little lasting harm. Under Claudius (reigned 41–54) the conquest of Britain was begun and Mauretania (now Morocco and Algeria) was occupied.

Problems of succession

A major weakness of the Augustan imperial system was that the succession was never formulated and when in AD 68 the last of his direct house, the unstable Nero, was killed, four rival candidates for emperor were put forward by different sections of the army. Following a terrible civil war, Vespasian (reigned 69–79) was successful and developed a system whereby each emperor "adopted" his successor, thus giving the empire stability.

Under Trajan (reigned 98–117) the frontiers of the empire were again extended and

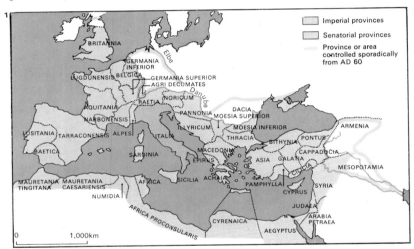

Imperial provinces
Senatorial provinces
Province or area controlled sporadically from AD 60

0   1,000km

1 The Roman Empire, at its height in the 2nd century AD, was theoretically divided into provinces controlled by the Senate and the emperor. In reality, the emperor had the power to intervene in senatorial provinces. Italy itself was ruled according to a modified version of the republican constitution.

3 Legionaries and Germans were often at war in the 1st century, as seen in a contemporary relief. The empire failed to find secure frontiers and barbarian invasions played a major part in its collapse. In the relatively remote east a client state in Armenia usually provided a reasonable buffer against the Persian and Parthian empires, but in the west, far closer to the heart of the empire, constant vigilance was needed. Of the 28 legions established by Augustus more than half were always stationed in the provinces bordering the Rhine and Danube. Augustus attempted to gain the more easily defensible Elbe-Danube river line, but failed to secure it.

2 The Roman Forum (now in ruins) was the centre of the government of the empire. Augustus and his successors symbolized their power in a series of impressive public buildings. The administration of all parts of the empire remained almost entirely in the hands of native Romans, who held both governorships and lesser posts until well into the 2nd century. Yet an increasing number of provincials succeeded in working their way up the administrative ladder, first from the western provinces and then from the eastern. By AD 200, 57% of the Senate were provincials.

4 The theatre at Palmyra in Syria is typical of the fine buildings – temples, amphitheatres, aqueducts and baths – that were built in all the provinces of the empire. Just as they tolerated other religions, the Romans took care not to interfere with the social customs of the peoples they conquered. The civilizations of Greece and the East continued to flourish, but provincials were inevitably influenced by the example of Roman culture.

Dacia, Armenia and Mesopotamia were added [8]; but the last two were abandoned by Hadrian (reigned 117–138). Hadrian concentrated on improving existing imperial defences – building a wall across northern England and a fortified line between the Danube and the Rhine – and travelled throughout the empire inspecting the imperial administration and legal systems.

Peace and prosperity

Under the Antonine emperors (so called because of the family name of Hadrian's successor Antoninus) the empire was at its most peaceful. A man could travel in safety from Britain to Arabia along superb roads and secure seaways [5]; trade flourished [6] and a single culture, two languages – Latin and Greek – and a single system of law and administration covered the whole empire. Great cities with fine public buildings grew up in the provinces [4], where the people strove to become Roman citizens.

There were, however, underlying weaknesses. Rome had grown rich in booty and taxation from the provinces and economic activity tended to be one-way, with wealth and produce flowing to Rome but little produced in return. This caused jealousy in the provinces and an increasing idleness within Italy [7], problems made worse by the widespread use of slaves for all productive labour and a steady decline in the population. It became more and more difficult to find native citizens to undertake the many administrative and military duties on which the government of the empire depended. As a result an increasing number of provincials reached high administrative positions.

Under Marcus Aurelius (reigned 161–180) [9] the peace ended. In the east a Parthian attack was defeated but the returning troops brought back a terrible plague which devastated the whole empire and further reduced its manpower. On the Rhine and Danube frontiers the barbarians were being forced forward by a massive migration of Goths in central Europe, and in 167 several tribes crossed the Danube and Alps and swept into Italy. Dacia was also overrun and Marcus spent the rest of his reign fighting to restore the frontiers.

KEY

The Emperor Augustus is shown here as Pontifex Maximus (High Priest). The state religion of the Roman Republic continued under the empire. The emperors were deified after their deaths and sometimes they were worshipped during their lifetimes by people in the Roman provinces. Other religions were tolerated as long as they accepted the divinity of the emperor, and many Eastern cults such as Mithraism or the worship of Isis, which provided greater mysticism and colour, became popular. Only Judaism and Christianity, because they denied the divinity of the emperor, were in disfavour, but there was little direct persecution.

5

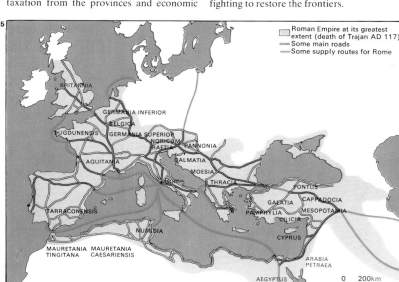

Roman Empire at its greatest extent (death of Trajan AD 117)
Some main roads
Some supply routes for Rome

BRITANNIA
GERMANIA INFERIOR
BELGICA
LUGDUNENSIS
GERMANIA SUPERIOR
NORICUM
RAETIA
PANNONIA
AQUITANIA
DALMATIA
MOESIA
THRACIA
PONTUS
GALATIA
CAPPADOCIA
PAMPHYLIA
MESOPOTAMIA
CILICIA
CYPRUS
TARRACONENSIS
NUMIDIA
MAURETANIA TINGITANA
MAURETANIA CAESARIENSIS
ARABIA PETRAEA
AEGYPTUS
0   200km

6

5 The network of roads and sea routes that held the empire together was built to enable troops, tax collectors and administrators to travel swiftly. But the great roads also allowed traders to cover the whole empire easily and the Roman world became an economic common market, with goods from one area available everywhere.

6 The loading of a Roman grain ship is shown in this picture. Egypt and North Africa were the granary of the empire and the importance of the grain trade, which provided a free ration for every citizen in the capital, was recognized by placing the trade directly under the personal control of the emperor.

7

8 Legionaries and their captives are shown here on Trajan's Column, a memorial in Rome to the Emperor's victories in Dacia. Trajan brought the empire to its largest extent with legions crossing the Danube into Dacia and down the Tigris and Euphrates to the Persian Gulf. But these areas were never fully pacified and their defence was a drain on the empire's military and financial resources.

8

9

7 Races – especially chariot races – and gladiatorial games became increasingly popular and important under the empire. Thus the emperor and rich patricians courted the popularity of the Roman mob by providing ever more lavish spectacles. The distribution of free bread and the frequency of free entertainments helped to insulate the Roman people from economic and political reality and played a major part in dissolving civic responsibility.

9 Marcus Aurelius, shown here addressing his troops, was a civilized, highly educated man and a staunch follower of the Stoic philosophy. He was fated to spend most of his reign struggling against the first great onrush of barbarians that threatened to overwhelm the empire. For the first time in centuries a foreign invader swept over the Alps into Italy and Aquileia was besieged. Only by conscripting every fit man, including gladiators and brigands, was Marcus able to drive back the barbarians, restore the River Danube line and begin the work of reconquering Dacia. His reign marked the end of the enduring imperial peace.

Roman rule in Britain

The Romans ruled Britain, or at least its southern half, for more than 300 years. The northern frontier was always a problem, but much of England enjoyed a peaceful and prosperous existence as one of the provinces of the Roman Empire. The people, particularly in the towns, adopted many of the social and religious practices of their occupiers. The final collapse of Roman power resulted as much from civil strife in Italy as from unrest among the Britons.

Patterns of conquest

A century separated the Romans' first tentative probes into Britain and their full-scale expedition of conquest. Britain fascinated the Romans as a strange romantic land in much the same way as the New World fascinated the Elizabethans. But the Romans also coveted its mineral wealth and realized its important strategic position. Julius Caesar (c. 100–44BC) was anxious to cut off the aid that the Britons were giving to the Gauls. His first landing, in 55 BC, achieved little of substance. In his second attempt the following year he marched north and probably captured the defenders' stronghold at Wheathampstead, Hertfordshire.

The Emperor Claudius [1], anxious to give some shine to his exceptionally dull image, ordered the conquest of Britain in AD 43. The troops of four legions (3,000 to 6,000 men in each), plus auxiliary regiments of provincials, landed and made their main base at Richborough. The invaders soon won a major victory on the River Medway. The main British force under Caractacus (died AD 54) withdrew and avoided further pitched battles. Claudius himself made a brief visit to lead his forces triumphantly into Colchester (Camulodunum).

The south and Midlands up to the Trent and Severn rivers were occupied without much opposition being met. But in Wales Caractacus led a stubborn resistance based on guerrilla tactics. When he was eventually inveigled into the open he was defeated, and was taken prisoner to Rome and exhibited in the triumph awarded to the successful Roman general. By AD 61 the Romans controlled England as far north as the River Humber. But in that year Boadicea (Boudicca) [3], Queen of the Iceni in East Anglia, led a revolt which spread so quickly that the governor, Suetonius Paulinus, was forced to abandon Colchester, London (Londinium) and St Albans (Verulamium). Inevitably the greater armed power, skill and discipline of the Romans prevailed and the rebellion was crushed with savage reprisals.

Consolidation and administration

By AD 80 the whole of Wales had been subdued. The new governor, Agricola (AD 37–93), pushed the frontier northwards to the Forth and Clyde, and at the Battle of Mons Graupius gained the whole of lowland Scotland. But as elsewhere in their empire, the Romans found it difficult to maintain their expanding frontiers securely. A period of consolidation led to the building of Hadrian's Wall [4] between the Tyne and the Solway during the AD 120s. Under the emperor Antoninus Pius further attempts were made to subdue Scotland, and the Antonine Wall was built linking the Firth of Forth and the Firth of Clyde. Hadrian's Wall fell into disuse for a period but by about AD

1 A bronze head of Claudius, Emperor of Rome from AD 41 to 54, was found in the River Alde in Suffolk. Claudius instigated and took a close personal interest in the conquest of Britain, which was important for his own prestige. During the 16 days that he spent campaigning in Britain he led his troops from the Thames and quickly captured Colchester. He also received the surrender of many tribes.

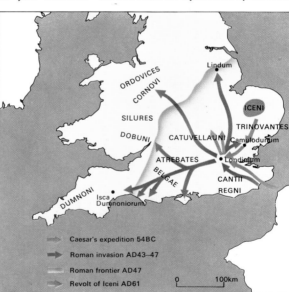

2 The two Roman invasions came about for different reasons. Julius Caesar saw the British campaign as a diversion from his main task of subduing Gaul, and retired after achieving his aim of defeating the British chief Cassivellaunus. By comparison Claudius hoped to confirm his own dubious power in Rome and at the same time saw the creation of a new colony in Britain as a means to balance the great power of the generals on the Rhine. Southeast Britain was the centre of both Celtic and Roman Britain at this time; the Fosse Way was the frontier of Roman Britain in AD 47, and served as a springboard for advances into south Wales in AD 49 and north Wales in AD 61.

Map labels: ORDOVICES, CORNOVI, SILURES, DOBUNI, DUMNONI (Isca Dumnoniorum), BELGAE, ATREBATES, CATUVELLAUNI (Camulodunum), Londinium, CANTII, REGNI, TRINOVANTES, ICENI, Lindum

Caesar's expedition 54BC
Roman invasion AD43–47
Roman frontier AD47
Revolt of Iceni AD61

0   100km

3 Boadicea, warrior queen of the Iceni in East Anglia, riding in her war chariot, symbolizes revolt against Roman oppression. This massive sculpture, executed by Thomas Thornycroft in 1902, stands nationalistically at the west end of Westminster Bridge, London.

4 A long stretch of Hadrian's Wall that is better preserved than most runs along the Whin Sill escarpment near the large fort of Housesteads. The wall runs for 117.5km (73.5 miles – 80 Roman miles) from Wallsend-on-Tyne to Bowness-on-Solway. It was built in the AD 120s, after a visit by the Emperor Hadrian, to help to defend the empire's most northerly frontier from invasions by the Picts. It was several times overrun and rebuilt, before the border was finally abandoned to the Picts in AD 383.

200 it again formed the northern frontier.

The first capital of the province of Britannia was Colchester, but London soon succeeded it as the centre of administration and trade. The bridge over the Thames there was the focal point of the road network and of the sea trade routes. London was burnt during Boadicea's rebellion but was quickly rebuilt and became one of the largest cities in the Roman Empire. The governor was assisted by an *iuridicus* who was responsible for justice in the civilian zone, and by many civil servants. Taxation and the treasury were under an independent *procurator*.

In the early third century Britain was divided into two provinces. London was the capital of Britannia Superior and York (Eburacum) became the capital of Britannia Inferior. Later, four provinces and then five were created. Communities had considerable control over their own affairs and the *coloniae* (important towns) [7] of Colchester, Lincoln (Lindum) and Gloucester (Glevum) were virtually self-governing.

Most Britons lived a settled, peaceful life under civil rule. But the territories between the Humber and Hadrian's Wall, much of Wales and some other areas, were under military rule and were controlled through a network of large garrison bases and smaller fortresses. The movement of troops was aided by an excellent road system. Roads and forts [6] always followed close on conquests. Important roads were Watling Street, from Dover (Dubrae) to Chester (Deva), Ermine Street, from London to York, and the Fosse Way, from Lincoln to near Axminster.

Withdrawal and abandonment

In the fourth century Saxon raiders became a serious threat to the southeast coast while the Picts and Scots increased pressure on the northern frontier, destroying Hadrian's Wall and overrunning much of the north in AD 367. The situation was temporarily restored but Roman forces began to withdraw, and by the early fifth century had completely abandoned the island. The last appeal for help was rejected by the authorities in Rome in AD 446 and the Anglo-Saxons began the process of effacing or transforming many of the signs of Roman occupation.

KEY

The arrogant attitude of the Romans towards the conquered Britons in the early years of their occupation is graphically illustrated by this scene of a Roman cavalryman trampling a Briton under his horse's hoofs. The horseman, named Longinus, came from Sofia, in modern Bulgaria, and was a *duplicarius* (junior officer) in a Thracian cavalry detachment. The scene was carved on a tombstone found on the site of Colchester. The troops of Boadicea (died AD 62) defaced the tomb, overthrew the headstone, and burnt Colchester as an act of vengeance after the Romans had violently taken over the territory of the Iceni, flogged Boadicea and raped her daughters.

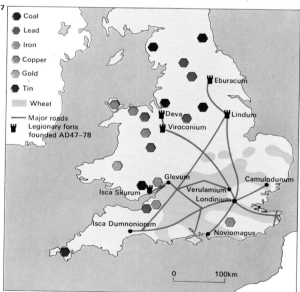

5 This reconstructed Roman boat is typical of the craft used by the Romans to bring troops and supplies from Gaul and up the main rivers of southern Britain. But communication by road was soon developed as the most efficient means of military control.

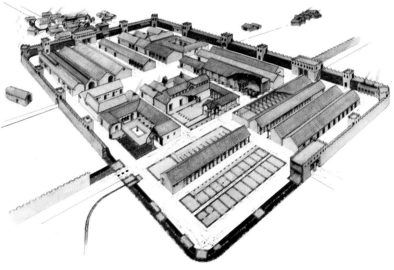

6 The typical Roman fort would hold 500 to 1,000 men. Each fort was strengthened by a wall with an earth rampart banked against it. From the main gate (*porta praetoria*) a main street (*via praetoria*) led to the headquarters building (*principia*). From there the other main street led to the side gateways, the *porta principalis dextra* and *porta principalis sinistra*. Within the fort there would usually be barracks for the troops and their officers, a workshop, granaries, stables, armouries, the regimental chapel and the commandant's house. Water would be supplied by an aqueduct running from a nearby river. Civilian settlements were often built and grew up outside the walls of the fort to provide a measure of comfort. Forts were built along Hadrian's Wall at regular intervals in order to repel aggressive barbarian tribesmen.

Coal
Lead
Iron
Copper
Gold
Tin
Wheat
— Major roads
⚑ Legionary forts founded AD47–78

Eburacum
Deva
Viroconium
Lindum
Glevum
Isca Silurum
Verulamium
Camulodunum
Londinium
Isca Dumnoniorum
Noviomagus

0   100km

7 The map of Roman Britain during the early decades of conquest and control shows roads, towns, frontiers, fort systems and industrial and agricultural regions. Throughout their empire, the Romans put the emphasis on urban rather than on rural life. The first of many towns they founded in Britain was at Colchester, while other towns developed from the settlements outside the military stations. The largest element of local government was the *civitas* (community), which was based on already existing tribal groupings. The most important towns received the title of *colonia* or *municipium*. Such towns were permitted to elect a town senate, which provided magistrates for administrative duties.

8 The Roman fort of Anderida at Pevensey in Sussex was built *c.* AD 280 to defend the south coast against Germanic invaders. A chain of these forts was built from Norfolk to the Isle of Wight to protect important harbours against the incursions of Saxon sea raiders. Others were built on the west coast. In AD 491 the Saxons stormed Anderida and slaughtered everyone in it. Today the fort stands several miles away from the shore because of the silting of the harbours.

Life in Roman Britain

In Celtic times Britain was already a prosperous country. The Romans rapidly increased its prosperity. Their colonial policy aimed at winning the approval of the upper classes by founding new centres such as Lincoln in the countryside, or by introducing Roman civilization to existing tribal centres. Their way of life affected the lower classes only slightly and gradually.

Life in the towns

At the centre of the typical Roman town [3] was the *forum*, round which were grouped buildings such as the *basilica* (hall of justice and public meeting-place), *curia* (council chamber), temples, baths and shops. The shape of each town was essentially that of Rome itself. The streets were paved and some had main drains beneath them to which house drains could be linked. Principal buildings were constructed of stone, with timber buildings surrounding them, and in the bigger towns wealthy Romans and Britons owned houses built of brick.

Labourers and craftsmen lived in simple rectangular houses made of wood, set with their longer dimension at right-angles to the street. They doubled as shop-front and workshop, the living rooms being set behind the working areas or on an upper floor and equipped with furniture and utensils that were usually homemade. Merchants' houses were larger, and made of brick, with an internal courtyard. They might have a hypocaust [7] (a system of underfloor central heating), a flush latrine (fed by rainwater collected in a lead tank on the roof), mosaic floors [Key] and brightly frescoed interior walls.

In Roman eyes the bath house was an essential feature of civilized life and even the smallest town had one, centrally placed and open to all. Amphitheatres were important, apart from the entertainment they provided, for the urban solidarity they demonstrated. The Romans enjoyed violent entertainments: as well as chariot races in which blood was shed accidentally, there were gladiatorial performances in which it was shed deliberately. Amphitheatres seating perhaps 5,000 spectators can still be seen at Silchester and Dorchester. Theatres existed in only the most important towns, such as St Albans [5].

Britons who lived in the towns, or visited them regularly, probably adopted the dress of the Romans. Those who, as Roman citizens, were entitled to wear a toga probably did so on formal occasions. Others wore tunics and, out of doors, hooded cloaks. Soldiers serving at Hadrian's Wall are known to have worn socks and underpants.

There was considerable interchange between Roman and Briton. Soldiers on the northern frontier, for instance, spoke a British dialect of Latin and drank not only imported wine but also "Celtic beer". Civilian settlements grew up beside forts and even in its most remote outposts the Roman army lived constantly in the company of the people it had conquered.

Agriculture and industry

Within Roman Britain the pattern of agricultural life seems to have changed little; even in their *coloniae*, or settlements of veteran soldiers, the Romans do not seem to have brought to bear their improved agricul-

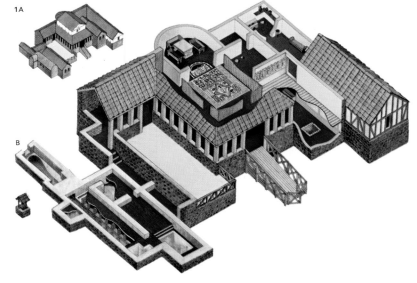

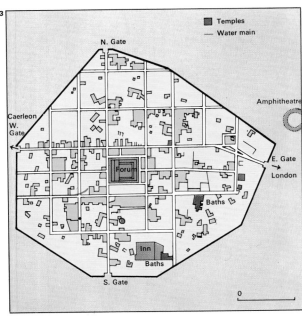

1 The Roman villa at Lullingstone, in Kent (shown here in model [A] and cutaway [B]) was built on an ancient site. The first stone house to appear on the site was probably built by a British farmer in the 1st century AD. By the end of that century it was occupied by a Roman and was greatly expanded by the addition of baths and other rooms. The house stood empty for most of the 3rd century until it was taken over by a prosperous Romano-British family. During the 4th century its celebrated mosaic floor was laid in a new reception room and another room was converted into a Christian chapel. The house was abandoned in about 400.

2 Pottery figurines fashioned to represent diners and reciters were excavated from a Roman child's grave at Colchester. The stocky build and general facial characteristics of the figurines suggest that they depict men from northern Italy. It is known that after AD 49, veteran soldiers, mostly Italian, increasingly settled at Colchester and raised families. Eventually there were several thousand members of this community.

3 Silchester, in Hampshire, is the only Roman town in Britain that has been completely excavated. The first town on the site, where seven main Roman roads met, is thought to have been founded about AD 45. It is a typically Roman town in its grid-iron arrangement, buildings such as baths and temples, its residential areas and its open spaces. The centre, as in Rome itself, was the *forum*, a large square surrounded by public buildings and shops. The most important building was the *basilica*, a long hall where business was transacted. The shops in the forum were more elaborate than the ordinary shops, which served also as houses and sometimes as workshops.

- Temples
— Water main

N. Gate

Caerleon W. Gate

Amphitheatre

E. Gate
London

Forum

Baths

Inn

Baths

S. Gate

4 St Martin's Church, Canterbury, is probably the oldest Christian church in Britain. It is also the only building in the country that was used for Christian worship in both Roman and Anglo-Saxon times. Before being adapted for Christ-ian use, the site had served as a pagan temple for Romano-Celtic religions. The Kentish queen Bertha worshipped there in the 6th century and St Augustine used it as his headquarters for his mission to re-Christianize England in 597. The western wall is Roman and was probably the wall of the church of Duro-vernum, built in the 3rd or 4th century. Few other buildings in Roman towns have been identified as specifically Christian edifices; usually small square rooms were set aside for worship.

tural technology. Most villas [1] were comparatively simple and even the most luxurious were primarily productive farms.

Agriculture was the means of subsistence of most people in Roman Britain [8]; after that came textiles, above all wool. Numerous varieties of weave have been recently unearthed at Vindolanda, near Hadrian's Wall, where remains of a tannery were also found. Wine, olive oil and works of skilful craftsmanship were imported into Britain, but in return Britain exported wool and, after native potters had learnt from the Romans, pottery such as Castor ware. Mineral deposits were worked by the Romans to a much greater degree than ever before.

The Roman ideals

The roads that fostered the urban way of life, and made Britain for the first time a living unity, were built primarily for the uses of the army and the administration. For those who used them there were *mansiones* (staging-posts) provided at regular intervals of ten Roman miles. A typical *mansio* had six guest rooms, a heated dining room, a kitchen and a small bath house, all of which, together with the staff, had to be maintained by the local community. Another significant way in which Britain was now unified was in the law, which probably did much to spread the use of Latin among the upper classes.

There was great variety in the practice of religion. Claudius had from the first introduced emperor worship, but this was a comparatively small part of a Roman's typical round of religious duties. The Romans had done their utmost to stamp out Druidism, for they violently disapproved of its human sacrifices, but they were otherwise tolerant either of the religions they had imported, such as Mithraism, originally from Persia, or of the native religion, of which little is known. The chief Celtic deity seems to have been a figure whom the Romans identified with Hercules [9]. Christianity made some converts, mostly in the fourth century [4].

When the Romans departed, the country quickly lapsed from its civilized ways. But the memory of Rome remained for centuries afterwards an ideal to which the emergent nation had continual recourse.

Mosaics, such as this 2nd-century depiction of autumn from a villa at Corinium (Cirencester), were put in many public and private buildings of Roman Britain. They often represented mythical scenes. The invaders tried to bring their Mediterranean style of life to Britain, adapting it as little as possible.

5 The theatre at Verulamium (St Albans), which was built in about AD 140, is one of only three found in Britain. It was used not only for plays but also for dancing, wrestling and bear-baiting. The audience sat on a semicircular bank of raised seats, with leading citizens occupying the front rows.

6 Romano-British farmhouses were generally single-storey buildings with rooms leading into one another rather than off a corridor. There were usually a number of buildings in addition to the living quarters.

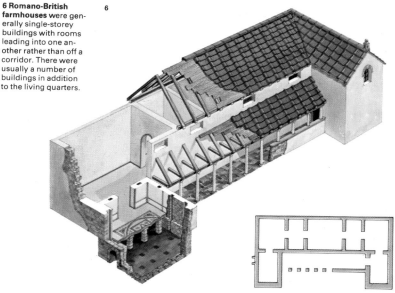

7 The central heating system (part of which is shown) at Chedworth Roman villa, near Cheltenham, worked by means of a hypocaust. In every Roman villa in Britain, one or more rooms were heated by the same means. The principle of the hypo- caust was simple but effective. It consisted basically of a furnace below ground-level, which was fed with charcoal by a stoke-hole located in an outside wall. The hot air from the furnace was drawn through channels under the floors (which were supported by tiled pillars) and from there up through small flues in the walls. In Mediterranean lands hypocausts were usually installed only for heating the baths, but the rigorous British climate demanded a more elaborate and effective system.

8 This bronze statuette of a British ploughman was found at Piercebridge in County Durham. Belgic immigrants from northeast Gaul before and after Caesar's invasions introduced the heavy *caruca,* which went deeper than the Celtic plough (shown here). The Romans used the Belgic plough, and Celtic farmers became confined to the areas of lighter soil.

9 The Cerne giant was cut in the chalk of a hillside north of Cerne Abbas in Dorset in the late Roman period. The figure stands 55m (180ft) tall and holds a club. It has been suggested that the giant is a British adaptation of the cult of Hercules, which was early introduced into Britain by the Romans and has been found in various parts of the country.

Roman remains in Britain

The remains of the Roman occupation of Britain that survive represent only a fragment of the civilization the Romans imported. So much has disappeared largely because the Anglo-Saxons had no use for the towns, roads, fortifications and villas the Romans had constructed, and, abandoning the Roman urban style of living, returned to the land. The remains lay unappreciated until the eighteenth century, and urban development from the medieval era to the present has all but obliterated many Roman towns.

Roman roads and towns
The most universal and unmistakable evidence of their existence left by the Romans is their road system, essential to the swift movement of troops and an efficient administration. Most of the more than 8,000km (5,000 miles) of sturdily constructed roads they laid down have been overlaid by modern highways, but are easily identifiable by the straightness of their alignment [2]. The finest stretches of intact road are found in the north, where the embankment (*agger*) of the road was often metalled with stone paving as,

for instance, at Blackstone Edge [Key]; in the south less durable gravel, chalk or small stones were used.

Not very much remains above ground of the towns that the network of roads was designed to link. In London fragments of the old Roman wall stand isolated, and there is a temple to Mithras similar to that found near Hadrian's Wall [8]. However, quite a good picture of life in the city can be built up from the numerous artefacts found in it [7]. At Silchester in Hampshire the complete circuit of the walls is intact, and the outline of the town's small amphitheatre is visible, as is the amphitheatre at Cirencester. At St Albans the southeasternmost of the four gates has been excavated: it was 30m (100ft) wide, with massive circular towers, and from it a street 12m (40ft) wide led into the city through a triumphal arch. Little of this remains, although there are stretches of wall and tessellated pavements (pavements of small cubical coloured stones) and hypocausts (underfloor heating systems) that testify to the wealth and high standard of living of the inhabitants. There is also there a

theatre which became the town's rubbish dump in medieval times.

Something of the former grandeur of provincial capitals can be seen at Bath [3], York, Lincoln and Caerwent in Gwent; this last was a religious centre where a temple and a bathhouse have been excavated. The Multangular Tower at York still rises amid intermittent stretches of wall; at Lincoln traffic still passes through the Newport Arch, formerly part of the great north gate.

Villas in Britain
Roman villas, ranging from enormous village-like settlements to simple farmhouses of five or six rooms, are most common in the milder and more Romanized south, particularly in the southeast, the Cotswolds and Hampshire. Only a few of the more than 500 known or suspected Roman villas in Britain have been properly excavated.

The largest and most splendid of these villas was only recently discovered at Fishbourne near Chichester; it was built in the first century AD and contained not only elaborate decoration but the only Roman

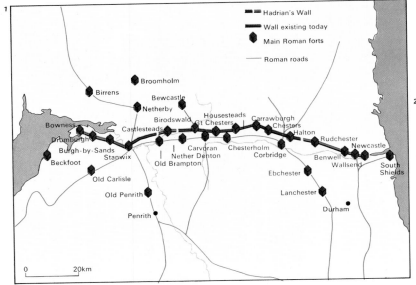

1 The vast complex of Hadrian's Wall is the largest Roman site in Britain. This map shows the course of the wall, linking the Tyne and the Solway, its many forts and the system of roads and towns that supplied it.

2 The Fosse Way, in an aerial view near Hinckley, a town in Leicestershire, is typically Roman in its straightness. Roman roadbuilders always took the shortest route between two places, crossing rather than

circling hills. Running from Lincoln towards Axminster in Devon, the Fosse Way, built after the first phase of conquest, was one of the earliest and straightest of all the roads constructed by the Romans in Britain.

3 The Great Bath at Aquae Sulis, modern Bath, is surrounded by a modern portico (in the Roman Doric order) but otherwise is substantially as it was in Roman times. The Great Bath is one of three plunge baths created in the 1st century AD, and filled by a natural hot spring with waters at 48.5°C (120°F) of renowned curative properties. Originally unroofed, it was later covered with a great barrel vault. Roman Bath was an elegant and sophisticated spa town with numerous luxurious villas in the surrounding country. It was visited not only by Britons but also by foreigners. Numbers of inscriptions found there are dedicated to the presiding goddess of healing, Minerva Sulis, from whom the town gained its name.

4 The most elegant mirror found in ancient Britain, this fine piece of Roman silverwork was once the property of a rich citizen of Viroconium, now Wroxeter. The town was known for its large *forum* and splendid baths. The mirror is 29cm (11.5 in) in diameter, and its skilful workmanship perhaps indicates that it was imported from Italy. The handle consists of two thick strands of interlocking silver wire, each of which bears two floral roundels. The mirror is now in Shrewsbury museum.

formal garden to have been found north of the Alps. A villa at Chedworth in Gloucestershire is particularly well preserved; both Chedworth and Fishbourne have museums attached, with a wide range of artefacts used by the inhabitants. Of other examples, Bignor Villa in Sussex is notable for its fine mosaics and the view it offers of Roman Stane Street climbing the Downs; and the villa at Lullingstone in Kent is interesting archaeologically because it is a small but elegant building that has undergone a number of structural alterations.

Around Hadrian's Wall

Something of what it cost to maintain and protect the Roman civilization can be seen from the straggling remnants of Hadrian's Wall [1]. The wall itself is 2m (6.5ft) high, and wide enough to walk upon; it has a ditch to the north 2.75m (9ft) deep and 8m (27ft) wide. Along its length are 17 forts of which the best preserved are Chesters, Housesteads and Greatchesters, but at every Roman mile (1,480m or 1,620 yards) there was a smaller bastion. A major rampart, 3m (10ft) deep and 6m (20ft) wide at the top, runs parallel to the wall 55 to 75m (60 to 82ft) to the south, and between the wall and the rampart runs a road. The Antonine Wall is also visible in Scotland, between the Forth and Clyde.

Numerous fortresses and civilian settlements surround Hadrian's Wall to the south, notably Vindolanda, where a detailed picture of everyday military and civilian life (and the earliest example of writing in Britain) has emerged from excavations. Corstopitum (Corbridge) [5] was once an important supply base and there is now a museum there with a good collection of Romano-British sculpture, including the "Corbridge Lion". Naturally, most military remains are in the frontier areas, to the north and the west of England, but in the southeast are the forts of the Saxon Shore, erected at the end of the third century AD to combat Saxon raiders. Reculver, Richborough, Pevensey, Portchester and, on the east coast, Burgh Castle are the best examples. Lastly, the Roman *pharos* (lighthouse) at Dover testifies to that port's position as the gateway to and from the heart of the empire.

At Blackstone Edge above Littleborough in Lancashire there is a long stretch of paved Roman road almost 5m (16ft) wide. It has a shallow trough built of shaped stones that runs down its centre. This may have been for drainage, but was more probably filled with turf to provide horses or oxen with a secure footing as they toiled uphill. Beneath the stone paving, so as to provide a firm base, was an embankment (*agger*) of earth or rubble, with a trench running on either side. This ensured that the road was drained. Not until the 18th century were roads of such a quality to be seen once more in Britain. They were built not for stimulating trade, but for the fast movement of troops and for administrative uses.

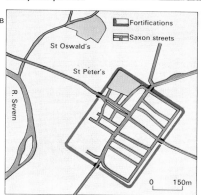

5 The Roman fort of Corstopitum, standing near Corbridge on the north bank of the River Tyne, is shown as it is now [A], and as it might have been [B]. The extensive 4th-century remains are typically square in basic plan. The town was both an important fort and supply base, and a wealthy civilian settlement. Its main buildings consist of two large granaries, a general storehouse in a courtyard, several temples and a fountain fed by an aqueduct.

6 These three plans show the development of the Roman city [A] of Gloucester, in late Saxon times [B], and in the Middle Ages [C]. Its development, in a pattern recurrent in Britain, reflects its defensible site near a bend in the Severn. Saxon Gloucester contained a shrunken community inside the Roman walls. But a growing population and burgeoning trade in medieval times caused expansion, especially towards the river and to the north.

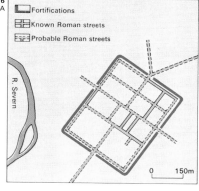

Fortifications
Known Roman streets
Probable Roman streets

R. Severn

0   150m

Fortifications
Saxon streets

St Oswald's

St Peter's

R. Severn

0   150m

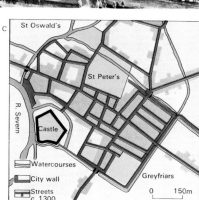

St Oswald's

St Peter's

R. Severn

Castle

Watercourses
City wall
Streets *c.* 1300

Greyfriars

0   150m

7 A Roman kitchen, reconstructed in the Museum of London, is fully equipped with Roman utensils found in the city. Food was cooked in copper pans over charcoal (stored in the arch below) on a stone hearth.

8 This Mithraeum, or temple to Mithras, the favourite god of the legionaries, is in the fort of Carrawburgh on Hadrian's Wall. At the head of a nave flanked by benches for the worshippers, stood three altars, on one of which Mithras is carved with a halo.

Roman art 27 BC–AD 117

Until recently, Roman art was regarded as the bridge between the art of Greece and that of the Middle Ages; largely decadent, it provided a continuity of tradition but because it derived almost entirely from Hellenic models, it offered few original contributions. Now it is widely accepted that, while assimilating Greek styles and techniques, Roman art in fact greatly modified and developed them. Rome remodelled Hellenic art to express imperial ideals – the power of the ruler and his role as patron, benefactor and protector of an ordered society.

Greek art and artists in Rome

The Romans themselves acknowledged the superiority of Greek art and architecture and from the time of the first contacts between the two cultures a steady flow of sculptors, architects and craftsmen in all fields entered the Roman Empire from the east in order to satisfy the increasing demands of her citizens. As early as 268 BC Greek artists had designed the Roman coinage, while as Rome conquered the Greek world huge quantities of booty, from small decorative objects to almost entire buildings, were taken to Rome. (In 187 BC, Marcus Fulvius Nubitor acquired 285 bronze and 230 marble statues during his conquests in Aetolia, while in 168 BC Aemilius Paulus seized no fewer than 250 wagon-loads of art treasures from Macedonia.) Greek artists were employed to copy and even to make pastiches of these works to meet an avid demand.

Different aesthetic attitudes and techniques separated Greek art from Roman. Greek art was produced by individuals whose names are recorded – Phidias' Parthenon, for example. Roman art was almost always anonymous, because the patron who commissioned it was of greater importance than the artisan who created it; so it is that we have the Arch of Constantine, the Baths of Caracalla and Hadrian's Wall.

By the age of Augustus (63 BC–AD 14) Greek (and to some extent Etruscan) influences had been assimilated and most architectural forms were established. Roman architecture offers an immense diversity of types and regional styles using a rich variety of local materials. Among its chief characteristics are the development of the columnar and trabeated (post and lintel) design of the Greeks and continuation, with modifications, of the three Greek orders (sometimes all in the same building, as in the Colosseum [6]). To them the Romans added the Tuscan, a variant of the Doric, and Composite, a combination of Ionic and Corinthian. All the orders were used decoratively and the elaborate Corinthian, the least popular with the Greeks, became the favourite under the Romans.

Similarly, interior decoration using marble and alabaster, paintings and mosaics was highly developed by the Romans, who placed a greater emphasis on interiors than did the Greeks. Added to this increased visual richness was a far greater variety of forms – Roman urban life prompted the demand for such structures as baths, theatres and amphitheatres, bridges, viaducts and aqueducts and multi-storied domestic buildings. Many of them were on a massive scale and were built rapidly, thanks to the Romans' introduction of concrete and brick.

The political emphasis of Roman art was

CONNECTIONS

See also
108 Roman art 117–550 BC
76 Greek art 1100–450 BC
78 Greek art 450–31 BC
98 Rome: the expansion of the empire

1 The unsurpassed richness of the reliefs of the Altar of Augustan Peace symbolize, through a synthesis of propaganda and art, the Augustan ideals of victory, piety and peace. They combine a decorative pattern with a narrative frieze, depicting Augustus, his family and officials.

2 This powerful representation of the ageless Augustus shows him in a cuirass decorated with images commemorating a victory over the Parthians. Although based on a Greek statue, its modified stance and the forcefulness of the head mark it as a truly Roman creation in concept and style.

3 In the Corinthian-style Maison Carrée, Nîmes, Hellenistic architecture is modified to accommodate the special requirements of the Roman religion, which placed emphasis on the interior rather than the exterior of a temple. The top two of a flight of steps form a threshold below the *cella* (the enclosed portion), the width of which equals the length of the portico. The colonnade is continued round the outside of the *cella* but, unlike that on a Greek temple, is embedded in solid walls. Modelled on prototypes in Rome it is one of the best preserved examples of Roman temple architecture of the Augustan age.

4 Augustan art was employed as a propaganda tool. In this cameo the ageless Augustus is represented as the revered messianic leader of a cultured and refined Roman society.

5 Roman portraiture developed out of the custom of producing wax death-masks for funerary processions. Both this patrician and the busts of his ancestors show identifiable individuals rather than types.

especially obvious under Augustas who, in keeping with his role as bringer of peace and prosperity to the empire, was himself portrayed as a pure, magnanimous, handsome and ageless ruler [4]; the serene formality and diverse imagery of Augustan art is summarized in the reliefs of the Ara Pacis (the Altar of Peace) [1]. In his programme of urban expansion he restored numerous buildings (82 temples in Rome alone) and built many others, including the magnificent temple of Mars Ultor, the Forum Augusti and the Maison Carrée, Nîmes [3].

Imperial patronage

The Augustan age had established the range of Roman architectural styles. Subsequent trends reflected the individual preferences of rulers and emphasized the intimacy of the link between the imperial patron and the objects he commissioned. Under Nero and the Flavians, for example, artists were given a freer hand and many extravagant works such as buildings with exotic octagonal, circular or oval rooms were constructed. Nero's sumptuous Golden Palace in Rome was a notable

example. After this flamboyant interlude the Trajan instigated a reaction in which utilitarian structures predominated.

Realism and portraiture

Portraiture, one of the major achievements of Roman art, developed throughout this period. It moved away from the idealism seen in the official propaganda portraits of Augustus toward a remarkable, almost brutal, realism under such later rulers as Claudius and Nero. Greek artists sought to transcend the individual in a quest for universal truths, while the Romans were concerned with accurate likenesses [2, 5]. The contrast between idealism and realism is seen particularly in portrait busts and also in the narrative reliefs on the Arch of Titus [7] and Trajan's Column.

Roman painting, known chiefly from examples at Pompeii [8], was largely derived from Greek, but developed along new lines. Urban Romans maintained their traditional connection with their rural background by favouring landscapes, which became increasingly theatrical and impressionistic.

A bronze head of **Augustus** first emperor of Rome, was found in Meröe, Sudan. Once part of a colossal statue, it characterizes the politically significant art of portraiture during the Roman Republic and after. In some ways idealized, it depicts Augustus in the prime of life. In fact it was made at the time of his death at the age of 76. It departs from Greek portraiture in conveying the strength of the man who symbolized the power and unity of the empire. Julian, in his letter to Theodorus, explained the correct attitude to imperial portraits: "He who loves the emperor delights to see the emperor's statues."

6 The Flavian amphitheatre in Rome – the Colosseum – was conceived by the Emperor Vespasian and constructed between AD 72 and 81. It was one of the greatest Roman architectural achievements and served as a model for amphitheatres throughout the empire. Its outer framework and basic interior structure were composed of gigantic blocks of travertine (a kind of hard limestone), the remainder being built of softer stone and concrete, largely faced with marble. Rising 45.7m (150ft) in four stories the arcaded façade included columns of all three orders – Doric, Ionic and Corinthian – with the upper part decorated with numerous statues. As many as 45,000 spectators, who gained admittance through 80 entrances, sat on tiers of marble seats arranged by class. They were protected from sun and rain by a movable awning.

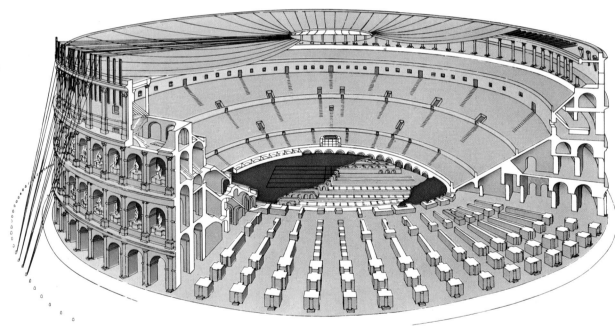

7 Titus' military achievements are commemorated on the reliefs of the Arch of Titus, which depicts his triumphal return from the sack of Jerusalem in AD 70. The arch was built by Domitian.

8 This Pompeian painting shows a mastery in the use of colour, representation of perspective, attention to detail and appreciation of architectural interiors. These were the hallmarks of the painter's art, in the first century AD.

Roman art 117–550

In the time of Trajan there was a revival of interest in the art of earlier periods which reached its apogee under his successor Hadrian (AD 76–138). The unprecedented economic prosperity of the empire, coupled with the emperor's personal patronage, led to a veritable Hellenic renaissance that persisted through Hadrian's reign and beyond, determining to a large degree the character of Roman art in the second century AD.

Hadrian's architectural revival

Hadrian's villa at Tivoli [3], described as "probably the most beautiful collection of ruins on Italian soil", was the most remarkable of all Roman architectural complexes. He made lavish use of multi-coloured marble and of diverse ornamental architectural features including the curved architraves that were popular in the Eastern Empire.

These and the innumerable Greek and Greek-influenced masterpieces of sculpture with which Hadrian surrounded himself introduced many new and important elements into the Roman artistic repertory. He created such emphatically Roman works as

the Pantheon – the most important domed structure of the age [1].

Under the Antonine emperors who succeeded Trajan, architecture was predominantly Greek in inspiration. Much use was made of the three orders and many massive works, such as large temple complexes, were built, especially in the Eastern Empire. A number of new architectural devices were employed, including ostentatious façades, exotically decorated (but detached and functionless) columns which visually fragmented interiors, niches in which works of sculpture were placed, and a profusion of surface decoration of all kinds.

During the rule of the Severan emperors in the third century, the rise of status and wealth of citizens in the Roman provinces in particular was asserted in a range of lavish architectural works. In an attempt to vie with the achievements of earlier emperors, such rulers as Caracalla (176–217) embarked on a number of grandiose building schemes that produced a wide range of structures. These were most notable for their impressive size and the pretentiousness of their decoration.

Among them one may include the Baths of Caracalla in Rome, the best preserved of all imperial baths. It accommodated 1,600 bathers and its central hall (*tepidarium*) was roofed by a vault soaring to a height of 54.8m (180ft), decorated with coloured marble and the most priceless statues. But in keeping with the Roman preference for interior rather than exterior elaborateness, it was largely undecorated on the outside.

The reigns of the later emperors of the third century and their preoccupation with the defence of the empire did not encourage major building programmes, other than those to do with military works. The fortified palace of Diocletian (245–313) at Spalato (Split) in Dalmatia (completed 305) stands out as the major military building of the late period and symbolizes the swing from decoration to pure functionalism.

Sculpture to the third century AD

Through Hadrian's influence, the earnest majesty of the sculpture from Trajan's era was replaced by Greek works in which sensual beauty predominates.

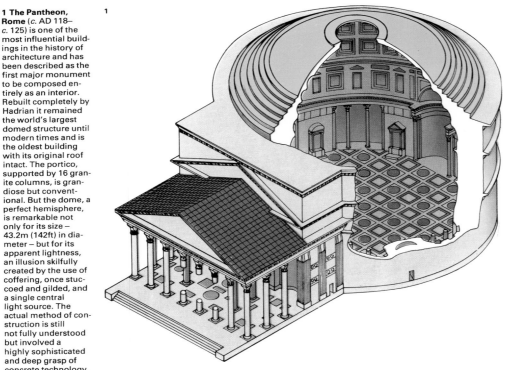

1 The Pantheon, Rome (*c.* AD 118–*c.* 125) is one of the most influential buildings in the history of architecture and has been described as the first major monument to be composed entirely as an interior. Rebuilt completely by Hadrian it remained the world's largest domed structure until modern times and is the oldest building with its original roof intact. The portico, supported by 16 granite columns, is grandiose but conventional. But the dome, a perfect hemisphere, is remarkable not only for its size – 43.2m (142ft) in diameter – but for its apparent lightness, an illusion skilfully created by the use of coffering, once stuccoed and gilded, and a single central light source. The actual method of construction is still not fully understood but involved a highly sophisticated and deep grasp of concrete technology.

2 The Portonaccio Battle Sarcophagus depicts Romans in a frantic struggle with barbarians and captures the moment when victory is unquestionably the Roman commander's.

The Emperor Hostilian (*d.* 251), detached from the mass of bodies, completely dominates the scene. The craftsmanship represents a theatrical example of mid-3rd century art.

3 Hadrian's Villa (*c.* AD 130), the most extravagant of all Roman country palaces, represents the eclectic tastes and restless architectural experimentation of a Roman ruler with unlimited resources. No single building is on a grand scale but the whole – an immense variety of courtyards, fountains, statues, pillars, pools, mosaics and individual buildings – covers 18km² (7sq miles) of countryside on the edge of the Roman Campagna and is skilfully blended into the landscape.

4 This bronze equestrian statue of Marcus Aurelius in Rome dates from the Antonine period when idealized figures began to displace the realism found under Trajan. Once gilded and with a figure of a barbarian under the horse's raised hoof, it was preserved while other Roman bronzes perished in the mistaken belief that it depicted the Christian Emperor, Constantine. The statue influenced the work of such Renaissance masters as Donatello (*c.* 1386–1466).

From *c.* AD 100, for reasons that are not fully understood, burial began to replace cremation as the customary method of disposing of the dead throughout the Roman Empire. From this time tombs became more elaborate and the art of making decorated sarcophagi was developed. In the late empire battle scenes became particularly popular on the sarcophagi – such as the Portonaccio Scarcophagus [2] – of military leaders.

In the Antonine period statues became more massive and elaborate. The use of chisels for carving stone was steadily superseded by the skilful use of the drill which enabled the production of more decorative effects in drapery and hair. Fine bronzes, such as the powerful equestrian statue of Marcus Aurelius (121–80) [4], were produced but few have survived.

Two particular types of relief co-existed from this period – a primitive, almost folk-art style, as shown on the column of Marcus Aurelius and the persistence of Greek classical elements – two streams of artistic endeavour that came together in the Arch of Constantine [5].

Under the Severan emperors massive statues and reliefs were often so elaborate that they virtually obscured entire architectural surfaces.

Sculpture of the turbulent third century was of variable quality – reliefs became cruder and there was a trend toward mannerism and abstract symmetry, which involved reintroduction of the chisel in place of the drill. This resulted in the angularity found in such stark and impressive works as the colossal head of Constantine [Key].

Painting and crafts

Painting did not apparently develop beyond the so-called Fourth Style identified at Pompeii. Mosaic, originally a flooring technique, developed rapidly in the second and third centuries and virtually replaced painting as a method of decorating walls [8].

Despite several imperial decrees against personal extravagance, fine jewellery, splendid ivory-inlaid furniture and a wide range of metalwork, such as the silverware found at Mildenhall, England [6], were popular throughout the empire.

This head from the **colossal statue** of the Emperor Constantine in Rome is the largest statue of its age – ten times life size. It was sculpted in *c.* AD 313 and was the peak of imperial art.

5 The Arch of Constantine is decorated with a rich variety of panel reliefs and statues, many of which were plundered from earlier public monuments. It thus summarizes the major artistic styles from Trajan to Constantine. It indicates a change towards painting and mosaics.

6 This great dish, part of the Roman silver hoard known as the Mildenhall Treasure (4th century AD), measures almost 60cm (2ft) across and weighs 8.3kg (18.3lb). The outer frieze represents the triumph of Bacchus, god of wine, over Hercules. Other mythological figures, including Neptune, also appear.

7 In this gold medallion minted at Thessalonica, the Emperor Constantine appears in a portrait that echoes those of Augustus. The restoration of order to the Roman Empire under Diocletian and Constantine prompted the standardization of Roman coinage. While achieving a high degree of craftsmanship, representations of rulers became stylized with religious overtones – a typical aspect of late Roman art, foreshadowing that of Byzantium.

8 The technique of mosaic, invented by the Greeks, was readily adopted by the Romans who perfected it and applied it on a large scale. Many local styles emerged in Italy and such provinces as Britain and Tunisia. Among the subjects employed are stories from classical mythology, interpretations of classical Greek paintings, landscapes, portraits, elaborate decorative motifs and particularly scenes of vigorous action – chariot racing, hunting and gladiatorial spectacles. By the 3rd century mosaic work had, in many instances, superseded the three-dimensional representations of gods and cult figures, as in mosaic portraits.

Christ and the Apostles

Jesus Christ was an orthodox Jew. Born in about 4 BC, he was crucified outside the walls of Jerusalem in about AD 30. He certainly gave Christianity its original impetus, but it is less certain that he meant to found a Church. His disciples came to think of him as the promised "Messiah" (the Hebrew form of the Greek "Christ"), and then gave to him the value and honour they reserved for the one God alone.

The figure of Jesus and his teaching still dominates Western imagination nearly 2,000 years after his death. Today more than 1,000 million people in many parts of the world describe themselves as Christians.

When Jesus lived there was in Judaism a variety of emphases and interpretations leading to controversies between parties. The evidence of the Dead Sea Scrolls [4] and other contemporary sources suggest that the teaching of Jesus had much in common with some of the sects current during his lifetime.

Responsibility for his death is still disputed. According to the Gospels he was crucified on the orders of the Roman governor at the insistence of the Jewish religious establishment. The Gospels suggest he was condemned for blasphemy (which was a capital offence) because he claimed to be the Messiah, although such a claim was not technically blasphemous in Jewish law.

The historical Jesus

The only detailed records we have of the life of Jesus are the Gospels, which were written by Christians for Christians. What are believed to be sayings of Jesus and stories about him circulated first by word of mouth. Collections of such sayings and stories were doubtless later written down, and it was from such sources that the four Gospels were compiled. They were written in Greek, although Jesus spoke in Aramaic. The first of the Gospels, St Mark's, was written in about AD 62.

The Gospels present us with highly interpreted history, and it has been impossible to disentangle fact from interpretation.

The earliest records (St Paul's Epistles and St Mark's Gospel) do not describe Jesus' birth. According to the later accounts of St Luke and St Matthew Jesus was born at Bethlehem of a virgin mother, Mary, his conception in her womb being due to the creative power of God's spirit. Jesus' home town was Nazareth in Galilee [1].

Jesus was baptized in about 26, when he was about 30, by his cousin John the Baptist [5]. Immediately after his baptism he went into spiritual retreat in the desert. Then he began teaching in Galilee, healing the sick in body and mind and proclaiming the imminent arrival of the kingdom (more accurately the kingly rule) of God.

Many have found its essence distilled in the Sermon on the Mount – a compilation of his sayings masterfully arranged. It emphasizes the need for spiritual rebirth and describes the heroic goodness that follows as a consequence. The teaching of Jesus had a challenging directness. He called men away from the letter of the Jewish law to its spirit. This led him to criticize severely the Jewish leaders of his day.

Arrest and crucifixion

Whatever the precise form of his criticisms, Jesus was marked down by the religious authorities as a public enemy. And this con-

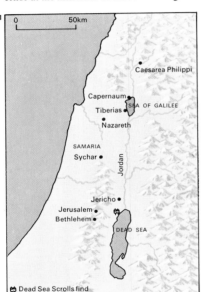

1 **Christianity began** in the northern Jewish province of Galilee, away from the religious centre of Jerusalem, and Jesus spent most of his ministry among simple country people. Traditionally, some Apostles travelled as far as Persia and India. The most widely travelled Apostle, Paul, was a Jew but a Roman citizen born at Tarsus in what is now Turkey. He took the Gospel both to Asia Minor and to Greece.

2 **Capernaum**, where the remains of this synagogue are still visible, is at the north end of the Sea of Galilee. It is mentioned several times in the Gospels in connection with Jesus' mission.

3 **Christ is transfigured** with Elias and Moses representing the Prophets and the Law, while John, Peter and James crouch before the light in this painting from a 12th-century church in Cyprus.

4 **The Dead Sea Scrolls** were discovered between 1947 and 1956 in caves to the northwestern end of the Dead Sea. They contain most of the books of the Old Testament and some other hitherto unknown Jewish writings in Hebrew and Aramaic from a monastic community at Qumran. Most of the manuscripts are made from leather and papyrus and are kept at the Israel Museum in Jerusalem.

cerned the civil (Roman) authorities because his wide popular support might involve disturbances. He probably realized that his life was threatened. On a visit to Jerusalem in about AD 30 to observe the Jewish Passover He provided his enemies with an occasion to arrest him by physically assaulting the money changers and traders in the Temple. He was eventually arrested by the Temple guard while he was praying in the Garden of Gethsemane – Judas Iscariot no doubt leading them to the place and betraying the identity of Jesus by kissing him. The Jewish leaders handed him over to the Roman governor, Pontius Pilate, who ordered his crucifixion (a Roman form of execution for slaves and low criminals) with two thieves at Golgotha.

The early Church

From the third day after the crucifixion the followers of Jesus became utterly convinced that he was alive. They were sure that he appeared and spoke to them as recognizably himself though now glorified. At a slightly later date the news was spread that certain women had gone to his tomb to anoint his dead body and had found the tomb empty, except for the graveclothes.

At the harvest festival of Pentecost, some five weeks after the crucifixion, the disciples believed themselves visited by the spirit of God sent down by the exalted Jesus. From then on they went about preaching that Jesus was the Messiah crucified and raised from the dead, through whom eternal salvation was offered to all.

Paul became the chief Apostle to the Gentiles (non-Jews) after a dramatic conversion to Jesus' teachings on the road to Damascus. Greek-speaking and a Roman citizen, he began the transformation of Christianity into a world religion for everybody. He travelled extensively, preaching in Asia Minor and Greece [8]. He was eventually arrested in Jerusalem, used his right as a Roman citizen to appeal to Caesar, and is said to have been executed in Rome at the same time as St Peter, about AD 60.

According to legend, St John, "the disciple whom Jesus loved", lived to an old age in exile. St John's Gospel is a profound meditation on the life and teaching of Jesus.

The image or name of Jesus Christ and the Gospel open at the words "I am the Light of the World" appear in many mosaics, including this 12th-century work on a ceiling in Palermo.

5 Jesus was baptized in the River Jordan by John the Baptist. In this sixth-century mosaic from Ravenna, the figure on the left represents the river god of the Jordan. Baptism is the method of entry into the Christian community.

6 Jesus celebrated his Last Supper with his disciples on the night before his death. This meal is the distinctive Christian ceremony called the Lord's Supper, Holy Communion, Eucharist, Liturgy and Mass. This sixth-century mosaic at Ravenna shows 11 disciples reclining with Christ. The two fish recall the miraculous feeding of the multitude.

7 A fifth-century ivory carving depicts the women who went to the rock-hewn sepulchre of Jesus at dawn on Easter Sunday and found that he had risen from the dead.

8 Paul went on his first preaching tour to Asia Minor. A return visit continued over into Greece and a third journey covered much the same ground. The final journey by sea went through Crete, Malta and Sicily to Rome.

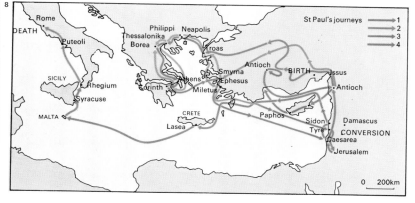

St Paul's journeys 1 2 3 4

Rome
DEATH
Puteoli
SICILY
Rhegium
Syracuse
MALTA
Philippi Neapolis
Thessalonika
Borea
Troas
Antioch
Smyrna
Athens Ephesus
Corinth Miletus
CRETE
Lasea
BIRTH
Jssus
Antioch
Paphos
Sidon
Tyre
Damascus
CONVERSION
Caesarea
Jerusalem

0 200km

Roman literature

In the first three centuries of Rome's development the influence of captive Greece on Roman literature was all pervasive. The Romans themselves regarded their literature as having been founded by a Greek – Livius Andronicus (c. 284–c. 204 BC). He wrote tragedies and comedies based on Greek originals and translated Homer's *Odyssey* using a native Italian metre the Saturnian.

The Hellenistic influence

The end of the First Punic War marked the beginning of literary work in Rome. Although the epic on the war by Naevius (c. 270–c. 201 BC) was in the Saturnian metre, Ennius' (239–169 BC) introduction of the hexameter – the metre of Greek epic – in his great epic poem the *Annales*, laid the foundations for subsequent Latin poetry.

However, the elder Cato (234–149 BC), to whom Ennius is said to have taught Greek, opposed this Hellenizing movement. The archetype of strict Roman morality, he wrote his works in as basic a Latin as possible.

The greatest writer of Latin comedy was Plautus (c. 255–c. 184 BC). He based his plays on Greek originals but added vigorous humour and brilliant handling of Latin. The refined style and deep interest in character of his successor Terence (P. Terentius Afer, c. 190–159 BC) [1], an African slave who won his freedom, were less successful on the stage. Drama never became established in Roman life and literature as it had been in Greece and Terence was the last significant Latin author to write for the theatre.

Satire, which the Romans considered to be the only genre they had not borrowed from the Greeks, was developed from Ennius' foundations by Lucilius (c. 180–102 BC). He established the hexameter as the regular metre for satire.

The greatest literary figures

The first century BC, the last era of the Roman Republic, produced some of the greatest figures in Roman literature. The huge output of Marcus Tullius Cicero (106–43 BC) [Key] included his speeches and works on rhetoric and philosophy. His friend and later enemy, the emperor Julius Caesar (c. 100–44 BC) [3] was also an outstanding orator. Caesar wrote an important grammar, but only his historical works survive.

Younger contemporaries of Caesar were the poets Lucretius (c. 99–55 BC) and Catullus (c. 84–c. 54 BC) [2, 4]. T. Lucretius Carus' hexameter technique in his didactic poem *De Rerum Natura* forms a link between Ennius and Virgil. G. Valerius Catullus was a member of a coterie that took Hellenistic poetry as its model for the cultivation of elegy and epigram, predominantly on the theme of love and in a variety of Greek metres. Sallust (G. Sallustius Crispus, 86–34 BC) raised the writing of history to a higher literary level in his monographs on the wars of Catiline and of Jugurtha, devising a taut, antiquarian style.

In Virgil (P. Vergilius Maro 70–19 BC), Latin literature at last acquired its Homer. His last and greatest work, the *Aeneid*, represents a fusion of the epic tradition with the newer, more subjective poetical ideas of the Hellenistic age. His friend Horace (Q. Horatius Flaccus, 65–8 BC) also rose from humble origins to enjoy the patronage of Augustus. His *Satires* lifted the genre to a high literary level, while his *Epistles* were an

1 A twelfth-century illustration shows Terence's comedy *Andria*: as Simo's slaves prepare for his son's wedding, he tells his freedman that the "wedding" is a trick being played on his son, who is in love with a girl from Andros.

2 Ariadne, daughter of Minos of Crete, wakes on the island of Naxos to find that she has been deserted by Theseus, son of the king of Athens, whom she has helped to enter the labyrinth and kill the Minotaur. The myth was a popular subject for wall paintings in the houses of the well-to-do of Pompeii. The myth of Ariadne forms a principal subject of Catullus' longest and finest poem, the *Peleus and Thetis*, and was later treated by Ovid.

3 Caesar is portrayed on this coin as "Perpetual Dictator" – the first Roman to have his image on coins. His accounts of the Gallic and Civil Wars are written in plain, elegant Latin.

4 Catullus recites his famous epigram on the death of his mistress Lesbia's pet sparrow. The picture is one of Lawrence Alma-Tadema's detailed evocations of Roman life.

original development from it. His *Odes* successfully adapted lyric poetry to Latin.

Livy (T. Livius, 59 BC–AD 17), one of the most notable historians of the age, extolled the glories of the departed republic in his *History of Rome*. Ovid (P. Ovidius Naso, 43 BC–AD 17) followed on from Catullus in the writing of love-elegy; his exuberant talent also produced the *Metamorphoses*.

Literary decline and transformation

At the end of the first century AD Latin literature in its classical form began to decline. The prose of L. Annaeus Seneca (*c.* 4 BC– AD 65) a Roman born in Spain, represents a reaction against Ciceronian fulsomeness in favour of brevity and point [5]. Together with the epics of his nephew Lucan (M. Annaeus Lucanus) (AD 39–65) and Statius (*c.* AD 45–*c.* 96) they typify a period in which both the Latin and Greek masters were imitated. However, among the more original poets and prose writers Petronius' (died *c.* AD 66) *Satyricon*, a discursive "novel", was written with immense verve and style. Another Roman born in Spain, Martial (M. Valerius Martialis) (*c.* AD 40–104), concentrated on epigram, in which his wit and humanity set the standard for all future attempts.

P. Cornelius Tacitus (*c.* 55–120), with his inimitable style and powerful personality, represents the peak of Roman historical writing. The polished letters of his colleague Pliny the Younger (G. Plinius Secundus *c.* 62–114) throw a vivid light on contemporary society, as also – less favourably – does the mordant wit and lurid hyperbole of the satirist Juvenal (*c.* 55–*c.* 140).

Prose authors of the second century tended to favour a style in which archaistic and colloquial elements are mixed. A striking case is the rhetorician and philosopher Apuleius (born *c.* 125), who is best known for his *Metamorphoses*.

From the third century onwards it is Christian authors such as Tertullian (*c.* 160–*c.* 230), St Augustine (354–430) and St Jerome (*c.* 347–420) who are most significant, and it was primarily as the language of the Roman Church that Latin was to remain alive right up to the Middle Ages.

KEY

Cicero rose from obscure origins to become a leading if ambivalent political figure in Rome, chiefly through the power of his matchless rhetoric, which was as effective on the political platform as in the law courts. His published works represent the summit of Latin prose.

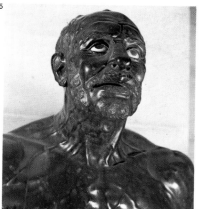

5 Seneca, statesman and philosopher, was rich, vain and often accused of hypocrisy for failing to live up to the ascetic ideals he professed. He attempted to realize Stoicism in practice by acting as Nero's adviser, but towards the end of his life, when his influence over Nero began to wane, he devoted himself to expounding the Stoic philosophy in numerous treatises and in his *Moral Letters*. His philosophy was derivative but influenced later Roman thought.

6 A palimpsest, a parchment reused after the original text has been removed, is the only surviving source of Cicero's *De Republica*, a dialogue concerning the best form of government. It was found in 1882. Cicero's text, which dates from *c.* AD 400, was partly obliterated when the parchment was used again for a commentary on the Psalms by St Augustine in the seventh century, but Cicero's text could still be read.

7 Virgil is seated between the Muses of Tragedy and History in this third-century mosaic discovered at Sousse, in Tunisia. He holds a papyrus roll of the *Aeneid*, showing the words from the first book, in which the poet invokes the Muses to tell him what caused Juno in her wrath to force Aeneas to suffer so much.

8 St Ambrose (*c.* 340–397), the German-born layman elected Bishop of Milan in 374, is enshrined in an early fifth-century mosaic at the basilica he founded in Milan which is named after him. A great preacher, he wrote a wide range of prose works and is regarded as the father of the Latin church hymn. His hymns, still in use, are written in a simple style to appeal to ordinary Christians, and in four-line verses.

Rome: soldier emperors to Constantine

After the murder of the Roman Emperor Commodus in AD 193 four rivals disputed the imperial succession. There was a costly civil war in which several major cities, including Antioch, Byzantium and Lyons, were sacked before Septimius Severus (reigned 193–211) was successful. Order was restored, but the military basis of imperial power then became more obvious.

Political anarchy and religious persecution

Septimius was succeeded by a family dynasty, but because a settled system of succession was lacking and because the legions increasingly realized that they had the power both to elect and to destroy emperors [1] there was continual unrest, with rivals being set up by local troops. In the 74 years from the death of Septimius to the accession of Diocletian (245–313) in 284 there were 27 emperors and many usurpers.

This political anarchy came at the worst possible time, for the barbarians were again pressing on the empire's frontiers. In 236 Alemanni and Franks crossed into Gaul and Goths poured over the Danube in 247

raiding the Balkans and killing the Emperor Decius (200–251). The Romans were forced either to allow the barbarians to settle within the frontiers or to buy them off.

In the east a new Persian dynasty, the Sassanids, invaded Syria and Asia Minor and then captured the Emperor Valerian (193–260) in 260 [2]. His son and co-regent Gallienus (reigned 253–68) had to put down five rivals before recovering the lost eastern provinces. Aurelian (reigned 270–75) finally abandoned Dacia but began restoring the Danube and Rhine frontiers. The work was completed by Probus (reigned 276–82).

The unrest was accompanied by a breakdown of civil order and a collapse of the economy. Inevitably, men searched for scapegoats and the most obvious ones were the Christians. The Romans had always been highly tolerant of religions provided they accepted the divinity of the emperor, and many Oriental cults, including the worship of Mithras and Isis, had become widely popular. But the Christians, who refused to sacrifice to the emperor, were an easy target. They were barbarously persecuted by Decius and even

more so by Valerian (reigned 253–60) [3].

Despite all the disasters, civilized life continued and the third century saw the work of some of the greatest commentators on Roman law – Papinian, Paulus and Ulpian – and considerable literary achievements.

Division of the empire

There was a desperate need for reorganization and in 285 Diocletian established a totally new governmental system [Key]. The empire was divided into Eastern and Western parts, each ruled by an "Augustus" with a "Caesar" as his deputy. The Augusti were to resign after 20 years and be succeeded by their Caesars. The imperial court moved away from Rome [4] and the provinces were replaced by dioceses ruled by a massive new bureaucracy and a reformed army. The whole system was supported by a new currency and heavy taxation and the emperors became absolute monarchs.

Diocletian's reforms did not solve the succession problem and when he and his co-Augustus Maximian (reigned 286–305) resigned in 305 chaos followed. In 312

CONNECTIONS

See also
142 The Byzantine Empire
136 The barbarian invasions
134 The rise of medieval Western Christendom
98 Rome the expansion of the empire
96 From the civil wars to Caesar's empire
92 Rome: the organization of the republic

1 The Praetorian Guard were the élite bodyguards of the emperors. They were the first body of Roman soldiers to realize they had the power to make emperors and, if need be, to break them too;

during the third century their example was followed by troops in the provinces. Anarchy ensued as local garrisons set up their own emperors and tried to dominate the empire.

2 The Emperor Valerian was forced to kneel to the Persian ruler in the worst disaster that befell Rome in the third century. The internal chaos of the empire coincided with renewed pressure

from barbarian tribes in the West and from the Persians in the East. The cohesion of the empire was further weakened as threatened areas took action independently of Rome to defend themselves.

3 Roman persecution of Christians [A], which reached its height under Decius and Valerian, contrasted with general religious toleration, limited only by the importance given to the unity of the empire. Provided the adherents of a religion were prepared to pay homage to the divine emperor, they were free to worship and make converts. Many Eastern cults such as Mithraism [B] flourished but the exclusiveness of Christians and their refusal to do homage caused them to be treated as a treasonable sect and to be deprived of citizenship.

Constantine (*c.* 285–337) emerged victorious in the Western Empire in a battle at the Milvian Bridge, during which his armies fought under the Christian cross. In gratitude he made Christianity the official religion of the empire. In 324, on the death of the Eastern Emperor Licinius (reigned 311–23), Constantine reunited the empire and moved his capital to Byzantium, which was rebuilt as a totally Christian city with the name Constantinople. Henceforth the Eastern Empire and Christianity were to be closely identified.

Constantine's death in 337 led to a division of the empire between his two sons; fighting followed until it was briefly reunited in 353. In 355, the Western Empire was placed under Julian, who briefly reunited the empire between 361 and 363. But thereafter the split became permanent.

The difference between the two halves was steadily being emphasized by the presence of barbarians in the West [6], as more tribes were settled within the frontiers, and by the development of a Christianity-dominated absolute empire in the East. In

the West the army was by then almost entirely recruited from barbarians; as a result, its resistance was severely reduced.

Major barbarian invasions

Towards the end of the fourth century came renewed major invasions; in 376 the Visigoths were allowed to cross the Danube to settle, but they were so badly treated that they revolted in 378 and killed the Emperor Valens (*c.* 328–78) at Adrianople.

Other groups also crossed the frontiers; the Vandals moved through France and Spain to set up an independent kingdom in North Africa; Jutes, Angles and Saxons occupied Britain; Franks and Burgundians settled in northern France and Ostrogoths in Italy. By the middle of the fifth century the Western Empire had been almost completely occupied by barbarians, although a Romanized administration and culture survived. It came as no great surprise when German troops in Italy elected Odoacer, the Ostrogoth (reigned 476–93), as king and he deposed Romulus Augustulus (reigned 475–6) ending emperors in the West.

Representing Diocletian's tetrarchy, this statue from St Mark's in Venice shows two Augusti clasping their Caesars, the deputies who would succeed them after 20 years. The chaos and disasters of the 3rd century forced Diocletian to impose major changes on the empire. His reforms, which were the first alterations to the system established by Augustus, can be seen as formalizing the practices of the years of anarchy when emperors reacted on an *ad hoc* basis to barbarian invasion and civil strife. The emperor and his deputies became full-time military leaders and the whole system was supported by stringent new laws.

5 Constantine the Great made the momentous decision that Christianity was to be the official religion of the empire. He also reunited the empire and moved the imperial capital from Rome to the strategically placed city of Byzantium (Constantinople), which he had rebuilt as a purely Christian city. Constantine identified with his new faith and put the whole weight of empire behind it.

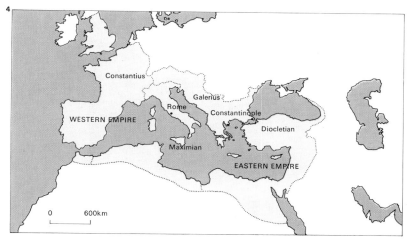

4 The reforms of Diocletian fundamentally altered the empire, splitting it into two almost independent halves. He and his Caesar Galerius were in charge of the crucial Danube and Eastern provinces, while Maximian and Constantius ruled the West. The emperors moved their headquarters nearer to the frontiers. Diocletian ruled from Nicomedia, near Constantinople.

6 This detail from a sarcophagus shows Roman soldiers subduing barbarians. Some fighting emperors during the third century were able to hold the frontiers and push back invaders but generally the barbarians were allowed to enter the empire, settle and infiltrate the army and administrative structure. During the fourth and early fifth centuries in the West the barbarian kingdoms took shape and their rise merged almost imperceptibly with the decline of the empire.

7 The influence of the Roman Empire continued as a civilizing force long after its fall. The massive grandeur of its buildings, such as these at Palmyra, remained as a visible reminder, while Rome's intellectual legacy was permanent.

115

Early Oriental and Western science

Modern scholarship has made it clear that in Roman times there was considerable cross-fertilization between different civilizations. The Roman Empire itself, with its emphasis on foreign trade, provided regular links between the diverse civilizations of Europe, Africa, India and even China.

The extent of Roman technology
The vastness of the empire posed immense problems in peace-keeping, in government and administration. On these counts the Romans excelled. They developed military technology, constructing large mechanized catapults, mechanical arrow ejectors and crossbows originally invented by the Chinese in the third century BC, elaborate battering-rams and wheeled siege towers, Assyrian inventions of the ninth century BC. The large standing army required not only weapons but clothing and food and here again the Romans used the most modern techniques, establishing in the south of France, for instance, water-operated multiple flour-grinding mills.

Roman roads stretched throughout the empire. Most were gravelled or stone set on concrete with curbstones and drainage channels. Since the empire extended into climates much colder and damper than that of Rome, forms of central heating and damp-proofing were common, at least in the houses built for Roman officials. Heated baths were also common in cities and army camps all over the empire. For their shipping, the Romans followed the example of Sostratos who, in the third century BC, built a huge lighthouse at Alexandria and constructed lighthouses at many other ports [1].

Yet, expert though the Romans were in the art of running an empire and in using up-to-date technology to help them, they made no progress in pure science. Such science as they knew they obtained from the Greeks. In the first century AD, Pliny (23–79) had written his *Naturalis Historia*, a vast compendium on all known science, but this was primarily a compilation of Greek science. Galen (*c.* 130–*c.* 200) [2], the greatest medical man of Roman times, was a Greek national. As a physician and surgeon to the gladiators, Galen obtained valuable knowledge about wounds and various internal organs and later he became personal physician to the Emperor Marcus Aurelius (121–180). He took many pupils and promoted his interpretation of the operation of the human body, which was based on the idea that the liver was the main organ of the venous and arterial systems.

Advances in other centres
It was not until the third century AD that there was a turn from the Greek concentration on geometry. Then, in Alexandria, the mathematicians Diophantus and Pappus recommended the study of numbers and evolved a kind of algebra. Their work was extended by later generations and particularly by the Muslim civilization, where algebra became highly developed. The "arabic" numerals [Key], which had originally been devised in India, were adopted.

Arithmetic and a form of algebra also characterized Chinese mathematics. It was in China that the abacus was invented and it became such a useful calculator that it is still common in Japan today.

In the West, some 1,800 years and more

1 A Roman lighthouse, erected at Dover, was one of many such aids to shipping, a typical use of technology by the vast and efficient Roman administration. Lighting was by means of a fire, usually of wood and tarry substances, contained at the top of the building. The lighthouse was built in a stepped form like the famous Pharos at Alexandria (3rd century BC).

2 Trajan's Column was erected by the Roman Emperor Trajan (*c.* 53–117) to celebrate his victories. Built of marble and more than 30m (100ft) high, it was set up in the Forum in Rome in 113. This section shows Roman legionaries being treated on the battlefield. It was here and in gladiatorial combat that Galen and other surgeons learned the basics of anatomy.

3 A crane is depicted on this section of Trajan's Column. The pulley block and ropes, and the men inside the treadmill at the bottom, can be seen clearly. Cranes such as this were found well into medieval times.

4 A euthytonon was a type of mechanical crossbow that was so large it had to be mounted on a stand and fired by a trigger mechanism. This improvement is usually credited to the mechanician Philo who worked at Byzantium (on the present-day site of Istanbul), sometime between 2000 and 250 BC. This engraving is taken from *Poliorceficon*, a book by Justus Lipsius (1547–1606), which was concerned with Roman armaments and was published in Antwerp in 1605.

5 Roman taximeters were operated by the wheels of the carriages. Worms and gear wheels reduced the rotations until dials and a counter could be run at a convenient speed. The dials had pointers that indicated the distance travelled by the carriage. The counting disc (at the top of the taximeter) allowed pebbles to drop into a holder at the bottom of the meter. By counting the number of pebbles in the holder, the fare could easily be worked out.

before the Roman conquest, astronomical observations were being made in Britain with great accuracy by using stone circles, of which Stonehenge is probably the greatest example. The Chaldeans (flourished 1000–540 BC) had discovered that eclipses occurred in cycles. Yet it was not until early in the second century AD that accurate eclipse predictions and studies of the earth and heavens were regularly made in China; however in the same period technological advances were accomplished. Chang Heng had invented a seismograph for recording earthquake shocks and the art of papermaking was established [7]. By the eighth and ninth centuries AD the Chinese had also invented gunpowder [6] and block printing, as well as the first clock ever to be made with an escapement – the large water clock in Sian [8]. Moreover Chinese technology brought in the first efficient form of horse harness, the sternpost rudder for ships, the manufacture of silk and most importantly, the compass which was in widespread use by the 1100s.

Alchemy, the forerunner of chemistry, was an ancient art of obscure origins that sought, among other things, to transform base metals such as lead into silver and gold. Carried westward, it was developed in Alexandria particularly by Zosimos in the fourth century AD, along with the important process of distillation.

Medicine and the Chinese influence

Medicine, too, was not free of mysticism and superstition, but herbal drugs were discovered and used by the Chinese, who in their pharmacopoeia had drugs against malaria and for bronchial diseases, while in India more than 500 drugs came into use, including an early form of tranquillizer. It was in China that the technique of acupuncture evolved.

After the Greeks the progress of science was slow and piecemeal, in the West at least, and seems virtually to have halted for some 600 years. Fortunately, though, the Muslims collected and collated all knowledge of Greek science and, with additions from India and the Far East, all this information filtered through to the West from the twelfth century AD onwards, thus paving the way for the great scientific revival 400 years later.

Pythagoras (right), using an abacus, seems to be competing against Boethius (475–524) [left] who is using arabic numerals.

6 Gunpowder, first invented by the Chinese, was originally used for fireworks and only later adapted for war. Shown here are early wheelbarrow-like rocket launchers. Gunpowder was first mentioned by Chinese Taoist alchemists about five centuries before it appeared in Europe (c. 13th century AD). The development of the gun is more obscure because at first gunpowder was used only in rockets and bombs.

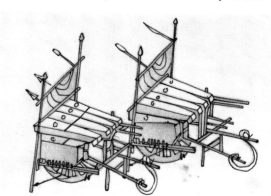

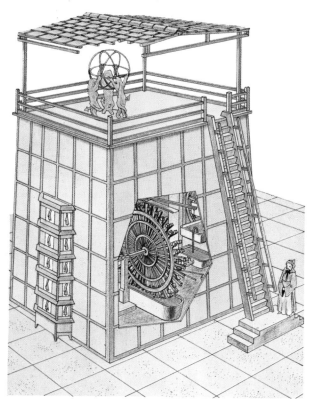

8 The first mechanical clock was Chinese, invented in the 8th century by I-Hsing. It was driven by an elaborately engineered waterwheel which acted as an escapement – the essence of all mechanical clocks. This clock was built by Su Sung about 1050. The clock was probably known in the West in the 9th century. The first European clocks with mechanical escapements were made in the 13th century.

7 Paper was another Chinese invention that only gradually reached the West over many centuries. Known in China in the 1st century AD, the necessary technique did not reach even the Muslim world until the 8th century, and it was 400 more years before knowledge of it penetrated to Spain and southern France. Once in the West, it still took 200 years more to reach Germany and 100 more to reach England.

9 Scientific chemistry developed slowly. Practical chemistry, on the other hand, was part of everyday life – as epitomized here in this copy of a 12th-century Arabic manuscript showing the preparation of perfumes. Attempts were made by the Greeks and the Muslims to classify natural substances and build chemistry into a science but success did not come until the 17th century.

117

India 300–1200

After centuries of political fragmentation and foreign domination northern India was once more united under the Gupta dynasty (c. AD 320–550), India's classical age. In southern India another great state gradually took shape under the Pallavas.

The Gupta dynasty and its empire
The Gupta kings, especially Samudra Gupta (reigned 330–75), Chandra Gupta II (reigned 375–415) and Kumara Gupta (reigned 415–55), founded and maintained, both by conquest and diplomacy, a great empire controlling nearly all of northern India. Good communications, security and relative prosperity created an atmosphere in which Indian culture attained unequalled heights. Thus the works of the poet Kalidasa (flourished fifth century AD) achieved such a degree of perfection that they were often imitated but never surpassed. In art and architecture, too, Indian genius revealed itself in its most accomplished form of refinement and symbolism, but without the overemphasis of detail that typifies much Indian art after about the seventh century.

The material prosperity of India in this period is emphasized in the accounts of a Chinese Buddhist pilgrim, Fa-hsien (flourished 399–414), who visited India in the fifth century, and by the discovery of many gold coins of the Gupta Empire [5].

At the beginning of the sixth century the Huns invaded India from the northwest and penetrated as far as central India. This invasion has often been described as the main cause of the downfall of the Guptas, but it can be argued that the Huns would never have succeeded if the Gupta Empire had not declined owing to internal factors.

The expulsion of the Huns from India
Although the Huns were expelled after 30 years, northern India became divided between rival powers in Surashtra, Uttar Pradesh and Bengal. There were important changes too in southern India in the present states of Madras and Kerala. A prosperous and cultured society, as reflected in classical Tamil literature, flourished in this area at least from the beginning of the Christian era. In the fourth century AD, the Pallavas made

Kanchi (Conjeevaram) the centre of a large kingdom. Although much smaller than the Gupta Empire in the north it was still of great importance. The Pallavas established a successful form of power-sharing between central and local government, which promoted political stability. The east coast of southern India remained under Pallava control until about 880, and from then until 1200 under that of the Cholas.

The Pallavas patronized the Brahmins who, in their turn, provided excellent educational facilities. In art and architecture a particular Dravidian style (named after the language spoken in central and southern India), culminating in the monolithic sanctuaries and rock reliefs of Mamallapuram (the "Seven Pagodas"), was developed [2]. The Pallavas contributed more than any other Indians to the expansion of Indian civilization into South-East Asia.

The influence of Harsha of Kanauj
Most of northern India was temporarily united by Harsha of Kanauj (606–47) whose career, admirably described by a Sanskrit

1 On the relief panel of this temple of the Gupta period, Vishnu is represented during his cosmic sleep on the coils of the seven-headed Naga. His consort, Lakshmi, is at his feet.

2 The greatest temple foundation by the Pallava dynasty of Kanchi, southern India, is the complex of Mamallapuram or Mahabalipuram, popularly known as the Seven Pagodas, south of Madras. The complex, built in 625–74, comprises a number of caves, a group of beautiful monolithic structures (the so-called *rathas*) and this splendid Shore Temple, dedicated to Shiva.

3 Frescoes depicting beautiful maidens are painted on the side of a huge rock at Sigiriya, Sri Lanka, where a fortified royal residence was built in the fifth century.

4 One of the most striking forms of the god Shiva is that of the four-armed Nataraja, dancing on top of a demon and surrounded by a halo with flames destroying the world at the end of an aeon. This is one of the finest bronzes of the Chola period (eleventh century).

5 The numerous gold coins of the Guptas (the Bayana hoard alone contains 1,021 specimens) are an important source for the history of the period. Their distribution gives an idea of the areas controlled by various Gupta kings and the frequency of the minting reflects economic activity. The representations show how the Gupta kings wished the world to see them. This king appears as a fearless hunter slaughtering a lion with a bow and arrow.

writer (Bana) and a Chinese pilgrim (Hsüan Tsang, in 630–43), reflects high standards of government and reasonable prosperity.

After the time of Harsha northern India showed progressive political fragmentation with larger states tending to split into smaller units which at first paid homage to the central authority but gradually became independent. Harsha's capital, Kanauj, was made the capital of the Pratihara dynasty in 750. The latter ruled paramount over the present states of Uttar Pradesh, Punjab and Rajasthan, but before the end of the ninth century their effective authority was limited to parts of the Punjab and Uttar Pradesh while different Rajput dynasties, originally of tribal descent, ruled in Rajasthan. Bihar and Bengal were under the Buddhist Pala dynasty (*c.* 750–1150) but from the tenth century they shared with minor dynasties.

During such divisions the Muslim Mahmud of Ghazni (Afghanistan) (971–1030) invaded and plundered northern India many times between 1000 and 1026 [7]. These were destructive raids, carried out mainly for booty. Although many Indian armies fought bravely, their resistance proved ineffective through internal rivalries and military miscalculations, such as over-reliance on elephants. Further political, but not cultural, decline led to new Muslim invasions and by the end of the twelfth century most of northern India had come under the control of the Muslim sultanate of Delhi.

Sanskrit literature of the post-Harsha period offers many excellent works, although few of the quality of the earlier periods. The most important historical text of ancient India, the *Kashmir Chronicle*, belongs to the twelfth century. In art and architecture some of the greatest achievements, such as the temples of Orissa and Khajuraho, belong to this late period.

There was no decline in southern India where the Cholas established one of the greatest Indian empires. Their kings invaded Sri Lanka and Bengal and even undertook a great maritime expedition to South-East Asia. While northern India suffered political fragmentation and Muslim invasions, the Chola kingdom established conditions in which Hinduism flourished.

Vishnu, one of the principal gods of Hinduism and the supreme deity for the Vaishnavas, some of whom find a close analogy between religious experience and sexual love, has revealed himself as a saviour of mankind in many different forms, in particular in ten descents (*avataras*) as a man or as an animal. His most celebrated *avatara* was as Krishna, the divine shepherd and king-philosopher. Of the animal *avataras*, the Boar (*varaha*), shown here with elaborate ornamentation, is most frequently represented. In Hindu mythology the god is believed to have descended in this form to rescue the earth, which had sunk in the ocean.

6

7

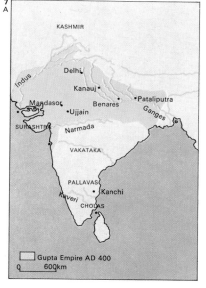

Gupta Empire AD 400
0 600km

Indian Empire of Mahmud of Ghazni 1030
0 600km

6 An older Shiva temple at Orissa shows the typical shape (*shikhara*) of a southern India temple. It is built on a platform in front of a pool that is used for ritual ablutions.

8 Puri is one of the great religious centres, attracting countless pilgrims from all over India during the annual cart festival of Jagannath (Juggernaut). The god used to be a tribal deity.

7 Although the Gupta Empire [A] represented the peak of classical Indian culture, its influence hardly reached the south. After its fall, the north was rarely united, while the Pallava and Chola dynasties brought continuity to the south. Despite the lack of cohesion, the north easily resisted the Muslims and lost only Sind until the growth of Mahmud of Ghazni's empire [B] as a powerful and aggressive neighbour.

Mahmud was not impelled by religious motives but Muslim raids continued until the establishment of the Muslim Sultanate of Delhi in 1206, marking the start of permanent Muslim influence in India.

8

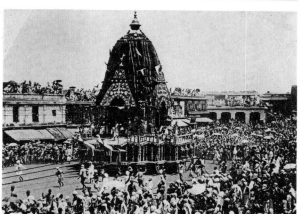

9

9 Although they were meant for monks, there is little trace of puritanism in the Buddhist caves of the western Deccan. This relief, probably of the sixth century, depicts Tārās, a kind of female saviour, who is performing a devotional dance.

Indian art to the Moguls

The beginning of Indian art can be traced to the Indus valley culture of Harappa and Mohenjo-daro (Pakistan) dating from about 2300 BC. At these two sites small examples of rather naturalistic as well as stylized sculpture were found that show a considerable understanding of the treatment of the human figure both in stone and in bronze. Clay figures in a somewhat primitive style were made and possibly represent popular art; in contrast, steatite seals showing animals and figures were cut in a tight, formalized design.

Mastery of sculptural technique

In the historical gap between this period and the Mauryan era (320–185 BC) it is clear that the Indian craftsman achieved complete control of his medium. The Ashokan pillar capitals, such as that from Sarnath adopted as the national emblem of India, show both stylized lions and a more naturalistic treatment of animals carried out in polished sandstone. This technique is a characteristic of Mauryan sculpture and it continued into a slightly later period. The *yakshi* [Key], although not a Mauryan piece, also has a polished stone sur-

face. This was probably intended for worship but it is remarkable for its unspiritual treatment of the female figure. This combination of the religious and the sensuous is found throughout Indian art. Similarly, it is not unusual for sculptured pairs of embracing figures (*maithuna*) to be found in a religious context in cave temples or in erotic postures adorning the outside of temples.

It is likely that major deities were not personalized but represented by symbols until the first two centuries AD an important development took place, as a result of the cult of *bhakti* (worship of a personal god), at that time images became the main icon. The change seems to have occurred in two places simultaneously, in Gandhara (now north-western Pakistan and northeastern Afghanistan) [5] and Mathura (southeastern Delhi), with some interaction.

Hellenistic influences on sculpture

In Gandhara, as a result of latent Hellenism remaining after the eastern extension of Greek culture, a style of sculpture arose that was deeply influenced by Hellenistic ideas.

These concerned not only interest in the human figure and the treatment of drapery but also the physical idealization of the deity. Whereas in Gandhara the religion served by sculpture was almost entirely Buddhist, at Mathura Hinduism and Buddhism existed side by side. Here also the style of the images was more Indian, and the stone used was frequently a mottled red or fawn sandstone instead of the grey schist in which most Gandhara stone sculpture was executed. The later (Gupta) Mathura Buddha figures [2], in which the robes appear to stick to the torso smoothly or in ridges as if they were wet, later became the exemplar of many other statues of Buddha that were made wherever Indian Buddhist culture spread.

Although brick and stone buildings were made, most of the remaining architecture from the first few centuries AD comprises caves and formalized Buddhist burial mounds (stupas) [3, 4]. The caves, such as those at Karli, Ellora and Ajanta, were cut from natural rock strata mainly between the second century BC and the ninth century AD for the worship of all three of the main Indian

1 The figures of tree-spirits on this 1st-century BC railing pillar from Bharut have a low relief and more formalized design than some of the later representations of tree-spirits. They made their contribution to later Indian sculpture, in which free-standing figures were not as popular as those that were carried out in high relief and made to be viewed from the front.

2 This elegant sandstone Buddha figure from Mathura was made in the Gupta period, about the 5th century AD. It expresses serene contemplation with a feeling for form and decorative style.

3 In the famous Ajanta caves every detail was cut from solid rock that forms part of a hillside. They were made between the 2nd century BC and the 9th century AD. The construction of the 7th-century Buddhist worship hall shown here can be compared with a church, having a nave, aisles and apse; the stupa, and seated Buddha (top, centre) make up the focal point of the whole design.

4 This low relief formerly decorated the lower part of a stupa similar to the one it illustrates. Made in about the 2nd century AD, it came from one of the many Buddhist stupas at Amaravati.

5 Late Hellenistic influence can be seen in the style of this 2nd-4th century AD image of a Bodhisattva (Buddha-to be) from Gandhara. Many conventional motifs in later Buddhist art originated in the sculpture of this period.

6 The sculpture of the Pala-Sena period (8th-12th century AD), such as this figure of Vishnu, shows a remarkable development both stylistically and in the deities portrayed. Some of the compositions were simple and restrained whereas others showed either figures in violent movement or in complex groups. Articles of dress, such as crowns, jewellery, robes and symbols, became more detailed. Sculpture was carved in grey sandstone or polished basalt.

religions, Hindu, Buddhist and Jain. Some were unadorned, but others were decorated with sculpture and painting.

Some impressive free-standing rock-cut temples were also made both in the western and southeastern Deccan. The shore temples at Mahabalipuram are the finest [10]. At the sites shared by more than one religion (as at Ellora) there is no difference in the artistic style adopted, which tends to be regional rather than sectarian. It is likely that much sculpture was painted in naturalistic colours, as in ancient Greece.

The theory and practice of sculpture and architecture were recorded in texts (*silpa sastras*) that probably had a wide distribution. In spite of this, Indian art is remarkable for the variety and sensitivity of its treatment of religious subjects. This is particularly noticeable if its bronze sculpture is compared with the stone sculpture of Orissa belonging to the eleventh and twelfth centuries [7]. In contrast, an elaborate and agitated style grew up in Mysore during the Hoyshala dynasty (late twelfth to fourteenth centuries) which is attractive in its unrestrained exuberance [9].

A little earlier, in Bengal and Bihar prior to the Muslim invasions, the art of the Pala-Sena period (eighth to twelfth centuries) flourished under generous royal patronage and religious fervour. Bronze and stone sculptures were produced in great variety and quantity [6] and had an important influence throughout eastern Asia. It was at that period that many of the forms of Indian deities, later accepted as conventional, were created in this part of the subcontinent and in the south.

Surviving examples of Indian painting

Painting also achieved high standards of skill as the wall paintings of the cave temples of the western Deccan (such as those at Ajanta) bear evidence. They suggest a long period of development and it is almost certain that there were other forms of painting that have now disappeared. Those on birch bark were especially susceptible to decay although a few examples have survived. Others, including illuminated palm-leaf manuscripts [8], remain and exhibit a mastery over line and brilliant colour that was to influence early Nepalese painting and the woodcuts of Tibet.

This fly-whisk bearer of polished sandstone comes from Didarganj (Bihar) and is of the 1st century AD. From its resemblance to figures at Sanchi and Bharut it is likely that this is a statue of a tree-spirit deity (*yaksha*). The fly-whisk is a symbol of power and is the only visual indication that this figure is not an ordinary mortal. It is in sharp contrast to later sculpture, showing figures that leave no doubt as to their status as deities. This image, although early, shows something of the significance with which female deities were eventually to be invested in Indian religion.

7 Orissa produced a characteristic local sculptural style, shared by all sects. It is illustrated by this 11th-century Jain figure of the mother-goddess Ambika that continues early traditions.

8 These palm leaf illuminated manuscripts of the 12th century are some of the earliest examples of Indian painting other than wall-painting. It is likely that there were earlier examples that have not survived, as these show evidence of mature development. Unlike later Indian miniatures, which developed from them, they did not necessarily illustrate the text but contributed to its sanctity as a religious object.

9 The Hindu art of Mysore in the 13th century is exemplified in this impressive sculpture. It originally formed part of an elaborate scheme decorating the outside of a temple. It is executed with characteristically fussy and meticulous treatment of details such as jewellery, costume and ornamental motifs. It shows a form of the goddess Durga overcoming the buffalo-demon Mahisha. Like many Indian deities she has several arms; they are necessary to hold the attributes ascribed to deities, partly to symbolize their powers, partly to distinguish them from other deities. Durga holds the symbols of several gods who combined to defeat the demon.

10 Free-standing temples were also carved in India out of natural rock outcrops, such as the Kailasanatha at Ellora. In southeastern India in the 7th and 8th centuries, a series of temples and huge reliefs, such as this one showing the descent of the Ganges, were hewn from granite rocks near the sea-shore at Mahabalipuram. They show a masterly control of form and design, especially when carried out in such a difficult medium.

China 1000 BC – AD 618

The Shang dynasty (c. 1600–c. 1030 BC), the first in the recorded history of China, was overthrown in about 1030 BC by a group of tribesmen from west China called the Chou. Their dynasty was to be China's longest, and its notable contribution to Chinese history was that it witnessed the birth and popular acceptance of Confucian philosophy.

The Chou and the Ch'in dynasties

The Chou period (c. 1030–221 BC) saw many important developments. The realm was extended to the sea in the east, the Yangtze River in the south and to the borders of Szechwan in the southwest. As this expansion continued, semi-independent states emerged which, although paying tribute to the emperor and his court, were more concerned with culture and religion than with political authority.

The delicate balance of power between the emperor's vassal states finally collapsed and the "Warring States" period began (475–221 BC). During this period of violent struggle, philosophical and moral thought flourished and a new, educated class arose.

Chief among the philosophers was Confucius (551–479 BC), whose teachings emphasized duties to the family and society rather than preoccupation with the dead. The influence of his thought signalled the decline of the old feudalism and began a tradition of close association between philosophical thought and political practice in China.

This was also a period of great technological change. Iron superseded the use of bronze, especially in weaponry; irrigation improved harvests; and the invention of the breast harness vastly improved the efficiency of the horse.

Gradually the smaller and weaker states were absorbed by the militarily and economically stronger states, until the chief contenders were the Chou in the south and the Ch'in in the west. Eventually the Ch'in became supreme rulers and in 221 BC China was for the first time unified under Shih Huang Ti (259–210 BC), the "First Emperor" [3]. He abolished the political system of the Chou and returned to the old feudal system, dividing the country into 36 provinces over which he set officials directly responsible to

himself. He completed and strengthened the Great Wall [1], today stretching 2,400km (1,500 miles) from southern Kansu province to the coast east of Peking. The written language was simplified and unified over the whole country. Weights, measures and coinage were standardized. Shih Huang Ti is remembered as a despotic but practical emperor who burned existing literature, exempting only works on agriculture, medicine, pharmacy and divination. After the First Emperor's death the structure soon collapsed under the feeble rule of the second emperor who was murdered in 207, bringing the Ch'i.. lynasty to an end.

The Han dynasty: education and wealth

Out of the chaos that followed there emerged a successful candidate for the throne, Liu Pang (247–195 BC), who founded the Han dynasty in 206 BC [2]. Initially the Han endeavoured to rule with the Ch'in system. But after about a century the principle of hereditary local power was curtailed and candidates for local government were selected by open examinations [4]. In 124 BC

1 **The Great Wall of China** was commenced under the Chou dynasty in the fourth and third centuries BC. It was designed to protect the Chinese people from attacks by the nomadic tribes who occupied the steppe lands in the far north and west. Building went on under ensuing dynasties and various sections of the wall were connected until, during the Ming dynasty (1368–1644), it extended for 2,400km (1,500 miles).

2 **China expanded under Han rule,** after unification by the Ch'in. The growing silk trade had to be made secure, but central Asia remained vulnerable to barbarian threats.

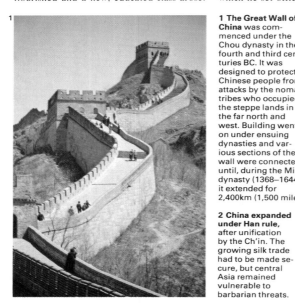

Han Empire 100 BC
Acquisitions by AD 100
Trade routes
Silk routes

ROMAN EMPIRE
HSUING NU
Great Wall
Antioch
Tyre
Alexandria
Seleucia
Samarkand
Kashgar
Khotan
Huang Ho
Lo Yang
Ch'ang-an
PARTHIA
Merv
Taxila
Yangtze
YUEH
Berenice
Indus
Ganges
Kattigara
Barbaricon
Barygaza
INDIA
Sabana
Muziris
Camara

3 **Standardization** of weights and measures was first ordered by Shih Huang Ti, First Emperor of all China. It was one of a number of measures to consolidate his rule, including standardization of the language, censuses and the construction of a defensive road system. After his death taxation and the forced labour system led to peasants' revolts and the eventual collapse of the dynasty.

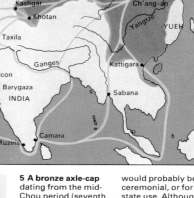

4 **A Han official** in his carriage is drawn by slaves. His attendant follows in the rear. During the Han dynasty a large bureaucracy was established to implement the growing powers of the state. The officials, usually nobles, were the product of the Confucian training of the official class. The Confucian ethical code required that the official class should possess wisdom, integrity, righteousness, conscientiousness, loyalty, altruism, love and humanity. Confucius insisted on the importance of education and training: "By nature men are pretty much alike; it is learning and practice that set them apart". The ruler's power could be forfeited.

5 **A bronze axle-cap** dating from the mid-Chou period (seventh century BC). The linchpin that held the wheel in place is decorated with a tiger's head. Chariots from which such axles came would probably be ceremonial, or for state use. Although chariots were also widely used in warfare, they were reduced to an auxiliary role as the raids by the mobile, northern horsemen increased.

an Imperial University was set up for the study of Confucian classics; its students were trained for government and rapidly increased until by the end of the dynasty they numbered nearly 30,000. Provincial schools were also established. Education and the growth of the civil service was greatly assisted by the invention of paper, and ink and brushes replaced sharp writing tools.

Under state patronage the arts revived and the early wealth of the Han dynasty can be seen from their rich tombs [6]. Most of the attainments of this period reflect the needs of a growing state bureaucracy – engineers developed irrigation methods and water clocks, and sundials and seismographs were also invented [9].

Until this time there had been little contact between China and the outside world, but under Han rule the empire was extended. Caravan routes were opened up, including the Old Silk Road which followed a chain of oases skirting the foothills of the Tarim basin. China sent ambassadors abroad along with its ever-popular silk, and products were exchanged as far afield as the outposts of the Greek world. Ideas travelled with the trade, most notable of which was Buddhism [10], introduced from India under the Han and which by the seventh century had become a major force in China.

Under the Ch'in and the Han, China for the first time became a great state. But the mandarins, so carefully picked by scholarly examinations, became corrupt and sided with the great landlords in their oppression of the peasants. The Han dynasty was brought to an end by widespread revolt in AD 220.

Disintegration of the empire
During the next three and a half centuries there was a succession of short-lived ruling dynasties. It was not until AD 581 that the country was at last reunited under the Sui dynasty. Prosperity increased, taxes were reduced, irrigation improved and public lands were distributed so that each family had some land of its own. But the extravagant second emperor increased taxation. In AD 618 he was assassinated by one of his officers, Li Shih-min (Emperor Li Yüan), who founded the Tang dynasty.

A rubbing from a Han stone relief shows a mounted barbarian archer at full gallop. The reins are looped on the horse's neck leaving the archer's hands free to loose off the arrow. Firing from the saddle he had great speed and mobility. Such raiders were a constant threat to the Han.

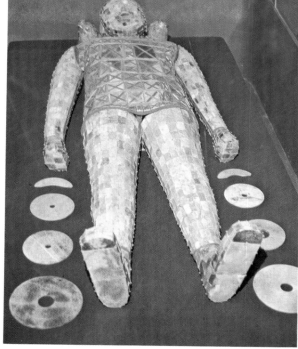

6 Princess Tou Wan was buried in this jade suit. She was consort of Prince Liu Sheng, who died some years before her, in 113 BC. Their tombs were accidentally discovered by some soldiers in 1968 in a cliff on the Ling mountain in the province of Hopei. The massive stone doors, which led to the burial chambers, were sealed by molten iron. The tombs were some 12.5m (40ft) square and beneath the collapsed jade suits lay some ash, all that was left of the royal couple. Jade was believed to have magical properties of preserving for eternity anything kept in it. The suits were made in 12 parts so that they totally encased the body.

7 A celestial horse of the Han dynasty was excavated in 1969 from a tomb at Wu-wei in Kansu province. Flying horses were a recurrent motif in Han art.

8 This bridge, at An-chi, built in AD 610, shows the remarkable sophistication of Chinese engineering, long pre-dating Western achievements. The building of canals and roads, needed for the transport of grain and the maintenance of peace, was fostered under Sui rule. The Grand Canal, built by forced labour, linked the Huang Ho and Yangtze rivers and connected the political centres of the north with the economically important Yangtze region.

9 The earliest known Chinese seismograph depicts eight dragons, each holding a ball in its mouth. Around the base of the vessel sit eight toads with open mouths. An earthquake at any point of the compass causes the dragon facing that direction to drop the ball it is holding into the mouth of the toad below, thus indicating the direction of the tremor. This instrument was invented during the Han period by a famous astronomer, mathematician, poet and writer, Chang Heng.

10 The Buddhist school of sculptors produced this white marble stele in the early part of the fifth century AD. It shows the Buddha Sakyamuni in a posture denoting him as the bestower of fearlessness. The Buddha is seated under some sal trees surrounded by his disciples and Bodhisattvas, including Ananda his favourite disciple, and Mahakasyapa, who became leader after his death. Above him float goddesses holding garlands of flowers.

Confucius and Confucianism

Thousands of oracle bones [1] which survive from the Shang dynasty (c. 1600–c. 1030 BC) give archaeologists clues to the form of religion in ancient China. From texts incised on these bones, a picture emerges of a world regulated by spirits of deceased kings (ti), ancestors, nature gods and guardian spirits. The Shang dynasty was overthrown by the Chou (c. 1030–221 BC) who believed that their dynasty had a mandate from heaven to rule the land. Heaven (T'ien) or the "supreme ancestor" (Shang Ti) was believed to govern the universe, fix the seasons, give fertility to men and animals and order the cycle of death and renewal. The emperor was also a priest who performed rituals to ensure the orderly succession of nature.

The life of Confucius

Documents have survived from the Chou period, which are quoted, and may even have been edited, by Confucius. They form part of the ancient tradition, which includes the complementary forces of yin and yang [Key] and reverence towards heaven and ancestors [5], that Confucius inherited. Elements of

Chou religion were transmitted by later Confucian teachers. Confucius is the Latinized form of K'ung Fu-tzu, Master K'ung, who was born in the city state of Lu in northern China in 551 BC and died in 479 BC. Confucius came from an aristocratic family but grew up in comparative poverty and, being disappointed in a political career, found his true work in training young men for public service [2].

Confucius founded his own private school, one of the first in China, and without claiming originality taught what he considered to be the best ancient wisdom. He discussed the arts of life in a city state, the study of old documents, and the Book of Poetry which included ritual hymns of early Chou rulers. But while claiming to preserve or restore earlier tradition, Confucius interpreted the documents in his own way and formulated an ethical and moral system that has influenced China ever since.

When he was about 50, Confucius was given office in the state council – some have claimed that he was prime minister of Lu – but he was dissatisfied with office and he

travelled to neighbouring states without success and meeting much hardship. He returned home a disappointed man and spent his last years in teaching and study.

The literature and teachings of Confucius

Confucius is traditionally credited with authorship, or at least editorship, of the five Confucian classics: the Book of Poetry, the Book of History, the Book of Changes, the Spring and Autumn Annals, and the Book of Rites, but few of these writings can be safely attributed to him. His true teachings are contained in the Analects (Lun-yu), a small book of his sayings recorded by his pupils. Modern specialists consider that some of these chapters are not authentic, but they are traditionally held to be the words of the Master. The Analects teach a way of goodness (jen) which includes courtesy, loyalty and unselfishness. Rulers should seek it, but it is almost a saintly quality. The ideal prince should rule by goodness and govern his conduct by ritual (li). This ritual is not confined to religious worship but is concerned with dress, good manners and personal morality. Confucius

1 Oracle bones were inscribed with questions to the spirits about the future. The bones were heated and the resultant cracks were "read".

2 This painting on silk from the Ming dynasty (1368–1644) shows Confucius as the ideal teacher, the "uncrowned king". During his life, he was a tutor to the sons of aristocrats and wandered from state to state, hoping to find some rulers who would put his teachings into practice. However, he met only with indifference and on occasion hostility. Indeed Confucius was unrecognized in his own lifetime as a moral teacher except by his small band of disciples.

3 The tomb of Confucius can still be seen at Chufu, in the province of Shantung. In front of the tomb stands a stone tablet and altar with candlesticks and incense vessels. The tablet bears a simple inscription: "Ancient Most Holy Teacher".

4 Court officials in China, such as this mandarin, even before the beginning of the Christian era were appointed to study the Confucian classics. Eventually the study of these works became universal among the educated classes and examinations in them had to be passed before service to the state could be undertaken or promotion gained; eventually a large Confucian-trained bureaucracy developed.

described this as the Way of the True Gentleman and it is his ideal. He also advocated filial piety (*hsiao*) to parents and ancestors and proposed a hierarchy of relationships: ruler and subject, father and son, older and younger brother, husband and wife, and friend with friend. Confucius believed in the Supreme Being, but held that service to God is meaningless if service to man is neglected.

Confucius and his contemporary Lao Tze (*c.* 604–531 BC) were teachers rather than founders of religions, but their supposed writings became the sacred scriptures of Confucianism and Taoism respectively, and part of the whole culture of China [8]. The teaching of Confucius was continued and extended a century later by Mencius (372–289 BC) and Hsun Tzu. Within a few centuries Confucian teachings became the orthodox doctrines of the state and the guiding lines of the official classes [4].

The role of Confucius in history

Confucius was neither a god nor a prophet, and it is often asserted that Confucianism is not a religion, but this statement should be qualified in order to be properly understood.

Before the Christian era, emperors offered sacrifice at the tomb of Confucius [3] – a practice that continued for many centuries. Yet Confucius was not a god with images but an ancestor or great sage, revered as the Teacher of Ten Thousand Generations. Nevertheless, the tradition of Chinese veneration of ancestors and Confucius' own emphasis on filial piety led to ancestral ceremonies being associated with Confucianism. The role of the emperor, and his performance of rituals on behalf of the people [6], further added to the complexities of Confucianism as did the moral and social teaching of Confucius himself.

In modern times Confucius has been alternately attacked as a feudal aristocrat and revered as the greatest teacher of ancient China. The dead are still venerated in China, and much time and money continue to be spent on preserving temples [7] and graves. Modern China, despite the shift away from tradition, brought about by the Marxist ideals practised by its rulers, still continues to be profoundly influenced by Confucianism.

P'an Ku, a mythological figure, here holds the symbols of yin and yang which appear in Confucian thought.

5 Bronze ritual vessels were used in sacrifices and ceremonies from ancient times. These vessels, grouped on or around the altar, bore stylized designs and masks. Their inscriptions describe royal or religious ceremonies, traditions that have influenced Confucian thought.

6 The offering of sacrifice and praising of the king for his laws is pictured in this illustration from the *Shih Ching*, a Confucian classic.

7 The Temple of Heaven in Peking, with its three roofs, gold-capped shrine and blue tiles, is one of the finest buildings in China. Here the emperor himself acted as high priest for the people. It is still a national monument today.

8 This fanciful picture illustrates the meeting of three ways of thought in ancient China. Confucius, left, as a scholar, may never have met Lao Tze, shown, right as an old man, and the Buddha, as a, centre, monk, may never have been to China. But Confucian morality and ceremony, Taoist nature mysticism, and Buddhist ascetism and devotion played formative parts in the traditional structure of Chinese religion, art and social life.

China 618–1368

The Emperor Li Yüan, who founded the T'ang dynasty (618–907), was followed by his son, the Emperor T'ai Tsung (reigned 627–49) under whose rule China became the most powerful and the largest empire on earth. The security China enjoyed in this position encouraged trade with the outside world and brought in a rich horde of goods. The trade also carried scientific ideas westwards beyond the borders of China [8].

Art, commerce and religion

Chinese arts flourished during the T'ang dynasty [3, 4] particularly poetry, and it produced such poets as Wang Wei, Li Po and Tu Fu. The earliest known printing commenced in this fruitful period and paper money was first issued [1]. Money-lenders thrived in the numerous markets and the growth of commerce brought prosperous trade with merchants from Japan, central Asia, Arabia, Turkey and the Mediterranean.

The emperor was as tolerant of the religions that the foreigners introduced as he was of the merchants themselves. Although a Taoist, he supported Confucianism for reasons of state and treated Buddhists with great respect. Zoroastrian temples and Nestorian Christian churches also existed in the capital Ch'ang-an, modern day Peking.

Wu Tse-t'ien – an efficient empress

The peace and prosperity that T'ai Tsung brought to the empire was continued by his former concubine, Wu Tse-t'ien, who came to the throne in 683 and ruled China with ruthless ability until she was forced to abdicate in 705 at the age of 82. She was a profound believer in Buddhism and was the first and only female "Son of Heaven". Much of the progress and stability of the country was due to the fact that civil service officials [2] were selected by examinations held under controlled procedures. The empress also permitted women to sit the examinations for government posts.

The main function of the government was the collection of revenue and the promotion of the agriculture on which it mainly depended. For this purpose the country was divided into districts controlled by magistrates. The people were divided into three groups mutually responsible for each other's conduct and for tax payments – encouraging a sense of collective responsibility that is still a feature of China today.

T'ang influence spread far afield, to such an extent that the Japanese capital of Nara was modelled on Ch'ang-an. In the west it clashed with Islam. Muslim armies had advanced, bringing their faith as far as Samarkand and Bokhara. Eventually they conquered central Asia, severing the overland route between China and the West. Trading continued by sea, but the power of China began to wane and as it weakened she became less tolerant of foreigners and their religions. In 845 all foreign religions were proscribed and a ban was placed on Buddhists and their rich but unproductive monasteries [9]. Disastrous revolts and invasions decimated the population and in 907 the T'ang dynasty ended in ruins.

Five dynasties and the Mongols

The T'ang dynasty was followed by a period called the Five Dynasties between 907 and 960, when, as a Chinese poet said, "States

1

**1 The *Diamond Sutra*, the world's oldest printed book, dates from AD 868, nearly six centuries before the first printing in Europe. With gunpowder and the magnetic compass, printing was one of the revolutionary inventions developed by China long before the West. The consequent growth in literacy meant that increasingly the civil service (for which, in theory, recruiting had always been democratic) was drawn from a wider circle of families. Printing facilitated the great expansion of the economy that characterized the Sung dynasty by making possible the introduction of paper money and credit notes.

2

2 This T'ang mandarin was one of the highly educated, privileged and wealthy élite who comprised the mandarinate or civil service. The continuity and resilience of the large state bureaucracy from earliest times is one of the more remarkable features of Chinese history. Its officials, selected by public examination, collected taxes, supervised state projects and the nationalized salt and iron industries (under state control since the Han dynasty) and administered local areas. They also supervised the merchant communities and foreign trade, a despised business largely in the hands of immigrants, but strictly controlled by the state.

3

3 A fine example of the elegant work that was produced in the classical period of Chinese civilization is this white T'ang porcelain spittoon of the late 9th century. It comes from Hsing-chou in the modern province of Hopei and exempli-fies the best of the high-fired ceramic ware produced there. Under the T'ang, Ch'ang-an was a thriving capital, one of the cultural centres of the East. Its wealth came partly from the prosperous western trade – Chinese goods were much in demand along the Silk Route. The T'ang is often seen as the artistic complement to the great scientific and technological achievements of the preceding Sui dynasty: the T'ang literary achievement in prose, as in verse, remains unsurpassed.

4 This silver wine flask of the T'ang dynasty is a typical piece of Chinese metalwork displaying strong foreign influence. Under this dynasty the capital, Ch'ang-an (now Peking), was probably the most cosmopolitan city in the world and as the empire expanded, merchandise arrived from all quarters. The flask is modelled on the leather water bottle widely used by travelling merchants.

4

5 Gunpowder was discovered by Taoist alchemists in about the 9th century and was first used strictly peacefully. By AD 1000 simple bombs, grenades and rockets were being made, but it was the Mongols who first exploited gunpowder for military ends. They probably used a type of cannon in their campaigns against the Sung troops and certainly employed Chinese engineers. The Mongols captured the Sung fleet which was armed with tre-buchets for firing bombs. Although gunpowder was a Chinese invention, it did not have the revolutionary effects on Chinese society that it had in Europe. Illustrated is a Chinese rocket of the Sung dynasty.

rose and fell as candles gutter in the wind". In 960 the Sung dynasty was founded by Chao Kuang-Yin. The war-weary country was at last glad to accept established rule and welcomed the new emperor, who took the title Sung T'ai Tsu. China was still threatened from the north and in 1044 an indemnified peace treaty was concluded with the Hsia, a former tributary kingdom. Gunpowder, which had been discovered during the T'ang period but had been used only for fireworks, was now used to produce the first military rockets in history [5]. The loss of the northern part of the country was partly offset by sea trade, which had been considerably helped by the Chinese discovery of the compass. Large ocean-going junks [6] carried cargoes of tea, silk, porcelain, paintings and other works of art to the East Indies, Africa and India. Another important invention was the abacus [Key], the first calculating machine and one still widely used.

The Sung dynasty ended in a similar manner to the previous empires. Corruption at court and discontent among the people permitted the ascendancy of the latest nomad

empire of the north, the Mongol nation. The new invasion started in the thirteenth century when Genghis Khan (1167–1227) invaded northern China [7]. By 1223 he had conquered most of the country north of the Yellow River and defeated the Hsia, killing about 90 per cent of the population [10]. However, it was not until 1264 that his grandson Kublai Khan (1215–94) was able to move his capital from Karakorum in Mongolia to Peking, and, in 1279 he overcame his former allies the southern Sung.

Mongol rule or, as it was officially known, the Yüan dynasty (1264–1368), was successful in uniting the Chinese and Mongol empires and under Kublai Khan Mongol power reached its peak. But the invaders were eventually overcome. Kublai Khan's successors did not have his ability and the oppression of the Chinese by hordes of foreign officials led to the formation of secret societies and to revolts. In 1356 Nanking fell to a peasant movement led by a monk. Chu Yüan Chang, who finally became first emperor of the new and purely Chinese dynasty, the Ming, in 1368.

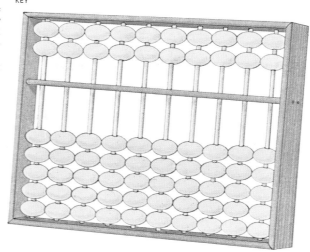

The abacus, introduced during the Sung dynasty, is still widely used today in all areas of commerce. Under the T'ang and Sung the economy underwent a rapid expansion, similar to that which occurred in 17th century Europe, but despite this Chinese society remained essentially feudal.

6 Maritime commerce expanded greatly during the Sung dynasty. Seagoing junks, such as the porcelain model here, carried cargoes of silk and porcelain to the East Indies, India and the east coast of Africa. Undoubtedly improvements in navigation (which was greatly aided by the invention of the floating compass in AD 1021) contributed to these ambitious trading expeditions.

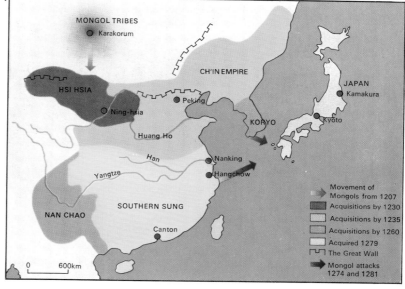

7 The Mongol conquest of China was finally completed in 1279. The Sung had previously allied with the Ch'in against the Ch'in in the north, but were themselves previously allied with the Mongols against

Movement of Mongols from 1207
Acquisitions by 1230
Acquisitions by 1235
Acquisitions by 1260
Acquired 1279
The Great Wall
Mongol attacks 1274 and 1281

8 This medical drawing from a Persian textbook of the 14th century is in fact of Chinese origin and shows the widespread influence of Chinese science.

9 Buddhist temples carved from rock caves were features of the Yun-kang and Lung-men periods (in the 5th and 6th centuries AD). The last of them (like the Fen-lai-feng cave temple near Hangchow, of which a detail is shown) were carved in the late 13th century (in the Yüan dynasty). Buddhism in China reached its peak in the middle of the T'ang dynasty, but thereafter the faith was gradually absorbed into traditional Chinese customs and philosophical beliefs.

10 Genghis Khan's army capturing a Chinese town is recorded in this painting. The Mongols came to rule almost all of eastern Asia as part of an empire stretching across Asia to Hungary and the Black Sea. After the conquest of the Sung in the south, China was for the first time ruled by foreign invaders. The Chinese-style Yüan dynasty ruled China for less than 100 years. Under the Mongols trade routes across Asia were safeguarded and a variety of religions, including Christianity, were permitted. Kublai Khan was the first Yüan emperor, but after his death Mongol rule was short-lived, ending in a series of rebellions and, finally, expulsion.

Chinese art to 1368

Neolithic Chinese society (*c.* 7000–*c.* 1600 BC) was centred on the Yellow River and organized in village communities. The potters made coil pots which they burnished and painted with black, white and red slip in striking animal, flower and abstract spiral designs. The shapes of later Neolithic pots were elegant and complicated; they appear to be related to the bronze vessels of the first historic dynasty (Shang *c.* 1600–*c.* 1030 BC).

Early art forms
The ritual vessels of the Shang slave society ruled by a god-king were magnificently designed and cast. The shapes varied and the rich decoration was based on animal motifs [1]. As society changed and the rituals lost their significance bronzes became more decorative, richer in appearance and often gilded. Decoration included floral motifs and was achieved by inlay and inserts of gold, silver and semi-precious stones.

Pottery had always been of great importance to the Chinese and through the centuries potters were developing the stoneware clay indigenous to the eastern areas of China, perfecting the throwing of the clay, glazing and firing to a high temperature. Lacquer painting was an early invention that also gradually evolved, particularly in the south of the country, to become an art form. With the invention of the brush and of paper, written calligraphy as opposed to the incised character became a major art from which painting evolved, at first for ritual purposes. By the second century BC, however, there is evidence of the use of painting as a medium of artistic expression.

The great Ch'in (221–206 BC) and Han (206 BC–AD 220) dynasties saw the unification of a country that encompassed many differing cultures. The newly evolved art of painting with colour on cloth or paper was used to express the many myths and legends of this rich heritage [3]. Some of the most lively painting came from the southern area, the old kingdom of Ch'u (present-day Hunan, Anhui area) and from the west in present-day Szechwan, where stamped brick and relief carvings were also used to portray everyday life and mythological scenes.

Buddhism was introduced to China during the first century AD but did not appear in the arts until the fourth century. However, between that time and the ninth century great temples were built or cut from the living rock and were decorated by rock carvings or massive wall paintings and hangings [4] and furnished with images of gilt bronze or stone. Unfortunately periods of anti-Buddhist destruction have obliterated much of this huge output, only the great rock temples of north China remaining.

The art of the T'ang dynasty (618–907)
The T'ang dynasty is famous as one of the finest periods of artistic work of all kinds in China. It was a time of cosmopolitan vitality and metropolitan splendour which encouraged all the minor arts; a taste for extravagance led workmen to produce colourful work ranging from jewellery to bronze mirrors inlaid with cornelian and gold, beautiful silks and exotic inlaid musical instruments. Great religious paintings are recorded although they have not been preserved. The two major artists were Wu Tao-tze (mid-eighth century), a majestic figure painter who

CONNECTIONS

See also
122 China 1000 BC–
AD 618
126 China 618–1368
44 China to 1000 BC

In other volumes
54 History and
Culture 2

1 Early Chinese bronze works, like this ritual vessel, a *chia*, of the fourteenth century BC (Shang dynasty), were cast from piece moulds, sections of mould bound together in such a way that they could be taken apart and re-used after casting. The decoration is arranged in bands round the vessel and is derived from animal motifs, which could be either realistic or mythical. Each ritual vessel had its function – this was used for libation.

2 This dagger handle, of the Warring States period (475–221 BC), is of cast gold. The lost wax method was used for casting such intricate "cut through" designs. Cast gold work of this period is extremely rare.

3 A painted banner from a tomb of the early Han dynasty depicts both upper and lower world mythology. In the centre is a portrait of the occupant, who was buried at Ma Wang Tui, Ch'ang-sha.

4 A meditating Buddha in Paradise is portrayed on this painted hanging, which shows the richness of T'ang painting. The influence of Buddhism in China reached its peak under the T'ang; the textiles, silver and pottery that were produced mark this as a classical period in Chinese art. The hanging comes from Tu-huang, a large Buddhist monastic centre in northwest China where hundreds of cave temples were decorated with wall paintings.

5 A winged lion of the Sung dynasty (960–1279), inlaid with silver and gold, was executed in the style of the late Chou dynasty (*c.* 1030–211 BC). It typifies one aspect of Sung taste. The decoration is similar to that of the Chou, but on a larger scale. In contrast to the preceding T'ang dynasty, when external influences such as Buddhism affected much of Chinese civilization, the Sung was a period of a selfconscious reassertion of traditional values in art and society.

used a flowing brush line expressing volume and vitality, and his contemporary, Wang Wei (699–759), a poetic landscape painter, the originator of ink painting of handscrolls in which the tonality of ink expressed depth, texture and atmosphere.

A tradition of landscape painting

It was landscape painting that was developed by major painters through the succeeding four centuries [Key]. The classical period of the tenth and eleventh centuries saw the painting of monumental mountain hanging scrolls, in which man is shown as a tiny inhabitant of a grand overpowering landscape. This concept gradually gave way to a study of the beauties of the smaller details – a single bird, a flower, or an incident on the banks of a river. Here the painter became interested in the style of his painting.

The scholarly taste of this period was expressed in all the crafts; the stoneware tradition of the celadons of Chekiang [7] and the court wares of Kaifeng jade carving, gilt bronze [5] and ceramics. At the same time however, particularly in the north of the country, a robust taste for flamboyant decoration was displayed in the decorated wares of Tzu Chou.

The great flowering of decoration on ceramics with the introduction of cobalt underglaze painting techniques and the painting of lacquer in gold and red to produce rich boxes and small furniture seems to mark a sharp change of taste but perhaps is more an underlining of the added riches and complementary character of the north and south of this huge country. During the early and mid-fourteenth century there was another great flowering of painting. Declining to serve the foreign Mongol court in the traditional role of bureaucrat, scholar-gentlemen gave their time to painting, comparable with the activity and gravity of production of the tenth and eleventh centuries. The Four Great Masters, Ni Tsan (1301–74), Wang Mêng (died 1385), Wu Chên (1280–c. 1354) and Huang Kung-wang (1269–1354) [9] painted in traditional but entirely personal styles to lead the way for the next three centuries in the unfolding of the honourable tradition of landscape painting.

"Travellers among Streams and Mountains" is a hanging scroll of the early eleventh century by Fan K'uan. Painted in ink and slight colour on silk, it is one of the early great masterpieces of landscape, and shows the classical style of construction. The foreground, in which tiny figures wend their way, is a little remote from the viewer. The horizon beyond them is defined by a cloud, a device used to express the space between the mid-ground and the towering mountain up which the eye travels to arrive at a third and fourth eye level. The painting is unselfconscious and has great dignity and serenity.

7 Fine quality grey stoneware with a thick blue-green glaze was produced by the kilns of the upper Tung River valley. It was known in Europe as celadon ware. This mallet-shaped vase is of L'ung Ch'uan celadon ware (late twelfth or early thirteenth century). Typically, it is simple in shape and decoration and depends for its quality on the exceptional texture of the glaze. Some of the world's most elegant ceramics were produced at that time, and this represents one of the classic wares of China, appealing to the refined Southern Sung taste and to the scholarly taste of later periods.

6 This handscroll of the thirteenth century was done in ink on paper by an anonymous artist. It depicts the preparations for the Spring Festival, when families gather to visit the graves of ancestors. It is a genre scene of great detail and interest, the style being complementary to the decorative romantic work of the same period.

8 Liang K'ai was a famous Ch'an (Zen) Buddhist priest-painter of the mid-thirteenth century. This hanging scroll of a walking priest is typical of his style. He used his ink and brush with the bravura of the academic Southern Sung masters, but added to this the Ch'an directness. The priest's personality is expressed with great simplicity.

9 One of the most revered paintings in China is "Fu Ch'un Mountains" by Huang Kung-wang. This ink on paper handscroll was painted over the space of about three years, when the artist, one of the Four Great Masters, was an old man. It typifies the character of the scholar-artist with its direct simplicity of technique which is based on the old masters, and the evocation of a quiet mountain landscape.

Japan 200 BC – AD 1185

By the first century AD successive waves of settlers from the Asian mainland, coming mostly through Korea, had brought three crucial skills to Japan. The casting of iron and bronze produced more effective tools and weapons. The potter's wheel speeded the production of earthenware. But more important than these was knowledge of rice and irrigation which replaced hunting and fishing with settled agriculture. The resulting Yayoi culture was based on farming villages [2], which had little or nothing by way of large-scale political organization.

Unification of Japan

In the third century AD a Chinese chronicle, *Wei Chih*, described Japan as a country of more than "100 communities" that had been unified by Queen Pimiko. Some clan (Uji) leaders had sufficient power to organize the construction of vast chambered tombs; one such family claimed descent from the sun goddess, and emerged as head of a loose confederation of powerful clans.

By the sixth century this embryonic Yamato Imperial House [Key] had organized Japanese intervention in Korean civil conflicts. This brought contact with Chinese ideas and skills which swept down the Korean peninsula, leading to a second and more radical transformation of Japanese life.

Buddhism [3], Confucianism, medicine, astronomy, Chinese-style architecture and the Chinese script all entered Japan in the sixth and seventh centuries [1]. Scholars travelled to China [5] and soon Japan's central rulers sought to model their state upon the bureaucracy of China's T'ang Empire. In 592 Shotoku Taishi [4] became regent and began a programme of spreading Buddhism and widening the power of the Yamato Imperial House.

In 646, measures known as the Taika Reforms included the imperial control of rice land, systematic taxation, and a nationwide network of imperial officials. At the centre of the new state was to be a Chinese-style capital with palaces, temples and broad straight avenues linking public buildings. China's system of civil service examinations was never reproduced, however. Powerful provincial families remained remarkably independent and a conscription system proved inefficient. Yet the legal codes of 702 were detailed and far-reaching and Heijo (Nara) contained impressive structures that still survive.

At the close of the eighth century the political influence of Buddhism had become so great that one priest, Dokyo, attempted to capture the imperial throne. Partly in response to this danger the court set up a new capital at Heian (Kyōto) in 794.

Rise of Fujiwara

After 50 years of stable administration, events at court produced new threats to imperial authority. The Fujiwara family had been loyal state servants throughout earlier centuries but now they used intermarriage and masterly intrigue to dominate palace appointments. In 857 Fujiwara Yoshifusa became grand minister. Soon after, his grandson was made child-emperor with himself as regent. Later Fujiwara remained regents after infant emperors reached maturity, and throughout the eleventh century they wielded overwhelming power.

CONNECTIONS

See also
122 China 1000 BC–AD 618
126 China 618–1368

In other volumes
54 History and Culture 2

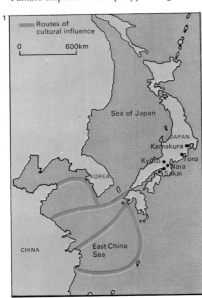

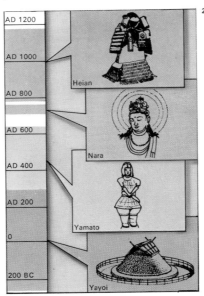

1 Routes by which Chinese culture and Buddhism entered Japan were established by the 9th and 10th centuries. Contacts dated from *c.* AD 400 when the King of Paekche, in Korea, sent scholars to Japan with Confucian writings. Koreans brought Buddhist writings and sculptures in the 6th century and Japan then began sending official embassies and students to T'ang China.

2 A house of the Yayoi period (250 BC – AD 300) was reconstructed at Toro in the suburbs of Shizuoka where the foundations exist of 11 houses, granaries and irrigation channels for paddies.

3 A mural painting of a Buddhist deity in the Kondo of Hōryūji temple, Nara, is in the style of contemporary T'ang painting of the 7th or 8th century and reveals the Indian origins of T'ang and Korean Buddhism. Only fragments of these murals remain, but in China no similar works survive at all.

4 Shotoku Taishi (571–621) made a profound study of Buddhism and founded such important temples as Hōryūji in Nara. He also tried to introduce Confucian ideas into the Japanese state and proclaimed a code of government in 604. The painting shows him in Chinese-style robes and is in the manner of a Chinese imperial portrait.

The Heian capital was the scene of outstanding cultural achievements. Whereas the dominant arts of Nara had been in the Chinese T'ang style, the new regime severed links with the continent and developed artistic styles that were authentic expressions of Japanese sensibilities. Architecture became less flamboyant and more refined. Vivid picture scrolls illustrated historical and literary themes. A new phonetic script supplemented Chinese characters, and permitted more supple forms of expression [7]. *The Tale of Genji* (c. 1010–20), Japan's most famous novel, was written by a court lady of the Heian capital.

Warrior families and court life

Parallel with the weakening of imperial authority came the rise of provincial families with new sources of power. To maintain law and order and combat northern aborigines, these lords increased their armies and became increasingly oblivious of imperial control. Their independent estates (*shoen*) paid little to the capital but stimulated the economic development of other territories.

These new centres of agriculture and organization produced leaders with a practical military ethic indifferent to many of the pretensions of court life.

In the eleventh century courtiers recognized the might of this new class and invited the powerful Taira and Minamoto families to aid them in suppressing dangerous rebellions. The Fujiwara may have hoped to control these robust warriors [6] but soon the Taira had replaced them as the effective masters of palace and throne. Taira Kiyomori used force and intrigue to overpower his rivals and in 1180 made his infant grandson emperor. After 20 years of dominance the Taira appeared unchallenged in the capital, but military power was now the only determinant of politics and the Minamoto rebelled against the new overlords of Heian Kyo.

From 1180 to 1185, these two families and their coalitions were embroiled in nationwide warfare. By 1184, the land forces of the Taira were annihilated and a year later their navy was destroyed. Now the Minamoto [8] were masters of Japan. Warriors ruled from their capital at Kamakura.

A *haniwa* is a hollow pottery figure, designed to house a spirit, which was often placed on the burial mounds of clan leaders and members of the imperial family. This figurine of an armoured warrior comes from the Yamato period (c. AD 300–c. 625); figures also exist in the form of animals, buildings and boats, as well as men and women. The idea of an anthropomorphic grave-statue was native to the Shinto tradition; when Buddhist craftsmen came to influence the Japanese artists, they brought refinements, but not the basic idea. Japanese *haniwa* differ from Chinese statues by having a hollow "eye" allegedly the entrance for the spirit within.

5

5 Naindaimon, an imposing Chinese-style structure, is the main gateway to Todaiji Temple. The style of the gateway reflects the architectural trend towards strength and simplicity that typified the Kamakura period and indicates the wide cultural links Japan developed with T'ang China.

6

6 Japanese armour of the Kamakura period shows the artistry associated with the late Heian period and the rise of a provincial military class that demanded very high skills of workmanship. Warriors often donated fine armour to important shrines. Their personal code emphasized simple dignity and courage.

7

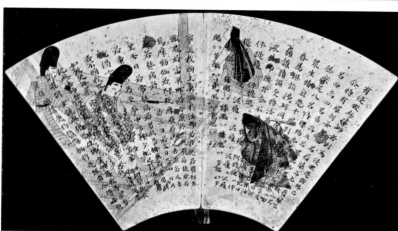

8

7 This hand-painted copy of a *sutra* (Buddhist scripture) dates from the late Heian period. It was believed that copying *sutras* by hand was one way to gain re-birth in Paradise. This was a feature of the Buddhist cult – the Jodo cult – which arose in the late 12th century and laid great stress on afterlife.

8 Minamoto Yoritomo (1147–99) led the armies that destroyed the power of the Taira family in 1185. This conflict, immortalized in *The Tale of Genji*, in-spired many important works of literature. Shown here in formal dress, Yoritomo set up his capital in his own territory, far from the imperial capital.

Japanese art

Japanese art since the introduction of Buddhism in the sixth century owes much to mainland China and Korea [1] and therefore it is sometimes dismissed as derivative. This is far from the truth, although the Japanese, always ready to absorb new influences, use imported techniques and ideas and in their own ways create new styles.

Early Japanese art

Until the sixth century Japanese art had been relatively simple but under the tuition of craftsmen from Korea, and later China, there was a burst of development in the arts and in government that led to the art of the Nara period (645–794). Much of its architecture and sculpture in wood, bronze and dry lacquer survives in temples to this day; in the eighth-century treasure-house in Nara, the Shōsō-in, quantities of lacquer, pottery and leather work, as well as painting and embroidery, give us a good picture of the arts and crafts of the period.

In the succeeding Heian period (794–1185) the Buddhist arts took on a more national flavour while the elegant court of the period, immortalized in the novel *The Tale of Genji* [4], is depicted in the famous scroll in Yamato-e style ("the Japanese style"). The Yamato-e style differs fundamentally from Chinese styles in its lack of interest in the brushstroke and in its use of flat areas of opaque colour.

In the thirteenth century there was a revival of the Nara styles in Buddhist sculpture led by the Kaikei school, and painting diversified into several groups according to the separate Buddhist sects. Secular art continued in the form of handscrolls depicting histories or satires while the handscroll format (originally a book) was also used by Buddhist artists.

Innovations from China and elsewhere

The arts flourished under the rule of the Ashikaga family during the Muromachi period (1333–1573) in spite of civil wars. Renewed contacts with China enabled painters from the Zen Buddhist monasteries to visit China and Korea and learn the art of ink-painting from Chinese painters [5]. At first the painters' academy was exclusively filled by monk-painters. As interest in mainland culture spread, two secular schools of painters arose, the Ami school and the more important Kanō school.

The Kanō school, which later became the "Classical" school in Japan, was a basically Chinese style but infused a decorative quality quite alien to Chinese scholar-painters' ideals. When in the Momoyama period (1573–1616) Kanō Eitoku [6] invented the use of gold leaf as a background to screen painting in opaque colour, a new unparalleled richness was introduced to Japanese art while retaining the brushstroke of Chinese convention but using Yamato-e colour. This decorative effect was ideally suited to the tastes of the new military leaders who succeeded the Ashikaga; where the Ashikaga rulers had encouraged the elegance and simplicity inherent in the aesthetics of the tea ceremony, Oda Nobunaga and his successor Toyotomi Hideyoshi wanted grandiose display. During this period the minor arts flourished as never before: lacquer, pottery and metalwork, particularly that associated with the sword and its fittings, all reached a

1 Carved in wood, this bodhisattva (future Buddha) is 7th century and in Korean style.

2 The Phoenix Hall of the Byōdō-in, near Kyōto (11th century), was so called because it is said to resemble a phoenix settling with outstretched wings. This beautiful building demonstrates how Japanese Buddhist architects interpreted Chinese ideas and retained Chinese construction methods.

3 This portrait sculpture in wood of the statesman Uesugi Shigefusa (14th century) is in a style which did not continue long. It does however demonstrate the economy of line and feature common to most later styles of Japanese portraiture. Portraits of laymen appear first in the 14th century, following posthumous Buddhist portraits of sages.

4 *The Tale of Genji*, a 10th-century novel, by Murasaki Shikibu, was illustrated in the Yamato-e style in the 12th century. As convention dictated, the view is from above with emphasis more on decoration than on the figures.

5 Sesshū (1420–1506) was a painter who had visited China and learned the art of ink painting from Chinese painters. This autumn landscape illustrates how he gave a distinctive Japanese flavour to the Chinese ink style.

peak of excellence. Military men prided themselves not only on their bravery and skill with the sword, but also on their ability to write verse and paint in ink.

Influences in recent times

Hideyoshi's successors, Tokugawa Ieyasu and his family, ruled Japan until 1868. The Edo or Tokugawa period was one of rigid exclusion or control of contact with the outside world in order to retain internal peace. Trade flourished within Japan, and brought with it prosperity for a new merchant class. This in turn brought new styles of painting and new ventures in the decorative arts.

The porcelain industry [9] began in Kyushu and in the middle of the seventeenth century started exporting to Europe and the Near East via the Dutch, who were allowed, with the Chinese, limited trade facilities.

The declining standards of the Kanō school assisted the rise of a new, popular school, Ukiyo-e, the school of the print artists, who produced printed books and broadsheets, theatre posters and ephemera for the dissolute world of the Edo: their sub-jects were the "pop heroes" of the day, courtesans, actors, dramatic moments of plays and erotica.

New contact with Chinese painting intro-duced the scholar's style, Nanga, and subse-quently two partial offshoots from this, the realist style of Maruyama Okyo and the con-trolled yet dashing Shijō style produced by Matsumura Goshun.

With the intrusions of the West into Japan in 1868 this activity changed abruptly. Western fashions became the rage, the court wore morning dress and top hats, painters studied oil-painting in Paris, and only the intervention of such enlightened men as Ernest Fenollosa (1853–1908) prevented neglect or even wholesale destruction of Japan's artistic and architectural heritage. In this century there has been not only a swing back to native ideals in many of the arts and crafts but also a serious expansion into Western materials and methods, so that Japanese artists are, for instance, among the leading print makers, while such architects as Kenzō Tange are among the best exponents of a wholly international style.

The Hōryūji temple, Nara, is a seventh-century five-storied pagoda similar in design to wooden pagodas of the T'ang dynasty in China.

6 These lion dogs are attributed to Kanō Eitoku (1543–90), one of the greatest painters of the Kanō school.

7 A screen of pine trees in a mist is one of a pair by Tōhaku (1539–1610), one of the contem-porary rivals of the Kanō artists. It shows how skill in ink was retained by artists who other-wise used brilliant colour and gold leaf.

8 These screens are called "Red and White Plum Trees" and were painted by Ogata Kōrin (1658–1716). They are a supreme example of the Rimpa or "Decora-tive" school of painting and they show the character-istic abbreviation of natural objects, treating them in a conventional way in a layout that is wholly original. Yamato-e was used for simplified yet decorative effects.

9 Porcelain was ex-ported to Europe via the Dutch from 1650 onwards. Enamelled wares like this Ka-kiemon vase of the late 17th century inspired Meissen, while blue and white wares influenced earthenware design in Delft and elsewhere.

10 "The Hollow of the Deep Sea Wave" is one of a series of 36 views of Fuji in woodblock colour prints by Katsushika Hokusai (1760–1849).

The rise of medieval Western Christendom

The barbarians who destroyed the Western Empires in the fifth and sixth centuries were either pagans or Arian heretics who denied the unity of God the Father and Son and violently rejected Roman Christianity. By the sixth century Europe needed to be reconverted. Missionary activity from Rome followed the growth of Benedictine monasticism and a rejuvenation of the papacy. St Benedict (c. 480–c. 547) lived in Italy when it was ruled by the Arian Ostrogoths. About the year 500 he had begun a hermetical life at Subiaco with emphasis on the performance of the liturgy and on a community life of moderate self-denial. In c. 529 he founded the Abbey of Monte Cassino, one of the bulwarks of Christianity and civilization in early medieval Europe. He also wrote a Rule for his monks that was humane and ensured the durability of his ideas and institutions [Key].

The Papacy – spiritual and temporal
The medieval papacy was founded by Pope Gregory the Great (c. 540–604) [2], who was a firm supporter and propagator of monasticism. He became pope in 590, a time

when the power of the Byzantine governors of Italy was rapidly declining and the Arian Lombards threatened to reduce the Papacy to little more than a Lombard bishopric. Gregory himself managed the Church estates in central and southern Italy, organized the defences of Rome, appointed governors of the leading Italian cities and in 592–3 made peace with the Lombards without reference to the Eastern Emperor. The Papacy henceforth existed as a temporal as well as a spiritual power in the West.

The Franks were the first of the barbarians to be converted when Clovis (465–511), founder of the Merovignian monarchy, was baptized, probably in 497.

Gregory dispatched St Augustine to England in 597 with monks from his own Roman monastery to begin the reconversion of Britain. But Britain was reconverted from two different directions because the Irish, who had remained Christian after the mission of St Patrick in 444 [7], had sent St Columba to Scotland c. 563 to found the monastery of Iona and convert the Picts. In c. 635 St Aidan went from Iona to Lindisfarne to convert the

English in Northumbria. There the Roman and the Irish traditions met and the result was the most advanced and flourishing culture of seventh- and eighth-century Europe [3].

Anglo-Irish influence
The Irish also penetrated deep into mainland Europe [4]. In 590 St Columban established monasteries at Anagratum and Luxeuil in the Vosges. Expelled from Burgundy for criticism of the behaviour of the court, he went to Italy to found the monastery of Bobbio, which set an example that was the most important impetus to the conversion of the Lombards. These and other Irish monasteries reintroduced the Catholic faith and brought with them their libraries, both classical and Christian, which had remained safe in Ireland during the migration period. The influence of the Irish on European culture can hardly be overemphasized.

Frisia and Germany were converted from Britain in the eighth century. Willibrord (c. 658–739), the apostle of the Frisians, was a Northumbrian who had joined a monastery in Ireland. Boniface (c. 680–754), an

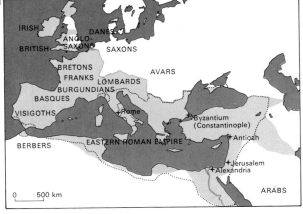

1 **Orthodox Christianity** in the later Roman world reached its greatest extent c. 600. North Africa, Rome and Ravenna were reconquered from the Arian Goths by Justinian in the mid-6th century. In the north, the Irish had been converted about 430, the Franks in 496, the Burgundians in 516 and the Visigoths in 589. The Anglo-Saxon missions began in 597. The new Christian unity of the Mediterranean, however, lasted only until the Arabs in the 7th century conquered an empire from Syria to Spain including three of the five original patriarchates, Antioch, Jerusalem and Alexandria.

Majority are Christian by 600
+ Patriarchates
······ Eastern Roman Empire

2 **Gregory I** (centre) was elected pope at a time when Rome was under strong pressure from the Arian Lombards. But his energy, dedication and grasp of administration enabled him to give the Roman Church a status it had never previously enjoyed. He established the principle of papal authority in temporal affairs both by his diplomatic initiatives and by asserting control of the "patrimony of Peter" which later grew into the papal states. By sending Augustine to England in 597, he made sure that the Church in Britain would look to Rome rather than Byzantium. Gregorian chant is named after him.

3 **The Ruthwell Cross**, possibly either a mass or preaching cross, is the finest monument of Northumbrian art during the British cultural renaissance of the 7th century.

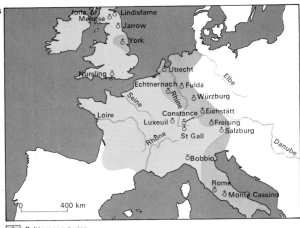

Celtic monasteries and their influence
Anglo-Saxon monasteries and their influence

4 **Many European monasteries** were founded by missionaries from Ireland where Christianity and classical learning had been preserved. Others were founded by their Anglo-Saxon converts.

5 **The Gatehouse of Lorsch Abbey**, in Hessen, West Germany, is an example of Charlemagne's impressive programme of new church building.

Englishman, continued the Frisian mission in 716, and in 719 was appointed by Gregory II to convert the Germans. He laid the foundations for the Carolingian Church and was martyred in Frisia in 754 [6].

The Arian Visigoths in Spain were converted to Christianity at the Third Council of Toledo in 589 (although Muslim invaders were soon to dominate the country). In northeastern Europe, conversions were delayed longer. Sweden and Denmark were only temporarily converted by St Anskar in the ninth century, Poland and Hungary in the late tenth and Norway at sword point by St Olav early in the eleventh century.

In the meantime the emperor of the Franks, Charlemagne (742–814), had imported Anglo-Irish missionaries to establish the basis of Carolingian Christianity, backed by energetic church building [5]. The late eighth century also saw an alliance between the pope and the Frankish emperors against the Lombards and the creation of the idea of the Holy Roman Empire. Anglo-Saxons, Germans and Franks all visited Rome, accepted the lead of the Papacy and

bought relics for their native dioceses [8].

The rise of the Holy Roman Empire saw a marked decline in the standards of the Church. There was widespread simony (the selling of ecclesiastical appointments), monasteries became rich and lax, and the Papacy itself was corrupt.

Church reform
Reform of the Church began at the house of Cluny, a Benedictine monastery in Burgundy [9]. Under Abbot Odo the Benedictine Rule was strictly enforced. The spirit of Cluniac reform permeated all aspects of Western Christianity, culminating in the pontificates of Leo IX [10] and Gregory VII (1073–85).

In 1054, the last year of Leo IX's pontificate, the Western Church broke with the Eastern. The Patriarch of Constantinople would not accept the universal supremacy of Rome, nor the people of the East the liturgy and practice of the Roman Church. Leo IX and the Emperor sought a political settlement between East and West to thwart their common enemy, the Normans, but the rift in religious practice was too wide to heal.

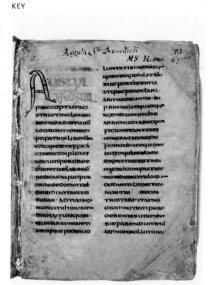

The Benedictine Rule was the cornerstone of the early medieval Church. It contained strict yet reasonable regulations for monastic life, in contrast to the ascetic excesses of Eastern monasticism. The Rule stressed the value of religious community life, humility, self-denial and the performance of the liturgy. For 600 years after Benedict's death in *c.* 547 there was no other monastic rule in the West. Its followers have included 20 popes and many pioneer missionaries. The learning and education of Benedictine monasteries in the Dark Ages also provided the only training for administrators faced with the increasingly complex problems of government.

6 **The story of the martyrdom** of St Ursula is the most famous of the martyrdom legends from the barbarian period. The saint, together with her companions, was murdered by the Huns in Cologne in 454 still protesting her virginity and her faith. Her triumphal funeral is shown in a painting by Vittore Carpaccio (1490). The cult of local martyrs became increasingly important during the reconversion period, at which time the supposed number of St Ursula's companions was increased to 11,000 by a clerical error. A notable missionary martyr was St Boniface, who went to Frisia and was murdered at Dokkum in the mid-8th century with 30 other monks.

7 **St Patrick, a Romano-Briton from Cumbria**, converted Ireland between 430 and 461. Ireland was the only country to escape the invasions of the 5th and 6th centuries. Christianity was preserved, along with many manuscripts containing both secular Latin and Christian literature. Irish missionary impetus was a prime factor in the re-education of Europe but its loosely ordered yet ascetic monasticism conflicted with the usages of the Roman Church after the reforms of Gregory the Great. This conflict was resolved at the Synod of Whitby (664) and led to period of peace which saw the creation of a British culture unrivalled in the rest of Europe.

8 **Relics** became an essential part of the furnishings of every church in the 9th century. They were housed in the greatest magnificence, as in the High Altar of St Ambrose, Milan, decorated by the German Volfinius about 835.

9 **The Abbey of Cluny** saw the start of the reform of the Benedictine system with a strict adherence to the Rule and stress on the splendour of the liturgy. This is the third church at Cluny dedicated to these ideals.

10 **Leo IX, who was pope from** 1049–54, was an ardent supporter of Cluniac reform. He began to improve papal standards.

The barbarian invasions

"Barbarian" was a term of abuse used by the Romans to describe anyone outside the Mediterranean civilization of Greece and Rome. Since the time of Caesar the frontiers of the Roman Empire had been menaced by invading Germanic peoples, many of whom were conquered, after which they settled as fairly peaceable – though armed – colonists.

The frontier that extended from the North Sea to the Black Sea via the Rhine and Danube was under renewed pressure in the fourth century from a fresh wave of hostile German peoples: Franks, Saxons, Burgundians, Visigoths, Ostrogoths, Sueves, Alans, Vandals and Gepids [Key]. Their societies were based on the clan and their tribes were relatively small in number, ranging from about 25,000 to 120,000; all able-bodied men were soldiers and farming was done by slaves. Their armies, supported by a powerful cavalry, proved too strong for the last of the Roman legions. The barbarians were converts to Arianism, a heresy abhorrent to orthodox Christians because it denied Christ's divinity. The westward pressure of these peoples at the end of the fourth century

was due to overpopulation and shortage of food as well as the arrival of the Huns, a fierce nomadic Mongol people from central Asia who came to pillage rather than to settle.

Eastern Roman Empire
The Eastern Roman Empire was threatened by barbarians on all fronts [2]. In Asia Minor it was under pressure first from the Sassanian Persians and later, after 622, from the rapid expansion of Islam. On the Danube and in the Balkans the threat of the Germans and Huns was ever present.

After the death of Theodosius the Great in 395 the Visigoths rose in Lower Moesia under Alaric (c. 370–410). The Eastern emperor Arcadius persuaded them to move west where Alaric was made Master of the Soldiers in Illyricum (coastal region of modern Yugoslavia); it was therefore as a Roman general that Alaric led a German invasion of Italy in 401–3. He invaded Italy again in 408 and in 410 captured Rome.

In 406 vast armies of Vandals, Sueves and Alans crossed the Rhine into Gaul which was already troubled by internal conflicts in the

imperial administration. External attack and internal conflict led to the withdrawal of the last legions from Britain where the Angles, Saxons and Jutes were now free to invade. The Vandals ravaged Gaul, moved into Spain in 408 and, in 429 under the leadership of Gaiseric (c. 390–477), accomplished the most crippling blow of all to the Western Empire – the invasion of Africa. In 442 Gaiseric was recognized as the independent ruler of North Africa on which Rome depended for the bulk of her food.

The first German empires
By 416 the Burgundians in Gaul had established the first German kingdoms within the old imperial frontiers and with the Franks scattered across northern France the pattern of post-Roman western Europe was beginning to emerge. But any hopes of peace were shattered by the Huns, who for nearly a century had been stable in what is now Hungary, threatening east and west alike and building up an enforced alliance of subject German peoples. In 450 the empire refused to buy off the Huns with any more gold so, led

CONNECTIONS

See also
134 The rise of medieval Western Christendom
138 Anglo-Saxon settlement
164 Anglo-Saxon art and society
180 Medieval Ireland
182 Scotland to the Battle of Bannockburn
20 The tools of history
114 Rome: soldier emperors to Constantine
142 The Byzantine Empire
152 Charlemagne and the Carolingian Empire
156 The Vikings

1 At the time of their invasion of Gaul in the 5th century, all freemen within the Frankish tribe were warriors. The Frankish army was mainly infantry rather than cavalry. Their most redoubtable weapon was the battleaxe, the *fran-* *cisca* [1], although they also fought with a cutting sword [2] and bows and arrows [3]. Because metal was scarce the round-shield [4] was made of wood covered with stretched hide. They wore close-fitting tunics [5] and plumed helmets [6].

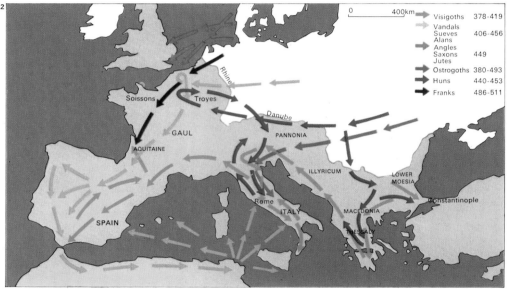

2 A chaotic wandering of barbarians within the confines of the Western Empire followed the collapse of the Rhine and Danube frontier. The barbarians, although quite un-co-ordinated, had four targets: Constantinople, Macedonia and Thrace; Italy; northeastern Roman Gaul; and North Africa, the granary of the Western Empire. Without any preconceived plan of attack, the citadels of the empire fell according to a "domino" pattern. Westward pressure was intensified as more tribes arrived from the east, in particular the Huns. The crucial blow to the empire was the Vandal invasion of Africa which, by 429, left Rome surrounded and the Vandals controlling the western Mediterranean.

3 The barbarians gradually formed independent kingdoms in the west. They expropriated some property but left much of the Roman landed aristocracy and administration intact. Roman and Teutonic societies existed side by side in an uneasy balance, the military strength of the invaders bringing the defensive capacity that defeated the Huns in 452. The Huns remained nomadic and never founded a lasting kingdom. The barbarian kingdoms were the origins of the national boundaries of medieval western Europe.

Kingdom of Visigoths
Kingdom of the Franks
Kingdom of the Burgundians
Kingdom of the Ostrogoths
Kingdom of the Vandals

Visigoths 378-419
Vandals Sueves Alans 406-456
Angles Saxons Jutes 449
Ostrogoths 380-493
Huns 440-453
Franks 486-511

by Attila (c. 406–53), they invaded Gaul. The Huns, with their powerful horses and incredible stamina, were considered almost invincible and were only halted near Troyes in 452 when the Franks and the Visigoths joined forces with the Romans. Attila turned south and menaced Rome, but died in 453 [7]. The destruction was widespread but the Germans had saved the legacy of Roman civilization in the west.

The Franks gradually consolidated their power in Gaul. Under Clovis I (465–511) they created a united and non-heretical Christian Gaul (Council of Orleans, 511) and Clovis was recognized by the Eastern emperor as the ruler of a country that roughly corresponds to modern France. He was baptised following his victory over the Alemanni.

After the deposition of Augustulus, the last Western emperor, in 476, the German soldier Odoacer (died 493) became the ruler of Italy. Meanwhile the Ostrogoths had left the Black Sea area and after the downfall of the Huns had moved into Pannonia and thence to Illyricum. Led by Theoderic from 471 they ravaged Macedonia and Thessaly

and in 487 marched on Constantinople. Theoderic was bought off by the emperor Zeno, who persuaded him to go to Italy and overthrow Odoacer. Theoderic defeated Odoacer on the River Adda in 490 and after a three-year siege of Ravenna Odoacer gave in. Theoderic assassinated him and established the Ostrogothic kingdom in Italy [8].

The Eastern Empire
The Eastern Empire averted most dangers from the Germans and the Huns by passing them over to the west, but the Balkans were ruined and depopulated and the Bulgars were able to threaten Constantinople in 493 and 499. In the east the emperor Justinian (c. 482–565) fought for 35 years against the Persian king Chosroes (reigned 531–79). The balance of power in the Mediterranean was temporarily reversed by Justinian's reconquest of Africa (533–4) and Italy (536–54); yet all this effort was wasted when, in the following century, the Arabs overran all imperial lands from the Middle East to North Africa, besieging Constantinople annually until their fleet's defeat there in 718.

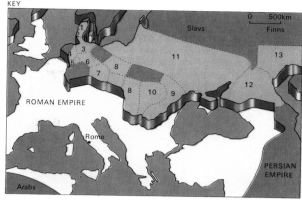

KEY

1 Jutes
2 Angles
3 Saxons
4 Franks
5 Burgundians
6 Thuringians
7 Sueves
8 Vandals
9 Ostrogoths
10 Visigoths
11 Gepids
12 Alans
13 Huns

Barbarian tribes c.395

Tribes not invading the Roman Empire 395-511

The positions of the barbarian tribes to the east of the Roman frontier along the Rhine and Danube rivers changed constantly. The pressure on Rome's increasingly ill-defended northeastern frontier was kept up by the arrival of new warring peoples who were impelled relentlessly westwards from central Asia by hunger and nomadic life style. The Huns invaded Italy in 452.

4 A B

4 The highest achievement of barbarian art was in its metalwork. The 6th-century silver-gilt dish from Sassanian Persia [B] was the equal of any contemporary east Roman metalwork. In the west, work ranged from the sternly Teutonic Visigothic crown [A] to the Anglo-Saxon Alfred Jewel [C]. This is made of continental cloisonné enamel and has classical motifs and a figure of Spring (or Christ).

5 The Mausoleum of Theoderic is the tomb of the first king of Ostrogothic Italy. It was built about 530 and is circular in shape, surmounted by a monolithic dome of Istrian stone weighing 477 tonnes.

6

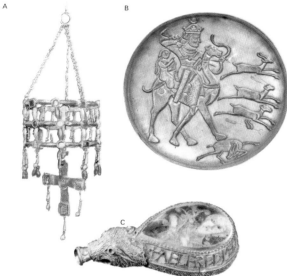

6 The Breviary of Alaric II, king of the Visigoths, an illuminated detail of which is shown here, is a collection of Roman law. Alaric's capital was Toulouse and his kingdom stretched into Spain. The breviary, which was completed in 506,

is an important source of late Roman law and is the result of an established Arian German barbarian king seeking to codify and reconcile his native tribal law with that of the lands of the old Roman Empire over which he found himself ruling.

7 Raphael's fresco of "The Repulse of Attila" (1513) shows Attila and his Huns confronted by Pope Leo I, protected only by the miraculous intervention of St Peter and St Paul. Attila, repulsed by

the Romans and Visigoths near Troyes in 452 and by Leo in Italy, retired to Pannonia and died on his wedding night of a burst blood vessel – four events which saved the legacy of Roman civilization.

8 German barbarians, when converted to Christianity, adopted the Arian heresy that denied the divinity of Christ. Theoderic, having established the capital of his Italian kingdom at Ravenna, built an

Arian cathedral and baptistery. The mosaics in the baptistery show the formalized presentation of the Baptism of Christ with the Apostles moving towards an altar-throne; a scene, the Etimasia,

taken from the fourth chapter of the Apocalypse. The existence of Arian and Catholic buildings side by side in the same city exemplifies the spirit of co-existence that prevailed in Theoderic's Italy.

7

8

Anglo-Saxon settlement

In the course of the fifth and sixth centuries the once-unified Roman province of Britain was split into a collection of petty kingdoms, barbarian Saxon in the south and east, Celtic in the west and north. The Venerable Bede (673–735) [Key], was clear about the origins of the barbarian invaders and the date of their arrival. They came, he said, in about 450 from three tribes, the Angles, Saxons and Jutes. Modern archaeological research has confirmed that the main regions from which distinct groups of settlers came were indeed Angeln, part of Schleswig and southern Denmark, and Lower Saxony, between the Elbe and Weser river mouths [2]. The identity and origins of the Jutes remains more problematic. Peoples not mentioned by Bede, including Frisians and Franks, also played some part. None of these peoples appears to have retained a separate cultural identity for long after their arrival.

The earliest Saxon settlements

Contrary to Bede's opinion, there were already Saxons living in Britain before the Roman withdrawal. Sea pirates, known generally as "Saxons", were raiding the eastern and southern coasts of Britain throughout the fourth century [1]. By the late fourth century it was common practice in the Roman army to employ mercenary bands under their own leaders, who fought, often against peoples related to themselves, in return for land and money.

There are large cemeteries in eastern England which contain thousands of cremated burials, many in pots similar to those at equivalent sites of the fourth and fifth centuries on the Continent [3]. Mucking, a village in Essex near Tilbury on the Thames, where occupation began early [7], is a site of some strategic importance, where Saxons would not have been able to settle so early without a position of some power, whether won by treaty or force.

When the regular Roman army withdrew from Britain in 410 the British leaders had to look after their own defence but continued the practice of employing mercenaries. Gildas, a Celtic monk who lived in the sixth century, wrote in graphic but perhaps exaggerated terms of the result of this policy: the barbarians rebelled against their employers and "fire and slaughter spread from sea to sea over the entire island". The story, recounted by Bede, of Hengist and Horsa, mercenaries who rose up against Vortigern who had hired them to protect Kent from the Picts, may have been a significant, although not the only, incident in this rebellion.

The survival of Romano-British society

Gildas was writing within living memory of the events he described, in a period of peace established after a British revival and defeat of the Anglo-Saxons, perhaps under the leadership of Arthur. Frequently the penetration must have been peaceful and gradual. As late as the mid-fifth century the native British had time to indulge in religious dispute, and St Germanus had to visit them twice from Gaul to combat the Pelagian heresy (which denied original sin). Despite the general decay of town life, he visited Verulamium (St Albans), a city in eastern Britain, which contained a functioning water system well into the fifth century.

During the invasions, many of the British

1 The "Saxon Shore" forts, shown here in a Roman map, were part of a system built in the 3rd century to protect the south and east coasts of Roman Britain from raids by barbarians. Some of these forts may have been manned by Saxons who stayed after the Roman withdrawal.

2 The 5th century invasions came from the far coasts of the North Sea. The Saxons were from Germany between the Elbe and Weser rivers, the Angles from Denmark and the Franks from the Rhineland. Other tribes came from Frisia. They had been driven from their homes by tribes from the east.

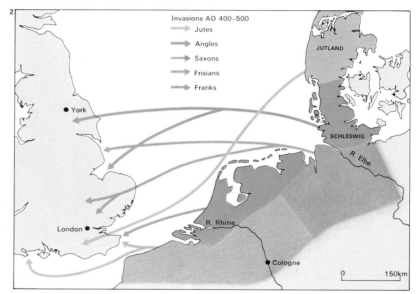

Invasions AD 400–500
→ Jutes
→ Angles
→ Saxons
→ Frisians
→ Franks

JUTLAND
SCHLESWIG
R. Elbe
York
London
R. Rhine
Cologne
0   150km

3 Early Anglo-Saxon pots survive in large numbers because they were used for the burial of cremated bodies. They were often clumsy and badly fired, although elaborately decorated. The Continental ancestors of the settlers also used these pots, so they can now aid in tracing the various tribes in Britain. The Continental Angles made pots with linear grooved patterns [A], while the Saxons preferred bizarre bossed and stamped decoration [B]. In England these styles became mixed, suggesting a quick loss of separate tribal identity. With the spread of Christianity, burial rather than cremation in pots became the rule. Use of the Romano-British potter's wheel died out by 500.

4 An Anglo-Saxon ship found at Sutton Hoo, Suffolk, in 1939, was the cenotaph or memorial of a great king, and contained much glorious jewellery and many domestic utensils. It is not certain which king the ship commemorated, but he may have been Raedwald of East Anglia. It dates from before the final acceptance of Christianity in that kingdom (c. 650). The invaders were sea-faring people, and ships played an important role in their mythology and religion. The Sutton Hoo ship, which was almost 29m (89ft) in length, is the best example of the boats they used. It was primarily a rowing boat, but may have had a sail. It probably had an elaborate carved figure head on the prow.

aristocracy may have emigrated or been killed, yet it is unlikely that all the existing population could have been slaughtered. Enslavement or absorption through marriage with the invaders were more likely. It is significant that the Anglo-Saxon language absorbed very few Celtic words other than names of natural features; but the Anglo-Saxon word for "Briton" also meant "slave". The barbarian invaders of France, by comparison, were not sufficiently numerous to eliminate the language, which became an amalgam of Germanic speech and Latin. But the distribution of placenames with a British origin suggests that the bulk of the Saxon settlers arrived and settled in the eastern half of the country. Anglo-Saxon penetration to the west was slow, and the invaders may have lived alongside the Britons, instead of displacing them, as was once thought.

The Anglo-Saxon kingdoms

The small warrior bands and groups of migrants, organized by family and clan, evolved gradually into petty kingdoms, of which seven developed into stable entities [9].

These kingdoms have given the name "Heptarchy" to Anglo-Saxon England from the sixth to the ninth century. Frontiers were not constant, kingship was personal rather than hereditary, and the balance of power was constantly changing. The most powerful king at any one time might be known as the "bretwalda", a name meaning perhaps "British leader" or "power wielder". The bretwalda had an informal power over the other kings. Raedwald (died 625), King of East Anglia, who was perhaps commemorated in the great ship burial at Sutton Hoo [4], may have been such a bretwalda.

Until the seventh century the Saxons were pagan. Their conversion to Christianity [6] resulted from two missions, the official Roman one by St Augustine (died 604), who arrived in Kent in 597, and the Celtic monastic one from Ireland [8], which won most favour in Northumbria. Differences in practice between these two, especially concerning the calculation of the date of Easter, led to dispute, which was resolved for Northumbria at the Synod of Whitby in 664 in favour of the Roman tradition.

The Venerable Bede was a monk who wrote the best surviving account of the early Anglo-Saxon period. Working in the early 8th century, he aimed to tell the story of the Church in England. He tried to be more critical of the sources he used (such as earlier chronicles as well as legends and hagiographies) than previous writers and is considered to have been the first English historian. Writing at Jarrow in his native kingdom of Northumbria, he tended to regard Celtic Christianity as heretical, but he admired its great men, especially St. Columba (521–97) who brought Irish Christianity to Scotland, and St Aidan (d. 651), bishop of Lindisfarne.

5 East Anglia was settled earlier and more densely than any other region, being the part of Britain closest to the invaders' homelands. From the pattern of 5th-century burial grounds in England it appears that the immigrants sailed up the rivers and settled along the valleys and Roman roads. Some invaders sailed up the Wash and moved both south and north. Cemeteries at North Elmham in Norfolk and Loveden Hill in Lincoln contain thousands of burials, and show that by the mid-5th century much of the population was already barbarian. The kingdom of East Anglia soon became wealthy as can be seen by the treasures found in the Sutton Hoo ship burial of c. 625.

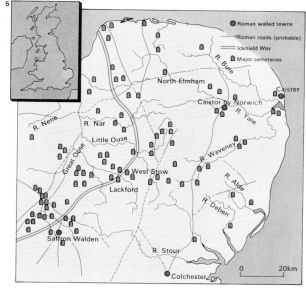

6 The Franks Casket, a small whalebone box, was made in Northumbria c. 700. Its decorative carvings include pagan and Christian myths, and its inscriptions are in both Latin script and Germanic runes. This panel shows the Adoration of the Magi and the story of Weland the Smith from Germanic mythology. The Anglo-Saxons often added Christian elements to their existing myths, as can be most clearly seen in the epic poem *Beowulf*, probably written after 700. But the Church was important in secular affairs by the 7th century, and many kings gave land and wealth to monasteries.

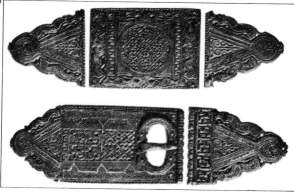

7 This bronze gilt belt-set was found in a grave in Mucking, Essex. It belongs to a type current in the late Roman army but is in a style found only in England, indicating there was a close relationship between the barbarians and the Roman army.

8 St Cuthbert (d. 687) was one of the best-loved Irish or Celtic churchmen of Anglo-Saxon England. He is seen here being presented with a book by King Athelstan. Although he had a Celtic training, he supported the decisions of the Synod of Whitby which ended the confusion over the date of Easter. A holy man known for his piety and humility, he was bishop of Lindisfarne. Miracles were said to occur at his tomb, now in Durham cathedral. His history was recorded by the Venerable Bede.

9 The kingdoms of the Heptarchy were often in conflict with one another and sometimes sought help from the Celtic kingdoms of Gwynnedd in Wales or Dalriada in Scotland. The centre of power altered quickly.

The Celts and Christianity

Prior to the Roman conquest, much of western and central Europe was dominated by Celtic tribes of Indo-European origin. Skilled in the use of iron, which they introduced into northern Europe, the Celts flourished from about 700 BC. They achieved a high level of material culture and also developed a highly complex religion and mythology which formed an integral part of their society. Celtic art reflected the importance of religion, and geometrical or magic symbols and cult animals are repeated on objects from all periods of Celtic history – pagan and Christian alike.

The finest phase of Celtic culture
Celtic culture reached its finest achievements during the La Tène period which began in about 500 BC and, in Europe, lasted until its absorption by Rome. Information about the Celts of this period is obtained both from archaeological evidence and also from Greek and Roman authors. They have recorded details of tribes, kings, nobles and place-names, all of which give important information about the Celtic languages, which have survived in a modified but recognizable form in Britain, Ireland and Brittany.

The Celts did not commit their religious beliefs and traditional learning to writing until after the arrival of Christianity in Ireland in the fifth century. The comments of the classical writers of the time when Rome's armies destroyed the Celtic world in Europe therefore have a unique importance. Almost fanatical in their religious fervour, the pagan Celts were dominated by powerful, highly aristocratic priests, the Druids, who often combined the role of king with their priesthood. Tribal in social organization, the Celts also possessed the oldest and most complex legal system in Europe. Every man had his rights, and crimes from murder to the smallest wrong were listed and categorized.

Celtic spiritual tradition
The Celts were highly conservative in matters of tradition. Non-literate, they had a complex oral tradition and were natural scholars with a deep admiration for intellectual power and a passionate love of words. The beauty of nature is depicted in some of their richest poems; its appeal to the Celts came at a time when their neighbours had little care for such intense spirituality.

It is understandable then that Christianity had such an early and widespread success in Ireland, the only country in western Europe to be totally untouched by Roman arms. Ireland fell in the fifth century, not to disciplined Roman soldiers, but to the equally disciplined Roman Church. All the Celtic fervour for religion was now transferred to the service of Christianity – a very Celtic type of Christianity, noted for its austere devotions and the selfless dedication of its clerics. The detailed and sophisticated laws were now transformed for Christian purposes; the glorious art once used to adorn the pagan warriors and their shrines and honour the gods now served to praise God in the form of superbly illuminated manuscripts [2, 3], the old pagan symbolism of spirals and circles taking on a new meaning.

In a Europe torn by invasion and disaster of every kind, Ireland remained a haven of peace and learning, far from the terrible ravages of the northern barbarians who from

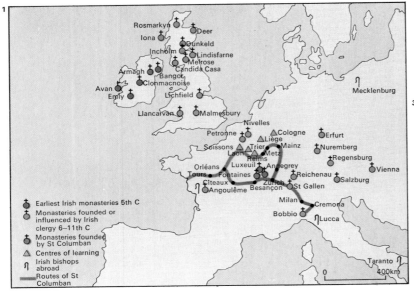

1 **St Columba** founded his monastery in Iona in AD 563. He was the first of numerous missionaries and scholars who, as shown on the map, established centres of sanctity and learning in Western Europe. The movement was represented by such figures as Aidan in England (635), Columba, in France, Switzerland and Italy (from 590), Feuillen in Belgium (c. 650), Kilian in central Germany (martyred c. 689) and Fearghal in Austria (mid-eighth century). Many of the Irish exiles were scholars, or monks wanting to evangelize the new pagan tribes who had overrun the Roman Empire.

2 **Illuminated manuscripts** are one of the great glories of European art, as this page from the seventh-century *Book of Durrow* shows. As Ireland remained untouched by the Roman Empire, its art retained its original style.

3 **The great period of Irish manuscript** illumination was from the late seventh to the early ninth centuries AD. In these unique works, vitality in the design is combined with austere representations of divine figures, showing by their elongated, sombre faces their long link with the Celtic past. In the *Book of Kells*, St Matthew is surrounded by decorated panels and motifs, which were familiar from the art of Ireland's pagan past.

406 were burning the churches and the towns of Europe and desecrating and destroying all that was sacred and beautiful. The early Irish Church favoured the monastic system and during the fifth and sixth centuries monasteries sprang up throughout Ireland.

The Age of Saints
This long period of tranquillity and learning was known as the Age of Saints. It was a time when the churchmen and their guests were occupied in studying the Gospels and illuminating the manuscripts, while the Christian scribes were busily committing to writing the old pagan oral traditions and poems of their country for, although Christians, they were also Celts and loyalty to the archaic traditions of their old and much-loved country was strong.

Only the coming of the Vikings at the end of the eighth century broke the spell that had made Ireland the cultural centre of the Western world. In the sixth and seventh centuries many Irish churchmen, such as St Columba (521–97), Aidan (died 651) and St Columban (543–615) travelled far over Europe, founding monasteries and churches, converting the heathen, teaching in the courts, establishing their own schools and inspiring all who came into contact with them by their austere devotion to their calling [1]. Following the Viking invasion, exiled monks continued to travel throughout Europe.

Their rich literature and that of their neighbours, the Welsh, profoundly influenced the evolution of medieval literature and provided new and thrilling themes for the enrichment of the troubadours' repertoires for the entertainment of the rich courts of later medieval Europe.

The Celtic story in Europe does not end in the romances of the courts of Eleanor of Aquitaine and her contemporaries in the twelfth century. In the eighteenth century the famous Ossianic controversy fascinated Europe. Although the poems of the legendary Gaelic bard, Ossian, were subsequently proved to be a mixture of traditional Gaelic folk poetry and poems attributed to James Macpherson (1736–96) they nevertheless inspired writers, painters, musicians and antiquarians with a fresh interest in the Celts.

This two-faced stone figure from Boa Island in County Fermanagh dates from the first century BC. It forms a link with the old pagan world of the Celtic past and the flowering of Celtic religious and artistic genius in the Christian era. It is impossible to say how long paganism lingered in Ireland after the arrival of Christianity. As Roman law and administration never interrupted the traditions of tribal life and religious awareness, the transition from pagan gods to the Christian God was complicated by the survival of some elements of Celtic pagan tradition. But by the fifth century, Christianity was firmly established.

4 The finest of all Irish brooches is the Tara brooch [A] (c. early eighth century). Both sides of the brooch are richly decorated, the back [B] being in a better state of preservation than the front.

5 The Ardagh Chalice [A] was found in Co Limerick in Ireland. The contrast of plain surface (silver) decorated with studs and gold filigree makes a striking impact. One of the more beautiful details is on the underside [B] of the foot of the bowl.

6 High crosses were free-standing monuments decorated with Christian or pagan symbols. One of the finest is the ninth-century South Cross at Castledermot, Co Kildare. These crosses presumably stood in the monastic precincts and were of varying heights with the wheel-shaped arcs joining the arms and shaft, and set on a substantial base.

7 Glendalough in Co Wicklow was a place of beauty and sanctity, sacred to St Kevin. Known as St Kevin's Kitchen (c. ninth century), the building has a vaulted ceiling which supports a corbelled roof in an ingenious manner. Small ecclesiastic buildings of this kind were probably widely distributed in early Christian Ireland, providing testimony to the spread of Christianity in that remote region, far from Rome.

The Byzantine Empire

In AD 293, the Emperor Diocletian decided, for military and administrative reasons, to shift the centre of the Roman Empire eastwards. From its origin as a new Rome, the Byzantine Empire became a vital trade and cultural link between Europe and Asia and a bastion within which Graeco-Roman civilization developed new and magnificent forms. Byzantium later found itself in doctrinal and political conflict with the Western popes and emperors but, as a Christian empire, it resisted Arab, Slav and Turkish invaders for more than 11 centuries.

The founding of Constantinople

Both Diocletian (245–313) and Constantine I (c. 285–337) sought a better base than Rome, closer to the troop-recruiting grounds of Anatolia and the Balkans. Constantine's choice fell upon a town on the Bosporus which had been the site of an ancient Greek city, Byzantium. Constantinople, as the new capital was called, had a fine harbour and was almost unassailable [Key]. The scale on which it was conceived surpassed anything in the ancient world. Constantine was deter-

mined to found an urban centre to which men throughout his empire could direct their loyalties. He believed a common religion could also provide a powerful cohesive force. He had been converted to Christianity in 312 and in 330 dedicated Constantinople to the Virgin Mary.

The close association of Church and state and the prestige and near impregnability of the capital gave the eastern sector of the Roman Empire remarkable unity after Theodosius (346–95) left the empire split into two. While the Western Empire was swept away in 476, the Eastern, or Byzantine, Empire was able to resist attacks on the Balkans by the Visigoths and Ostrogoths. In the fourth and fifth centuries an intellectual élite created the science of theology from Greek logic and the Christian revelation.

Although theology helped to strengthen the people's faith, it also led to religious discord. Many of the Byzantine Empire's later problems can be attributed to factional struggles in which theological differences hardened into political ones that could be exploited by invaders. The Christological con-

troversy, which came to a head in the fifth century, was concerned with the relationship between the human and divine aspects of Christ's nature. The followers of Nestorius, Patriarch of Constantinople, stressed the human side while the Monophysites, based at Alexandria, stressed the divine. An equally bitter controversy, which reached its height in the eighth century, was iconoclasm – opposition to the worship of images.

The Justinian era

The empire reached its apogee under Justinian the Great (c. 482–565), a brilliant administrator with wide-reaching military ambitions [6]. His general, Belisarius, reasserted Christian-Roman authority over large areas of the former Western Empire. Justinian greatly expanded the capital and built Hagia Sophia [8] which was intended to provide a centre of worship for all Christendom. Perhaps Justinian's greatest achievement was the codification of Roman law. His *Codex Justinianus* remained in the Middle Ages the main legal source book in Europe.

Byzantium nevertheless became increas-

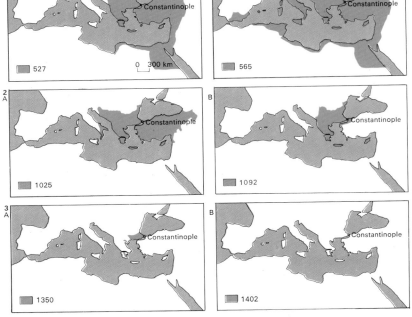

1 The Byzantine Empire under Justinian I grew from an exclusively Eastern power in 527 [A] to an empire controlling by 565 many former territories of imperial Rome [B]. Germanic invaders were ousted from many areas of the Mediterranean.

2 The recovery of many areas which the empire had lost to the Slavs, Germans and Arabs in the 7th and 8th centuries was completed by the conquests of Basil II ("the Bulgar slayer") [A]. But Normans and Turks had made large inroads by 1092 [B].

3 Dismembered by Turkish attacks and by internal feuds, the empire had shrunk in 1350 to a corner of the Balkans and some land in Greece [A]. By 1402, even the Balkan territory was lost [B] and Constantinople was soon to fall.

4 A patrician couple of the 6th century had a similar social status to their traditional Roman counterparts. As a middle class emerged, mobility between plebeian and patrician classes was higher than in Rome.

5 Byzantine coins reflected a change to a predominantly Greek culture after the 7th century. The Latin inscription on the gold solidus [A] gave way to one in the new official language of Greek [B].

6 Justinian I ("reigned" 527–65) is the central figure in a glowing mosaic from the Church of S. Vitale, Ravenna. Justinian made a determined effort to reunite the old Roman Empire under Christianity. Byzantine churches, such as at Ravenna were built after Justinian's general, Belisarius, overran Italy as far north as Milan in the years following 535. Justinian's military, cultural and administrative achievements earned him the title of "The Great". His consort, Theodora, daughter of an animal keeper, was influential in Byzantine court politics.

7 The dromon was a Byzantine development of the traditional Greek galley. Much Byzantine trade was carried by sea and the empire kept large and efficient mercantile and naval fleets. The dockyards along the Marmara coast were the finest found in Europe until the 12th century.

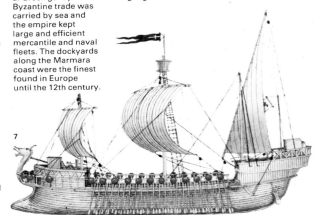

ingly Greek in character after the reign of Justinian and Greek replaced Latin as the official language [5]. The conquests in the west were short-lived and the ravages of plague weakened the empire's ability to resist Persian attacks in the east. Aided by dissident Monophysites, the Persians had occupied most of Egypt, Syria and Palestine by 615. A greater menace appeared in 637 when the Arabs, five years after the death of the prophet Mohammed, overran Syria and Palestine. Later they took North Africa, Sicily and the important grain lands of Egypt while their fleets secured Cyprus, Rhodes and several other islands.

Revival and decline (867–1453)
As its boundaries shrank, the empire regained its ethnic unity and acquired renewed strength. From 867, under a Macedonian dynasty, it took the initiative against the Muslims and by the time of the death of Basil II (958–1025) its borders reached from the Danube to Crete and from southern Italy to Syria [2]. During this last period of greatness, trade flourished [9] and

missionaries spread Christianity throughout the Balkans and into Russia.

The Turks were soon to bring the empire to its knees, however. The defeat of Romanus IV by the Seljuks in 1071 and subsequent capture of Anatolia were the beginning of the end. From then until 1453 the empire was steadily eroded by intrigues, attacks and religious conflicts [3]. The Crusaders, whose help was enlisted against the Seljuks, fell out with the Byzantines and exploited a quarrel to take Constantinople in 1204 and set up a number of semi-independent Latin states. Religious schisms and trade rivalries prevented any concerted western effort against the Ottoman Turks who made Byzantium a vassal state in 1371. Although only Constantinople and a few outposts along the Sea of Marmara remained by 1453, it took the Sultan Mehmet II (1431–81) [10] nearly two months to capture the great city itself. With his final victory the old Byzantine world from the Balkans to Palestine was once again united, this time under the Ottoman state, which dominated the Mediterranean for the next 200 years.

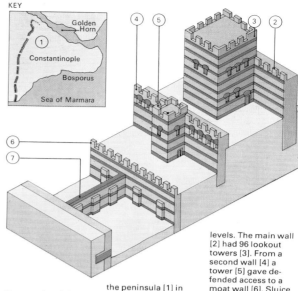

Constantinople's walls, built across the peninsula [1] in the 5th century, were defensible at varying levels. The main wall [2] had 96 lookout towers [3]. From a second wall [4] a tower [5] gave defended access to a moat wall [6]. Sluice gates [7] controlled water in the moat.

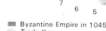

8 Hagia Sophia (Church of Holy Wisdom), built during the reign of Justinian I, was completed in only five years. Intended to provide a spiritual centre for the empire, it is the largest Christian church in the Eastern world and is exceeded in splendour only by St Peter's in Rome. The most famous of many architects who worked on the project were Anthemius of Tralles and Isidorus of Miletus. The overall design shows little classical influence, although Justinian despoiled classical buildings in Athens, Ephesus, Rome and Baalbek for marble. Technically, the most striking feature is the massive central dome [1] which measures 31m (100ft) across. Its thrust is borne by four arches [2], joined by pendentives [3] which separate the semi-domes [4 and 5]. Lesser semi-domes [6] flank the main piers [7]. The thrust from [5] is taken at the west by an arch [8] supported on piers [9]. Vaulting [10] transfers the outward thrust to a series of flying buttresses [11]. Buttresses [12] support the dome's north and south thrust. By building domes on arches [13 and 14], large areas could be spanned. The exterior brickwork is plastered. Brick domes and semi-domes are lead-covered. Interior walls, piers and floors are clad in various marbles, and vaults and domes in rich mosaics. The church was used as a mosque after 1453 and has been a museum since 1933.

9 Major trade routes of the 11th century reached a natural junction at Constantinople and the city became a great east-west market. A duty of 10% was levied on imports reaching Hieron from the Black Sea, Abydos from the Mediterranean, Trebizond from Asia, and Salonika from the Balkans. Byzantine craftsmen were famed for their working of gold, silver, amber, ivory and all kinds of precious stones.

Byzantine Empire in 1045
Trade flow
Customs houses

1	Amber	9	Timber
2	Ivory	10	Spices
3	Gems	11	Salt
4	Minerals	12	Weapons
5	Gold	13	Slaves
6	Textiles	14	Wax
7	Cotton	15	Furs
8	Silk	16	Dried fish

10 Sultan Mehmet II gave the Ottoman Empire a European outlook when he took Constantinople in 1453 and made it a centre of learning and religious tolerance. Although autocratic, he was a gifted administrator. Gentile Bellini, who painted this portrait, was one of several Italian artists he patronized.

Mohammed and Islam

Mohammed (Mahomet) was born in Mecca in west-central Arabia in *c.* 570. He had an unhappy childhood: his father, mother and grandfather died before he was eight and left him in the care of an uncle. At the age of 25 he married a wealthy widow, Khadija, who bore him six children, and for 24 years they lived happily together. Only after her death in 619 did Mohammed take other wives, to strengthen ties with important families and to seek – unsuccessfully – a male heir.

The visions of Mohammed

When he was 40 years old Mohammed, who loved solitude, was in a cave on Mount Hira outside Mecca when he had visions of the angel Gabriel calling him to "recite" in the name of God the creator. He received revelations that were to become the first parts of the Koran ("recitation"). Mohammed conveyed these teachings to a group of friends who believed with him in the unity of God.

At first the little group, which met for prayers to God (Allah), was ignored or scorned, but as their numbers grew they were persecuted. Some took refuge for a time in Christian Ethiopia. They were called Muslims (Moslems) – "surrendered men" – in the religion of Islam they had "surrendered" or submitted to the one God. Early converts were made from Yathrib, a town 200km (120 miles) north of Mecca, and Mohammed was invited to go there. In 622 the *hegira* (migration) took place and Mohammed and his followers moved from Mecca to Yathrib, henceforth called Medina the "city" of the prophet. The Muslim year is dated from the *hegira.*

At Medina, Mohammed built a mosque and a house, and sent his followers on raids to provide funds and ensure protection against armies from Mecca. There were battles at Badr and Uhud and finally Mohammed's armies and influence grew so that in 630 he was able to capture the city of Mecca almost without loss [1]. He rode around the Kaaba shrine [4] and had its idols destroyed.

After the death of the prophet

The death of the prophet in 632 was sudden, but after some hesitation his friend, the elderly Abu Bakr (573–634), was appointed as caliph – successor to the prophet and vice-regent of God. Arabian tribes that had been bound to Mohammed by oath began to break away but Abu Bakr sent armies to establish Muslim rule. They were so successful that his forces broke out of Arabia into the rest of southwestern Asia. Abu Bakr died two years after his appointment, but under his successors, the caliphs Omar and Othman, Arab armies rapidly conquered Mesopotamia and entered Persia, while others entered Syria. Jerusalem surrendered to them and Omar visited the Christian churches there and the site of the ancient Jewish temple where later a great shrine was built, incorrectly called the Mosque of Omar but more properly the Dome of the Rock, one of the most holy places of Islam. Arab armies went to Egypt, where Alexandria surrendered; then after some delay they travelled along North Africa and in 711 crossed into Spain at Gibraltar. They even penetrated into the heart of France, where between Tours and Poitiers the Muslim armies met Frankish forces under Charles Martel in 732. After seven days of fierce skirmishes the Arabs were forced to retreat southwards.

1 This painting is from a copy of *Siyar-i-Nabi* (Life of the Prophet) and shows Mohammed and Abu Bakr on their way to Mecca from Medina. The prophet is traditionally depicted faceless with a flaming halo. The paintings of Islamic art are unique in that the historical events depicted take a secondary place to the intensity of the religious feeling. This work is in the 16th-century Ottoman court style, yet it has an almost contemporary realism.

2 The Prophet Mohammed was a "warner", calling men to turn from idols to worship the one true God (Allah). There are no contemporary pictures of the prophet but later artists have pictured him as an Eastern holy man.

3 The Koran is the sacred book of Islam. All Muslims know some of its verses by heart and use them in daily prayers. Illuminated copies of the Koran in gold and bright colours were written by hand by skilled scribes.

They remained in southern France for some years but Spain was the limit of their rule and here they remained for centuries until the fall of Granada in 1492, after which Muslims and Jews were expelled from Spain. In the east, Persia came completely under Muslim rule as well as a large part of India. The Muslims preserved much of the cultures they encountered and transmitted them, taking Indian numerals as Arabic numerals to Europe and preserving Greek medicine, astronomy and philosophy during the period of the Dark Ages in Europe.

The importance of the Koran

Mohammed, who is said to have been illiterate, passed the Koran, the divine Word, to his followers and it was written down by scribes at his recitation or from memory [3]. The final official version was completed under Othman, about 20 years after the death of Mohammed. The Koran is in Arabic, in 112 *suras* (chapters) the first of which is always recited in daily prayers. Most of the early chapters are long and deal with religious and social matters, while the later ones are short and challenging in content.

The Koran teaches faith in God, the coming judgement against unbelief, and the ideas of heaven and hell; it also sets out duties appropriate to marriage, the family and social life. Many stories in it are parallel to some in the Old and New Testaments and Adam, Abraham, Moses and Jesus appear as prophets. The religious duties of Islam are taught in Five Pillars: confession of faith in one God and Mohammed as his apostle, prayer five times a day [5, 7], alms-giving of a proportion of one's income, fasting from all food and drink during the hours of day throughout the whole of Ramadan (the ninth month of the year) and pilgrimage to Mecca at least once in a lifetime.

All men and women are bound to perform these religious and social duties, with exemptions for the young, sick and old. Islam is an international religion, with perhaps 500 million followers, mostly in Asia and Africa. The Arabic language prevails in the southern Mediterranean and many Near Eastern countries, and is used by all Muslims to recite the Koran and in formal prayers.

"Praise be to Allah", the first words of the Koran, is a favourite text inscribed on the walls of mosques.

4 In the centre of Mecca stands the Kaaba (cube), a stone building covered with a black cloth, towards which Muslims turn in prayer and round which they go at times of pilgrimage.

5 Five times every day Muslims are called to prayer from a mosque or minaret tower. The *muezzin* calls that "God is most great, there is no god but God, Mohammed is the Apostle of God, come to prayers, come to salvation".

6 The Royal Mosque Masjid-i-Shah, in Isfahan, Iran, was built in the 17th century for Shah Abbas the Great. It is composed of porches, halls, domes and minarets covered with blue tiles and mosaics. Long friezes of elegant lettering proclaim the glory of God.

7 Prayer rugs, such as this 18th-century Persian rug, are famous products of Muslim craftsmen. Large ones cover the floors of mosques.

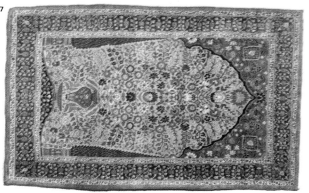

Byzantine art

The small proportion of Byzantine art to survive is mostly "official" art – the aesthetic visualization of Christian beliefs, or of imperial authority, or a combination of both as the Christian Roman state. Such art was the dominant production of this society, and secular art was almost non-existent. Byzantine art, therefore, begins with the foundation of Constantinople in 330, and ends with its fall to the Turks in 1453.

"Official" art
The new capital on the site of Byzantium brought Roman art back into the geographical setting where many of its traditions, forms and media had originated. Yet Byzantine art was no mere continuation of Graeco-Roman art in the service of the Orthodox Church. When fourth-century Constantinople emulated Old Rome by erecting two cochleate (spiral) columns, their sculptured reliefs (known from drawings) were developed far beyond Trajan's classical forms; they abound not merely with Christian symbols, but convey imperial triumph with an endless procession of stiff

figures. The style reflects the hierarchy of the state. This new art developed in the rich cities of the eastern Mediterranean, for example in the cupola mosaics of St George in Thessaloniki (450), where a superficially Hellenistic style portrays the place of Christian martyrs in the heavens.

The first great achievements of Byzantine art belong to the reign of Justinian (527–65), whose patronage ranged from St Sophia in the capital [2] to the remote mosaics of Mt Sinai [1]. The Dark Ages after Justinian, when Byzantium was under constant threat of invasion, saw a rise in the production of icons or, rather, in the superstitious cult of images or saints as the mediators for personal salvation. The Iconoclast emperors from 726 until 843 tried to impose a non-figurative religious art, comparable to medieval Muslim and Jewish societies, but the Byzantium managed to survive the Dark Ages and so did Christian figurative art.

After Iconoclasm, the Macedonian emperors (867–1065) celebrated the Triumph of Orthodoxy with the redecoration of churches, starting with the apse of St

Sophia. Progress in Byzantium customarily meant regress to past ages, and this period recreated the style of Justinian.

New patrons
More survives from the Comnenian dynasty (1081-1185) than the Macedonian. Under the Comnenes, the wealth of Byzantium significantly shifted from the emperors and bishoprics into the hands of aristocratic landowners and monasteries.

Taste for gold and saturated colours stimulated an unrivalled mosaic expertise. Cubes of glass (coloured or fused with gold or silver leaf), marble, even precious stones were pressed one by one into lime plaster beds on church vaults and each could be tilted to reflect the light. Vast mosaic figures in an ethereal golden glow surrounded the Byzantine worshippers. Fresco was a cheaper, but less permanent substitute.

The twelfth century was a period of refinement and sophistication of earlier forms, accompanied by a growing self-consciousness: artists began to sign their works in this century. Two major mosaic

1 This mosaic medallion of St John the Baptist is from the basilica founded around 550 by Justinian on the Burning Bush site on Sinai (now St Catherine's Monastery). Its great expressiveness suggests that the decorators were from Constantinople and were able to transform a classical tragic mask into mosaic.

2 The mosaic of the "Enthroned Madonna and Child" in the apse of St Sophia, Istanbul, was probably unveiled by Patriarch Photius in March 867. The Patriarch claimed in a celebrated sermon that the mosaic seemed so lifelike that Mary was "not incapable of speaking".

3 David fighting a lion is one of seven author portraits opening the *Paris Psalter*. The atmospheric setting and female personification assisting David are devices of classical art, suggesting that the artist was an antiquarian.

4 This "Crucifixion" is in the monastery of St Luke of Stiris, central Greece. Luke's relics attracted pilgrims, and so financed the 11th-century church. The saints here express monastic piety, and the scenes have a direct style.

5 The mosaics of the "Crucifixion" in the Monastery of Daphni, near Athens (c. 1100), suffered extensive restoration in the 1890s. The style is more narrative than at St Luke's Monastery and is also more emotional.

decorations in Greece illuminate the middle centuries of Byzantium. The "Crucifixion" in the Monastery of St Luke [4], of the Macedonian period, delineates Christ's death as a stark episode in Church history followed by the "Resurrection". The Comnenian "Crucifixion" at Daphni [5] still forms one unit in a dogmatic cycle, but the artist emphasizes the human aspects.

Individuality within conformity
Since Comnenian artists normally decorated a standard architecture (the cross-in-a-square cupola church) with a standard repertory (life of Christ and Mary), this may paradoxically have stimulated individuality within conformity. In Monreale the mosaicists brought to Sicily had to adapt their repertory to a different structure and larger scale [6]. The challenge was to their technical expertise rather than to their inventiveness with subject-matter.

The Latin occupation of Constantinople (1204–61) did permanent damage to the Byzantine economy, and Palaeologan art (1261–1453) is on a reduced scale. The last great mosaic decoration was that of the Kariye Camii, where another aristocrat provided himself with a mortuary chapel [7, 8]. Icon production increased in this period, especially for church sanctuary screens. Palaeologan artists absorbed earlier styles and made further developments. Late Byzantine art influenced Western European art; although it is harder to see what it derived from Western contacts. Whereas pictorial perspective was being developed in Italy, Byzantine compositions used architecture to punctuate the action rather than to unify the picture space. Did Byzantium stimulate, reject, or fail to understand the Western developments? Late Byzantine compositions are increasingly complicated – clarity of expression was seldom valued.

After the middle of the fourteenth century there was little money left for art in Constantinople, and many artists left to find work in other Orthodox societies, such as Russia. Although the consequence was to disseminate the Byzantine tradition, the destruction of imperial Constantinople in 1453 terminated creative Byzantine art.

The frontispiece of a luxurious psalter painted in Constantinople between 1018 and 1025 commemorates the victory of Basil II over the Bulgarians. The tri-umphant emperor, crowned by Christ, armed by archangels and assisted by military saints, tramples his conquered enemies underfoot. The picture both declares the power of the Byzantines and expresses their conception of sovereignty, with their emperor as representative of Jesus Christ on earth.

6 Monreale Cathedral [A] was founded by William II (r. 1166–89) as a monastic cathedral to rival the archbishopric of Palermo. It is a Latin basilica, with a Greek cross-in-square sanctuary.

The polychrome intricate east façade dominates the hill seen from Palermo. Christ Pantocrator is the focus of the vast mosaic interior [B]. The dramatic style was the current fashion in Constantinople.

7 The Kariye Camii, Istanbul, with minaret belonging to its Turkish conversion into a mosque, is now a museum. This was the 11th-century Chora Monastery which was refounded in 1315–21.

8 The Chora (Kariye Camii) was decorated by its 14th-century refounder Theodore Metochites with mosaics, except in the south chapel, which held his tomb in a frescoed setting. The apse contained the "Anastasis" (shown here), Christ at Easter rescuing Adam and Eve from Hell. Metochites justified his greed by his use of wealth, and this decoration alleviated his guilt by promising life after death.

9 The "Annunciation" is one side of a double-sided icon in the National Museum, Ohrid, Yugoslavia. The icon also portrays the Virgin, Saviour of Souls. A furnishing of the Peribleptos Church (now St Clement's), the icon was perhaps sent from Constantinople and was probably painted around 1300.

10 In this 14th-century miniature mosaic icon of the Annunciation the minute glass cubes are set in beeswax. Such portable mosaics became fashionable among the few rich families of Palaeologan Constantinople and were collected by Renaissance popes. This was virtually the only form of Byzantine art appreciated in the West.

147

Arabs and the rise of Islam

Arab expansion was a tribal conquest of civilization. Such conquests are commonplace in history, but the Arabs are the only people who started a new civilization as a result of their conquest. This came about by a combination of two special circumstances. First, on the side of the conquerors there was a unique fusion of religious conviction [2] and tribal military force. The Arabs conquered the Middle East [4] in the name of a monotheism that sanctified their tribal heritage over and against the conquered civilizations and their conquests were so successful that they had no need to come to terms with those civilizations anyway. Thus the Arabs could avoid being culturally absorbed by the peoples they had conquered.

Second, on the side of the conquered peoples there prevailed in Egypt, Syria and Iraq a unique type of provincial culture. Having lost their own civilizations some 1,400 years before when they first came under the rule of alien empires, these provinces had not yet been fully assimilated by Byzantium and Iran. As a result, they were less committed to these civilizations and the

Arabs were exposed only to a culture filtered through a provincial milieu and language. These provincial cultures, pale versions of their imperial civilizations, could thus be reshaped by the Arabs.

The strength of the Arab position as against the weakness of provincial culture is the keynote of early Islamic history politically, culturally and ethnically.

The political influence

Politically, the strength of the Arab position determined the evolution of the Arab conquest society. With the Umayyad dynasty [4] (661–750) the capital was moved from Medina to Syria where a tribal confederacy formed the basis of the caliphs' power; in the provinces the tribal armies were placed under tribal leaders in a system of indirect rule. Within some 40 years kinship ties had been eroded and the tribal armies gave way to professional soldiers and civilians.

Normally, the loss of the tribal organization means that the conquerors must borrow the political organization of their subjects or suffer political disintegration; either way the

conquered civilizations eventually win. But among the Arabs the sanctity of the tribal past meant that neither of these eventualities came to pass and the obsolete organization was retained until the third civil war (744), which was followed by the Abbasid revolution. It was thus that the Abbasids (750–1258) [3] had to govern an Islamic empire as opposed to an Arab conquest society without losing the link with the tribal past, a problem that eventually proved insoluble. Although the Arabs attempted in various ways to foster an imperial ideal and aristocracy within Islam on the Iranian model, the fourth civil war (811–13) meant the failure of such attempts, the adoption of slave armies [5] as the instrument of government and, soon afterwards, the dissolution of the unitary state. The delay, however, meant that meanwhile Islamic civilization had developed sufficiently to survive.

Islamic culture and learning

Culturally, Arab strength accounts for the character of Islamic learning. The core of Islam is a revealed law, actually created in

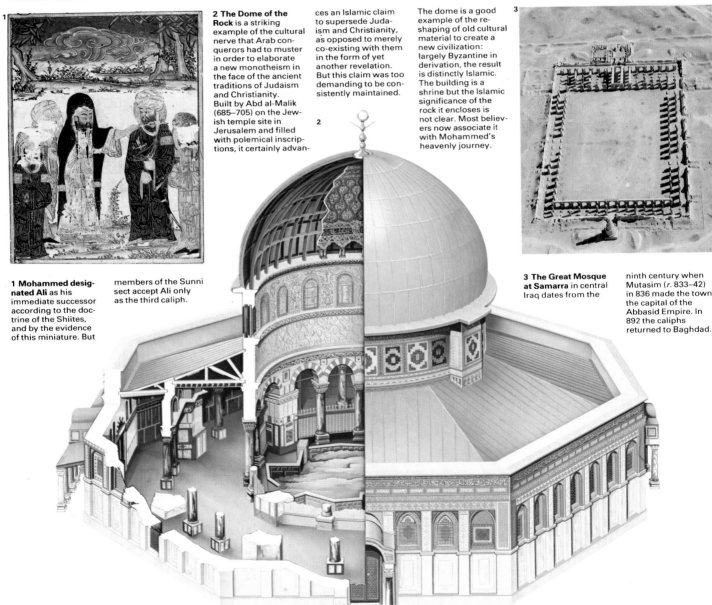

2 The Dome of the Rock is a striking example of the cultural nerve that Arab conquerors had to muster in order to elaborate a new monotheism in the face of the ancient traditions of Judaism and Christianity. Built by Abd al-Malik (685–705) on the Jewish temple site in Jerusalem and filled with polemical inscriptions, it certainly advan-ces an Islamic claim to supersede Judaism and Christianity, as opposed to merely co-existing with them in the form of yet another revelation. But this claim was too demanding to be consistently maintained.

The dome is a good example of the re-shaping of old cultural material to create a new civilization: largely Byzantine in derivation, the result is distinctly Islamic. The building is a shrine but the Islamic significance of the rock it encloses is not clear. Most believers now associate it with Mohammed's heavenly journey.

1 Mohammed designated Ali as his immediate successor according to the doctrine of the Shiites, and by the evidence of this miniature. But members of the Sunni sect accept Ali only as the third caliph.

3 The Great Mosque at Samarra in central Iraq dates from the ninth century when Mutasim (r. 833–42) in 836 made the town the capital of the Abbasid Empire. In 892 the caliphs returned to Baghdad.

Iraq in the eighth and ninth centuries from a variety of foreign materials, but in theory based exclusively on the Koran and the Prophet's works. It was the learned laity studying Islamic law who came to be seen as the heirs to Mohammed's preaching.

It follows that it was not difficult to create an Islamic scholarship, overwhelmingly Arab in orientation; but it was not so easy to create an Islamic philosophy and science. The Arabs did inherit Greek philosophy from the conquered provinces, but being neither Arab nor Islamic, such teaching was regarded as ungodly wisdom. Although philosophy continued to be cultivated, it was gradually relegated to marginal and heretical circles.

Ethnic development

Ethnically, the weakness of provincial culture explains the overwhelming Arab influence on the Middle East. Islam began as a religion for Arabs, but could not remain so when the non-Arabs began to convert to it. From being an ethnic faith on the Judaic model it had to become a universal belief on the Christian model, but the transition was

never quite completed. Mohammed was an Arab who, unlike Jesus, had never ceased to be honoured in his own community and in his name the Arabs had conquered a kingdom that was very much of this world [Key]. The notion that a Muslim was in some sense an Arab, and Islamic civilization in some sense Arabian, therefore proved extremely tenacious. In the ninth and tenth centuries non-Arab converts, especially Iranians, attempted to disentangle Islam from its Arab origins, insisting that Islam was a faith that could be combined with any identity and culture; but the success of these so-called *shuubis* was limited. The three provinces – Egypt, Syria and Iraq – all became Arab countries, while Iran, not a province but an empire, retained its Iranian identity but largely lost its Iranian civilization. Only where Islam has spread peacefully, as in parts of black Africa and Java, has it proved flexible enough to combine with a local culture. In the Middle East the ancient intransigence of Islam *vis-à-vis* such cultures has a modern sequel in the preference for Arab as opposed to Egyptian, Syrian or Iraqi nationalism.

Camel-breeding Bedouin tribesmen roamed over most of Arabia before the days of oil. South Arabia had sufficient internal resources to maintain stable state structures, while in the north external resources were often available in the form of commercial revenues or imperial subsidies. But most of the peninsula was too poor to support a non-tribal organization. Here Mohammed was the first, but not the last, to create a state in the name of a religious doctrine. By uniting the tribesmen as believers and calling on them to wage war against the unbelievers, he provided a rationale for a conquest of the fertile lands that culminated in the formation of a vast empire.

4 The unification of Arabia is traditionally credited to Mohammed and was accomplished before his death in 632. Egypt and Syria were taken from the Byzantines and Iraq, and the Iranian plateau from the Sasanids be-

tween 632 and 656, when civil war broke out. The first Umayyad caliph, Muawiya (r. 661–80), resumed Arab expansion in North Africa and eastern Iran, while a second thrust under Walid I (r. 705–15)

pushed the Arabs into Spain and India; but a last attempt to conquer Constantinople in 715–16 failed. At the end of the Umayyad period (750) the limits of Arab expansion were the Pyrenees, the Sahara, the Cau-

casus and Turkestan. Further expansion, whether Arab or Islamic, became the work of local dynasties, militant fraternities and missionaries and merchants, and such expansion continued far beyond 950.

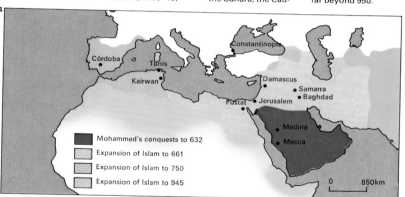

Mohammed's conquests to 632	
Expansion of Islam to 661	
Expansion of Islam to 750	
Expansion of Islam to 945	

0   850km

6 A hoard of Arab coins from a tenth-century Viking grave in Sweden represents payment for the slaves, fur and honey that were exported to the Muslims by the Vikings who colonized Russia.

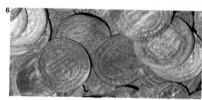

7 Harun ar-Rashid is one of the few caliphs to have fired Western imagination, thanks to *The Arabian Nights' Entertainments*. His reign (786-809) fell in the period following the transfer of the capital from Syria to the culturally richer Iraq, but politically his days were troubled. Faced with revolts and sectarian discontent, he divided his lands between his sons; this provoked a fourth civil war which, disastrously for royal power, was won not by the caliphate but by a provincial army.

8 Kairwan, in Tunisia, was founded as a garrison city for the settlement of Arab soldiers. It soon became a centre of learning and orthodoxy in North Africa, which was at that time inhabited overwhelmingly by Berbers. On conversion, the Berbers repeatedly made use of the programme of tribal state formation and conquest enshrined in the Prophet's career, in the name of a doctrine that was sometimes reformist, sometimes heretical. Although North Africa is extensively Arabized today, it still has a substantial Berber population. But Arab or Berber, Islam among North African tribesmen is highly distinctive, centring on holy men often identified as descendants of the Prophet.

5 A mounted archer (detail from a Palmyra fresco) typifies the kind of Turkish slave soldier from Transoxania who swelled the ranks of the Islamic armies. Transoxania, which was once the eastern frontier of Iran, is now wholly Turkish. Divided into well-entrenched principalities, it was a difficult place for the Arabs to conquer. As a result, Iranian culture survived and it was in this marginal province that the revival of Iranian literature in Islam took place from the tenth century onwards. Transoxania was exposed to Turkish tribes and after several invasions was overrun, eventually to become Turkish.

Islamic art

Architecture and its decoration best illustrates what is known as Islamic art. But the glamour of the great religious buildings has also tended to obscure the importance of Islamic secular art, which expressed itself most readily in painting, pottery, metalwork and textiles – hitherto the "minor arts"; under Islam they achieved major status.

The extent of Islamic art

Geographically Islamic art extends from Spain in the west to Indonesia in the east. It flourished from the seventh to the seventeenth centuries and thereafter gradually declined. Despite this wide range in space and time it has an instantly recognizable *cachet*. This is most apparent in Islamic decoration, which displays unequalled resource and virtuosity. The ban on the representation of living figures in a religious context – a ban that did not apply in secular art – directed the imaginative energy of the artist towards geometric, floral and epigraphic ornament such as the ubiquitous and aptly named arabesque (winding stems and leaves) and numerous varieties of Arabic

script. In religious buildings Koranic inscriptions predominated and calligraphy was given special reverence [Key].

In architecture the characteristic Islamic building is the mosque [1]. In its primitive form it derived from Mohammed's house in Medina, which had an open enclosed courtyard with a roofed area facing Mecca. Many mosques had a minaret – a tall tower for the call to prayer. Other characteristic Islamic buildings include the caravanserai for accommodating travellers, the *madrasa* or theological college and the *ribat*, a kind of fortified monastery. Palaces and tombs [5] abound. These various building types differ considerably from one country to the next, but they all have a number of domes; lengthy arcades with pointed or horseshoe arches; complex vaulting; open courtyards as an integral part of the design; and large expanses of flat wall surfaces decorated with carved stucco, polychrome tilework or inlaid marble.

Early Islamic art was produced mainly in Syria under the Umayyad dynasty (661–750). Umayyad religious buildings clothed Byzantine forms in glittering

mosaics, while palaces used chiefly Persian motifs within a basically Roman structure. Such eclecticism remained one of the constant features of Islamic art.

Stylistic variations in Islamic art

By the ninth century a classical Islamic art had returned under the Abbasid caliphate in Iraq and spread widely throughout Islam [7]. Its prime feature was carved or moulded stucco using geometric and rigorously stylized vegetal motifs, endlessly repeated.

As the power of the Abbasid caliphs waned, new political groupings generated five distinctive regional styles: Hispano-Moorish, Syro–Egyptian, Turkish, Persian and Indo-Muslim. In Spain the establishment of an anti-Abbasid Umayyad caliphate gave Syrian art a new lease of life, inspiring a distinctive style that spread through northwest Africa and was marked by an extreme, mannered delicacy of ornament applied to basically simple structures [8].

The Syro-Egyptian tradition produced a notable variety of religious buildings with carved stone exteriors and rather cramped

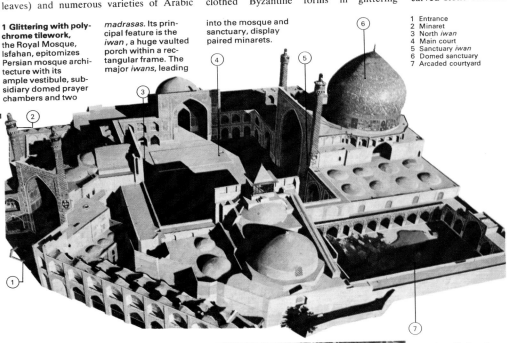

1 Glittering with polychrome tilework, the Royal Mosque, Isfahan, epitomizes Persian mosque architecture with its ample vestibule, subsidiary domed prayer chambers and two *madrasas*. Its principal feature is the *iwan*, a huge vaulted porch within a rectangular frame. The major *iwans*, leading into the mosque and sanctuary, display paired minarets.

1 Entrance
2 Minaret
3 North *iwan*
4 Main court
5 Sanctuary *iwan*
6 Domed sanctuary
7 Arcaded courtyard

2 The Ardabil Carpet (1539–40), signed by Maqsud of Kashan, is 61.85m² (637 sq ft), giving ample room for its complex design.

3 Painted in Iran in 1296, this bestiary page reflects the rule of the Mongols. Through their pan-Asiatic empire, numerous Chinese motifs were introduced into Persian painting – in this case the phoenix and the conventions used for feathers, plants and border.

4 This ceramic bowl with lustre-painted decoration (*c.* 1300) depicts a Coptic priest swinging a censer. The design was scratched with a stick and is essentially a pattern of rhythmical reciprocating curves. The double-fired technique and decoration reflects Mesopotamian influence.

interiors. Turkish architects, at first greatly influenced by the neighbouring Arab and Persian styles, later responded to the challenge of Hagia Sophia in Constantinople, the greatest of Byzantine churches, by perfecting the Ottoman type of mosque. This has a great central dome, visually (but not structurally) shored up by tiers of half-domes, and slender, pencil-shaped minarets at the corners.

Pottery, metalwork and painting

Among the minor arts, which occasionally borrowed from China and also from Europe, Islamic pottery displayed technical virtuosity of a high order, especially in lustre ware. The accent was always on colour and decoration (which could be figural, epigraphic, geometric or vegetal) rather than on shape or body [4]. The major centres of these arts were successively Mesopotamia, Egypt, Persia and Turkey.

Islamic metalwork, mainly in bronze, was technically extremely diverse, used a variety of shapes and favoured scenes of courtly life framed by bands of stately inscriptions. Its heyday was in Persia and the Arab Near East

from the twelfth to the fourteenth centuries [9]. Metalwork was usually cast or engraved and niello, a black metallic composition, was used to fill engraved lines.

Islamic textiles fall naturally into two groups. Silks were used mainly for ceremonial and funerary purposes; most date from the ninth to the twelfth centuries and bear heraldic beasts and inscriptions. For carpets [2] the golden age flowered in sixteenth- and seventeenth-century Persia. Hunting scenes and floral motifs of a complex symbolism are commonplace.

Islam has never had a tradition of easel painting, and frescoes and mosaics are also rare, but book painting has always been popular [6]. Arab painting delights in animal fables, scientific treatises and genre scenes of an unexpectedly humorous quality. The mature Persian tradition [3], which influenced both India and Turkey, favoured scenes from narrative poetry. Indian painting modified this by a more naturalistic approach (a rarity in Islamic art), as its portraits and crowd scenes show, while Turkish artists stressed historical subjects.

KEY

These drawings of angular inscriptions dating between 790 and 1543 are part of the rich repertory available to the Islamic designer. Often several scripts were used in a monument for added contrast. Extraneous ornament makes some virtually illegible, thus emphasizing their decorative function.

5 At Gunbad-i Qabus this tomb of a minor prince, whose sarcophagus was apparently suspended from the roof, is the first and greatest of a series of tomb towers built for rulers or saints and found throughout northern Iran and Anatolia from the 11th to the 15th centuries. Their form probably originated in central Asia and was later modified by the influence of nomad tents and Caucasian churches.

6 This late 15th-century portrait of a painter in Turkish costume was painted by Kamal Al-Din Bihzad (c. 1460– c. 1533). His use of colour and compositional sense profoundly influenced Islamic artists.

7 The Mosque of Ibn Tulun, Cairo (876–9) with its huge scale, enclosed and empty precinct, crenalated walls, spiral minaret, brick piers and abstract stucco ornament, carried into Egypt the style of Abbasid Iraq. The arcaded courtyard with a deep sanctuary is a constant of early Arab mosques. The mosque's uncluttered spaces are in deliberate contrast to the busy city outside.

8 The frequent extensions to the interior of the Great Mosque of Córdoba, Spain (begun 785–6), finally resulted in a sanctuary disproportionately deep in relation to the courtyard. Endless diminishing vistas of horseshoe arches open on every side. A pitched roof covered each arched bay and to gain the requisite height the architect either placed columns one above the other or built special piers over them. This was a classical device.

9 Inlaid with gold and silver, this bronze basin, the "Baptistère de Saint Louis" (Egypt or Syria c. 1300), celebrates the technical virtuosity and iconographical resources of early Mameluke metalworkers. Externally narrow animal friezes frame a broad central band with monumental scenes of courtly life; insignia of rank and heraldic emblems identify the main notables.

151

Charlemagne and the Carolingian Empire

The Carolingians, a dynasty named after Charles Martel ("the Hammer") and his grandson Charlemagne, became the leading aristocratic family among the Franks in the seventh century. The family's power and prestige were greatly increased during the rule of Charles Martel (c. 688–741) who united the Frankish kingdom, halted the advance of the Arabs at the Battle of Poitiers in 732 [1] and began a political relationship with the Papacy that led to the foundation of the Holy Roman Empire. Charles Martel's son Pepin (c. 715–68), the father of Charlemagne, ruled from 747 and was anointed king with papal approval in 754. In 753 Pope Stephen sought the aid of the Franks against the Lombards who, having taken Italy, were threatening Rome. The king of the Franks became a regular ally of the pope, and the Carolingian house was invited to divide Italy with the Papacy.

The Frankish dynasty

The Franks traditionally considered their kings as being of divine origin and saw the tribe as the possession of its royal family. The Frankish state as such only existed under strong kings such as Clovis (465–511) and Pepin who eliminated their rivals. On Pepin's death his sons Charlemagne (742–814) and Carloman succeeded. When Carloman died in 771 Charlemagne seized full control [Key]. In 773 he answered the Papacy's call and defeated the Lombards. From 774 he ruled Italy by conquest and swore an oath of assistance with the pope.

The reign of Charlemagne

The Papacy was now as frightened of the Franks as it had been of the Lombards and sought a way of restraining Charlemagne. The solution came in 800 when Leo III crowned Charlemagne [4], initiating the Holy Roman Empire and giving rise to the claim that the emperor held his power from God bestowed upon him by the pope.

Charlemagne was a warrior king. Wars were fought against the Lombards in Italy and against the Agilolfing dukes of Bavaria; in addition there was a constant succession of campaigns against the barbarians on the borders of the kingdom: Arabs, Avars, Slavs, Saxons and Danes. War was carried out not only for political reasons but also for plunder. Charlemagne was often poor in the early years of his reign and a Frankish king's power rested on his ability to reward his followers. Booty taken in war was his single most important revenue.

Charlemagne fought the Arabs in Spain in 778, a campaign that ended in the ignominious defeat of Count Roland at Roncesvalles at the hands of the Basques [7]. In the south he defeated the Lombards and overthrew Tassilo of Bavaria in 788, making Bavaria for the first time an integral part of the Frankish Empire [8]. From 772 to 804 he waged a bloody and almost continual war against the Saxons, led until 785 by Widukind, and he proceeded to conquer Bohemia in 805–6.

Charlemagne's military prowess recreated a centralized European government that needed an administration more complex than any known to the Franks. Charlemagne created a new court at Aachen [3], with a palace and cathedral built on an imperial model. He employed a circle of

1 **A battle of crucial importance** for the future of Europe was fought at Poitiers in 732 when Charles Martel and the Franks finally put a stop to the advance of the Arabs, who threatened to destroy the Christian West completely. The Franks had already been successful in defeating the German tribes east of the Rhine. Frankish expansion under Charlemagne was therefore based on 300 years of Frankish strength.

2 **Charlemagne's military leadership** was the basis of his power. Frankish custom, based on a tribal levy, made every freeman liable for military service and for equipping and feeding himself at war. Later, middle-class freemen gave up their lands to local lords and fought in the lords' retinues, saving the small man expense but destroying the unity of the Frankish army on which Charlemagne had built his power.

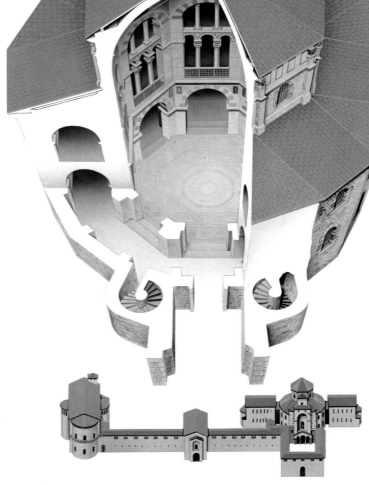

3 **Charlemagne's Palace Chapel** at Aachen was the architectural masterpiece of Carolingian Europe and a symbol of imperial power. It was based on the design of St Vitale in Ravenna, capital of the empire in Italy after the fall of Rome, and was designed as a chapel to house Charlemagne's throne.

scholars, including Alcuin from Northumbria and Theodulf from Spain, to educate a new literate class of administrator, to produce a new and legible script, to reform the practice and liturgy of the Church and to produce a theory of empire to accompany the reality of imperial power. Alcuin, above all, formulated the role and responsibilities of the Christian emperor, thus justifying the imperial side of the relationship with the Papacy.

Organization of Frankish society

Charlemagne ruled his lands through local counts, of whom there were more than 200. Many were of royal blood, and their appointment by Charlemagne created the beginnings of an international aristocracy that long outlived the Carolingian Empire. He employed stewards to carry out the business of government and special travelling agents, the *missi dominici*, to keep the counts in line with imperial policy and to raise troops when necessary. The royal will was expressed in a series of imperial charters that were lucid and authoritative. The Church played a central role in administration both through the ser-

vices of educated bishops and clerics and the unification of doctrine and practice.

Frankish society had earlier been based solely on personal loyalty to the king. Gradually Charlemagne insisted on a new oath of fidelity, initially only in times of crisis. These oaths were the beginning of a feudal monarchy based on the sworn allegiance of a landed nobility, a completely different concept from the personal loyalty of a tribe to its chief, for example.

Charlemagne intended to leave his empire divided between his sons, but Louis' reign, which began in 814, was a chapter of bitter family rivalry and the Church was increasing its control over secular affairs. A new wave of external attack from the Arabs, Bretons, Vikings and Normans further weakened the empire. Louis died in 840 and after three more years of feuding the Treaty of Verdun divided the empire in three.

Carolingian rule had disintegrated by the end of the ninth century but Europe had been given an imperial ideal, an international landed aristocracy and a series of significant social bonds that were soon to be revitalized.

Charlemagne was the most powerful ruler in early medieval Europe. Standing 193cm (6ft 3.5in) tall with broad shoulders, he was physically impressive. His character was enigmatic and his personal religion erratic, although he oversaw the consolidation of Christianity throughout his realm. He was politically ambitious, appointed able ministers and understood the importance of education. He unified Western Europe and recreated an equivalent of the old Roman Empire. But he regarded his lands as private property and willed them to his sons. It is hard to know whether he saw himself as an international leader or simply as an unusually successful Frankish tribal chief.

4 The coronation of Charlemagne by Pope Leo III in St Peter's on Christmas Day 800 was depicted in a 15th-century miniature by Jean Fouquet. Charlemagne, who had just restored Rome to the pope after a revolt, needed a sacred seal on his *de facto* position as emperor. The coronation made him legally heir to the Western Roman emperors. It was more the culmination of Carolingian expansion than an expression of a papal claim to select temporal rule.

5 The Lothar Crystal, a solid piece of rock crystal delicately engraved with biblical scenes, made in the 9th century, was owned by Lothair II of Lorraine (r. 855–69). The quality of the carving demonstrates the continuing artistic achievements of the Carolingian Empire despite the political decline of the 9th century. The classical motifs show how Rome was being used as an example by the Franks.

6 This page from the Sacramentary of Charles the Bald (823–77) shows the coronation of a Frankish prince – possibly Charles himself. He is flanked by two clerics and appears to be being crowned by God in person, handing a crown down from heaven; the Church's role in supporting the throne and the royal sense of divine mission are thus illustrated together. The Sacramentary was written and illustrated in 869–70 and saw the height of achievement of the last great school of Carolingian illumination which had developed around the Court School at Aachen and spread to Reims and Tours before it reached St Denis.

7 The death of Roland at the Battle of Roncesvalles in 778 gave rise to one of the great epic poems, the *Song of Roland*. It epitomized Charlemagne's knights as chivalrous defenders of Christianity against the Saracens. In literature, as in politics, the Carolingians thus laid a foundation of medieval lore. On his immediate retreat across the Pyrenees to put down a Saxon rising in the north, the rearguard of his army was in fact annihilated by a Basque force at Roncesvalles.

8 Carolingian Empire 814
States tributary to Charlemagne
Byzantine Empire

Partition of Carolingian Empire at Treaty of Verdun 843:
(1) To Charles
(2) To Lothair
(3) To Louis

8 The size of Charlemagne's empire on his death in 814 makes a sharp contrast with the partitions ratified at Verdun in 843 after 30 years of squabbling among his successors. The poet Theodulf wrote: "The wall, so firm and artistically decorated in the days of my youth is showing cracks.... Everything sweet has fled from the ageing world and nothing is left of its former strength."

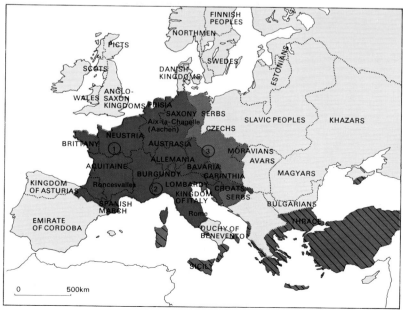

The Anglo-Saxon monarchy

Between the seventh and eleventh centuries, Anglo-Saxon England became a strong and stable nation, organized around a solid and powerful monarchy. This institution developed from that of the warrior chieftain, and throughout the period it continued to rely on the personal relationship of a king with his thegns [1], to whom he gave lands in exchange for companionship in peace and war, and who had a role, uncertain and fluctuating according to the strength of the previous king, in electing and advising a new monarch. Gradually the king acquired institutional authority through the patronage of the Church and the organization of the currency [5] and, despite the informal nature of much of his authority, he was the hub of the stability and prosperity of the kingdom.

The growth of a unified monarchy

In the eighth century kings of Mercia, in the Midlands, dominated England. The most famous of these was Offa (reigned 757–96) [2], the first king to have personal authority over most of southern England. He wrote a code of laws, set up a stable currency and was a figure of international standing, dealing with Charlemagne on equal terms.

During the early ninth century power passed from Mercia to Wessex, but the first Viking raids [6], in 793 on the east coast, threatened the stability of England. Sporadic Viking attacks continued until 865, when the "Great Army" chose not to return with its loot to Scandinavia for the winter but instead divided up the land of East Anglia, the East Midlands and Yorkshire, and began to farm it. The English rallied under Alfred (reigned 871–99), king of Wessex, who defeated the Danish commander Guthrum [9] decisively at Edington in 878, and made a treaty confining the Danes to the region east of Watling Street, known as Danelaw [7]. Alfred's son, Edward the Elder (reigned 899–925), reconquered this district by 920, but the substantial numbers of Scandinavian settlers left a permanent mark on social institutions and placenames, notably in the suffix "-by", found, for instance, in "Derby". Alfred's daughter, Ethelfleda (died 918) Lady of the Mercians, led campaigns against a new enemy, Norwegian pirates, who attacked the northwest coast, settled in Lancashire, and set up a kingdom in York in 919.

Alfred had organized his defence around burhs (forts), a practice continued in the tenth century. Some of the sites chosen were already towns. The burhs (later called boroughs) became the centres of trade and administration of taxation, and the site of mints [5], in the late Anglo-Saxon period.

The revival of the monasteries

The Viking attacks had been particularly severe on the wealthy and unprotected monasteries, and had brought the Church to a low ebb by the beginning of the tenth century. A similar decline on the Continent was combated by the reform movement started at the monastery of Cluny in 910. Its ideas were brought to England, about 940, by some English monks who had spent time in the reformed monasteries. Notable among the monks were St Dunstan (c. 925–88) [11], who revitalized the monastery of Glastonbury, St Ethelwold (c. 908–84), Bishop of Winchester and St Oswald (died 992), Bishop of Worcester. Edgar (944–75), who

1 The king and his witan (council) were the ultimate source of justice. The witan comprised the leading churchmen and thegns, or nobles, who were bound to the king by ties of personal loyalty. These formal assemblies developed out of the king's council of the period of Mercian supremacy. They brought problems to the attention of the king and assented to his most important ecclesiastical appointments and land-grants.

2 A standardized coinage, based on the silver penny, was introduced by Offa of Mercia, whose name is shown on this coin. It was used for most transactions and was accepted on the Continent. Other coins of Offa copy Arabic models. Offa effectively ended the old heptarchy, or division of England into seven separate kingdoms, even though his immediate successors were unable to maintain the unity that he had created.

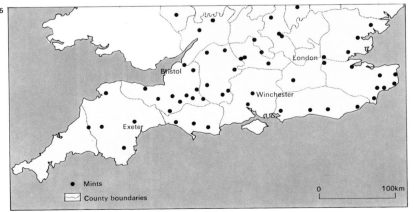

3 Offa's Dyke, running along much of the frontier between Mercia and Wales, testifies to the power of the 8th-century Mercian kings. It was intended more to mark the border than for defence.

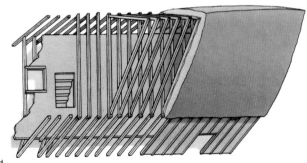

4 The Royal Hall of Cheddar is typical of the halls that were the centre of royal power and were the first permanent royal residences. The witan enjoyed the king's hospitality and advised him in such halls.

- Mints
- County boundaries

0   100km

5 Mints were set up in many towns in the later Anglo-Saxon period. As many as 44 places issued coins during the ten-month reign of Harold in 1066 alone. The king kept strict control over the currency, maintaining the purity of the metal and regularly recalling the entire coinage for reissue, at one time every six years. Although the name of the mint appeared on the coins, the dies were centrally issued. Penalties for counterfeiting included mutilation and death.

6 The Vikings raided Lindisfarne monastery in 793, and raids on the east coast for loot continued for 50 years. Most of the surviving accounts of the raids were written by monks, so the Vikings have acquired an image of savagery that ignores the constructive side of their settlement.

became king in 959, was an ardent supporter of the monastic revival and was the first king to be crowned with a ceremony that underlined the religious associations of the monarchy [8]. The service, written by St Dunstan, remains the basis for coronation services. The alliance of monasticism with the king gave the ruler a new agency for asserting his authority in the country.

Edgar's son Ethelred II (reigned 978–1016), was again faced with attacks by aggressive Scandinavians. Large taxes, known as Danegeld, were collected to buy off these warbands, so that most of the late Saxon coins still in existence are in the collections of Scandinavian museums. The Danegeld did not prevent the eventual victory of the Danes under Sweyn Forkbeard (died 1014), whose son Canute (994–1035) [10] became king of England in 1016, and of his home country Denmark in 1018.

The nature of Anglo-Saxon kingship

The monarchy changed considerably in the course of the period 500–1050. There were, however, some constant themes. The importance of royal blood is emphasized in the early period by genealogies, whose chief function was to demonstrate the ancestry of the royal houses, and to relate them through mythical rulers to pagan gods and, in Christian times, through biblical characters to Adam and Eve. Later, with the growing regularization of rules of descent, the question of the right of the king to the throne took on a religious aspect.

The king did not rule absolutely, but was advised by his council of thegns, the witan [1] and observed accepted codes of law and custom. These might be issued by each king in accordance with established usage. There was a complex system of taxation; coinage was maintained at a high standard of purity and constantly changed. Local justice was organized through the courts of the shire and hundred, and the defence of each county by earls, appointed by the king. These institutions were strengthened, rather than weakened, by the desire to oppose the invasions of the Scandinavians. Norman administration was grafted onto an already existing complex and efficient system.

The *Anglo-Saxon Chronicle,* shown here in the version copied at Peterborough in 1122, is a vital source for knowledge of political history before the Norman Conquest. A collection of genealogies and chronicles prepared by monks, it appeared in a number of different versions. As a result, it is possible to see a regional view of events, as in the case of the crisis of 1051–2, when the authority of Edward the Confessor (r. 1042–66) was challenged by the sons of Earl Godwin (d. 1053). The *Chronicle* is said to have been founded by King Alfred as part of his encouragement of English culture.

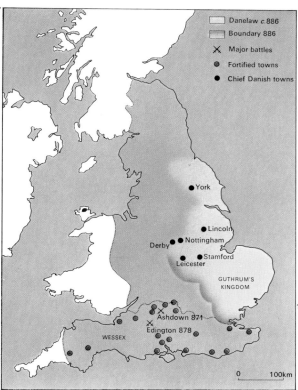

7 The Danelaw was recognized by King Alfred in the treaty of 886. In the east of the country the Danish traditions of independence and local autonomy continued well into the medieval period. A Viking kingdom was soon established in Lancashire which prevented the Danish expansion to the north.

Danelaw c.886
Boundary 886
X Major battles
● Fortified towns
● Chief Danish towns

York
Lincoln
Derby Nottingham
Stamford
Leicester
GUTHRUM'S KINGDOM
Ashdown 871
Edington 878
WESSEX
0 100km

8 Edgar the Peaceable was a powerful and efficient king who is said to have shown his political authority over the whole of England by being rowed on the River Dee by six sub-kings. He encouraged the monastic revival and this manuscript records his coronation, which took place in 973, two years before he died.

9 Guthrum (d. 890) is said to have been the first Scandinavian to settle in England. In the treaty of 878 with Alfred he agreed to convert the Danes to Christianity and took the name Athelstan after his own baptism.

10 Canute, or Cnut, set up an empire that took in much of the coastlands of the North Sea. As king of England he encouraged the established institutions (here he grants a charter to the New Minster at Winchester with his wife).

11 St Dunstan, shown here prostrate at the feet of Christ, was Archbishop of Canterbury from 959 to 988. He promoted the growth of the new monasticism. He founded abbeys at Peterborough and Ely.

The Vikings

In the late eighth century the pagan Scandinavians, known as the Vikings, burst upon the rich kingdoms of Western Europe. Improvements in ship design, and overpopulation at home were two reasons behind Viking expansions at this time. They attacked rich and undefended monasteries, pillaging their treasures and killing and enslaving the priests. The attack in 793 on St Cuthbert's monastery at Lindisfarne in Northumberland [2] was the most shocking of these occurrences because it was the first.

Colonization of Britain

Some of those who came from Scandinavia turned their attention to colonization. Graves in western Scotland, the Hebrides, Shetland, Orkney and Isle of Man, furnished after the pagan fashions of their Scandinavian homelands with weapons and household goods, tell of the gradual settlement of these areas [1]. In England, raids gave way to settlement in the 860s and, after a series of campaigns against the English kingdoms, the Scandinavians – by treaty with Alfred the Great in 878 – settled as conquerors to the

east and north of a line from the River Lea to the Dee. Here they established at least one kingdom (based on York) as well as other political groupings in East Anglia and the Midlands (where their power was based on the "Five Boroughs" – Leicester, Stamford, Lincoln, Derby and Nottingham). By the middle of the tenth century the English had reconquered these areas, although in the early eleventh century the whole of England was again conquered by the Scandinavians, and a Dane, Canute (994–1035), became king of England for a short period.

The Scandinavian settlements in Scotland and the Isle of Man [8] continued until the thirteenth century (in Orkney and Shetland islands until the fifteenth century). The Scandinavians in Ireland did not attempt to conquer the whole country, but founded a series of towns (of which Dublin was the most important) through which they could influence the trade of the Western European seaboard. Goods from France and Spain were exchanged for slaves, furs, ivory and other products of the north.

In other Western European countries the

Scandinavians were less successful as colonists or conquerors. Only in Normandy did they succeed. There in 911 a Scandinavian, Rollo, was given the right to settle and govern much of that corner of France.

The Scandinavians also moved into the North Atlantic looking for plunder and farmland. They settled the virtually unknown Faroe Islands and Iceland and reached Greenland, where they founded two major settlements. They also appear to have made landings on the coast of North America.

Eastern Europe

The Scandinavians were also influential in Eastern Europe. From the Roman period onwards they traded spasmodically with the eastern Mediteranean, travelling by way of the Polish and Russian rivers. In the eighth century a larger and more organized commercial traffic developed. The Swedes founded trading stations and collected tribute from Finno-Ugric, Balt and Slav tribes in the east Baltic. In the ninth century the Scandinavians contributed to the growth of Russian trading towns at Staraja Ladoga,

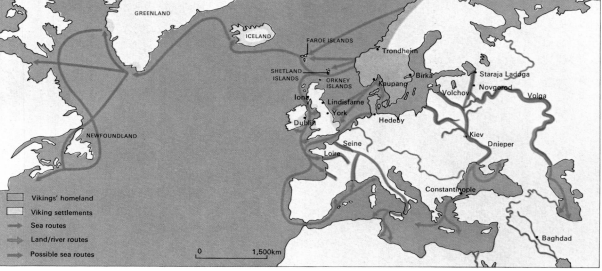

1 The Scandinavians of the Viking age used these main routes in their search for wealth and land to settle. In the east are the riverside towns they established as centres, such as Kiev and Novgorod. From these, along rivers, the Scandinavians reached the rich trade and wealth of the Orient and the Byzantine Empire. In the west the chief areas of settlement were the northern and western islands of Britain, northern and eastern England, northern France and the Irish ports (which they also founded). By the western sea routes they reached America, but failed to establish any settlements there.

Vikings' homeland
Viking settlements
Sea routes
Land/river routes
Possible sea routes

0 1,500km

2 These two views from a stone on Lindisfarne commemorate the Viking invasion. An inscription from it reads: "793. In this year dire portents appeared over Northumbria... the ravages of heathen men destroyed God's church on Lindisfarne".

3 This bronze figure of the Buddha was imported from Afghanistan or India to Sweden in the early Viking age, presumably as a souvenir. Other objects imported into Scandinavia and found during excavation include Arabic, Byzantine, English, French and German coins, as well as silks and foreign animal skins, metalwork from Britain and pottery and glass from northern Germany.

4 Thor, the Scandinavian god of thunder, agriculture and war, is usually symbolized by a hammer. In this 6.7cm (2.5in) bronze, his beard develops into a hammer. The symbol was often used as an amulet to protect the wearer against evil.

House walls of turf
Pit with fire-scarred stones
Edge of terrace
Hearth
Post holes

0 10
metres

5 This plan of a house excavated at L'Anse-aux-Meadows in Newfoundland can be dated to the 11th century. It has many of the structural features of houses built by the Vikings in their colonial settlements in Iceland and Greenland.

Novgorod and Kiev. The river route along the Volchov and the Dnieper (which led from Lake Ladoga to the Black Sea and Constantinople, now called Istanbul) was largely controlled by the Scandinavians from the ninth to the eleventh centuries.

Trade and economic organization

The main exports of the Scandinavians were furs, honey and slaves; the main imports silver, spices and other luxury goods. The volume of this eastern trade is demonstrated by the fact that (at a conservative estimate) some 40,000 Arabic coins have been found in Viking age sites in Gotland and Sweden.

These eastern adventures are also reflected in long inscriptions carved, in runic characters, on Swedish stones. Some refer to merchant expeditions and imply a fairly peaceful Scandinavian presence in Russia. However some tell of more warlike episodes – a military expedition to Arabia, for example, or of Scandinavian mercenaries in Russia or Byzantium: Greek literary sources attest to the presence of Scandinavians serving in the bodyguard of the emperor.

The raids, colonization, trade and other activities of the Vikings, which have so firmly established them in European history, were based on a settled domestic economy. This economy had its foundation in agriculture but was expanded in the Viking age by the development of marginal land in Scandinavia itself and by the foundation of towns which functioned as major trading stations – towns such as Birka in Sweden and Hedeby in South Jutland [7]. The international contacts of the Scandinavians are clearly seen in these towns, both in excavated material and in the written descriptions of travellers. The Scandinavian of the Viking age was, however, basically a farmer – even traders and pirates were apt to travel extensively abroad, mainly in summer.

The image of the Vikings as marauders and pirates is influenced by the chronicles of priests who identified them with pillaged monasteries. The fact that the Scandinavians were pagan compounded any offence. But in the tenth century the Scandinavians were gradually converted to Christianity and so their image improved.

KEY

This richly carved ship's prow is part of a complete Viking longship found in a woman's grave of the 9th century at Oseberg in Norway. The ship itself, built of oak, was 23m (75ft) long, and was basically a coastal sailing vessel which could also be propelled by oars. One of the larger examples of Viking longships, it was probably used for raiding and longer voyages of discovery and colonization. Much smaller ships were used for coastal and river warfare and trading. The prow here is decorated with intertwining animals, and is shaped in the form of a serpent, a style that is characteristic of the final pre-Christian period of Viking art.

6

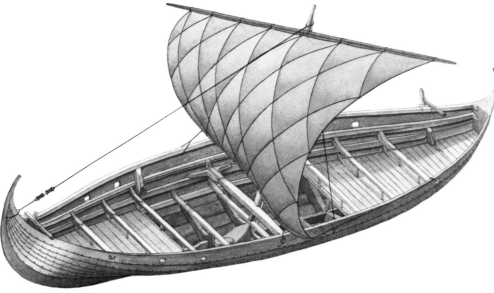

6 The hull of a small cargo vessel was raised from the bed of Roskilde fjord in Denmark and reconstructed. It is 13.5m (44ft) long. There is decking fore and aft with a hold in the centre. The mast is seated on the keel and the one large sail is supported by a single transverse spar. When not in use the spar could be lowered and stowed away. The vessel could also be rowed by means of sweeps inserted through the square holes that are clearly seen in the gunwale plank. Steering was by means of a paddle attached aft on the starboard quarter (starboard being an old Norse word for "steering side").

7

8

9

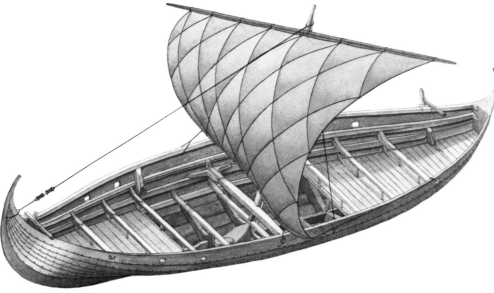

7 The open fire in the centre of this house, a reconstruction from the town of Hedeby in South Jutland, provided heat for the room and for cooking. The family would sit on the earth benches at the sides of the room and eat off low tables. The walls are of wattle and daub and may have been covered with hangings. There was a room at each end.

8 The annual ceremony at Tynwald on the Isle of Man comes from the assembly established there by the Scandinavians in the Viking age (the Norwegians finally surrendered Man in 1266). Such assemblies acted as a combination of town meeting, law court and fair, and were held in various places throughout the Viking world.

9 The craftsmanship of the Vikings is well demonstrated by this early 11th-century weather vane. Made of gilt bronze, it may originally have graced a Viking ship. The Scandinavians were fascinated by contorted animal ornament.

Throughout the Viking age, their craftsmen produced distinguished objects decorated with animal ornament largely free from the influence of contemporary European art. Objects like this vane demonstrate the brilliance of the ornamentation.

Western European economy 800–1000

Western Europe in the ninth and tenth centuries was the poor relation of both the Arab and the Byzantine worlds. It was poor in the exploitation of its natural resources, in technical ability, in political cohesion and cultural achievement. Europe's social and economic backwardness was further threatened by Magyar hordes from the east and incessant raids by Vikings who swept southwards from the Scandinavian countries.

The growth of self-sufficiency

The temporary political unity and economic regulations of the Carolingians (Charlemagne and his successors) throw little light on actual conditions of life in the ninth century. Charlemagne attempted to stabilize and centralize the coinage throughout his empire, centred on France, to avoid exploitation in times of scarcity by regulating prices, to facilitate trade by keeping down internal customs dues and to introduce a universal system of weights and measures. But the regularity with which his instructions were issued indicates their ineffectiveness; the actual unit of production and consumption remained the great estate [1], a legacy of the Roman villa.

The economy of the Roman world in the invasion period was based on the great estates of the old landowning aristocracy. The decline of easy communications and trade forced these social units to become more self-sufficient. Alongside the estates, some of which passed into the hands of the invaders, there remained a free peasantry that still existed in the ninth century. But its decline was hastened by the centralization policies of Charlemagne and the obligations imposed by the government on all free men. It became preferable to surrender individual land rights to the local lord in order to escape the fiscal liabilities of freedom.

The Church, trade and commerce

The decline in personal freedom benefited the great estates of the aristocracy and the Church. Most available records of ninth-century estates are monastic and show a pattern of organized growth indicating that the Church played a crucial role in maintaining economic stability in difficult times.

The growth of European towns was based both on the old Roman cities (especially in Italy and the south) and on the garrisons and royal residences of the Carolingians and the Merovingians, their predecessors, such as Aachen, Nijmegen, Worms, Frankfurt and Ratisbon. Other cities, and mercantile suburbs of older cities such as Paris, grew round the monasteries which were the major producers of surplus goods for sale.

The uncertainty of statistics of trade and commerce after the fall of the Western Empire has given rise to much theorizing by historians. It fostered especially the belief that the economy of the Western world survived the Germanic invasions, but was brought to a halt by the Islamic conquests of the seventh century. The loss of the Mediterranean, it was argued, destroyed the movement of merchandise both between East and West and also internally within the West, reducing trade to mere barter. In fact trade continued, although on a much reduced scale. The centres of commerce initially switched from Provence and Languedoc to the north of Europe – Frisia, the Low Coun-

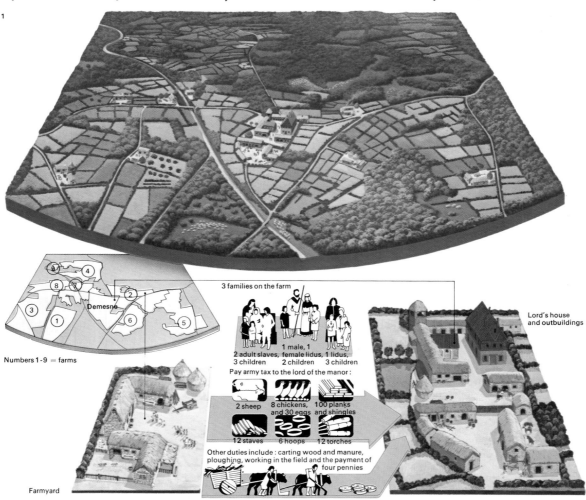

Numbers 1-9 = farms

Demesne

Farmyard

3 families on the farm

1 male, 1 female lidus, 3 children
2 adult slaves, 2 children
1 lidus, 3 children

Pay army tax to the lord of the manor:

2 sheep | 8 chickens, and 30 eggs | 100 planks and shingles
12 staves | 6 hoops | 12 torches

Other duties include: carting wood and manure, ploughing, working in the field and the payment of four pennies

Lord's house and outbuildings

1 A typical "villa" or rural social unit in Carolingian Europe is here illustrated, and its population and composition shown. This organization was the result of the fusion of the old Roman "villa" with the customs of the invading Germanic peoples. The status of the peasants varied: some were free, some owed various degrees and types of service to the lord. An estate was divided into the demesne land, the property of the landlord cultivated by forced labour, and the land cultivated by his tenants as "mansi". The return for their tenancy was calculated in various burdens of service. These could be increased if the tenant exchanged his military obligations for service to the lord. Carolingian peasants were divided into "coloni", free men still obliged to till their own land, "lidi", half-freemen owing certain legal obligations, and "servi" who had no legal rights at all. But although outside the law, "servi" did hold property. The total population of the estate was known as the "familia" and was supervised by the landlord or his agent. Slaves in principle worked the demesne land for three days a week, while "coloni" and "lidi" performed set tasks throughout the year. Estates such as the one illustrated were centralized units of production on land redeemed from the surrounding forests. They were also early centres of rudimentary industrial organization for weaving and cloth production. Finally, they were the basic unit of the rural feudal society that provided both the stability and the opportunity for oppression in early medieval society.

tries, northern France and the Rhineland. The old route to the north from Ostia (the port of Rome) to Provence and thence by land gave way to a land route over the Alps. Italian ports such as Amalfi [4] and Gaeta maintained a precarious commercial liaison with the East and in the tenth century the emergence of Venice [3] foreshadowed a new era in Mediterranean trade.

Charlemagne sought to create a stable coinage, but never succeeded in making the minting of money a royal monopoly which was one possible way of avoiding constant debasement. Charlemagne did understand the need for a coinage to be used as a medium of exchange, rather than barter, and his silver denarius, or penny, became one of the standard coins of the medieval West. But Carolingian coinage supplied only the small change of the West, whereas the international trading currency of the Middle Ages became the Byzantine gold nomisma and the Arab gold dinar [2].

The population of Western Europe expanded very slowly after the ravages of the invasion years. Stagnant economies could not support cities of 50,000 to 100,000 people such as those of the Arab world at Cairo, Antioch and in Spain at Córdoba. Under the manorial system, towns had lost their importance and many were abandoned.

Industry and agriculture

Industry, such as mining and weaving, was only in its infancy in the ninth century and was mainly centred on monastic properties. Comprehensive rotational systems of agriculture barely survived on the old Roman estates. Agricultural technology gradually improved with the introduction of heavy wheeled ploughs for northern soils. The harrow and the flail were probably introduced at that time and the decline of slavery made the extended use of watermills for grinding corn an economic necessity. Yet production was too low to support any great increase in population and trade was too ill-organized to offset local effects of crop failure and plague [Key]. The technological knowledge of the Romans was never wholly lost, however, and new inventions were gradually developed.

KEY

Early medieval life was primarily agricultural. Harvest time was the crucial part of the year. Cereals were the staple of life, meat being a luxury. Limited productivity and trade meant that a successful harvest was the only insurance against famine and its constant attendant, plague.

2 Medieval currency problems were two-fold: to preserve a coinage acceptable to all trading partners and also to enable transactions to be performed through a monetary medium rather than by barter. The Clovis II coin [A] shows the Merovingian attempt to maintain a prestige form of Roman coinage, but debased both in design and metallic content. Coins such as the gold solidus [B] of Louis the Pious (778–840) were minted in many towns, although under the control of the king. No Western economic system at this time could support such reliable coinages as the Byzantine nomisma [C] of Justinian II or the Arab gold dinar [D]. These coins were the basic tender of medieval international trade until the Florentine florin was minted in the 14th century.

3 Trade in the Mediterranean was sharply curtailed by the pirates who followed in the wake of Arab conquests. Yet gradually the need for raw materials from the West and for luxury goods and spices from the East ensured that increasing numbers of vessels, as this one portrayed in a Venetian mosaic, plied the Mediterranean.

4 Amalfi, with its natural harbour, was one of the key commercial links between the Western world and Byzantium before the rise of Venice. The revival of western Mediterranean trade came from Italian towns such as Amalfi, Naples, Salerno and Ravenna that had the closest links with the Byzantine world.

The Holy Roman Empire

The Holy Roman Empire sought to re-create a united Christian West such as had existed in theory during the last years of the Roman Empire. It was holy because it was based on the theory that the pope was supreme in ecclesiastical affairs, the emperor being the secular arm and the defender of the Church. It was Roman – despite the fact that the emperor was first a Frank and then a German – because Rome had for so long been the political centre of the world.

Power struggle

The theory that the pope represented the spiritual authority of God on earth and the emperor temporal authority was rarely a reality. In practice there were strong emperors and strong popes, and power fluctuated. From the mid-eighth century the Papacy had looked increasingly to the Franks for protection against the Lombards, a dependence which culminated in 800 when Leo III crowned Charlemagne as emperor in Rome. Although the term was not used until the twelfth century, the papal recognition of Charlemagne may be seen as mark-ing the beginning of the Holy Roman Empire. The tenth century was dominated by the political success of the Ottonian emperors (962–1024), a Saxon royal house [Key 1, 2, 3]. Otto the Great (912–73) was enthroned as king of the Germans at Aachen in 936 and crowned as roman Emperor in 962 [4].

The Salian dynasty (1024–1125) saw the height of imperial power at a time when the Papacy was also powerful. The Salians were a Frankish family. Their empire under Conrad II (990–1039) included Germany, Burgundy and Italy, and Conrad saw himself as ruler of the city of Rome.

The Investiture Contest

But a new reforming spirit within the Papacy made a major issue over the problems of investiture: who should appoint bishops, the pope or the secular ruler? The quarrel was essentially about the great wealth and power of the benefices at stake. No ruler could afford to relinquish control over appointments to the wealthiest positions in the land.

The so-called Investiture Contest culmi-nated in a complete breakdown of imperial-papal relations which damaged the theoretical justification and the effective power of the Holy Roman Empire. The contest reached its height in the reign of Henry IV (1050–1106), who found that there was no possible compromise on this issue with Pope Gregory VII. Accordingly he called an imperial synod at Worms in 1076 and deposed Gregory. Gregory retaliated, deposed and excommunicated Henry, and began to build up a strong anti-imperial coalition. Henry, realizing the seriousness of his position, spent three days in a hair shirt at Canossa [6] awaiting papal forgiveness. Gregory then withdrew the excommunication but not the deposition. Henry had his revenge in 1084 when he marched on Rome and drove Gregory to a bitter death in exile. The Investiture Contest was finally solved by a compromise in which the emperor was allowed to invest a bishop with his sceptre before consecration by the ecclesiastical authorities.

The religious and political pretensions of the empire reached their height under the Hohenstaufens (1138–1254). Frederick I

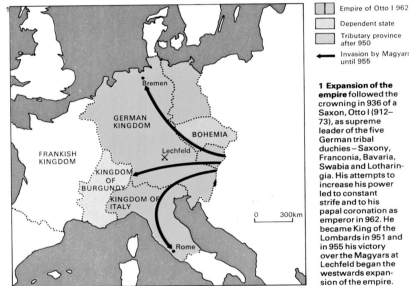

Empire of Otto I 962
Dependent state
Tributary province after 950
Invasion by Magyars until 955

1 Expansion of the empire followed the crowning in 936 of a Saxon, Otto I (912–73), as supreme leader of the five German tribal duchies – Saxony, Franconia, Bavaria, Swabia and Lotharingia. His attempts to increase his power led to constant strife and to his papal coronation as emperor in 962. He became King of the Lombards in 951 and in 955 his victory over the Magyars at Lechfeld began the westwards expansion of the empire.

2 St Michael's at Hildesheim in Saxony, destroyed in World War II, was built in the early 11th century under the supervision of Bishop Bernward. It was later decorated with fine bronze ornaments and a magnificent flat, painted ceiling entirely covered with a representation of the Tree of Jesse. The plan of the church referred back both to early Christian and to Carolingian architecture but its internal rhythms were made subtler by two transepts and an apse.

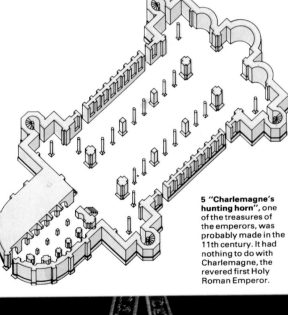

3 The Holy Lance, which pierced the side of Christ, was politically the most important relic. In legend given by St Helena to Constantine, it passed to Otto I, who bore it in battle.

4 The imperial crown of Otto I was made for his papal investiture in 962. With the Christian cross in front, it is octagonal to represent the heavenly Jerusalem, with a semicircular strip above to represent world dominion. Two panels back and front each contain 12 stones representing the tribes of Israel. On the sides are biblical kings. The crown symbolizes the functions of the wearer as temporal ruler and regent of Christ on earth.

5 "Charlemagne's hunting horn", one of the treasures of the emperors, was probably made in the 11th century. It had nothing to do with Charlemagne, the revered first Holy Roman Emperor.

"Barbarossa" (1123–90) and Henry VI (1165–97) saw themselves as the holy leaders of a God-given German Empire. Frederick I's power was based on an alliance with Henry the Lion of Saxony against the Normans and the Romans, and the opposition of the pope in 1159 led him to appoint his own pope, Victor IV. The Papacy thereafter backed the Lombard League – an association of north Italian cities that opposed Frederick and forced him to make peace after the Battle of Legano in 1176.

A dream shattered

Henry VI claimed all the lands of the Normans, especially Sicily [9], and dreamed of capturing Tunis and even Constantinople. He built the first imperial fleet, and defeated a hostile coalition between northern Germany and Britain when he made Richard Coeur de Lion his prisoner. His son Frederick II (1194–1250) had been brought up in Sicily and was more Mediterranean than German. Ambitious but sceptical, he was the mortal enemy of the Papacy and achieved the greatest expansion of imperial influence [7].

Sicily under the Hohenstaufens was culturally a mixture of Italian, Arabic, Greek, Norman and Germanic influences, and was part of a larger culture that spread from Mesopotamia through Moorish Spain. Frederick saw politics as an art, government as a bureaucratic skill. He patronized the arts and sciences and founded the university of Naples to rival Bologna. He saw the empire as a German federation and in 1231 was prepared to recognize the territorial claims of the princes. In 1235, he published the Landfrieden of Mainz, the first German law written in German which defined the empire as a league of princes within a monarchic framework. His negotiations with the Muslims scandalized the Christian West but enabled him to be crowned at Jerusalem without fighting a crusade. At the Council of Lyons in 1245 he was condemned and deposed. When, on his death, the Papacy broke the power of his family, the future of the international empire of his dreams was shattered. The empire survived, but as a spectral organization, until Napoleon Bonaparte's arrival in Germany in 1806.

KEY

The imperial seal of Otto III (r. 980–1002) epitomized his ambition with the inscription *Otto Imperator Augustus Renovatio Imperii Romanorum* – Emperor Otto Augustus, the Renewal of the Empire of the Romans. The grandson of Otto I who founded the dynasty, Otto III was crowned aged three. His education was designed to produce an emperor who combined German strength and tradition with the new Arab, Greek and Latin learning. To identify himself with Charlemagne, Otto opened his tomb at Aachen and stole a tooth, nail clippings and clothes. On feast days his clothes were decorated with lions, eagles and dragons and hung with bells.

6 Canossa was the most humiliating episode in the history of the empire.

Although the political results ultimately favoured Henry IV, the spectacle of a penitent emperor begging for the intercession of Abbot Hugh of Cluny and Matilda of Tuscany in seeking the pope's pardon was one that injured German pride so much that even Bismarck's 19th-century conflict with the Papacy was based on a desire "never to go to Canossa again".

7 The territorial achievements of Frederick II (1194–1250), culminating in the Battle of Bouvines (1214) at which he defeated his last remaining opponents, were the fulfilment of German imperial ambition. Frederick inherited his Sicilian kingdom and his southern ambitions from his father, and his lands completely surrounded the papal state. He achieved the federation of Germanic countries and an administration imbued with Mediterranean culture that had been the dream of Otto III.

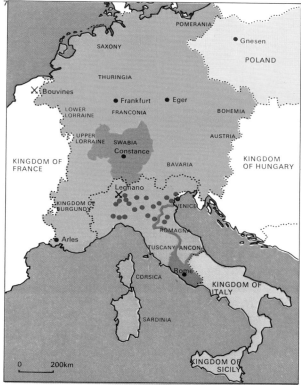

Sole Hohenstaufen duchy 1152
Empire of Frederick I
Allied with Empire 1184
Tributary under Frederick I
Ruled by Henry VI and Frederick II
Papal states 1152
Papal states from 1213
● Towns of Leagues of Lombardy and Verona

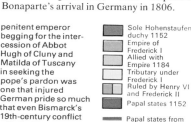

8 The final humiliation of Henry IV, portrayed in this contemporary illustration, came in 1105 when he was forced to surrender his imperial regalia to his son, Henry V. Some of Henry IV's family had disagreed with his deposition of Gregory VII and installation of a new pope. Henry V joined them after becoming co-ruler in 1099 and soon overthrew the emperor.

9 Peter of Eboli's illumination of the siege of Naples (1191) shows the army of Henry VI attempting to enforce his wife's claim to the south Italian Norman kingdom. Despite his early successes, Naples, aided by the failure of the imperial naval allies and by plague in the besieging army, did not fall. Henry VI was crowned King of Sicily only in 1194.

10 Castel del Monte, in southern Italy, is the finest of a chain of strategic fortifications designed in a novel and functional manner by Frederick II. He built up in Sicily and Apulia a system of defence and administration that was the most thorough and rational in Europe. As fortifications the castles protected Frederick from attack by the Papacy. As architecture, they were part of the flowering of the arts during Frederick's reign.

European expansion to the east

In the centuries before AD 1000, while a lively, enterprising society was emerging in the villages and towns of Western Europe, Eastern Europe was a land of sparsely occupied forests, grasslands and low, easily crossed mountains through which passed at random a variety of peoples. The roaming shepherds of the Carpathians, the nomads from the south Russian steppe who settled in the Danube basin, and the forest peoples of the Vistula (ancestors respectively of the modern Romanians, Hungarians and Poles) travelled up and down the region between the Baltic and the Danube, the Pripet Marshes and the Elbe. They were untouched either by the political or the religious allegiances of the settled lands to the west and south and were impervious even to the cultures of each other.

The forest peoples, Polish and Czech Slavs, seem to have been divided into numerous tribes and many of the *grody* – fortified settlements on hills or marsh islands which were their capitals – have been uncovered. The grasslands of the middle Danube were the haunts of successive nomadic peoples who lived by raiding neighbouring lands. The last ones, the fierce Magyars of mixed Finnish and Turkish origin, reached the area by the late ninth century.

The spread of German influence

To the west of these peoples were the Germans, whose national identity was also recent. Their relations with the Slavs were close, sometimes hostile but often peaceful; their society of well-organized communities and their Christian civilization, contained within the Holy Roman Empire, exercised a constant and fruitful pressure. Under the powerful Ottonian kings (919–1024; emperors from 962) the German advance was rapid and for the first time there is evidence that the tribes, in their turn, were beginning to unite: by 1000 the Magyars (whose raids were checked by the German victory at the Lechfeld in 955), the Bohemians and the Poles [1] had each united under single, independent dynasties. The earliest sign of the evolution was the advance of Christianity [7]. Already in the ninth century Byzantine missionaries converted the Moravians, whose mushroom empire was annihilated by the Magyars; in its place, under German influence, there grew up the Czech duchy of Bohemia, whose first ruler, St Wenceslas (*c.* 907–29), became a Christian and a tributary of the German king.

Bohemia under German auspices was the focus for the rapid development of all Eastern Europe. Duke Boleslav I (reigned 929–63) was the father-in-law of the first Polish duke, Mieszko I (reigned 963–92), who received Christianity at his hands; while the influence of St Adalbert, Bishop of Prague [3], caused the Hungarian Duke Stephen (reigned 997–1038) to be baptized. But German influence was ambivalent. In AD 1000 the Polish duke Boleslav the Mighty (*c.* 996–1025) received a crown from the emperor and Stephen, who ended the influence of Eastern Christianity in Hungary, received a crown from the pope.

The emergence of national identities

In the eleventh century the Bohemians, Poles and Hungarians held their own and they acquired a sense of national identity under

1 Boleslav the Mighty was the founder of the first Polish monarchy. His reign resulted in 30 years of internal consolidation for his country, the building of a national Polish Christian Church and significant expansion abroad. All of these developments changed Poland from an alliance of Slav tribes into a powerful, centralized monarchy. Alliances with Bohemia, Hungary and Kiev enabled the king to make the Oder and the Vistula virtually Polish rivers and also allowed him to declare war on Emperor Henry II (973–1024) between 1004 and 1008. The emperor finally recognized the integrity of the Polish state and Boleslav was crowned in 1024.

2 St Wenceslas, Duke of Bohemia and an enthusiastic Christian convert, became the patron saint of Hungary, Poland and Bohemia. As duke he failed to resist the aggression of the German king Henry I (*r.* 919–36), who meant to subdue the Wends and Slavs as well as the Bohemians. Bohemia became a German fief (owing nominal allegiance to the emperor) and Wenceslas, who was blamed for the defeat, was murdered by his brother, Boleslav (who succeeded him as Boleslav I).

3 The cathedral of Gniezno (Gnesen) was the centre of the Christian religion of the new Polish state. It housed the relics of St Adalbert of Prague, one of the apostles of Eastern Europe. Gniezno, an ancient Polish centre, was the most important of the castle towns of Boleslav the Mighty. These towns were thriving garrisons which gradually fostered local trade and finally became centres for the export of corn to the West.

native dynasties. The Premyslids of Bohemia did not receive the royal title until 1198, but although they were included in the German Empire they retained intact their Slavonic language and customs.

The Polish dynasty of Piast was more aggressively anti-German and their court at Gniezno became the focus of resistance to the ambitions of the Salian emperors. But the unity of their vast territories was superficial and from 1079 to the end of the thirteenth century the country was ruled jointly by, and then divided among, several Piast princes. Throughout this period the unity and identity of the Poles was maintained only by a national Church and a common culture.

The Hungarian house of Arpad was more fortunate in maintaining its unity in close relation with the German emperors and Hungary flourished as a bridge between Byzantium, Russia and western Europe.

Later German migration
The eastward advance of the Germanic peoples had been checked after 1002 short of Pomerania, Poland and Hungary [Key]; after 1100 it resumed in force. Slavonic and Magyar rulers welcomed the Germans as cultivators of the sparsely inhabited soil. First came merchants, to swell the Wendish, Polish and Bohemian towns, especially Lübeck and Danzig on the Baltic. A massive migration of farmers followed in the twelfth century: they occupied the fertile lands of Silesia, spread throughout Bohemia and parts of Hungary and pioneered the opening up of Transylvania. Finally came the knights, ostensibly to convert the pagan Lithuanians to Christianity, but also to carve out new territories along the Baltic.

Several military orders were formed which in 1237 united as the Order of the Teutonic Knights of Livonia. This heralded German domination of the Baltic and the maritime communities soon founded an association, or Hansa [4]. The Hansa played the predominant role culturally and politically in northeastern Europe. But German rule did not accompany the migration; apart from Polish friction with the Teutonic Knights, relations with the new settlers were generally friendly and mutually profitable.

KEY

4 Bremen, on the Weser, is one of the great north German trading ports. Founded in the ninth century by Charlemagne, its merchants set up the city of Riga in 1158 and in 1358 joined the Hansa for protection.

5 Marienberg, with its famous castle, was the capital of the Teutonic Knights in Prussia. The knights, founded as a noble military, charitable and missionary organization in 1190, abandoned work in the Holy Land and settled on the Baltic coast to enforce the conversion of the pagan Prussians. In 1309 they established their headquarters at Marienburg.

6 Pope Sylvester III gave St Stephen the upper part of the crown of St. Stephen, symbol of Hungarian nationhood, in 1000. The circlet was given by the Byzantine emperor Michael VII 75 years later.

7 The advance of Christianity in Eastern Europe (1000–1250), and the conversion of the Slavs and Magyars, was quick on a superficial level, but thorough Christ-tianization took centuries. First came the German or Byzantine mission-aries, whose real purpose was to convert the rulers and estab-lish bishoprics; only later, in the twelfth cen-tury, did Christianity begin to reach the rural communities.

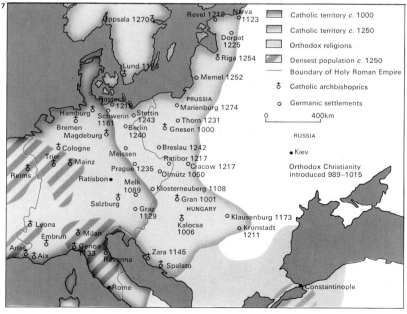

Anglo-Saxon art and society

Most modern knowledge about the society of the early Anglo-Saxon period comes from artefacts buried with the dead. They include weapons and jewellery, glass and pottery, as well as many small items of household and personal equipment [3], their range of wealth reflecting the hierarchical nature of the society. For later periods, cultural achievements revealing the nature of society include poetry surviving from a great tradition of oral literature; a magnificent range of visual arts, including manuscript illuminations, sculpture and architecture; law codes and wills; and, perhaps most enduringly, a structure of local government based on shire, hundred and parish, that has been in use to modern times.

Kingship in the age of *Beowulf*

The epic poem *Beowulf*, written sometime after 700, presents a probably accurate picture of the warrior king, with a band of followers. The exchange of gifts for hospitality was intrinsic to the personal bonds of this society, and its effect can be seen in the wealth of the burial of a seventh-century king at Sutton Hoo, Suffolk. This contained rich gold and garnet jewellery, weapons, bronze bowls and many other objects, some imported, some of local manufacture. The helmet [1] found at Sutton Hoo probably resembled the gold inlaid one said to be worn by Beowulf, and the monster he fought is perhaps an alternative expression of the elaborate distortions of nature shown in the animals on the gold buckle [Key].

Accompanying the king were the thegns or nobles, who owned at least five hides – very roughly 250 acres of good land. The ordinary ceorl (freeman) owned one hide or about 50 acres, freehold, and was buried with only a spear and sometimes a shield. There were also slaves, often defeated enemies, who could own nothing.

The graves of women, as well as the poetry, indicate the high status of women in Anglo-Saxon society; a thegn's wife might be buried with elaborate brooches and perhaps with a weaving baton, a symbol of her authority as *hlafdige* or lady. Women certainly could dispose of their own property, but no woman was queen in her own right [6]. In sixth-century Kent jewellery was worn similar to that found amongst the Franks on the Continent, suggesting close cultural links between the areas. Ethelbert (*c.* 552–616), King of Kent, married a Christian Frankish princess, Bertha, and St Augustine (died 604) began his mission to convert the Anglo-Saxons under her patronage. Christianity did not bring any sudden change to the nature of society, but it did alter its material culture. The vision previously incorporated in intricate metalwork was now brought to equally complex manuscript illuminations [2].

Christian art and architecture

There were new skills also. Benedict Biscop (*c.* 628–90) used glaziers and masons from Gaul to build the monasteries he founded at Monkwearmouth and Jarrow in Northumbria. Travellers brought books, pictures and relics from throughout Europe to enrich the new churches and monasteries. The Irish monks added Celtic traditions to the art style that emerged in northern England in the eighth century. The manuscripts and the great stone carved crosses of this period reflect Celtic, Saxon and Mediterranean

1 The helmet from Sutton Hoo, now reconstructed on a leather base, was made of bronze with silver wire inlays and garnets lining the eyebrows. The nose and moustache are of gilt-bronze. The goods contained in the burial, which included objects that the dead man might need in the afterlife as well as his personal treasures, suggest that pagan Anglo-Saxon religion was still flourishing in spirit despite the nominal introduction of Christianity. Raedwald, who may have been the king celebrated by the burial (although no body was found in the ship – which may have been more of a cenotaph), changed his religion several times.

2 The *Lindisfarne Gospels* were written and illustrated *c.* 698, by Eadfrith, Bishop of Lindisfarne in Northumbria from 698 to 721. They represent a fortunate amalgam of many styles, but owe most to the decorative traditions of Christian Irish art, as shown in the *Book of Kells* (late 8th century). Other influences visible in the *Lindisfarne Gospels* include Coptic patterns from Egypt and Italianate portraits of the four evangelists. Christianity was relatively easily introduced to the Anglo-Saxons; and many kings granted land and wealth to the monasteries and saw the bishops as spiritual companions, equivalent to the thegns in the military sphere.

3 An early 6th-century glass beaker from a grave in Mucking, Essex, suggests the importance given to feasting by the Anglo-Saxons. Epic poetry such as *Beowulf* might have been recited at such feasts.

4 The modern statue of the warrior Bryhtnoth recalls the heroism of Anglo-Saxon warfare. *The Battle of Maldon*, possibly the only survivor of a large group of battle poems, tells of the earl's defence of Essex against Danish raiders in 991: although it was clear that the defenders could not win the fight, all of Bryhtnoth's thegns died fighting at his side. To outlive the leader in battle was thought a disgrace. Almost 400 years of Christianity did little to change Saxon martial ideals.

5 Aelfric, Abbot of Eynsham after 1002, wrote a number of books in both Latin and English, including this *Grammar Book*, which was intended for the education of novice monks and the sons of the nobility. As a leading figure in the movement for Church reform, he wrote two sets of sermons to be used by parish priests who might otherwise have been unable to fulfil their pastoral functions. These sermons, known as the *Catholic Homilies*, were in the West Saxon dialect, the basis of modern English.

The use of English as a written as well as a spoken language was stimulated by King Alfred in the late 9th century. Although most of those able to read and write were clerics, a higher proportion of laymen were literate than in later medieval times.

stylistic inspiration. In contrast to the open shrines of the pagan era, churches were built, at first in wood and later in stone [7]. Above all, a literate class developed, conscious of its English heritage. Bede wrote the history of the English Church and King Alfred (848–99) translated Latin texts into English and encouraged native learning.

Despite the Viking incursions, in the tenth century, there was a revival both in religion and art, but it was centred on Wessex rather than Northumbria. A great variety of literature, both secular and religious [5] survives from this period, including many poems, elegiac as well as heroic. A new art style, the Winchester school, revealed a new interest in naturalism and dynamism [8]. It is found mostly in manuscripts, but also in some sculptures, most notably an angel set into the wall of Bradford-on-Avon church.

Law and local government

Early justice was based on the extended family and the vendetta. Later this form of violent retribution became formalized with a complex system of payments known as wer-

gild, based on the severity of the offence and the social standing of the victim. From the period of the invasions, administration of land was based on the unit of the hide (which varied in size from time to time, and according to the quality of the land, but was theoretically enough land to support a single family). Shires were divided into hundreds (each in theory representing 100 hides); the men of each hundred or shire met regularly in moots to see justice done. There was a detailed structure of common law often with harsh penalties: there were no prisons, and death, exile, outlawry, mutilation or enslavement were the alternatives to fines which many could probably not pay.

As well as being the focus of local government, the shire became an important unit of defence. This was in the hands of the ealdorman, a royal official in some ways corresponding to the sheriff, who, although often not a native of the shire, organized the fyrd, or militia. Eventually the office became hereditary and the same man might acquire authority over two or more shires; the ealdorman gradually merged with the earl.

The great gold buckle from Sutton Hoo is one of the finest examples of early Anglo-Saxon metalwork. Although it seems to be a massive object, it is in fact hollow and might have held some aromatic or precious substance. The intricate and complicated patterns are a stylized representation of two animal figures. They have long narrow bodies, rudimentary eyes and heads, and legs that become almost indistinguishable from vine tendrils. Animal ornament was a favourite motif of the Saxon metalworkers, but after the 7th century naturalism became common. A delight in flowing patterns is characteristic of Anglo-Saxon art in all its forms.

6 The queen attended the councils of Anglo-Saxon kings, as in this illustration of c. 1000. Women were shown greater respect, both socially and legally, in Anglo-Saxon times than after the Norman Conquest. They could buy and sell property, and might make separate wills from their husbands'. One woman is even recorded as having disinherited her son in favour of a kinswoman. Even in early Anglo-Saxon times women played an important part in the spread of Christianity. St Hilda (614–80), who was related to King Edwin of Northumbria, founded the twin male and female monastery (following an Irish model) at Whitby in 657, and stayed as abbess until she died.

7 The church tower at Earls Barton in Northamptonshire is one of the best preserved pre-Conquest buildings. Most Saxon churches were made of timber, so few have survived (although the stave church at Greensted in Essex may be one such, dating from the early 11th century). The stone Roman remains were regarded with awe as the work of giants. The pilaster strips on this tower may derive from patterns of timber frames. Most of the parish boundaries were created by 1000, and many churches of later date incorporate Saxon elements, often hidden by plaster. Sometimes a stone cross took the place of a church.

8 *St Ethelwold's Benedictional*, named after a 10th-century bishop of Winchester, the capital of the Anglo-Saxon kings, is a fine example of the late style of manuscript illumination known as the Winchester school. Its most striking character-istics are the lively figures in flowing draperies and curled acanthus leaves. Carolingian manuscripts probably were the inspiration for the Winchester school, which is lighter and more delicate than the Continental styles that preceded it.

9 Westminster Abbey was begun by Edward the Confessor (c. 1002–66), but he was too ill to attend its consecration and died soon afterwards. This scene from the Bayeux tapestry depicts his burial in the abbey. The largest known Saxon building, it dwarfed the royal palace of Cheddar. The size of the abbey attests to the wealth of the country, the power of the king and Church, and the new importance of London, which after the Conquest would take over the role of leading city of England.

The end of Anglo-Saxon England

Edward the Confessor (*c.* 1002–66) lived in exile in Normandy for 25 years during the reigns of Canute and his sons and succeeded to the English throne in 1042. His mother, Emma (died 1052), was a Norman and although there may not have been much love lost between Edward and Emma, he probably thought of himself as at least half Norman. He brought Norman followers with him to England, including Robert of Jumièges (died *c.* 1055) whom Edward made Archbishop of Canterbury in 1051 despite opposition from the earls. In a sense, therefore, Edward began the process, completed by William, of a transition from an English to a Norman society. Even royal castles, so much a part of of William's conquest, were introduced in a primitive form by Edward.

The rise of Godwin and Harold
Edward's reign was constantly over-shadowed by the power of the family of Godwin (died 1053). Under Canute England had been divided into a few large earldoms, not unlike the old Saxon kingdoms. This consolidation produced a new and dangerous political situation, with a few earls controlling great territorial interests [2]. Much of the history of Edward's reign concerns quarrels between himself and these earls, and between members of the Godwin family, which controlled most of southern England.

Edward married Edith the daughter of Godwin, in 1045, and it seemed at first that Godwin controlled the country. But in 1051 the earl refused to obey Edward's order to subdue Dover, a town in Godwin's own earldom where disturbances had taken place. Open conflict was averted by Siward and Leofric, the other important earls, who supported the king. Edward exiled Godwin's entire family, but the following year Godwin returned, secure in his authority within his earldom, where his territorial control was much greater than that of the king. The mutual dependence of king and earl had been made apparent.

Towards the end of his reign Edward turned increasingly towards thoughts of piety and he left much of the detail of government to Harold (*c.* 1022–66), Godwin's son. At some time Harold visited France and fell into the power of William of Normandy (1027–87), who extracted from him an oath of support for his own claim to the English throne. Edward and Edith had no children (the king's reputation for piety derived from a legend that he never consummated his marriage), but towards the end of Edward's life an English candidate for the throne appeared, Edward the Aethling, son of the Confessor's brother. The Aethling died soon after his return from exile (*c.* 1059), leaving a son, Edgar (*c.* 1050–*c.* 1130), too young to be a serious candidate for kingship, and a daughter who later became St Margaret of Scotland (died 1093). So when Edward died in January 1066, Harold [4] proclaimed himself king and was crowned in Westminster, supported by Edward's council.

The Norman invasion
Since the death of Canute, Norwegian and Danish rulers had been awaiting an opportunity to renew their claim to England. In 1066, after defeating the Danes, Harald Hardrada (1015–66) of Norway set sail for England, with the support of Tostig, the son of

CONNECTIONS

See also
168 Norman and Angevin England
164 Anglo-Saxon art and society
190 Western European economy 1000–1250
172 Romanesque art of the 11th century

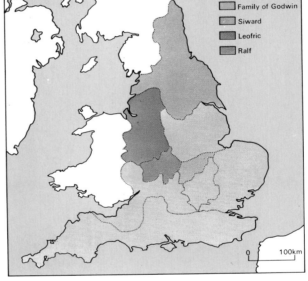

1 **Edward the Confessor** was the last truly English king of England, because Harold was Danish by descent. The son of Ethelred II, he ruled over a court of Danes, Englishmen and Normans. This cultural mixture arose from the wealth and strength of the English throne in this period, which encouraged the ambitious Norsemen to invade the country. It was only the simultaneous attack on England on two fronts in 1066 that brought the downfall of the Anglo-Saxon kingdom. Although not such a great landowner as William became, Edward still commanded great authority through the veneration in which the monarchy was held.

| English earldoms 1046 |
| Family of Godwin |
| Siward |
| Leofric |
| Ralf |

100km

2 **The great earldoms of Edward's reign** were a result of Canute's policy of land distribution. Godwin himself had been an obscure soldier who rose to prominence in the service of the Danish king. In many areas, particularly Sussex and the west, he held considerably more land than the king, and the local thegns showed as much loyalty to him as to the king. Leofric and Siward, the other leading earls of Edward's reign, were unwilling to see their countryman Godwin acquiring unlimited power over the king. The distribution of earldoms on the map shows the power of the House of Godwin at its peak in 1051.

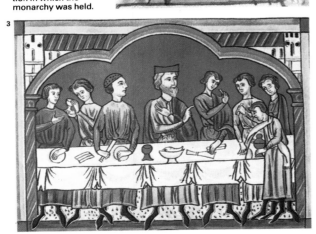

4 **Harold, the son of Godwin,** was accepted as king on Edward's death, although Edward had probably offered the throne to William of Normandy, and Harold had sworn to support him.

5 **Stigand,** an infamous pluralist, was Archbishop of Canterbury in 1066, but William, an ardent supporter of Cluniac reform, replaced him with Lanfranc, who introduced feudalism to the Church.

3 **Edward (centre) quarrelled with Godwin** (right foreground) in 1051 over the latter's apparent disregard for his son Sweyn's irresponsibility in 1046–9. But in 1052, the Earl Godwin was able to return virtually unopposed because Edward's policy of encouraging Normans at his court was unpopular (it was at this time that Edward is said to have promised the throne to William of Normandy). Although Godwin died in 1053, Harold was able to take advantage of this sympathy and take virtual control of the country until the death of Edward.

Godwin. The English king decisively defeated his northern enemies at Stamford Bridge but then had to march from York to Hastings to meet the second enemy, William, who had landed at Pevensey on 28 September to assert his own claim to the throne. William brought new military skills to England, using ranks of lancers on horseback against the traditional English foot-soldiers fighting with axe and sword. Nevertheless, the Norman victory at the Battle of Hastings, fought on 14 October, was not a foregone conclusion and owed much to the exhaustion of the English.

England after the Conquest
The Battle of Hastings is depicted on the Bayeux tapestry [Key, 5, 7], a long embroidery completed *c.* 1080, perhaps commissioned by Bishop Odo of Bayeux (*c.* 1036–97). William refused to be crowned by Stigand, Archbishop of Canterbury (died 1072), whom he replaced in 1070 with Lanfranc (*c.* 1005–89), an Italian monk. After a rebellion in Exeter in 1067, and in the north in 1069, drastic measures were taken to bring

the country under control [7]. There were a few pockets of resistance [8] but much of the English nobility had died in battle; their heiresses often married Norman barons.

A uniform system of land tenure was imposed, all land being distributed in return for specific military services, which facilitated long-term garrisoning of the castles that defended Norman rule [6]. In order to establish precisely the obligations of his tenants, William ordered a survey to be made of his new possessions – the Domesday Survey, completed in 1086.

Much of the Anglo-Saxon way of life disappeared with the ending of free land tenure for the thegn and ceorl. Some legal institutions continued in a debased form, but art and literature, which had been flourishing at that time, died out. Fifty years after the Conquest, the distinction between English and Norman had been lost. The Conquest gave new vitality to the towns, reorganized the Church and brought England closer to the culture of Europe; in many ways it can therefore be considered the beginning, as much as the end, of an era of English history.

Harold was killed at Hastings either by an arrow in his eye or, as is more likely, by being cut down by a Norman horseman. His army of about 7,000 men consisted of both the traditional Anglo-Saxon king's band of followers, and the fyrd, or local militia comprising less well-trained soldiers who were called out to deal with emergencies. After the battle, William advanced swiftly to London before any significant new opposition could be organized around the last remaining member of the English royal family, Edgar the Aethling. William founded a monastery on the site of the battle, a symbol of the unity of political power and religious duties for a Norman.

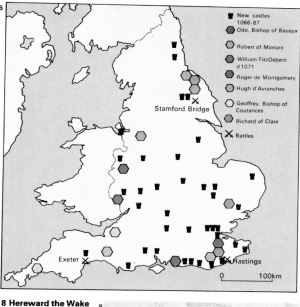

6 William allocated the land of England in such a way as to allow his invading forces to control the country. He built castles in the towns, and gave estates to his followers, who were spread throughout the kingdom to prevent any of them from gaining local power as Godwin had done. The south of England, the most strategically vital area, remained in the hands of a few trusted friends. But he gave the lands of more than 2,000 thegns to fewer than 200 Norman barons, and so radically centralized the power structure of England. William kept large estates for his own use. The introduction of feudalism was perhaps the most dramatic effect of his rule.

Map legend:
- New castles 1066-87
- Odo, Bishop of Bayeux
- Robert of Mortain
- William FitzOsbern d 1071
- Roger de Montgomery
- Hugh d'Avranches
- Geoffrey, Bishop of Coutances
- Richard of Clare
- Battles

Stamford Bridge
Exeter
Hastings
0   100km

VR: HIC:EST:VVAD AR

7 The revolt of the north, in 1069, was punished by William with such severity that the Domesday Survey, of 20 years later, shows the most affected areas – Yorkshire and Lancashire – as still far less prosperous in agriculture than the rest of the country. Such brutality, and the force with which the Normans cleared the centres of towns to build castles to overawe the populace, meant that many Englishmen hated the invaders. Although the Normans ended the Anglo-Saxon institution of slavery, they reduced freemen to the level of serfs and cut the authority of the common law.

8 Hereward the Wake was a semi-legendary English thegn who carried on opposition to William from a hideout in the Fens of East Anglia. Before the Conquest he had held significant estates in Lincolnshire. But opposition to the Normans as violent as shown in this 19th-century illustration was not common; most English accepted the new king and by the early 12th century it was common for a man of English descent to become a royal official such as a sheriff. At the end of the story of Hereward, it is said that he was pardoned by the king. Only two Englishmen survived the Conquest as important landowners; others became managers of their old estates.

9 The late Anglo-Saxon stone carving at Kilpeck Church, Herefordshire, is one of the few examples of an English cultural survival after the Conquest. In most other respects, a sharp change occurred after 1066. Norman-French and Latin became the official languages, and vernacular literature died out, as learning became more completely the province of the Church, and secular literacy became less common. But the *Anglo-Saxon Chronicle* was still written until the early 12th century. The Church itself was revitalized from the torpor into which it had fallen since 1020. Stone architecture on a massive Romanesque scale became common for churches and castles.

Norman and Angevin England

Between 1050 and 1100 Europe was transformed by the conquests of a warrior people from a section of the north French coastlands: the Normans. They overran England and southern Italy, and settled in Scotland, Wales, the Byzantine Empire and (after the first Crusade) the Levant. These extraordinarily successful people were a closely interrelated group of families, several of them related to the Viking-descended dukes of Normandy. They took the art of cavalry warfare further than any of their contemporaries and crystallized as a distinct group under the leadership of Duke William I (1027–87), the foremost general of his age.

The conquest of England

The Normans' greatest achievement was the conquest of England, which because of its distinctive civilization, advanced organization and great wealth was the richest prize in Europe for soldiers of fortune. Normans had begun to settle there before 1066; but the transformation of England into a Norman kingdom required the ambition of William, an adventurer like the rest. The invasion of 1066 was a corporate enterprise, ostensibly to establish William as the heir of King Edward the Confessor, but really to win the Normans new fortunes. The narrow victory of Hastings, the slow advance from Canterbury to York, and the rewarding of his companions with wide estates was only just successful, for his success created in the midst of a living society an uneasy circle of adventurers [1], in some ways less civilized than the surrounding people.

Although William hoped to rule like an Anglo-Saxon king, his enemies – King Philip of France, Malcolm of Scotland, Canute of Denmark, in league with dissident Englishmen and rebellious Normans – imposed a state of virtual siege on his dominions and forced him to original expedients, critical for the future. These included the elaborate system of military service in return for land, out of which the striking solidarity of the Anglo-Norman state was born [4], the series of mighty castles – at the Tower of London, Wallingford and Colchester, in which the grandeur of Norman ambitions can still be seen [5] – and the detailed survey of English wealth made by his order in 1086: the Domesday Book [2]. The Normans transformed England, not least by reordering the diocesan structure of the Church.

Under the Norman Empire

When William died, in 1087, England and Normandy were ruled separately by his sons William Rufus (r. 1087–1100) and Robert of Normandy, but so strongly were their societies united that within a few years they were again governed by a single ruler.

The Norman Empire, ruled by Rufus's successor and Robert's supplanter Henry I (1068–1135) and his descendants, held together until 1204. For the great Norman families, England was a colonial El Dorado but Normandy was "home". Rarely have English and continental history been so intertwined, with marked effect on English civilization: Englishmen such as Adelard of Bath and John of Salisbury contributed to the European revival of learning and science; the great English cathedrals, from Durham at the beginning of the twelfth century to Canterbury at the end, were designed by the most

1 The Conqueror and his companions, as portrayed in the Bayeux tapestry, demonstrate the *esprit de corps* of the Norman leaders, which was one of the secrets of their success. Bishop Odo of Bayeux (on the Conqueror's right), half-brother to William, was probably highlighted because he commissioned the tapestry.

2 The Domesday Book was an astonishing monument to masterful and inventive rule. The survey was ordered by William in 1085 and completed in six months. Its precise purpose is uncertain, but it constituted a record of the wealth of the king and his principal subjects, of stock, of the land and the condition of the peasants who cultivated it. Teams of commissioners gathered the necessary information from jurors. This particular page describes a manor in Somerset.

Map 3 (legend)

SCOTLAND
IRELAND
WALES ENGLAND
NORMANDY
Paris
BRITTANY
MAINE
ANJOU
TOURAINE
AQUITAINE
TOULOUSE
GASCONY

- Maternal inheritance
- Paternal inheritance
- Acquired through marriage with Eleanor of Aquitaine
- Owed suzerainty to Henry
- Conquered territory
- Kingdom of France

0   250km

3 The Norman Empire spread farthest under Henry II, partly because of the union, via Matilda, of Norman and Angevin domains, but largely through his marriage to Eleanor of Aquitaine.

4 The network of Norman power was gradually extended by linking powerful families in marriage alliances, although civil war was needed to bring the Angevins to the English throne. England alone was acquired through conquest.

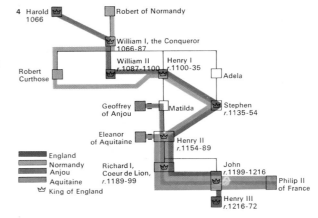

4 Harold 1066 — Robert of Normandy
William I, the Conqueror 1066-87
Robert Curthose — William II r. 1087-1100 — Henry I r. 1100-35 — Adela
Geoffrey of Anjou — Matilda — Stephen r. 1135-54
Eleanor of Aquitaine — Henry II r. 1154-89
Richard I, Coeur de Lion, r. 1189-99 — John r. 1199-1216 — Philip II of France
Henry III r. 1216-72

- England
- Normandy
- Anjou
- Aquitaine
- King of England

advanced French builders; an English scholar, Nicholas Breakspear, was the first and last Englishman to become pope (as Adrian IV from 1154 to 1159); and Norman French remained for three centuries the language of the English courts and society.

England's growing prosperity

It was also a period of unprecedented prosperity for England, apparently based on sheep: England was the main supplier of wool to the Flemish textile industry. Many landlords, especially the monasteries, owned vast estates. They were progressive farmers, opening up the fens and the Yorkshire uplands. English towns [6] such as Norwich, Oxford and Salisbury developed rapidly, usually around existing episcopal seats and castles. Flourishing markets for agricultural products needed special protection and for the first time there was a definite movement towards self-governing guilds.

Norman England was also remarkably advanced in the arts of government and a series of careful monarchs (broken only by the disputed succession between Henry I's daughter Matilda and his nephew Stephen of Blois) sought to marshal England's resources rationally. On Stephen's death, Matilda's son Henry II of Anjou (reigned 1154–89) governed with professional skill, using men who were prepared to experiment with legal forms, establish precedents for consistent action and keep records as a matter of routine. The exchequer, the jury and the common law courts are institutions that began their organized life in his reign [7].

Their significance, however, lay in the future. At the time there were many signs that the now unwieldy empire was breaking up [3]. Its French inhabitants increasingly looked to the king of France, and the last years of Henry II's reign were disrupted by rebellions, in which his sons joined. After his death the empire was held together by the sheer military genius of his son Richard I (reigned 1189–99); but Richard's brother John, who was no general, lost it forever in only five years [8]. By 1204, England had once more become a separate kingdom, enriched and transformed by a century and a half of Norman and Angevin rule.

Orford Castle in Suffolk, built by Henry II, was typical of the many stone castles that were constructed in every part of England. Control of these castles was the first requirement of effective authority; they could be captured only with the expenditure of much time and effort. The earliest was the Conqueror's wooden prefabricated castle set up on the beach at Hastings before the battle. Henry II showed characteristic inventiveness in the design of Orford, built between 1166 and 1172 to control the Alde estuary and the Suffolk hinterland. Its keep is polygonal outside and cylindrical within; its services are in the outer shell.

5

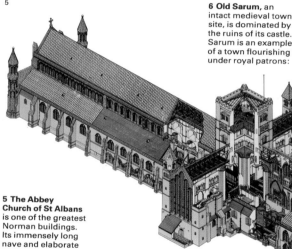

5 The Abbey Church of St Albans is one of the greatest Norman buildings. Its immensely long nave and elaborate east end, which was later rebuilt, are characteristic of the Anglo-Norman style. The Norman part (the central tower and most of the nave) was built, 1077–93, under the direction of Abbot Paul. Although Norman in inspiration, many of the details are Anglo-Saxon since rebuilding had been planned before the Conquest.

6 Old Sarum, an intact medieval town site, is dominated by the ruins of its castle. Sarum is an example of a town flourishing under royal patrons: Norman kings often resided there and William I made it a bishopric. It declined from 1222 when the bishop moved to nearby Salisbury.

6

7

7 The Great Seal of Henry II, despite its use in the orderly life of the realm, bears a warlike image. With the aid of capable ministers, Henry transformed the existing customs into a systematic body of law and an efficient administrative procedure.

8

8 King John was buried at Worcester Cathedral after a reign that provoked hatred and distrust among most of his subjects, principally because he presided over the dissolution of the Angevin Empire. After 1204, when he lost most of his continental possessions, he devoted himself to their recovery. Most of the troubles of his reign were caused by the need for heavy taxation when prices were rising sharply. Abortive campaigns ended with the rebellion of the barons and the imposition on him of the Great Charter at Runnymede in 1215. After the sealing of the charter the Norman Empire had to make way for the emergent English nation.

Islam in Europe

The overthrow of the Visigothic state of Iberia in AD 711 by Berber forces recently converted to Islam (further penetration into France was stopped by Charles Martel in 732) was an event unforeseen by the Arabs of the East. The distance of Spain from Damascus was such that these forces could not have expected to enjoy the full fruits of their conquest.

Independence from the East
In the early years the governors of Al-Andalus, as the Muslim-controlled area of the Iberian Peninsula was called, were nominated, however, by the caliphs in the East and revenue from taxation did find its way across the Mediterranean. No such benefits were forthcoming after AD 756 when the Umayyad family achieved pre-eminence among the small number of Arabs (probably fewer than 20,000) who had settled in the peninsula. This family retained supremacy in Córdoba [2] for nearly 300 years, maintaining from the outset political independence from the East.

Although there was an increasing inci-dence of conversion to Islam among the indigenous population during that period, the Umayyads of Córdoba could seldom command the loyalty of all Muslims in the peninsula. It was only in the tenth century, under Abd ar-Rahman III, who declared himself caliph in AD 929, that hitherto semi-autonomous regions acknowledged, sometimes after protracted conflicts, the supremacy of the Umayyads. The ensuing unity gave Al-Andalus a strength that enabled the state to brush aside sporadic forays by the kingdoms of the north and even, on occasions, to arbitrate in the dynastic disputes of these kingdoms. Profitable treaties were struck with some of the small North African dynasties and trade was established with the German and Byzantine empires.

Importance of Córdoba
Córdoba became the undisputed capital of western Islam and a magnet for scholars [9], poets and craftsmen who assembled there from throughout the Islamic world. A palace was built in the cooler foothills of the nearby Sierra Nevada where the caliph resided and conducted affairs of state, while in Córdoba itself successive additions imparted increasing splendour to the Great Mosque.

Unity in Al-Andalus proved, however, to be difficult to maintain and as a result the caliphate was formally dissolved in AD 1031. The subsequent fragmentation of Al-Andalus into some 30 city states, all jealous of their own independence and covetous of the territories of their neighbours, occurred in the eleventh century when Christians from beyond the Pyrenees, actively encouraged by the pope, became involved in the internal affairs of the northern Spanish kingdoms.

The reconquest of Muslim Spain
The reconquest of Muslim Spain may be considered in two phases, before and after the capture of Toledo in 1085 [6]. Before this date there is scant evidence from Latin or Arabic chronicles that the gradual occupation of territories to their south by the Christian states was either premeditated or concerted. This settling of sparsely populated areas indicated a colonial intention rather than any attempt to wrest territory away from

1 **Attributed to Alfonso the Wise (1221–84)** the *Cantigas de Santa María* is a collection of 400 songs recounting a variety of miraculous and legendary incidents connected with the Virgin Mary. This miniature from a contemporary manuscript depicts the salvation from a storm of merchants bound for Acre, their successful business transactions and the homage they paid to a Marian shrine.

2 **The Great Mosque of Córdoba,** begun in the 8th century, was later enlarged and embellished and is a lasting testimony to the magnificence of Islamic civilization in Spain. By 1031 the city had been devastated by civil strife and lost its place of eminence. The mosque survived, was consecrated to Christianity after the reconquest (1236) and had a church built in its centre in the 16th century.

3 **A commentary on the Apocalypse** of St John the Apostle by Beatus of Liebana shows one remarkable side of the culture of Islamic Spain. Christian manuscripts were still copied and studied in monasteries, but their style shows an enormous Islamic influence. Thus in this illumination to the text (which was done at Gerona in about 975) Jerusalem becomes a Mozarabic (Spanish-Muslim) city with the horseshoe arches to the first-floor windows that are the hallmark of the style.

4 **Rodrigo Diaz de Vivar** (*c.* 1043–99) *left,* known as "El Cid", was a Castilian knight estranged from the king, Alfonso VI. His exploits were widely celebrated and resemble those of a present-day mercenary. Supported by a force of faithful Castilians he moved freely in Muslim-held territory. His crowning achievement was the capture of Valencia in 1094, which effected his reconciliation with the Castilian king.

5 **El Cid was born** at Vivar, Burgos, where these monuments mark the site of his ancestral home. The Spanish epic *Poema de Mío Cid* portrays him as a noble Christian hero, loyal to his king despite being banished from Castile. Arabic sources emphasize his cruelty. Many ballads celebrate his achievements as an invincible knight and it has become difficult to distinguish the Cid of history from the Cid of legend.

Muslim control and it achieved very little.

Lack of any effective political cohesion among the Christian states was the main obstacle to territorial expansion at the expense of the Muslims. The Duero and Ebro valleys remained the approximate boundaries until Alfonso VI's definitive occupation of Toledo in 1085 altered the map of the peninsula by placing a permanent Christian wedge in Al-Andalus. From this time the reconquest gathered momentum and the religious factor, hitherto largely dormant, now emerged. The confrontation between Santiago and the Bible for the Christians and Mohammed and the Koran for the Muslims was evident in the clashes between the Almoravids and Almohads, Berber tribes from northern Africa, on the one hand, and the forces of Castile and Aragon on the other.

After suffering reverses, notably at the battle of Alarcos in 1195, the forces of Castile, Aragon and Portugal, usually acting independently of each other, reduced the power of Al-Andalus to some 400km (250 miles) of coastline from Gibraltar eastwards. The Nasrid kingdom with its capital at

Granada lasted 250 years by dint of shrewd diplomacy, judicious alliances contracted from time to time with both Castilians and Muslims from North Africa, trading links with Genoese and Catalan merchants and geographical barriers, such as the Sierra Nevada, that discouraged assaults. The Alhambra, built in the fourteenth century, is the major monument of this bastion of Islamic civilization in a Spain whose political orientation was by then the same as that of the other Western European powers.

Hostility towards the Muslims and reverence for the Islamic cultural tradition were not incompatible in the new Spanish state. While wars were being waged in frontier zones, Toledo, like Norman Sicily [8], became a centre from which Greek and Arabic learning was transmitted to Western scholars. Churches in ornate styles were built by Muslim craftsmen; kings were familiar with Arabic and yet the power of Islam in Spain and Sicily was on the wane. The reconquest was a major political achievement, but the vestiges of nearly 700 years of Islamic presence in Spain were indelible.

The castle of Manzanares el Real, to the north of Madrid, was constructed near the site of an earlier fortress during the second half of the 15th century in the elaborate Mudejar and late Gothic styles. It was the residence of the Mendoza family, whose members were granted the marquisate of Santillana and the dukedom of Infantado for distinguished political and military service. This brought them material enrichment and increasing prominence in Castile's affairs.

Christian territory

Muslim territory

······· Internal boundaries

6 The boundaries between Christian and Muslim Spain altered little until the capture of Toledo (1085). The reconquest achieved its greatest momentum in the 12th and 13th centuries, when the crusading zeal was at its height. The small kingdom of Granada, however, survived until 1492.

6 A
CAROLINGIAN EMPIRE
KINGDOM OF ASTURIAS
UMMAYAD EMIRATE OF CORDOBA
814

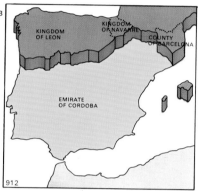
B
KINGDOM OF LEON
KINGDOM OF NAVARRE
COUNTY OF BARCELONA
EMIRATE OF CORDOBA
912

C
KINGDOM OF NAVARRE
KINGDOM OF ARAGON
KINGDOMS OF LEON AND CASTILE
COUNTY OF BARCELONA
INDEPENDENT MOORISH STATES
MOORISH STATES
1037

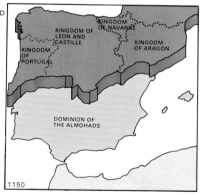
D
KINGDOM OF NAVARRE
KINGDOM OF LEON AND CASTILLE
KINGDOM OF ARAGON
KINGDOM OF PORTUGAL
DOMINION OF THE ALMOHADS
1150

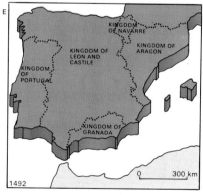
E
KINGDOM OF NAVARRE
KINGDOM OF LEON AND CASTILE
KINGDOM OF ARAGON
KINGDOM OF PORTUGAL
KINGDOM OF GRANADA
0   300 km
1492

7 St James the Elder (died *c.* 44) is the patron saint of Spain. His body was miraculously discovered in the 9th century in the remote province of Galicia where the city of Santiago de Compostela now stands. The shrine became a centre of pilgrimage for Christians, thus opening Spain to European influences, and the Spanish Christians acquired a warrior saint who would lead them into battle against the Muslims.

7

8

8 Byzantine and Arabic features are prominent in Monreale Cathedral, built by the Norman William II (1154–89) who challenged Muslim political and religious supremacy in the Mediterranean.

9 During Raymond of Toledo's archbishopric (1125–51), large numbers of Arabic works were translated into Latin by scholars from all over Europe. Scientific and philosophical treatises were thus introduced to western Christendom.

9

Romanesque art of the 11th century

The term "Romanesque" was borrowed from literary sources by nineteenth-century antiquaries to describe pre-Gothic architecture, implying that it reflected that of Rome. Today it is applied to the art, and more especially the architecture, of the eleventh and twelfth centuries. The political and economic instability of Western Europe after the abandonment of the Western Roman Empire accelerated the decline of widespread patronage of the arts. Despite the glories of the artistic revival under the emperor Charlemagne and his successors, and personal patronage in more localized centres, it was only the re-establishment of strong governments and the parallel reforms within the Church c. AD 1000 that caused the arts to revive throughout Europe.

Patronage of the arts by the Church
Land was the primary source of income in the Middle Ages, and the power of landowners was supreme. The Church readily accepted gifts of land and became a great feudal landowner, owning as much as a third of France in the eleventh century. The other source of income was the offerings of the faithful at saints' shrines [7], and pilgrimages to them were common for people of all classes.

Although Rome and Santiago de Compostela in Spain were the major destinations, visits to many other centres were encouraged. The shrines contained holy relics which were believed by many to have great healing powers. To display them in safety was an incentive for rebuilding a church. The creation of monastic orders – particularly the Cluniacs – to protect these relics and the founding of monasteries to administer both spiritual and temporal affairs also encouraged building.

The Church therefore became the greatest patron of the arts, building religious houses, decorating them with paintings [5] and hangings and filling them with altars, candelabra and screens made of wood or iron. As religious services grew more elaborate, more manuscripts and religious artefacts were acquired. Stone castles and houses from this period are rare, which may indicate that they were not built in great numbers or, more likely, were soon replaced by better and more comfortable structures. The creation of a church for both pilgrim and monk was the builder's main problem.

Romanesque churches of two forms
Two basic church types were current in Western Europe: the wooden-roofed, aisled basilica, and the vaulted small-scale martyrium (erected on a site of religious significance). The former had been taken up by the early Christians in Rome and continued to flourish with little development in central Italy up to the Gothic period. When Charlemagne and his successors sought to revive the Roman Empire from c. AD 800, it was this type of church they copied in Germany.

These churches had apsidal choirs, sometimes at both ends, with west fronts forming multi-storied masses, called *westwerks*, and they were crowned at each end with towers. Linking these complex elements was the long nave, with expanses of flat wall between the ground-floor arcades and upper windows. These areas were painted or hung with embroideries. All Romanesque churches were a mass of colour, now lost.

1 The nave of St Michael's, Hildesheim (1000–33) (looking west) has a 12th-century wooden ceiling and the original suspended candelabrum to light the eastern crossing. The church has no vaults, in the Carolingian tradition, and so is wide and well lit. The solid walls of the elevation are now dull and monotonous, but they were once brightly painted, probably to emphasize the width.

2 The west façade of the monastic church at Sta Maria at Ripoll near Gerona was rebuilt between 1020 and 1032 by Abbot Olivia. This style of church can be seen throughout southern Europe and is distinguished by the use of small rubble building material, the covering of all interiors with vaults and, particularly, by the arched corbel-table running around the summit of all external walls – seen here on the towers and gable.

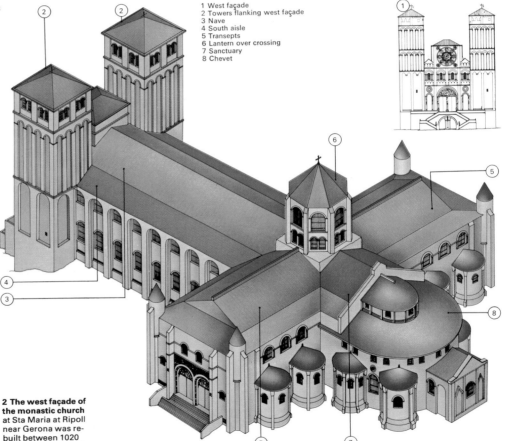

1 West façade
2 Towers flanking west façade
3 Nave
4 South aisle
5 Transepts
6 Lantern over crossing
7 Sanctuary
8 Chevet

3 The cathedral of Santiago de Compostela in Spain is the reputed burial place of the apostle James and was the goal of vast numbers of pilgrims from all over Europe during the Middle Ages. Four main "pilgrimage roads" from France fed a single route through northern Spain, each road being marked by an important shrine. The five greatest churches built between 1050 and 1100 were at Tours (St Martin); Limoges (St Martial); Conques (Ste-Foy); Toulouse (St Sernin); and Compostela. They have similar plans and were influential in spreading the concept of the great Romanesque church – a large, cross-shaped building with a towered main western entrance [1, 2], galleries running around the interior at an upper level and emphasis on the east end with many protruding apses for altars. The nave [3] had altars, but served mainly as a congregating area. Each altar was dedicated to a different saint and had a reliquary on or within it. Candelabra, screens and hangings were placed around the altar. (It was this profusion of fittings that fed accidental fires.) This is the third church on the Compostela site, probably built by French masons and finished c. 1125, although many alterations have since been made.

In central Italy, no such evolution took place. Marble panels were often used instead of wall-paintings in Tuscany, as at San Miniato al Monte, Florence, and bell-towers and baptisteries were usually completely detached, as is seen most clearly at Pisa.

The martyrium type is not so easily identified as the Roman, although its distribution corresponds to the extent of the Roman Empire in the fifth century – that is, north and east Spain, the southern half of France, north Italy, Yugoslavia and Asia Minor. Its roots lay in the provincial Roman buildings, built of small local stone or brick rather than the faced stone blocks of imperial monuments. Although limiting size, this material facilitated the vaulting of the narrower spans, in contrast to the open wood roofs of the wide basilica-church.

As a consequence, however, the mason became a craftsman in stone rather than a sculptor, and any architectural sculpture, such as capitals, was often crude and carved in relief rather than deeply cut into the stone. Apart from the materials used, the decorative system of thin pilasters (ornamental

rectangular columns) rising to a corbel-table (a line of projections running along under the eaves) clearly distinguishes this work.

Romanesque churches in provincial France
These two church types merged, and after much experiment (especially in France), two basic plans evolved, which were modified in each region to suit local conditions. The great pilgrimage churches [3] are typical of the "ambulatory" plan, which allowed pilgrims to see the relics behind the high altar – or beneath it in a crypt – without disturbing the monks at their worship [Key].

Churches associated with the Cluniac order, especially in Burgundy, used parallel apses, that is, a series of semicircular or octagonal recesses of varying heights and depths flanking the main monk's choir. Unlike the ambulatory layout, this plan probably encouraged masons to vault the main areas, as more solid walls were available for support. Lighting such vaulted spaces was a difficulty that was not satisfactorily overcome until the widespread use of cut-stone (ashlar) in the second half of the twelfth century.

KEY

This ground plan of Sainte-Foy, Conques, France (*c.* 1050–1110) allowed for the needs of pilgrim and monk alike. The former would enter at the west [1], and go along the aisles [2] to the ambulatory [3] to look at the principal shrine in the sanctuary [4]. The monks would enter the church by the transept [5] from the cloister (around which the monastery was grouped), and then go in procession to their choir behind screens or to minor altars [6].

4 The nave at St Etienne, Nevers (*c.* 1070), looking east towards the high altar, is made narrow and dark by the need to support the stone barrel vaults. Behind the middle range of arches a half-barrel links the main elevation, helping to support the main span and allowing for small upper windows. The wall shafts divide the elevation into vertical rather than horizontal bay units.

5 In the church of S Angelo in Formis, near Capua (rebuilt after 1072), this picture shows Christ touching the eyes of the blind man, who then washes at the well and sees again. Spots on the cheeks and angular drapery patterns show the Byzantine influence. The paintings illustrate scenes from the Old and New Testaments and cover the whole interior of the church. Such a complete cycle is rare today.

6 The marble lintels at St Genis-des-Fontaines, French Pyrenees (1020–21), show the enthroned Christ [B] supported by angels. Saints [A] flank the figure. It is one of the earliest known pieces of monumental medieval stone sculpture. The layout and the light chip carving indicate that the sculptors had studied earlier stone reliefs and contemporary metalwork such as the Basle altar-frontal.

7 The gold-covered cult-figure of Ste-Foy was originally made in the 10th century, but pious visitors to her shrine at Conques have, over the centuries, encrusted this reliquary with precious stones, cameos and crystals. Although other figures were made, a statue was considered idolatrous and most reliquaries are casket or shrine shaped and contain the saint's relics. Sometimes a representation of a particular part of the holy body is on top.

8 This large altar-frontal of beaten gold was given to Basle Cathedral by the German emperor Henry II (*c.* 1019). The central figure of Christ is flanked by the archangels Gabriel and Raphael on the left, and SS Michael and Benedict on the right. Henry and the Empress Kunigunde kneel at Christ's feet. Few large precious-metal objects have survived from the early periods. Their value as metal tempted later generations to melt them down.

Romanesque art of the 12th century

The huge churches built towards the end of the eleventh century placed great emphasis on the east end, where the main altars and relics were sited. Outside, the layered walls and roofs of the apsidal chapels, ambulatory, clerestory and perhaps galleries, culminated in mighty towers; inside, the increased use of sculptural decoration and sumptuous church furniture emphasized the respect owed to the shrines in the sanctuary. On entering, usually at the west front, the pilgrim often had to pass under evocative, carved tympana, dominated by a central Christ figure surrounded by the Heavenly Host, or vigorous apocalyptic scenes from the Last Judgment.

Interior decoration by secular artists

The increased use of ashlar had led to greater skill in stone-cutting. Because the walls were now divided into arcades and piers, the didactic and decorative functions of the large-scale wall-paintings were transferred to sculptured capitals and portals (which were then painted) or, in some regions, to stained glass. These capital designs had to follow the lines of the particular structure but a tym-panum over a door, or a panel inserted into the wall, had no structural function, so only the shape was determined by architectural factors. Animal and human figures, there-fore, had to be reinterpreted and traditional iconography amended to fit these new shapes. For subject matter, the sculptor looked to the older arts of metalwork and manuscripts which supplied an endless variety of suitable iconography. But for figurative techniques, surviving Roman sculpture was undoubtedly influential.

This divergence from the original, and the different treatment of stone, help to explain the many styles that emerged, often with little connection to regional variations. It must be remembered that workshops were composed of itinerant masons, sculptors and painters hired by ecclesiastical patrons, and not of resident monks. Some even autographed their work, like Gislebertus at Autun [7], or Wiligelmus at Modena. Although a man might specialize in designing secular buildings such as castles and town houses, his workmen would probably also have built and embellished abbeys, cathedrals or parish churches. Decorative details are therefore common to all types of structure from centres of worship to dwelling-places.

Reactions to ostentation

An enormous increase in the demand for church furnishings led to a concentration of lay craftsmen in places where patronage or raw materials were most readily available. Products from centres such as Cologne or Liège (metalwork), or Limoges (enamels) [4], and sometimes master craftsmen, too, were sent throughout Europe. Similar methods probably applied to the production of illuminated manuscripts [5], particularly the lavish "picture books" for the personal use of abbots and bishops. Everyday service books, however, were most likely copied out within the cloisters of major monasteries. Such ostentation produced inevitable reactions. In particular, the Cistercian order, led by St Bernard of Clairvaux (1090–1153), returned to the monastic ideals of poverty and purity which they felt had been betrayed by painted sculpture and the liberal use of precious metals and jewels.

1 The nave of Durham Cathedral (finished in 1133) is a tentative application of rib-vaults to a main span, but clearly demonstrates the advantages that such a co-ordinated vault type could bring. The emphasis is still on strength and solidity and the cylindrical piers cannot hide their size behind the simple surface patterns. The chevron or zigzag was a popular motif with Norman builders in 12th-century England.

2 The medieval builder depended as much on resources of wood as on stone because huge scaffolds were needed for the construction of vaults and arches. The arch stones (voussoirs) were placed on wooden centring, the last keystone locking the radiating voussoirs together. Once the mortar had set, the centring was lowered and moved on. Exact workshop practices are, however, difficult to ascertain.

3 This is the massive central keep of the Tower of London (1078–97), one of the earliest Norman castles in Britain. Few stone castles existed in western Europe before 1000, but by the end of the 12th century hundreds had been constructed to house noblemen's families and defend their dependants in times of attack. The Crusades gave rise to more sophisticated and expensive techniques.

4 The life of St Valerie, the patron saint of Limoges, is illustrated on this enamelled casket (c. 1170). According to legend, she was engaged to a prince, who found on his return from war that she had dedicated her life to God and so could not marry. The prince then had her beheaded (right). On the lid, St Valerie can be seen, head in hand, walking to mass escorted by angels. This is one of many ornately embellished copper casket reliquaries with attached panels decorated with figures enamelled in blue, red and turquoise on golden backgrounds. Most surviving medieval enamelwork is on church accessories, the medium lending itself to delicate and glittering small-scale work.

5 The Bury Bible is an illustrated English manuscript of about 1140. On the left, a large painted panel depicts the scene of the prophet Jeremiah lamenting the imminent fall of Jerusalem. On the right, a decorated initial heads the text with armed men and dragons inhabiting the vine-like foliage. The rich palette, especially the solid background colouring, and the carefully drawn faces are evidence of the high quality of the manuscript.

In order to display these treasures to their best advantage, the vogue throughout Europe for stone-vaulted churches had to be reconciled with another factor, that of natural, direct lighting. Dark, mysterious interiors were rivalled by open, well-lit spans, the light sometimes being filtered through stained glass. In barrel-vaulted churches the half-cylinder of stone that makes up the roof exerts an even, outwards thrust along its entire length, requiring continuous support. Building windows into these walls therefore compromised the stability of the vault.

In Poitou, the barrel-vault rises directly from the top of an arcade of very tall columns. Large windows in the aisle wall flood the central space with light. However, this was only efficient in narrow, single-aisled buildings, and left the upper half of the church dark. A more practical system shed light from both aisles and clerestory: to achieve this, vaulted galleries above the aisles were used to buttress the main elevation, just below the "springing" level of the main span. This assured stability and allowed small clerestory windows to be made. However,

the problem could only be solved by dispensing with continuous support, and this meant new concepts of vault design.

The development of new vault types

In western France, about 1100, great domes were built on huge arches over the main span, which concentrated pressure on the four corners. These were then heavily buttressed. This solution was admirable for the traditionally aisleless churches of that area, but elsewhere, especially in the Anglo-Norman region, aisles surmounted by open galleries as high as the main arcade were considered necessary. Over aisles and around ambulatories, "groined" vaults (formed by two simple tunnel vaults intersecting at right angles) had been used. Again, the four corners provided support so that the walls between could be pierced for windows.

In northern France, from about 1140, such a system evolved, later to be called Gothic. But elsewhere, Romanesque wall architecture continued until the thirteenth century, when the other arts had begun to look for more naturalistic inspiration.

In this detail of a capital from the narthex at Vézelay, St Paul (on the left) is praying for the redemption of the world. The function of the capital is to make a smooth transition between the round shaft below and the square or rectangular-shaped masonry above. The Romans used a number of capital types, employing stylized acanthus leaves and scrolls. When figurative scenes were carved out of capitals the sculpture had to conform to the architectural shape and not interrupt the gently splaying lines. Here, St Paul's leg, back and head continue the capital's contours, although the actual pose of the figure appears unnatural.

6 Originally the west doors at Vézelay fronted the hilltop church, but a large towered narthex or porch was added in about 1150, and this vulnerable sculpture was protected from the weather. It is one of the greatest ensembles of Romanesque carving, full of energetic figures in ample and swirling draperies.

7 The existence of many Roman monuments in Autun influenced the architectural decoration of the 12th-century cathedral there; flat fluted pilasters, for example, were used rather than the usual round shafts. The wealth of sculpture typifies the decorative concepts of a developed Romanesque church.

8 The painted apse of the chapel at Berzé-la-Ville is thought to have been commissioned by Abbot Hugh of Cluny before his death in 1109, and to be a smaller version of the main apse painting of his great abbey church at Cluny. Directly above the altar, Christ sits in glory within an almond-shaped panel surrounded by the Heavenly Host. He is giving a scroll to St Peter, who is holding the keys to heaven. Below the windows there is a range of saints and around the base of the apse wall, simulated drapery. Originally all great churches had such scenes over the altar, often illustrating the martyrdom and life of the saint to whom the altar was dedicated.

9 The west façade of Notre-Dame-la-Grande, Poitiers (c. 1140), is considered to be one of the finest in western France, an area noted for its copious use of external sculpture. Also typical of the area are the corner turrets faced by shafts and topped with scale-like stone roofs. Although there is only one central entrance, the custom of inserting three or more deep decorated portals within a grandiose façade was popular in the Middle Ages.

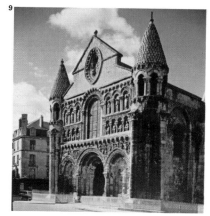

The Crusades

The immediate cause of the First Crusade – and the starting-point for a major stage of European expansion – was the threat to the Byzantine Empire created by the Seljuks. These Muslim tribesmen had conquered the empire's richest province, Anatolia, and farther east had dispossessed the caliphs of Antioch, Tripoli and Jerusalem. An attack on Constantinople seemed inevitable.

In 1095 the emperor, Alexius I Comnenus (reigned 1081–1118), an able soldier and diplomat, therefore asked Pope Urban II (1042–99) for assistance, baiting his request for a contingent of mercenaries to retake Anatolia with the suggestion that they could then travel on to liberate Jerusalem. Urban responded [1] and for the first time the Papacy sanctioned a "holy war". It promised that whoever undertook this pilgrimage would be freed from all penances due.

The First and Second Crusades

The four separate armies of the First Crusade [3] converged on Constantinople in the winter of 1096–7: men from Lorraine, the Norman kingdom of Apulia, Provence, Brit-tany, Normandy and Flanders. The Franks took Nicaea and Dorylaeum; Antioch fell after a lengthy siege, and the Crusaders stormed Jerusalem in July 1099 [2].

With the holy places conquered, territorial motive now became paramount. Four Frankish states [5], defended by the castles and garrisons of the Templars and Hospitallers [6], survived increasing Muslim pressure until 1144, when Edessa fell. A crusade was called for by Pope Eugenius III (died 1153). Emperor Conrad III (1093–1152) and King Louis VII of France (1121–80) incompetently led the armies of the Second Crusade (1147–9) to starvation and disaster in Anatolia.

Jerusalem falls to the Turks

A further revival of Islam and the empire's final loss of Anatolia to the Seljuks in 1176 left the Latin states in danger. Saladin (Salah ad-Din, 1137–93), the brilliant Kurdish vizier of Egypt, united Islam from the Nile to the Tigris and in 1187 invaded the Latin Kingdom of Jerusalem, and overran the Frankish states.

The armies of the Third Crusade (1189–91) came to their aid and Frederick I (Barbarossa) (1123–90), the Holy Roman Emperor, took the Seljuk capital of Iconium. Phillip II Augustus of France (1165–1223) and Richard I of England (1157–99) joined the ex-King of Jerusalem, Guy of Lusignan (1140–94), in besieging Acre, which surrendered after a two-year siege. Richard then set out for Jaffa, the port of Jerusalem, and although he won the coast from Tyre to Jaffa for the Christians, was prevented from attacking Jerusalem.

The Fourth Crusade (1202–04), supported by Pope Innocent III (c. 1160–1216) to restore the kingdom of Jerusalem, resulted in the debasement of the ideal: war was now made against fellow-Christians for gain. Venice, which largely controlled the eastern Mediterranean, forced the army to accept a price for transport to Egypt which it could not pay. The Doge of Venice agreed to remit the debt only if the troops were diverted to repossess Zara on the Adriatic, a former Venetian city taken by the Magyars in 1186. The army was then persuaded to intervene in

1 **Pope Urban II's appeal** to the Council of Clermont (1095) launched the First Crusade and was an attempt to reconcile Church and state. The reply to his call, a shout of "*Deus vult*" (God wills it), later became the battle-cry of Crusader-knights in the Holy Land. His appeal led to the spontaneous and ill-disciplined People's Crusade (1096). The Pope's appeal to biblical images of Jerusalem, the heavenly city, made the idea of freeing the earthly city one of great splendour and power. The Pope reinforced his appeal with calculated references to Western overcrowding and famine.

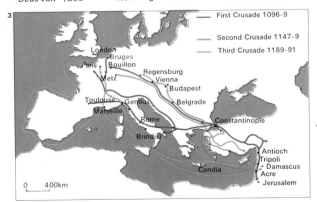

First Crusade 1096-9
Second Crusade 1147-9
Third Crusade 1189-91

London
Bruges
Paris
Bouillon
Metz
Regensburg
Vienna
Toulouse
Genoa
Budapest
Marseille
Belgrade
Rome
Constantinople
Brindisi
Antioch
Tripoli
Candia
Damascus
Acre
Jerusalem

0   400km

3 **The recovery of the holy places** and the protection of the subsequently established Frankish states were the aims of the early Crusaders. The separate armies of the First Crusade (1096–9), from France, Provence, Normandy, Flanders and Apulia, joined at meeting-points throughout Europe, marched through Magyar territory to Constantinople and fought their way across Asia Minor. Those of the Second Crusade (1147–9), led by the kings of France and Germany, also went overland, but their refusal to adapt to conditions of Eastern warfare led to their destruction by the Seljuks of Asia Minor. By the Third Crusade Western naval strength had improved and Richard I chose a sea route to Acre.

2 **Crusaders besieging Jerusalem** in June and July 1099, faced fortifications more complex than any in northern Europe. A quick assault was necessary as the defenders had poisoned the wells for 10km (6 miles) around the city. Wood had to be fetched by sea for the scaling ladders, mangonels (beams that hurled boulders), giant catapults and trebuchets (slings worked by counterweights). Three wooden "castles" on wheels, to make possible attacks on the upper levels of the walls, were hung with hides to ward off arrows and "Greek fire" (a blazing naphtha-based mixture extinguishable only by vinegar). The knights cared for their horses before they looked after themselves: heavy warhorses, trained to charge home against infantry and able to carry a man wearing a third of his own weight again in armour, were irreplaceable in the East. Loss of his horse reduced the knight to the ranks of the foot soldiers. The heat claimed many lives.

Fourth Crusade 1202-04 Venice-Constantinople
Fifth Crusade 1218-21
Sixth and Seventh Crusades 1228-9 and 1248-50

Aigues-Mortes
Venice
Brindisi
Constantinople
Tunis
Candia
Tripoli
Acre
Damietta

0   400km

4 **Directly concerned with saving Jerusalem**, the 4th–7th Crusades, aroused far less enthusiasm in the West than had their predecessors. Greed and hatred for fellow-Christians in the East made the army of the Fourth Crusade easy prey for the manipulations of Venice. Personal magnetism and negotiating skill brought about what successes were later achieved by Emperor Frederick II and King Louis IX of France.

a Byzantine dynastic quarrel and besiege Constantinople itself. The presence of a Frankish army brought to a head hatred between Greek and Latin Christians, long fostered by mutual blame for successive crusading disasters. In April 1204 the Crusaders seized and looted the city.

The Crusaders finally defeated

After the failure of the Fifth Crusade, 1218–21, the last in which the Papacy was actively involved, the Pope's Hohenstaufen enemy in southern Italy, the Holy Roman Emperor Frederick II (1194–1250), conquered Jerusalem by political means. He claimed its throne through his wife and sailed on the Sixth Crusade (1228–9) while actually excommunicate. Supported by the Teutonic Knights, he negotiated a ten-year truce that restored Jerusalem (except for the Temple and Muslim holy places) to the Franks. In 1229 Frederick crowned himself King of Jerusalem but, after the truce, quarrels over territory between Templars and Hospitallers so weakened the kingdom that it fell in 1244 to the onslaught of mercenary Turks.

Louis IX of France (St Louis, 1215–70) made another attack on Cairo in the Seventh Crusade of 1248–50, but was taken prisoner. Freed by ransom, he rebuilt Jaffa and Acre and conciliated Muslim leaders. His return to France, however, left the kingdom of Jerusalem crumbling because of the renewed rivalry of the military orders. Baibars, Mameluke Sultan of Egypt from 1260 to 1277, took advantage of this division; in 1268 he seized Jaffa and Antioch, and in 1271 the castle of Krak des Chevaliers [Key]. Louis IX, mortally ill, set out again on the Eighth Crusade (1270) but died in Tunis; Prince Edward (later Edward I of England, 1239–1307) reached Acre in 1271 and negotiated an 11-year truce, but in 1289 Tripoli fell to the Mamelukes and in 1291 they captured the last stronghold, Acre.

The territorial and spiritual triumphs of the Crusades were short-lived. Urban's vision of a united Christendom degenerated into papal autocracy; the division between Latin and Orthodox Christians became absolute. The West's chief defence against Islam, the Byzantine Empire, was fatally weakened.

Krak des Chevaliers, best-preserved of Crusader castles, guarded the north-west flank of the County of Tripoli. Begun in 1142 by the early Crusaders, it defied 12 sieges, falling to Baibars in 1271 when its garrison of 2,000 was reduced to under 200. Frankish forts were first built by the Templars to protect the pilgrim route to Jerusalem and later grew into chains of castles guarding the frontier and ports.

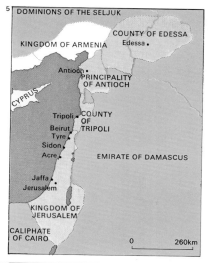

□ Muslim territory
□ Extent of Crusader States, about 1140

5 The four Crusader states comprised the County of Edessa (founded 1098), the Principality of Antioch (1098), the Kingdom of Jerusalem (1099), which claimed the overlordship, and the County of Tripoli (1109). The last, the smallest and weakest, had an unexpected ally in the Assassins, an heretical Muslim sect which supported the Franks against Damascus for 200 years. The states were feudal, had their own coinage and traded widely with Europe and the Levant.

Hospitaller   Teutonic Knight   Templar

6 The military orders' offer of a life of religion and battle brought recruits from England, France, Portugal, Spain, and Italy. The Hospitallers began in the 11th century as guardians of the pilgrim hospice of St John in Jerusalem. The first Templars, given their rule by Bernard of Clairvaux in 1128, lived in the palace of the King of Jerusalem, the "Temple". In contrast the Order of Teutonic Knights, founded in 1198, was confined to German nationals only. The orders were answerable only to the pope and held the only standing armies in the East.

7 The fortified cathedral of Albi in southern France, founded in 1277, was a deliberate symbol of the Church Militant's triumph over the Cathar heresy, which denied the reality of Christ's incarnation. The Cathars were brutally supressed by Simon de Montfort l'Amaury (1160–1218) in a crusade of 1209, inspired by Innocent III, who introduced the idea of mounting crusades to crush Christian heretics and enemies of the Papacy.

9 Saladin, a chivalrous and courteous enemy, was taken as an ideal by many Frankish knights. This 13th-century drawing from the *Chronica Majora* of Matthew Paris shows him wresting a relic of the True Cross from Guy de Lusignan, King of Jerusalem, at the battle of Hattin (1187).

8 The sixth-century Quadriga of Lysippus, four splendid bronze horses taken during the sack of Constantinople (1204), still stand over the portal of St Mark's, Venice. A less tangible result of the Fourth Crusade was the flow into Western Europe of Greek scholars seeking refuge from the Turks, which gave impetus to the Renaissance. Earlier Crusades imported to the West new goods: damask, muslin carpets, rice, sugar, lemons, spices, dyes; and the Arabic numerals which revolutionized mathematics.

10 Crusaders' tombs in English churches usually show a knight lying peacefully, his legs crossed. But in Dorchester Abbey Sir John Holcombe, who died on the Third Crusade, is shown in effigy struggling to draw his sword.

The king and the barons

The institutions of English government were born between 1100 and 1400; only the monarchy itself is older. Kings before Henry I had taxed and dispensed justice, but the permanent institutions of the Exchequer and the King's Court on Justice took shape in his reign. They were the foundations of the vast structure of government that developed in the twelfth century: the Chancery, the Privy Seal, the courts of King's Bench and Common Pleas. The workings of this great machine penetrated into every corner of the kingdom, and its powers and limits were the underlying political question of the age.

The growth of the leading families

Naturally, those with most to gain or lose from the king's government were his greatest subjects, whom medieval chronicles and documents such as Magna Carta call the "barons". During the twelfth century, the English nobility constituted a formidable order. Until then, the members of great noble families had been adventurers, and few survived more than two generations. From Henry I's reign onwards the natural life of a baronial family was much longer, as political co-operation and the development of landed estates became more attractive.

In defence of the principle of inheritance, the families made a determined front, and subsequent kings could not ignore the power and the steadily accumulating wealth of successive generations of Lacys, Clares, Bigods and Mortimers. Far from being opponents of central government, they usually wished to participate in it, and the "common law" administered in the king's courts, with its sensitivity to questions of land and inheritance, was developed to meet their needs.

The twelfth century was the period of formation for both the baronage and the king's government. Henry I (reigned 1100–35) [Key] raised many new families from obscurity, which threatened old-established barons, but it was reasonably possible in his reign for a landed family to hold on to its property and to prosper. But the civil wars of Stephen's reign (1135–54) [1], which followed Henry's death, was the critical time for all landed families, old or new. The succession dispute, far from pro-

viding opportunities for them, made their tenures radically insecure. The Treaty of Winchester (1153), which assured the succession of Henry II, also ensured the survival of the great landed estates, and most of the barons gave Henry their loyal support in return for the profitable "incidents" (wardships, or marriage to heiresses in the king's protection) or offices (such as the shrievalty of a county) created by renewed royal power.

Magna Carta and after

The loss of the European dominions by King John (reigned 1199–1216) [2], and his attempt to regain them forced him to exploit his authority and patronage, and "amerce" or take a fine from the barons for every favour he did them and every real or imagined offence he "overlooked".

Many were brought to the point of ruin. The inevitable reaction resulted in Magna Carta (1215) [4], in which the barons tried to enumerate and forbid the various abuses of royal power. Its reassertion of fundamental rights was not a revolutionary idea, but it marked the resumption of the leadership

1 Henry II crossed to England from France in 1153, as this contemporary miniature shows. By 1150 it was clear that King Stephen could not eject the Empress Matilda or her son Henry from Normandy, nor could they expel the king from England. Henry's invasion threatened another bloody stalemate, but Stephen was made to disinherit his son. Henry II and his magnates co-operated in restoring order, devising a more just and effective law, and introducing trial by jury and assize courts. These courts, which relied on travelling judges dispensing royal justice, impinged to some extent on the barons' rights to hold feudal courts on their own estates.

2 The Great Seal of King John (c. 1210) symbolized the beginnings of a momentous change in methods of government. In John's reign the Chancery formalized its procedure and began to keep systematic records of the grants and royal acts made under the Great Seal. The two series of Patent Rolls (open letters) and Close Rolls (written to individuals) show the birth of a professional civil service with its own interests.

3 John gives the traditional kiss of peace to Philip II Augustus, King of France (r. 1180–1223), his feudal superior. John, who as the brother of Richard I (r. 1189–99), had occasionally been an accomplice of Richard's enemy, King Philip, was far less successful at holding together the Angevin Empire than Richard. By 1204, the French king's armies had swept through the fortresses of Normandy; the attempt to regain them put a strain on English resources until the Treaty of Paris in 1259. The loss of Normandy brought to an end long wars in France, but in the subsequent recriminations it also had the effect of tarnishing John's image, and from then on he was regarded as an incompetent king.

4 A copy of Magna Carta (one of only four official copies to survive) symbolizes the acknowledgment of the fact that the monarchy is subject to the law of the land. This recognition was extracted from an unwilling King John at Runnymede by a league of barons, churchmen and townspeople. Magna Carta is a detailed schedule of specific articles that guaranteed many kinds of inherited rights and privileges and limited the king's prerogatives.

5 The Battle of Evesham (1265), which ended the "Barons' Wars", saw the defeat and death of Simon de Montfort. Despite this, his reforms of government reflected the growing authority of the classes below the baronial one, were largely accepted.

of the community, by the united baronage.

Most of the subsequent conflicts between king and barons were the result of the Plantagenets' foreign ambitions and the strain this put upon their subjects. In each case the barons acted merely as the spokesmen of a now entrenched and articulate landed class. The ambition of Henry III (reigned 1216–72) to intervene in Italian politics led to a spirited contest with Simon de Montfort (c. 1208–65) at the head of a strong popular movement. The baronial programme, enshrined in the Provisions of Oxford of 1258, put control of the administration into the joint hands of the king and the barons to prevent the Crown developing interests against those of the nobility and, by implication, those of the country. De Montfort's cause was finally overthrown [5], but he had sufficiently shown the impossibility of foreign wars without a broad consent.

The barons and external wars
Edward I (reigned 1272–1307) was more masterful than his father, Henry III: but the strain put on the economy by his Scottish wars – taxation, the king's right to "purveyance" (seizure of supplies for his campaigns), and the hard military service – alienated his barons in the absence of decisive victory. They vented their frustration on his weaker son Edward II (reigned 1307–27). At first only his favourite, Piers Gaveston (died 1312), was under attack, but the barons demanded in the "Ordinances" of 1311 control over the king's household and his powers of patronage.

Edward III (reigned 1327–77), however, showed that confrontation could be avoided. He reconciled the baronial factions in the great national enterprise of the French wars and opened up for them prospects of new wealth at French expense, at a time of falling profits from land. Richard II (reigned 1377–99) [10], because of his reliance on a private army and household government and Henry VI (reigned 1422–61; 1470–1), because of his insanity, were both driven from their thrones; but a fundamentally stronger national government came into existence in the fourteenth century through the combined efforts of kings and barons.

The nightmare of Henry I was narrated in an illustrated 12th-century chronicle by John of Worcester. According to this, Henry dreamed that he was confronted first with a mob of infuriated peasants, then with a group of armed knights brandishing their swords, and finally with a company of aggrieved prelates. The dream symbolizes the contemporary idea that society was divided, as King Alfred said, into "Men who fight, men who work, and men who pray". Henry's extortions from the propertied classes had given him a reputation as a rapacious ruler. He also encouraged centralization of power, a concept that aroused opposition from the entrenched nobles and their followers.

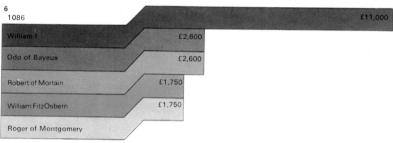

6
1086

William I	£2,600
Odo of Bayeux	£2,600
Robert of Mortain	£1,750
William FitzOsbern	£1,750
Roger of Montgomery	

£11,000

1436

	£5,900
Henry VI	£3,400
Richard, Duke of York	£3,100
Richard, Earl of Warwick	£2,800
Humphrey, Earl of Stafford	£2,250
Humphrey, Duke of Gloucester	

6 The king's income in 1086 and 1436, compared well with that of noblemen for the same period. The king's land in 1436 was mostly Lancaster (ie private) estates but his income was greatly boosted by taxes and customs (which are not shown here). By 1436 the king's advantage in landed wealth had obviously diminished.

7 The Court of King's Bench in about 1250 administered laws and functioned by means of already established procedure. The judges, clerks and attorneys were all professional lawyers. The court was one of the most valuable institutions to the Crown in terms of prestige in the country and income from fines.

8 The wheel of fortune was a popular symbol in the 1300s and 1400s. It testified to the preoccupation with the acquisition of wealth and advancement via patronage. Many medieval illustrations stressed the impermanence of authority, and new balances of power had constantly to be made. The opportunities open to able, ambitious but poor young men increased in the 1300s. Fortunes could be made from war, and time spent in the service of the king or a great nobleman could also be profitable.

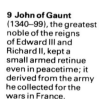
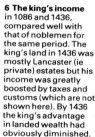

9 John of Gaunt (1340–99), the greatest noble of the reigns of Edward III and Richard II, kept a small armed retinue even in peacetime; it derived from the army he collected for the wars in France.

10 Richard II handed over his sceptre when challenged by his exiled cousin, Henry Bolingbroke (Henry IV). The unreliable Richard had made too many of his subjects feel insecure to attract their permanent loyalty.

Medieval Ireland

The continual feuding of the petty Irish kings during and after the Viking period eventually brought about English intervention in 1169, and the Norman barons who settled in the country soon controlled the greater part of it. Throughout the period there remained in the west a distinctly Gaelic culture, by which the Anglo-Irish lords of the Midlands in the fourteenth century came to be affected; the east maintained its ties with England. The great Midland families gradually increased their power and independence from the English king and the eastern English colony, and because of internal strife the English kings were in no position to reassert their control until the end of the fifteenth century.

The Vikings and the English

By 900 the Vikings had already gained footholds along the Irish coast [1, 2], and Dublin had been founded as a trading colony as early as 841. But the Vikings had not shaken the rule of the Irish kings in the interior. The strongest of these proved to be the northern Ui Néill dynasty who defeated the Eóganacht kings of Munster in 908. The Eóganacht power in Munster then gradually passed not to the Vikings in Waterford and Limerick but to the Dál Cais of north Munster. The most famous personality of this *sept* (Irish clan) was Brian Boru (941–1014), who succeeded in becoming king of all Ireland by 1002, and finally destroyed the Viking threat at the Battle of Clontarf in 1014 [3].

Despite the reputation the Vikings had earned as ruthless plunderers, they contributed much to Irish life. Many of the country's most important maritime towns were founded by them, and from the Vikings the Irish learnt seafaring, and commerce based on coinage and market-places.

By the eleventh century the great Irish monasteries were in decline, and at a number of synods, culminating in the synod of Kells in 1152, reforms were introduced to bring them into conformity with the rest of Europe. Many older monasteries were not revived by the new division of Ireland into dioceses, but others gained new purpose by adopting Augustinian rule. The foundation of Mellifont by the Cistercians in 1142 kindled a new enthusiasm for monastic life, which was maintained and fostered by the Dominicans and, particularly in the fifteenth century, by the Franciscans.

Invasion and resurgence

With the extinction of Viking power and the death of Brian Boru, the Irish chieftains were free to pursue unresolved wars for 150 years. In 1166 the ruler of Connaught, Rory O'Connor (1116–98), became the first king of all Ireland since 1014, but his glory was brief. Dermot McMurrough (1110–71), whom he had ousted from Leinster, sought English aid and the English king, Henry II, sent over from Wales a contingent of Norman barons. But the conflict of interests soon ended any arrangement between the Irish and the invaders, despite the brief peace of 1171 and a second compromise, the Treaty of Windsor of 1175. The Irish kings were outraged by Henry's repeatedly granting land to his vassals, but what the English king made over on parchment, the barons were determined to conquer by the sword. One region after another – although not the

1 **A ship drawn on wood,** dating from about 1100, was discovered during excavations in Dublin. Showing clear Viking influence, it offers a rare insight into the type of vessel used by traders of the period.

3 **"Brian Boru's harp",** traditionally said to have belonged to the Irish high-king, was in fact made about three centuries after his death in 1014. It was Brian who finally subjugated the Viking invaders in Ireland. In 978 Brian inherited his brother's kingdom of Munster and began his avowed conquest of all Ireland. By 984 he controlled Leinster and was supreme over southern Ireland; in 1005 Brian became the first King of Ireland. In 1013 the opposition to his rule – the men of Leinster and the Vikings with support from the Orkneys – united in armed revolt. Brian Boru's army met them at Clontarf in 1014 and emerged victorious, although Brian was slain by the fleeing enemy.

2 **The many Viking descents upon Ireland** began in the fateful year of 795. During the next 50 years, Norwegian Vikings made sporadic raids down the west and east coasts, plundering as they went. They also sailed up the Shannon from Limerick in an attempt to gain a foothold in the area of Clonmacnoise and Lough Ree. They established a permanent settlement with the founding of Dublin in 841. In the century after the Danes appeared (c. 850), Viking settlement was extended south of Dublin along the east and south coasts, and towns with Viking names such as Wicklow, Waterford and Wexford were founded.

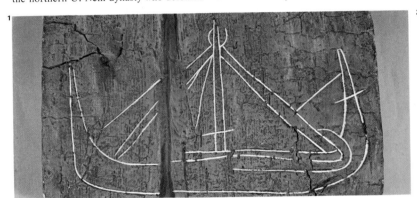

First Viking attack 795

Viking invasions 9th century

Viking fleets

Viking settlements

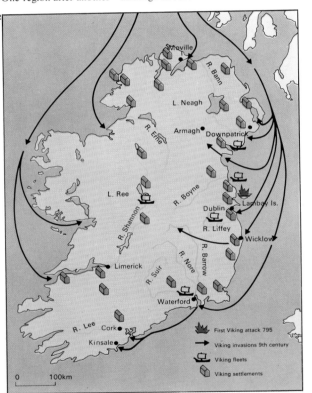

4 **Dunguaire Castle,** on Galway Bay, is one of many surviving tower-houses built by native Irish chieftains between 1450 and 1650. The fortifications of these structures were much smaller than those of the 13th-century Norman castles.

northwest – fell to the Normans [6], who were better equipped and better disciplined than the Irish. The invasion of Connaught in 1235 marked the greatest extent of Norman power. The Normans consolidated their newly won territories by building castles [Key], and settlers founded and fortified many inland towns.

Despite the Norman Conquest, Gaelic Ireland survived. It not only retained its laws, language, traditional institutions and culture, but from 1260 began to fight back. The English presence in Ireland was severely shaken by the invasion in 1315 of the brother of the king of Scotland, Edward Bruce (c. 1275–1318), who saw himself as a liberator, and lost only his final battle in 1318. The colonists were further reduced by the onset of the Black Death in 1348. It was in the west, the stronghold of Gaelic power and culture, that the de Burgo family in Connaught began the trend among the Anglo-Irish lords of adopting Irish customs and even language. In 1361 Lionel, Duke of Clarence, the second son of Edward III, having failed to reduce the whole country to

his rule, enacted the famous and divisive Statutes of Kilkenny (1366). These in effect outlawed all who adopted "the manners, fashion and language of the Irish enemies".

Virtual independence

Two visits by Richard II in the 1390s were no more than superficially successful [7]. The "Pale", the area remaining English and loyal, became smaller in the fifteenth century, while the virtually independent Anglo-Irish lords of the Midlands expanded their power. In the west, Gaelic chieftains had wrested the land from absentee landlords. First the earls of Ormond, then the earls of Desmond and finally the earls of Kildare were the greatest power in the land; and the greatest of these was Garret More Fitzgerald, the Great Earl of Kildare, who when he died in 1513 was the uncrowned king of Ireland. A parliament at Drogheda in 1460 had declared that Ireland, should be bound only by laws passed by itself. The 1494 parliament under Edward Poynings (1459–1521), the viceroy of Henry VII, revoked this and heralded the Tudor reconquest [8].

Trim Castle is the largest stone castle in Ireland. Completed by a Norman baron named De Lacey (c. 1200), Trim consists of a central tower in the middle of a free-standing area defended by a curtain wall, originally surrounded by water. Shortly after the Normans first landed in Ireland, in 1169–70, they built strong fortifications to enforce their claims. At first they built mottes – earth mounds shaped like plum puddings – with a wooden tower on top. The massive stone castles were built later.

5 The Normans defeated the native Irish armies, at first, both by superior tactics and by better weapons and armour. This superiority is epitomized in the splendid effigy of a knight at Kilfane, County Kilkenny, shown here. He is clad from head to toe in mail with his body covered by a surcoat (a sleeveless outer garment, worn on top of chain mail). A sheathed sword hangs from a loosely buckled belt slung round his waist, and he wears the early type of rowel spurs, which helps to date the effigy at around 1320. The coat of arms on his triangular shield proclaims him to be a Cantwell, probably "Long Cantwell", who described himself as old in 1319. This was a period when Norman power had already begun to decline in Ireland, and shortly before the Normans changed over to the use of plate armour.

- ▲ Mottes
- ♜ Major castles
- 🏰 Walled towns

6 The Normans arrived in Ireland in 1169–70, and by 1172 were masters not only in Dublin, Wexford and Waterford, but also of much of Leinster. By submitting to King Henry II of England, many Irish rulers opened the way for Norman colonization and the founding of towns in their lands. Subsequently, Norman barons conquered Meath, eastern Ulster and large parts of Munster. By the early 13th century, they had overrun most of the areas east of the Shannon. In 1235 they established their dominion over the western province of Connaught, and 15 years later over Co Clare. By 1250, the greater part of Ireland was in their hands, the northwest alone succeeded in retaining its independence. But by 1400, the Gaelic chieftains had regained many of their lost territories, so that by the 15th century, Anglo-Norman domination was largely confined to the provinces of Leinster and Munster.

7 Art McMurrough Kavanagh (right), King of Leinster confronts the Earl of Gloucester. The Earl had been sent by Richard II to demand McMurrough's submission. But he averred that he was Ireland's king.

8 Henry VII (r. 1485–1509), seen here in effigy in Westminster Abbey imprisoned his viceroy in Ireland, Garret More Fitzgerald, in London for a year while the Irish parliament passed Poyning's Law, in 1494, reasserting English control.

Scotland to the Battle of Bannockburn

The Romans attempted to protect their British province by extending their power amongst the Pictish tribes beyond the frontier. For a while, in the second century AD, they occupied the south of what was later called Scotland and built the defensive Antonine Wall across the narrow waist between the Forth and the Clyde [1].

The foundation of Scotland
In the sixth century "Scots" from Ireland crossed the narrow sea and began to settle among the scattered tribal communities of Picts, founding the kingdom of Dalriada in what is now Argyll. Eventually in the ninth century, by a mixture of royal intermarriage and conquest, the Scottish king Kenneth mac Alpin (died 858) became king of both Scots and Picts, and his successors extended this kingdom to include the area of Strathclyde, occupied by Britons, and of Lothian, where there were also Anglo-Saxon settlements. The Picts disappeared from history and the Scots spread their language, Gaelic, over most of their new kingdom, from then on to be called Scotland.

The kingdom was Christian, for the Scots had come from Christian Ireland and the Picts had been converted, partly by the mission of St Columba (521–97) [3]. It did not include all of modern Scotland, for the northern and western islands, Caithness and parts of the southwest were held by a new group of invaders, the Vikings from Scandinavia. These men had started raiding in the late eighth century, and made settlements in the ninth. Their mark in Scotland survives in placenames, and ruined buildings. It was not until the mid-1200s that a Scottish king, Alexander III (reigned 1249–86), defeated a king of Norway and took over the western isles. The northern isles remained under Scandinavian rule until the fifteenth century.

The monarchy and feudalism
The monarchy gradually developed its own institutions. For a time the succession moved between two related lines of kings, but with the succession of Duncan I (died 1040) in 1034 an attempt was made to confine it to a single direct line. Duncan was defeated by his rival, Macbeth in 1040, but his son Malcolm

III (c. 1031–93) regained the throne in 1058 and established direct succession. Malcolm married Margaret (c. 1045–93), an English princess, initiating two centuries of frequent royal marriages between the two countries. Their youngest son, David (1084–1153) [4], who became king in 1124, had spent much of his youth in England and held the earldom of Huntingdon. As King of Scotland he brought some Norman barons and friends from England and settled them in fiefs in southern Scotland. His long reign is important not only for this but also for the start of medieval monasticism in the country.

Succession was still not entirely assured, so to ensure peaceful inheritance David, when his only son died, took his 11-year-old grandson Malcolm round the kingdom to be proclaimed as his heir [Key]. Malcolm IV (reigned 1153–65) and, following him, his brother William the Lion (reigned 1165–1214) took David's policy of feudalization further, and William spread it to central and northern Scotland. Feudalism provided the kings with mounted knights in time of war and established clearly defined

1 The Antonine Wall, the northernmost boundary of Roman rule in Britain, was built in the 2nd century AD. But even before the end of the century it had been abandoned by Roman forces under pressure from invading Pictish tribes. This monument, from where the wall meets the North Sea, shows a Roman horseman riding down four naked tribesmen armed with spears, swords and shields.

2 This defensive tower or broch, built of flat stones without cement, still stands more than 12m (40ft) high on the island of Mousa in the Shetland Isles. Its thick walls offered a safe refuge from raiders and contained dwelling rooms and galleries. There are more than 500 known examples of brochs scattered along the coast of north Scotland, some dating back to the 1st century AD.

3 Iona church stands on the former site of St Columba's monastery (founded in 563). It was from here that Christianity was first carried by Columba to the Scottish mainland.

4 The impulse that David's founding of burghs gave to Scottish economic life is symbolized by the fact that his were the first coins to be minted by a Scottish king. Under his administration, trade and commerce flourished.

obligations. The new office of sheriff gave them administrators for a wide number of tasks. The founding of monasteries brought Scotland into the important religious movement of the twelfth century.

In spite of close marriage ties and cultural links with England, and the fact that some Scottish barons held land in England too, peace between England and Scotland was not firmly established. The border between the two long remained unstable; under David I, Scotland controlled Northumberland, Cumberland and Westmorland. Its claim to these countries was formally abandoned only in 1237. Successive kings on one side or the other of the border attempted to enlarge their kingdoms by conquest.

Rebellions and the War of Independence

The male Scottish royal line ended with the sudden death of Alexander III in 1286. It was arranged that his Norwegian granddaughter should succeed and be married to the son of King Edward I of England (reigned 1272–1307). On her death in 1290, the Scots left the decision about the succession to Edward. He chose John Balliol (reigned 1292–96), the best claimant by feudal law, but attempted to dictate to the new king. This led to an unsuccessful rebellion in 1296 [5], invasion by Edward [6], the resignation of Balliol and the removal of the Stone of Scone – on which Scottish kings were enthroned – to Westminster. For a time Scotland appeared to be conquered. But in 1297 a knight, William Wallace (c. 1270–1305), launched another and more serious national rebellion, which was not suppressed for several years. In 1306 came a third rising, ultimately successful, by Robert Bruce (1306–29) [7], whose grandfather had been a claimant against Balliol in 1291. In the long War of Independence Bruce taught the Scots how to avoid pitched battles, dismantle castles and rely on the inability of the English to sustain an army for long so far from home. Finally he defeated Edward II (reigned 1307–27) in the Battle of Bannockburn in June 1314 [8]. English claims to Scotland were not ended, and the two countries were often at war during the next three centuries, but Scottish independence was proved at Bannockburn.

David I (left) and his grandson Malcolm IV, who succeeded him as a young boy, are shown on this 12th-century abbey charter. With David's death the country had been transformed through the widespread introduction of feudal, Anglo-Norman ways of life, a policy continued under Malcolm. David had to have Malcolm proclaimed as his rightful heir because there were other claimants to the throne by earlier rules of succession.

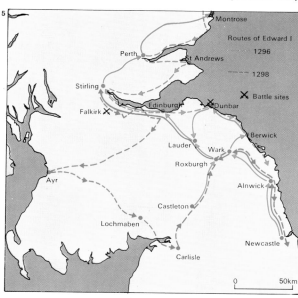

5 In 1296, Edward I invaded Scotland to put down Balliol's rebellion. After sacking Berwick and Dunbar he easily took other castles and towns, but resistance continued to the end of his reign in 1307.

7 Robert Bruce was crowned King of the Scots in 1306, but within three months he had been defeated by an army sent by Edward I. It was not until 1309 that Bruce had gained enough power to hold a parliament – then his guerrilla tactics were gaining the upper hand against the English forces.

8 The Battle of Bannockburn was a resounding victory for the Scottish forces under Robert Bruce. The two sides met as the English troops attempted to relieve Stirling Castle. But Edward II, unlike his father, was neither a good general nor a man of determination, and the Scots, outnumbered three to one outmanoeuvred the English.

6 The "Ragman Roll" records those Scottish nobles and landholders who swore loyalty to Edward in his campaign of 1296, recognizing him as their king. Robert Bruce appears on it but not William Wallace. Later Bruce was to join Wallace in his unsuccessful rebellion, submit again to Edward and finally rebel again and emerge victorious at the fateful Battle of Bannockburn.

Wales to the Act of Union

The Welsh language and nation came into recognizable being with the expulsion of the Goidelic (Irish) Celts in the fifth century AD. The Welsh were slow to establish stable political and social order, but by the tenth century there had emerged uniquely Welsh institutions, codified by Hywel Dda or Hywel the Good (died 950) [3]. Threatened by the Saxons, encroached upon by the Normans and finally conquered by Edward I, Welsh nationalism smouldered and flared, notably in the rebellion of Owain Glyn Dŵr, or Owen Glendower (1359–1416).

The creation of a Welsh kingdom
Evidence from Goat's Hole at Paviland in Gower proves that Wales was inhabited in Palaeolithic times by cave-dwelling hunters. After 8000 BC and until about 1000 BC Iberian migrants settled along the coast. Between 1000 and 500 BC the Celts penetrated Wales, bringing with them Druidism and the basis of the Welsh language.

The Celts remained the power in Wales until the arrival of the Romans in AD 43. Roman rule was largely military [1], and

their chief interest lay in the Welsh mineral deposits, but they established a number of urban communities which exerted a civilizing influence [2]. With the withdrawal of the Romans at the end of the fourth century, Wales was rent by the struggle for power between the Brythonic (native) Celts and incoming Goidelic (Irish) Celts. Largely through the efforts of Cunedda (flourished early fifth century), who founded a number of dynasties and the new kingdom of Gwynedd [Key], the Irish were expelled. In the fifth and sixth centuries Christianity came to Wales, brought by such men as St Illtyd and St David.

Anglo-Saxon incursions during the seventh and eighth centuries greatly hindered attempts to achieve political unity in Wales. The limits of Saxon colonization of the Welsh borders are marked by Offa's Dyke, constructed in the mid-eighth century. The reigns of Rhodri the Great (died 877) and Hywel Dda established a sound political and social framework, but the greatest measure of success was achieved by Gruffydd ap Llywelyn (died 1063) who fleet-

ingly held sway over the whole of Wales.

Towards the end of the eleventh century, Norman invaders seized control of large parts of Wales and consolidated their conquest with powerful castles [4] built at strategic points in the March of Wales (central and eastern Wales). But the princes of Gwynedd resisted their incursions and in the twelfth century presented a strong challenge to the English Crown. Llywelyn the Great (died 1240) began to weld the Welsh territories into a feudal state, and his grandson Llywelyn II ap Gruffydd (died 1282) [5] was able to assume the title Prince of Wales and was recognized in it by Henry III of England in 1267. Henry's successor, Edward I, was less easily defied. He conquered Gwynedd in 1277, humbling Llywelyn and restricting his domain to west of Conway. Following Llywelyn's death in 1282 Welsh political independence came to an end.

English rule and Welsh discontent
The Edwardian conquest destroyed the social system of five orders established by Hywel Dda. The *brenin*, or king, with the

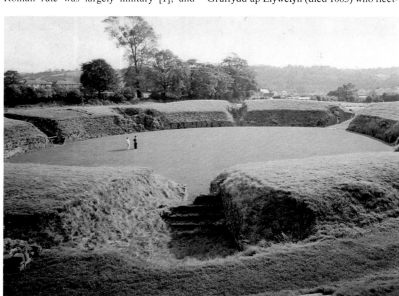

1

2

1 **Caerleon-upon-Usk (Roman Isca) was,** with Chester (Deva), the largest legionary base in or near Wales. The view shows the stone amphitheatre at which soldiers of the 2nd Augustan legion might watch entertainments or military demonstrations. The amphitheatre stands outside the fortress, built in AD 73.

2 **The tiny 4th-century Christian chapel,** here shown reconstructed, at Caerwent (Roman Venta Silurum) is the sole archaeological evidence of Christianity in Roman Wales. Venta Silurum was a tribal capital situated in the better developed southeast, and it was in urban communities of this kind that Romano-British Christianity had modest popularity.

3

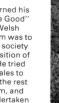

4 **Pembroke Castle is** typical of the Norman castles built in Wales. It was constructed originally as an earthwork motte, or mound, topped by a wooden tower, by Arnulf de Montgomery (*fl.* 1093–1102). In the revolt of

1094, when the Welsh severely defeated the Normans, it survived two sieges, and was almost the only castle in Wales not to fall. It subsequently became an important base for the reconquest of the country. It was rebuilt

with a round keep of stone by the powerful William Marshal, 1st Earl of Pembroke (*d.* 1219). Extended in the 13th century, it was much damaged in the English Civil War and greatly restored in the late 19th century.

5

4

3 **This silver penny** was minted by Hywel Dda, the first Welsh king to issue his own coinage. His coins, like this one, bore the words "Hopael Rex", or King Hywel, thus asserting his claim to rule all Wales and, by its Latin, demonstrating his civilizing intention.

Hywel Dda earned his sobriquet "the Good" by codifying Welsh law; his system was to govern Welsh society until the imposition of English rule. He tried to organize Wales to conform with the rest of Christendom, and *c.* 930 had undertaken a pilgrimage to Rome.

5 **The head of the last independent Prince of Wales,** Llywelyn II ap Gruffydd, King of Gwynedd, was borne through the streets of London after he was killed in a skirmish in 1282. Llywelyn was a first-rate soldier and a clever diplomat; he had behaved and been

treated as a sovereign both by Henry III and by Edward I. The revolt in which he died was not his doing, but he was bound to support it for reasons of kinship and pride. At his death "Wales was cast to the ground", as one chronicler wrote, and swiftly subdued.

uchelwyr (nobles) and *priodorion* (freemen) had owned and ruled the land; subservient to them had been the *taeog*, serfs with some privileges, and the *caethion*, bondsmen with none. Kinship was the primary social tie, and because inheritances were divided between all the sons, smallholdings were the norm. Fast-moving social and economic changes after 1284 swept the traditional way of life away and produced a pattern of substantial landed estates and farms consolidated under a single ownership.

With these agrarian changes, and plague, slump and, not least, alien rule, there was discontent. From 1284 to 1536 Wales was divided into the English Principality (comprising the old Welsh kingdoms) and the Welsh March [6]. In the Principality, English administration was harsh, Welsh law was disregarded, and commercial privileges were conferred mostly on English burgesses. There was at this time opposition throughout Britain to the Lancastrian regime, and, aggravated by local conditions, there were a number of rebellions in Wales, the most important being the Glyn Dŵr revolt of 1400.

Triggered by a dispute between Owain Glyn Dŵr, Lord of Glyn Dyfrdwy [8] and his neighbour Reginald Grey of Ruthin, the revolt soon spread nationally. Although it brought no political amelioration, the revolt stands high in Welsh national sentiment.

Henry Tudor raises hopes

The collapse of the rebellion brought in its train penal laws reducing Welshmen to second-class citizens, but hostility to the English rule could find an outlet in the fifteenth century only in prophetic poetry. Prophetic bards assured Welshmen that a son of destiny would arrive to reassert Welsh supremacy in their land. During the Wars of the Roses the bards threw in their lot variously with the House of York or the House of Lancaster until, when the Lancastrian claim passed to Henry Tudor, they urged his support unanimously. Henry's victory at Bosworth in 1485 was hailed as a Welsh triumph. But the new king was an opportunist, and the call by the Welsh gentry for improved law and order and for equal rights with the English was not yet answered.

KEY

- - - - Kingdom boundaries 7th & 8th centuries
— — Offa's Dyke
· · · · Wat's Dyke

The map shows the Welsh kingdoms in the 7th and 8th centuries AD. The largest of them, Gwynedd, was roughly the area that Cunedda had seized in the 5th century. It was to be the last outpost of Welsh independence. Powys was the kingdom next in importance and in the south Dyfed was sometimes united with Seisyllwg. The southeastern kingdoms were frailer, so that Rhodri the Great and Hywel Dda managed more easily to subdue them in their attempts to unite Wales in the 9th and 10th centuries. In the east the boundary of Wales was defined for all time by Offa's Dyke, constructed to mark the limits of the Kingdom of Mercia *c*. 790. Wat's Dyke was built to defend Mercia against Welsh raiders 80 or 90 years earlier.

6 There was a new dispensation after the Edwardian conquest. The Welsh kingdom of Gwynedd was entirely dismembered and now became the shires of Anglesey, Merioneth and Caernarvonshire. With these, Cardiganshire, Carmarthenshire and Flint were held under the direct rule of the Crown, and these six shires were known as the Principality. The rest of the country, of which the possession had long been Norman, passing only occasionally into Welsh hands, was held as before by Marcher lords. The statute of Rhuddlan, enacted in 1284, provided the judicial and administrative framework by which the Principality was to be governed until 1536. But it made little difference to Welsh culture.

7 Harlech Castle was considered at the time to be the best fortified of all the powerful castles constructed by Edward I. It was built on a rocky site with a series of concentric defences. Its mighty gatehouse was buttressed by several strong doors and portcullises, and it had no equal in Wales. The work, which cost over £8,000, was undertaken in 1283 by about 900 labourers, carpenters and stonemasons, and was completed within six years. The master mason was James of St George, who had gained his experience of castle-building in Savoy. His wage is recorded as three shillings a day, more than any other builder in the realm could command.

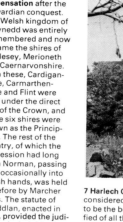

8 Owain Glyn Dŵr is a national Welsh hero (this statue stands in Cardiff City Hall), but ill-recorded as a man. His rebellion in 1400 drew enormous support from the peasantry but was mainly sustained by gentlemen and squires. Once he was established as a self-styled Prince of Wales, Glyn Dŵr had the enterprise to form a constructive policy: he established an alliance with France, and proposed to create two universities and an episcopacy independent of Canterbury. He governed through the Welsh parliament he had summoned. When the tide turned against him after 1409, Glyn Dŵr went in and out of hiding; he was active in 1415 for the last time. With this mysterious end, he was an inspiration to the bards who kept nationalism alive.

9 Henry Tudor was acclaimed in Wales as the first Welshman to be ensconced on the English throne. The bards jubilantly greeted Henry's triumph in late August 1485 as a victory for Wales and fulfilment of the hoary prophecy that a descendant of the Ancient Britons would oust the Saxon oppressors of their land. When he landed at Dale in Pembrokeshire with 2,000 Frenchmen in August 1485 and promised to deliver the Welsh from their "miserable servitudes", he did indeed attract a great many recruits. But although Henry rewarded his supporters and made some attempt to render the Welsh equal subjects with the English, he disappointed many expectations. In 1489 he revived the title of Prince of Wales for his young son Arthur.

Triumph of the Church

The years from the accession of Gregory VII in 1073 to the removal of the Papacy to Avignon in 1309 saw the highest achievements of the medieval Christian Church in all its spheres of activity. Western Christendom could genuinely be seen as a unity with about 500 bishoprics all working as part of an international papal system that became ever more organized and powerful. Roman Christianity laid the ground rules for everyone in European society, rules from which only the Jews were exempt. The twelfth and thirteenth centuries also saw the triumph of Gothic architecture and decoration, an achievement paralleled by the complete codification of law, knowledge and philosophy in a Christian theological context by the scholastic philosophers, St Thomas Aquinas (1225–74) in particular.

The power of the Papacy
The eleventh-century Papacy had been inspired by the example of Cluniac reform, demonstrated best in the Lenten Synod at Rome in 1074 which laid down strict rules for the appointment and behaviour of the clergy.

The following year Gregory VII formally prohibited lay investiture (the appointment of bishops by a secular ruler), which not only antagonized Emperor Henry IV (reigned 1056–1106), but generally incurred violent hostility throughout Germany, France and England [8]. The resulting battle made it necessary for the Papacy to indulge in extensive legal justification of its position and led eventually to an institutionalized Papacy that would have been abhorrent to the original Cluniac reform.

The pope claimed to be the supreme judge and moral arbiter in temporal disputes. Only in this period could Emperor Henry IV have been humbled at Canossa by Gregory VII; could so powerful an English king as Henry II have appeared at Canterbury in a hair shirt to atone for the murder of Thomas à Becket; could King John (reigned 1199–1216) have been forced to submit himself (and the English crown) as a vassal of Pope Innocent III [Key]; or could Innocent IV have been able to annihilate the Hohenstaufen line of emperors in 1250.

The most imposing instrument of papal

government was the Ecumenical Council, a decision-making body of all the bishops of Western Christendom summoned to attend at the request and direction of the pope. There were no fewer than seven such councils during the years 1123 to 1312.

Papal administration
The growth of the Curia, the administrative focal point of the Vatican, also led to its increasing use as an international court of law. By the thirteenth century the ecclesiastical nature of the Papacy was at times overshadowed and popes such as Innocent IV (reigned 1243–54) were lawyers and administrators rather than religious leaders. The Papacy became increasingly embroiled in international politics, a development which culminated in its forcible transference to Avignon by the French monarchy from 1309 to 1377 [10].

In contrast to the growing complexity of papal administration there was a wave of reforms within the religious orders, all of which were based on piety, simplicity and austerity. Despite the Cluniac reforms, the

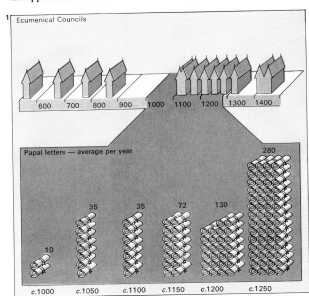

1 Ecumenical councils, meetings of representatives of the whole Church, reached a peak in the 1100s and 1200s. As the Roman Church established its role as religious authority and arbiter of political events its administrative function developed: the ecumenical council grew and litigation at the Roman courts increased, as did the correspondence of the Curia, the juridical, legislative and administrative offices of the Papacy. Papal correspondence, a record of ecclesiasical intervention, grew from 130 letters during Adrian IV's Papacy (1154–9) to 3,646 during John XXII's (1316–34).

2 This miniature of St Bernard of Clairvaux (1090–1153) being nourished by milk from the breast of the Virgin Mary is taken from a 15th-century Venetian Cistercian breviary. It exemplifies one of the chief tenets of St Bernard's teaching, an almost sentimental devotion to the mother of God. The reverence of Mary was one of the striking elements of popular piety in the 12th century.

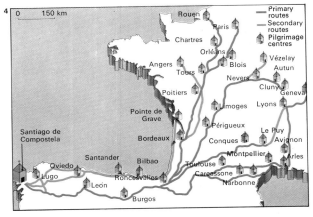

3 The *Decretals* of Gregory IX, the manuscript from which this miniature is taken, show the importance attached to the 13th century to the doctrine of transubstantiation. This was made the official teaching of the Church by the 4th Lateran Council (1215) which also prescribed annual confession. Gregory IX patronized the new monastic orders and canonized Dominic, Francis and Anthony.

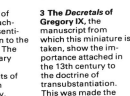

4 The pilgrimage was the "package tour" of the Middle Ages; pious Muslims went to Mecca just as devout Christians sought to go to Rome, Compostela or Canterbury. The pilgrimage was at once the fulfilment of an active religion and also the experience of a lifetime. In the early Middle Ages the focus of Christian pilgrimages was Rome; but the gradual appearance of famous relics around the Western world led to a proliferation of provincial shrines, the most famous of which was that of the body of the apostle James at Compostela in Spain. The pilgrimage route to Compostela became a thoroughfare along which many famous Romanesque churches were constructed.

5 The Dance of Death or *danse macabre* shows death in the role in which he appeared in late medieval popular mythology. He was seen to be an immediate tangible threat, a maleficent joker. Although many of the precepts of Christianity which are emphasized today appear to have been ignored in the early Middle Ages, it was the common religion of all in the West, and death for the Christian was the consummation of life. Death was therefore prominent in contemporary hagiography and was personified in the popular religious mystery plays. Awareness of death was intensified by the various catastrophes of the 14th century, especially the Great Plague of 1348 which depopulated Europe.

Benedictine Order had lost its pious intensity by the early twelfth century. The constitution of the Cistercians, the *Charter of Love*, written by St Stephen Harding in 1119, was an adaptation of the Benedictine Rule based on a denial of ostentation and an austere life located away from centres of population. The response to this rigorous life of prayer was overwhelming and 530 Cistercian houses had been established by 1200. Another offshoot of the Benedictine system were the Carthusians, founded by St Bruno at the monastery of the Grande Chartreuse in 1084.

There were many other new orders in the twelfth century – Augustinian canons, hermits, Hospitallers, crusading orders, Templars and organizations for the visitation of the sick and the imprisoned. The ultimate in clerical reform was the foundation of the Friars [7]. St Francis of Assisi (1181–1226) and St Dominic (c. 1170–1221) both stressed that their followers should take vows of complete poverty, and that neither individuals nor the order as a whole should possess any property. Friars were by definition to live by the work of their hands and, if necessary, by begging.

They were to preach and exhort and to set a visible example of self-denial. By the fourteenth century, however, the orders denied their founders' intentions, held communal property and became part of the structure of the Church.

The twelfth and thirteenth centuries were ages of secular piety and involvement with the Church as well as clerical reform. The Crusades were central to religious activity as were pilgrimages to Rome. Compostela in Spain [4], Canterbury and local shrines.

Persecution of heretics

The ugly side of the universality of papal pretensions and the increasing extremes of the religious life was the persecution of heretics [6]. The Jews, as non-believers, were largely left alone at this time, but heretics, as deserters from the true faith, had to be reconverted, by force if necessary. Among those heretics were the Albigensians, who held all material things and social life to be evil, thus presenting a challenge both to the Church and to the stability of society. The Albigensians were destroyed in the 1200s.

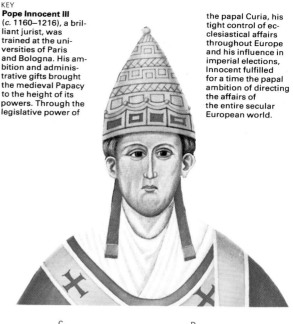
Pope Innocent III (c. 1160–1216), a brilliant jurist, was trained at the universities of Paris and Bologna. His ambition and administrative gifts brought the medieval Papacy to the height of its powers. Through the legislative power of the papal Curia, his tight control of ecclesiastical affairs throughout Europe and his influence in imperial elections, Innocent fulfilled for a time the papal ambition of directing the affairs of the entire secular European world.

6 Some followers of Amalric of Chartres were burnt at the stake in the early 12th century for believing that those in love with God partake in His perfection and are proof against sin. Traditionally it was the business of the Church to preserve orthodoxy; the ecclesiastical authorities, the Dominicans and Franciscans in particular, investigated heresy, the secular powers punished it.

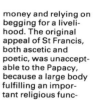

7 The religious orders increased greatly in number as a result of popular piety and the growing complexity of Church organization. The Cluniac reformers of the 10th century merely tightened the Benedictine Rule. The Carthusians [A], founded by St Bruno in 1084, and the Cistercians, the "white monks" [B], lived according to new rules that emphasized the virtues of self-denial and the strictest religious observance. The orders of friars, the Dominicans [C] and Franciscans [D], were a new kind of wandering priest who travelled Europe preaching, possessing neither property nor money and relying on begging for a livelihood. The original appeal of St Francis, both ascetic and poetic, was unacceptable to the Papacy, because a large body fulfilling an important religious function needed corporate property if it was not to become socially subversive. Despite opposition from loyal followers, the friars soon obtained property and became an integral part of the Church.

8 The question of who should invest bishops caused a rift between popes and secular rulers. Great local wealth and political power went with the office of bishop, the sort of wealth with which any ruler would want to reward or bribe a useful servant. This bronze depicting Emperor Otto I investing Bishop Adalbert of Magdeburg in 962 demonstrates the imperial position at its strongest. The new bishop is seen receiving his authority from the secular ruler, not the pope.

9 This fresco illustrates the triumph of the Church as seen through the eyes of the late 14th-century Florentine artist Andrea di Buonaiuto. It is in the Spanish Chapel of the church of Sta Maria Novella in Florence and is a contemporary commentary on the development of the structure of the Church into a vast international organization with a role in administration and religious teaching. Sta Maria Novella is a Dominican church and the picture emphasizes the role of the Dominican friars, many of whom had been made permanent commissioners for the extermination of heresy, the origins of the Holy Inquisition, under Gregory IX. Dominican friars are here shown as the *domini canes*, "hounds of the Lord".

10 The Papal Palace in Avignon was a more efficient centre of papal government than Rome, and it is testimony to the wealth, success and culture of the papal court in captivity. After two centuries of great power, the Papacy entered a period of submission to temporal rulers when from 1309 to 1377 it was forcibly removed to Avignon. Ostensibly the move was to resolve the conflicts it had suffered in Rome; but the influence of the French monarchy over the Papacy increased.

187

European learning in the 13th century

The medieval papacy reached the height of its power, wealth and cohesion in the thirteenth century, at the same time as St Thomas Aquinas (c. 1225–74) was finishing his *Summa Theologica* (1266–73), the greatest work of scholastic philosophy.

The genesis of Scholasticism
The highest achievement of medieval Christian thought, Scholasticism was an intellectual system that employed the logic, or dialectic [Key], of Aristotle. In the course of the twelfth century, as Aristotle's works became available to Western scholars in Latin translation, his analytic method was applied in all fields of inquiry. The result was a revolution in the methods and content of learning which produced a coherent system of philosophy and theology and a new science of jurisprudence and which laid the first fragile foundations of experimental science.

As a growing range of Greek and Arabic philosophical and scientific texts became available, centres of learning shifted away from the monasteries and episcopal schools to a new kind of academic organization – the university [1]. Universities were associations or corporations of masters or students that arose in a number of urban centres where the new learning was available. The earliest of them appeared early in the thirteenth century at Bologna [5], Paris, Montpelier, Naples, Oxford, Cambridge, Toulouse and Salamanca. Such schools gave Scholasticism its name (the teaching of the "Schoolmen").

Training at a medieval university was long and arduous. It usually took six years to become a Master of Arts, after studying logic, grammar, rhetoric, arithmetic, geometry, astronomy and music. The doctorate of theology, the highest academic achievement, required about eight more years of study.

This was the age of translation, both from the original Greek and from Arabic and Jewish texts and interpretations. Much classical cosmology, mathematics and philosophy was revealed through the writings of Arabic scholars such as Al Farabi (died 950), Avicenna (980–1037) and Averrhöes (1126–98). Gradually through the works of Aristotle and Ptolemy (90–168), Averrhöes and Avicenna, the knowledge of Greek cosmology and medicine became available, but it was above all the ideas of Aristotle [2] that influenced the thought of the early universities. Aristotelian thought was opposed to the emphasis on divine grace in early Scholastic thought and as such incurred the wrath of the conservatives. Paris in particular was the seat of new Aristotelian learning and, although the teaching of Aristotle's *Metaphysics* and *Physics* was initially banned in the schools during the thirteenth century, the prohibition was widely ignored.

Reconciling faith and reason
The crux of the Scholastic controversy was the relationship between revelation and reason. Aristotle's writings challenged the primacy of theology by asserting that rationality was the basis of knowledge, and the greatest works of this period were devoted to reconciling this apparent contradiction between faith and reason. A group of arts masters at Paris, the leader of whom was Siger de Brabant (c. 1235–81), adopted Averrhöes' view that reason and philosophy were superior to faith and knowledge from faith.

CONNECTIONS

See also
170 Islam in Europe
134 The rise of medieval Western Christendom
194 The Church in medieval England

In other volumes
24 Science and The Universe
230 Man and Society

1 Cino da' Sinibaldi was a famous Tuscan lawyer and poet who died in 1336. His tomb is one of many Italian funerary monuments for academics showing the university classrooms of the later Middle Ages. The magnificence of this tomb is an indication of the importance that Italians were beginning to accord to teachers of civil law in the Italian cities. Cino had studied at the great university of Bologna and had worked at Pistoia, Rome, Treviso, Florence, Siena and Perugia. He was a link between the Scholastic academics of the 13th century and the humanist scholars who were to establish the intellectual framework for the Renaissance period from the 14th century.

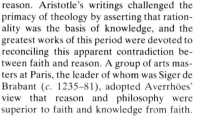

2 This 13th-century French miniature shows Aristotle teaching the Emperor Alexander as a child. The illustration is from the *Treasury* of Brunetto Latini (c. 1210–c. 1295), a compendium of history, philosophy and legend completed in 1265. Instruction in philosophy was thought indispensable in the training of a ruler.

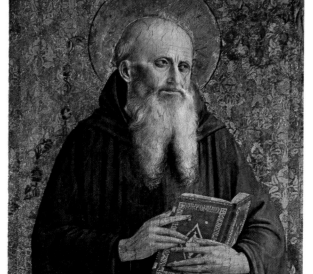

3 St Bonaventure, minister-general of the Franciscan order and cardinal bishop of Albano, taught at Paris with St Thomas Aquinas. First and foremost a theologian, his most widely read work was *The Journey of the Soul to God.*

4 This 15th-century miniature of Roger Bacon shows him in scholarly pose. His intellectual curiosity seemed almost without limit, but the novelty of his interests and ideas led in the end to his imprisonment.

This provoked a fierce controversy, in the course of which Aquinas composed his famous apologia for revealed religion – the *Summa contra Gentiles.*

A more conservative defender of theology against the rationalist philosophers was St Bonaventure (1221–74) [3] who conceded that some knowledge could be gained from philosophy alone without the need of faith. Like St Thomas, Bonaventure held that a philosophical proof of God's existence could be found in the need for an original cause or motion to begin the chain of events in the universe. However, he believed that philosophy could not explain the details of revelation nor could it provide a moral framework for life.

Roger Bacon's (*c.* 1214–92) [4] work showed an interest in empirical observation that epitomized the growing belief in man's powers of investigation and understanding. Following the Scholastic tradition, he saw in the behaviour of material things, including the commonsense behaviour of man, a "natural law" that manifested divine insight. Consequently, Bacon believed that science

was a natural and harmonious basis for religion. But he was also a firm believer in alchemy [9], the "science" of turning base metal into gold, and in astrology, the art of telling the future through the stars.

The development of Aquinas' thought

Albertus Magnus (*c.*1200–80) incorporated the new knowledge of Aristotle within an encyclopedic formula that embraced Platonism, Neo-Platonism and Arabic theology. Thomas Aquinas, his pupil, inherited this breadth of reference and consciously set out to provide a logical Aristotelian substructure for Christian thought. Aquinas took faith as the starting point and argued that everything discovered from reason could be interpreted in the light of it. Apparent conflict between reason and faith meant that the reasoning was wrong. He conceded that human knowledge depended on sense perception, and in his theory of "natural law" – a body of universal moral principles that can be ascertained by human reason alone – he provided the basis for a purely rational system of ethics.

This 15th-century figure is an allegory of dialectic. Christian vices and virtues, portrayed in human form throughout medieval times, were joined in the 13th century by figures representing the disciplines of the philosophers. Dialectics, the branch of logic concerned with the rules of reasoning, was first formulated in Greece by Aristotle (384–322 BC).

5 Bologna, one of the earliest and most famous of European universities, was noted for its law schools. The 12th century saw the emergence of student guilds which gradually became independent of the law of the city. By the 14th century, Bologna University had an organized collegiate system.

7 Mathematics, like logic, played a part in the curriculum of the Scholastic philosopher, revealing through the human intellect the divine plan for the universe. Medieval mathematics consisted almost completely of a rediscovery of Greek knowledge in the field, preserved and transmitted by Islamic scholars. These Pythagorean theorems are from a 13th-century treatise which is typical of the manuals of the time.

6 St Thomas Aquinas was the greatest doctor of the medieval Church and the greatest exponent of Scholasticism. His *Summa Theologica,* the best known philosophical treatise of his time, attempted a comprehensive synthesis of reason and revelation that established him as the Church's foremost theologian. Whereas initially his Aristotelian analysis was criticized, his statements have come to be regarded as a wholly Christian and rather overbearing interpretation of man's being and actions.

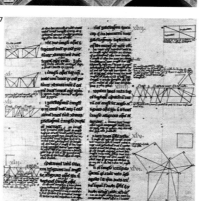
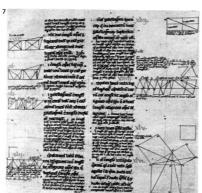

8 Notations on this 13th-century manuscript of the Roman writer, Lactantius, indicate textual criticisms common to a generation aware both of the ambiguities of classical texts and the textual corruptions that had developed over the centuries. Greek and Latin works had been translated into Syriac, thence into Arabic and back into medieval Latin. Careless transcription created more inaccuracy, but greater care ushered in a new era of classical scholarship.

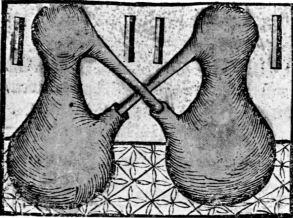

9 Alchemy was a pseudo-science of the Middle Ages. Although much effort went into improbable experiments (such as this one, which uses a *double* alembic) and into vain attempts to transmute base metals into gold, many alchemists were able chemists whose work laid the foundations of later knowledge. Like astrology, alchemy had its roots in classical antiquity and flourished in the 13th century alongside more fruitful scientific work on mathematics, optics and astronomy.

Western European economy 1000–1250

Between 1000 and 1250 the economic foundation of Europe shifted to the north. The new wealth came from a rapidly developing agriculture in the northern European plain [2] between the Loire and the Elbe. Under the Romans this had been a sparsely populated forest area; now as a result of expansion, it was the Mediterranean that became a frontier zone.

An era of expansion
In this new world all the energy and enterprise of the pioneer was yoked to clearing the forests, establishing villages and marking out the routes, largely independent of the Roman system, that bound the plain together. Here and there the directing force of some powerful family or monastic community could be detected, but almost certainly most of the work was done by small family groups and peasant communities. In the progressive draining of the marshes of Lincolnshire, Flanders and the Po basin – among the most remarkable achievements of these centuries – the lead may have been given by abbeys such as Ramsey or Les Dunes, but the fiercely independent spirit of the subsequent peasant proprietors suggests that many took matters into their own hands and proceeded to drain the land themselves.

Intensive settlement of the soil implies an increasing population. In this period, without records of taxation, there is no direct evidence to measure population growth; but an indication that numbers increased rapidly comes from the twelfth-century migrations of both peasants and knights from France, Germany and Flanders to Eastern Europe, Spain, Sicily and Palestine.

Within the bounds of Western Europe the few surviving monastic estate documents give glimpses of the slow development of a farming expertise geared specifically to northern climates, such as the use of the heavy-wheeled plough (*carruca*), the harnessing of draught animals in columns and the development of the watermill for grinding corn, which gradually came into use between 1000 and 1250. By 1200, horses were shod and harnessed with a shoulder collar. More had, by then, been learnt about spring and autumn crops: wheat and rye for standard autumn-sown cereals, but other varieties for the more rigorous climates. The use of spring-sown oats and barley for poorer soils was understood by progressive estate managers of the thirteenth century, such as the agricultural writer Walter of Henley.

Social and economic organization
The social structure of this developing agrarian society is difficult to determine and probably varied dramatically from area to area. Where seigneurial and monastic landlords took an interest in developing their lands (and many did, especially the new order of Cistercian monks), it was common to attract landless labourers with the promise of heritable tenures, on condition that they bound themselves and their heirs to the soil. Serfdom was one way of organizing labour for co-operative work, but it is a mistake to see it as universal. The peasant "communes" of the open-field system in France and England, with their periodic redistribution of holdings, the independent holdings (*allods*) of Aquitaine and the farmsteads (*casalia*) of Italy all resulted from peculiarities of local

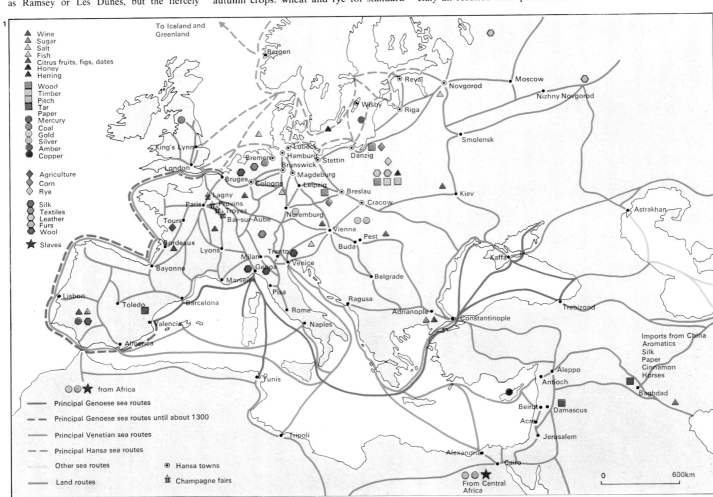

1 The number of trade routes, by both land and sea, increased during the High Middle Ages. The most popular form of bulk transport in the early Middle Ages was by river. In the 11th and 12th centuries land transport be- came much easier, partly because the routes were safer, partly because of the increased use of pack animals. The trade of the carrier, often a native of the high Alps, had by 1250 become an integral link in Italian commerce with northern Europe. The routes primarily linked importing centres such as the Italian and Hansa towns with the centres of production and exchange – the fairs of northern Europe and Flemish towns. But links were also forged through Poland and Hungary with the ancient Asiatic routes. Sea routes developed dramatically during the 12th century. Mediterranean ships of the period were small, rarely more than 100 tonnes. They usually sailed in convoy and with enterprising crews from Venice, Pisa and Genoa to most Mediterranean ports. Some Levant towns had Italian quarters as early as 1110. Subsequently the Venetians established themselves at Alexandria and their attempts to control the trade of Constantinople culminated in their conquest of that city in 1204. The northern seas had their own Viking tradition of seamanship. The trade in furs originally carried by Scandinavians from Novgorod to the Baltic fell into the hands of merchants from the north German towns in the 12th century. The merchants soon organized themselves into a protective league or *Hansa*.

custom and circumstance. What is reasonably certain is that the number of such communities increased sharply during the eleventh and twelfth centuries.

The rise of towns

The other consequence of population increase was the growing importance of towns as centres both of exchange and production. The origins of towns varied, but most of them grew up to serve local needs, as strongpoints (the *burg*) to which agricultural markets were attached, or as ports importing and exporting goods. The more intensive the settlement of the land, the larger and more numerous were the towns. The number of cities in Flanders and northeastern France – Bruges, Arras, Valenciennes, and others – reflects a high agricultural population. The burgeoning cloth production of Bruges in the twelfth century, in part depended on the farming of sheep in the Flemish salt marshes. The towns' specialized way of life needed independent institutions: the earliest guild of merchants known, that of Tiel in Holland, existed earlier than 1000, but the merchant guild as an instrument of communal government became common only in the twelfth century. The towns flourished, however, because of the vast consumer market that had developed by 1200.

By 1000 Western Europe was in sporadic but persistent contact with the East through the Baltic and Russia, and with the decline of Saracen piracy in the Mediterranean the Italian ports developed strong links with Egypt and (after 1098) the Crusading states, which enabled Oriental silks and spices to reach the West. Mediterranean trade [1], necessarily large-scale and risky since goods bought abroad were never pre-sold, gradually led Venetian merchants to develop a "capitalist" system of investment [3].

The High Middle Ages was a period of almost universal expansion in Western Europe, an underdeveloped continent ready for sustained exploitation. As conditions changed, however, the warning signs of overpopulation and overuse of the land went unnoticed until persistent famine and plague swept over the region brought a general recession in the fourteenth century.

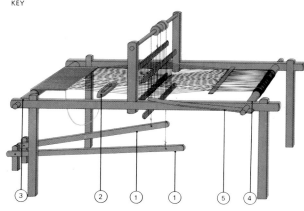

Horizonal treadle looms, introduced in the 13th century, stimulated textile manufacture, the medieval industry that employed most urban craftsmen. The less efficient vertical loom was replaced by a machine using treadles [1] to raise or lower the lengthwise threads as the shuttle [2] passed through them. Cloth was wound on to the cloth beam [3] as thread was released from the warp beam [4] by depressing a lever [5]. About two-thirds of all trade was in cloth or raw materials required for it. Most weaving was done in the Low Countries, partly by urban industry and partly by cottage labour.

2 Drainage and land reclamation was one of the most spectacular aspects of the clearing of the northern European plain in the 11th and 12th centuries. An area near St Omer in France shows how, in the early Middle Ages, polders were diked, marshy land made fit for grazing and a system of waterways constructed. Cultivation of new lands both fed a larger population and gave greater freedom to farmers, who paid for pioneer lands in fees rather than in services: two vital factors for the commercial and industrial expansion of the Low Countries during this period.

3 The Fondaco dei Turchi (Turkish warehouse) on the Grand Canal in Venice was built in the late Romanesque style in the 13th century. Although it takes its name from a later period of the city's mercantile development, the building exemplifies Venetian power in the first era of its commercial success. From the 11th century, Venice dominated European trade with the East and its arsenal and dockyard comprised the biggest industrial unit of the age.

4 Illustrated calendars of seasonal activities provide the best visual evidence of medieval costume and methods of agriculture. Shown here are May: a shepherd and his flock [A]; June: cutting wood [B] and July: haymaking [C].

5 The grape harvest depicted in an Italian calendar shows the method of pressing the grapes with the feet, a tradition that survived in parts of rural Italy until after 1945. Wine was produced in quantity throughout southern Europe, in most of France and in the Rhine and Moselle areas of Germany. Production methods had changed little since the days of the great Roman villas and viticulture was, throughout the Middle Ages, still the economic prerogative of the big landowner or entrepreneur. By the 13th century a genuinely international wine trade had developed along a north-south axis with large quantities of wine being shipped in barrels by water transport.

Town life in medieval England

Town life in England re-emerged in the tenth and eleventh centuries after being practically non-existent since the departure of the Romans in the early fifth century. The medieval towns, especially London, were important as centres of trade and production and were also the cradles of a sophisticated non-feudal political consciousness.

The earliest towns

There were a number of towns in England before 1066 which formed part of the tenth-century kings' coherent plan of local government and defence. They were often planned towns: excavations at Winchester have revealed the grid plan of the tenth-century streets, and it seems that Oxford was consciously populated in about 910 by drafting villagers in from nearby communities. Although they were really fortified villages rather than towns, with agriculture far more important than industry, towns were nevertheless the means by which a measure of peace and government was imposed.

Some 52 "towns" were described in the Domesday Book (1086). In the twelfth century, growing international trade and a rising standard of living combined to multiply their number and to give birth to new crafts. The *negotiatores* (general traders) of the past gave way to specialized occupations organized in guilds. The purveyors of luxury goods, the vintners, goldsmiths and pepperers, were the earliest specialized trades in London. By 1191 representatives of a score of different trades witnessed a charter at Oxford, only a middle-sized town. Guilds [7] enabled the towns to maintain their supplies and markets: smaller towns began with a single "guild merchant" of all the trades, but during the thirteenth century humbler craft guilds [6], of weavers, bakers, fishmongers or smiths also emerged to claim their share in city government. These craft guilds looked after their members' interests whereas the guilds merchant organized the fundamental trading and market privileges of the towns [3].

Life in the towns

Crowded sites soon made necessary a new style of life: closed street frontages, houses with their narrow ends to the street stretching far back into their gardens, created the town house and the modern townscape [Key, 1]. Warehouses, counting-houses, shops and living-rooms shared many cramped sites in burgeoning towns such as Southampton. The streets themselves were narrow, because space was valuable: ordinances frequently forbade householders to encroach upon them. Open gutters in the middle of streets carried much refuse, although nearly every house had a cesspit as well. By 1300 town authorities were beginning to improve sanitary conditions: paviours were engaged to restore the surface of the streets, and butchers were generally forbidden to slaughter livestock except in "shambles" situated safely beyond the walls [9].

The new communities where men could suddenly rise to great wealth and which had relative political freedom were naturally conscious of their difference from the more static rural society. At first they were dominated by small oligarchies made up of men in the luxury trades who had great influence at court. In their confident hands London and Oxford tried in 1191 to become communes

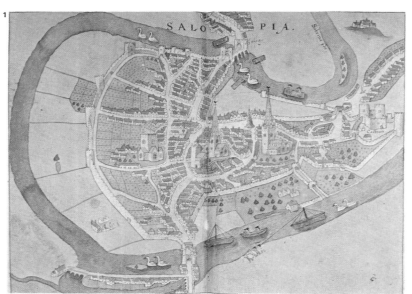

1 Shrewsbury, in a 16th-century map, appears crowded round its market square. The castle was first built in 1070 and the town obtained its first charter in 1189. In the great period of medieval town life, before 1350, many towns outgrew their walls. The walls in most English towns were not built until the middle of the 12th century.

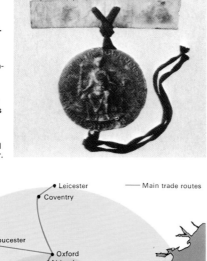

2 John granted a charter of incorporation to London, England's largest town, in 1215. From the mid-12th century, London's citizens, who claimed even the right to elect the king, were seeking and winning privileges of self-government. John's charter confirmed the citizens' liberties and permitted the election of a mayor.

3 Markets were the life-blood of a medieval town, and they provided an outlet for peasants from the surrounding countryside to sell their own goods, as well as being the centre for the inland distributive trade in more luxury products. They were situated in regular places in each town, often marked by a market cross or a covered hall. Markets could be violent places, and there were strict regulations governing them, particularly on the standards of the goods sold. Towns hired officials such as the assayers of ale and bread, and offenders suffered heavy fines. A successful market could be an important source of profit for the town or the baron who set it up. The great period for the creation of market rights was the late 12th century.

4 Southampton was an important town for both imports and exports in the Middle Ages. The network of roads was used to take these goods to inland towns. Goods from Southampton were found as far away as Leicester, 220km (137 miles) to the north. Only valuable and easily carried goods were taken so far. The quality of the roads, particularly in winter, was not good, and although the trading towns tried to ensure that bridges on the vital routes were kept in good repair, responsibility for road upkeep was in the hands of each parish. As a result, many goods were carried by water whenever possible.

Furthest distribution of:
Dyestuffs
Wine
Household goods, iron, fish
Tile, stone, slate, coal
— Main trade routes

Leicester
Coventry
Gloucester
Oxford
Abingdon
Bristol
Newbury
Reading
London
Andover
Basingstoke
Guildford
Salisbury
Winchester
Southampton
Exeter
Honiton
0   80km

independent of the surrounding shires, and many towns had mayors, aldermen, a common council and a town clerk by the middle of the thirteenth century [2].

But just as remarkable as the emergence of these towns was their vitality and tenacity in the face of radical social change. In the late thirteenth century the great urban oligarchies in London, Lincoln, York and other towns were swamped by the practitioners of the lesser guilds, many of whom were immigrants from the country. But although they displaced the old families, sometimes violently, these new families everywhere maintained and developed the old privileges, traditions and organizations of the towns.

Social unrest in the towns
Urban upheavals were common, especially after 1300. Many towns such as Louth and Stamford, dependent on their production of cloth, declined into small market towns when the introduction of the fulling mill made it more practical to make woollen cloth in the villages. The Black Death (1348–50) hit especially hard at the densely packed towns.

A general movement of wealth and population from the Midlands to East Anglia and the south-west left behind the shells of once prosperous communities: at Oxford falling house rents after 1325 attest the declining population [8]. English towns were spared the violence of fourteenth-century Ghent or Florence, but the rising of the peasants in 1381 found allies in York, Norwich, and especially in London. And thieves at all times could find a ready refuge in the slums outside the city walls.

Only London could muster a mob with the violent potential of the urban masses in Paris or other European towns, but on several occasions in the late fourteenth century that London mob proved able to frustrate the government. For a few days in 1381, in association with the peasants led by Wat Tyler (died 1381), it managed to render the government completely powerless. Urban life in England was precarious, yet it remained attractive. Trade, law, and government itself could not function without towns, and by 1500 they had become a vital element in all other English institutions.

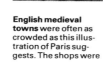

English medieval towns were often as crowded as this illustration of Paris suggests. The shops were little more than market stalls, and each trade was concentrated into a specific street. Most stable towns grew out of markets; others, relying on single industries, were liable suddenly to expand or disappear.

5 The export of wool was vital to the economy of many towns in the early Middle Ages, because the Flemish cloth industry came to rely on English wool in the 13th century. But when the English cloth industry itself grew up, based largely on the villages, many towns contracted. In the 15th century several towns, such as Bradford in Yorkshire and Devizes in Somerset, prospered on cloth, but most of the towns that had grown in the 13th century now stagnated. Their political importance declined as the economy faded after 1400 and political hegemony reverted to the nobility. Henry VII tried to revitalize town life after 1485 by stimulating trade overseas.

6 The craft guilds held examinations for apprentices before they were admitted to the guild as master craftsmen. Apprentices were required to learn their craft for seven years, and after qualifying they served as journeymen, travelling the country until they could afford to set up in business permanently. One of the most important functions of the craft guilds in the early 14th century was to maintain the standard of workmanship, and they adopted the concept of examinations from the universities. During the 15th century, when many towns were in decline, guilds generally tried to limit the number of apprentices by restricting entry to the sons of existing craftsmen.

7 The Guildhall of the Merchant Adventurers at York was built in about 1360 and served as a meeting place for the guild. The Merchant Adventurers became one of the greatest guilds in York as the trade in cloth overtook that in wool in the later Middle Ages. Cloth exports, unlike wool, were not bound to any staple and a merchant community emerged, ousting the foreign traders from their control of the business. Like most other guilds, the Merchant Adventurers was not purely economic in function, but had a partly religious and charitable purpose, because it undertook to support any of its members who became destitute. By the 15th century, the guild had many branches throughout the Low Countries.

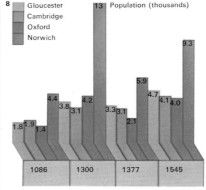

8 The population of medieval towns is hard to determine exactly, but it is clear that most towns lost between a third and a half of their population in the Black Death. This fall was only slowly reversed in the later Middle Ages.

| | Gloucester | Cambridge | Oxford | Norwich | Population (thousands) |

| 1086 | 1300 | 1377 | 1545 |

Values: 1.8, 1.9, 1.4; 4.4, 3.8, 3.1, 4.2; 3.3, 3.1, 2.1, 13; 5.9, 4.7, 4.1, 4.0, 9.3

9 The Shambles, York, is one of the few remaining medieval streets in England. It was originally devoted to butchers' shops. Despite legislation that butchers should only work outside the towns, the regulations, as here, were often ignored.

The Church in medieval England

The medieval English Church, like the Church elsewhere in Europe, had little spiritual or organizational unity, but was a tangled association of virtually independent bodies. It included monasteries and cathedral chapters, bishops and their officials and individual clergy, great international orders such as the friars, all catering more or less directly to the spiritual welfare of the country, and all under the general jurisdiction of a distant pope. It was a tinder-box of conflicting rights and privileges, careerism and spirituality.

Shrines and monasteries
To the layman in the Middle Ages the Church was a collection of shrines such as Canterbury, which had relics of St Dunstan and St Thomas Becket [1], Bury, with the body of St Edmund, and St Paul's, London, where the image of St Wilgeflot ("St Uncumber") was invoked by women who were tired of their husbands. Powerful monasteries [8] or cathedral churches grew up round these shrines, served by clerks or monks jealous of their privileges and profiting from the pilgrimages made to them.

Church authorities respected these popular cults but also attempted to impose order upon them and to raise the spiritual quality and education of the clergy. The first attempt, in the tenth century, was monastic. By 1200 the older monasteries were no longer the powerhouses of spiritual renewal, but they remained the most distinctive and often the best-loved of English religious institutions. Although varying greatly in size, they all shared a strong vested interest in their landed rights and, as a result, kept a weather-eye on political affairs. They had a slow, agrarian pace of life, and a conservative dislike of religious enthusiasm and novelty. In the twelfth century, many austere houses including those of the Premonstratensians, Carthusians and especially Cistercian orders were founded to regenerate spiritual life [3].

At first the monasteries maintained their corporate discipline. The monk's day began before dawn, and was devoted primarily to prayer, with labour in the fields taking up the rest of the time. According to the chronicler William of Malmesbury (c. 1095–c. 1143), monks abstained from meat, ate only one meal a day in winter, never wore fur or linen garments, and were sparing in speech. The abbot observed the same austere customs as his monks, although he dined separately from them, "with the strangers and the poor". In the later Middle Ages many orders of monks acquired a reputation for avarice, and although monks were rarely guilty of the debauchery portrayed by Protestant and anticlerical propaganda, comfort gradually replaced austerity after 1200.

The bishops and reform
The second attempt to organize the Church for pastoral work was episcopal. It began in the late eleventh century in the atmosphere of reform engendered by Pope Gregory VII (reigned 1073–85). Archbishop Lanfranc (c. 1005–89) of Canterbury and his successor St Anselm (1033–1109) set up a regular organization of the English dioceses, and by 1150 the forms of diocesan organization, were clearly established. The beginnings of the canon law of the Church made possible stricter control of the various offices.

A spate of parish church building be-

CONNECTIONS

See also
186 Triumph of the Church
178 The king and the barons
188 Thirteenth-century European learning
154 The Anglo-Saxon monarchy
174 Romanesque art of the 12th century
200 Gothic art of the 13th century
216 Gothic art of the 14th century
218 Gothic art 1400–1550
208 European society 1250–1450

1 Thomas à Becket (c. 118–70) was killed by followers of his former friend Henry II (r. 1154–89). As Archbishop of Canterbury, he defied the king's attempt to set up a centralized legal system that would cover members of the Church as well as the laity. Becket had served previously as chancellor to Henry; throughout the period the rulers of Church and state were often linked personally, even though their interests were kept carefully separated. Henry underwent many penances for the murder of Becket, who was canonized in 1173, and whose tomb in Canterbury Cathedral became the object of pilgrimages until the Reformation.

2 The friars came to England in the 1220s with the intention of revitalizing spiritual life by preaching and living in poverty. They were bound to no particular convent (although friaries were set up soon in many towns) and stressed pastoral work. Encouraged by Bishop Grosseteste, they were pioneers of philosophy and theology at Oxford University and included two of the finest medieval philosophers, Duns Scotus (c. 1264–1308) and William of Ockham (c. 1300–47). They also transformed the art of preaching and used a vivid outspoken style to attract large congregations. They were prepared to attack injustice and defend the interests of the poor.

3 Rievaulx Abbey was a Cistercian monastery founded in 1131. These monasteries were set far from civilization to encourage austerity, but they grew rich on the profits of wool production, which financed their great building programmes.

In the later Middle Ages the Cistercians had become the principal butt of anticlerical propaganda aimed at the monasteries; they were seen as an overtly political and wealth-seeking organization, whose original puritanism had become entirely lost.

4 Almshouses, such as this one at Burford, Oxfordshire, became common in the 14th century, when the rural economy ceased to grow and urban poverty became a palpable problem. Charity had always been a charge upon both monasteries and bishops, and now hospitals, almshouses and religious fraternities were set up specifically to care for the victims of destitution and disease. The founders were often secular and ranged from the high nobility to permanent guilds of prominent townsmen.

tween 1100 and 1250 [Key] provided most communities with at least the rudiments of religion, and in the early thirteenth century the first attempt was made to regulate the religious life of the laity. Stephen Langton, Archbishop of Canterbury (1207–28) and Robert Grosseteste (c. 1175–1253), Bishop of Lincoln (1235–53), proclaimed that the laity should take common and private confession at least once a year.

An ambitious and wide-ranging attempt to provide an educated clergy was necessary: the ignorance of many parish priests was an evil often bewailed by Grosseteste. The new universities at Oxford and Cambridge became, under Grosseteste's influence, the breeding ground of a small group of highly educated churchmen [2]. These men naturally came to occupy the commanding heights of the Church; by 1400 nearly all the bishops had attended a university and many parish priests, too, were educated there. The stream of educated graduates came to occupy posts of responsibility in the state as well as the Church and throughout the Middle Ages high promotion in the Church often

depended upon previous service to the Crown [9], and vice versa. As a result, the question of who had ultimate authority to appoint the higher officers of the Church, whether it was the king or pope, was a major source of contention, as was secular infringement on clerical jurisdictions.

Wycliffe and Lollardy
Many of the graduate clergy, as well as laymen, were uneasy about ecclesiastical wealth [5]. The most determined of the later movements of reform was the work of the fiery Oxford theologian and royal clerk John Wycliffe (c. 1328–84). Wycliffe saw the organized Church as a creation of Antichrist and proposed a return to "simple" Christianity, free of both hierarchy and superstition, dependent on the Bible, and led by the Crown. His followers, known as the "Lollards", attracted much influential support, and their translation of the Bible (c. 1388), although officially condemned, was widely read. Their attack on established traditions, however, eventually brought down on them all the force of the Church.

Henry III (r. 1216–72) organized the rebuilding of Westminster Abbey. Church-building was an important activity in the Middle Ages, uniting all classes in a common endeavour.

The results of this building programme are still obvious. The great period of parish church building began after the Norman Conquest and continued into the 13th century. Most of the

"decorated" or "perpendicular" churches of the later Middle Ages were alterations or additions to older buildings. Local guilds or landowners often bore the building and maintenance costs.

5 This image of a "greedy friar" occurs in an early manuscript of the poem *Piers Plowman* (1362–92). There were many attacks on the Church for its corruption for many parish priests were badly educated.

6 Sacraments such as marriage brought the Church to the centre of the life of many people. The priest was often the only available doctor, marital adviser, teacher and lawyer for the village poor.

7 Tithe-barns, such as this one in Tisbury, Wiltshire, were built to receive the tithes or taxes payable to the Church, comprising ten per cent of each parishioner's gross earnings. Lawyers insisted that the poor, like the rich, had to pay tithes, but they added that priests were equally bound in charity to reprieve the very poor. At tithe-gathering, the village haywards used to ring the church bells and then go from house to house; but tithes were hard to enforce, and excommunications for nonpayment were common. It was generally assumed that everyone owed a debt to the Church for protecting their spiritual interests, but not to be spent on worldly ostentation.

8 Monasteries were found in almost every part of England. These institutions, which were central to medieval Christianity, were often established as shrines and as centres for the dispensation of charity, as well as being land-owning corporations. Each monastery was more or less independent of any superior authority, and the main problem of any prospective reformer such as Grosseteste was to battle with the abbots and cathedral chapters who were jealous of their privileges and local autonomy. As a result, much of the history of the English medieval Church is littered with obscure legalistic wrangles, with moral issues submerged. Benedictine monasteries were the oldest, dating from the Anglo-Saxons.

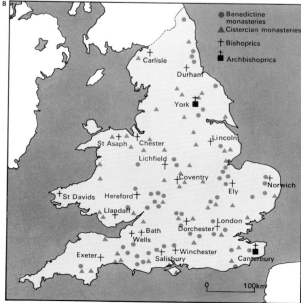

● Benedictine monasteries
▲ Cistercian monasteries
† Bishoprics
■ Archbishoprics

Carlisle
Durham
York
Lincoln
St Asaph Chester
Lichfield
Coventry
St Davids Hereford Ely
Llandaff Norwich
London
Bath Dorchester
Wells
Exeter Winchester
Salisbury Canterbury

0   100km

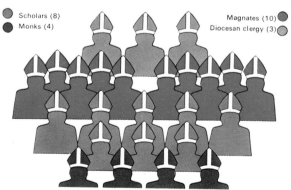

Scholars (8)
Monks (4)
Magnates (10)
Diocesan clergy (3)

9 The episcopacy in England in 1275 comprised men from different backgrounds, but there was a higher preponderance of scholars and university-educated bishops than at any other time in the Middle Ages. During the next 50 years,

the number of civil servants, or royal administrators, sharply increased, reflecting a more secular outlook in the Church. At the same time, the rise in factionalism among the nobility made the Church – an important source of patronage and power – a more

attractive object of baronial ambition, and many bishoprics came under the direct control of the barons. After 1350 encroachment of secular politics on those of the Church (as opposed to a secular attack on clerical privilege) was common.

Rural life in medieval England

The structure of the English village originated in a deliberately chosen pattern of settlement. Brought to England probably by the Anglo-Saxons, it was characteristic of the open landscape of northern Europe. Except in the west and north, where Anglo-Saxon settlement was less complete, it generally replaced the isolated farmsteads of the Celts, and was well suited to the cultivation of the open or "champion" country of the Midlands and southern England.

The medieval English village

In the eleventh century villages acquired the characteristic shape and size that survives today, clustered around a church and often a green. Little essential has changed since *Domesday Book* (1086), in which were recorded almost all the settlements and village structures that now exist: the large village greens typical of north Yorkshire and many other areas, the straggling villages of much of East Anglia, the hamlets of Dorset. By 1100 the idea of several families living in close proximity and co-operating in the agricultural tasks had wholly taken root.

The peasant's house usually stood in its own vegetable plot. Many peasants also owned a pig or a few fowls. The whole family took part in the struggle for survival: the women looked after the garden and made the normal leather clothes (and, after 1350, the woollen clothes that slowly replaced them) while the men worked in the fields. The primitive implements included the ox-drawn plough, but backbreaking work from sunrise to sunset was still necessary [1] both in the peasants' own fields and in those of the lord to whom they owed labour services. A monotonous diet of bread and ale (not yet flavoured with hops) was only partly mitigated by vegetables and fruit, although the plentiful livestock of most parts of England made meat-eating reasonably frequent for all classes.

The social system of the village

In any village there were men of many different social degrees. In the eleventh century there were *geneats* (a lord's free retainers and messengers), *cottars* (who in theory owned five acres but owed labour services to the lord) and *geburs* (who had more onerous services than cottars); landless labourers and slaves also existed in some areas. By 1200 these various distinctions had been simplified into *freemen*, who could sell their land and leave the village, and *villeins* (bondmen), who were hereditarily tied to their lord. These classes, however, were not exclusive or separate. The extensive records of Ely show that free peasants, villeins and labourers intermarried, and by the early fourteenth century some villeins could buy and sell property without hindrance and amass considerable personal wealth. Whatever the importance to landowners of the legal status of peasantry, it was certainly less central to peasant society itself than the ties of kinship and membership of the village community.

The village community demanded organization, and most villages of the thirteenth century were "manors", or estates owned by a lord (whether resident or not) and composed partly of *demesne* (fields directly cultivated for him) and partly of fields held by peasants in return for rent or

1 Most of the tasks of the agricultural year were done by hand. The agricultural writer, Walter of Henley (*fl.*1260) advised that seed should always be brought from other manors at Michaelmas, because it grew better than the home-produced product; and similar lore surrounded other jobs. Agricultural output was at its medieval peak in the 13th century when many improvements to estates were introduced and virgin land was cultivated.

3 Hunting was an overriding passion for King John (*r.*1199–1216), as well as many other kings. William II (*r.*1087–1100) died in a hunting accident and the itineraries of many rulers' journeys through the realm were tours of their hunting lodges, some of which, such as Woodstock in Oxfordshire, developed into sizeable towns. Forests were richly stocked with bear, wolves, boar and all kinds of deer. Falconry was also practised.

Royal forests c. 1250

0   100km

2 Royal forests took up a significant proportion of the land of England in the 13th century. Woodland was as important to the medieval economy as open country for it yielded not only timber and charcoal, but served as a reserve for game and provided pasturage for pigs. After the Norman Conquest the Crown began to assert its authority to gain control of the woodland, designating areas (and sometimes whole counties) as royal forests in which rights of hunting, farming and gathering were severely restricted. These controls were often resented by local inhabitants, and disputes over encroachments by farmers seeking to extend their estates were common throughout the Middle Ages.

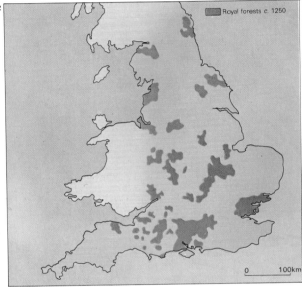

4 The reeve was the local agent of the landowner, and he was often a native of the village in which he officiated. Despite this he was frequently depicted as a tyrant. Here a reeve is shown directing peasants as they reap on the lord's demesne.

5 The postmill, an elaborate version of the windmill, was introduced into England in the 1180s and quickly became common. The whole mill turned to face the wind. A monopoly of milling was an important part of a feudal lord's authority in his manor.

6 The birth-rate in the 13th century was higher than today, but infant mortality varied between 15% and 20%. Once a man reached the age of 20, he might expect to live to 50. The Black Death, however, severely reduced life expectancy, even in the countryside.

week-work [7]. Labour services varied widely, but they generally included up to six days' ploughing or reaping [4] for a few weeks of every year (perhaps two days at other times) and additional "boon-work" at harvest time, as well as carting, tending the lord's garden, and marling (fertilizing).

Declining feudal obligations
These obligations were often taken lightly; in theory milling at the manorial mill [5] or baking in the manorial oven, for a small payment, were heavy duties upon unfree peasants, but in practice the absence of many landlords could render them obsolete. In some instances, as at Wigston Magna, Leicester, the village was divided between two manors, and the peasants ploughed together the common fields of both manors; feudal obligations there can have amounted to little more than a few customary dues. From the twelfth century, labour services were increasingly commuted to money rents. The manorial lord controlled justice amongst his villeins, but villagers could apply to hundred or shire courts, in which

they might hope for a greater degree of impartiality in cases involving their lords.

By the fourteenth century the life of the peasant was changing rapidly. The pressure of population before 1300 [6] led to great distress, exacerbated by the great famine of 1315–17; but the Black Death of 1348–50 and its recurrences in the next half-century dramatically reduced the number of mouths to feed. It brought about a shortage of labour and so forced up wages, despite attempts to hold them down. While prices remained stable, the wage for a farm labourer doubled between 1350 and 1415. At the same time the great estates were less often directly farmed, and instead the demesne was rented out. Thus the distinction between free and unfree came to be irrelevant in practice, and the yeoman farmer, with consolidated fields and wage-earning farm labourers, began to dominate the rural scene. Numerous solid and prosperous fifteenth-century yeoman farmhouses separate from the village are a feature of much of the English countryside, a symbol of the decline of the manorial system.

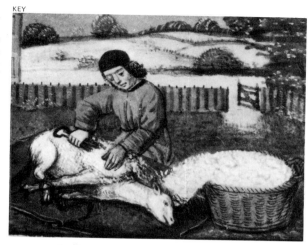

KEY

Wool provided one of the most important sources of English wealth in the Middle Ages. The Flemish and, later, the English cloth industries were almost insatiable, and after the 12th century when the numbers of sheep as farm livestock overtook pigs, sheep-farming brought great profits to many areas. In the late Middle Ages, peasant-owned flocks became as important as those belonging to the lords, as peasants' estates grew in size.

7

7 Peasant farming generally took place in open fields. Signs of these fields can still be seen in some places, as here in Crimscote, Warwickshire, where the patterns of medieval furrows, undivided by hedges or fences, can be detected by aerial photography. Fields were divided into strips, which were measured by the amount that could be ploughed in a day. They were distributed by lot, annually at Michaelmas, so that a man's fields changed each year. A primitive system of crop rotation, with one field out of two or three lying fallow while wheat alternated with oats or barley, was usually operated; but without adequate manure, the yield was only half that of modern agriculture.

8

8 Stokesay Castle, Shropshire, was built between 1260 and 1300 and was a fortified manor house in the characteristically expansive and lavish style of the age of "high farming" and booming agricultural rents. Its wide halls without aisles, and its tall traceried windows, seem to represent a new standard of comfort for the provincial landowner. The solar wing with small chambers indicates a growing inclination for privacy. But the polygonal tower and gatehouse serve as reminders that Stokesay was in the unruly Marches, or border areas with Wales; the whole enclosure could provide protection for the village from raiders.

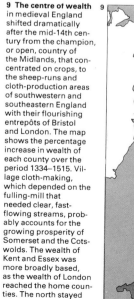

9 The centre of wealth in medieval England shifted dramatically after the mid-14th century from the champion, or open, country of the Midlands, that concentrated on crops, to the sheep-runs and cloth-production areas of southwestern and southeastern England with their flourishing entrepôts of Bristol and London. The map shows the percentage increase in wealth of each county over the period 1334–1515. Village cloth-making, which depended on the fulling-mill that needed clear, fastflowing streams, probably accounts for the growing prosperity of Somerset and the Cotswolds. The wealth of Kent and Essex was more broadly based, as the wealth of London reached the home counties. The north stayed relatively backward.

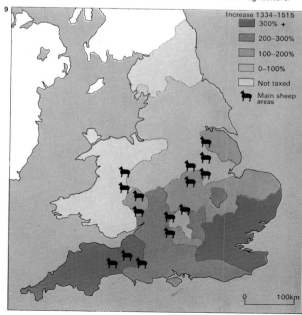

9

Increase 1334–1515
- 300% +
- 200–300%
- 100–200%
- 0–100%
- Not taxed
- Main sheep areas

10

0   100km

10 Lavenham, Suffolk, was one of a series of East Anglian towns that won great prosperity in the 15th century, arising from the production of wool and a certain amount of cloth. As well as boasting an extravagant "wool-church" financed by this industry, Lavenham's wealth was seen in the elaborate timber-framed houses, which have been preserved almost intact. In such houses the art of domestic living was perfected, unassailed by the pressures of town life that prevailed in more urbanized boroughs. The old central hall was divided into several units, including a solar or small hall, and probably a kitchen. The fireplace was an original part of such houses, and was situated on a wall instead of in the centre of the hall. Glazed windows were another luxury.

Gothic art 1140-1200

The term "Gothic" was coined during the Italian Renaissance for a style of architecture then considered so barbaric that it could be ascribed only to the Goths, the fifth-century ravagers of Classical Rome. Gradually "Gothic" has ceased to be pejorative. Its application to sculpture and the graphic arts is a fairly recent development.

Beginnings of Gothic in France
In the early Gothic period, as in the early Middle Ages, the principal artistic enterprises were the great abbey and cathedral churches and their decoration. No new form of church was evolved by early Gothic masons, but both architectural style and building technique underwent a complete transformation. Indeed, the Gothic style is, of all architectural styles before the twentieth century, the one that can claim to be most completely independent of the classical traditions of antiquity.

For the first 30 years of its existence, Gothic architecture was an exclusively French development, but in the late twelfth century it spread beyond France and by about 1250 it had displaced the local Romanesque styles of most regions of western Europe. This expansion may well have been due to the aesthetic and technical advantages of the style, but it must also have been connected with the rise of France as a major political force and her widely acknowledged superiority in cultural matters. It is therefore highly appropriate that the first mature example of Gothic architecture should be the choir of the Abbey of St Denis [Key], which lies just to the north of Paris, and was to become the principal mausoleum of the kings of France. The Abbot of St Denis, Suger (c. 1081–1151), was a passionate believer in the natural superiority of all things French.

The choir of St Denis (begun in 1140 and consecrated in 1144) is a building whose extraordinary originality is matched by perfect consistency of execution. The qualities that distinguish it from previous buildings, however, all stem from a single technical innovation, namely the combination of rib-vaulting with pointed arches. Both of these features were derived from earlier buildings; rib-vaulting from Norman churches such as Lessay (begun c. 1100) and pointed arches from the Burgundian Abbey of Cluny (begun c. 1090) or a derivative.

Rib-vaults and weight-bearing buttresses
Rib-vaults had several advantages over other types of vault [4]. Their diagonally crossed pairs of arches (ribs) provided a framework allowing the cells of the vault to be filled in afterwards. Centring (curved scaffolding on which vaults were built) was thus required only for the ribs, not for the whole vault. Rib-vaults were also relatively light, being made of carefully shaped stones rather than rubble embedded in mortar, and the use of geometrically regular arches made them easier to fit over variously shaped spaces. The disadvantage of semicircular ribs as used at Lessay and elsewhere was that their height (radius) was bound to be half their span (diameter). This meant that in an oblong vault diagonal ribs and wall ribs required some rather clumsy adjustments to make them equal in height to transverse ribs.

By combining rib-vaults with pointed

1 Senlis Cathedral is the earliest surviving French cathedral (begun in 1153) based on St Denis. Pointed rib-vaulting is used throughout, mostly carried on slim columns with foliage capitals. This view looking north-east, shows the ambulatory with one of the semicircular chapels in the right distance.

2 Laon Cathedral remains largely unaltered except that the ambulatory was replaced in the early 13th century by a flat east end lit by a rose window. In the nave there was a brief attempt to extend the alternating rhythm of the vaults downwards by adding extra shafts to every other column of the arcade.

4 Vaults under semicircular ribs appeared in England and Normandy about 1090. The main problem with semicircular ribs was that the wide diagonal ribs had to be depressed and the narrow wall-ribs raised as vertical pieces in order to rise to the same height as the transverse ribs. Gothic architects solved this problem by using pointed ribs. These are struck from two centres, which means that their height is no longer dictated by their width. Pointed arches were adopted for aesthetic rather than structural reasons. Sexpartite vaults (so called because they have six compartments) were used over paired rectangular bays at Laon in the late 12th century.

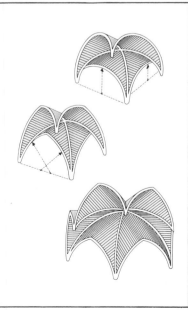

3 A simplified version of Gothic architecture had been evolved by 1160 in the Cistercian abbeys of northern England, but the adoption of Gothic started with the rebuilding of Canterbury choir c. 1175 by the French architect William of Sens. The design has several features not usual in France, notably the dark marble shafts. These became standard in English Gothic for about a century afterwards.

5 The jamb figures at Chartres Cathedral (central west portal) probably represent Old Testament personages. The Christ in Majesty above the doors has a calm nobility different from the terrifying Judging Christs of the Romanesque portals.

6 The subject of this sculpture at Senlis Cathedral (west portal) is the Coronation of the Virgin, one of several scenes that became popular in the 12th century with the advent of the Virgin cult. Traces of original paint remain.

arches, whose height can be kept constant whatever the span, the St Denis mason put vaults of even height over a complicated series of spaces of various sizes and shapes. He emphasized the lightness of the vaults by setting them on single, slender columns. He also extended the skeleton principle of rib-vaults to the outer wall of the choir. Realizing that rib-vaults did not exert a continuous outward thrust that needed to be absorbed by massive walls, he placed deeply projecting buttresses only at those points where the thrust was concentrated. The walls, relieved of their load-bearing function, opened up as an almost continuous band of windows. It was the brilliant lighting of the St Denis choir that most impressed Abbot Suger.

Laon Cathedral

St Denis immediately became the inspiration for a whole series of northern French cathedrals, of which Laon [2], begun c. 1160, is in many respects the classic example. The internal elevation is of four storeys, including an ample gallery or tribune. The vault is in six sections with shafts arranged alternately

singly and in threes. The narrow proportions of churches of this period, regarded as typically Gothic, are no less than those of some major Romanesque churches, such as Cluny or the churches that flank the pilgrim route on the way to Santiago.

The smooth exterior contours of twelfth-century Gothic churches were generally enlivened by towers, ranging from the usual two at the west end to the seven projected at Laon. West fronts were pierced by large portals in which high-relief sculpture was used not only on the tympana (the area between the top of the door and the arch above it), as in the Romanesque period, but also on the framing arches (archivolts) and on the jambs (the sides of door or archway). These large-scale jamb figures [5] became a hallmark of the Gothic style.

Apart from foliage capitals, early Gothic interiors contain little carved decoration beyond the moulded arches, ribs and shafts of the buildings themselves. Their austerity was to some extent offset by the lavish use of stained glasswork and a restricted use of painted colour.

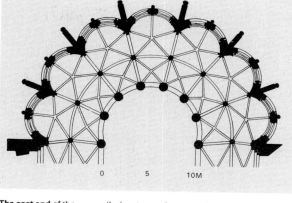

0   5   10M

The east end of the 8th-century church of St Denis was demolished and replaced in the 12th century by this complex structure in which the semi-circular processional path or "ambulatory" enabled visitors and pilgrims to see the shrine of St Denis without disturbing the services there. Opening out of the ambulatory are seven chapels partly contained in shallow round apses. Their windows form a continuous band separated only by small stretches of wall and thin, deep buttresses. Crossed lines indicate the position of the vaulting ribs and crosses, the sites of the altars.

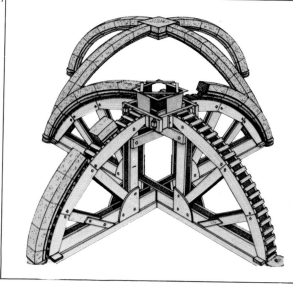

7 The south choir aisle at Exeter Cathedral (c. 1300) was bombed in 1942 and rebuilt in 1949. This view shows the wooden arched scaffolding or "centring" used to build up the ribs. Some of the ribs are already in position but none of the sculptured keystones ("bosses") at the intersections have yet been replaced. After the completion of the ribs the next stage is to fill the curved triangular spaces in between with horizontal courses of stone. No scaffolding is required for this process because the ribs serve the purpose adequately. By this stage the wooden protective roof would already have been built.

8 In the Jesse tree window at Chartres Cathedral, Jesse lies in the lowest panel. Out of his loins grows a tree with Christ and His ancestors, including the Virgin seated in the branches.

9 The windows round St Thomas à Becket's shrine at Canterbury Cathedral were filled with stained glass illustrating the miracles worked by his intervention. They recall late 12th-century illumination.

10 Church metalwork was among the most esteemed forms of art in the Middle Ages. Abbot Sugar of St Denis wrote long descriptions of the treasures which he collected for his abbey, but made no mention of the stone sculptures of the new west portals which are now regarded as the starting point of Gothic sculpture. The Klosterneuburg altarpiece (1181) was originally made as a pulpit, which explains why more precious materials were not used. Most unusually the artist, Nicholas of Verdun, was allowed to sign his name. These classical figures possess a three-dimensional solidity and weight that has nothing to do with the linear exaggerations of most Romanesque art.

11 Like metalwork, manuscript illumination was a major art form. Many fine manuscripts were produced in monastic scriptoria but there are good grounds for thinking that the leading artists – of whom this Winchester master was one – were professional laymen. The two figures in this initial to the Book of Isaiah (c. 1170) in the Winchester Bible achieve in one way what Nicholas of Verdun did in another, that is the effect of solidity and a respect for the proportions and articulation of the human figure. The Winchester style derives from recent Sicilian mosaics of provincial Byzantine style. The same artist may have executed some wall paintings in Winchester Cathedral.

199

Gothic art of the 13th century

In their plans and in the disposition of their internal spaces the twelfth-century Gothic cathedrals adhered largely to Romanesque precedent. But in a number of churches begun towards the end of the century, of which the most important was Chartres Cathedral, the middle storey (the gallery or "tribune") was omitted to make room for high arcades and clerestory, separated only by a low triforium (a passage along the wall).

The immense enlargement of the clerestory windows necessitated extra support to resist the thrust of the high vaults and this was provided by flying buttresses, which henceforth became a feature of French cathedral design. Externally, flying buttresses effaced the simple contours of the early Gothic churches and substituted the complex, restless silhouette now regarded as characteristically Gothic. The grandeur and simplicity of the interior at Chartres was enhanced by the plain quadripartite rib-vaulting, supported by piers consisting of a column with four attached shafts, thereby bringing the vertical lines of the vault shafts down to the ground.

Chartres, like St Denis 50 years earlier, became the prototype of a whole series of northern French cathedrals [7]. Reims, begun in 1210, followed Chartres closely but made an important advance in the window design. Instead of the groups of three openings, two pointed and one circular, that made up the Chartres clerestory, the Reims mason treated all windows as one unit, but with subdivisions built up as an interdependent series. This innovation, known as "bar tracery" or simply "tracery", was taken up by other masons and became an enduring characteristic of Gothic architecture.

European developments

In the reign of Philip II Augustus (1165–1223), France was established as the leading power in Western Europe and in the reign of his grandson Louis IX (1215–70) it became the arbiter of European taste. As a result, there are major churches outside France, such as León Cathedral in Castile (begun c. 1250) and Cologne Cathedral (begun 1248), that are practically indistinguishable from French work. The same cannot be said of Westminster Abbey (begun 1245), even though it has more French features than any other thirteenth-century English church.

Among the most impressive examples of this so-called "Early English" style is Lincoln Cathedral (begun 1192). The proportions are broad and the walls are of Romanesque thickness, although any effect of heaviness is offset by the use of fine mouldings on the arches and marble shafts on the walls and supports. This English predilection for complicated linear pattern is remote from the austerity and single-mindedness of contemporary French designs. During the second half of the thirteenth century the English masons learned all they wanted to know of French Gothic, but their evident enthusiasm for elaborate architectural decoration remained undiminished.

Master masons

Unfortunately, little is known of the organization of thirteenth-century masons because of the scarcity of contemporary documents. But it seems clear that master masons performed much the same role as

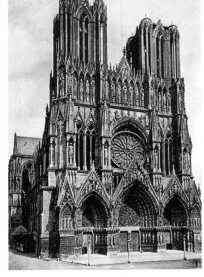

1 All the important Gothic churches in northern France, beginning with St Denis, were intended to have two-tower west façades. Here at Reims as well as at Laon and Chartres towers were also intended over the transept façades but were never completed. The rose window is a usual feature of 13th-century west fronts, but the portals are exceptional in substituting rose windows for sculpted tympana.

2 The influence of Chartres on the cathedral at Soissons is apparent in the high arches and clerestory. However the overall proportions are taller than at Chartres and the piers are simple, round columns with only one shaft attached, as opposed to the four at Chartres. The extreme lightness of the interior is due to the destruction of the original stained-glass windows.

3 The foliage on the chapter-house door (late 13th century) at Southwell Minster (left to right: buttercup, vine, oak) achieves a fine balance between the natural and the ideal while still respecting the architectural function of the capitals (the tops of columns). Sculptors at Reims in about 1230 were the first to carve leaves that corresponded closely to natural forms.

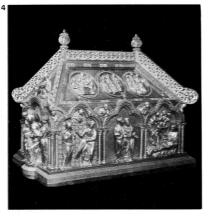

4 The Reliquary of Our Lady at Tournai Cathedral has been attributed to Nicholas of Verdun who finished the Klosterneuburg altarpiece in 1181. The figures have ample, fluid draperies anticipating those of some of the stone sculptures at Reims. The reliefs show the Annunciation, the Visitation and the Nativity. More important shrines like St Thomas à Becket's at Canterbury had the same house-like shape but stood on tall stone bases.

5 The sole surviving large exterior sculpture at Notre Dame, Paris, this Virgin stands on the *trumeau* or central pillar of the north transept portal. Her pose, with the weight on the left leg, serves to draw attention to the infant Christ on her arm and is accentuated by pointed, hanging folds. Her air of mannered refinement and swaying posture became typical of 14th-century sculpture. Like most Gothic sculpture this figure was originally coloured.

6 The broad proportions and elaborate decoration at Lincoln Cathedral are in complete contrast to contemporary French cathedrals. The arcades are so wide that the aisles merge with the central space. Most available surfaces are enlivened by brownish-grey polished shafts from Purbeck (Dorset). The vault is the earliest in which additional decorative ribs are introduced to blur the divisions between bays and create an effect like a series of branching palm fronds.

architects in later periods although, unlike architects, they were promoted from the ranks of the working masons and not trained purely as designers [Key]. Master masons were generally well paid and often the owners of extensive property. The working masons were divided into "cutters", who carved the architectural components, and "setters" who actually put up the building. Figure sculpture was usually the province of highly skilled specialists.

The masons' tools were extremely simple. Stone was given various degrees of finish by using finer or coarser chisels. Simple lifting tackle was used and a primitive wooden crane remained in position at Cologne Cathedral until the building was finished in the nineteenth century. The vast front of Cologne was completed then with the aid of the original designs on parchment, which were probably made as presentation drawings for the approval of the design by the cathedral authorities. Working drawings were usually drawn full-size on boards or on a plaster "tracing floor" in the masons' lodge. A surviving example of such a floor at York

has many layers of partly erased drawings underneath the most recent set, which is late fourteenth century in origin.

Decorative features

Stained glass remained the principal form of internal decoration, but by about 1250 the deep blues and reds of the early thirteenth century gave way to a wider and less predictable range of colour.

Thirteenth-century sculpted portals retained the basic twelfth-century format but, like the cathedrals they adorned, were much larger than those of the preceding century. There was an increased emphasis on the Virgin and even in Last Judgment scenes she and St John appear interceding with Christ on behalf of mankind. A certain humanizing of religious subject-matter was achieved in the opening decades of the century by the increasing depiction of character types. By about 1250 this process had gone further and figures such as that of Joseph on the west front of Reims were endowed with the worldly grace that was to characterize much courtly art of the succeeding century.

Hugues Libergier was master mason of the important church of St Nicaise in Reims which was destroyed in the French Revolution. He appears holding a model of his church with his square and dividers by his feet. His dress is indicative of the professional status of masons in this period. The Dominican friar Nicholas de Biard compared bishops with master masons who, he said, gave orders to others wearing gloves and holding a measuring rod and receiving higher pay although they themselves never set their hands to the work. Often master masons were commemorated by inscriptions placed in "their" buildings by patrons.

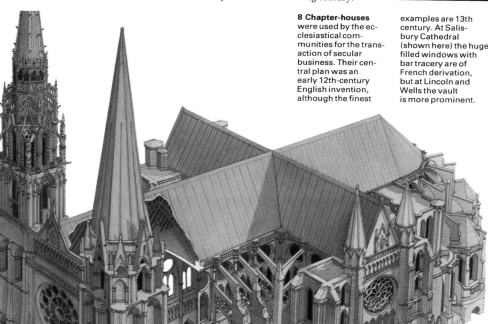

8 Chapter-houses were used by the ecclesiastical communities for the transaction of secular business. Their central plan was an early 12th-century English invention, although the finest examples are 13th century. At Salisbury Cathedral (shown here) the huge filled windows with bar tracery are of French derivation, but at Lincoln and Wells the vault is more prominent.

9 This detail at Westminster Abbey shows St Peter and is one of 13 compartments of a wooden panel whose original purpose is unknown. It may have been placed behind or in front of an altar. Despite severe damage, it is still recognizable as the finest panel painting surviving from the 13th century. The thin features and mannered gestures recur in wall paintings at Westminster.

7 The west towers of Chartres Cathedral, with the portals and windows, were added to an 11th-century, wooden-roofed basilica in the period 1130–50 and retained after the basilica was burned in 1194. The new Gothic church has the same width as the old basilica but the increased height may be gauged from the rose window added above the west windows. Smaller roses occur in the clerestory windows and more complex ones in the transept fronts. Even the lower tiers of flying buttresses are linked by radiating arches to make them look like segments of rose windows. The niches on the buttresses are the first examples of what became a favourite device of the Gothic designers. The austere interior displays to perfection the medieval stained-glass windows that have been preserved in their entirety at Chartres, alone among French cathedrals. After the decision to retain the old west portals, the transept fronts were decorated with as much sculpture as if they had been west fronts; the similarity would have been complete if their twin towers had been carried up. The number of towers planned was nine, two more than Laon. In this, Chartres set new standards of magnificence.

The emergence of France

The beginnings of the French nation can be traced to the medieval house of Capet. Hugh Capet, a feudal lord whose lands centred on the middle Seine, was elected king of the west Frankish domains in 987 and reigned until 996 (superseding the Carolingians). France was then only a part of Gaul, which stretched from the Pyrenees to the Rhine and had been, since the Frankish invasions of the fifth century, a conglomeration of Germanic, Celtic and Romance elements. The French-speaking peoples came to the fore and were united only with the slow extension of Capetian authority. The Capetians initially controlled only a small area around Paris and Orléans and for 200 years were given only nominal allegiance by the many more powerful feudal lords in France.

The Capetian kings
The Capetian kings had almost no contact with Aquitaine and the south, while in the north powerful dynasties of the nobility in Normandy, Anjou, Flanders and Burgundy were their equals in wealth and influence. But the real source of Capetian strength was

the popular veneration they were increasingly accorded. The kings, although at first powerless, gradually emerged as sacred figures, consecrated by the holy oil first used by St Remy (c. 438–533), the "Apostle of the Franks", in c. 497 at the baptism of Clovis. This oil was thereafter kept at Reims in an ampoule that never emptied [Key]. The royal touch was also believed to cure scrofula (lymphatic tuberculosis). In 1297 the kings added a halo to their inheritance with the canonization of Louis IX, whose spirit became the focus of both religion and patriotism.

Gradually the French kings built up their political power too. This had a solid base in the growing prosperity of France – in the thriving city communes, which were actively encouraged by the twelfth-century kings, and which became foci of loyalty. A dramatic expansion of authority beyond the royal patrimony took place in the reign of Philip II Augustus (1165–1223); he took advantage of the unpopularity of King John of England to regain English-ruled Normandy and bring his own vast territories extending to the Pyrenees under direct royal rule. His son,

Louis VIII (1187–1226), introduced royal power into the heartland of southern France (Languedoc), after the Albigensians there had been crushed by a papal crusade against their heretical view of the world as a creation of the Devil. The power and prestige of Louis IX (St Louis) (1215–70) were so great that he was able to act as the arbiter of all Europe. The apogee came in the reign of Philip IV (1268–1314), who advanced the frontier far to the east and subdued even the Papacy.

The change in government
With this advance came a new kind of government. The thirteenth-century kings attracted formerly independent lords into their service and, with the help of educated men from the University of Paris [5], began to establish a "civil service" of local officers who judged cases and collected revenue in the provinces. At the centre were the *parlement* of Paris, the highest court of the kingdom, and the Chambre des Comptes, a financial department, staffed by lawyers of high calibre who made French royal justice widely sought. By 1314 the French king was

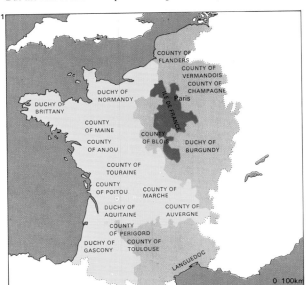

1 **By 1180 the map of France** included parts of modern Belgium but excluded Franche Comté, Dauphiné and Provence, which were ruled by the emperor. But even within the confines of the Capetian dynasty's formal authority, many provinces were almost wholly beyond royal control. The real centre of the king's power was between Paris and the Loire. To the west lay the dominions of his Plantagenet rivals. Almost all of these were gained by Philip II Augustus (r. 1180–1223).

- Royal demesne 1180
- Fiefs held by English king 1180
- Other fiefs 1180

2 **By 1328 France had expanded** beyond her borders of 1180, and in 1349 she gained the Dauphiné. Internally the Capetian dynasty had established its hold on the former Plantagenet lands, and

all the feudatories in Burgundy, Brittany and Languedoc acknowledged it, as did the English king in Guyenne. But the many enclaves and noble houses indicate how shallow were the roots of royal author-

ity, which often merely confirmed the positions of local magnates. In ensuing wars, many nobles played off the English and French crowns against each other, to gain almost complete independence.

- Royal demesne 1328
- Territories of royal princes 1328
- Fiefs held by English king 1328
- Other fiefs 1328

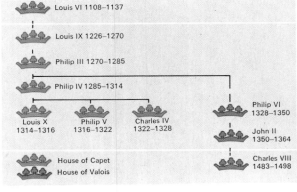

Louis VI 1108–1137	
Louis IX 1226–1270	
Philip III 1270–1285	
Philip IV 1285–1314	Philip VI 1328–1350
Louis X 1314–1316 / Philip V 1316–1322 / Charles IV 1322–1328	John II 1350–1364
House of Capet	Charles VIII 1483–1498
House of Valois	

3 **The French monarchy** descended in the male line from 987 to 1848. The direct succession of son to father was carefully preserved until 1316. The elder Capetian line died out altogether in 1328. The Valois line then ruled until 1589.

4 **The ambulatory** of the Abbey of Saint-Denis was one of the earliest inspirations of Gothic architecture and it represented the cradle of French national sentiment. It was built in the mid-12th century during the reign of Louis VII.

5 **The seal of the University of Paris** was made in 1215. The university began as a corporation of scholars in the 12th century and in the 13th century trained many of the most useful servants of the Crown as well as great philosophers.

richer, more respected and better served than any other European ruler.

That his power was nevertheless limited was shown by the crises of the fourteenth century. Up to 1314, by a remarkable chance, power had passed uninterruptedly from father to son; but the direct line of Capetians ended in 1328 and the crown passed to their Valois cousins [3]. The Valois claim was challenged by Edward III of England (1312–77), to whose court flocked all the dissidents of France: in Flanders, in Brittany and above all in the south, local noblemen hoped to increase their patrimony by playing off Philip VI (1293–1350) against his rival. Edward crushed the French arms at Crécy (1346), while France became the prey of war-bands who made their fortune from the profits of ransom, pillage and terror. At Poitiers (1356) [7] King John II (1319–64) was himself captured by Edward's son, the Black Prince, and released only for an enormous ransom. English successes were not continuous, but during the reign of Charles the Mad (1368–1422) a murderous factional struggle between the rival houses of Burgundy and Orléans exposed France to a renewed attack from England. With the help of the Duke of Burgundy, the Lancastrian King Henry V was able to conquer Normandy, occupy Paris and induce Charles VI to disinherit his son in his favour.

Joan of Arc, the Maid of Lorraine

At the lowest ebb of the Valois fortunes, when the English and Burgundians ruled northern France, the popular and religious aspect of kingship reasserted itself in the extraordinary events involving Joan of Arc (c. 1412–31) [6]. Joan, a humble girl from Lorraine, went to the court of the disowned heir, Charles VII (1403–61), [10] and by claiming the miraculous intervention of the saints in her cause endowed it with a popular fervour. She herself was burnt by the English but a reconciliation between Burgundy and Charles VII at the Congress of Arras (1435) made the English position hopeless.

As the English withdrew from France, the monarchy found itself firmly established in the affections of its subjects. The unity of the nation was never again in doubt.

The Coronation Chalice at Reims Cathedral is the symbol of a sacred kingship. The French monarchy made up for its lack of physical power by the prestige of its religious sanction. Kings were anointed with the oil said to have been used at the baptism of Clovis, the first king of the Franks; the oil was kept in a phial miraculously refilled for each coronation. The king was regarded by many as a religious figure. The lilies on his shield were said to have first appeared supernaturally on the shield of Clovis, and his banner, the oriflamme, was said to be the mythical flaming lance of Charlemagne, King of the Franks.

6 Joan of Arc was captured at the battle of Compiègne in 1430. She had appeared at the court of Charles VII when his fortunes were at their lowest in 1429. Her adoption of the dress and manners of the mercenaries shocked some but inspired many more and she eventually turned events in Charles's favour. She was tried by the English in 1431 as a heretic and burnt at the stake.

7 The Battle of Poitiers (1356) (from a 15th-century manuscript) was a great defeat for the French army. The English, led by Edward, "the Black Prince", were heavily outnumbered, but at the height of the battle they launched an attack from behind the French lines and the French king, John II, fled, only to be captured with his son Philip. He was taken to England, where he died.

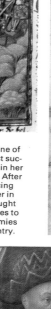

8 Charles V of France is seen here entertaining Emperor Charles IV on the latter's state visit to Paris in 1377. Charles V (1338–80), who came to power at the age of 19, became one of France's most successful kings in her darkest days. After heavily reducing English power in France he sought European allies to drive his enemies from his country.

9 A miniature by Jean Fouquet (c. 1420–80) shows the trial of John, Duke of Alençon, for treason in 1458. The duke fought with Joan of Arc and was loyal to her even during the court intrigues of the time. However, as an outspoken rebel who had made no secret of his loathing for the king, he was still given a fair and formal trial and did not die until 1476. French judicial institutions, which enjoyed a high prestige throughout Europe, had been perfected in the 13th century. They were based on Roman law.

10 Charles VII was painted by Jean Fouquet in 1445. This much-abused monarch is the least understood of French rulers. He overcame appalling misfortunes which began with his disinheritance by his father Charles VI in 1420 in favour of Henry V of England, the victor of the Battle of Agincourt. Yet he became the focus for all enemies of English rule in France and gradually succeeded over 30 years in expelling his opponents from the realm. However, revolts of the nobles disturbed the last years of his reign.

203

Feudalism

In a civilization like Europe's in the early Middle Ages, with no civil service, police, or legal profession, the only stable institutions were the family group and especially – since wealth and power could result from it – the warrior band with its lord. The bond between lord and warriors was the basis of "feudal" society, in the Germany, England and France of the eleventh and twelfth centuries. The ethic of the medieval warrior demanded fierce loyalty from the retainer and unstinting generosity from his lord.

Land tenure

The personal bond between the warrior and his lord, or later the tenant and the king, which was expressed in fealty, vassalage and homage (increasing grades of obligation), became translated into the sphere of property relations. Absolute ownership of land was unknown in the Middle Ages; instead the king granted estates or fiefs, in return for definite services, often military, which were supported by oaths. The king's tenants-in-chief, too, granted out land to knights in return for oaths of loyalty to themselves. The

ceremony of homage [1] did not necessarily imply either a permanent or an exclusive tie, and it was common for a knight to owe homage for an estate to several lords, or even to a social inferior. Feudal terminology was therefore simply a way of describing the complexities of land tenure: society was not in any hierarchical "feudal pyramid".

Nevertheless, the period from about 950 to 1250 is properly called "feudal" because of the dominant position of a warrior aristocracy in Western Europe and in Crusader Palestine. This aristocracy shared a common training and fighting technique, that of the mounted knight, and a common code of conduct, the ideal of "chivalry". Its basis was the technological superiority of mounted men over foot soldiers, which emerged about the time of Charlemagne. In antiquity the stirrup and the horseshoe were unknown; introduced from central Asia about 750, stirrups, in particular, had revolutionary consequences: they made it possible to charge effectively with a lance carrying the full impetus of a galloping horse without the rider being unhorsed on impact with his target [2].

This placed great value on mounted warriors [Key], but before about 950, horses were scarce; the Franks' relative wealth in them partly explains Charlemagne's successes.

Rise of the knights

There is evidence that the number of horses increased rapidly after 950 and in the next century mounted men everywhere decided the fortunes of war. Their numbers were small, their equipment expensive and their training long. But this increased their pride and prestige and by 1100 mounted warriors would symbolize their corporate spirit by the initiation ceremonies of knighthood.

Local society was thus dominated by the knightly classes. They served in various capacities: many, without a permanent master, sold their services as mercenaries to the highest bidder. Others, especially at the start of their careers, took service as the household knights of a great lord. Still others, particularly the Normans, were bound by the conditions of their tenure to fight at specified seasons; "knight-service" was the most important obligation for landowners.

1 The ceremony of homage is shown in a metaphorical context in this 12th-century illustration of Theophilus paying homage to the Devil. The Devil carries a charter or written record. But the original ritual of homage was designed to register the contract in public in a memorable way without the need of documentary record. Because few men were literate in the earliest period of feudalism, it was essential to make legal contracts before witnesses whose memories could be relied on. Submission was represented by the lord taking the hands of his man between his own; afterwards they kissed to symbolize friendship.

2 A 14th-century knight, Sir Geoffrey Luttrell, receives his helmet and lance (from the Luttrell Psalter). Knights were regarded as heroes and their combats were invested with glamour. The impetus of their charge depended on the innovation of stirrups; the high saddle also acted as a lever. The lance, intended to unseat opposing horsemen, could be used properly only with these aids. Few lords had many horses but evidence exists that careful breeding was increasing the number. An important stud at Corvey, Saxony, supplied the German imperial armies.

3 Functional armour was characteristic of Norman knights. As the sword [G] and axe [F] became longer, protection had to be increased with heavy shields [D, E] and the 10th-century helmet [A] gave way to a helmet with visor [B] and a fitted coif of chain mail [C].

4 Langeais Castle was one of 13 built by Fulk Nerra, Count of Anjou (987–1040); from them the House of Anjou, which later ruled England, began its formidable rise to power. In each, a castellan and garrison controlled roughly as much country as could be traversed in a day. Although fortified towns had existed before, the stone keep, which was the oldest type of castle, was an invention of the tenth century. Keeps, such as that at Langeais, stood on an artificial mound with a stockade. Internally they might have had only one big room and a store, but they were focal points of feudal power and by 1100 had spread throughout Western Europe.

The most formidable of knightly warlike skills was probably the charge in close formation, which was used devastatingly against the Turks in the First Crusade. But as professionals, knights also learnt defensive skills, the most spectacular of which was the development of the castle, from the simple eleventh-century keep to the elaborate bastioned castles of the 1200s [5].

Chivalry and landowning
The knightly ideal of chivalry demanded that knights should fight to avenge the oppressed, to vindicate the honour of ladies – the theme of romantic love first appears in knightly circles in southern France in the twelfth century – and to advance the Christian religion against the Muslim Saracens [7, 8]. The Templars and Hospitallers combined knightly prowess with monastic chastity in the defence of the Holy Land.

A knight's fighting career, unless he became a lord's military official, a marshal or a constable, was comparatively short, and his ambition was usually to acquire and cultivate an estate. Once on the land most of them lived simply in keeps with one first-floor room for living and sleeping and, below, a storehouse. Although knights had customary rights over the tenants who held plots by their grant, villagers did not live entirely in the shadow of their lords. The spread of a "three-field system" of farming with its complex organization, and the appearance of village communes in France, suggests that many villages [6] while paying dues had an independent corporate life, to which the knight contributed by providing leadership and protection and often by taking the lead in cultivating neighbouring wastelands.

Feudalism thus maintained a series of obligations which linked the peasant with his immediate lord and, indirectly, with royal power. Although the dominance of a warrior caste was crumbling by the fourteenth century through the competition of other groups of a less military nature, the disappearance of feudalism was gradual and the knight would constitute a social ideal for centuries. The material remains are with us still, as are the vestiges of chivalry as a guiding principle of Western European civilization.

KEY

The knight, armed and mounted, was at the centre of feudal society. Only an élite could afford the costly equipment.

5 Caernarvon Castle, Wales (1283–92), was one of the massive castles built by Edward I after he conquered Wales. By 1300 the primitive keep had developed into a fortified community large enough to house, and strong enough to protect, a provincial government behind technically innovative bastions.

6 An English village [centre, left], dating from feudal times, was created on uncultivated land at Chelmerton, Derbyshire. It shows a pattern implying a planned settlement, either through the enterprise of a lord or simply peasant co-operation. Each house has a garden and narrow strip extending into the waste land.

7 Ekkhard and Uta, a thirteenth century Crusader and his lady sculpted in Naumburg Cathedral, represent the highest ideal of European nobility – the Christian warrior – towards the end of the feudal period. From the First Crusade (1096-9), warfare found an idealized form in the defence of Christendom against the Saracens. Great numbers took part in expeditions to Jerusalem and enthusiasts united in the military orders of the Templars and Hospitallers. Besides the great international Crusades, many knights went individually to win a reputation or expiate an offence, to fight in Palestine, in Spain, or in pagan Lithuania.

8 Chivalry is exemplified by St Louis IX, King of France (1226–70), rescuing a Saracen lady and her child in battle. The feudal knight evolved a code of conduct transcending even his Christian allegiance, and chivalry was a central theme.

Medieval literature and drama

Latin was the common literary language of medieval Europe and was extensively used in works of philosophy, theology, law, history and even storytelling and romance. Yet the fascination of medieval imaginative literature lies in the evolution of the use of vernacular, or local, language. The invaders of the Western Roman Empire brought new cultures, languages and mythic traditions which in some cases superseded, and in others fused with, the ancient literary traditions and language of Rome.

From epic to romance

The myth, epic and folklore of the Germans [3] and Norsemen was initially passed on by word of mouth from recitations in the banqueting halls of the great; it was rarely written, which makes any accurate assessment of its growth and origins incomplete. Contemporary history was sometimes turned into instant legend, as in the Anglo-Saxon poem *The Battle of Maldon* recording a famous battle with the Danes in 991. In contrast *Beowulf* (*c.* 700–1000) was the written version of an ancient epic of the kil-

ling of the dragon Grendel, and the Norse *Eddas* are epics from the same tradition.

By the eleventh century such native ancestral traditions were mingled with other literary interests; lyrical, allegorical, amatory and narrative themes were subjected to Graeco-Roman classical revivals of content and form with periodic romantic yearnings for a non-classical past. The Arthurian legend pervaded medieval literature although it was not until the fifteenth century that Thomas Malory (died 1471) recorded in full, in his *Morte d'Arthur*, the legendary exploits of King Arthur and his Knights, a romantic fable of Romano-Celtic Britain.

At the beginning of the twelfth century an influential literary model was the French *chanson de geste* such as *The Song of Roland*, an epic poem in Old French based on a heroic interpretation of Charlemagne's defeat at the Battle of Roncesvalles. As with *Morte d'Arthur*, it involved changing history into legend and legend into literary epic. Another similar epic was the Spanish *El Cid*.

Epic and narrative poetry of the early Middle Ages gave way to the poetry of

courtly love, the most famous exponent of which was the French poet Chrétièn de Troyes (flourished 1165–80). *Roman de la Rose* [2], a vast thirteenth-century poem of allegory, was more a cynical variation on Latin love poetry, while the Middle English *Gawain and the Green Knight* is an example of a chivalric romance of the fourteenth century. Meanwhile the troubadour poets of Provence, such as Bertran de Born (*c.* 1135–1207), and the German Minnesänger, such as Wolfram von Eschenbach (*c.* 1170– *c.* 1220) and Walther von der Vogelweide (*c.* 1170–1230), produced a new sophisticated poetry of courtly romance which was recited by wandering minstrels and poets [5] and marked the high point of medieval courtly culture.

Development of music and drama

Troubadour song also marked a meeting point between developments in medieval literature and music. In the early Middle Ages music had been almost entirely in the service of the liturgy of the Church and had been dominated by Gregorian chant, or

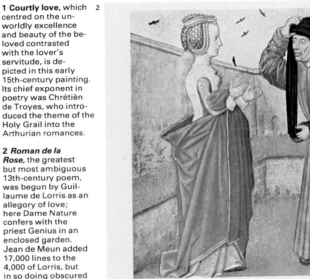

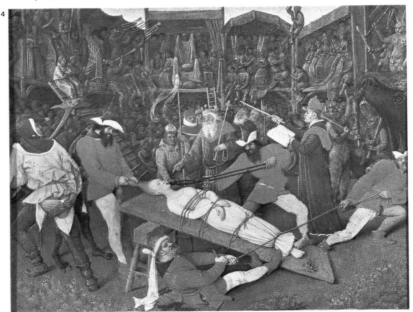

1 Courtly love, which centred on the unworldly excellence and beauty of the beloved contrasted with the lover's servitude, is depicted in this early 15th-century painting. Its chief exponent in poetry was Chrétièn de Troyes, who introduced the theme of the Holy Grail into the Arthurian romances.

2 *Roman de la Rose,* the greatest but most ambiguous 13th-century poem, was begun by Guillaume de Lorris as an allegory of love; here Dame Nature confers with the priest Genius in an enclosed garden. Jean de Meun added 17,000 lines to the 4,000 of Lorris, but in so doing obscured the original theme.

3 The capture of the magic ring by Brünnhilde, a Valkyrie maiden, on the funeral pyre of Siegfried, is the dramatic Wagnerian finale of *The Ring of the Nibelungen,* the most emotive epic of the German race. The story is Burgundian-Frankish in origin.

4 A medieval mystery play is recorded pictorially in a unique miniature by Jean Fouquet (*c.* 1420–80). It shows a circular theatre surrounded by galleries or scaffolds, with players above and spectators below. The drama has been removed from the church interior and although the action, a scene of martydom, takes place on the raised central area, the angels, throne and audience intermingle.

plainsong. The great musical innovation of the eleventh century was polyphony – the juxtaposition of two or more voice lines harmoniously intertwined. It was exemplified by the musical school of Notre Dame in Paris which in the twelfth century produced composers such as Léonin and Pérotin.

Church music also led into dramatic representations of the Easter and Christmas Gospels, performed at Winchester as early as 970. Gradually other religious cycles were dramatized, at first within the Church and later outside; thus began the mystery [4] and morality plays of the late Middle Ages.

Triumph of the vernacular
Medieval song culminated in the motet in the thirteenth century, the expression of sung verse in a polyphonic context that gave musical form to troubadour and Minnesänger poetry. In Italy the lyric flourished at the Sicilian court of Frederick II, where Provençal language and themes were merged into the new Italian language. A parallel movement occurred in Galicia and Portugal in the thirteenth and fourteenth centuries, and in the

north of France with Rutebeuf (active 1250–80), Eustache Deschamps (1346–c. 1406) and Charles d'Orléans (c. 1384–1465).

The fourteenth century marked the true literary emergence of the languages of modern Europe. Renaissance rediscovery and re-evaluation of Latin and Greek as imaginative literature began at the same time and poets such as Francesco Petrarch (1304–74) wrote equally in Italian and Latin. His idealized love poetry and his wholly modern appreciation of nature were expressed in Italian lyric verse.

Geoffrey Chaucer (c. 1345–1400) in *The Canterbury Tales* [7], and Giovanni Boccaccio (1313–75) in *The Decameron* [8], developed the art of storytelling in the manner of the Italian collections of the *Novellino*, the French *Bestiaries*, and the French novel *Aucassin et Nicolette.* But both of them added a new breadth of humour and experience born of vivid observation. The masterpiece of late medieval literature was the *Divine Comedy* [9] by Dante Alighieri (1265–1321), an imaginary personal voyage through Hell, Purgatory and Paradise.

The *jongleurs*, jugglers or tumblers, in this ornamented medieval manuscript initial were travelling entertainers who sang, danced, clowned and, originally, recited the narrative poems of popular medieval poetry. They brought diversion to the far-flung courts of medieval Europe and carried with them the traditions of popular music, poetry, epic and drama. Their musical role was gradually taken over by minstrels and troubadours. *Jongleurs*, however, remained throughout Europe as part of a tradition of itinerant entertainers, ranging from bear-keepers and street buskers to courtly clowns, that was common from Roman times to the 19th century.

5 Adenez, "King of the minstrels", is shown in this manuscript reciting the *Roman de Cleomadès* to the Queen of France. Minstrels, wandering singers and musicians such as troubadours sometimes attained exalted status; Adenez was the employee of Henry, Duke of Flanders. Throughout the Middle Ages they passed on orally the great epic romances of the ancient heroes such as Roland.

6 The Exultet ("Let the angelic choirs of heaven now rejoice") was one of the great Easter hymns. In its present form it was probably written by St Ambrose of Milan (c. 340–97), creator of Ambrosian plainsong which was the forerunner of the Gregorian chant. The words of the Exultet were written on special scrolls, often highly illustrated with scenes of contemporary life as in this 11th-century Italian example.

7 The Canterbury pilgrims are shown setting forth in this illustration from *The Canterbury Tales.* Each pilgrim entertained the others with a story on the journey. In this first major work of English literature, Chaucer incorporated ancient tales in a new, medieval setting and also produced a complete and highly entertaining account of 14th-century manners, customs and morals.

8 Giovanni Boccaccio ranks with Dante and Petrarch as one of the great 14th-century poets. His major achievement was the narrative realism of *The Decameron* (c. 1348–53), a loosely knit series of engrossing and often bawdy tales of contemporary manners and morals. But writing in Italian was for him an indulgence and his weightiest works, such as *De claris mulieribus* (c. 1360–74), were in Latin. He was a mixture of medieval writer and Renaissance theorist.

9 The topography of Hell into which Dante descended in his *Divine Comedy* was drawn by Bartolomeo de Fruosino about 1420. Although it had a high theological and philosophical theme, the poem gave the most graphic account of the tortures undergone by sinners, particularly those Dante himself most disliked. Its genius lay in a progression from the descriptive, in Hell, to the lyrical, in Purgatory, and the mystically rapturous, in Paradise.

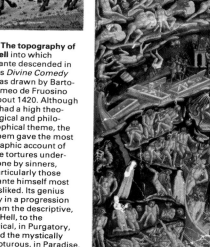

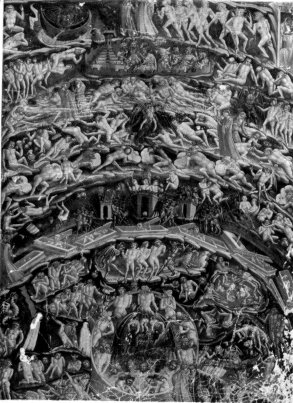

European society 1250–1450

By 1250, the material basis of Europe as it is now known was established; most of the towns and villages that exist today were already inhabited. Although forests were more extensive than now, the area under cultivation had passed the period of most rapid expansion. The next 200 years were characterized by crises of population and production, alternating boom and slump, and by the development of the techniques of finance [Key, 3], business and trade. The underlying crisis was probably the high population of the thirteenth century. Tax records show that in southern England and Provence the number of inhabitants doubled during the century. Older towns such as Florence [2] pushed out beyond their walls and multiplied their parish churches, while new towns were founded in every part of Europe.

The increase in urban populations
In 1250 the population of Europe had reached about 70 million and evidence from several sources suggests that the rate of increase was about one per cent a year. In consequence the food resources of the conti-

nent were stretched. Migration from Germany into central and eastern Europe reached its height in the thirteenth century, while the poor soils and marginal lands of the alpine and Apennine uplands came under the plough [1] for the first time. A higher expectation of life combined with a higher birth-rate to push European society into a slowly maturing demographic crisis.

This process created new social forms, above all the large towns that profoundly modified European civilization. Since classical times, probably no town with more than 40,000 inhabitants had existed; now Venice Naples, Barcelona, Bruges, Paris and some others far exceeded this limit. Immigrants, often from distant places, crowded into them, to work in the textiles of Bruges or Florence, or the more diversified trades of London and Paris. Two great ports, Venice [4] and Genoa, provided Europe with the products of the East, and among the first Europeans to visit China was the Venetian merchant Marco Polo. The Baltic trade was monopolized by the Hanse – a powerful league of North German cities.

In the urban centres, more specialized ways of life could develop. The skills of the accountant and the banker (developed first in Pisa and Florence) and the practice of marine insurance (a Venetian speciality) made possible the first international companies; the Bardi of Florence, through their agents in London and Bruges, were the greatest creditors of the English crown, for instance. Courtiers, civic patrons of art and people of fashion flourished in the midst of a new, degrading poverty. The friars adapted religion for the urban masses, while poverty and disease were mitigated by burgeoning charities. Whatever the crises, the great cities continued to grow.

A century of catastrophe
Undernourishment left the teeming population of the thirteenth century a prey to a worsening climate, hunger and disease. In 1315 persistent rain inaugurated a series of harvest failures and two years of famine throughout northern Europe. The high prices and booming business of the previous century now proved unstable; the economy had

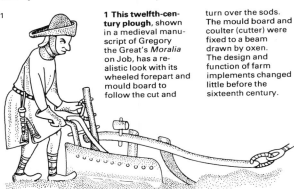

1 This twelfth-century plough, shown in a medieval manuscript of Gregory the Great's *Moralia* on Job, has a realistic look with its wheeled forepart and mould board to follow the cut and turn over the sods. The mould board and coulter (cutter) were fixed to a beam drawn by oxen. The design and function of farm implements changed little before the sixteenth century.

2 The fortified Roman town of Florence was reduced during the stagnation of the Dark Ages but grew rapidly in the twelfth and thirteenth centuries. Its expanding size and power were marked by the wider walls of 1172 and 1284–1330. In spite of internal factional struggles the city's position as a centre of cloth-making and the wool trade led to an era of prosperity and cultural achievement, particularly after the rise of the Medici merchant family in the early 1400s. Florentine bankers were the most influential in Europe and the florin, minted in the thirteenth century, became a monetary standard.

Walled city in Byzantine period
Walled city in 1172
Walled city in 1330

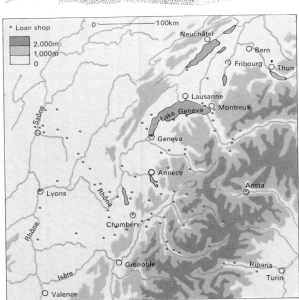

3 Loan shops or *casane* of Lombard moneylenders were a feature of the alpine valleys of Savoy in the thirteenth century. In many parts of Europe country life was transformed by money at this time, especially in Italy.

The rise in the population made necessary the rapid development of different soils. Peasants obtained the required capital to a large extent from loans. Moneylenders operated from shops set up in rural areas.

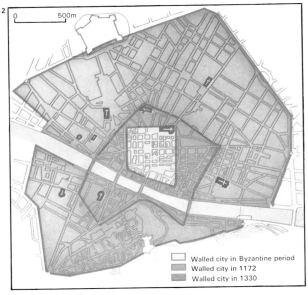

4 Venice at the time of Marco Polo's departure for Asia in 1271 (depicted imaginatively in this book illustration) was the busiest port in the world. The city was also a centre of ship-building and a leader in ship design.

advanced too fast. The boldest venturers, the Italian bankers, were hit hardest of all and in 1343 repudiation of the English royal debt precipitated a crash in which the house of Bardi and many others collapsed.

In 1348–50 came the terrible Black Death, the first of the recurrent bubonic and pneumonic plagues that were to beset Europe until the 1730s. Carried by both rats and humans in ships from the Crimea, the plague was immediately fatal and spread rapidly from southern to northern Europe. Its effects were catastrophic: probably about 50 per cent of the total population died, but some towns in Provence lost four-fifths of their inhabitants and many villages were finally deserted. In many parts of Europe there was some recovery, but the plague returned in 1361, 1369 and 1379, and by 1400 the population had shrunk to about a half or two-thirds of its total a century before. This was reflected in the all-pervading theme of death in late medieval art [6].

One of the effects of these calamities was social disorder, in towns and in the country. For the first time peasants and townsmen took violent action against bad government, in the Jacquerie rising [7] of northern France (1358) and peasant rebellions in England (1381) and Catalonia (1409–13). Governments everywhere were fearful of such movements and tightened their control wherever they could. These revolts were led by men of enterprise, peasants, farmers or artisans who resented the legal limits on their activities. The fall in population led to a demand for labour and labourers demanded the right to sell their services to the highest bidder.

The standard of living
The revolts reflected the rising expectations of the peasantry; in England, wage rates doubled while prices remained stable between 1350 and 1415. They were a sign of the vitality and independence of humbler men [8]; for them, the demographic decline meant a higher standard of living and the opportunity to turn peasant holdings into small farms. After a century of catastrophe, the fifteenth-century peasant followed the bankers and merchants of the pre-plague age into a measure of prosperity.

Coins proliferated in the late Middle Ages with the development of economies based on trading. A gold coinage, almost unknown since the seventh century, was restored in the thirteenth. Pioneers of the new money were the republics of Venice with the ducat and Florence with the florin. These coins are of roughly standard appearance and were issued in 1357–67 by Pedro I of Castile [A]; in 1399–1413 by Henry IV of England [B]; in 1419–34 by Conrad III of Mainz [C]; in 1368 by the republic of Florence [D]; in 1350–64 by John II of France [E]; and in about 1420 by Tommaso Mocenigo representing the Venetian republic [F].

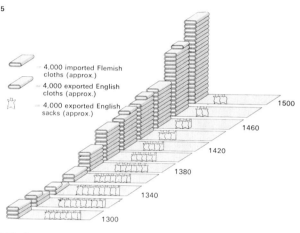

5 The English wool trade expanded rapidly after 1200. Raw wool was at first sent for processing to the established cloth makers of Bruges, Florence and other centres in France, Flanders and Italy but their monopoly was broken by the growing export of English-made cloths.

- = 4,000 imported Flemish cloths (approx.)
- = 4,000 exported English cloths (approx.)
- = 4,000 exported English sacks (approx.)

1500
1460
1420
1380
1340
1300

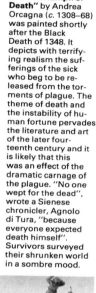

6 "The Triumph of Death" by Andrea Orcagna (c. 1308–68) was painted shortly after the Black Death of 1348. It depicts with terrifying realism the sufferings of the sick who beg to be released from the torments of plague. The theme of death and the instability of human fortune pervades the literature and art of the later fourteenth century and it is likely that this was an effect of the dramatic carnage of the plague. "No one wept for the dead", wrote a Sienese chronicler, Agnolo di Tura, "because everyone expected death himself". Survivors surveyed their shrunken world in a sombre mood.

7 The Jacquerie, a peasant rising which engulfed the Paris region in 1358, resulted from the disorder in France after the English captured King John II in 1356. The main grievance arose from attempts by landowners to keep down wages in a time of labour scarcity. All the fourteenth-century peasant rebellions had broadly similar causes.

8 The shepherds in "Nativity" by Hugo van der Goes (c. 1440–82), clearly drawn from peasant life, show a marked individuality characteristic of the confident peasant culture of the age and exemplified also in the popular art of the woodcut. Compared with the animal skins of three centuries earlier, the clothes worn here are luxurious.

The origins of Parliament

The English Parliament has no definite origin; but it clearly came into existence during the thirteenth century. At that time, the word "Parliament" meant a special session of the king's Council with the judges present to hear petitions to the king from his subjects. Early "parliaments" were largely occasions for judicial business in the presence of the king's councillors or barons, where great matters could be decided more solemnly than could be done by the judges alone. In the reign of Henry III (reigned 1216–72), whose judges actively opposed baronial privileges, this was a widely popular precaution; and the greater the number of earls and barons, bishops and abbots present, the more definitive the settlement of the business. In this sense Parliament was a variation of the ancient king's Council.

Parliament and taxation

Parliament was not only judicial in function: it also provided an occasion for the king's subjects to give their consent to taxation and their advice on important matters. This function implied representation of the various interests and communities by men with power to bind their constituents to their decisions. Taxation with the consent of such representatives was the general rule in the "Parliaments" called by Simon de Montfort (c. 1208–65) [1] to show the broad support for his rebellion of 1258–65.

Edward I (reigned 1272–1307) [2] defeated de Montfort's revolt in 1265, but during his reign all the various functions of Parliament came together: a session of the Council with as many councillors, barons and bishops present as possible; and representation of the shires by two knights, of the boroughs by two burgesses, and the clergy of the northern and southern provinces by proctors, all armed with plenipotentiary powers to grant taxes.

Advising the king

Although Edward I summoned representatives to only a few of his early parliaments, their presence became increasingly common as his need for funds grew, for without their consent taxes proved difficult to collect. By 1307, it was normal for parliaments individu-ally summoned lords and representatives of the "Commons" to meet each year. Parliament was the creation of the government for its own purposes, but it was also a call to the propertied classes to participate in the processes of government. This call was accepted with reluctance by shires and boroughs that were preoccupied with local affairs. Nevertheless their representatives came, and in a fourteenth-century parliament the "Commons" were generally represented by men who had experience on the countless local commissions of the peace, for the collection and assessment of taxes and so on, that shouldered the executive work of government. For instance, a certain John Morteyn, who was a large landowner, several times serving in the wars of Edward II (reigned 1307–27), a frequent member of commissions, and a strong partisan who evidently made enemies, acted nine times as Member of Parliament for Bedfordshire. Such men naturally had independent views, and their voice, joined with that of the lords, became increasingly influential.

Parliaments continued to hear petitions

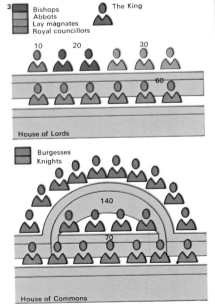

1 Simon de Montfort, whose shield is shown here, is sometimes said to have invented the English Parliament. To bolster his revolt against his brother-in-law Henry III he summoned a representative "parliament" in 1264. He genuinely believed in corporate, legal government, with frequent sessions of parliament to act as a bulwark against untrammelled royal or bureaucratic power. His ideas were later taken up by Edward I.

2 Edward I sits in Parliament with the kings of Scotland and Wales in this contemporary illustration, but in fact they never attended together. Behind them are royal princes and the archbishops. Barons and the other bishops sit down both sides, and in the centre are the judges, the councillors and the legal advisers of the Crown. The Commons and the proctors of the clergy (who must have outnumbered the rest) stand before the king.

3 The structure of Parliament was fairly settled after 1350. After a peer had been summoned once, he was usually automatically called to later parliaments, and many knights and burgesses were chosen regularly. Some shires and boroughs did not bother to send representatives; as a result, the size of the Commons might vary. The royal councillors were a reminder of Parliament's original function.

Bishops
Abbots
Lay magnates
Royal councillors

The King

10   20   30

60

House of Lords

Burgesses
Knights

140

70

House of Commons

4 The Chapter House at Westminster was the venue when the Commons first sat separately from the Lords. This was during the "Good Parliament" of 1376 when impeachment was first used against royal ministers for allegedly embezzling taxes for the war. The step marked a clear realization that the Commons had a different role from the Lords. By this time they had already evolved a formal procedure and elected a speaker. The Chapter House was used by the Commons until the mid-16th century.

5 Thomas Hungerford (d. 1398) was one of the first recorded speakers of the Commons. A client of John of Gaunt, one of the main objects of the Good Parliament's attacks in 1376, Hungerford represented the Commons the next year when Gaunt managed to reverse many of that Parliament's decisions. The speaker was elected by the Commons to be their spokesman, and also served as chairman of their debates. The institution of Speaker crystallized the rising confidence of the Commons: in the 1400s it felt able to oppose the king himself, asserting that Parliament was more competent in administrative matters than the king.

and act as courts of justice, and to make grants of taxation to the Crown. But many parliaments did neither and it seems that it was their advice and on occasion their consent to legislation that was sought by Edward II and Edward III (reigned 1327–77). By 1370 statutes had to be promulgated by the king in Parliament, and the kind of business on which they legislated broadened in scope: the Statute of Labourers, 1351, arose from petitions by landowners to Parliament to control labourers' wages.

The Lords and the Commons

During the period 1370–1450 Parliament evolved its classical structure and procedure. The representatives of the lower clergy broke away to form their own "Convocation", and by 1376 Lords and Commons met as separate bodies. Both Lords and Commons developed as political bodies able to put pressure on the king's ministers. Their weapon was "impeachment", a judicial process in which the Commons as a body "appealed" or accused a minister before the Lords who acted as judges of the case.

Under Richard II (reigned 1377–99) and Henry IV (reigned 1399–1413), parliaments were keen to experiment with ways of influencing the processes of government: in 1377 they appointed special "war-treasurers" to supervise taxes granted for military purposes; in 1401 they proposed to grant no taxes before redress of grievances. In these demands, the Commons were led by a new kind of member: the national figure, often himself a member of the Council, independent, outspoken and politically experienced, such as Arnold Savage, Speaker in 1401 and 1404. In the hands of such national "front-bench" politicians Parliament was ready to assume a central role in the politics of the fifteenth century. Its elections became the arenas for factions to assert their authority, and many leading noblemen acquired sizeable groups of supporters in the Commons. Henry VII (reigned 1485–1509) used Parliament to legitimize his dubious claim to the throne [7], but tried to reduce its incursions on the royal authority by cutting his expenditure so that Parliament would not have to be called so often.

The three estates, the division of society into knights, priests and labourers, was basic to the concept of Parliament, but was not directly expressed in the structure of the Lords and Commons.

6 Westminster Hall, an 11th-century building, was reconstructed between 1394 and 1402. As the central hall of the king's palace, it was the usual place of formal sessions of Parliament in the Middle Ages. But sometimes, as in 1388, Parliament was held in the "White Hall".

7 Henry VII formally declared his claim to the throne in Parliament in 1485, and recorded it in the Statute Roll, in the already antique "Norman-French" of legal documents. Unlike the long acts of Richard III and Edward IV declaring their titles upon intricate legal arguments, Henry's title rested simply on the declaration of Parliament in this quite short bill. It marks not only the succession of an entirely new dynasty, but also the recognition of the sovereign efficacy of Parliament. This foreshadowed the Parliamentary declaration of royal supremacy in 1534.

8 Henry VII opened Parliament in 1485 and tried to ensure that it acted as an ally of the Crown. His policies were well suited to the interests of the merchants who were prominent in the Commons; he significantly reduced the power of the Hanseatic League in England and in 1496 passed the *Magnus Intercursus* to improve trading relations with The Netherlands. But when Henry had consolidated his income, he called Parliament only in cases of abnormal expenditure.

The Hundred Years War

The conflict between the Plantagenet and Valois dynasties, lasting from 1337 to 1453, was marked by short campaigns, longer truces and periods of stalemate. Known as the Hundred Years War, it dominated the history of both England and France in the fourteenth and fifteenth centuries. Even after 1500 the threat of its resumption was a potent weapon in the hands of the English.

The start of the war

The origin of the war lay in the Norman Conquest itself: an Anglo-Norman empire was a reality up to 1204 and its shadow was pursued with energy by Henry III (reigned 1216–72). Gascony, which the Treaty of Paris (1259) left in English hands but effectively in remote dependence on the French Crown, proved a source of constant dispute. It would have remained merely a legal issue had Edward III (reigned 1327–77) not laid claim to the French throne [1] after the death of Charles IV (reigned 1322–8).

Although Edward's mother was a sister of Charles IV, the French nobles ruled out succession through females and decided that

Philip of Valois (1293–1350), as a cousin of Charles, had the best claim. He therefore became king as Philip VI. Despite this setback, Edward did not press his claim until public opinion had been sufficiently aroused by constant pinpricks over Gascony and French support for Scottish independence.

After war began in 1337 and the English won a naval victory at Sluys (1340) Edward relied on the traditional method of recruiting huge armies through subsidies to the princes of Germany and the Low Countries, which cost the English taxpayer enormous sums. Although his clumsy invasion of France ended in disaster through the deceit of his continental allies and shortage of money, the adaptable Edward learned quickly.

After 1340 he employed what could almost be called guerrilla tactics. He acquired footholds on the French coast, strongly fortified, which served as bases for devastating raids on the countryside. This had the advantage of requiring smaller forces than usual. More important, the profits of prisoners' ransoms and the plunder of the country [4] made the expeditions largely self-supporting and

attracted the nobility and ambitious but impoverished knights.

His tactics relied on speed and flexibility against the more numerous but rigidly arrayed mounted knights of France. His archers [7], although less accurate than the French crossbowmen, were highly effective from carefully chosen positions. At Crécy (1346) [Key] Edward inflicted a crushing defeat with a rain of arrows; at Poitiers his son, Edward the Black Prince (1330–76) [6], routed the French cavalry and captured John II (reigned 1350–64). Perhaps even more important was the capture of a base at Calais (1347) which remained as an English colony for two hundred years. By 1360 both sides were exhausted, and in return for peace at Brétigny the French agreed to abandon Gascony to Edward. From the English point of view the war seemed to have ended in victory.

Fighting for fortunes

Peace, however, proved illusory. The war was not really a conflict between nations: it had become a highly organized business

1 The fleur-de-lys was incorporated in the newly designed royal arms of England when Edward III assumed the title of King of France in 1340. He acted only after securing the support of the Flemish and discontented subjects of France. The arms and the title "King of France and England" lasted to 1801.

2 The English campaigns in Gascony and France consisted of plundering raids on most of western and northern France. Except for a few captured strongholds, from which ambitious and profitable expeditions could be launched, there was little attempt to conquer French territory until Henry V invaded Normandy early in 1417.

Map labels: Calais; Agincourt 1415; Crécy 1346; Harfleur; Amiens; Formigny 1450; Caen; Rouen; Rheims; Paris; Patay 1429; Troyes; Tours; Orleans; Chinon; Amboise; Poitiers 1356; Bordeaux; Castillon 1453; Dax; Bayonne; Vittoria; Narbonne; Najera 1367; Pamplona

English possessions
- in 1360
- in 1429
- 1360–1453
- Edward III 1346–7
- Black Prince 1356 & 1367
- Henry V 1415
- Joan of Arc 1429–30
- ✕ Main battles

0   100km

3 Control of the Channel gave a decisive advantage in war and up to 1340 the French, with the help of the Genoese admiral Barbanera, had been able to raid Southampton and show their flag at the mouth of the Thames. The decisive Battle of Sluys in which English vessels won the first important naval engagement in their history, marked the turning point. At the Battle of La Rochelle (shown here), fought in 1372, the Castilians, who were allied to the French, surprised an English fleet taking reinforcements to Gascony. This defeat ended the English domination of the Channel and was followed by defeat for the army in Gascony, marking a decline in the military fortunes of Edward III.

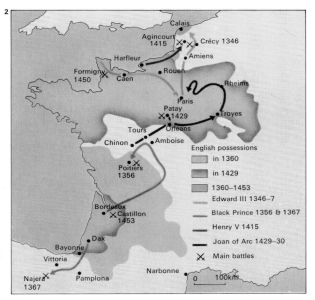

4 Plunder was one of the main reasons why ordinary soldiers took part in the war, regularly supplementing their wages. It could be acquired in two ways: by ransom or by capture of a place by storm. When a whole village or town was taken the usual practice was to plunder it and subject the inhabitants to fire and the sword. When the Black Prince took Limoges in 1370 he not only permitted looting but put the entire garrison to death, and after Calais fell it was said that "every Englishwoman was wearing its booty". The profits of war, especially for the victors, were great. When Henry V wanted to gain the good will of captured towns he managed to restrain his men from looting, but this was exceptional for the times.

enterprise for the exploitation of France. Individual captains such as Walter Mauny (died 1372) and John Chandos (died 1370) recruited soldiers by "indenture" or contract with pay for a limited period; they themselves signed similar contracts with the king which committed them to bring a specified number of soldiers [9].

Expeditions or cavalry raids such as that of the Black Prince in 1355–7 or of John of Gaunt (1340–99) in 1373 were designed to make a profit for the participants through plunder, levy of tribute from the countryside, and especially through prisoners' ransoms. Such ransoms were divided between the king and the captain who took the prisoner, and they could amount to huge sums. The ransom for King John of France was set at £500,000 (but was never fully paid) and even the humbler prisoners of Mauny were sold by the captain to Edward for £8,000.

The resumption of war

As the war was fought on French soil the English made far more profit than their enemies, and they resumed hostilities in 1369. English victories were less common in the years before a "twenty-eight years' truce" was agreed in 1396.

The most spectacular phase of the war followed the accession of Henry V (reigned 1413–22). Henry was a military genius who could organize and discipline a large army for years on end, and his renewal of the war was immensely popular. His victory against superior forces at Agincourt (1415), followed by the systematic occupation of Normandy in 1417, was made possible by the feud between the Armagnac and Burgundian factions, and Henry's alliance with Burgundy. By the Treaty of Troyes (1420) Henry married Catherine, the daughter of Charles VI, and was made his heir with the intention of inaugurating a single Anglo-French monarchy. Although he died before this was fulfilled, Henry controlled the whole of northern France with his Burgundian allies, and his infant son, Henry VI (reigned 1422–60; 1470–71) held sway in Paris until 1435. By 1453 the revival of the House of Valois finally deprived the English of Normandy and even Gascony.

The Battle of Crécy was the first great English success of the land war, fought when the superior and larger French army overtook Edward's forces who were making for Flanders. English archers checked the French cavalry raid with concentrated fire and the French were almost totally destroyed.

5 Sieges dominated warfare in the later Middle Ages and were accompanied by strict conventions. Formal siege began when a herald of the besieging party demanded entry in his master's name. Once this was done the besieger could not retreat without dishonour.

The Duke of Burgundy excused his failure to take Calais in 1436 by claiming that he had never formally besieged it. If there was no hope of relief for a besieged castle or town, the honourable practice was for the captain to surrender. Mining the walls was the most common form of assault; by 1400 cannons had largely replaced siege engines. The command of a captured garrison was one of the greatest rewards that could be offered to a benefactor of the king, and, like other positions of responsibility, might be highly profitable.

6 Edward, the Black Prince, the most notable knight of the age, and victor of Poitiers (1356), took his name from the black armour he wore. This effigy lies on top of his tomb in Canterbury Cathedral.

7 The skill of the English archers paved the way for several victories. The crossbow [right] was an effective weapon that had been used to deadly effect in the Crusades. It was more accurate than the longbow over short distances and had greater force, but it could not fire more than one arrow every 30 seconds. The longbow, 1.83m (6ft) long, could fire as many as six arrows in the same time.

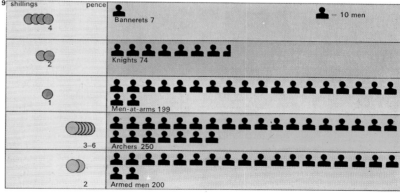

shillings		pence		
◯◯◯ 4			Bannerets 7	♟ = 10 men
◯◯ 2			Knights 74	
◯ 1			Men-at-arms 199	
◯ 3–6			Archers 250	
◯ 2			Armed men 200	

8 Caister Castle was built by Sir John Fastolf, the successful captain of Henry V and VI, about 1432–5. Made of brick, crenellated and surrounded by a moat, it cost several thousand pounds and indicates the wealth that Fastolf won in war. After the aggression of Henry V, English policy became more defensive, to preserve the profitable estates and towns from being recaptured. Charges of governmental mismanagement of the war became common in the 15th century.

9 The armies of the Hundred Years War comprised companies recruited and trained by indentured captains who were noblemen or soldiers distinguished by their military ability. Captains contracted with the king to muster their men and fight in return for wages. They usually led their men in specific military operations, designated in the contract. Captains subcontracted for men-at-arms. The foot soldiers and archers were often volunteers, but many were recruited by "commissions of array" which were appointed for each county. The offer of good wages and free pardons stimulated recruitment, and often there were more volunteers than the king could employ. This diagram shows the retinue of the Earl of Northampton in 1341. The system proved more flexible than the old system of feudal obligations; the short-term contracts were soon applied to peacetime relationships between lord and servant, and were instrumental in the growth of "bastard feudalism".

213

The age of Chaucer

Thanks to the poet Geoffrey Chaucer (*c.* 1345–1400) [Key] the late fourteenth century is more vivid than any other period of medieval English history. In his *Canterbury Tales* (*c.* 1385–1400), he describes the pilgrimage, which was a social event that united all classes, and through his immortal sketches of the various pilgrims and the tales they tell along the way he unfolds an ironical, dispassionate but profoundly sympathetic vision of English Society [3].

Chaucer's chequered career

Chaucer himself, with his many-sided experience of life, was perhaps uniquely placed to interpret his age. The son of a London vintner, he learnt the world of the court through his service with Lionel, Duke of Clarence (1338–68), and later with John of Gaunt, Duke of Lancaster (1340–99). He served in the French wars and as a diplomatic agent of Edward III (reigned 1327–77); he was a Justice of the Peace in Kent, sat in parliament for the shire, and was a controller in the port of London. He was several times both creditor and debtor, suffered robbery from highwaymen and was perhaps guilty of rape. London and the court were his special province, and Chaucer the poet probably had in mind a sophisticated court audience. This did not limit his appeal because he was widely known by all classes in his own lifetime.

The growth of a court culture

The age of Chaucer was distinguished by the emergence under Edward III and Richard II (reigned 1377–99) [4] of a "court", in the sense of a luxurious and fashionable cultural centre. It was sustained by London's new role as the unchallenged metropolis of England, and imbued with the delicate tastes of the rival "court" in Paris. The transformation of a somewhat provincial aristocracy into worldly, elegant courtiers can be observed in many innovations of Chaucer's lifetime: the luxurious appointments of Edward III's new palaces at Eltham and Sheen; the changing fashions of court ladies; the new, subtle style of cuisine; the taste for witty songs and elegant dances, and for poems and stories of delicate refinement, such as Chaucer's *Book of the Duchess* (1369), a lament for the Duchess of Lancaster, or with a touch of irony, such as his philosophical tradgedy *Troilus and Criseyde* (*c.* 1385) [Key].

Court culture, however, was not simply exquisite or exclusive. At its heart was the broad experience of the world given to men of all classes by the Hundred Years War. This tough but liberating experience gave the nobility an extrovert interest in the ordinary life and character of the ploughman as well as the prince. Chaucer's *Canterbury Tales* with their rich, earthy realism, or the scenes of London low life by Thomas Hoccleve [2], were not only for the nobility but also a much wider audience – the now literate middling townsmen and countrymen.

The Peasants' Revolt

The period of Chaucer's maturity, from the resumption of the war with France in 1369 to the deposition of Richard II in 1399, was the testing-ground of a new popular attention to the activities of government. It was punctuated with eruptions of opposition, sometimes in parliamentary form as in 1376 and 1386, sometimes in a violent defiance of

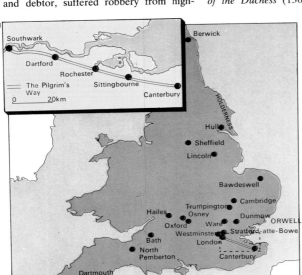

1 The England of the *Canterbury Tales* was mainly confined to the south-eastern part of the country, and was dominated by London and, of course, Canterbury, to which the pilgrims were journeying to visit the tomb of St Thomas à Becket (*c.* 1118–70). Even though the population of every other town in the country had fallen since 1330 because of the ravages of the Black Death, London's continued to rise because of constant immigration, and it stood at more than 50,000 in 1400. South-east England was the richest and most populous area at that time and the dialect of the south-eastern Midlands was becoming accepted as standard English in the court.

2 Thomas Hoccleve (*c.* 1370–*c.* 1450) is shown presenting his book, *Regement of Princes* to the Prince of Wales, who later became Henry V (*r.* 1413–22). As in this case, a "patron" was often merely a potential benefactor of the poet who hoped to gain favour by presenting the book. Patronage from a notable member of court was necessary to win a poet recognition, and perhaps a job to enable him to live. Hoccleve was a clerk of the Privy Seal, where his duty was to clothe royal orders and proclamations in elegant prose. His poetry is notable for its realistic and acute observations of life in the London taverns, and centres on his own involvement in that life.

3 Chaucer's pilgrims were depicted on the Ellesmere manuscript, an authoritative illuminated text of *c.* 1400–10, which may have been written for Chaucer's son Thomas (*c.* 1367–1434). Shown here are [A] the merchant "with a forked berd, in motelee". As a customs officer, Chaucer was familiar with such great international merchants. The Nun's Priest [B] is not described in the Prologue, but the tale of the Chantecleer reveals a distinctly witty cleric, even if riding a "jade", a horse "both foul and lene". The Wife of Bath [D] is one of Chaucer's most vivid portraits; a middle-aged Venus with a huge amorous appetite who had had five husbands, she was outrageously dressed. The brawny miller [F] was, like most of his type, enterprising and thrusting, benefiting from his lord's monopoly of milling; and the squire [C] was an apprentice in the arts of war and love. Chaucer himself had been a squire in the household of Lionel Duke of Clarence. The Prioress [E], "ful simple and coy" was an exquisitely ironic sketch of an over-refined lady.

4 The portrait of Richard II on the Wilton Diptych, probably painted for the king himself *c.* 1395, is by far the best portrait of a medieval English king, and exemplifies the finest achievements of the court culture of his reign. On the back of the diptych the king's emblem – the white hart – is depicted with refinement, probably by John Siferwas, and on the front Richard is seen being ushered into heaven by his patrons Edward the Confessor and Edmund the Martyr. Its painter is unknown.

royal authority, as in 1387–8, when the "Lords Appellant" attacked the court favourites, and as in 1399, when Henry Bolingbroke (1367–1413) overthrew Richard II. Religious controversy, too, was heard at court, stimulated by John Wycliffe (*c.* 1328–84) and supported by John of Gaunt and the so-called Lollard knights.

The greatest explosion of all came from the humbler folk of town and country, in the "Peasants' Revolt" of 1381. The name is misleading, for it was really a movement of substantial farmers and tradesmen with a following of farm labourers, infuriated by the constant war-taxation and the arrogance of the officials who collected it.

This followed the social unrest caused by the Black Death [5] and its consequent recession and upheavals. The Statute of Labourers (1351) tried to restrict the rise in wages that followed the plague. The unrest was expressed by William Langland (*c.* 1332–1400) in his poem *Piers Plowman* (*c.* 1362).

Beginning in Essex in June 1381, when one village erupted and killed a tax-collector, the revolt spread as other villages heard of

the event [7]. It became a more serious threat when the men of Kent, led by a veteran of the wars, Wat Tyler (died 1381), marched upon London. The 14-year-old Richard II parleyed with them, first at Mile End and then at Smithfield, and promised to grant their demands – an end to serfdom and the dispossession of the nobility and the Church. At Smithfield, however, the Mayor of London, William Walworth, killed Wat Tyler [9], and Richard, with presence of mind and courage, forthwith persuaded the insurgents to disperse. Reprisals began once the rebels had returned home.

Although the revolt was against a government hated by most people, the excesses of the peasants and their demand for social emancipation alienated the landed classes; Parliament insisted on severe punishment.

Local riots were frequent in subsequent years; the government never again dared to tax the ordinary villager so heavily, and serfdom gradually disappeared in the following generations. In the increasingly complex society of the time, the peasants had shown that they too had interests to protect.

Geoffrey Chaucer regularly read his poems to his audience in the royal court. *Troilus and Criseyde*, which he is reading here, is a fine example of his narrative style, its plot being based on a story by the Italian poet Giovanni Boccaccio (1313–75). Such borrowing from a foreign culture was typical of the international atmosphere of the court; Chaucer may have met Boccaccio on a diplomatic visit to Italy in the 1370s. The theme of the poem is the traditional one of courtly love, treated with a new attention to personality and emotion. The seduction of highborn ladies was a favourite pastime of Chaucer's audience, and Edward III set an example with his mistress Alice Perrers.

5 The Black Death 1348–50, killed probably a third of the population and recurred throughout the next 50 years. The shortage of labour thus created led the peasants to more demands for higher wages.

6 John Ball (*d.* 1381) was an ex-priest who had been imprisoned for heresy in 1360. He voiced the most radical ideas of the rebels, demanding an end to ancient distinctions between lord and serf.

8 The rebels burned the Savoy Palace of John of Gaunt, and the house of the Chancellor, Sudbury. After the king spoke to the rebels at Mile End, they entered the Tower unopposed and executed Sudbury.

9 Wat Tyler was killed by the Mayor of London two days after the rebels entered the city. His death ended such unity of purpose as they had possessed. He was described as "endowed with much sense".

7 The Peasant's Revolt occurred mainly in the south-east; the rebels in Essex were in touch with those in Kent, from which the main march on London began, and townsmen were involved as much as countrymen. St Albans, Bury St Edmunds and towns in Norfolk all suffered violent movements: a Norfolk man even proclaimed himself king. Isolated incidents occurred as far away as York and Bridgewater, Somerset. In general, risings took place in the prosperous areas where legal restraints such as villeinage no longer had any economic function and merely restricted the virtually independent farmer from pursuing his own fortunes. In addition the London mob was a particularly volatile element in English society, as it would be for many centuries.

Gothic art of the 14th century

Building and the decorative arts from the end of the thirteenth century to the end of the fourteenth century combine grace with naturalism. The source was French Rayonnant, the radiating style, best seen in rose windows, but by the end of the period a new style – International Gothic – had developed.

At the beginning of the period, buildings became larger and more elaborate. In England the rebuilding, on a grand scale, of the east end of Old St Paul's, London (1256–c. 1328) inspired the square-ended Angel Choir of Lincoln Cathedral (1256–80) and the new Exeter Cathedral (c. 1280–1375). Towers with spires became popular – for instance Salisbury (1330), by Richard Farley. Capitals (heads of columns), bosses (ornamental knobs) and corbels (stones that support wall shafts) were richly ornamented: hence the name "decorated" for this style.

The use of wood
Many impressive effects were achieved by using wood instead of stone. The nave of York Minster (1291–1345) was given a wood vault that may have been the model for later Continental net vaults. The idea may have come from the Chapter House at York, which did not include a central column – an innovation taken further at Ely, near Cambridge [Key].

Gradually wood was recognized as a building material in its own right. Elaborate roofs were built, such as that of Westminster Hall (1390s), although even here stone construction includes details such as the tracery "panelling" above the arches. Tracery was first applied in France about 1250, but was not taken up in England until the building of St Stephen's Chapel, Palace of Westminster in about 1310. English "panel" tracery then developed and fitted well with the simplified lines of Perpendicular architecture as exemplified by York choir and Canterbury nave [3]. It was even applied to vaults: probably first at Tewkesbury [1], then as fan vaulting at Gloucester (after 1357).

European architecture followed the Rayonnant style of the thirteenth century, increasing in scale: Beauvais – the tallest cathedral of all – had a choir vault of 48m (157ft), which collapsed in 1275. Cologne incorporated five aisles, and Strasbourg was designed with two western spires, the second of which was never built.

After the middle of the fourteenth century, France did not rival the inventiveness of Germany and Bohemia. The dominant German architects were the Parléř family, the eldest of whom had worked at Cologne. His design for Schwäbisch Gmünd [6] became a model for the great town churches of southern Germany in the later Middle Ages in which, with nave and aisles of equal width and height, greater prominence was given to the windows and vault. Such constructions are known as hall-churches.

Developments in Italy
Italian architects tried to graft northern Gothic ideas of scale and ornament onto a Romanesque tradition. The results were picturesque, but often structurally overambitious. Work on the new Florence Cathedral, begun in 1296, had reached the crossing in 1367 when it was realized that no one knew how to vault such a space. (Brunelleschi finally solved the problem in the 1420s

1 **The additions at Tewkesbury Abbey** in the early 1300s transformed a plain Norman choir into a sumptuous burial chapel for members of the Despenser family. The church was also vaulted throughout.

2 **The stained-glass programme** in the clerestory windows of Tewkesbury included a series of armoured knights; these were the counterparts of the effigies on the canopied Despenser tombs in the chancel.

3 **The nave of Canterbury Cathedral** (1378–1405) followed the new fashion for a greater delicacy of line. Originally it also had stained glass throughout, but little is left of its splendour.

4 **A new interest in nature** was shown in the graceful decoration (largely by Italian artists) of the Papal Palace, Avignon. The papacy had retreated there from Rome in 1309, transforming a fortress into a series of elegant apartments.

5 **The Angers tapestries** were used for interior decoration. Their subject, however, was religious, for they were based on an illustrated apocalypse manuscript owned by the French king Charles V, who commissioned seven tapestries in 1377.

6 **At Schwäbisch Gmünd in Germany,** the Church of the Holy Cross was built as a hall-church by Heinrich Parler. Although influential, the design was not original, being based on that of St Elizabeth, Marburg. The choir was begun 1351.

and 1430s.) Siena Cathedral reflects the aspirations of its citizens [7,10]; at Bologna, the San Petronio remained unfinished; and Milan Cathedral, begun in 1387, was finished only in the nineteenth century.

A greater contribution was made by Italy in the field of sculpture. Although everywhere this remained closely connected with architecture, figures were conceived more and more in the round. In the thirteenth century Nicola (c. 1225–c. 1284) and Giovanni Pisano (c. 1250–c. 1320) and Arnolfo di Cambio began this trend towards sculpture as a work of art in its own right.

Developments in painting

In contrast to architecture, little northern European painting survives. Some English panel paintings are of high quality, but whether they were done by English or French artists is uncertain. For example, the faces on the retable (altar shelf) in Westminster Abbey (late thirteenth century), the Wilton Diptych and the portrait of Richard II in Westminster Abbey (both late fourteenth century) are all of uncertain origin.

In contrast, illuminated manuscripts of high quality survive in large numbers. Perhaps the first to combine naturalism and grace was the *Douce Apocalypse* of the early 1270s. Religious texts began to show scenes from everyday life as well as traditional subjects. For instance, the late thirteenth-century Windmill Psalter is full of accurate portraits of birds. Marginal illustrations to the Luttrell Psalter (c. 1340) show scenes of country and farming life.

Italy has been much luckier in the survival of its pictures, which show a similar progression. Coppo di Marcovaldo's Madonnas at Siena (1260) and Orvieto (c. 1265), still in the Byzantine tradition, are now seen as seated human beings. Cimabue's Santa Croce crucifix (1283) [8] shows an understanding of anatomy and a delight in transparent drapery, while in his frescoes at Assisi (c. 1280) he experimented with perspective for the first time since the Roman era. An understanding of depth and form is clear in the works of Ambrogio Lorenzetti (flourished 1319–48), as in the fresco at the Chapter House of San Francesco, Siena [11].

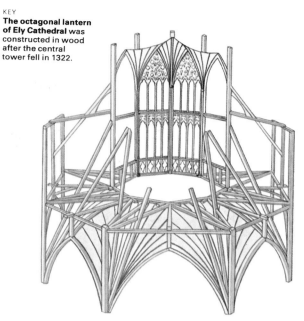

The octagonal lantern of Ely Cathedral was constructed in wood after the central tower fell in 1322.

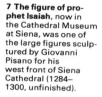

7 The figure of prophet Isaiah, now in the Cathedral Museum at Siena, was one of the large figures sculptured by Giovanni Pisano for his west front of Siena Cathedral (1284–1300, unfinished).

8 The realism of Cimabue's Sta Croce painted crucifix provided a challenge for contemporary painters. It marked a departure from the Byzantine tradition and brought a new freedom to Italian painting.

10 Siena Cathedral had later Gothic additions to the west front. Aisle and clerestory windows received tracery, the vault was raised and the choir built out over the precipice (with a baptistery below). Finally a gigantic new nave was to have been constructed southward from the hexagonal crossing, for which the existing nave and choir would have been merely transepts. Insufficient buttressing made this unsafe and the project was abandoned, but the new aisle walls and vaults remain embedded in later buildings.

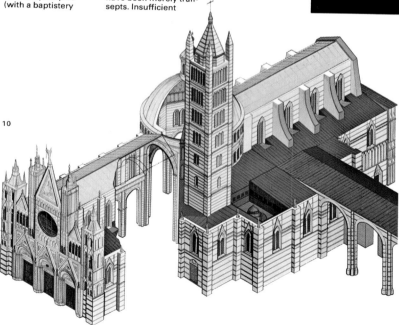

10

11

9 The Arena chapel at Padua was entirely decorated with frescoes by Giotto (1266–1337) about 1305. Enrico Scrovegni a moneylender (here presenting a model of the chapel), paid for the work.

11 Ambrogia Lorenzetti brought new realism to portraiture in his frescoes in the church of S Francesco, Siena (c. 1329). Except for scenes from the life of St Francis, as in the basilica at Assisi, near-contemporary events did not usually appear in frescoes. This scene shows St Louis of Toulouse (canonized 1317) being received at Avignon by the Pope, who is seated to the left, while worldly-wise cardinals and courtiers look on.

217

Gothic art 1400–1550

Art in Europe during the late Middle Ages can be seen as a further development of the International Gothic style. Delight in things colourful, courtly and complex is found in all branches of art and architecture, although obviously there are regional variations.

Gothic art and patronage
In Italy the love of Gothic decoration and rich materials persisted in the work of Gentile da Fabriano (c. 1370–c. 1427) and Benozzo Gozzoli (c. 1420–97). In the north of Europe, International Gothic had no rival. On the whole, there was a redoubling of decorative detail in parallel with increasing realism. For example, the animals, birds and plants that surround the principal figures in the tapestries of the "Lady with the Unicorn" [5] stand out in lifelike detail. This is probably helped by the novel use of red instead of the usual blue for the ground. As a result, blue and green could be used for modelling without spoiling the overall effect.

Almost everywhere, panel paintings are known to have been a popular art form for use as altarpieces, but few have survived other than Flemish or German ones. In contrast, illuminated manuscripts have suffered much less. Books of Hours [1, 8] seem to have been extremely popular. These were service-books for wealthy laymen, and especially for ladies, and they appear to have been objects of prestige and pleasure. The *Bedford Hours*, for example, was given to King Henry VI for Christmas when he was a boy.

By no means all patronage came from the aristocracy, however; the "Lady with the Unicorn" tapestries were commissioned by a family of the high bourgeoisie rather than of the nobility. The patronage of this section of society was extremely important during the period. Many outstanding English manor houses were owned by merchants; for example, Hever Castle in Kent was bought by Sir Geoffrey Boleyn, a mercer, in 1462. The house of Jacques Coeur in Bourges [3] is a French example of property being accumulated by the merchant classes. Merchants founded charitable institutions like Browne's Hospital in Stamford, Lincolnshire (around 1490). They also financed the building of many great "wool-churches" in East Anglia.

In Europe generally, and especially in central Europe, this was the great age of the parish church, paid for by the local trading community but often built on a cathedral scale.

Late Gothic architecture
In Europe, late Gothic architecture tends towards the exotic. For the French version, with its flame-like tracery rippling over walls as well as in windows, the term "flamboyant" is appropriately used. Examples can be seen at Rouen (St Maclou), Caudebec in Normandy and Notre Dame de l'Epine near Nancy. In the Iberian Peninsula plant-like decoration was popular. In Germany and Bohemia architects became adept at designing ingenious new patterns for vaults. Net-vaults were extremely popular. Instead of the ribs splaying out from the supports and crossing each other at various angles, the whole area was covered by a fretwork of parallel ribs. Sometimes nave and aisles are treated together; at other there is a different pattern for each area, and at Nördlingen in Bavaria there are three distinct patterns for the nave alone.

1 **Mehun-sur-Yèvre**, the favourite castle of the Duc de Berry, was made even more fantastical in this illustration to his manuscript, the *Tres Riches Heures*, by the Limbourgs, than it was in reality.

2 **The ruins of the Château** of La Ferté Millon, built for a cousin of the Duc de Berry, give some idea of how fantasy castles really looked, even though it has now lost the greater part of its ornamentation.

3 **The house of Jacques Coeur** in Bourges (mid-15th century) was also a warehouse. Coeur was an immensely successful merchant who built a small palace from which to carry on his business. He paid for the restoration of Bourges Cathedral, in return for which the pope made his son an archbishop. He took the unusual step of having these statues of himself and his wife Macée, leaning from behind stone balconies, added to the façade of the building. It was at this time that merchants throughout western Europe were having more influence on the arts as their purchasing power increased with wealth.

4 **In St Stephen's, Vienna**, the architectural canopies above figures of the Hapsburg emperors were designed c. 1400 as a three-light unit with convincing perspective to give an illusion of depth.

5 **The six French tapestries** of the "Lady with the Unicorn" in Paris, were made about 1500 for the Le Viste family and illustrate the five senses. The sense of taste is shown here.

6 **Little is known of medieval masons** and builders, and portraits are rare. Adam Kraft (c. 1450–c. 1508) carved himself holding a mallet and chisel as one of the "Atlas" supports in the Sacrament House of the Church of St Lorenz, Nuremberg (1493). In sculpting this church he took the opportunity to invent architectural forms that did not have to bear any weight and which would writhe right up to the vault. He introduced a new restraint into late Gothic sculpture.

An alternative to the net-vault is to make the ribs curve sinuously across the surface of the vault. This makes flower-like effects as at Wasserburg in Germany, and invites such excesses as making the rib appear to jump diagonally across the space of a window, as at Kutna Hora in Czechoslovakia.

Similar artistry was lavished on the small structures that make up church furnishings: fonts, pulpits and especially containers for the sacrament. These architectural motifs were repeated in the stained glass, which was often provided to add colour to the interiors [9], and especially in the wooden altarpieces, which are such outstanding works of art of the period. Tilman Riemenschneider and Veit Stoss are justifiably the most famous carvers and sculptors of this period: their figures are so lifelike as to appear portraits, yet they are stylized in order to fit naturally into the narrative subjects.

English architecture
English architects created a style known as Perpendicular, lavishing their efforts on vaults, with elaborate wooden angel-roofs as

a substitute in smaller structures. Many vaults are variations of the lierne design (a system of smaller connecting ribs) of Canterbury: for instance, York Minster choir (c. 1400). Others like Bath Abbey, Sherborne Abbey (c. 1430 to c. 1500 in stages), St George's Chapel, Windsor (begun 1481), and Henry VII's Chapel at Westminster, develop the idea of "fan" vaults [Key] first intitiated in the cloister at Gloucester during the previous century.

In the late Gothic period, windows generally became larger and included brilliant displays of stained glass. But in most decorative forms of the fifteenth century there is less variety than in the fourteenth, a trend that led, for example, to numerous instances of the same cartoon being used for stained glass in different churches. An example is the figure of St John the Evangelist in Sherborne Almshouse, Dorset, which is repeated at East Brent in Somerset. Such repetition, both in art and architecture, reduced the original freshness and inventiveness of the Gothic style, permitting a readier acceptance of the Renaissance in northern Europe.

KEY

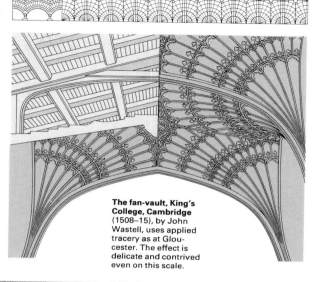

The fan-vault, King's College, Cambridge (1508–15), by John Wastell, uses applied tracery as at Gloucester. The effect is delicate and contrived even on this scale.

7 The Royal Pew in St Vitus' Cathedral, Prague, demonstrates a different illusionism from that of the Church of St Lorenz. At Prague the normal stone rib construction is carved to imitate wood with its knots and forks. Benedict Ried's Vladislav Hall (1487–1502) and stairs in Hradčany Castle nearby have ribs that intersect and overshoot each other as if they were single branches.

8 The *Hastings Hours* (c. 1480), which includes a portrait of Edward V, contains detailed studies of wild and garden flowers and this picture of the royal barge on the River Thames.

9 The Beauchamp Chapel (c. 1447) of St Mary's Church, Warwick, was a religious enterprise financed by lay patronage. No expense was spared over the sculpture of the Warwick tombs or the stained glass that fills the windows on three sides of the chapel. It includes a series of angels holding a continuous scroll of the Te Deum and magnificent figures in the east windows.

10 The Great Hall of Hampton Court Palace was built by Henry VIII in 1531–6 after the fall of Cardinal Wolsey. As at Warwick, all details were planned together to produce a rich effect. The hammer-beam roof, with elaborate pendants, was carved by Richard Rydge of London. Armorial glass filled the windows and imported tapestries lined the walls of this domestic interior.

11 The cones of ribs in the vault of Oxford University Divinity School (1483) are designed to hang free as pendants. On a small scale this is the principle of the fanvault in Henry VII's Chapel at Westminster Abbey (1503–19). The panels of tracery "ribs" there are merely carved on the surface of the stone slabs of which the vault is constructed. This is a measure of the development of vaulting – that the load-bearers could be treated as ornament.

The Ottoman Empire to 1600

The Ottoman Turks or Osmanlis (so called after their founder Osman, who died in 1326) began as Muslim warriors who patrolled the eastern borders of the Byzantine world. Osman's military genius raised them from nomadic tribesmen lacking any political institutions or national consciousness to become the formidable potential masters of a great empire by the mid-fourteenth century.

Osman's son and successor, Orkhan (ruled 1326–c. 1360) defeated a Byzantine army sent against him in 1329 and went on to capture Nicaea (Iznik) and other Greek cities. He annexed the neighbouring Turkish principality of Karasi. But his success in the holy war against the Christians attracted numerous other Turkoman warriors voluntarily to his lucrative service. Before 1453, the emerging Ottoman state held more lands of greater extent in eastern Europe than its considerable provinces in Asia Minor.

An era of expansion
By the time Orkhan's son Suleiman had set himself up in Gallipoli in 1354, the Ottomans commanded a large enough army to begin

major campaigns against Europe. Murad I, Suleiman's brother and Orkhan's successor, took Adrianople and by 1365 had made it his European capital, thus establishing a pattern of conquest whereby the Turks took over a Greek capital and with it the machinery both of government and Church administration; they employed the local clergy as tax collectors and held them responsible for the behaviour of their charges. The average peasant was allowed as much freedom as he had enjoyed under Greek rule, if not more. Ottoman society was very flexible and the success of the Turks was in part due to the greater opportunities they offered to the peasant class.

The move westwards was halted momentarily during the latter part of the fourteenth century by the incursions into Anatolia of the Tatar ruler Timur (Tamerlane). Timur set up independent Turkoman emirates and although these did not survive long, their existence demonstrated the weakness of the empire. As long as the Ottoman objective remained the conquest of European territory (and the militarist structure of the state made

constant expansion a necessity) Anatolia, so vital to the survival of the empire, would be vulnerable to attack and internal revolt.

The golden age
The golden age of Ottoman power occurred under Murad II (ruled 1421–51) and his son Mohammed II (ruled 1451–81). Murad was responsible for the creation of the Janissaries [Key], a corps of troops and administrators conscripted from among the Christians of the Balkans and raised to unquestioning obedience. This levy, called the *devşirme*, created a new social class whose fortunes were identified with those of the sultanate.

Mohammed II, called "The Conqueror", was responsible for the demise of the Byzantine Empire. In 1453, after a prolonged siege, he took Constantinople (now Istanbul) [3, 4], thus giving the empire the cultural and administrative centre it had lacked. Mohammed's achievement was to reunite the old Eastern Empire under a single sovereign. The translation of the sultan and his court to the new capital finally brought about the triumph of the *devşirme* faction over the old

CONNECTIONS

See also
148 Arabs and the rise of Islam
176 The Crusades
234 The age of exploration
150 Islamic art

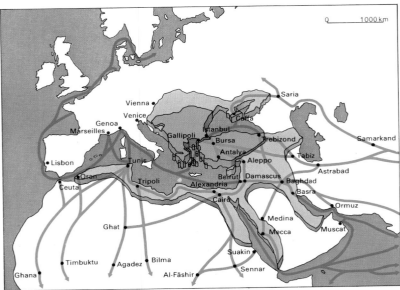

1 The major east-west trade routes were taken over by the Ottoman Empire as it absorbed the old Greek world. Trading stations in the Peloponnese, along the Sea of Marmara and also in Cyprus, came under attack. But commercial interest was only one motive for Turkish aggression and trade profits were sometimes sacrificed to the overriding economic need for military expansion by the Ottoman state into the non-Islamic realms of Europe.

→ Sea trade routes 15th century
← Land trade routes

▨ Ottoman Empire 1480
▨ Ottoman Empire 1600

2 Caravanserais were built as staging inns for the camel caravans that carried trade for the Ottoman Empire on a vast network of overland routes to and from the East. These routes declined when Portugal traded with India by sea.

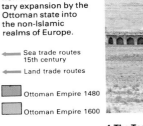

The traditional caravanserai, following the Persian model, consisted of a two-tiered building with a lower floor that consisted of stables built round an open courtyard. The oldest of these buildings still standing dates back to about 1080.

3 When Constantinople (seen here on a 15th-century map) fell in 1453 a new phase opened in Ottoman history. The great city had long been the focus of Turkish ambition, but repeated attempts to over-

come it had failed. When at last the old Byzantine capital was taken it became the Ottoman Empire's cultural and administrative centre.

4 The Topkapi Sarayi or Old Palace was built by Mohammed II, the conqueror of Constantinople, on the site of the old Acropolis. One of

the earliest Ottoman buildings in the new capital, it was the sultan's official residence and also housed the harem. It was built round a series of courtyards and was conceived on a grand scale.

Stone frame around glazed brickwork

Glazed tiles

Stairs

Arcade

Turkoman nobility, but it also removed the centre of power away from Anatolia, which was to weaken still further the all-important eastern frontier.

The next major period of expansion occurred under Suleiman I, "The Magnificent" (ruled 1520–66) [5, 6], who took the empire still farther into Europe. The main force of the attack fell on Hungary, which in 1526 became a vassal state of the sultan. But the need to maintain a strong presence in Anatolia and the problems of supply and transport over such vast areas meant that no farther westward expansion was feasible. The siege of Vienna in 1529 [6] was a failure. The struggle by sea, however, continued, for the shipbuilding yards at Constantinople had made the Ottomans a major sea power. It was the Battle of Lepanto in 1571 [8] that reduced Turkish naval power and drove the Turks back into the eastern Mediterranean.

Decline and fall

Throughout the late fifteenth and the sixteenth centuries the two superpowers, the Ottomans and the Hapsburgs, faced each other menacingly. But by 1600 Ottoman power began to decline as a result of internal discord, factional struggles and harem politics at the centre, as well as constant pressure on the eastern frontiers. A revival occurred in the mid-seventeenth century but the second siege of Vienna failed in 1683.

The Ottoman Empire remained throughout its long history essentially tribal in structure. The divan, the sultan's administrative body, had only slight powers and although after the reign of Suleiman the grand vizier came increasingly to rule the empire, his position was always tenuous and provided no means for an easy succession on his death or loss of favour. The distribution of land in 'timars, quasi-feudal grants, never created a landed nobility that could identify itself with the sultan and although the *devşirme* gave the sultan a strong military power base, they also alienated him from the Turkoman nobility and became themselves in time a threat to his security. Despite its internal weaknesses the Ottoman Empire succeeded in knitting into a single race a group of scattered nomadic tribesmen.

The Janissaries were recruited from Balkan Christians taken at the capture of Adrianople and reared as Muslims with unquestioning obedience to the sultan. They were not only the army's best soldiers but, after they had received *timars* (land grants), they formed a social class. Because they had no links with any of the traditional tribal groups, they became the sultan's chief defence against the Turkoman nobles who resented attempts to curtail their autonomy. After the 1600s, the sultanate declined and the Janissaries, like the barbarians in the Roman army in the 3rd century, came to manipulate rather than uphold the government.

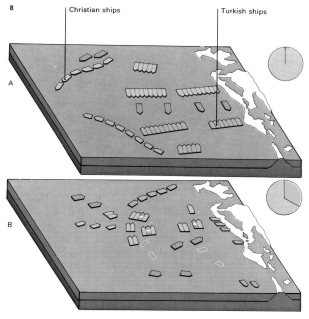

5 Suleiman I, called "The Magnificent", brought the empire to the height of its power. Here his workmen are restoring a castle in Egypt – a mark of the extent of his military power and authority.

6 The first siege of Vienna, undertaken by Suleiman the Magnificent in 1529, marked the limit of Ottoman expansion in the west. The Ottoman supply lines were already stretched too far from Istanbul.

7 The Cathedral at Famagusta symbolized the magnificence of Cyprus in the 1500s. Under the Venetians it had become the wealthiest island in the Mediterranean. The Ottoman conquest of the island in 1570 was the main event in the reign of Sultan Selim II, an otherwise unworthy successor of Suleiman the Magnificent. It marked the beginning of a new phase of hostilities between Christians in the west and the Turks. The war was fought out largely at sea and culminated a year later (1571) in the Battle of Lepanto.

8 The Battle of Lepanto, between the allies (Spain, Venice and the papacy) and the combined Ottoman fleet was fought off the Greek coast. At the start [A], both navies were grouped in two advance lines and one rearguard. The Christians had both galleasses (dark blue) and galleys (light blue). Four hours later [B] the Turkish fleet lay scattered with most of its ships beached, sunk (white outlines) or boarded (white-barred vessels between blue).

9 A miniature painting of Suleiman from the mid-1500s shows the borrowing of foreign styles typical of Turkish art. The Turks adapted the skills of their subjects and there is little original Turkish art except for carpets and other textiles.

Christian ships

Turkish ships

A

B

The world Europe set out to explore

At the end of the fifteenth century Vasco da Gama (c. 1469–1525) sailed round the Cape of Good Hope to India and Christopher Columbus (1451–1506) stumbled upon the Americas. So began an age of discovery in which Europeans were to navigate the seven seas, make their landfall in most of the inhabited regions of the globe and come to think of the world as a whole. The voyages east, however, were simply new ways of going to places already, if imperfectly, known.

Alexander the Great had marched the Greeks through Persia into India; Rome had bought Asia's silks and peppers and had bequeathed Ptolemy's geography to medieval Europe; Byzantium had long been a bridge between Europe and Asia. But the hostile crescent of Islam had hemmed in Christian Europe, until the extensive conquests of Genghis Khan (1167–1227) [4] gave it a brief respite from Islamic pressure.

The achievement of Marco Polo
The Mongol empires, stretching from Russia to China, straddled the land routes between Europe and Asia and allowed the two conti-

nents to trade directly with each other. In 1271 Marco Polo (1254–1324) travelled via Bokhara to Kublai Khan's court at Peking [2], and in the mid-fourteenth century an Italian handbook for merchants, *la Practica della Mercatura*, described the 140-day journey from the Black Sea to China [3] and listed no fewer than 288 spices and drugs that could be bought in the markets of Asia. But these tenuous contacts of traders and also missionaries were once again snapped by the hordes of Tamerlane (1336–1405) and the dynasties that emerged out of the wreckage of the Tatar empires. That is why discovering a route to the East by sea was so important for European commerce and trade.

The importance of China
At the end of the long journey east lay China; where the native Ming dynasty (1368–1644) expelled the foreign Mongols, cultivated its own empire, restored its economy and refined its bureaucracy. Threatened by offshore rivals and by the scourge of Japanese piracy, the Ming withdrew into partial isolation, broke off relations with some of

China's old tributaries, forbade its people to travel overseas, threw out foreign traders and prohibited private foreign trade. But the first Ming emperor had established relations with 17 different neighbouring states, and in 1502 more than 150 self-styled rulers from central Asia traded with China under the cloak of tribute relations. The maritime expeditions of Cheng Ho [6] hinted at a vast Chinese potential for seaborne expansion that was never to be realized.

By its self-denying ordinance Ming China did not fully exploit the valuable interport trade of the Indian Ocean, with its hub in the archipelago. This was left to the merchant principalities of the East Indies and to a motley crew of traders – Arabs, Persians and Indians. Ever since the time of the Cholas, India's mainland empires, expanding from their bases in Hindustan, were more concerned to defend their northern frontiers and to acquire territory in the south than to probe overseas. Babur, the first of the Moguls, began his conquest of India soon after Vasco da Gama reached Calicut. But the empire he founded remained land-based, and was

1 This Venetian map of 1448 shows how the typical medieval "wheel" map of the world was beginning to change. The medieval *mappamundi* was biblically inspired, with Jerusalem at the centre, the terrestrial paradise at the top and systematically disposed continents. By the mid-15th century the conventional medieval map was being influenced by contemporary marine charts and enriched by incorporating some of the information from travellers. It was also modified by the recovery of classical writing about the outside world, in particular the geography of Ptolemy and Fra Mauro, as evidenced by this map.

2 Marco Polo travelled to Peking in 1275 and served the Mongols in the East for 17 years. Returning to Venice in 1295, Polo gave Europe its first detailed account of China and its neighbours in *The Description of the World*. It was the most comprehensive account of the East produced before 1550, full of hard details about cities, canals, rivers, ports and industries. A practical administrator and merchant, Polo had little eye for religion and civilization but his accounts are factual and relatively free from the fantasies that formed the staple of contemporary travel literature. He re-created Asia for the European mind.

3 In 1260 the Polos, Nicolo and Maffeo, travelled from the Crimea via Bokhara to Peking. Returning to Europe, they set out once again in 1271 with Marco Polo. His return journey, from 1292–1295, was along the Malay Peninsula, Sumatra and India. These journeys revealed how the Pax Mongolica had helped to connect Europe with Asia and enabled Europeans to learn about the East.

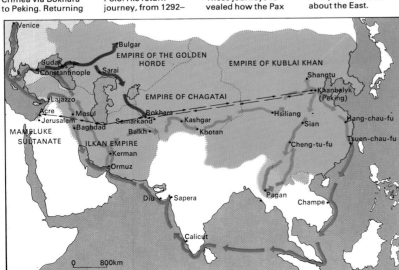

Mongol Empire late 13th century
Mongol tributaries late 13th century
Christian world late 13th century
Polo's known first journey 1260-69
Unknown portion of first journey.
Polo's second journey 1271–95

4 Genghis Khan the Tatar conqueror, rode roughshod over the Chinese civilization, selling many people into slavery. The Chinese resented the Khan; and they did not appreciate the links that their conqueror forged for them with the outside would. For China, the imposition of Mongol rule meant the breaking point in the continuity of its tradition. The Mongols were looked upon as foreign overlords, and were finally expelled by the native Ming dynasty in the 14th century.

never to possess a deep-sea fleet of its own.

Europe's deadliest enemies were also its closest neighbours – the Ottoman Turks. By capturing Constantinople in 1453 [5], the Ottomans held the gateway between Europe and Asia; by taking Mameluke Egypt (the Mamelukes were originally Turkish prisoners of Genghis Khan who seized control in 1250) and Syria (1516–17), they severed the European trade route east and wrested much of the profits of the Eastern trade from the Venetians and Genoese.

The Ottomans move into Europe

Perhaps the most fateful decision in modern times was the Ottoman resolve to push westwards into Europe. This took the Ottoman armies through the plains of Hungary to the gates of Vienna and held their navy in the Mediterranean. By establishing their empire throughout the Balkans, the Black Sea region and the Levant, the Ottomans sealed off these areas from European expansion and gave the Iberian powers the incentive and the opportunity to find new outlets, whether by creeping round the African coast to the East

or by making their landfall in America.

In Asia, Europe once again came into contact with great Oriental despotisms, mainland empires, many of which were Muslim, which it did not dare to challenge and which it could not hope to penetrate. In getting to the East by the new sea routes, Europeans merely touched upon the western and eastern edges of Africa, a continent whose northern territories had long been influenced by Islam but whose interior was long to remain unknown to the outside world.

In the New World the story was different. Here the Europeans actually discovered a continent that was out of touch with the rest of the world. There were remote, isolated, civilizations that had wondrous monuments and strange customs, but whose technology lagged far behind that of Europe or Asia. America's discovery by Europe opened her swiftly to the full blast of European influence: conquest and exploitation, disease and religion. Here the result was a clash of two wholly different cultures, which had come into contact with each other for the first time in recorded history.

Until Marco Polo's return to Venice, Europe learned little new about Asia. In legend it was still a continent of monsters and demons. In the 12th century Christians debated whether the dog-headed men of India might be converted, and even as late as the 14th century Western manuscripts showed Indians with dog heads, fantasies which the discovery of direct sea routes to the East were at last effectively to dispel.

5 The siege of Constantinople in 1453 vividly illustrates how Europe in the 15th century was contracting, not expanding. The Ottomans had triumphed against the last Crusade from Europe. The capture of Constantinople (today known as Istanbul) meant that the Ottomans were in Europe to stay.

6 Cheng Ho's seven maritime expeditions, beginning in 1405, and visiting over 30 countries, were remarkable feats of seamanship. The largest of his ships was 121m (400ft) long and 54m (180ft) wide, with four decks and watertight compartments. The 62 ships of his voyage to India carried 28,000 men.

7 Nicolas Deslien's "upside-down" map of the world of 1567, based on lost Portuguese originals, shows how radically the European view of the world was changing as exploration continued. Whereas maps in the Ptolemaic tradition had often been drawn with east at the left and Jerusalem at the centre, this has north at the bottom and, as nearly as possible, France at the centre. Its accurate description of known parts of the world reflects the growing precision of navigational techniques, but when describing unknown parts – the land mass to the south of Java, for example – it is still seriously inaccurate.

8 By the end of the 15th century, the vessels known as carracks were the largest merchant ships, at the other end of the scale from the small caravels. Portuguese carracks could be from 600-1,000 tonnes and were heavily built, with large castles, commonly three-masted with square rigging on fore and main and with lateen mizzens. The castle structures became more elaborate (as this 16th-century picture shows) and they were more often incorporated into the hull. These were the ships whose size, capacity for goods and men and solid construction made them the characteristic vessel used by the Portuguese in their Eastern reconaissance and trade, even though their bulk made them less well suited to carrying out the more detailed tasks of exploration.

Asian empires of the Mongols

The Mongol Empire, at the height of its power in the thirteenth century, was the largest land empire in history. It stretched from the Yellow Sea in the east to the Danube in the west and included areas of present-day Russia, China and Iran.

The origins of the Mongols are obscure. They were traditionally nomadic tribes who lived in felt tents called *yurts*, and followed their herds of horses, cattle, camels and sheep on an annual round of pasturage in the areas that are now Manchuria, Mongolia and Siberia. The numerous, loosely organized and constantly feuding Mongol tribes were first brought together as a unified nation under Genghis Khan (1167–1227) who became ruler of all the Mongols in 1206 [2].

Genghis Khan and the Mongol Empire
Genghis Khan's first move was to reorganize the major tribal groups in Mongolia as well as those on the Siberian borders. He then turned his attention to China. In 1211 he launched an attack on the Ch'in Empire, in northern China, an attack that continued until the whole of China finally came under

Mongol domination in 1279.

But China was not the only target: the hitherto unknown Mongols also raided the west. In 1219–20 Genghis Khan defeated Mohammed Shah of Khwarizm, and as a result acquired Transoxiana and Persia. Two of his generals defeated successively the Georgians, the Kuman-Turks, on the Volga-Don steppe – later to provide the manpower reserve of the Mongol Golden Horde that dominated Russia – and the Russian armies themselves, on the Dnieper. The Mongol armies then withdrew to central Asia.

Genghis Khan's successors
Ogodei (1185–1241), who succeeded to the khanate after the death of Genghis in 1227, resumed the western offensive and Persia, Georgia and Armenia were overrun as far as the Black Sea. At the same time the eastern campaign led to the defeat of the Ch'in in 1234 and pressure against the Sung dynasty in South China increased.

Batu (died 1255), Genghis's grandson, drove north of the Caspian and the Caucasus into Europe, defeating the Bulgars on the

Volga and capturing many cities, including Kiev, before splitting his army into two and initiating attacks on Poland and Hungary. Cracow and Breslau were captured and a German and Polish army defeated at Legnica in 1241. Batu himself devastated the Hungarian army at the Sajo, captured the towns of Pesth and Gran, and then led his forces to the Adriatic. No attempt was made to hold Hungary or Poland, but a Mongol base was established on the lower Volga to supervise the Russians and the Kumans.

Another grandson, Hulagu (1217–65) [6], the founder of the Persian Ilkhan dynasty (the title recognized subordination to the Great Khan), campaigned to the southwest. The Isma'ili sect, the Assassins, lost their great stronghold of Maymundiz in 1256; Baghdad fell in 1258 when the Abbasid caliph was killed. The conquest of Aleppo and Damascus followed, but in 1260 the Mongols met their first defeat, at Ain Jalut, at the hands of the Mamelukes under Baibars (1233–77), whose army had been battle-hardened against the Crusaders.

The Mongol army was commanded in this

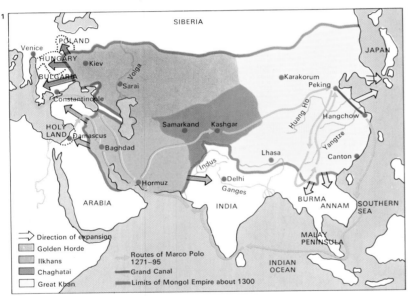

1 Despite the terror which the Mongols inspired, their domination of large areas of Asia and parts of Europe led to the development of trade routes used by traders of many nations in the 13th century. The Venetian Marco Polo set out for Peking in 1271. It was more than 20 years before he returned to Europe to describe the wealth and splendour of the Khan's court.

2 Genghis Khan united the Mongol tribes in 1206 under his leadership. After his death in 1227, the growing Mongol Empire was divided into four among his descendants, with Ogodei as chief.

3 Genghis Khan, portrayed here by a Chinese artist, was a politician as well as a warrior. He skilfully used patronage and alliances to further his aims.

4 A Yuan empress (from the dynasty founded by Kublai Khan) wears a medieval Mongolian headdress. Similar headdresses were still worn in this century.

5 Mongol troops are shown attacking a town with the aid of siege engines. These engines, called mangonels, were made by a German in China in 1273.

battle by the Christian Kitbogha, which illustrates the wide range of religions accepted by the Mongols. Shamanism, Buddhism, Islam and Nestorian Christianity were all practised, but in general the Mongols themselves were Shamanists. This tolerance explains the presence of various Christian priests at the Mongol court, and it encouraged and enabled Marco Polo (1254–1324) to travel across Asia to China. In fact the Mongol khans appear to have had a genuine intellectual interest in religion, and debates between experts of different faiths are reported by observers [8].

Kublai Khan conquers China

In 1260 Kublai (1216–94) became Great Khan, moving his capital from Karakorum to the site of present-day Peking in 1264. From there he conducted the campaigns that led to the annexation of all China, an area where the Mongol terror tactic was not general.

Although the campaign did not finish until 1279, the Yuan dynasty that Kublai founded is generally reckoned to run from the foundation of the new capital until the

last ruler fled before the Ming armies to seek refuge in the Mongol homeland in 1368. But the empire of which Kublai was the last ruler had broken up much earlier, for by 1295 the western Khans had accepted Islam and were no longer willing to submit to the overlordship of a non-Muslim Great Khan.

Nor did all Kublai's campaigns prove successful. In the north he was never able to subdue Kaidu (died 1301), the grandson of Ogodei. To the south, Burma surrendered but was not occupied, while in northern Vietnam disease forced a withdrawal. A seaborne campaign against Java was defeated, as was an attempted invasion of Japan. After Kublai's death there were nine rulers up to 1368. The dissolute rule of the last emperor, Togan Timur, saw revolts and chaos develop into open rebellion in 1348 and eventual defeat by the Chinese insurgents.

In the Middle East and in Russia Mongol dynasties continued to rule until 1502, although by that time they were little more than nominally Mongol because their various conquests had diluted the original Mongol groups to a considerable extent.

KEY

The horsemen who formed the élite of the Mongol armies were the key to their military success. They were trained to use the bow or sword while at full gallop.

6 Hulagu, invaded Iran in 1256, captured Baghdad and later defeated the Assassins, but was himself defeated by the Mamelukes in Syria. This was a turning point in Mongol history.

7 The Gur Emir, the tomb of Tamerlane (1336–1405) at Samarkand, was finished in 1434. This is one of the many magnificent buildings which this Mongol ruler, who was a great patron of the arts as well as a warrior, had constructed in Turkestan, his favourite region, and elsewhere. With his Tatar supporters, he temporarily reunited the empire of Genghis Khan.

8 The prophet Jeremiah is illustrated in Rashid ad-Din's *History of the World* (1306). The author, a physician in the court of Abaga-Khan, the Mongol ruler of Persia, included biblical themes in his world history – an indication of Mongol tolerance of alien religions. This illustration also shows Chinese cultural influences – another indication of Mongol absorption of foreign ideas.

The empires of South-East Asia

The lands that lie along the maritime route between the Indian subcontinent and China [1] have been strongly influenced by both of these regions. Except in Vietnam, the major cultural influence has been from India, but for most of the Christian era the kingdoms of South-East Asia have recognized, to a greater or lesser degree, the ultimate political suzerainty of the emperors of China.

The first centuries AD
The involvement of India and China in the affairs of this complex region seems to have been the largely accidental result of a need to find an alternative route between them when the land journey was made difficult by political instability in central Asia. But the various parts of South-East Asia had already achieved considerable technological, economic and political development by the time they came under the influence of their larger neighbours in the first centuries AD.

Lin-i, with its capital near Hue, and Fu-nan in the Mekong Delta, are two of the best known states that existed to the south of China in the early Christian era. Lin-i

became the kingdom of Champa, which dominated central Vietnam and parts of the south until the fourteenth century. Fu-nan grew into a substantial empire that dominated the greater part of the northern and eastern shores of the Gulf of Siam and their hinterland until the centre of power shifted, in about the middle of the sixth century, to a former vassal state, Chen-la, probably in the vicinity of the Tonle Sap. From this kingdom the Khmer Empire of Cambodia [Key] developed from the beginning of the ninth century onwards. To the west, in the seventh century, lay Dvaravati, near present-day Bangkok, and farther west again, in Burma, lay the Pyu kingdom of Shrikshetra, with its capital at Prome.

Southwards in the Indonesian archipelago, and on the Malaysian peninsula (parts of which seem to have been dominated by Fu-nan), a number of small kingdoms flourished, due in part at least to the development of trading routes between China and the West. These routes brought Buddhist pilgrims through the region and traders whose posts seem to have attracted teachers of Hin-

duism as well. These religions, originating in India, became the state faiths of the kingdoms of South-East Asia, a role that Buddhism has retained, but Hinduism [5, 7], except in Bali, has almost disappeared.

Buddhist and Hindu influences
By about the seventh century AD a Chinese Buddhist traveller, I-ching, was advising his fellows to spend some time in Sumatra studying Sanskrit and Buddhism before going on to India. He himself spent almost a decade there translating Buddhist texts into Chinese. The rise of this centre in western Indonesia, the beginnings of a state known as Shrivijaya which exercised commercial control in western South-East Asia for several centuries, followed a shift of power from the coast to the interior on the mainland. Meanwhile, elsewhere in the archipelago, in west Java (Taruma) and Borneo, Indian influences began to be detectable and a major dynasty, the Shailendras, Lords of the Mountain, who may have had links with Fu-nan, came to power in central Java. There, from about the eighth to the ninth centuries,

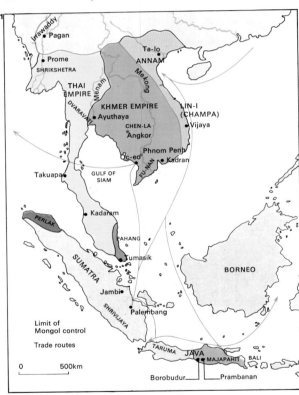

1 **The geographic position** of South-East Asia, lying between India and China at the centre of a monsoon system that facilitated sailing to and from both these countries, explains much of its cultural development and historical importance. The thriving commercial trade, carried inland along the great river systems, also brought a diversity of religious, political and cultural influences to the area.

2 **Chandi Plaosan** in central Java is a large Buddhist complex built in about the mid-9th century. Two apparently symmetrical groups have central shrines framed by rectangles of temples. Each main building is two-storied and houses a pantheon of Buddhas and Bodhisattvas. Inscriptions say that the images "shine forth the Doctrine" and windows were evidently arranged to create a radiant effect.

Irrawaddy
Pagan
Prome
SHRIKSHETRA
Ta-lo
ANNAM
THAI EMPIRE
DVARAVATI
KHMER EMPIRE
Ayuthaya
LIN-I (CHAMPA)
CHEN-LA
Angkor
Vijaya
Phnom Penh
Oc-eo
Kadran
Takuapa
GULF OF SIAM
FU-NAN
Kadaram
PERLAK
PAHANG
SUMATRA
Tumasik
BORNEO
Jambi
SHRIVIJAYA
Palembang
Limit of Mongol control
Trade routes
0   500km
TARUMA   JAVA
MAJAPAHIT   BALI
Borobudur   Prambanan

3 **Borobudur**, one of the world's greatest Buddhist shrines, was built in about the middle of the 9th century to a unique plan involving colossal resources – 570,000 cubic metres (two million cubic feet) of stone were moved from a river bed, dressed, positioned and carved with countless spouts, urns and other embellishments. The walls are covered with reliefs relating to Buddhist doctrine and there are altogether 504 shrines with seated Buddhas.

4 **Chandi Mendut**, a small temple of the Borobudur group, probably served as an antechapel. This relief, on the north wall of the porch, shows Kuvera, god of wealth, often associated with the merchant class who supported Buddhism.

Buddhism [2] appears to have flourished, its culmination being seen in the shrine of Borobudur [3, 4]. This, with its miles of reliefs expounding the faith, is one of the world's greatest religious monuments.

Hinduism was not neglected, however, and it was perhaps from this setting that Jayavarman II (c. 770–850), "returning from Java" as an inscription says, established in Cambodia the kingdom that dominated the central mainland from about the ninth to the fourteenth centuries. Hinduism, centred upon a lingam (phallic) cult located in a temple at the centre of the capital, was the state religion. The temples were of ever-increasing complexity, culminating in the magnificent structure of Angkor Wat [8] and the enigmatic Bayon in the centre of Angkor Thom. The economic strain of these ostentatious building programmes possibly contributed to the fall of the Khmer Empire under attacks from both the Thai, newly established as a power to the west, and the Vietnamese on the eastern borders.

Eastern Java, also perhaps for socio-economic reasons, saw the rise of the kingdom of Majapahit – a state whose maritime power enabled her to repel a Chinese fleet in 1293. Its influence extended as far west as central Sumatra and its blending of Hinduism and Buddhism was the culmination of a trend that can be detected in central Java as early as 782. The end of this kingdom seems to have been linked with the coming of Islam which, already established in northern Sumatra at the time of Marco Polo's visit in 1291, became important on the coast of Java a century or so later, although Majapahit's fall is usually dated to 1480.

Developments on the mainland

On the mainland, the Mongols, although unsuccessful in Java, had intervened with limited results in Vietnam. In Burma, the kingdom centred upon Pagan on the Irrawaddy, where some thousands of temples built over a period of two centuries testify to the power of Buddhism. It fell to the Mongols in 1287. At about the same time Rama Khamhaeng consolidated Thai power in what had been the western Khmer Empire to found the present state of Thailand.

Jayavarman VII (c. 1120–1215) became king of Cambodia in 1181 after driving out Cham invaders. Following his father, Suryavarman II, who built Angkor Wat, he embarked on an enormous building programme. In addition to temples and associated buildings he created hospitals and rest-houses for travellers and improved roads, with many stone bridges still in use today. Most of his predecessors were Hindu, identifying themselves with Hindu gods; Jayavarman was a Buddhist who seems to have had a special relationship with the god Lokeshvara (shown here), whose carved head dominates the towers and gateways of his buildings.

5

6 The Lake Pavilion at the Summer Palace, Bangkok, with its elaborate carving and gilding, is a reminder that bamboo and wood have been used as the materials for most buildings in South-East Asia during the past millennium (as they still are today). Even shrines and their images were often wooden, so that much of the past has been lost from the archaeological record.

6

5 A terracotta head from southern Thailand depicts a manifestation of Shiva's wrath – a creature that ate its own body to satisfy its hunger after the demon it was born to eat had been pardoned.

7 Garuda, vehicle of the god Vishnu, was a magic bird and enemy of snakes. In South-East Asia it became a divinity in its own right and was the centre of salvationist cults of various kinds.

7

8

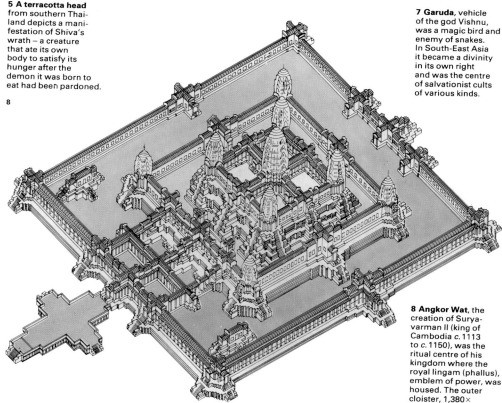

8 Angkor Wat, the creation of Suryavarman II (king of Cambodia c. 1113 to c. 1150), was the ritual centre of his kingdom where the royal lingam (phallus), emblem of power, was housed. The outer cloister, 1,380 × 1,150m (4,529 ×3,733ft), enclosed a complex of buildings, the main group arranged as a square of four at the corners and one in the centre. All were covered in exquisite low relief with divine dancers, plants, birds and animals. The many towers housed images. The walls of the central group were covered with reliefs depicting Hindu stories and battle scenes. In two, the king was shown. It was to be his shrine after death when he became identified with the god Vishnu.

African empires 500–1500

The most obvious sign of Africa's emergence from the primitive status of a "prehistorical continent" was the growth of political states [1]. The 1,000 years from 500 to 1500 saw the gradual emergence and then the great flowering of the black kingdoms that created such a rich and varied culture. By the beginning of the sixteenth century, much of the continent had entered this stage, evolving organized political societies with rulers, a soldiery under their direct command, and an administrative class. These were economically supported by the tribute that could be exacted from the mainly agricultural peoples considered to be either directly or indirectly the subjects of these states. Rulers often also controlled important trade routes.

The first black empires
The most ancient black African states were the empires of Kush/Meroë in the middle Nile valley (c. 800 BC–c. AD 400) and Axum in northeastern Ethiopia (first to fifth centuries). These were shaped by the influence of Egypt and south Arabia respectively and, although they had considerable effects upon

later developments farther south, were rather special cases. These empires apart, the main African empires were in the Sudanic belt, that is in the area to the south of the Sahara and north of the tropical forests.

The earliest of these Sudanese empires was Ghana, which was founded by the West African Soninke people. The first reference to Ghana comes from a North African writer in AD 773, and by 800 it had emerged as a powerful trading state, ruling the whole of the country between the Senegal and upper Niger rivers. The prosperity of Ghana was based largely upon its control of the gold trade. The gold fields of West Africa lay well to the south and Ghanaian traders obtained the precious metal by a strange process known as dumb barter in which the gold producers never met these traders face to face. The Ghanaians then sold the gold to North African merchants, who gathered in the southernmost oases. These oasis communities on the edge of the Sudanic belt served as the termini for the caravans that braved the routes across the Sahara [3].

Sometimes the fierce Berber nomads who

usually guided these caravans turned upon the settled trading empires. In the middle of the eleventh century, the Almoravids, a Berber confederation, led a Muslim holy war out of the desert to the north and to the south. In 1056 they invaded Morocco (later conquering southern Spain) and in 1076–7 seized the capital of Ghana.

The growth of the desert empires
Although the Ghanaian Empire fell, many smaller kingdoms survived and one of these grew into the spectacular empire of Mali [2]. Three great kings (who ruled between c. 1230 and c. 1340), Sundiata, Mansa Uli and Mansa Musa, so expanded Mali that it became one of the greatest empires in the world. It covered much of the western Sudan, and included the famous city of Timbuktu. The rulers had become Muslim (the religion travelled across the desert trading routes), and in 1324 Mansa Musa made the pilgrimage to Mecca, taking so much gold that he upset the Cairo money market en route. The successor state to Mali was Songhay, which had its centre on the middle Niger. The

1 The empires of Africa

The languages of Africa

HAMITO-SEMITIC
NILO-SAHARAN
NIGER-CONGO
BANTU
KHOISAN

Axum: 5th cent AD
Ethiopia: 14th cent
Ghana: 11th cent
Mali Empire: 14th cent
Songhay Empire: 15th cent
Berber dynasties: 11th–13th cent Almoravid c. 1050–1140 Almohad c. 1125–1269
Hausa states: 14th cent
Oyo and Benin: 15th cent
Monomatapa Empire: 15th cent
Kanem-Bornu 14th cent

1 The population of subsaharan Africa is mostly Negroid, although the Khoisan, in the south, are smaller peoples of different origins. The inhabitants of Africa north of the Sahara are paler – often Caucasoid in origin. Three main features distinguished the development of African languages: the long evolution of western African languages; the fairly rapid spread, after about 2000 BC, of an offshoot of them, the Bantu languages, over all of Africa south of the Equator; and the imposition of Arabic on the much more ancient Semitic languages, such as Berber, still spoken in northern Africa.

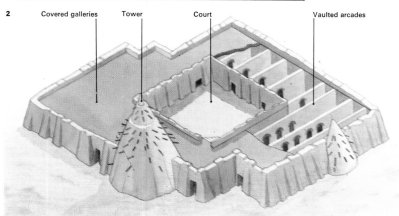

2 Covered galleries Tower Court Vaulted arcades

2 The great mosque at Timbuktu was designed in the 14th century by As-Saheli, one of the Egyptians brought back to Mali by the emperor Mansa Musa after his pilgrimage to Mecca in 1324. Timbuktu grew to be an important centre of commerce, religion and learning, producing many fine Muslim scholars.

3 Trade routes across the Sahara had developed in Greek and Roman times, but first came into their own with the introduction of the camel to Africa around AD 100 and the growth of the Islamic states six centuries later. This 13th-century picture shows a Muslim merchant of the kind that engaged in this ancient and hazardous commerce. He would have traded in West African gold, ivory, kola nuts, slaves and leather wares in exchange for salt, weapons and luxury goods. Control of the southern end of these trade routes made rich the great Sudanic states of Ghana, Mali, Songhay, the Hausa states and Bornu.

great rulers of Songhay at the height of its power were Sonni Ali (reigned 1464–92) and Askia the Great (reigned 1493–1528).

To the east of Songhay were the Hausa states, such as Kano and Katsina, whose origins are traditionally traced back to the eleventh century. By the fourteenth century they had become the domains of powerful kings and prosperous merchants, centres of population, crafts and trade. They were famous for their leather work, which was exported north across the desert. Europeans obtained it in North Africa, and knew it as Moroccan leather. In the central Sudan, on either side of Lake Chad, was the great state of Kanem-Bornu. From as early as the eleventh century its rulers had been Muslims. Kanem-Bornu was one of the oldest and largest African states, retaining its independent existence (although with changes of ruling dynasty) until it was overthrown by the European invaders at the end of the nineteenth century. In the mountains at the eastern end of the Sudanic belt, the Christian empire of Ethiopia [6] became the successor of ancient Axum.

In the woodland and forest areas to the south of the Sudan, kingdoms made a somewhat later appearance [4], but many, including Benin and the Oyo empires of Yorubaland (both of which produced some of the world's great sculptures), were in existence before the coming of the first Europeans in the fifteenth century.

The kingdoms of the Bantu nations
In other parts of Africa, especially in the vast regions south of the Equator over which the Bantu language family had spread rapidly during a few thousand years, states were beginning to be established. A cluster of kingdoms came into existence between the great lakes of East Africa, including Ruanda and Buganda. Another group, the Luba-Lunda kingdoms, grew up south of the Congo (Zaïre) forests, and the Kongo kingdom also emerged south of the river estuary that in colonial times bore its name. Much farther to the south were Great Zimbabwe and the empire of Monomatapa [7], on the Zimbabwe/Rhodesia plateau, which traded gold to the Muslims on the East African coast.

This bronze head with ivory headdress portrays an *Oba* (King of Benin). The splendour of the great African states was epitomized in the persons of their rulers. The headdress is carved with pictures showing the power of the *Oba*, which was, in the case of most African rulers, circumscribed; they were seldom absolute monarchs, being regarded instead as fathers of their people and personally responsible for their welfare. Many African kings were considered to be divine – their function in the world being to mediate between man and the gods. This high concept of leadership did not, of course, prevent corruption in ambitious rulers.

4 Powerful kingdoms had grown up in the forest regions of West Africa by the 1400s and were in trading contact with the older Sudanic states to the north. The former had a rich and ancient artistic tradition, especially in sculpture – Ife terracotta and bronze heads, and Benin bronzes have achieved world renown. This detail of a bronze head from Benin lacks the finesse of very early Benin workmanship but, although heavier, still exhibits considerable skill.

5 Knowledge of iron technology for producing tools and weapons was important in the history of African cultures and political systems, contributing to the ascendancy of many kingdoms that later became great empires. The spread of ironworking occurred over West Africa before the end of the first millennium BC and is linked south of the Equator with the rapid expansion of the Bantu. This picture shows the successive processes of smelting, forging and trading.

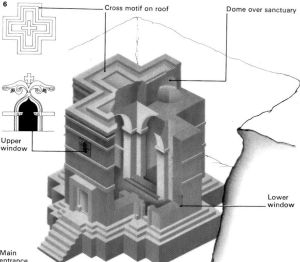

6   Cross motif on roof   Dome over sanctuary   7

Upper window

Lower window

Main entrance

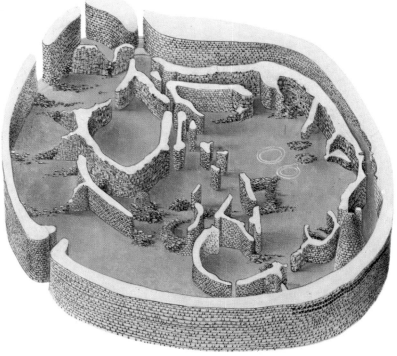

6 This church at Lalibela in Ethiopia was one of several hewn out of solid rock during the 13th century. The Middle Ages were a time of great church building and of general revival and expansion for this ancient Christian empire.

7 The Great Enclosure at Zimbabwe was built mainly in the 14th and 15th centuries on a site used for ritual purposes since *c.* AD 1000. The plateau area of Zimbabwe, the modern Rhodesia, supplied gold to Arab traders at Sofala (a coastal outpost of the rich trading city of Kilwa in East Africa). To the north of Great Zimbabwe, the Monomatapa kingdom was formed, probably in the 15th century, and by 1500 the Portuguese were supplanting the Arab trading links.

Mesoamerica AD 300–1521

During the greater part of the first 1,500 years after the birth of Christ there was a definable "Mesoamerican" civilization. This term was coined by the scholar Paul Kirchhoff in 1943 to describe the very similar cultures that occupied what are now southern Mexico, Belize, Guatemala, Honduras and El Salvador, extending eastwards through Nicaragua into Costa Rica between AD 1000 and 1500. These cultures shared temple-pyramids as religious centres; the sacred ball-game called *pok-ta-pok* by the Maya and *tlachtli* by the Aztec; a pantheon of deities including sun, wind and rain gods; and in the latter part of the period, especially, an iconography with a grisly emphasis on death.

Classic and Preclassic cultures

Many of these features had first appeared in the Preclassic cultures, and in the succeeding Postclassic period (from AD 900 to the Spanish Conquest in 1519) they reached their most complex form and widest distribution.

The first of the Classic cultures to attain great importance was that of Teotihuacán, based on the great city of that name in the Valley of Mexico. From AD 300–600 the city was at its maximum size, estimated to have contained about 200,000 people.

One of the surrounding cultures was that of the Maya. These people occupied the Yucatán Peninsula and the adjacent highlands of Guatemala and their area can be divided into three contrasting regions: the southern, consisting of the volcanic highlands and the short steep slope down to the Pacific shore; the northern, the arid scrub plateau of northern Yucatán; and the central, the jungle-filled basins of the Usumacinta, Belize and Hondo rivers. It was in this central area, that Classic Maya civilization emerged and flourished.

The appearance of this civilization is defined by the erection of stone stelae (upright slabs) with inscriptions in the complex calendar known as the Long Count, which combined three different calendars in one; by the building of massive stone temples and other public buildings in civic and religious complexes normally described as "ceremonial centres"; and by beautifully decorated polychrome pottery in new forms.

Oaxaca, a highland valley in southern Mexico, has one of the longest histories in Mesoamerica. The apogee of the culture of the Zapotec Indians who lived there came between AD 300 and 600 and was focused on the great hilltop city and ceremonial centre of Monte Albán [3].

The site has one enormous plaza laid out by levelling the hilltop, lined on all four sides by large buildings approached by broad sweeps of steps. The working of precious stones and metals was one of the most notable characteristics of the Zapotec and of their successors as rulers of Oaxaca, the Mixtec. After Classic Monte Albán was abandoned in the seventh century, some of the tombs were re-used by the Mixtec nobility [7]. Mixtec expertise also extended to architecture and their capital at Mitla [6] has walls decorated with long, repetitive mosaics, of thousands of rectangular pieces of stone.

The Gulf Coast

No such spectacular manifestation of architectural brilliance existed along the Gulf Coast, the home of the Olmec. But the cul-

CONNECTIONS

See also
232 Colombia and Peru 300–1534
46 Preclassic America to AD 300
236 Americas: conquest and settlement

1 The Pyramid of the Moon at Teotihuacán is one of two massive structures that dominate the heart of this great city, the first major planned settlement in Mesoamerica. It flour-ished from about 100 BC until AD 700 in a small valley branching off from the Valley of Mexico. The pyramid and the great plaza in front close the northern end of the Street of the Dead.

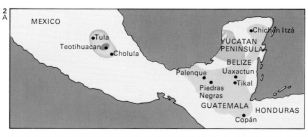

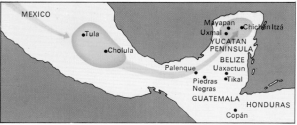

2 There were three stages in the prehistory of Mesoamerica. [A] The cities of Teotihuacán and Cholula dominated highland Mexico (AD 500). [B] Toltec influence from Tula reached Chi-chén Itzá in the Maya lowlands. The Maya civilization began to collapse in about AD 1000. [C] The Aztecs ruled the highlands, and the Yucatán Maya sites were eventually abandoned (AD 1500).

A
☐ Maya Empire c.300-630
☐ Maya Empire by 960
■ Teotihuacán c.300
☐ Toltecs by 960

B
☐ Maya Empire 960-1200
☐ Toltecs 960-1200
→ Toltec expansion 10th-12th centuries
→ Aztec migration 12th century

C
☐ Maya Empire 1200-c.1450
■ Aztec Empire by 1519
⊙ Aztec city alliance by 14th century

3 The Great Plaza forms the core of the main Zapotec ceremonial centre of Monte Albán, on the hills overlooking the Valley of Oaxaca. The building in the foreground, Mound J, lies on a different orientation from the rest of the site, and has been identified as an astronomical observatory; it is also adorned with carved panels depicting the towns conquered by the lords of Monte Albán. A large population lived nearby.

tures of Veracruz, and to the north that of the Huasteca, had their own distinctive characteristics. Veracruz sculpture was marked by panels filled with complex designs, many of them concerned with sacrifice, the ball-game, or both. The best-known site is El Tajín [Key], where five ball-courts have been uncovered. These panels are close in conception to those on the great ball-court of Chichén Itzá in Yucatán.

Chichén Itzá is a Maya site in origin. Its most spectacular ruins mark the occupation of Chichén and the domination of northern Yucatán by the Toltec, a warrior people from highland Mexico north of Mexico City. Many of the buildings at Chichén are derived from the architecture of the Toltec capital Tula. The most spectacular are the Castillo, a massive temple-pyramid with steps on all four sides, the Temple of the Warriors, and the *tzompantli* or skull-rack, where the heads of sacrificed victims were displayed [5].

The most impressive feature of the site is natural – the great circular Cenote (well) of Sacrifice, more than 18m (60ft) deep, into which victims were flung to bring rain.

The civilization of the Aztec was in full flower when it was destroyed by the European invaders and was the only Mesoamerican high culture to be observed and recorded as a living entity.

The cities of the Aztec

The Aztec had taken Teotihuacán [1] and Tula as models in making the central highlands of Mexico their base and like them had extended their way down into the coastal lowlands. Their capital, Tenochtitlán, was sited on islands in Lake Texcoco and had a population estimated at 300,000. It comprised two cities, the second being Tlatelolco to the north of Tenochtitlán, which acted as the commercial centre. Part of the marketplace and temples of this city have been excavated and restored, whereas the major part was demolished by the Spaniards during the conquest; the centre of the colonial capital was built on its ruins. Most of what remains of Aztec culture consists of grim lava sculpture, delicate turquoise mosaic work, and a number of manuscripts in picture writing, often with marginal notes in European script.

The Pyramid of the Niches at El Tajín is a large ceremonial centre on the Gulf of Mexico and one of the best-known sites of the Classic Veracruz civilization. It flourished through the first millen-

nium AD contemporary with the Maya to the southeast and Teotihuacán to the south. There are 365 niches on the pyramid which have been interpreted as reflecting the days of the solar year in another aspect of

the Mesoamerican obsession with the calendar and the passage of time. El Tajín also had a number of ball-courts, decorated with sculptures showing sacrifice, possibly of an earlier date than Chichén Itzá.

4 Famous for its stucco sculptures is the temple at the western lowland Maya site of Palenque. This was one of the first Maya sites to be explored in the eighteenth and nineteenth centuries,

and work has continued there since. The most spectacular discovery came in 1952 when a stairway was found leading to a buried vault where a great stone sarcophagus contained the jade-

laden body of a ruler. Recent study of the hieroglyphic tablets that give this Temple of the Inscriptions its name identify him as Pacal, first and greatest ruler of Palenque.

5 The Great Ball-court at Chichén Itzá was the largest in Mesoamerica. This huge structure, about 83m (270ft) long, was erected by the Toltec conquerors of Yucatán in AD 1000. Stone rings

in each wall were targets for the ball and sculptures along the base of the walls depict the decapitation of a ball-player, perhaps the captain of the losing team. A small temple stands at the end.

6 A room in the Mixtec palace at Mitla shows the complex stone mosaic decoration based mainly on the step-fret motif and built up of thousands of individually shaped stone blocks. The

rooms, which were roofed in timber, lie round a series of closed courtyards accessible through narrow passages. There is also a pillared hall and tombs that lie below courtyard level.

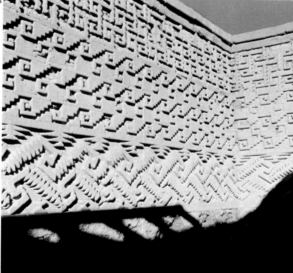

7 This head is one of the superb pieces of Mixtec gold work discovered in 1932 in the excavation of Tomb 7 at Monte Albán. This was a Zapotec tomb of an earlier period that had been re-used for the burial of a Mixtec lord. The tomb also contained carved bones, rock crystal, turquoise, mosaics and jades. This piece was probably worn on the chest and reveals a complex symbolism. The gold-working technique came from South America.

Colombia and Peru 300–1534

In South America one area has always led the rise to civilization – that of the Andes from Lake Titicaca northwards to Panama, and their flanking Pacific and Amazon slopes. There the earliest pottery, the first signs of agriculture and the first settlements and public buildings have been found. From AD 300 onwards, after 1,500 years of increasing momentum, a number of regional cultures of a diverse character arose. They were so strongly regionalized in their art as to suggest separate political units, with strongholds attesting to warfare between them.

The arts of Colombia

In the far north of the South American continent, in present-day Colombia, a gold-working tradition of great technical competence and artistic originality emerged [7]. Pendants and nose ornaments of sheet gold with added detail commonly survive, but there are also magnificent gold vessels for holding the drug coca and unique items such as the model raft with a god and his attendants in the Bogotá Museo del Oro. Techniques extended from the simple hammering of sheet metal to granulation, lost-wax casting and the creation of a gilt surface on a gold-copper alloy by removing the copper with acids from plants.

In southern Colombia the monumental art of San Agustín [1] flourished. The mounds in the area contain stone megalithic chambers, apparently both tombs and shrines, entered by stone tunnels decorated with painted designs. Huge blocks of the volcanic rock andesite were worked into box-like sarcophagi, and shafts of rock turned into menacing statues of warriors and demons. Some of them were double figures, with an animal alter ego looming over the man's head; others represented birds of prey wrestling with serpents. In the nearby Tierradentro region there are rock-cut tombs fashioned with domed roofs and equipped with stepped entrance shafts.

At the time of the Spanish Conquest, Colombia was occupied by large populations living in defended and palisaded settlements. Religion centred on sun-worship and the economy was based on potatoes and maize, and trade in salt, gold and emeralds.

The principal centre in northern Peru was the basin of the River Moche, the seat of Mochica, Chimú, Inca and Spanish colonial rulers. Moche civilization dates from about AD 200 to about 700 and the resources its rulers could command are demonstrated by the colossal Temple of the Sun [2], a terraced pyramid 228m by 136m (741ft by 442ft) and 41m (133ft) high, made of mud-adobe bricks, and the nearby Temple of the Moon, a palace complex adorned with wall paintings, as well as vast irrigation canals cutting across the desert. Some Moche cemeteries have been properly excavated, many others looted. The most notable furnishings are the stirrup-spouted jars [4], some with painted scenes of warfare, hunting and daily life, others modelled into three-dimensional portrait heads.

Palaces for life and death

From about AD 1000 onwards, the Moche valley held the capital of the kingdom of Chimor, the great city of Chan Chan, which covers more than 15.5 square kilometres (6 square miles) [8, 9]. The centre of the city

CONNECTIONS

See also
230 Mesoamerica 300–1521
46 Preclassic America to AD 300
236 Americas: conquest and settlement

In other volumes
208 The Physical Earth

1

1 Enigmatic and massive carvings are scattered on the hills round the present town of San Agustín in southern Colombia. Many depict men with spirit alter egos in the form of animals sitting on their heads and others combine human and feline features, with long fangs. There are megalithic chambers – possibly tombs – which are not unlike those of prehistoric Europe, with rock-cut basins and carved boulders.

2 Millions of mud bricks were used to build the Temple of the Sun (or Sun Pyramid) and the neighbouring Temple of the Moon at Moche, on the north coast of Peru. The Temple of the Sun formed the ritual heart of the Mochica state during the first thousand years AD. Both palaces have suffered greatly from erosion and looting and the original form of the Sun Pyramid is now hard to discern.

2

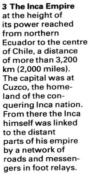

3 The Inca Empire at the height of its power reached from northern Ecuador to the centre of Chile, a distance of more than 3,200 km (2,000 miles). The capital was at Cuzco, the homeland of the conquering Inca nation. From there the Inca himself was linked to the distant parts of his empire by a network of roads and messengers in foot relays.

ECUADOR

PERU

Cajamarca

Moche Chavin

Paracas Machu Picchu
Nazca Cuzco
Lake Titicaca
Tiahuanaco

BOLIVIA

CHILE

ARGENTINA

=== Inca roads
⊙ Inca Empire *c.* 1200–1400
☐ Acquisitions 1438–71
☐ Acquisitions 1471–81
☐ Acquisitions 1493–1525

0   650km

4 Stirrup-spouted pottery vessels were characteristic of the Mochica period in the first thousand years AD. These vessels were made by specialist potters. One type of vessel took the form of portrait heads, while another (an example of which is shown here) had plain, smooth bodies with painted designs. These often took the form of scenes from Mochica life and warriors in action. The spout shape changed subtly as centuries passed.

4

5

5 Nazca potters, renowned for the strong blocks of colour that give their work a cartoon-like effect, made this lifelike figure of a drummer. Nazca culture flourished in Peru at the same time as Mochica in the north.

consists of ten basically similar walled enclosures, each with a maze of rooms and open courtyards and once rich but now looted tombs. It has been suggested that these were the successive palaces of the Chimú rulers, the new king building a new compound while his predecessor's was maintained in perpetuity as a funerary shrine. The tombs are known to have contained gold and silver vessels and jewellery, pottery and textiles.

On the central coast of Peru the "Lima style" of pottery suggests a state similar to Moche with ceremonial centres such as Aramburu, near present-day Lima. But the most spectacular pottery style was undoubtedly that of Nazca on the south coast, with polychrome vessels [5], often in four or five colours, and a dominant motif of a cat demon.

The most influential culture in southern Peru at that time, however, was based on Tiahuanaco, south of Lake Titicaca. Vast areas of ridged fields have been identified along the lake, indicating a large agricultural potential and the labour to exploit it. Tiahuanaco is noted for its monolithic stone carving, including the famous Gate of the Sun [6], which lies on one side of a large enclosure, and tall column statues in a stiff but detailed relief style. Tiahuanaco was preceded in the Andes by Pucará as a cultural centre and passed on some of its features to Huari, north of the lake.

Rise and fall of the Incas

Inca grandeur lasted for less than a century, from 1476 to 1534. Inca expansion began under the ruler Pachacuti Inca Yupanqui (reigned 1438–71) and continued under his successor Topa Inca. At its peak it reached from Ecuador in the north, into what are now Chile, Bolivia and Argentina, running along the cordilleras of the Andes and the coast for more than 3,200km (2,000 miles) [3], which left the Inca military over-extended. Much is known of Inca social and political organization – the system of recording information on knotted string *quipus* [10], the professional army, the communications by relays of runners, and the supreme authority of the Inca himself – the source of strength and ultimately the downfall of the empire.

Inca stone walls in the Peruvian Andes were constructed with quite remarkable skill, the hard stone blocks fitting closely together even though their outlines are irregular. One block in a Cuzco street has 12 angles, but such intricacy is not unusual. In Cuzco itself many Inca structures survive, including part of the Temple of the Sun. It now forms a section of the Dominican monastery. On the hills above Cuzco the great fortress of Sacsayhuaman presents its multiple ramparts to an enemy. Similar stonework is known at the mountaintop city of Machu Picchu and a version was used for agricultural terracing.

6

6 The Gate of the Sun, Tiahuanaco, Bolivia, is carved from a single block of stone and formed part of a great ceremonial enclosure. It is adorned with low-relief carving.

7 The art of gold ornamentation was highly developed in South America. The main centres of innovation were in Colombia and Ecuador, spreading northwards from there into Mexico.

7

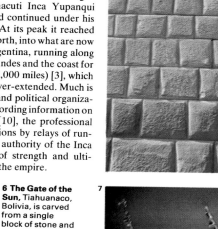

8 A restored panel of decoration in adobe – mud brick – forms part of the outlying temple of El Dragón at the site of Chan Chan, the ancient city north of Trujillo in the Moche Valley of northern Peru. Moche was an earlier focus of civilization in this valley, as the colonial and modern city of Trujillo was after Chan Chan. The site consists of a series of great walled enclosures, altogether covering some 28 sq km (11 square miles).

9

8

9 Moulded adobe relief decorates the interior of a room at Chan Chan. The friezes are repetitive and consist of birds, fish or abstract designs. The rooms have niches in the walls and some have a U-shaped structure called an *audiencia*, which is thought to have been the seat of a clerk, who checked goods in and out.

10

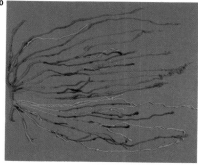

10 A *quipu*, a series of knotted cords, was used by the Inca as a counting and memory device. The knots were of different colours to denote different kinds of numerical information and are an eloquent testimony to the bureaucratic structure of the Inca Empire. For recording the constant payments of incoming tribute they were indispensable.

The age of exploration

The later fifteenth and early sixteenth centuries make up one of the most momentous periods in the history of Europe. In 1492 Christopher Columbus (1451–1506) sailed west across the Sea of Darkness and discovered the Americas. In 1497 Vasco da Gama (c. 1496–1525) embarked from Lisbon, sailed down the West African coast and around the Cape of Good Hope, up to Mozambique and Mombasa and then across to Calicut in India. In 1519 Ferdinand Magellan (c. 1480–1521) [3], seeking the route to the Orient that Columbus had failed to find, led a Spanish expedition around the southern extremity of South America and across the Pacific to the East Indies. Magellan himself was killed, but the survivors of his expedition, returning by way of the Cape of Good Hope, circumnavigated the globe for the first time.

The voyage of the Portuguese
Vasco da Gama's expedition [2] eastwards was less a voyage of discovery and more an armed embassy determined to open up Portuguese commerce with the East. It was the culmination of almost a century of tentative, hesitant exploration by the Portuguese in which their frail caravels (light Mediterranean sailing ships) had groped their way along the West African coast and finally rounded the Cape. Lured by the gold and ivory of Africa, and by the prize of Eastern trade which awaited Europeans who could reach India by sea, the captains of Henry the Navigator (1394–1460) [1] paved the way for da Gama by voyages to Madeira (1418) and the Azores (1431); by rounding Cape Bojeador (1434); the discovery of the mouth of the Senegal (1444); the sighting of Sierra Leone (1460) and eventually the discovery of the Cape of Good Hope by Bartholomew Diaz (c. 1450–1500) in 1487.

The motives of the explorers
The motives behind these early explorations, whether Portuguese or Spanish, were a combination of acquisitiveness and religious zeal. Barred by the Italians from the large prize of Mediterranean trade, deprived of the profits of the luxury commerce in Eastern goods which Muslims and Venetians together controlled, the Iberians sought new routes to the sources of supply. At the same time they were spurred on by their crusading zeal.

But not all the voyages during this period were by Iberians. Under Spanish patronage the Genoan Christopher Columbus sailed in search of a westward sea route to the Indies trade, and in October 1492 landed in the Bahamas, believing them to be an Asiatic archipelago. By 1504 he had made three more voyages to the Caribbean, but had come no nearer to proving that Asia had been found. Meanwhile, from England, the voyages of John Cabot (c. 1450–c. 1500) along the northeast coast of America, and the explorations of the Florentine Amerigo Vespucci (1454–1512) along the north coast of South America and Brazil on behalf of the Spanish, led to the belief that there was an uncharted land mass to the west between Europe and Asia, a New World [4].

The Portuguese had many advantages over other European rivals in the exploration of Africa and Asia. The Italians were bottled up in the Mediterranean; the Spanish were not united into one kingdom until 1479 (the

1 **Prince Henry of Portugal**, "the Navigator", was the most important of the precursors of the age of exploration. Placing gentlemen of his own household in command of his ships, Henry developed a systematic if intermittent programme of exploration beyond Cape Bojeador. By the time of his death his ships had advanced south by 2,415km (1,500 miles).

2 **Vasco da Gama's voyage** by which he reached Calicut in 1498 was an event of great significance, but his achievement cannot be ranked with that of Magellan or Columbus. In sailing east he completed what others had begun – Diaz had rounded the Cape of Good Hope in 1488 and in fact accompanied da Gama part of the way on his Indian voyage.

3 **Ferdinand Magellan**, a Portuguese employed by the King of Spain, embarked in 1519 from Seville on the voyage in which he sailed through the strait that bears his name. He then crossed the Pacific to the Philippines where he was killed. Del Cano completed the circumnavigation.

4 **Westward voyages**, until Magellan's circumnavigation, were a gradual process of realization that an uncharted continent lay between Europe and the Indies. While the Portuguese sailed eastward to the spice trade, Spain and England were anxious to find a quicker, westward route.

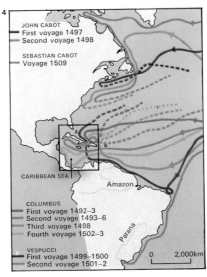

JOHN CABOT
First voyage 1497
Second voyage 1498

SEBASTIAN CABOT
Voyage 1509

CARIBBEAN SEA

Amazon

COLUMBUS
First voyage 1492–3
Second voyage 1493–6
Third voyage 1498
Fourth voyage 1502–3

VESPUCCI
First voyage 1499–1500
Second voyage 1501–2

Parana

0   2,000km

expulsion of the Moors was not complete until 1492). By then the Portuguese had taken the lead.

A small seafaring nation, Portugal possessed a large fleet of ships, a seafaring population trained on ocean fishing, a well-organized system of marine insurance and investment and a royal family ready to back these maritime enterprises. Moreover, in the fifteenth century the design of European ships had developed rapidly and so had the necessary navigational aids.

Portuguese dominance in the spice trade

By finding a direct route to India by sea the Portuguese were able to gain an advantage in the spice trade. The architect of Portuguese supremacy in the Orient was Alfonso d'Albuquerque (1453–1515). In 1510 he seized the island of Goa; in 1511, Malacca. From the East Indies Portuguese ships went to China and sailed annually between China and Japan. Their twin commercial aims were to monopolize the spice trade with Europe and to get as large a share as possible of the inter-port trade of the Indian Ocean.

But in fact the Portuguese commercial empire of the sixteenth century – the result of the age of exploration – achieved less than its architects had hoped. The Portuguese succeeded in overawing but not in controlling their Asian competitors at sea. Until the coming of the Dutch they retained a monopoly of the sea route around the Cape which they had pioneered. But they never achieved a monopoly of the spice trade between Europe and Asia. The Venetians, supplied by the old land routes, continued to sell some spices in Europe, while the Portuguese did not achieve control of all the spice islands. Their Estado da India [7], a set of fortified trading-posts clinging sometimes precariously to the coast or to islands, never penetrated and certainly did not dare to challenge the empires of mainland Asia.

The explorations in the East were initially more important to Europe than for the countries explored. They indicate a shift in the centre of gravity of European trade from the Italian states to the Atlantic. The new routes to the East by sea permanently changed the mercantile map of the world.

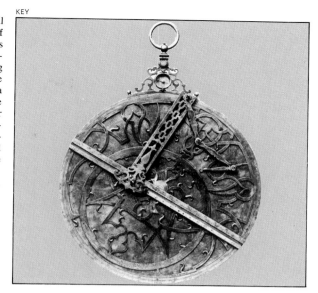

The astrolabe, together with the quadrant, was one of the chief navigational aids that made exploration possible.

5 Lisbon in the late 16th century, with a population of about 100,000, was the largest city in Portugal. It was the nerve centre of her seaborne empire where the spices of Asia were redistributed to the Mediterranean and Atlantic world in exchange for their goods. In the next century Portugal was to lose her commercial dominance in Asia and her political independence in Europe.

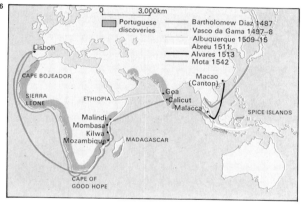

6 The route to the Indies took the Portuguese 50 years to develop from the time that Diaz rounded the Cape of Good Hope in 1488. The greatest single step was da Gama's crossing of the Indian Ocean, but it was the later venturers who gained for Portugal her central position in the trade of the Far East.

7 The Portuguese empire in the East consisted of a string of fortified trading-posts all the way from Sofala in east Africa to Macao in the China Sea. The headquarters of the Estado da India was Goa on the western coast of India. The grand forts, such as this one, that the Portuguese built in Goa were meant to overawe rivals from the sea and to dissuade attack from the hinterland.

8 The Dutch began to sail eastwards in 1595. Their innovation in getting to the East Indies was to leave India on their flank and sail direct from the Cape to the Sunda Strait and then turn north to Java. This new route had the advantage of by-passing Goa and Malacca, but it also meant that they had to carry bullion or European goods direct to Indonesia, without the support of local Asian trade.

Americas: conquest and settlement

By the early sixteenth century the Spaniards had established colonies in the Antilles and the Isthmus of Panama. The mineral resources of these islands, however, proved to be slender and, despite the name Golden Castile, the area had little to offer. Trading expeditions along the coast of Yucatán revealed a likelihood of greater wealth in the interior. In 1519, after two preliminary expeditions, the governor of Cuba, Diego de Velázquez (1599–1660), sent a fleet under Cortés to settle the region [1A].

The major conquests

Hernán Cortés (1485–1547) [2] and his expedition of 550 men landed on the coast of Mexico and founded the settlement of Veracruz. Cortés then threw over Velázquez's authority and placed himself directly under the crown. By a series of adroit diplomatic moves and superior military technology, Cortés took formal possession of the Aztec capital of Tenochtitlán. He was welcomed by its ruler, Montezuma, whom he promptly imprisoned. According to Cortés, Montezuma made a willing donation of his empire to Charles V in the mistaken belief that Cortés was the emissary of the god Quetzalcoatl. This widely circulated story is almost certainly a fable.

The conquest of Peru in 1532 by the adventurer Francisco Pizarro (c. 1471–1541) [4] was the result of the search for gold. When the Spaniards arrived on the Peruvian coast at Túmbez, the Inca Empire was divided by a civil war. Atahualpa emerged victorious shortly before Pizarro caught up with him at the fortress of Cajamarca [3]. Here the Spaniards succeeded in killing most of Atahualpa's retinue and capturing the Inca himself. Atahualpa offered to fill his cell with gold in exchange for his freedom. Pizarro accepted, but although Atahualpa kept his part of the bargain he was not released. The value of his ransom has been estimated at about 20 million dollars. It included the treasures of the Temple of the Sun and was all melted down for bullion.

Pizarro lacked Cortés's powers of leadership and after the subjugation of the Inca Empire, the conquerors began to fight among themselves. The civil war lasted until the death of Pizarro in 1541 and the execution of his brother Gonzalo in 1548.

With the conquest of Peru, the attention of the European explorers turned elsewhere. Expeditions were sent to Texas and Florida and some settlements were established. The colonization of New Mexico was slightly more determined but the area soon became little more than a military outpost.

Administration

The Spanish crown, fearful lest the more successful of the *conquistadores* (conquerors) should set up independent feudatories, rapidly took control of the government of the new colonies. The administration was complex. At the top was the viceroy, responsible only to the crown [5]. The distance between Spain and the Indies inevitably placed great power in his hands.

To maintain the balance of power within the colony, a separate court of appeal, the *audiencia*, was established. This had the right of direct representation before the crown and could suspend crown officials from their duties; it also served to represent the Indians

Aztec Empire
— Route of Cortés 1519–20
····· Route of Cortés 1521
▲ Tribes who allied with Cortés
🌿 Marshland

4,500 metres
4,000
3,500
3,000
2,500
2,000
0

1 Cortés landed at Veracruz [A] on 22 April 1519. By forming alliances with the peoples of Cempoala, Jalapa and Tlaxcala, he maintained his supply line with the coast [B]. After some skirmishes, he persuaded the powerful independent "state" of Tlaxcala to join him, thus obtaining the much-needed base from which to launch his attack on the Aztec capital of Tenochtitlán. After the retreat of the "Sorrowful Night" on 30 June 1520, Cortés fled to Tlaxcala. With the help of local workmen he built a fleet of brigantines to harass the defenders of Tenochtitlán from the lake. These were carried in pieces and assembled on the lake shore, while the army, composed largely of Tlaxcalans, attacked the towns that seemed sympathetic to the beleaguered Aztecs.

2 Hernán Cortés was born in the province of Estremadura in Spain. He sailed for Hispaniola in 1504 and as Velázquez's lieutenant, he took part in the conquest of Cuba. His career subsequent to the conquest of Mexico was, like that of so many *conquistadores*, spent largely in an effort to secure from the crown due recognition of his achievements. He died in Spain in 1547.

Inca Empire at its greatest extent in 1525
— Inca roads
— Pizarro's expeditions 1524–8
- - - Pizarro's expedition 1531–3
▣ Spanish settlements
⚜ Capital of Inca Empire

3 A superb network of roads helped Pizarro to conquer Peru. Reaching Túmbez in 1531, he then marched inland to Cajamarca to find the Inca ruler Atahualpa. Following a surprise attack, the city fell, and Atahualpa himself was later executed. Pizarro founded a new capital at Lima, thus shifting the centre of Inca society from Cuzco to the coast.

4 Charles V gave his support in 1528 to Pizarro's third expedition to Peru. Pizarro had accompanied Balboa on his march across the isthmus of Panama in 1513 and later formed a partnership with Diego de Almagro and Fernando de Luque to settle Peru and seek out the riches of the Incas. The first expeditions (1524–8) were unsuccessful. In 1531, Pizarro and his brothers left Panama for Peru, joined later by Almagro. Pizarro lacked the diplomatic skills of Cortés and was unable to curtail the ambitions of his followers. Shortly after the subjugation of the Incas, Almagro broke away and during the ensuing civil war Pizarro was murdered.

before their overlords and to administer justice. The crown also imposed the system of *residencia*, whereby a crown official was examined at the end of his term of office and any misconduct punished. These cumbersome institutions were often ill-equipped to deal with the American situation.

The Indians were "granted" to settlers under the *encomienda* system. Theoretically this provided the colony with a salaried labour force; each settler was given a number of Indians (but not their lands) from whom he took tribute or labour. The position of the Indian was therefore rather like that of the European serf and the power that this appeared to give to the colonists forced the crown to make several efforts to abolish the system. By the Laws of Burgos of 1512–13 and subsequent royal decrees, the number of Indians held in *encomienda* was limited and their duties were to some extent lightened.

The wealth of the Americas

The great wealth of America lay in its silver deposits. The mines of Potosí in Peru and of Zacatecas, Guanajuato and the other Potosí

in Mexico were all discovered during the 1540s and their output soon came to dominate the European silver market. The colonists benefited little, however, and were forced to subsist on agriculture and trade in lesser commodities, such as silks, with the Philippines. Precious metals were shipped via Seville which enjoyed a monopoly of America's trade and where a clearing house, the *Casa de la Contratación*, had been set up as early as 1503. The Americas had always been a Castilian venture and the Castilian crown benefited directly [6].

The impact of Spanish colonization upon the New World was considerable [9]. The introduction of European crops and livestock destroyed the domestic life of the Indians and altered the ecological balance of central Mexico. The disruption of old tribal divisions and of the hierarchy of Aztec society produced an alienated, enfeebled community which rapidly succumbed to European diseases so that the population declined rapidly during the sixteenth century. Latin America thereafter was a mixed culture dominated by that of the Spanish settlers.

This 17th-century Inca wooden beaker shows Peruvian Indians and a European together. At first, the Spaniards tried to preserve the Indian social structure, using local chieftains as justices and tax-collectors. However, over the course of the 17th century, the native nobility intermarried and the Spaniards took Indian women as concubines, producing a mixed group known as mestizos. The Indian culture declined and was supplanted by that of a white, or near-white society.

5 The Spanish crown rapidly replaced the *conquistadores* with its own officials, who set up a complicated government machine providing checks and balances to prevent any single individual from growing too powerful. There was also the *audiencia*, a court of appeal to which every citizen had recourse and which was responsible directly to Spain. This former governor's palace is in what is now Morelia, central Mexico.

6 Silver bullion from the Indies provided Castile especially with great wealth during the 16th century, but the influx slowed as the mines were worked out in the 17th. The massive inflation of 17th-century Europe may have derived from these imports of American silver.

Spanish territories
Other European territories

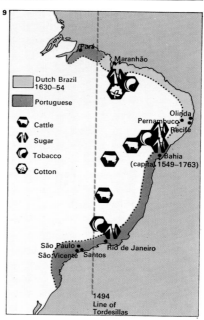

Revenue in millions of Venetian ducats

7 Spanish cruelty towards the Indians was notorious. *Brief Relation of the Destructions of the Indies* by Bartolomé de las Casas, which became a best-seller during the 16th century, helped to foster and disseminate this "black legend". Its popularity, however, was due more to fear and hatred of Spain than to love of the Indians. This illustration by De Bry appeared in The Netherlands as anti-Spanish propaganda.

8 The conquest of America by Europeans in the 16th century meant that the indigenous cultures were almost wiped out. Because of the Spaniards' desire to convert the Amerindians to Christianity, the traditional features of the native way of life were absorbed into the dominant culture, although this process was uneven. The tribes that survived were the most primitive and they never developed, but the civilization of the Aztecs and Incas wholly disappeared.

9 By the Treaty of Tordesillas of 1494 the Spaniards and the Portuguese – at the instigation of Pope Alexander VI – agreed to a demarcation line between their territorial claims. When Brazil was discovered in 1500 by Cabral, Portugal naturally claimed it but did not seriously colonize it until 1530. Spain and Portugal were united between 1580 and 1640, so that Brazil was open to attack by Spain's enemies. In 1630, the Dutch seized a rich area in the north, which they kept until 1654 when they were driven out by wealthy landowners. Throughout the 17th century, the Brazilian economy developed rapidly, based on sugar and tobacco together with cotton and cattle.

Dutch Brazil 1630–54
Portuguese
Cattle
Sugar
Tobacco
Cotton

Pará
Maranhão
Olinda
Pernambuco
Recife
Bahia (capital 1549–1763)
São Paulo
Rio de Janeiro
São Vicente Santos
1494 Line of Tordesillas

The rise of banking

Between the twelfth and the fifteenth centuries Italy exercised a European supremacy in commerce and finance and was responsible for most innovations in business and banking. By the sixteenth century that supremacy was being eroded. In twelfth-century Genoa, banks accepted deposits, exchanged foreign coins for local currency and engaged in bullion dealing. Later the Florentines became the leading Italian bankers. Throughout this period banking and trade were invariably combined and the commercial contacts of the Italians throughout Europe necessitated the regular international transfer of money. The simple but commercially important bill of exchange [Key] was evolved in the fourteenth century and avoided the problems and dangers of transporting gold and silver coins.

Activities of medieval bankers
Largely as a consequence of the Church's ban on lending money for interest (usury), medieval bankers made exchange operations a key activity. Although the bill of exchange did involve a credit element until the money was paid out at a later date, it was not classified as a loan.

The banker's profit came from the exchange rate – the Medici averaged 15 per cent. The prohibition on charging interest may also have impeded the evolution of the basic banking activities of borrowing and lending at interest, although deposit banks did exist, mainly in Italy, but also in Bruges [1]. They offered non-interest bearing current accounts and the facility of making transfers between customers' accounts.

From the fourteenth century onwards some towns allowed written orders (the distant ancestor of the cheque) to make payments or transfers. These orders replaced the original requirement that such orders had to be made orally by the depositor in person.

A decline in private banking in the Low Countries and Italy in the fifteenth century is illustrated by the fall in numbers of large Florentine banks from 80 to eight between 1330 and 1526. The hazards of medieval banking are revealed in an assertion made in 1585 that over the years 96 out of 103 Venetian banks had failed, often through

incautious loans to merchants and governments. Deposit banking recovered in the late sixteenth century, often in the form of municipally controlled public banks, as in Venice and later in Amsterdam [9], both of which accepted deposits and transferred money between customers' accounts.

Small banks and money-lenders
More widespread than the handful of great Italian merchant banks such as the Medici [6] were small local banks. These shared many of the functions of their large counterparts, accepting deposits, making loans on pledges and changing money, but avoided exchange and transfer operations.

Throughout Europe there were also the reviled but necessary money-lenders and pawnbrokers who made small, short-term loans to the poor [3]. Even in a largely subsistence peasant economy, money was needed to pay taxes and the poorest might seek a loan in hard times. Some of these money-lenders were Christians, but most were Jews to whom the usury laws did not apply and who could therefore charge interest. Some, such as

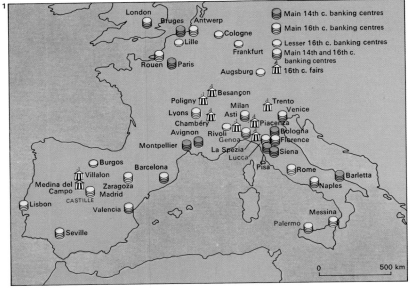

1 **Bruges,** whose financial activities were dominated by Italians, was the only permanent medieval banking centre outside Italy. The late 15th century witnessed the rise of the south German houses. The major 16th-century centres were Antwerp, Lyons and the seasonal fairs in Castille and at Besançon. The Genoese controlled the Besançon fairs which soon left and reconvened in Piacenza by 1579.

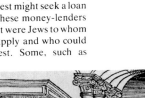

2 **The Papacy received dues** and taxes from all over Europe. Heedless of the usury prohibition, the Papacy also borrowed at interest from bankers. The great basilica of St Peter's, begun in the early 16th century, was partly financed by the proceeds from the sale of indulgences conveyed by the Fuggers, who had broken the Italian monopoly.

3 **Aaron,** a noted Jewish usurer, built this house in Lincoln in about 1170-80. He conducted business with the greatest families of England and financed part of the building of Canterbury Cathedral. On his death a special department of the Exchequer had to wind up his huge financial assets. The Jews were expelled from England in 1290, as the king hoped to take over their assets and debts from the nobility.

4 **Jacques Coeur** (*c.* 1395–1456), French merchant, money-lender and financier, built this palace in Bruges during his spectacular career. He lost his fortune in 1453 and was exiled until his death.

5 **This painting by Quentin Massys** (*c.* 1465–1530) shows the money-changer and his wife at work. For a small fee they would exchange gold for silver coins or the coins of one currency into another.

those in Florence in 1437, were officially invited to set up licensed pawnshops so that Christian souls should not be jeopardized through usury, although Christian money-lenders were not above charging disguised interest.

In mid-fifteenth-century Italy churchmen established non-Jewish pawnshops, the *Monti di Pietà*, to meet the needs of the poor. The charitable purpose of these pawnshops was perverted, however, when governments and the rich turned to them for loans.

Sixteenth-century banking

The sixteenth century saw no major innovations in banking practices but, as the European economy expanded, the scale of operations increased, credit became more extensive and other nationalities, especially the south Germans, adopted Italian methods. London remained a second-rank financial centre until the late seventeenth century. Methods of obtaining credit became more flexible as a result of the increasing use of endorsement and discounting, although until the seventeenth century these were rarely applied to bills of exchange. Interest rates fell and some Protestant states adopted a more liberal attitude towards usury. Expanding trade was facilitated by the evolution of an unofficial system of settling international trading debts and balances during the great seasonal European trade fairs. This lasted until the late sixteenth century when the fairs began to lose their importance.

The most spectacular development was the mid-sixteenth-century boom in government borrowing to meet the escalating costs of war. The French kings raised money in Lyons, the English and Spanish [8] rulers in Antwerp, the greatest centre for long-distance trade of the era. Fortunes were lost as the kings of France and Spain defaulted on their debts, just as Edward III of England had ruined two Florentine banks in the 1340s. Ordinary individuals as well as great banks such as the south German Fuggers [7] were hit. Although the Fuggers survived on a reduced scale, by the end of the sixteenth century the era of the international financier was over, although more mundane forms of banking continued to prosper and develop.

The form of the bill of exchange altered little from its evolution in the 14th century. To avoid the Church's prohibition on usury it had to instruct the recipient in another country to make payment in another currency to the person specified. Usually the rate of exchange was given. This bill, signed by Anton Fugger, required the Fugger agents in Rome to pay 2,700 ducats to the Dean of Elwagen.

6 The Medici family were the greatest of all 15th-century Italian bankers with branches in France, England, the Low Countries and Italy. Under Cosimo de' Medici (1389–1464), like his successors a patron of the arts, they reached the peak of their influence and wealth. Already in serious financial trouble, the firm finally foundered when the Medici were expelled from Florence in 1494.

7 In the early 16th century the Europe-wide commercial and financial empire of the Augsburg family, Fugger was the largest the Western world had ever seen. It rested on lucrative mining monopolies, extensive trade and loans on a vast scale. The most spectacular period occurred under the guidance of Jakob Fugger the Rich (1459–1525) (here painted by Albrecht Dürer). He learnt the banking trade in Italy.

8 Jakob Fugger's most famous loan was made to the Hapsburgs: 543,585 florins that helped to buy the votes of the Imperial Electors who made Charles V of Spain Holy Roman Emperor in 1519. The fortunes of the Fuggers and the Hapsburgs were thereafter closely linked. Charles V had to borrow unprecedentedly large sums from foreign bankers living in Antwerp to pay for his wars with the French, the Turks and the German Protestants.

9 The Amsterdam Exchange Bank (1609) was one of the most famous and successful of the municipally supervised banks set up after the financial recovery of the late 16th century. It received cash deposits of over 300 florins, transferred money between customers' accounts, traded in bullion and handled all bills of exchange over 600 florins. Like other municipal banks it did not make loans, grant overdrafts or discount bills nor did it issue bank-notes.

The politics of Europe 1450–1600

Medieval Europe was the *respublica christiana*, the universal world of Christendom ruled by two swords – the spiritual wielded by the pope, the secular by the emperor. But the sixteenth century saw the disintegration of this ideal and the growth of monarchical states whose triumph was at the expense of the Church and the nobility. The process was at its clearest in France [Key] where the crown gradually succeeded in converting lands held under the old feudal system into lands held by the crown, through a succession of accommodations between the crown, the papacy and the nobility.

Popes and kings

The disputed claim to the papal succession from 1378 to 1417 damaged the position of the pope as head of Christendom and the rival candidates became the puppets of secular rulers. Later popes retreated to play Italian politics while the princes of Europe exploited them to control the Church.

Monarchs of the sixteenth century attempted to increase their control over the nobility. The medieval king's real power rested ultimately on his wealth and the military strength that he could command. Towards the end of the Middle Ages a few noble families in many parts of Europe greatly extended their own power and fortunes. The Percy family, for example, had as much power in the north of England as kings. At times of weak kingship this situation produced civil war in France, Spain and England (the Wars of the Roses, 1455–85). But the disorder of war threatened the privileged classes as a whole and they looked to a strong king to restore stability. The princes set about re-erecting the former effective monarchical rule based on financial resources, powerful adherents, the centralization of the administration and the neutralizing of the power of the independent nobles.

In England the Lancastrian Henry VII (1457-1509), who won his crown in battle, attempted to establish personal control of government. Henry enjoyed the forfeited lands of the vanquished and was financially stable. He administered the affairs of the realm from his own household and was able to discipline a nobility weakened by wars and

to reduce the numbers of their military retainers. By marrying into the House of York he minimized the possibilities of dynastic rivalry. Henry rebuilt a strong personal monarchy in a realm that had seen nothing like it since Edward I (1239–1307). His son, Henry VIII (reigned 1509–47) [5], continued to strengthen the crown: he incorporated Wales within the kingdom and brought the wealth and authority of the Church into his hands.

Francis I and Charles V

Francis I (1494–1547) came to the throne of France facing a difficult situation. France had been rent by war, nobles such as the Bourbon family were powerful and the kingdom was large and divided. Francis exploited the royal rights of direct taxation, centralized the treasury and attempted to impose uniform Roman law codes on all the provinces.

There seemed some hope for a degree of unity and centralization in Spain when the kingdoms of Aragon and Castile were both inherited by the Hapsburg Charles V (1500–58) in 1516 [1]. Charles suppressed

1 The empire of Charles V was the product of dynastic accident. He succeeded his father Philip as Duke of Burgundy in 1506, and his grandfather Ferdinand as King of Aragon and of Castile in 1516. The territories of the Holy Roman Empire came with his election to the title in 1519. There was no logic to the empire and problems of communication made imperial rule almost impossible. Charles resolved these difficulties by handing The Netherlands (1555) and Spain (1556) to his son Philip II and the rest of his empire to his brother Ferdinand (1556).

4 The "Field of the Cloth of Gold" is a painting that commemorates the historic meeting of King Henry VIII of England and King Francis I of France in June 1520 to conclude a long awaited peace. The encounter was marked by great festivity and elaborate preparation.

2 The warring Hapsburgs and Valois dominated early 16th-century politics. Charles V, of the Hapsburgs, inherited disputed claims over parts of Burgundy and Italy [A]. Francis I, consolidating Valois rule in France, feared encirclement by the Hapsburgs. The conflict centred on Italy, important for its wealth and a vital hub of communications for the Hapsburg possessions in Spain and the empire. The route through the Italian

Henry had built a temporary palace, all tents were embossed with gold and velvet and the royal interviews were occasions for jousting and other contests of chivalry, ceremonial banquets and firework displays. As each king tried to outdo the other

passes was the only alternative to the Channel route, which was not under Charles's control. Most of the disputed areas came under imperial control by the Peace of Cateau-Cambrésis (1559) [B].

in pomp and splendour, the occasion turned into a trial of strength. No alliance resulted and open war had broken out by 1522.

Legend for map 2:
Disputed areas in Italy 1519-59
Held by Francis I of France
Under Spanish (Hapsburg) rule
States under Spanish influence 1559

3 The life of Niccolò Machiavelli (1469–1527) coincided with the rise of the nation state and the beginning of the art of statecraft. Machiavelli outlined a guide for princes in his book *The Prince* (1532), and his *Discourses* on the first ten books of Livy, the Roman historian. *The Prince* shocked Christian Europe and earned for its author a reputation for devious and treacherous statesmanship. He became a leading figure of the Renaissance.

the "comuneros" revolt and sent royal officials (*corregidores*) to enforce his will. But the nobility remained powerful and jealous of their privileges and the two kingdoms remained separate. Charles was distracted by his imperial duties (he was also Holy Roman Emperor) from consolidating Spain. In order to obtain Castilian money, he recognized the privileges of the nobles. Philip II (1527–98) continued to rely on Castilian bullion for his campaigns and neglected Aragon, which voted little money to the crown. The kingdoms of Spain were never centralized and in the 1600s Aragon and Catalonia revolted against the crown.

When he became Emperor in 1519, Charles V tried to centralize his lands through imperial institutions and constant visits, but the task was impossible. The empire had no unity; large cities such as Nuremberg were proud of their independence, the princes were trying to assert themselves and powerful families such as the Wittelsbachs were anxious to stop the Hapsburgs from becoming too strong. The Reformation [8] finally divided the empire.

In the late sixteenth century the Hapsburgs gained control over their hereditary lands in order to provoke resistance in Bohemia.

Attempts by both Charles V and Francis I of France to increase control of their kingdoms led them into more than 40 years of war over disputed claims in Italy [2]. Instead of becoming more unified, Italy was completely fragmented by the war.

The need for a strong monarchy

Although more powerful monarchies emerged in Western Europe during the sixteenth century, there were serious limitations on their control. Without a paid civil service kings still depended on the nobility. Men thought of Catalonia or Provence before they thought of Spain or France; allegiance to family, locality or lord could still overcome allegiance to the king. In the late sixteenth century religious divisions threatened the monarchy in England and encouraged a return to separatism in France. But the chaos of war once again alerted the ordinary nobles to the need for a strong king as the guardian of law and order.

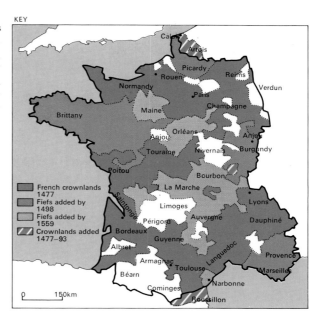

KEY

	French crownlands 1477
	Fiefs added by 1498
	Fiefs added by 1559
	Crownlands added 1477–93

0 150km

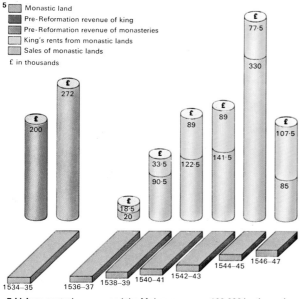

5
- Monastic land
- Pre-Reformation revenue of king
- Pre-Reformation revenue of monasteries
- King's rents from monastic lands
- Sales of monastic lands

£ in thousands

£200 £272 £18·5 / 20 £33·5 / 90·5 £89 / 122·5 £89 / 141·5 £77·5 / 330 £107·5 / 85

1534–35   1536–37   1538–39   1540–41   1542–43   1544–45   1546–47

5 Henry VIII's increased revenue from his dissolution of the monasteries, with the proportion received from the rent and sale of the lands, is shown here. Henry took this action four years after his break with Rome and claimed that he did it to put an end to corruption. But his real motive was to acquire an extra income for wars to further his dynastic ambitions in Europe.

6 Christ Church (Cardinal's College), at Oxford was founded by Cardinal Wolsey (c. 1475–1530) in the 16th century. Its purpose was to train young scholars in philosophy, rhetoric, humanities and civil law. The educational foundations of the period owe much to the humanist belief in the relationship between education and the nobility of service to the realm.

7 Lisbon, central market for the spices of Asia, was the heart of a Portuguese empire that expanded east to the Indian Ocean and the Moluccas and west to Brazil. The population of this cosmopolitan city increased from about 50,000 in 1500 to more than 100,000 by the end of the sixteenth century. Similar developments in Amsterdam, Madrid, Antwerp, London and Paris were integral to the rise of the modern states. Capital cities, already centres of trade, also became the base of government and of court life.

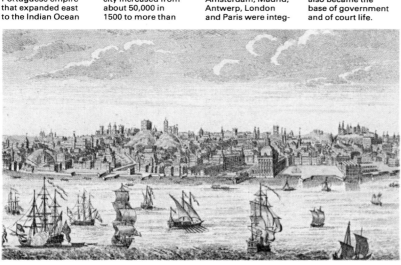

8 Political propaganda in Reformation Germany is typified by this woodcut of a figure brandishing the banner of freedom. The development of printing, the growing respectability of vernacular languages and a period of religious and political conflict gave rise in the early sixteenth century to cartoons and broadsheets as a means of persuasion. Pictorial representation had the widest appeal in an age of limited literacy. Monks were depicted as wolves and the enemies of Luther were caricatured. The method of contrasting woodcuts in juxtaposition was employed frequently. In one the Ascension of Christ was shown next to the descent of the pope into hell.

England in the 15th century

The period from the deposition in 1399 of Richard II (in favour of his cousin, Henry of Lancaster) to the death of Richard III on the field of Bosworth in 1485 is among the most turbulent in English political history. The murders of Richard II, Henry VI and Edward V, and the violent death of Richard III, illustrate the inability of the Crown to establish the line of succession clearly enough and rule firmly enough to rise above faction.

An insecure throne
The claim of Henry IV (reigned 1399–1413) was powerfully supported by conquest and parliamentary title. Henry V's overwhelming popularity put the question of succession into abeyance while he lived and during his son's minority, but the feebleness of the adult Henry VI (reigned 1422–60; 1470–1) raised it anew. Richard, Duke of York (1411–60), popular as the enemy of the court favourites, had a strong hereditary claim to the throne. After his death his son succeeded in driving out Henry and reigning as Edward IV (reigned 1461–83) [9] – although not securely until he had extinguished the Lan-

castrian dynasty after its short-lived restoration [7, 8] at the hands of the Earl of Warwick, "the Kingmaker" [Key], in 1470–1.

Factional dispute arose again during the minority of Edward's son, Edward V (1470–83), in 1483 and he was quickly imprisoned and almost certainly murdered as one of the "princes in the Tower" by his uncle Richard, Duke of Gloucester (1452–85), who reigned as Richard III (1483–5). The invasion of Henry VI's half-nephew, Henry Tudor (1457–1509), rallied the numerous enemies of the king but it was many years before Henry VII, his weak hereditary claim bolstered by marriage to the sister of Edward V, would sit securely on the throne.

Areas of consolidation
This political turbulence, although it bred insecurity, did not prevent solid progress and development of English wealth [1] and institutions. Parliament, by 1400 the essential organ of royal co-operation with the landed classes of the shires began by 1500 to command increasing respect. The courts of justice [2] were sustained by a vocal and pow-

erful body of common lawyers. The traditional bodies that governed the counties and boroughs generally increased their autonomy: the counties in the hands of local gentry as justices of the peace (JPs) and the boroughs under their new charters of incorporation as independent legal entities. For the Church it was a time of peaceful enjoyment of traditional revenues.

The letters of the Paston family of Norfolk reveal the intricate web of local and national institutions, secular and ecclesiastical, and the need for patrons, connections and friends at court to use them effectively. Threatened by powerful neighbours in his inheritance from John Fastolf (c. 1378–1459), John Paston (1421–66) spent several months in London in 1465 seeking remedy from the law and help from patrons. The Wars of the Roses [6] did not seriously disturb the fortunes of such people.

The limits of ambition
Historians once attributed many of the political disturbances of the century to the overwhelming power of the nobility and their

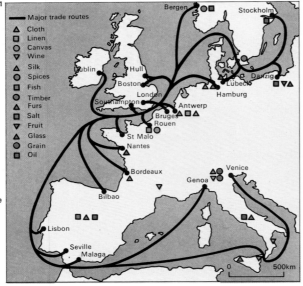

1 Trade in northwest Europe increased markedly during the 15th century and England participated fully in the boom. The nation imported furs and corn from the Baltic, wine from southern Europe and spices from the East. Cloth was the main English export. The Hanseatic League of north German towns won trade concessions in several English ports in the 1370s and it dominated London's trade with the Baltic, where the League had a virtual monopoly. The Genoese dominated the trade with the south that was centred on Southampton. But English merchant adventurers gradually increased their share of the trade during the 15th century.

Major trade routes
Cloth
Linen
Canvas
Wine
Silk
Spices
Fish
Timber
Furs
Salt
Fruit
Glass
Grain
Oil

2 At the Court of Common Pleas, the plaintiff spoke in a kneeling position. Although courts were subject to intimidation, the increasing influence of the legal profession and the independence of judges such as Chief Justice Fortescue suggests that on many occasions they refused to be corrupted.

3 A yeoman's cottage at Didbrook, Gloucestershire, is typical of the stone or half-timbered houses put up in the 15th century which reflected the growing prosperity of the independent rural landowner. From the mid-14th century, landlords increasingly rented or leased out the demesne since the fall in rural population made it hard to farm in the old way; the consequent new class of small farmers, freed from feudal obligations, strengthened their hold on the land and provided a new basis for rural prosperity.

4 William Caxton (c. 1422–91) printed *The Game and Playe of Chesse* in Bruges shortly before he moved to London in 1476. Caxton was the first English printer and his books reached a far wider audience than could afford to buy manuscripts.

5 The Lollards regarded the Bible as the sole authority of religious truth and inveighed against the wealth and worldliness of the Church. They were the disciples of John Wycliffe (c. 1328–84) who had denounced clerical authority and the worship of images. After a "rising" in 1414 the Lollards were subjected again to persecution and public burning. This picture shows the prison chamber in which Richard Hunne, a rich merchant accused of holding unofficial religious services, was found dead in 1514.

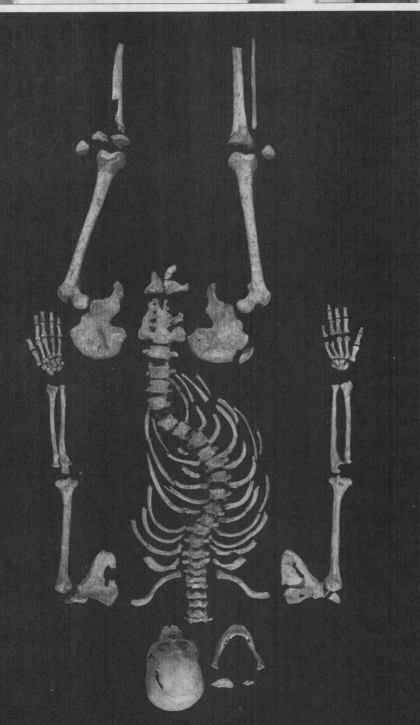

saves a city's sight

Shapps brand

Glasgow's grey skies meant the eclipse was visible only briefly during breaks in the cloud, in contrast to Edinburgh, below left, where – properly equipped – viewers could watch it in all its glory

torso or stomach in which to digest it, the sun would fall out the bottom of his neck and continue to illuminate all our lives.

Then there was the tale of Christopher Columbus who, apparently, bullied an indigenous tribe into providing food and water for his crew after insisting that if they refused he would darken the sun, which – as he was aware of an impending eclipse on 29 February 1504 – he duly made "happen", terrifying the natives into submission.

While I can see how terrifying the phenomenon may have been for the ancient

IMAGINE being a
Grant Shapps's ta
appears at the fro
the room and in
himself: "Hi everyone
Michael Green. I'm g
talk to you about bei
how I can help YOU
go digital. Except I ca
about being me. I car
full rundown on my
and my qualification
not me, I'm someone
we won't mention th

One political nerd
corner nudges his b
guy looks familiar. H
bit like… no, it can't
there is a striking res
to that Tory party ch
bloke. Maybe they're

For while having a
is one thing, pretend
be someone else is a
kettle of fish altoget

hired henchmen, recruited to serve them for pay or political favour, without the moral or social obligations of earlier feudalism. "Bastard feudalism" – the short-term recruitment of military retinues and local agents from among the gentry by formidable political figures – was certainly practised, but these "empires" of patronage were never very solid. The great nobility might indulge in affrays like the Battle of Nibley Green (1469) between the Berkeley and Talbot families, but so did respectable gentlemen such as the Pastons. Moreover, "bastard feudalism" restrained potential trouble-makers by associating them with great political factions.

For most of the century the nobility co-operated remarkably well to govern the country. Powerful captains such as Richard Beauchamp, Earl of Warwick (1382–1438), hoped the Lancastrian kings would lead them in victorious and profitable Continental war, and until the English were finally expelled in 1453 men such as Warwick had more to gain by exploiting the French than by civil war.

During Henry VI's long minority

(1422–42) co-operative rule by the king's Council proved a successful and generally popular experiment, despite the rivalry of Henry Beaufort and Humphrey, Duke of Gloucester (1391–1447). The financial weakness of the Crown became an increasing problem however, and this, combined with the unpopularity of the queen and Henry's lapse into madness allowed the dormant claims of York to acquire genuine support during the 1450s.

The social progress and cultural achievements of the century were considerable. The rise of the independent yeoman farmer [3], the cloth-maker, the wool exporter and the small-town merchant encouraged a multitude of local enterprises, to the general profit. Such men were usually literate, increasingly so as printed books became more readily available [4]. English architecture, both ecclesiastical as in the magnificent King's College Chapel, Cambridge, or domestic as in Tattershall Castle, reached the zenith of its brilliantly distinctive Perpendicular phase. Whatever the troubles of the monarchy these achievements remained.

Richard Neville, Earl of Warwick (1428–71), called "the King-maker", supported the York claim to the throne. Upon the Yorkists' victory in battle he received vast tracts of land and important posts of state. At first a dominant influence on Edward IV, he then went out of favour, engineered the brief restoration of Henry VI and was killed fighting at Barnet.

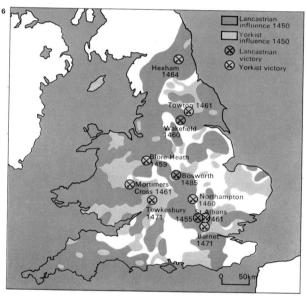

6 The Wars of the Roses were fought between descendants of Edward III. There was little consistent support from any particular area for either Lancastrian or Yorkist claimant to the throne, although from 1461 the Lancastrians relied on the support of the "reivers" of the Scottish borders.

7 The confused Henry VI was taken from the Tower and restored to the throne in a ceremony at St Paul's in 1470. His wife, Margaret of Anjou, was tenacious in keeping the Lancastrian cause alive in spite of his madness. Henry had become king in 1422, aged nine months. He was finally murdered in 1471 after the return of Edward IV.

8 At the battle of Barnet (April 1471), Edward IV defeated his former ally Warwick and cleared the way for his own victory the following month over the forces of Margaret of Anjou and her son, the Lancastrian heir, who was killed fighting.

Although many of the leaders of the defeated side might lose their heads after such a battle, it was not impossible for their supporters from the gentry to survive by offering their services to their victors. The wars hardly affected the lives of the poor.

9 Edward IV had political as well as military talents, and despite his light-hearted manner he was an intelligent, determined king. His centralizing policies, particularly in the financial sphere, laid the ground for those of the Tudors.

10 The execution of Edward Beaufort, Duke of Somerset, followed his capture in the Battle of Tewkesbury (1471), the final Yorkist victory that restored Edward IV. Many Lancastrians were executed for treason after summary trial by military law.

243

Scotland 1314–1560

Early in the War of Independence (1306–14) against England, the Scots had made the "auld alliance" with France – an agreement between the two countries to act together against England. This alliance was renewed at various times during the next three centuries although it is doubtful whether it brought advantages to Scotland.

Conflict with England and the clans

Peace was eventually made some years after 1314, but as Edward III (reigned 1327–77), the young English king, began to assert himself it ceased to be kept [Key]. In 1332 the English king allowed Edward (died 1363) the son of John Balliol (reigned 1292–6), to try to reclaim Scotland from David II (reigned 1329–71), the son of Robert Bruce. Defeat forced David to flee to France and the conflict became absorbed in the Hundred Years War between England and France. At various times in the 14th and 15th centuries Scotland became involved in this conflict.

In spite of frequent wars, which particularly damaged the trade of the principal Scottish towns – those in southern Scotland – the economy developed in this period, although the characteristic features of a primitive structure still adhered to it. From 1349 to 1350 the Black Death struck Scotland as it struck the rest of Europe and this plague returned several times, but because it became mainly an urban pestilence and Scottish towns remained very small, the country did not suffer greatly.

A more serious long-term development in the fourteenth century was the building up of the two branches of the great baronial family of Douglas, descended from Bruce's second-in-command, James Douglas (1286–c. 1330). This family became powerfully esconced in the Border area, where at times it carried on negotiations or wars with English Border families as if it were an independent power. The Lordship of the Isles under the Macdonalds held a similar almost independent position in the Western Highlands and islands. Great independent aristocratic families could often supplant the king in the collection of his revenue and coerce and corrupt the local church, placing members of their own family into the better-paid ecclesiastical positions. Nepotism was not confined to the nobles, many kings also used the Church to provide regular incomes for their bastard offspring.

The power of the Crown

James I (reigned 1406–37) was the first in a series of vigorous, often impetuous, short-lived kings who had a significant effect on the institutions of their country in their turbulent reigns. His reign saw the creation of Scotland's first university, at St Andrews, and the use of Parliament by the king as a means of gaining support from lesser folk in his struggle with the overmighty nobility [3]. The king also began the process by which a professional central court, instead of the King's Council, was developed to hear cases. Under James V (reigned 1513–42), this Court of Session obtained a more regular supply of money for the Crown by increased taxation. This provided funds for the Court – an important development towards professionalism in Scottish law.

James II (reigned 1437–60) reduced the power of the Douglases by the extermination

1 **Scottish troops and their French** allies are shown here attacking Wark Castle in 1385, one of the lesser English Border castles. The "auld alliance" meant that Scottish forces frequently became involved in the Hundred Years War. This illumination is from Froissart's *Chronicles* of the 15th century.

2 **The Black Douglas family's conflict** with a succession of Scottish kings well illustrates the struggle between the nobility and the Crown in medieval Scotland. During the 15th century the power of the Douglas family in the Border area posed a growing threat to the Crown until they were finally suppressed by James II.

Centres of Douglas fiefdoms
Black Douglas allies
Douglas castles
James II allies
Royalist castles
Battles between the Crown and the Douglases 1448–84

3 **James I** succeeded to the Scottish throne in 1406 at the age of 12, when he was held captive by the English. He remained a prisoner until 1424 when he was released in return for a large ransom. He returned, determined to assert the power of the Crown and to give the Scottish Parliament an importance and authority similar to that of the English Parliament. Under his reign it introduced taxes, passed laws and reforms and supervised trade. A period of vigorous reform began that touched many aspects of daily life, extending the Crown's control. In 1426 a court for civil cases was created that sat three times a year to settle disputes.

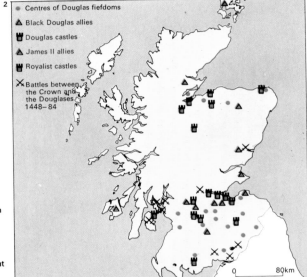

4 **James III's reign ended** with an armed rising of the nobility, his defeat at Sauchieburn near Stirling in 1488, and his murder after the battle. This 19th-century print shows the battle with Stirling Castle, where James had in vain sought refuge, in the background. The nobles' rebellion was probably motivated more by James's personality than by his policies: he was suspicious, avaricious and unconcerned with affairs of state. The relatively short reigns of Scottish monarchs, often ending violently amidst intrigue and conspiracy, resulted in a series of long minorities, beginning with James I, that undermined royal authority. For 200 years no Scottish ruler succeeded to the throne as an adult.

of the leaders [2]. James IV (reigned 1488–1513) in 1493 attempted to increase royal authority in the Highlands by destroying the Lordship of the Isles. He was unable to replace this power with effective central rule, and the main result was to leave the Highlands broken into separate and often warring clans. Efforts by successive Scottish kings to have courts appropriate to Renaissance princes led to the cultivation of the arts by James III (reigned 1460–88) [4] – who thereby annoyed his nobility – some splendid poetry in the reign of James IV [5] and great spending by James V on royal palaces.

Protestantism established
James IV's marriage in 1503 to Margaret Tudor (1489–1541) [7], elder daughter of the English King, Henry VII (reigned 1485–1509), was of great long-term significance to both countries, for it led to the union of the crowns in 1603 and, even before that, in the later sixteenth century, to peace between them. But peace was not achieved at first. James IV was drawn into war with Henry VIII (reigned 1509–47) in support of

his renewed alliance with France. James invaded northern England and was killed in an overwhelming Scottish defeat at Flodden Edge in 1513.

In spite of this disaster the Scots clung to the French alliance, partly because of the ruthless attempt of Henry VIII to dominate their country [7]. In particular, Henry hoped to marry his son Edward (1537–53) to James V's daughter and successor, Mary. But James's marriage allied him to the rising family of Guise, who came to represent the Catholic party in France. Meanwhile the weaknesses and corruption of the Catholic Church and the political needs of the government encouraged the promotion of a powerful group of Protestant nobles.

In 1560 the joint issues of external alliance and religion led to a brief war between the Catholic and French regent, Mary of Guise, and the Protestant lords in alliance with the English. The victory went to Protestantism and alliance with England, and this was confirmed in the Reformation Parliament of 1560, when the authority of the pope in Scotland was abolished.

David II (left) was captured by the English at the Battle of Neville's Cross in 1346 which effectively ended his invasion attempt. The English king Edward III (right) held David prisoner for 11 years before freeing him in return for a large ransom. This proved to be a crippling burden for Scotland and taxes were greatly increased to meet the debt. The Scottish economy, already undermined by wars with England, was weakened still further, although over half the ransom money was finally paid. David also debased the coinage by minting more coins to meet the debt, a precedent followed by his successors – the first Stewart kings.

5 One of the more attractive personalities among the early Scottish kings, James IV styled himself in the role of a Renaissance monarch. He built new royal palaces, notably at Falkland and Linlithgow, and encouraged learning and the arts. But he was not so successful in foreign affairs: in support of France, James invaded England in 1513 but his army was defeated at Flodden Edge and he was slain.

6 Margaret Tudor lacked the political drive and skill of her brother, Henry VIII, while sharing similar marriage difficulties. Thus, following the death of her husband, James IV at Flodden, her personal life was a considerable source of instability during the minority of James V. The man shown with her here [left] is probably the Duke of Albany (c. 1484–1536), who acted as regent until 1524.

7 The formidable fortress at Stirling has played a major role in Scottish history. Probably built in the 12th century, it was the birthplace and residence of several monarchs, rivalling the capital of Edinburgh.

8 The Augustinian Border abbey of Jedburgh was one of the richest foundations in Scotland. During the 16th century it was burnt three times by the invading armies of Henry VIII.

The Renaissance: humanism

It is impossible to ascribe precise dates to the Renaissance. The traditional picture of a rebirth of creative thought and learning after the dark years of the Middle Ages has rightly been discarded. The new learning was not wholly original, and there was no dearth of creative thought in the Middle Ages. But this modification does not detract from the importance of the Renaissance, which from the fourteenth to the late sixteenth centuries produced the growth of a more secular spirit, a renewed interest in classical civilization and an increased respect for literature.

The classical revival
Interest in the classics was promoted by Francesco Petrarch (1304–74) who worked on classical texts and wrote poetry in Italian and classical Latin. The prosperous towns of fifteenth-century Italy became the centres of classical scholarship, which was pursued outside the universities and under the patronage of urban patricians interested in the civilization of the pre-Christian world. Through the residence of a Byzantine scholar, Manuel Chrysoloras (c. 1350–1415), Florence

became famous for Greek studies and Marsilio Ficino (1433–99) established an academy for the study of Plato's philosophy. The concentration on classical texts owed something to the medieval interest in Arabic translations of the ancients, but the textual scholarship of men such as Lorenzo Valla (1407–57), who used philological investigation to discover pure texts of the classics, was completely original.

In Italy, burgher patronage and the casual acceptance of a worldly Church meant that the new learning was predominantly secular. Patrons and scholars were fascinated by the achievements of man and came to believe that on his own he had real power to shape his own destiny. They looked back to a time in which human achievement had reached its peak – to the past of classical Rome. The new learning was named *humanitas* (humanism) by Leonardo Bruni (1370–1444), who wrote a *History of the Florentine People* inspired by classical models.

Since Italy was the centre of the Mediterranean trade routes the movement soon spread. Those fighting in the Italian

wars took back books to the court of Francis I (1494–1547), who encouraged the new scholarship. By the mid-sixteenth century the influence of Italian humanism on French scholarship was evident. French humanists such as Lefèvre d'Etaples (1455–1536) and Guillaume Budé (1468–1540) [5] embarked upon etymological and philological studies of Greek and Roman texts and of their own institutions and codes of law.

The growth of printing
The printing press was another factor encouraging the rapid northward spread of Italian scholarship. Venice became a major centre for the printing of Greek and Latin classics and patristics. In conjunction with the scholars of Germany and The Netherlands, printers produced new editions of the classics [7] and works by the Italian humanists.

In The Netherlands and Germany, however, the influence of Italian humanism was tempered by the new devotional movements such as the *devotia moderna*. Gradually new ideas were admitted into the structure of medieval scholasticism and the Christian

1 The woodcut shows a lecture being delivered by Cristoforo Landino. Landino was one of the most widely read commentators on classical literature and a professor at Florence. His principal work was a study of Virgil and Dante; he also translated the works of Aristotle. Landino was much interested in politics. One of the circle of Lorenzo de Medici, he believed in the freedom of the citizens. In his *Disputationes Canaldulenses* (1475), he discussed one of the central questions of philosophical speculation; namely the relative advantages of the active as distinct from the contemplative life-style.

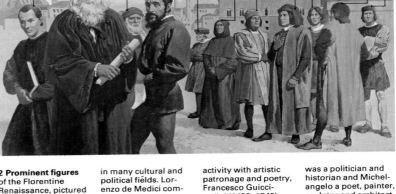

2 Prominent figures of the Florentine Renaissance, pictured here, often worked

Machiavelli
Leonardo da Vinci
Michelangelo
Verrochio
Amerigo Vespucci
Savonarola
Donatello
Francesco Guicciardini
Angelo Ambrogini
Lorenzo de Medici
Botticelli

in many cultural and political fields. Lorenzo de Medici combined diplomatic

activity with artistic patronage and poetry, Francesco Guicciardini (1483–1540)

was a politician and historian and Michelangelo a poet, painter, sculptor and architect.

3 The rich library of Lorenzo the Magnificent (1449–92), grandson of Cosimo de Medici (who built the Bibliotheca Marciana in St Marco Florence), became a public library in 1571, housed in a building by Michelangelo.

4 The new humanist learning of the centuries of the Renaissance would not have had so much impact but for the invention of printing in the mid-15th century. The map shows those European cities that had sizeable printing presses between 1454

and 1494. By 1500 Italy had 73, Germany 50, France 45 and England only 4. Although the printing press was extremely important, secular literacy was still not very extensive until the spread of education during the Reformation.

Invention of printing 1454
Printing begun 1454-64
1465-74
1475-84
1485-94

Stockholm
Odense
Oxford
Leiden
Zwolle
London
Delft
Deventer
Alost
Utrecht
Bruges
Gouda
Louvain
Bamburg
Paris
Mainz
Pilsen
Cracow
Strasbourg
Basel
Vienna
Lyons
Budapest
Venice
Salamanca
Bologna
Rieka
Lisbon
Saragossa
Lerida
Tortosa
Barcelona
Subiaco
Rome
Valencia
Valdemosa
Seville
Constantinople

philosophy of St Thomas Aquinas. The interests of the northern humanists were less secular than those of their Italian counterparts – although their concern with a pre-Christian culture and more personal religion led to some clashes with the Church. Their philosophy is best exemplified in the work of Desiderius Erasmus (*c.* 1466–1536) of Rotterdam [6B]. Erasmus applied the new textual scholarship to the production of a pure Greek text of the Scriptures [6A] and by 1516 he had published a translation of the New Testament, based on the Greek texts that he had collected. If Erasmus shared some of the Italians' beliefs in the self-sufficiency of man, his humanism was diluted by mystical piety.

In England, Humphrey, Duke of Gloucester (1391–1447), was a patron of the new learning. After 1490 the teaching of William Grocyn (*c.* 1446–1519) made Oxford a centre of Greek studies and Thomas Wolsey (*c.* 1475–1530) built a college at Oxford (now Christ Church) to be devoted to the new humanism.

In large part the new learning was sponsored by laymen and by the princes of Europe, who were growing in power and building brilliant courts as monuments to their prestige and authority. It is not surprising that many of the questions with which scholars concerned themselves were those of the ideal state, the model ruler, the best laws.

Politics and humanism

The humanists believed that men could improve their condition and so encouraged speculation about politics, a speculation evident even in Erasmus' writings and more obvious in those of Thomas More (*c.* 1478–1535) [8], Claude de Seyssel of France and Niccolò Machiavelli (1469–1527). Scholars who regarded the classical past with interest and who employed new critical skills in evaluating evidence, looked to the new learning to build the future. As they adapted the models of the classics to a new and more vital vernacular literature, so they hoped to find in the past the blueprint for a better society. It was that spirit of optimism, a belief in the potential achievements of men, that gave the Renaissance its unique vitality.

Renaissance intellectual life was characterized by a desire to emulate the scholars of Greece who had embraced all philosophical and rhetorical knowledge. The religious and devotional movements of The Netherlands. To return to the Church of the Apostles he worked to produce a pure Greek text of the New Testament [A] in the hope of ending religious dispute.

Florentine academies were often modelled on those of Plato and Aristotle, who are pictured here engaged in animated discussion in a relief by Giotto (1266–1337) from the campanile of the Duomo in Florence.

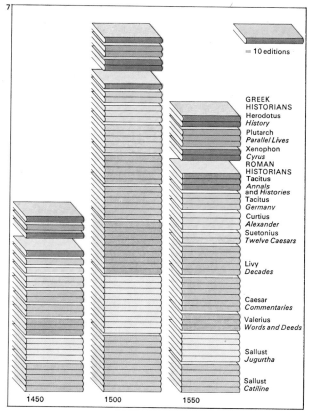

5 Guillaume Budé was the father of Greek studies in France and introduced the new sciences of philology and etymology. His critical study of Justinian's *Digest*, a compendium of Roman civil jurisprudence, showed how the text had become corrupted over the centuries. Budé demonstrated to the French humanists how language and law, as all institutions, were affected by time and change.

6 Erasmus of Rotterdam [B] more than any other scholar represents the humanist scholarship of northern Europe. He combined the Italian interest in the classics and textual scholarship with the

7 Editions of classical texts were produced in great numbers following the re-emergence of interest in the classics and the invention of printing. The diagram shows a survey of classical book production, 1450–1600. Beginning in Italy, the editing of texts spread to France, Germany and England by the 16th century. That century's interest in Livy and Caesar gave way in the 17th century to a greater interest in Florus and Tacitus. Perhaps more than two million copies of classical texts circulated in Europe between the invention of printing and the year 1700.

= 10 editions

GREEK
HISTORIANS
Herodotus
History
Plutarch
Parallel Lives
Xenophon
Cyrus
ROMAN
HISTORIANS
Tacitus
Annals
and *Histories*
Tacitus
Germany
Curtius
Alexander
Suetonius
Twelve Caesars

Livy
Decades

Caesar
Commentaries
Valerius
Words and Deeds

Sallust
Jugurtha

Sallust
Catiline

1450   1500   1550

8 A blueprint for the perfect republic –Utopia – was presented by Sir Thomas More in 1516. It described the communal ownership of land, the education of men and women alike and religious toleration. As a humanist More believed that all could by their own endeavours improve the quality of their lives. In particular humanists thought that learning would teach men how to govern and serve the commonwealth. Often they advised rulers: at the court of Henry VIII the humanist "commonwealthmen" advised the government on social policy.

Renaissance music

The Renaissance in music was very largely the work of Flemish composers – some six generations of them in all, between 1400 and 1570 – who spurred musical development in France, Italy, Germany, Spain and England.

Rise and spread of Flemish music

It is surprising that such a small country should have produced such an enormous amount of talent, but at a time when its larger neighbours were engaged in long and disruptive wars or suffering the effects of plague, Flanders – part of the dukedom of Burgundy – had a stable, highly developed and thriving middle-class civilization which was relatively free, fond of the arts, and ready to patronize its artists. Musical education in the cathedral choir schools of Flanders and northern France was one of the chief factors that contributed to this wealth of talent. Musical composition was taught as well as singing, and every chorister therefore had the opportunity to emerge as a fully fledged composer.

For this reason Flemish singers were highly sought after throughout Europe. The career of Orlando di Lasso (c. 1532–94) [6] illustrates extremely well the possiblities for choristers. As a child he had a beautiful voice, and was taken into the service of the viceroy of Sicily, with whom he travelled to Milan and Sicily. As a result of his talent as a composer he was given a patent of nobility by Emperor Maximilian in 1570, and made a Knight of the Gold Spur by the pope in 1574.

Orlando's career demonstrates vividly two important aspects of Renaissance music. First, the social status of the musician had changed dramatically. In the new appreciation of art of all kinds that the Renaissance brought, creative talent was at a premium. A musician was no longer simply a servant or artisan to be employed. If his music won favour and esteem, there was virtually no limit to what he might aspire to. Second, but more far-reaching in its effects, the Renaissance saw a secularization of music. Men such as Lasso, who emerged from a tradition that had hitherto been centred on the cathedrals and monasteries, were now acquired to enhance the music establishments of kings and princes, popes and bishops, much as works of art were acquired.

Inevitably this new patronage had its effect on written music. Gregorian chant had often been the thematic base for writing music of the mass, but now secular tunes were used by such composers as Josquin des Prés (c. 1450–1521), chapel master to Lorenzo de' Medici, Louis XII and Maximilian I, who took the tune "L'Homme armé", and John Taverner (c. 1495–1545) in England who used the "Western Wynde". Gradually composers began to turn to secular forms, and this was of crucial importance for the evolution of Western music. Composers turned to poets for lyrics, the madrigal was born, and the *chanson* (which corresponded in France to the madrigal) received a further valuable stimulus to its development.

Vocal and instrumental music

According to Baldassare Castiglione's manual for courtiers *Il Cortegiano*, singing was permitted as an aristocratic accomplishment for the perfect gentleman. Castiglione went on: "Harmonious too are all keyed instruments; and with ease many things can be performed on them which fill the soul with

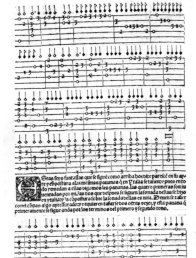

2 John Day (1522–84) published his *Psalter* in 1560 containing this illustration of musical instruction in the home. The *Psalter* was issued after the requirement of 1559 that music in the reformed Church be "a modest and distinct song . . . as plain understanded as if it were read without singing". Day was an exile during the reign of Mary Tudor and, while abroad, studied printing.

3 Jean d'Ockeghem (c. 1430–95) was one of the most important Flemish composers. His 36-part motet was held to be one of the wonders of the world. He was master of the king's chapel in Paris in the reign of Charles VII (1403–61), the choir of which he is directing here. Among the pupils in his new music school was Josquin des Près.

4 Music for the lute was usually written in tablature, instead of staff notation, between the 14th and 17th centuries. In tablature there is no indication of pitch, either between one note and another or between the strings themselves, although there was a conventional tuning of the instrument. The horizontal lines show the string courses of the instrument, the figures (sometimes letters were used) indicate which finger is to be used for stopping the string, and the notes at the top indicate rhythm. Another form of tablature existed for keyboard music. The piece shown here is from a collection of the music of John Dowland (1563–1626), one of the greatest exponents of the lute. Lutenist to Christian IV of Denmark and then to James I of England, Dowland is chiefly remembered as the composer of exquisite songs with lute accompaniment.

1 Royal and noble courts offered regular employment to musicians, whose duties included taking part in entertainments staged for visiting dignitaries. The court ballets or masques, of which the entertainments often consisted, were of Italian origin and included singing, dancing, mime and elaborate staging, usually on an allegorical theme. One such was the *Ballet des Polonais*, represented in this tapestry. It was given at the French court in 1573 in honour of Polish ambassadors, and the music was written by Orlando di Lasso. A group of musicians is shown here seated on a stage machine. They are playing an antique lyre (top), cornett, lute and viols. Violins were also used and three singers took part.

5 The racket was a double reed instrument that preceded the bassoon. It was an attempt to compress the necessary length of tube for an instrument of the bassoon's compass into the smallest possible space; but it gave a muffled sound.

the sweetness of music". It is not surprising that one of the stringed instruments of the time, a *lira da braccio* [10C], made by Giovanni d'Andrea of Verona in 1511, has a Greek humanistic inscription on its back: "Song is doctor to the pains of Man".

Side by side with these developments in vocal music went developments in instrumental music. Lute, harpsichord, organ, viols [7], and many kinds of wind instrument [5] were in common use, and with aristocratic encouragement began to be further exploited. Wind instruments were thought unsuitable for gentlemen to play, for as Castiglione wrote, they were instruments "disdained by Minerva", although even that was to change. Wind bands were, however, an essential feature of ceremonies and festivities in aristocratic circles, and professional musicians were retained to provide them.

Italian and English Renaissance

The Flemish migrants' domination of the places to which they went lasted only a few years, but they acted as stimulants to the native musicians. In Italy, for example, the Flemings can be said to have founded the Roman classical tradition, the Venetian choral tradition and the madrigal, but it was the Italians themselves who brought them to their full development.

For one idyllic moment, however, there was a kind of balance between the north and south of Europe; between polyphony – several voices sounding in independent musical parts at the same time – and melody; between words and music; between what had gone before and what was to come. That golden age was the blossoming of the Renaissance. The advent of music printing [2] gave it impetus, and the upheaval of Reformation and Counter-Reformation influenced its development into paths that it might not otherwise have followed.

In England the Renaissance arrived late, although this did not prevent an English composer, William Byrd (1543–1623), from being one of the pioneers of keyboard music, and in many respects England was especially lucky, since the accession of Elizabeth I gave the country peace and stability in order to unite and assimilate its new nature.

Giovanni Pierluigi da Palestrina (1525–94) summarized in his work the achievements of polyphonic music – the art of combining two or more voices in harmony with one another – and brought to a climax a tradition that had begun in the 9th century and subsequently never improved. He also represented the rise of native Italian talent in Rome, which had previously been dominated by Flemish musicians. Indeed, almost the entire musical foundation of St Peter's was, at one stage, Flemish. Palestrina wrote about 100 masses; one of the most beautiful is that dedicated to Pope Marcellus II, seen here receiving the music.

6 Orlando di Lasso, (left, foreground) was born in Mons, travelled as a young man to Sicily and Milan and was choirmaster at St John Lateran, Rome. He returned home and settled in Antwerp in 1555. Soon afterwards he entered the service of Duke Albrecht of Bavaria and spent the rest of his life there and in the service of his successor, Duke Wilhelm. He is seen here directing a concert at the Bavarian court. He contributed to all the sacred and secular vocal forms of his time, and ranks with Palestrina and Byrd for the quality, variety and sheer quantity of his output. As such, he is undoubtedly one of the finest composers and practising musicians of his day.

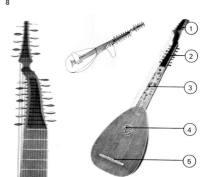

8 The theorbo was a bass lute, shown here sideways, in the playing position and front view. Five pairs of bass strings were fitted to the extra pegbox [1] and seven to the main one [2]. There were frets [3] spaced down the fingerboard, an elaborately carved soundhole or rose [4], and a bridge [5] to which the strings were attached. The performer was unable to play the theorbo at any great speed so it was more suited to accompaniment work than for use as a solo instrument.

7 The viol family, of which this bass [A], tenor [B] and treble [C] are examples, succeeded the medieval fiddles. Viols emerged during the 15th century and survived into the 18th, when the family strikes the string. Because of its small size and intimate sound the clavichord long remained a favourite domestic instrument, although it was replaced by the harpsichord and eventually by the piano. as a whole was displaced by the violin family. The viols had movable frets of gut on the fingerboard, and usually six strings, although the number varied. The most important difference from the violin family was the way the bow was held underhand resulting in a rather soft, unaccented sound. The much more incisive quality of the violin and related instruments rendered the viol obsolete.

9 The strings of the clavichord are struck, as on the piano – but unlike the harpsichord, whose strings are plucked. The key [1] is depressed; it pivots [2]; and the metal tangent [3]

9

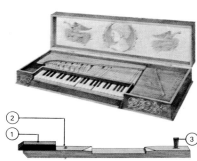

10A

10 The cornett [A] was made of wood or ivory and was a cross between a woodwind and a brass instrument, because although it had fingerholes it also had a cupped mouthpiece which was vibrated, "reed" fashion. The Renaissance fiddle [B] and the *lira da braccio* [C] were two of the forerunners of the violin. The fiddle usually had five strings, as did the *lira*, but in addition the *lira* had two open strings off the fingerboard, which could be bowed or plucked, as required.

Florentine art 1400–1460

The opening years of the fifteenth century in Florence were marked by costly and alarming wars, which the city survived somewhat luckily, and by political upheaval. These circumstances led to high taxation, the ruin or exile of several prosperous families and a great decline in the city's absolute wealth; but they also provoked a resurgence of patriotism, a consciousness of the city's cultural traditions and supposed Roman origin, and public competition between patrons.

The cathedral dome
The building of the dome [1] of the Florentine cathedral by Filippo Brunelleschi (1377–1446) symbolized the new condition of the city; it seemed the realization of an impossible civic dream, it was achieved by the study of medieval and Roman engineering, and inevitably it was compared with the greatest triumphs of Antique architecture. The dome is not classical in detail, except in its latest elements, the lantern and the four semicircular exedrae which act as buttresses round its drum; and it is characteristic that even these, formed as they are out of classical

pilasters, capitals, entablatures, scrolls and shell-niches, are not classical in function. Brunelleschi solved with extraordinary confidence the problem that remained fundamental to most later Renaissance architecture of applying the Antique formal vocabulary to building types of a new age. To some extent he was helped by the example of Florentine Romanesque architecture, notably the Baptistery and SS. Apostoli. His determined use of traditional local materials, a dark grey stone known as *pietra serena*, brick and whitewashed stucco, even coloured-marble facing, also meant that in their colours his buildings were much more closely related to the vernacular than to antiquity. The strong contrasts inherent in these materials gave his buildings very distinct articulation of their parts [8].

Florentine sculpture
Clarity of structure and evidence of intensive study of the Antique also characterize the mature sculpture of Donatello (1386–1466), whose "Saint Mark" (1412) [Key] made the first forceful statement *all'antica* (in imita-

tion of the Antique) – a statement about the nobility and gravity of man such as Cicero might have made, and in a language which revived that of Roman sculpture [2]. It was followed by a long series of saints and prophets which peopled the public buildings of Florence with heroic, alert and intensely personal beings, each an embodiment of a particular virtue.

Donatello's work is not dependent for its expressiveness on the beauty of craftsmanship and rich materials; in this respect he exemplifies an aesthetic revolution, enunciated in Alberti's essay *On Painting* (1435), which advocated a revival of the Antique standards of simplicity, even austerity, and eloquence. But Donatello's contemporary Ghiberti (1378–1455), a much less radical character, found an elegant compromise between Gothic ideals of beauty and the new style. A matchless craftsman, he won for this reason the competition (1401–2) for the first set of gilt bronze doors for the Baptistery; their design and details owed practically nothing to antiquity. But the second set – the "Paradise Doors" – are much more

1 **The Cathedral, Florence,** was begun by Arnolfo di Cambio *c.* 1300. By the early fifteenth century construction had reached the octagon. Advice was sought in all directions for covering the 40m (130ft) opening before, in 1420, Brunelleschi's solution was accepted: a dome of double herringbone brick shells and stone ribs, which could be built without scaffolding. The dome was completed in 1436, and the lantern on top begun in 1446.

2 **"David" by Donatello** (*c.* 1435) was made for the Medici. Sensual and pagan, it was the first convincing revival of the canons of the classical nude. The head derives from a Roman "Antinous".

3 **"Tribute Money" by Masaccio** (*c.* 1427) is the dominant scene in a cycle of frescoes in the chapel of the Brancacci, S. Maria del Carmine, Florence, detailing the story of Peter. The series was begun *c.* 1424 by an older artist, Masolino. The subject was interpreted by theologians as one of the main proofs of the supremacy of the pope as Vicar of Christ.

4 **"Tarquinia Madonna" by Filippo Lippi** (1437) was an early work, painted in tempera on panel, after a period of close imitation of Masaccio and a journey to north Italy. A knowledge of Flemish painting accounts for the informal, domestic treatment of the subject and the dark, rich colours.

lucid in design and participate in the revival in many ways; and in the conspicuous use of mathematical perspective, effectively a rediscovery of Brunelleschi's, they show where his new allegiance lay [7].

Florentine painting

Perspective was very quickly exploited by the youngest of the pioneers, the painter Masaccio (1401–28). It allowed him to place objects in apparently rational relationships in space; but he saw that the visual laws it expressed, when applied consistently, controlled not only the boundaries and intervals of space but also the drawing of forms to suggest three dimensions. The result was realism in a new sense, the realism of structure. But the same way of thinking, about space and form as unities, also led him to the first true pictorial light, unified and rational to the extent that shadow was its inevitable companion. In his "Tribute Money" [3] light as much as linear perspective produces a convincing reality. Like Brunelleschi, he was inspired by Tuscan tradition, particularly Giotto, and by antiquity, in his case a reper-

tory of realistic and expressive forms. His art has the same tone of austerity and moral seriousness as Donatello's.

No painter changed the course of his art in so short a life as Masaccio did. He simplified formal problems to make them amenable to laws, and some of his followers, such as Paolo Uccello (1397–1475), continued in this way; most, however, applied his discoveries to complexities more in line with both Gothic traditions and natural appearances. Fra Angelico (c. 1387–1455), indeed, owed as much to the example of late Gothic painters such as Lorenzo Monaco (died 1424). Filippo Lippi (c. 1406–69) [4] and Domenico Veneziano (c. 1410–61) further modified the new style towards a descriptive naturalism, the former by responding to Flemish art, the latter by an intenser study of reality and especially of light. Domenico's "Saint Lucy Madonna" [9] represents the most advanced stage reached in Florentine painting by the mid-fifteenth century, and it exemplifies the dominant type of altarpiece – in a single, rectangular, window-like frame – then current.

"Saint Mark", by Donatello (c. 1412) was made for the Weavers' Guild for their niche on the guild hall, Orsanmichele. Donatello imitates Antique statues not only in details, such as drapery folds, but also in the whole design. The balanced movement of the figure's structure (contrapposto) is expressed with clarity through and by the superimposed drapery, and there is no trace of the swinging posture and elegantly curved forms characteristic of late Gothic sculpture. The solemnity and intensity of expression is typical of Donatello, but his rugged simplicity is now exaggerated by the loss of gilt details and weathering.

7 Solomon and Sheba by Lorenzo Ghiberti, from the "Paradise Doors" of the Baptistery (1425–47) (gilt bronze) is one of ten Old Testament reliefs from the third set of doors (his second), all commissioned by the guild of businessmen who had responsibility for the Baptistery. The reliefs were modelled in wax and cast separately,

chased to a very high finish and then set into a massive framework. A rich leafy border frames the opening. Ghiberti, one of the most notable figures of the early Renaissance, left Florence in 1400 but was recalled a year later to take part in a competition that led to his casting the Baptistery's second pair of doors.

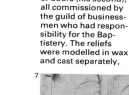

5 The Palazzo Medici-Riccardi was built by Michelozzo c. 1444–60 for Cosimo de' Medici who had rejected a design by Brunelleschi as being too ostentatious. Michelozzo's design conforms superficially on its exterior to the traditional Florentine merchant's house, although a heavy classical cornice relaces the projecting eaves. Inside it has a splendid Renaissance arcaded court.

6 San Lorenzo by Brunelleschi was begun c. 1418 with the Medici Sacristy, attached to the left transept. The church was largely funded by the Medici family (Cosimo is buried under the central dome). Although the façade was never completed, Michelangelo added the New Sacristy to the right transept.

8 The Pazzi Chapel, Santa Croce, Florence, was designed by Brunelleschi c. 1430 to be both the private chapel of a wealthy family and a chapter-house. The plan, a lateral domed rectangle with a smaller domed chancel, is a development of Brunelleschi's earlier Sacristy in San Lorenzo (1420–28). But the elevations articulated characteristically in dark pietra serena on white stucco are so much more logically composed that it is generally considered his most perfect building. The exterior is only partially to his design.

9 "Saint Lucy Madonna" by Domenico Veneziano (c. 1445) is an altarpiece of the type known as a sacra conversazione, which implies a close and informal relationship, spatially and psychologically, between the Madonna and the saints. The single panel generally supplants the polyptych, but its hierarchical division is often preserved, as here, by a subtle design of painted architecture. The perspective construction is extremely sophisticated and the colour and light are exceptionally unified. A predella with five small narratives of the lives of the saints was formerly attached.

The Flemish Renaissance

In the years just before 1400 the two most vital artistic centres in northern Europe were far apart, in Paris and Prague, but the distance made strangely little difference because of a widespread style in court art known by two unsatisfactory but descriptive titles, International Gothic, or the Soft Style, which was also to be found in London and Lombardy. A new beginning sprang from a dynastic event, the union under Philip the Bold of the Duchy of Burgundy with Flanders. This north-south (Bruges-Dijon) axis brought together artists from Flanders, Holland and Germany to form a school that is conventionally called Flemish.

From naturalism to realism

The first phase of the patronage of the dukes of Burgundy was centred on Dijon. Philip the Bold (1342–1404) brought the painter Melchoir Broederlam (died c. 1410) from Ypres and the sculptor Claus Sluter (died c. 1406) from Haarlem and Brussels. The main surviving work by Broederlam is the painted part of a very large altarpiece at Dijon. It is of a kind found all over northern Europe in the fifteenth century, from Burgundy to Poland, and quite distinct from Italian traditions: the main part is polychrome and gilded wood sculpture with a highly wrought Gothic frame, which is enclosed by wings with painted narratives.

Sluter's greatest work was his last, the so-called "Well of Moses" (a Calvary group) in the Charterhouse of Dijon [1]; he developed a strikingly massive and lifelike style, which was to be a characteristic of Burgundian sculpture for a generation.

Philip also employed the most brilliant miniaturists of the time, the Limbourg brothers (from Nimwegen, trained in Paris); but it was for Philip's brother, Jean de France, Duc de Berry, that they made their most celebrated manuscript, the *Très Riches Heures*. The narratives in this book reflect the cosmopolitan, especially Italianate, nature of French art at the end of the fourteenth century, but the calendar pages have a startling naturalism that is quite new. It is upon a firm tradition of panel painting, but newly inspired by sculptors and miniaturists, that the "Flemish" style is based.

The second phase of Burgundian patronage is centred on Bruges, the commercial and banking capital of the north, where Philip the Good (duke from 1419) and his chancellor Rolin employed Jan van Eyck (died 1441) [2] from 1426, the year in which Jan's brother Hubert died. The enormous painted polyptych "The Adoration of the Lamb" in the Cathedral of St Bavon, Ghent, seems to have been begun by Hubert and finished by Jan in 1432, and it was commissioned by a merchant; Jan also painted for churchmen, the bourgeoisie, courtiers and visiting Italians. His style, still more naturalistic than the Limbourgs' or Sluter's, was technically the most sophisticated and luxurious of its period [3]; while not the first to use oil, he put it to great effect in obtaining strength and transparency of colour, closely observed descriptions of light and texture – as in the sheen on a pearl – and minuscule detail. A contemporary artist of almost equal importance, Robert Campin (c. 1375–1444), worked mainly in Tournai in a tougher, less refined style with strong, expressive and realistic effect; and this style was developed

1 **Claus Sluter's** "The Prophet Zachaiah" (c. 1401) is one of six prophets round the base, commonly called the "Well of Moses", of a Calvary group of which only fragments survive, begun in 1395 for Philip the Bold. The figures were painted by Jean Malouel; such collaboration between major painters and sculptors was common in the 15th century, when unpainted sculpture was the exception to the rule.

2 **Jan van Eyck's** "The Madonna of Chancellor Rolin" (c. 1435) was in the Cathedral of Autun, the home town of Nicholas Rolin, who was also one of the principal patrons of Rogier van der Weyden.

3 **"The Arnolfini Wedding"** (1434) is by Jan van Eyck. The signature states "Johannes de Eyck fuit hic", meaning "Jan was here" – that is, was witness to what is clearly a marriage. It is believed that the couple are Giovanni Arnolfini, a merchant from Lucca, and Giovanna Cenami. The picture is filled with appropriate symbolism: the marriage-candle a reminder of Christ's presence, the mirror, signifying purity, and the dog fidelity. The statuette on the back of the chair is of St Margaret, the patron saint of childbirth.

by his pupil Rogier van der Weyden (1400–64) [5], who established in Brussels the most important mid-century workshop.

Enrichment of tradition

Exchange of artistic ideas between Italy and the Low Countries in the fifteenth century left the Flemings richly in credit. The three founders of the Flemish school were followed by a host of painters operating in the Low Countries in the second half of the century, who enriched the tradition and maintained its prestige; but the Flemish "rebirth" also revitalized German painting, which for the first half of the century was in general an independent development from the Soft Style. The decisive change here came through the influence of Rogier, particularly upon Martin Schongauer (c. 1450–91) who worked mainly in the upper Rhine around Colmar. Schongauer was not only, as painter and draughtsman, the principal inspiration for Dürer, but also the first great engraver (from c. 1470) [Key]; through this medium his influence was international.

German sculpture, by the end of the cen-tury the liveliest of northern schools, was also transformed by Dutch influence, mainly through the activity in several German centres of Nicholas Gerhaert of Leyden (flourished 1463–73), who brought to the Gothic style a richer variety of surface through strong undercutting. The masters of this final quasi-Baroque stage of Gothic sculpture were the exuberant Veit Stoss [4] (c. 1440–1533, born Nuremberg but 1477–96 mainly in Cracow) who, typically, worked as much in limewood and boxwood as in stone, and the more restrained Tilman Riemenschneider (c. 1460–1531).

Virtuosity in architecture

Architecture was far from conservative in this period. Builders pursued virtuosity and enormous scale in their work, as in a number of Netherlandish town halls, notably Brussels (begun 1402). The most inventive results were in spectacular vaulting systems, above all in England, especially by delicate fan-vaults [7], sometimes with pendants, and in Germany by ever more complex geometry of rib patterns.

Martin Schongauer's engraving of St Michael (c. 1475) is closely related in style to contempor-ary German wood sculpture. Schongauer regularly signed his work and is in fact the earliest northern engraver known by name. His work formed one of the major channels of northern influence on Italian painters such as Giorgione, while one of the earliest works by Michel-angelo, the "Tempta-tion of St Anthony" (c. 1490), was a painted copy of a similarly spiky, fantastic design. Schongauer published more than 100 prints, covering a wide range of sacred and profane subjects; their com-bination of rich inven-tion and disciplined clarity makes them particularly attractive.

4 Veit Stoss's "The Archangel Raphael and Tobias" (1516–18) was com-missioned for the Car-melite church in Nürnberg by a Flo-rentine silk mer-chant, who undoubt-edly chose the sub-ject because in Florence Raphael was popularly the guardian angel of tra-vellers. In this sculpture the natural beauty of the material is revealed, and not covered by the usual polychromy and gold. It is one of the latest, most vivacious works of Veit Stoss.

5 Rogier van der Weyden's "The Mag-dalen" (c. 1451) is the right wing of a triptych, with Christ in the century flanked by the Virgin and St John, painted for Jean de Braque. The trip-tych was made imme-diately after Ro-gier's return from Italy and the sim-plicity of style and sense of order may reflect his experiences there.

6 Hans Memling (c. 1440–49), a German pupil of Rogier, was from 1465 the most suc-cessful painter in Bruges, where he en-joyed an interna-tional clientele. Several of his por-traits were of Italians, such as this one – "Portrait of an Italian" (c. 1480) – and these pictures exerted considerable influence back in Italy. His land-scape style was much imitated there. The sitter here may be a medallist, but is more probably a collector.

7 King's College Chapel, Cambridge, was begun for Henry VI in 1446; after interruptions the fabric was completed in 1515. The fan-vault is the work of Henry VII's mason, John Wastell. Of all English fan-vaults this is the most majestic and the most happily balanced between clarity and finesse of design. Although of massive construc-tion its taut curv-ature and brilliantly resolved geometrical web of ribs, carved in relief, give an illusion of weight-lessness. The screen dates from the 1530s.

Italian art 1450–1490

The stylistic revolution in Italy in the second and third decades of the fifteenth century had results that were in two crucial respects different from those that followed Duccio's and Giotto's achievements in the fourteenth century. First, the revolution's effects were quickly felt in almost all centres of any artistic importance; second, a profusion of artists of talent and imagination ensured that what followed was not reassessment, nor merely exploitation, but a continuing ferment of activity with a sustained momentum of exploration and invention.

Increase in secular patronage
One reason for this situation was the wider opportunity offered to artists in the second half of the century by a multiplicity of centres of patronage. These may be divided crudely into two types: on the one hand were the old nominal republics such as Florence and Venice, where a prosperous bourgeoisie with a patrician upper crust offered at least as much patronage as the state and, on the other, the petty principalities, such as Milan, Mantua, Ferrara, Rimini, Urbino and

Naples, where a confused mixture of despotism and ancient aristocracy tended to focus patronage on a centre such as a court, often with pretensions, as that of the dukes of Burgundy. A long period of comparative peace made these courts – among them the special case of Rome – much more significant artistically after the mid-century, although they were never entirely separate.

A change in status
These opportunities for artists were augmented by a slowly, and subtly changing attitude to the status and use of works of art. This was the period in which the battle to promote the visual arts to the level music held among the liberal arts was in general won. And independently of any philosophical change of this kind there was more lavish investment in culture for culture's sake; so if artists were not painting more fresco cycles in churches than before, they were producing more paintings and sculpture for private enjoyment. The quantity of artistic objects in domestic settings was significantly higher at the end of the century than at the beginning.

The result of these changes in demand is in harmony with new possibilities of supply that arose from technical and expressive progress; in general terms there was a wider diversity in the forms and functions of works of art in the second half of the century. The bronze statuette and the portrait-bust, for example, became firmly established in this period. These two types illustrate a general phenomenon – the meeting of new and mostly secular demands by the adaptation of earlier and mostly religious forms. The small bronze of mythological or allegorical subjects followed a long tradition of statuettes of saints and the composition of the sculptured portraits is best explained by reference to the continuing manufacture of reliquary busts. The evolution of each type was stimulated by the collecting of similar objects from antiquity. Similarly, the new large-scale secular figurative paintings, which decorated princes' palaces and patricians' houses were made possible by the painters' experience of ecclesiastical art. Thus one is reminded in Andrea Mantegna's Camera degli Sposi in the Ducal Palace, Mantua (1473–4), of

1 Antonio Pollaiuolo's "Hercules and Antaeus", a bronze statuette c. 1470–80, is similar in design to a small painting by Pollaiuolo (1429–98) on the same subject (Virtue) done for the Medici; the bronze was also most probably in the Medici collection. Although it has a triangular base it is only intelligible from one viewpoint; the all-round view was first achieved by Verrocchio.

2 In Andrea Mantegna's fresco "The Gonzaga Court" (c. 1470–74), Ludovico Gonzaga sits, surrounded by his family and court, receiving a letter. There are similar scenes on a second wall, and the ceiling is painted as an illusion with a circular opening to blue sky.

4 The courtyard, Palazzo Ducale, Urbino, was built 1464–72 by Luciano da Laurana (fl. 1468–82) for Federico da Montefeltro (1422–82), warrior, statesman, scholar and patron of the arts, later first Duke of Urbino. The broad proportions and subtlety of rhythm distinguish it from Florentine courtyards; the whole palace is spacious, most refined in its sculptural decoration and placed so as to command a stupendous view.

3 Villa Medici, Poggio a Caiano (Florence), was begun c. 1482 by Giuliano da Sangallo for Lorenzo de' Medici and completed for Leo X c. 1520. The curved external staircase replaced the original one which comprised two simple ramps. Close to the roads into Florence from the north, the villa was intended for receptions as well as the usual rural and summer pursuits; hence, probably, its unusual size and formality.

religious narrative and in Sandro Botticelli's "Allegory of Spring" [5] of a madonna and saints. More rarely – although both Mantegna (c. 1431–1506) and Botticelli (1445–1510) provide important examples – secular compositions by artists during this period are inspired by classical sculpture.

Styles of architecture

Architecture shows that diversification by cross-fertilization works in both directions. Giuliano da Sangallo (1445–1516) and Lorenzo de' Medici (1449–92) designed at Poggio a Caiano [3] the first of the great Renaissance villas, for a life-style inherited from the Middle Ages and fortified by re-reading the ancients. Similarly, Giuliano's building is a blend of traditional big farmhouse and selected classical elements – externally, the arched basement (recalling an amenity Pliny described in one of his villas) and the applied temple-front, a dignifying feature with a great future in domestic architecture. Conversely, the principal source of inspiration for the church designs of Leon Battista Alberti (1404–72) was

a secular form, the Roman triumphal arch.

These architectural examples also illustrate the adjustment of newly enriched styles to what were enduring problems and it is only too easy to overstate the degree of change. Churches, tombs, altarpieces and so on were of course made as prolifically as before. The diversity of styles was as great as that of art forms and showed similar vigorous continuity and change, so that generalization is unusually difficult. But in architecture it may be said that style was characterized by a more exact, complex and literate knowledge of antiquity; in sculpture and painting, on the other hand, typically in the work of the greatest sculptor of the period, Andrea del Verrocchio (c. 1434–88), or of the greatest painter, Giovanni Bellini (c. 1430–1516) [7], the antique was not the overwhelming inspiration it had been for their predecessors Donatello and Masaccio, but was replaced by the imitation of nature. Simplicity was superseded by complexity, austerity by refinement, and in these ways late fifteenth-century Italian art reflected the structure and values of its market.

Filippo Strozzi (1428–91), the most important Florentine patron after the Medici, built a palace grander than theirs. This marble bust of him (c. 1485) is by Benedetto da Maiano.

5 Sandro Botticelli's "Allegory of Spring" (c. 1478) was painted to be fixed above a couch in the town house of the young cousins of Lorenzo de' Medici. The earliest secular paintings are generally either wall decoration or on furniture.

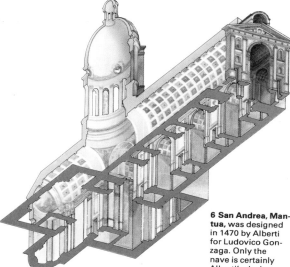

6 San Andrea, Mantua, was designed in 1470 by Alberti for Ludovico Gonzaga. Only the nave is certainly Alberti's design.

7 The "Madonna and Saints" of Bellini in S. Maria dei Frari, Venice (1488), is characteristic of the artist's dignity and restraint, with especially refined description of light and resonant colour. It was painted when he was turning, in his later work, to simpler more monumental forms. The classically shaped frame is original.

8 Andrea Verrocchio's Colleoni Monument commemorates Bartolomeo Colleoni who was *condottiere* of the Venetians. Verrocchio made his competition model for the monument in Florence and moved to Venice in 1483 to execute it. It was cast, after his death (1488), by Leopardi, who also designed the base.

Florence and Rome: the High Renaissance

Historical events around 1500 seem in general once again to concentrate artistic activity in Florence, Rome and Venice and reverse the earlier tendency to diffusion among many centres. Foreign invasions and power struggles within Italy did clearly have a disastrous effect upon patronage in courts such as Naples, Urbino and Milan. Indeed, the presence of Leonardo (1452–1519) and Bramante (1444–1514) in Milan, and the ambitious patronage of Lodovico Sforza (1451–1508), might have made that city one of the artistic capitals of Europe, but for the French conquest of 1499. Leonardo then returned to Florence, where for six vital years he and Michelangelo (1475–1564) were in direct rivalry; they were joined in 1504 by Raphael (1483–1520) [6, 7] and successively by several younger men, notably Andrea del Sarto (1486–1530).

In Florence, too, there had been a political earthquake, but the expulsion of the Medici in 1494 and the formal revival of the republic (until 1512) in fact left a prosperous oligarchy in power and the new state institutions provided some of the greatest oppor-tunities any Florentine artist had so far been offered, especially in the commissioning of two enormous battle-pieces from Leonardo and Michelangelo for the new council chamber (neither was finished).

Bramante and St Peter's

After 1499 Bramante went to Rome. His great opportunity came in 1506, when the energetic Julius II (pope 1503–13) [2] chose him as architect for New St Peter's; this building is unavoidably a symbol of the power and centralization of the spiritual Church, and for Julius its rebuilding was not so much a practical necessity as a gesture of regeneration. Bramante's plan [Key] expresses these ideals with majestic clarity, and the piers of the crossing that he con-structed established the colossal scale of the present building.

Michelangelo had already been in Rome; he returned to work for Julius, first (1506) on a projected tomb of a massive size and pro-lixity of sculpture undreamed of since the mausolea of antiquity [4], and next (1508–12) on the painting of the ceiling of the Sistine Chapel [1]. Before the Sistine ceiling was complete Raphael had decorated for Julius the first of the suite of rooms in the Vatican Palace, the Stanze; and the Stanza della Segnatura, painted with ideal examples of the faculties of the mind, was immediately followed by his Stanza d'Eliodoro (1511–44), decorated very differently with dramatic narratives from Church history.

Pope Julius made Rome an artistic centre as important as Florence. He was succeeded by Leo X (pope 1513–21), son of Lorenzo de' Medici; his taste for the arts was more personal, more refined than visionary, but many projects such as St Peter's and the decoration of the Vatican Palace and Sistine Chapel were continued. And artists enjoyed the patronage of cultivated members of the papal court and their bankers on a scarcely less lavish scale and far above fifteenth-century levels; this was a great period of palace building and urban redesign.

The architecture of the High Renaissance was not only Bramante's; the work of the Sangallo family is scarcely less important and Bramante's position at St Peter's, and the

2 Raphael's painting of Pope Julius II was executed be-tween 1511 and 1513.

1 Michelangelo's painting of the ceiling of the Sistine Chapel in the Vatican was commissioned by Pope Julius II, who asked Michelangelo to paint the 12 Apostles. The artist, if we be-lieve his account, said this was a poor scheme and was given a free hand. The prin-cipal figures, on the supports of the dome, are 7 prophets and 5 sibyls, representing revelation to Jews and Gentiles. Down the centre is a sequence of 9 Old Testament subjects, from the Creation (shown here) over the altar to the Drunken-ness of Noah, illustrat-ing an epoch of religious history not previously represent-ed in the chapel. The first half was unveiled in August 1511; this second part was completed a year later.

3 Bramante's Tempietto in S. Pietro in Montorio, Rome, was built as a shrine over the traditional site of St Peter's martyrdom. Bramante intended that it should be tightly en-closed in a circular courtyard. The de-sign departs from clas-sical circular temples in its high drum with windows between colonnades and dome.

4 Michelangelo's "Moses" (1513–16), in S. Pietro in Vincoli, Rome, was made for the tomb of Julius II. According to the second plan (1513) it was to be set fairly high up on the tomb. Its present setting, at floor level, is in a much reduced monu-ment – a compromise reached, after many frustrations, during the 1540s.

leadership that this almost inevitably entailed, were inherited by Raphael; Michelangelo's career as architect began about 1516. All of these, as was normal in the Renaissance, were trained in some other art and this fact is yet another cause for variety.

New intensity of feeling

A High Renaissance architectural style does exist, however: it is distinguished by total mastery, now, of the language of antiquity, by clean and measured design, by solemnity, mass and – increasingly – richness of materials. The units tend to be larger in proportion to the whole, the forms denser and more plastic than in the earlier Renaissance, and the same characteristics are found in painting and sculpture. Here, however, because figurative art offers a direct comparison with the norm of nature, the effect is, by artifice, superhuman. Figures move, with violence or deliberation, in postures that far exceed natural behaviour; in their heroic and idealized forms they appear a race of supermen. This style is eloquent, rhetorical, with a new intensity of feeling.

In painting, largely through Leonardo's example, a new range of dramatic effects, of action and of physical presence of form, was made possible by descriptive discoveries about light and atmosphere and especially by powerful and unified effects of tonal contrast (chiaroscuro). In all three arts drawing had a new role (again, this was largely Leonardo's doing); the quantity of preparation in drawing was greatly increased and exploration in drawing much more conscious, and these changes express the artists' ambition towards statements that are at once more considered and more novel.

The artist's status

The four principal artists were each personalities such as to change definitively the social position of the artist (Raphael lived in a palace designed by Bramante for a cardinal). And the maturity, the apparent perfection of their forms and the inventiveness of their compositions gave their works the status of classics. These artists earned the title Divine, hitherto reserved for poets, and their style was to be labelled the Grand Manner.

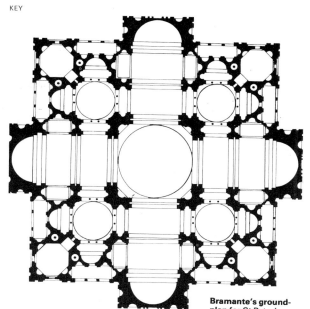

Bramante's ground-plan for St Peter's.

5 Leonardo da Vinci's "Virgin and Child with Saint Anne" (c.1510) first appeared in a cartoon (now lost) shown in Florence in 1501. The composition was slightly different from this version.

6 Raphael's "Paul Preaching at Athens" (1515–16) was one of a set of ten coloured cartoons (of which seven survive) for tapestries, made in Brussels, with which Leo X intended to complete the decoration of the Sistine Chapel. The weaving technique required that each scene, from the lives of Saints Peter and Paul, be designed in reverse. The surviving tapestries are in the Vatican.

7 Villa Madama, Rome, was designed by Raphael in 1517 for Cardinal Giulio de' Medici. This garden loggia, decorated after Raphael's death by his pupils, particularly the brilliant decorator Giovanni da Udine, is a rare example of form, function and decoration all imitating the classical.

8 Leonardo's "Heads of Warriors" (1503–4) is a detailed study of the expression of horsemen for the "Battle of Anghiari", a mural commissioned by the Florentine republic and never completed. It is representative of the new drawing techniques that were invented by Leonardo as part of an intensive preparatory process.

257

The High Renaissance in north Italy

Many of the opportunities and interests of Roman and Florentine artists around 1500 were shared by their contemporaries in northern Italy; but there too it makes no sense to think of the High Renaissance in terms of generations. Giorgione (c. 1475–1510), with whom a new phase of Venetian painting seems to begin, was outlived by the much older Giovanni Bellini.

Transfusions of style
The legacy of Leonardo's most active periods in Lombardy (1483–99, 1506–8) was an assurance that north Italian artists would be as dedicated as any Florentine to forceful presentation and animation of all subject types, and there were continued migrations, in both directions, that were direct and even intended causes of style transfusions. The feeling – correct, in the medium term – that a Roman golden age had passed by the early 1520s dispersed Raphael's pupils, among them Giulio Romano (1499–1546), who fulfilled the ambitions of Federico Gonzaga [1] to make of Mantua a new Rome. And then the catastrophe of the sack of Rome (1527)

benefited several north Italian cities, notably Venice which acquired Jacopo Sansovino (1486–1570) [Key].

The Gonzaga family in Mantua and the Este family in Ferrara escaped the momentary eclipse that most Italian courts suffered around 1500 and continued their tradition of enlightened, if strong-willed, patronage. In such centres there was much local talent, but it is characteristic of a new situation that the major artists were employed at a distance – Titian (c. 1487–1576) in Venice or Antonio Correggio (c. 1490–1534) in Parma. This was briefly an artistic centre of the first importance not because of peculiarly favourable, temporary patronal conditions but rather because of the residence there of Correggio and Parmigianino (1503–40). Andrea Palladio (1508–80) in Vicenza exemplifies best of all the phenomenon of the great "provincial" artist. None of these, however, was provincial in the sense of artistic isolation, partly because distances are short across the roughly triangular area between the Apennines, the Alps and the Adriatic. In the sixteenth century this area was characterized

artistically by a large number of minor centres of activity – such as Bergamo or Bologna – whose fortunes fluctuated, and by the continuous dominance of Venice.

Venice offered civic and corporate patronage on an enormous scale in this period, and in all the visual arts; in architecture the new library and mint commissioned from Sansovino [7] are outstanding examples. These new buildings, together with the old Doge's Palace, were decorated lavishly by painters and sculptors.

Sumptuous decoration
In general the style of Venetian decoration is individual in its sumptuousness and harmoniously interwoven complexity, festive as Venetian music of the same period [4]. Personalities have rarely been more important in defining the character of a city's art. Giorgione seems to have worked, immune to market pressures, within a restricted range of his own choice; and by focusing on particular problems, such as transient states of nature in the "Tempesta" [2] or the description of sentiment in portraits, enlarged rather as

1 Federico Gonzaga, painted by Titian c. 1528, was the son of Isabella d'Este and spent his youth at the court of Julius II. He was an important patron of Titian's and introduced him to the emperor Charles V, whom Titian later painted. This portrait, in the relatively new three-quarter length format, set a fashion for including dogs: this one is probably a symbol of fidelity.

2 Giorgione's "Tempesta" (c. 1507) is one of the few attributions to Giorgione sustained by evidence. The subject has given rise to much speculation: perhaps Giorgione, inspired by the legend of the Greek painter Apelles painting the unpaintable – a thunderstorm – took this challenge as his main theme. The painting ranks among the earliest Western landscapes.

3 Palladio's Villa Rotunda was begun about 1550 in Vicenza. It is exceptional among villas for its perfect symmetry, designed to take advantage of the view from the hilltop site. The dome, a new feature in domestic buildings, surmounts a circular hall. The classical portico had appeared before at Poggio a Caiano, Florence, the Medici villa designed by Guiliano da Sangallo and completed in 1520 for Pope Leo X. The six-columned porticoes on all four sides are an elaboration of Palladio's basic design.

4 Paolo Veronese painted the fresco decoration (c.1562) in the Villa Barbaro, Maser (near Treviso), designed by Palladio. The villa is Palladio's most perfect blend of palace and farmhouse. Veronese was his ideal decorator. His illusionism takes amusing but slight liberties with the real architecture. Veronese's brilliant colour and realistic observations of landscape make the frescoes quite remarkable. Some have religious subjects – such as the "Marriage of Saint Catherine" seen here. Some are landscapes and others contain portraits of the owner's family, friends and servants.

Leonardo did the whole notion of what a work of art might be. Then Titian – very different, especially through the universality of his art – with a succession of masterpieces over a very long career kept Venetian painting concentrated on problems such as those of atmosphere and colour, dramatic expression and characterization [6], which tended to be neglected in central Italy. Titian, largely through patronage, became the most esteemed painter in Europe.

Correggio [5], whom Titian admired, was in some respects a yet more original artist and of all sixteenth-century painters he came nearest to a baroque style. His were the most advanced experiments with movement and emotion and – again following Leonardo's lead – he achieved through naturalism and immediacy a newly intimate relationship between works of art and their spectators.

Palaces and villas

The other great "provincial", Palladio, has European stature for different reasons. Building activity generally gathered pace in north Italy from about 1525, especially in civic and domestic fields. Giulio Romano, Sansovino and Sanmicheli (c. 1484–1559), another transmitter of Roman experience, especially to Verona, built palaces and villas, and Palladio was much indebted for his designs to their example.

The growth of villa-building in the period was extraordinary but natural; the patriciate, especially from Venice, tended to put their capital increasingly into land and wanted to live on their estates. Palladio's genius lay in a rare combination of theoretical equipment and flexible, common sense. Very practical by temperament and also imaginative, he could provide for a new clientele with widely varied aspirations and needs, a range of building types that became (partly through publication) almost definitive. They ranged from a thinly disguised farmhouse, the wings of which were barns, to an impressively spreading rural palace [3, 4]; and they fulfilled a social need no less perfectly by their evident use of classical building styles as a mark of social distinction. The architectural needs of gentlemen in Europe (and America) were met by two centuries of Palladianism.

Jacopo Sansovino's "Neptune" typified Venetian sea power.

5 Correggio's "Adoration of the Shepherds" (c.1528–30) is the outstanding example of the artist's illusionistic painting of light. The informal, eccentric design of this scene is calculated to make the spectator feel he completes the group.

7 Jacopo Sansovino's Library of S. Marco, Venice, was begun in 1536. The site required a building filling the whole length of the Piazzetta, incorporating an arcade (as a public amenity, and to accommodate shops), reflecting that of the Doge's Palace which it faces.

6 Titian's "The Rape of Europa" (1559–62) is one of a set of mythologies painted for Philip II, King of Spain and Emperor, Titian's principal patron in the later part of his career. The artist seems to have been free to choose his subjects.

8 Palladio's S. Giorgio Maggiore, Venice, was begun in 1566. This was Palladio's first church. Like many architects working in Venice he was inspired by St Mark's to use columns in large numbers, but the interior is exceptionally light in colour, well-lit and unadorned.

The German and Netherlandish Renaissance

The tides of stylistic influence flowed northwards in the sixteenth century as the prestige of Italian culture compensated for Italy's loss of political power. Albrecht Dürer of Nuremberg (1471–1528) [1] was one of the first northern artists to digest the Italian Renaissance achievement and the Antique itself, which he saw at first hand, mainly in Venice. The greater part of his work was in woodcuts and engravings [Key], to which he brought immense refinement of technique and capacities for narrative invention that made him the most imitated artist in Europe. The religious series issued for the popular market were more Germanic, more realistic in style, while the prints of humanist subjects, and the later panel paintings (from *c.* 1506), were more clearly of the Renaissance.

Grünewald and Holbein

A painter and draughtsman of equal imaginative power, but narrower range, was Mathias Grünewald (*c.* 1480–1528), court artist to the archbishops of Mainz and Brandenburg, who is best known for his prodigious altarpiece for the hospital of Isenheim, now Colmar (*c.* 1512–16) [4]. By expressive distortion, brilliant colour and technique, and extraordinary invention Grünewald painted with emotive power.

A cooler, and much more cosmopolitan character was Hans Holbein the Younger (1497–1543), who was born in Augsburg, matured in Basel [2] and later worked in London. His thoroughly Germanic beginnings were transformed by knowledge, probably first-hand, of north Italian High Renaissance art, and also of French and Flemish art. His later work, especially in England, is largely portraiture, where he combines an Italian sense of form and composition with dispassionate observation of detail and texture. In England, too, he worked as court artist to Henry VIII and established a tradition of miniature portrait painting; his principal native follower was Nicholas Hilliard (*c.* 1547–1619). German painting at the beginning of the century had another focus along the Danube between Bavaria and Austria. The major master of the Danube school was Albrecht Altdorfer (*c.* 1480–1538) who began, at about the same time as Giorgione (1475–1510) in Venice, to make paintings of which the main subject is landscape.

Painting in Flanders was centred on Antwerp, which supplanted Bruges as the commercial capital from about 1490. The founder of the Antwerp school was Quentin Massys (*c.* 1465–1530) who, like his German contemporaries, assimilated native traditions with Italian ideas. His collaborator Joachim Patenier (died *c.* 1525) was a landscape artist whose panoramic view of the world initiated a long Netherlandish tradition.

Rejection of local traditions

Massys had a group of followers whose complicated, neo-Gothic figurative painting has the dubious label "Antwerp Mannerist". But in Flanders, and still more in Holland, a contrary current impelled by a rejection of local traditions (except of technique) in favour of Italian is labelled more sensibly Romanist; chief among these were Scorel (1495–1562, Utrecht and Haarlem) and Heemskerck (1498–1574), and out of this school came a flourishing group of genre painters.

Netherlandish painting in the sixteenth

1 "SS. Paul and Mark" by Albrecht Dürer (1526) was one of a pair of paintings representing the Four Temperaments, which the artist presented to the city of Nuremberg. The other painting showed SS. Peter and John. Dürer combined the oil painting tradition of the north, emphasizing realism, with Italian monumental figure style. He was influenced by Bellini, whom he had seen in Venice, and by more recent studies of Raphael (probably via engravings), with whom Dürer corresponded. They also exchanged works.

2 Hans Holbein the Younger's "The Meyer Madonna" (1526) was painted for the burgomaster Meyer's castle near Basel. It was completed immediately before Holbein's first trip to England. The painting was later modified (about 1530) when Holbein added Meyer's first wife. Holbein, one of the most accomplished portrait painters of his time, was a close friend of Erasmus and painted three portraits of him. While in Basel, Holbein also worked on designs for wood engravings and illustrated the Luther Bible.

3 "Derich Born" by Holbein was painted in 1533 on the artist's second London visit. Derich Born was a Cologne merchant resident in London. He was one of the many Germans Holbein painted for the Merchants of the Steelyard before his royal service.

4 Mathias Grünewald's "Resurrection" (*c.* 1512–16) is a wing of an enormous altarpiece from the Antonite monastery of Isenheim which shows, when closed, the Crucifixion. On the first opening the Annunciation, Nativity and Resurrection are revealed; on the second opening the hermits Paul and Anthony and the Temptation of Anthony. The wings flank three carved wooden figures of saints. Grünewald rejected Renaissance ideas. The remarkable emotional intensity in his work is expressed in Gothic imagery.

century already has that character so familiar in the seventeenth of a bewildering quantity of minor masters but few of real stature. Two, however, stand out: Hieronymus Bosch (*c.* 1450–1516) and Pieter Bruegel the Elder (*c.* 1520–69). Bosch, who worked in apparent isolation in Brabant, was active well before 1500, but about that date began to enjoy European patronage (such as that of Philip, Archduke of Austria). His peculiar fantasies, based on popular proverbs and vernacular visual imagery, are moralistic comments on the human condition [6]; and to him is to be traced a fluid, transparent panel-painting technique continuous down to Rubens. Bruegel was one of the links in this tradition, but he looked still harder at nature and its seasons [5], at social classes, and enriched the content of secular painting.

Sculpture and architecture
A Renaissance style in sculpture began in south Germany around 1520, especially at Innsbruck and Augsburg; the exceptional Apollo Fountain at Nuremberg [8] by Pieter Flötner (*c.* 1493–1546) is clearly part of the

legacy of Dürer's Renaissance and begins a series of remarkable south German fountains ending with those of Hubert Gerhard (*c.* 1550–1620). Antwerp was the other main centre of northern sculpture and the training ground of Gianbologna (1529–1608), who spent most of his career in Florence. Buildings in a true Renaissance style are isolated freaks: they include the Fugger Chapel at Augsburg (1509–18), the Chapel of Sigismund at Cracow (1517–33) and the Castle of Breda (begun 1536), the last two by Italians. Clumsy imitations or superficial applications of Italianate detail are more common. A distinctively northern decorative style, spreading from England to Poland, is based upon the Mannerist engravings of the Antwerp school (particularly Cornelis Floris [1514–75]), with strapwork, scrolls and grotesque masks. But the most interesting architectural developments were in large, functional and stylistically unselfconscious buildings with many windows among the civic architecture of German towns such as the Rathaus at Bremen [7] and houses such as Longleat and Hardwick in England.

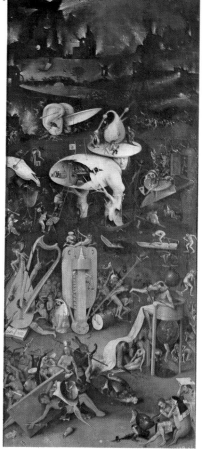

Albrecht Dürer's engraving "Portrait of Erasmus" was done in 1526. Desiderius Erasmus (1466–1536) was the most important figure in north European Renaissance humanism.

5 Pieter Bruegel the Elder's "The Cornharvest (August)" (1565) is one of an incomplete set of Months, of which three more are in Vienna and one in Prague. They were apparently painted as part of the decoration of a house in Antwerp. Rubens was later to paint this type of landscape.

6 Hieronymus Bosch's "Hell" (*c.* 1500) is the right wing of the "Garden of Delights" triptych. The left panel shows the Creation. With the wings closed one sees, in near monochrome, the Flood. Bosch's macabre fantasies seem to be meditations on the follies of mankind and their results.

7 Bremen Rathaus was built in the 15th century and is mainly Gothic in style. Its façade was later re-built, between 1609 and 1614, by a Flemish architect and it includes many characteristic features of Renaissance civic architecture. The high-pitched roof and the stepped gables show a strong Flemish influence, while the round-headed arcade and the decorative panels are typically Italian. Civic building during the 17th century flourished, illustrating the wealth and pride of the expanding commercial centres in northern Europe.

8 In the Apollo Fountain (1532) by Pieter Flötner the figure is about half life-size and at this early date is unique in northern sculpture for its classicism. This is partly based on Dürer's research into the ideal proportions of the human body (as in his "Adam and Eve" engraving, 1504), and partly (presumably) upon Flötner's own experience in Italy, where he may have seen the Apollo Belvedere in Rome. Flötner's work precedes all the known Italian examples of this pyramidal type of fountain, with sea monsters and putti positioned around the base.

The French Renaissance

The origins of French painting and sculpture of the fifteenth century have much in common with those of Burgundian and Netherlandish art; indeed, artists such as the Dutchmen Jean Malouel (died 1415) and the Limbourgs worked for both French and Burgundian courts. French painting, however, was rooted in a lively tradition of miniature painting, especially in Paris, which was also one of the main centres of the International Gothic style [1].

Naturalism and simplicity

Inspiration for a new beginning came from two directions, Italy and Flanders, and the most important artist concerned was Jean Fouquet (*c.* 1420–80) [2], who was in fact in Rome *c.* 1446–50. The combination of the new naturalism of the north with the simplification and lucidity of form and space of the south is without parallel in the mid-century. Fouquet, who appears to have been the principal artist of Charles VII (1403–61) and Louis XI (1423–83), worked as a panel painter and miniaturist in Tours and Bourges; and other artists connected with the court, such as the Master of Moulins (Jean Hey) [3], worked in this part of the Loire valley, then favoured by the royal household. There was also, however, a very active school in Provence from about 1440, the painters being sometimes more inclined to a Flemish style (Nicolas Froment [*c.* 1450–90] for example) and sometimes to an Italian (Enguerrand Quarton).

French sculpture of the fifteenth century was mainly the product of regional schools pursuing the Gothic style into its last naturalistic stages, and the chief new stimulus was provided by Claus Sluter (*fl.* 1390–1406) in Dijon. However, Michel Colombe (*c.* 1430–*c.* 1512), who shared royal patronage and even commissions with Fouquet, was perhaps as outstanding a figure; almost all his works have been destroyed and those that survive, such as the transitional Gothic-Renaissance tomb of François of Brittany at Nantes (1499–1507, partly by Italian assistants), are amongst his last works. Architecture is generally most original in secular genres, châteaux and great town houses such as the Hôtel Jacques-Coeur at Bourges

(1444–51) or the Hôtel de Cluny in Paris (begun 1485); in such buildings there appear elements of planning that were to recur in the great châteaux of the next century, broad courtyards, external staircases, round towers, steep roofs and rudimentary galleries.

Culture from conflict

The reinvigoration of French art from about 1500 was mainly the result of the successes of Charles VIII and Louis XII in military raids into Italy in 1494 and 1499; culturally the French became subject to Italy, overwhelmed by the cultivated Renaissance they met in Florence, Naples and Milan. The beginnings of an Italianate court were made under these kings. The earliest well-documented (but imperfectly surviving) product is the Château de Gaillon (1501–10), built and decorated, in part by Italians, for the Cardinal-Minister d'Amboise. But this Italianism became a programme under Francis I (reigned 1515–47), who visited and admired Italy and attracted to his court Italian poets, musicians and artists (among them Leonardo). Later it was secular

1 "The Visitation" from the *Boucicaut Hours* (*c.* 1405) is a narrative scene surrounded by a foliate border. The soft, convoluted drapery typifies International Gothic, but the landscape looks forward to Van Eyck's naturalism.

2 Jean Fouquet's "Etienne Chevalier and St Stephen" (*c.* 1452) is the left half of a diptych that stood over the tomb of Chevalier at Melun. The right half of the work shows a madonna with angels; the donor's patron saint was thus seen to be recommending him to the Virgin's mercy. Chevalier was one of the most important of Charles VII's courtiers and for him Fouquet made a book of hours. This work also reveals the virtuosity of his perspective and classical pilastered backgrounds – clearly the fruits of his recent Italian journey. But few contemporary Italians were so advanced, and Fouquet's realism derives more strongly from his roots in Flemish art.

3 Master of Moulins' "Madonna" (*c.* 1490) was influenced by the Fleming Hugo van der Goes as much as by Fouquet. The artist was probably the Dutchman Jean Hey, court painter to the Bourbons in Moulins, hence his title.

building that dominated French art for a century, as the monarchy and its dependants provided themselves with settings in the new style, adapted, naturally to social and climatic conditions quite different from those in Italy and retaining features characteristically French, such as the gallery.

The earlier of Francis's own surviving buildings – a wing at Blois and the Château de Chambord [6] – are transitional and Italian detail is grafted on to local forms, sometimes with bizarre results, as among the half-flamboyant, half-Renaissance chimneys and dormers on the fantastic roof of Chambord. But all traces of Gothic have disappeared in the new buildings at Fontainebleau, begun for Francis in 1528 by Gilles Le Breton [7] and completed in the 1560s by Primaticcio. The result cannot, of course, be mistaken for an Italian building – its steep roofs and dormers preclude that – but by about 1530 a genuinely French Renaissance style is established, by far the most mature outside Italy. French architecture maintained this position under the leadership in mid-century of Philibert de l'Orme (c. 1510–70) [4] and

Pierre Lescot (c. 1510–78) [5]; the chaste, cerebral style of the former seems from the standpoint of later centuries the foundation of the French classical tradition.

The School of Fontainebleau

The royal court was so much the focus of activity in painting and sculpture that "the School of Fontainebleau" is a title that adequately covers the most striking developments in these arts. Again the presence of Italians summoned by Francis I was crucial: Rosso Fiorentino (1495–1540) and Francesco Primaticcio (1504–c. 1570) among the painters, Benvenuto Cellini (1500–71) among the sculptors. As it happened there was far more native talent among the sculptors than among the painters; Jean Goujon (c. 1515–62) and Germain Pilon (c. 1535–90) [Key] treat the human figure with grace and competence to match any Italian, save Michelangelo alone. The School of Fontainebleau, which was also strong in engraving and metalwork, was to become, from 1530, one of the most influential centres of Mannerism in the North.

"Diana", attributed to Pilon (c. 1550), is from a fountain of the Château of Anet.

4 The tomb of Francis I in St Denis, Paris, was begun in 1547 by Philibert de l'Orme. The form of a triumphal arch symbolizes the triumph of life over death, as in Italian tombs. The couple are presented within the tomb as dead and on top of it as alive, praying. The sculptor was Pierre Bontemps.

5 Pierre Lescot's Cour Carrée of the Louvre was begun in 1546. Francis I's rebuilding of the old Louvre (from 1527) was a sign of the return of central power to Paris after a century's absence. Lescot collaborated here and elsewhere with the sculptor Goujon, for whose reliefs there are Italian precedents.

6 Château de Chambord was begun in 1519 and its architect was probably Domenica da Cortona, although the only real traces of Italian planning are in the cruciform keep. Otherwise the overall design is characteristic of French medieval châteaux. It was built for Francis I.

7 The Cour Ovale of the Château de Fontainebleau (begun c. 1530) was added by Gilles Le Breton to the medieval hunting lodge of Francis I.

The upper triple-arched opening on the left was originally approached by an exterior staircase. The domed gateway was added c. 1640.

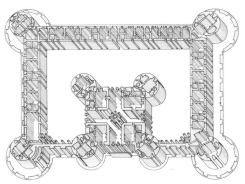

Mannerism

The High Renaissance style in Florence and Rome had been in many ways artificial, most obviously in its grace, idealism and tendency to exaggerated posturing. The achievement of the major artists was a perfect style, or so it seemed; and the pursuit of perfect style was to become an end in itself. The phase so characterized is called Mannerism (from *maniera*, Italian for "style").

Qualities of style

Artifice was not the only quality of style felt worthy of further cultivation; novelty, licence, variety and the esoteric were appreciated in the sixteenth century – in the criticism of Giorgio Vasari (1511–74), for example – as never before and the demonstration of artistic genius was thought a legitimate object in a work of art. This demonstration could become the work's true subject. Mannerism was a late Renaissance style and it was important that all the beliefs on which it was based could be justified by selective reading of ancient authorities. But the surviving monuments of antiquity were studied in a new spirit, paying particular attention to licence, complexity and grace. In northern countries the appreciation of similar qualities in late Gothic art facilitated the reception of the new style, but Gothic as such was held in contempt and what was supplied was of classical breeding.

The phenomenon of Mannerism did not occur everywhere – Venice, for example, was almost immune – and its arrival was never marked by a sharp break but by a shift of emphasis. Qualities such as grace appear in many stages in Renaissance art and were generally subject to inflation across the period; in some centres, as art looked inward at itself and not outward at nature, an extra twist of the spiral produced hyperinflation. This first happened in Rome; and perhaps it had already happened by 1520 in the work of Raphael and Michelangelo.

Artistic conditions in Rome soon after 1520 were ideal for the growth of Mannerism. A brilliant group of young artists, the pupils of Raphael, Florentines such as G. B. Rosso (Rosso Fiorentino, 1494–1540) and Benvenuto Cellini (1500–71), and north Italians such as Girolamo Parmigianino (1503–40), were impelled towards the display of virtuosity by their own competitive situation, by the qualities of the last works of Raphael (died 1520), and by the current work of Michelangelo (1475–1564). The latter's drawings were rich in invention, his sculpture suave and refined away from the natural, his architecture increasingly more marked by licence than by rule [2, 3]. The dispersal of this group had begun before the sack of Rome in 1527, but that event was directly effective in the spread of Mannerism, through Perino del Vaga (*c.* 1500–47) to Genoa, through Parmigianino to Emilia, and through Rosso, eventually, to Fontainebleau. Furthermore the Mannerist style was unusually accessible because so much of it appeared in engravings.

Medici patronage

In Italy the most impressive Mannerist in the 1530s, if Michelangelo is excepted, was certainly Parmigianino, first in Bologna and then in Parma. His "Madonna with the Long Neck" [4] is one of a number of religious works whose spiritual message is less obvious

1 **Rosso Fiorentino's** section of the Galerie François Ier (1535–9) in the Château de Fontainebleau is one of 12 bays in the gallery, each different. In general, the subjects appear to be allegories of the reign of Francis I. The central fresco here – one of those most likely to be by Rosso's own hand – is the "Death of Adonis", which may refer to the death of the dauphin in 1536. The salamander above is the king's personal emblem. A similar mixture of painting and stucco – Roman and Florentine ideas – was used by Primaticcio.

4 **Parmigianino's "Madonna with the Long Neck"** (1534–6) was painted for Elena Baiardi, for a church in Parma. The subject refers to St Helen's vision of the Cross, represented in the crystal vase. All the figures react to this vision – the Virgin by seeing in the child across her knees a prefiguration of the Passion. The column is a symbol of the Virgin's purity. The content is comparable to that of an altarpiece by Correggio (*c.* 1494–1534), but the ecstasy seems more stylish than expressive, and strikingly elegant in form.

2 **Michelangelo's Porta Pia, Rome,** begun in 1561, is a fine example of Mannerist licence. There are no true capitals and the pseudo-capitals are fragments of an architrave topped by the guttae of a frieze.

3 **Michelangelo's "Victory"** (*c.* 1527–8) was designed to stand in a niche in the lowest zone of the tomb of Julius II to be flanked by a pair of slaves. The continuous spiral movement of the figures is known as the *figura serpentinata* (snake-like figure) which Michelangelo is said to have thought most perfect.

than their elegant forms, rhythmic grace, refinement of detail and airless, unreal colour. After 1540 Florence became the most vital Italian centre and remained so until the early seventeenth century under the patronage of the Medici dukes. Cosimo I (1519–74) assembled a team of highly talented artists: the painters Salviati (1510–63) and Il Bronzino (1502–72), the sculptors Cellini [Key] and Bartolommeo Ammanati (1511–92), and the architect Buontalenti (1536–1608). To this court and its new nobility was attracted, in about 1560, the one really great artist who worked wholly in the Mannerist style – the Fleming Gianbologna (1529–1608). No sculptor's work has ever been so suavely complex, nor made so purposefully to be admired for its art. He brought the bronze statuette to perfection but, partly in emulation of Michelangelo, worked also on a large scale in marble [5, 8].

It is natural that a style that tended to preciosity and rested upon technical accomplishment should flourish in metalwork and jewellery; Cellini, whose gold saltcellar was completed for Francis I, was the most famous

goldsmith of the period, but the richest concentration of talent was in south Germany in the second half of the century [6].

Northern centres

Centres of Mannerism in the north were mainly, but not exclusively, at courts, and the first of these was at Fontainebleau. There, from about 1530, the decoration of Francis's new château was emphatically Mannerist, and very influential (for example, in England). The principal artists there were Italian – Rosso and Primaticcio – and their main work was the Galerie François Ier, an exuberant mixture of stylized allegorical painting and complex stucco relief [1] where first appeared the universal currency of Mannerist decoration, strapwork (ornamental strap-like bands). Around 1600 the last vital centres of Mannerism were in Holland (the school of Haarlem) and at the court of Rudolph II at Prague, where the Flemish painter Bartholomäus Spranger (1546–1611) and the Dutch sculptor Adriaen de Vries (c. 1548–1626) produced decadent, erotic art of the highest order.

Cellini's bronze bust of Duke Cosimo I of Florence was made between 1545 and 1547. Cosimo was a Medici who was created Grand Duke of Tuscany in 1569. He was a great art-lover and Cellini's patron.

5 Gianbologna's "Astronomy" (c. 1573) is one of the most highly finished of this artist's statuettes and was made to be turned round in the hand so that the brilliance and variety of its composition could be admired. From any angle the forms make a balanced and complicated design, each one different. In his work Gianbologna took a characteristic aspect of the 16th century – making art for connoisseurs – to its logical conclusion. He believed that the subject was of no particular significance, the figures being chosen "to give scope to the science and accomplishment of art".

6 Wenzel Jamnitzer's ewer (c. 1570) is made of silver gilt, enamelled and inlaid. Vases of surprising fantasy, in bronze and silver, survived from the Roman Empire. Thereafter, the creation of more witty and complex tableware was a main Mannerist preoccupation. Jamnitzer (1508–85) used casts of real animals.

7 Bartholomäus Spranger's "Venus and Adonis" (c. 1590) was characteristic of the taste of Rudolph II and his court at Prague, where there was an unusual interest in all things strange and improbable. Spranger was a cosmopolitan artist, trained in Antwerp, Paris, Parma and Rome, who exerted great influence in the last stages of Mannerism through his engraver Goltzius. In general he gives the impression, as here, of finding his style as absurd as his subjects.

8 Gianbologna's "The Rape of the Sabine" (1582) seems to have been made without the artist having in view a subject, a site or a patron, but only the ambition to show the world that, contrary to rumour, he could work as well on a colossal scale in marble as in small bronzes (on which he made his reputation). He believed that he had made his task as difficult as possible by multiplying the figures and by varying their age and sex. The piece's title was arrived at when the work was complete.

265

The Tudors and the new nation state

When Henry Tudor (1457–1509) [1, 2] took the English Crown from the last Plantagenet king on 22 August 1485, he became master of a rich kingdom. Its fertility and its mineral wealth appeared to a Venetian observer of 1500 to be "greater than those of any other country of Europe"; its trade in the northern and western seas was increasing rapidly, and its population was slowly growing. Yet the monarchy itself had minimal prestige, ruined by the mental incapacity of Henry VI (reigned 1422–61; 1470–1) and the confusion of the Wars of the Roses.

The establishment of the Tudor dynasty

In 1485 few estimated that Henry VII's new dynasty would last long; Henry had to earn respect. There were potentially dangerous challengers to his throne: Simnel Lambert (c. 1475–1525), who pretended to be the imprisoned Earl of Warwick (the last male Plantagenet) and who momentarily won over Ireland; and Perkin Warbeck (c. 1474–99) whose claim to be Richard of York, the younger of the princes murdered in the Tower, was taken up by many European

rulers. Henry survived partly because of his prudent statecraft, but primarily through the desire of all classes for firm leadership and an end to political uncertainty.

Henry is said to have introduced a "new monarchy", but actually it was no more than the new life breathed into traditional institutions by a strong personality. Although parsimonious, he understood the value of ceremony [2] and pageantry [4] and his court impressed foreigners. Behind this façade lay the increasingly solid support of most of his subjects. The old nobility, weakened but not extinguished by the civil wars, was happier with a subordinate role than before. Henry treated most of these families with respect: he passed the Statute of Livery and Maintenance in 1487 against keeping private armies, but it did not seriously affect their retinues or their local influence.

The king rested his power upon the traditional rulers of the counties, the gentry as justices of the peace; he gave them stronger authority than before and framed his policies with a careful eye on what they would accept. In particular, he emulated Edward IV

(reigned 1641–70; 1471–83) in a policy of peace and therefore low expenditure: he avoided straining the loyalty of the gentry with extensive taxation [3]. The main achievement of Henry VII was to nurse the monarchy back to its traditional role of leadership by avoiding serious confrontation with any group of subjects.

The court of Henry VIII

Henry VIII (reigned 1509–47) [Key] was the first Prince of Wales to succeed peacefully since 1422. Far more obviously than his father, the second Tudor king was a typical Renaissance prince: well educated, polyglot, author of a sprightly theological work against Luther, accomplished on lute and harp, a jouster and a bowman, combining magnificence with intellectual power and political will exactly as the age demanded of its rulers. His court was a centre for the new learning: his ministers Richard Fox (c. 1448–1528) and Thomas Wolsey (c. 1475–1530) [6] were great patrons, and his later chancellor, Thomas More (c. 1478–1535), embodied more than any contemporary Englishman the

1 The striking of medals was a characteristic form of propaganda in the 15th and 16th centuries. Henry VII used them together with other projections of the royal image in pageantry and artifice to bolster his weak hereditary claims to the throne. His marriage with Elizabeth of York (1465–1503), shown here with Henry, was thought to give him a much better title to the throne than the right of conquest or the Act of Parliament upholding
it, both of which could be reversed. His use of the "Tudor Rose", combining both the red and white roses, symbols of the two factions in the Wars of the Roses, exemplified his desire to appear as unifier of the nation.

3 Henry VII's revenue grew dramatically in his reign, based on an extremely careful management of Crown estates. Henry aimed to "live of his own" and tried to make even war profitable, by keeping costs low and suing for peace, thus
staying independent of both nobility and Parliament. In this way he hoped to establish a strong monarchy, which he regarded as a vital prerequisite for the establishment of English political authority in Western Europe.

£3.0 (7%)
£20.0 (47%)
£11.0 (14%)
1487
£19.5 (46%)
£30.0 (34.6%)
£42.0 (27.8%)
1495
£37.0 (51.4%)
£38.0 (25.2%)
1505
£81.0 (47%)

Crown lands
Customs
Other

Figures in £000
Percentage total revenue in brackets

2 Henry VII and his queen with their children are depicted with St George and the dragon on this altarpiece, probably made for the royal palace at Sheen. The prominence given to the queen in early Tudor art was new. This painting commemorated Henry's patronage of the Order of the Garter, which honoured equally the old nobility and new soldiers of fortune. Henry did not intend to crush the nobility into submission but rather to forge a new alliance with them. But his unpopular taxation seemed to show that Henry had autocratic intentions. His ministers Richard Empson (d. 1510) and Edmund Dudley (1462–1510) were condemned for treason after the king's death.

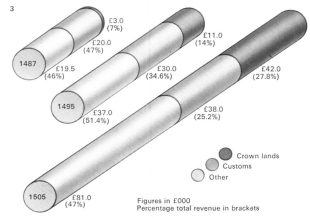

4 The baptism of Arthur (1486–1502) Henry's eldest son, was an occasion for a typical royal procession. The name Arthur was chosen to echo that of the semi-mythical British hero whose cult was at its height in the 15th century
and to stress the Welsh origins of the Tudors. Similar events were organized at the start of his reign for each main town, to affirm the glory of his rule. Henry also planned a major addition to Westminster Abbey, begun in 1509.

cultivated practical intellect of the Renaissance. Henry's cruelty in the interest of state, dramatically evidenced in the fate of four of his six wives (two of whom were executed and two divorced), only underlines the brilliant, thunderous atmosphere of his court.

Whether wittingly or not, Henry's use of the revived authority of the Crown had, in the form of the Reformation, revolutionary effects on England. He was initially content to work through powerful ministers, the first of whom, Thomas Wolsey, cut a figure as a cardinal hardly less magnificent than his master's. Wolsey established a position of unprecedented dominance in Church and State. Even before the Reformation he planned a wide-ranging reform of the Church, but it was unachieved at his fall from power in 1529, and his legacy was a hierarchy henceforth subservient to the king alone.

Cromwell and the bureaucracy
By 1534 Henry had found in Wolsey's former servant Thomas Cromwell (c. 1485–1540) [7] a minister to carry through all the implications of the break with Rome, and

realize the new potential power of the monarchy. A hard, practical man, Cromwell used the growing authority of Parliament. In his popular reformation of the Church, he was supported by the gentry, whom he manipulated with his system of patronage and an unprecedented network of informers [9]. Yet even Cromwell was only an instrument of government, dismissed and executed by the king in 1540 when his enemies won the ear of Henry.

Henry VIII was by then the master of a terrifying royal power, sustained by an efficient and much reformed bureaucratic machine, and able to impose his own middle way between the proponents of a radical reformation and the adherents of the old, Catholic ways. His achievement, a remarkable display of statecraft, was to have understood and given a lead to the strongest strands in public opinion, from which, as institutionalized in Parliament, he never moved very far. By the time of his death in 1547, the Tudor monarchs had become the natural expression of the growing national confidence of sixteenth-century England.

Henry VIII was an imperious man whose will dominated the political history of his reign. His court was a focus for all kinds of artistic patronage and with John Skelton (1460–1529), the poet, as his tutor he was probably the first English king to know Italian and Spanish. He took advantage of the shrewd diplomacy of his father to make England a respected state, fighting many costly, although rarely glorious, wars. In this policy he claimed to embody the new assertiveness and aggression of Englishmen overseas that also found expression in the steady erosion of the Hanseatic League's monopoly of English trade. The culmination of this nationalism was the Church's break with Rome.

5 The *Henry Grace à Dieu*, a carrack of 1,000 tonnes was built in 1510–14 and became Henry VIII's flagship. By 1510 Henry had won a strong position, holding the balance of power between the Hapsburgs and Valois, but he was never able effectively to assert his authority in Europe although he defeated the French in 1513.

6 Thomas Wolsey, who had ambitions to become pope and to dominate European politics, was the chief architect of Henry VIII's foreign policy before 1529. As papal legate he had great power in the English Church, and seemed almost independent of the king's jurisdiction.

8 Walmer Castle, Kent, was one of the fortresses built by Henry VIII in 1539 as part of a system of coastal defence during a brief alliance of his enemies the Emperor Charles V and Francis I of France. These castles used the most modern style of fortification.

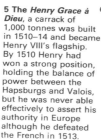

9 The Pilgrimage of Grace was a rebellion of the north in 1536 in opposition to the Oath of Supremacy demonastery. It was typical of Cromwell's rule that men could be executed for opinions expressed in private. Marshall, abbot of Colchester, who verbally supported the rebellion, uttered treasonable ideas in private and who was determined to resist the suppression of his monastery. It was typical of Cromwell's rule that men could be executed for opinions expressed in private.

7 Thomas Cromwell was the architect of the English Reformation. His ruthless determination, born of soldiering in the Italian wars and in business experience in The Netherlands, served him in good stead in Wolsey's service and later in the king's. A genuine supporter of reform, he believed that the king should be the head of the English Church. He excelled as parliamentary manager, marshalling the broad support that the Tudor monarchy enjoyed, to create a great series of Acts establishing the royal supremacy and suppressing the monasteries. His extraordinary executive ability helped him to enforce them through his system of informers and active control of the JPs. He made many enemies, particularly amongst the nobility, and in 1540 he was executed for treason.

The Reformation

When Martin Luther (1483–1546) posted his *Ninety-Five Theses* on the door of the Castle Church, Wittenberg, Saxony, on 31 October 1517, he was initiating a religious debate in the traditional fashion. But the consequences of his protest were revolutionary and heralded a new historical era: the end of the dominance of a single European Catholic Church in the Middle Ages; the creation of "reformed" or Protestant Churches; a century and a half of "religious" wars that convulsed the emerging nation states of Europe; and a new understanding of the Christian faith that has personally affected millions of men and women throughout the world. All these developments are historically part of "the Reformation".

The original intention

The term Reformation precisely describes Luther's original intentions. His *Ninety-Five Theses* [Key] were an attack on abuses inside the Roman Catholic Church. He denounced the frivolous uses to which the pope put his vast wealth and in particular the way in which he was raising money by the sale of "indulgences" [1], documents that the pardoner Tetzel claimed gave any purchaser automatic remission of his sins.

The history of the Catholic Church had for centuries, however, been a sequence of "re-formations", and for more than a decade before 1517 John Colet (*c.* 1467–1519), Erasmus [3] and Thomas More (*c.* 1478–1535) had been working to cleanse the Church of its follies. What made Luther more than merely a traditional reformer was his deliberate personal conviction coupled with the accidental political situation at that moment in Europe's development.

Luther's complaint against indulgences went beyond distaste of the money involved to a rejection of the whole concept of spiritual book-keeping with God. Luther felt a sense of infinite personal unworthiness that no amount of good works could ever overcome. As an Augustinian monk, a doctor of theology and then professor at the University of Wittenberg he had mortified himself mercilessly but still felt impure, repulsive, sinful – unworthy even to approach God's presence. Only blind faith could save him.

From this conviction was born the doctrine of "justification by faith", which was to inspire the spiritual side of the Reformation. It involved great emphasis on the single gesture of faith – "conversion" – and rejected reliance on the priest as a middleman between the believer and God.

The effect on Germany

Luther's ideas, broadcast by the medium of the printing press, had a dynamic impact throughout Germany and were given immeasurable help in taking root by a fatal four-year delay on the part of the Catholic authorities. Pope Leo X (1475–1521), who hoped to strengthen his personal power by manipulating the imminent election of a new Holy Roman Emperor, adopted the traditional papal tactic of supporting a weak candidate for this position that carried so much influence in Germany. He had already selected as his protégé Frederick the Wise of Saxony (1482–1556), Luther's own ruler and protector. Thus Luther's teachings were not officially condemned until 1521 at the Diet (congress) of Worms.

1 The sale of indulgences was to raise money for the rebuilding of St Peter's in Rome. An indulgence was originally a remission from punishment for a sin and was conditional on the sinner's sincere repentance. But in the Middle Ages, indulgences came to be granted to sinners who did good works, such as going on crusades or contributing money to the building of a church. This led to the idea that forgiveness could be purchased as a commodity.

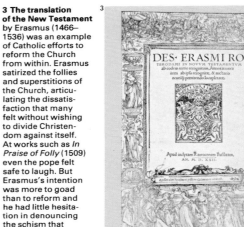

2 The religious orders of monks, nuns and friars provoked hostility. To ordinary people and reformers they seemed idle and excessively wealthy, spending their money on their own enjoyment – not in the service of God. They were envied by rulers for their valuable town properties and country estates.

4 Encouraged by Luther's success and sharing some of his ideals, the peasants of the southwest and central German states rose against their rulers in 1524. But Luther denounced their complaints – mainly political social and economic – and in the face of stern repression the revolts collapsed.

3 The translation of the New Testament by Erasmus (1466–1536) was an example of Catholic efforts to reform the Church from within. Erasmus satirized the follies and superstitions of the Church, articulating the dissatisfaction that many felt without wishing to divide Christendom against itself. At works such as *In Praise of Folly* (1509) even the pope felt safe to laugh. But Erasmus's intention was more to goad than to reform and he had little hesitation in denouncing the schism that Luther created.

5 The Holy Roman Empire, in theory the political expression of the Catholic Church, was in practice a grouping of rival German states. The Reformation provided the chance many of them had been waiting for to defy the power of the emperor. The Protestant rulers formed themselves into the Schmalkaldic League in 1531 and the emperor replied with the Nuremberg League in 1538. War seemed imminent but a compromise was reached in 1539.

Holy Roman Empire *c.*1500
Church lands *c.*1500
Hapsburg lands *c.*1500
Nuremberg League (1538)
(S) Schmalkaldic League (1531-47)

Rostock · Pomerania (S)
Hanover (S)
Einbeck (S) · Magdeburg (S)
Göttingen (S)
Schmalkalden (S)
Luxembourg
Worms
Bohemia
Bavaria
Isny (S)
Constance (S)
Burgundy
Styria

0   200 km

Frederick the Wise then sent Luther into hiding for a year and the new emperor, Charles V [6], had neither the time nor the power to control this defiance to his authority. In 1520 Luther had urged princes to throw off the unbiblical authority Rome claimed over the Church in their states and rulers were not slow to profit from this invitation. In 1529 all the Lutheran states and towns in Germany put their names to a "Protestation" against the emperor – the origin of the term Protestant – and after more than 25 years of conflict the Peace of Augsburg in 1555 recognized the right of a ruler to determine his country's religion.

England and Switzerland

The principle of the rights of a ruler had been exercised by Henry VIII (1491–1547) in his Reformation in England, which was confirmed by the Act of Supremacy of 1534. It was essentially a forcible transfer of the supreme power over the Church from the pope in Rome to the English sovereign. As in Sweden in 1527 and Denmark and Norway in 1536-9, there was little pretence that

Thomas Cromwell's dissolution of the monasteries (1536–40) was any more than a confiscation of the Church's wealth.

Only in Switzerland was the "purification" of the old Church's images and decorations a matter of genuine Christian austerity. In Zurich in the 1520s Huldreich Zwingli (1484–1531) had the church organs destroyed because their sound was profane while the citizens as a whole were banded into a new democracy of the faithful.

In Geneva [9] after 1541 Jean Calvin (1509–64) created a holy city that gave shelter to more than 6,000 refugees fleeing from persecution in France, Italy, Spain and, during Mary's reign, from England as well. Calvin's book *Institutes of the Christian Religion* (1536) enshrined the other great Reformation theme of "predestination" – that God chooses His own elect. It was the buoyancy of those who felt the assurance of their own election that inspired the Calvinists in the Netherlands as well as the Huguenots in France, the Presbyterians in Scotland and the Puritans in England and later in the New World.

KEY

Martin Luther, an Augustinian monk, was the first great inspirer of the Reformation. On a visit to Rome in 1511 he was appalled by the wealth and spiritual emptiness of the Catholic Church so different from the ideals of primitive Christianity. These ideals became his watchword as he argued against the Church's corruption in his *Ninety-Five Theses* (1517). Luther was not a revolutionary, for he retained vestments and certain Catholic ceremonies, but in his reformed Church the communion cup was given to the congregation, saints were no longer objects of special prayers and Church wealth went to pastoral and educational activities.

6 The Emperor Charles V (1500–58) inherited the lands belonging to the Austrian and Burgundian Hapsburgs and also the kingdoms of Aragon and Castile. This provoked the fear of France, which fought throughout his rule to prevent encirclement, and encouraged German Protestant princes to defy his authority. Defeated in his ambition to reunite Christendom, he gave his possessions to his brother and son.

7 Fountains Abbey was among the finest religious houses destroyed during the dissolution of the monasteries in England (1536–40). The first act (1536) based dissolution on whether a monastery enjoyed an annual income of less than £200. The Act for the Dissolution of the Greater Monasteries (1539) completed the process. Pensions were paid to monks refusing to join the secular clergy and the Crown took the rest. A few lands were kept by the Crown, a few given away, but most were sold to gentleman farmers, thus creating a new class of landowners with the strongest reasons for staying loyal to the new order.

8 The Bible replaced the priest as the ordinary man's spiritual authority. The scriptures were translated from the Latin and Greek and offered to the people in their own languages – notably German in a translation by Luther, and English in the King James version of 1611.

9 Under John Calvin the Swiss city state of Geneva became the most influential single centre of the Reformation. Its citizens lived under the strict moral rule of the Calvinist Church, which readily burned opponents such as the anti-Trinitarian "heretic" Michael Servetus in 1553.

Catholic
Lutheran
Calvinist
Anglican
Greek Orthodox
Religious centres

Canterbury
Wittenberg
Augsburg
Geneva
Rome

0   500 km

10 By 1560 Lutheranism had spread from north Germany to Scandinavia, Poland and Hungary along German trade routes and among the scattered communities in the trading cities. After 25 years of religious wars the Peace of Augsburg (1555) had given the Lutheran princes and cities the right to freedom of worship, in effect strengthening the hands of both Catholic and Protestant rulers within their own territories. But it gave no privileges to the followers of Zwingli and Calvin, who were the dominant religious groups in The Netherlands, Scotland and most Swiss cantons, and who formed a substantial Swiss minority in France, Hungary and southern Germany.

269

The English Reformation

The English Reformation had as its particular event Henry VIII's desire for an annulment of his marriage; this the pope was unwilling to grant him. The more general discontent was related to that which brought religious turmoil to the Holy Roman Empire, France and The Netherlands in the early sixteenth century. The Church's taxes drained money from the country to finance the pope's political ambitions; ordained priests could seek exemption from secular penalties; and the pope claimed to appoint higher clergy and bishops. The Church's superstitious practices, too, came under attack. There was a tradition of opposition to the pope that had grown up with John Wycliffe (c. 1328–84) and which had never completely died out, and a mood of questioning old assumptions in the light of Renaissance humanism.

Nationalism and the Reformation

The Reformation evolved from national opposition to external interference. Wycliffe's ideas, although mainly dealing with doctrinal matters, implied a revolt against papal authority. He translated the Bible into English and suggested that the laity should make their own evaluations about religion. Such ideas anticipated those of Martin Luther (1483–1546); but by the early sixteenth century opposition by a nation-state to the universal claims of the papacy was now conceivable. Henry VII (reigned 1485–1509) had built up a monarchy strong enough to win the respect of the powers of Europe. Henry VIII (reigned 1509–47) [Key] was determined to be an ideal Renaissance prince, supreme at home and ostentatiously magnificent abroad.

Henry feared that he would be unable to have a male heir by his first wife Catherine of Aragon (1485–1536). The pope influenced by the Holy Roman Emperor Charles V, Catherine's nephew, was reluctant to annul the marriage. In 1533, Henry's mistress, Anne Boleyn [1], became pregnant, and Thomas Cromwell (c. 1485–1540) said that the problem could be speedily resolved by claiming that the king was not answerable to the pope. The marriage was annulled by Thomas Cranmer (1489–1556) [5], who was made Archbishop of Canterbury to deal with the divorce issue, and in 1533 Catherine was forbidden to appeal to Rome. The Reformation Parliament (1529–36) was directed at every stage of the break with Rome to enact the necessary legislation. In 1534 it passed the crucial Act of Supremacy, declaring Henry to be "Supreme Head" of the English Church, and then sanctioned the dissolution of the monasteries by the Crown.

The introduction of Protestantism

Henry VIII wished to be head of the English Church, but wished to dissociate it from the Lutheran ideas that were spreading through the Holy Roman Empire. Cromwell and Cranmer, however, urged him to adopt a more Protestant attitude, and in 1538 an English translation of the Bible was ordered to be placed in every church.

The reign of Edward VI (1547–53) saw Protestant doctrine more fully accepted [2]. Edward was a minor, and the Protector Somerset (1506–52) relaxed the laws against heretics and welcomed radical foreign preachers to the country. The clergy were allowed to marry and the new prayer books

1 Anne Boleyn (c. 1507–36), was secretly married to Henry in January 1533 to ensure the legitimacy of the child that she was carrying. In the event, however, she gave birth to a daughter, Elizabeth, and not the male heir for which Henry had been hoping. During Anne's later pregnancy, Henry fell in love with Jane Seymour (c. 1509–37), accused Anne of adultery and incest, and had her executed in 1536.

2 Henry on his deathbed is shown giving pious instructions to his young son, Edward VI, to ensure the stability of the Reformation. Also shown are members of Edward's council who carried out many of the most extreme Protestant policies to be adopted in England in the 16th century. Among these was an order against "superstitious" images in churches, which led to much iconoclasm. Chantries – chapels used specifically for the purpose of praying for the souls of those recently dead and thought to be in purgatory – were also abolished in 1547. Their foundations had often included provision for schools, and so many schools were set up under Edward.

3 The prayer book of 1549 introduced a national liturgy in place of the various local "uses" practised previously. A rebellion in the West Country objected to the use of the vernacular in the service, claiming that it was like "a Christmas game". Doctrinally the book was a compromise and was followed by a more Protestant version three years later.

4 Mary I refused to repudiate her mother's faith during Henry's reign, and after her succession she fortified her restoration of Catholicism by marrying Philip II of Spain, leader of the Counter-Reformation. Her ruthless determination to end Protestantism in England earned her the title of "Bloody Mary", but in private affairs she was gentle.

5 Thomas Cranmer ennobled the Reformation with the majestic rhythms of his translations of the Bible and liturgy, but politically he was weak and pliant. On occasion he went against his personal convictions to defer to royal authority. He was the theological inspiration behind the innovations of Edward's reign. Mary made him recant his Protestant faith, but later he renounced the recantation itself, and thrust the hand that signed it into the flames first, when burnt at the stake.

of 1549 [3] and 1552 showed a distinct turn towards Protestant doctrine.

Mary (reigned 1553–8) [4], the daughter of Catherine of Aragon, introduced a severe Catholic reaction; many Protestants were driven into exile, and nearly 300 were burnt at the stake as heretics [6]. But her repression was unpopular: Henry VIII had sold to the gentry many of the lands acquired by the dissolution of the monasteries. As a result, this class had a vested interest in opposing a full restoration of Catholicism, and welcomed the accession of Elizabeth, who, as daughter of Anne Boleyn, seemed certain to restore Protestantism.

The Anglican compromise

Elizabeth I (1533–1603) acceded to the throne in 1558, and sought a middle way between the policies of the previous two reigns. But the return of many of the Marian exiles from Calvinist Geneva and a strongly Protestant Parliament forced her to adopt a more extreme attitude than she intended. She took the title of "Supreme Governor" of the Church, hoping thereby not to offend the Catholics; the Thirty-nine Articles of belief, to which all priests had to swear in 1563, contained ambiguous assertions acceptable to men of all current persuasions. And she also rejected all suggestions that the episcopacy should be abandoned.

In these ways, Elizabeth hoped to gain loyalty through compromise rather than to satisfy conscience, and despite constant pressure in Parliament and elsewhere from the Puritans (as the politically radical Calvinists came to be known) to modify this settlement, she assisted the Anglican Church to gain a positive identity. Some Puritans tried after the 1570s to be allowed to worship outside the Established Church. Eventually John Whitgift (c. 1530–1604) [7], drove underground all Puritan opposition throughout the country.

Fear of Catholic plots however, continued throughout Elizabeth's reign, allowing the Puritans to pose as good patriots. The Catholic threat remained constant, fuelled from abroad by Jesuit priests and support from Spain, and assisted by a few families in England itself.

Henry VIII, assisted by Cranmer, tramples the pope underfoot while monks wring their hands in despair, in this contemporary cartoon. Henry was determined to make his authority absolute in his kingdom; he claimed that this was the wish of the people.

6 Foxe's *Book of Martyres* (1563) is an account of the persecution of English Protestants since the 14th century. Written by John Foxe (1516–87) in Strasburg and Basel where he was in exile during Mary's reign, its emotive language and illustrations won great popularity and strengthened English hatred of Catholicism for many years. Here a victim of Mary's persecution is dressed as the Antichrist as he is led away to his execution.

7 John Whitgift was made Archbishop of Canterbury in 1583, and although Calvinist in his personal views, he strictly enforced the laws against dissenters. His use of the Court of High Commission, the supreme ecclesiastical court of the country, was challenged as unconstitutional, but he ensured that the Puritans were broken as a political force. They re-emerged strongly, however, after the accession of James I.

8 A Puritan church, shown here in a 17th-century illustration, was sparsely furnished and austere. The most extreme Puritans, led by Thomas Cartwright (c. 1535–1603) attacked the bishops, hoping to return to the principles of the early Church, accepting only biblical authority. Brownists, led by Robert Browne (c. 1550–1633) were a radical wing hoping to run the Church by a loose federation of ministers elected by their congregation.

9 Fear of "popish plots" and invasions haunted Elizabeth's reign. This illustration of the various insurrections of the Catholics since 1550 forms part of a broadsheet on which a ballad recounting the same events was printed for sale to professional balladeers. In 1570 the pope excommunicated Elizabeth and urged Catholics to depose her. A crisis of conscience ensued for previously unpolitical Catholics, who thereafter were regarded with suspicion if not actually persecuted. There was little danger of a Catholic rebellion since the Revolt of the North of 1569 had proved abortive; but several isolated plots against the queen's life continued throughout her reign, culminating in the infamous Gunpowder Plot of 1605 against James I.

The Counter-Reformation

The Reformation in the early 1500s divided Europe into the Catholic countries of the south and the Protestant countries of the north. Under the protection of the German princes Protestantism became established and the Holy Roman Empire existed in an uneasy state of religious cold war. In the sixteenth century it was believed that two religions could not co-exist in one political community. Heterodoxy, championed by the princes, was an assault upon the secular as well as the spiritual order.

Within the Catholic Church, reformist cardinals such as Gasparo Contarini (1483–1542) advocated a conciliatory policy and reform of abuses. But discussions between Catholics and Protestants at the Diet of Regensburg in 1541 revealed only the impossibility of a compromise and blasted hopes of a revived Christian unity.

Since the Lutheran Reformation a new, formidable Protestantism had assaulted the Catholic Church – the organized Church of John Calvin. As Protestants gained ground in Germany and Switzerland, and missionaries went out from Geneva to France and the Low Countries, the Catholic Church realized the need for action. Between 1536 and 1545 Pope Paul III repeatedly called for a general council, but was forced to postpone it for political reasons. The council finally met in Trent on 15 December 1545. Ostensibly this was a council for all Christendom. But the location in the Italian Alps and Pope Paul's successful coup in determining votes by representative rather than by nation ensured an Italian, papal dominance.

The new orthodoxy

Contarini's followers still hoped for reconciliation with the Protestants. Cardinal Gian Petro Caraffa (1476–1559) and his party were in no such mood. Meeting in three sessions, (1545–7, 1551–2, 1562–3) the council answered the Lutheran challenge for the first time. On all major points of theological controversy, concerning salvation, transubstantiation and the authority of the fathers, the Catholic position was reasserted with a clarity that late medieval theology had lacked. The council made token reforms of the glaring abuses. Most important, the Council of Trent reinforced the authorities of the Church from the bishops to the pope, whose supremacy was not questioned. The defenders of a new orthodoxy were already armed with the Inquisition (established in 1233) and the papal Index of prohibited books. Trent made the Roman Catholic Church one denomination among many – but a denomination that was ready to fight. Even before the council met, Ignatius Loyola (1491–1556) [2], who was appointed general of the Society of Jesus in 1541, a year after the pope's approval, had provided an army of missionaries – the Jesuits.

Beginnings of a religious war

The end of a policy of conciliation meant religious war. Where nobles and princes had emerged as the patrons of the Protestant faiths it meant civil war. In France the rival noble factions of Bourbon and Guise adopting the cause of Calvinism and Tridentine Catholicism (Roman Catholicism as reformed at the Council of Trent) began 30 years of conflict. Political ambition was often the handmaiden of religious zeal.

1 **The Counter-Reformation** in the century following the Council of Trent won back much of the territory lost to Protestantism. It was most successful in achieving a popular reconversion in Poland, where the enactment of religious toleration in 1573 led to a sudden growth in Jesuit activity in the field of education until the support of Sigismund III (r. 1587–1632) for the Jesuit cause led to the complete suppression of Protestantism. The Counter-Reformation was often marked by the banning of "heretical" books. Catholic control of higher education, supported by royal and noble families, and the denial of Protestant rights. Its success was often achieved by war.

2 **Ignatius de Loyola** was a Spanish nobleman born in 1491. He underwent a religious experience during a period of convalescence after being wounded in battle. After studying theology at the Sorbonne in Paris in 1534 he formed a group of devout men who swore to serve Christ and his vicar on earth, the pope. His religious order, the Society of Jesus, was strictly organized, its members being subject to rigorous discipline. The theology outlined by Loyola in his *Spiritual Exercises* has been described as a "shock tactic spiritual gymnastic". Originally, membership was confined to 60 Jesuits, but in 1540 the pope authorized the order to increase its membership without limit. By 1556 the order was firmly established in Europe and it grew greatly in size until the mid-17th century, acting as the principal agent of the Counter-Reformation. It conducted widespread missionary work.

3 **William of Orange** (1533–84), "the Silent", was the largest landowner in The Netherlands. He led the opposition to Philip II's erosion of the aristocratic and constitutional liberties of The Netherlands. During the war William tried to unify the provinces, which were jealous of their rights and promoted religious toleration to prevent a rift between Protestants and Catholics in their defence of common constitutional rights. Dutch independence was recognized in 1648.

4 **The Massacre of St Bartholomew's Day** – 24 Aug 1572 – was a slaughter of French Protestants (Huguenots). The massacre spread throughout France leading to the outbreak of civil war. Catherine de' Medici (1519–89), the French regent, alarmed at the growing Huguenot political power, plotted against the leaders and instigated the massacre.

5 **The rebellious Netherlands** were helped in their struggle against Spain in the 1560s and 1570s by Elizabeth I of England. It became increasingly evident to Philip II that the Low Countries would not be settled while the rebels received foreign aid. In 1580 Philip successfully pressed a claim to the crown of Portugal and the acquisition of a powerful fleet, together with the increasing revenue from the Indies, perhaps persuaded him to launch the Armada. In league with leading French Catholics, Philip hoped to put in at French ports, while his regent for The Netherlands, the Duke of Parma, was to clear towns along the coast to meet the Spanish ships. Philip may have intended the conquest of England (we cannot be sure) or more probably have planned to contain the English while Parma militarily reasserted Spanish control in The Netherlands. The Armada was defeated in 1588.

The popes looked to the Catholic princes of Europe to be the secular arm of Tridentine decrees. In particular they looked to the principal heir of Charles V – Philip II of Spain (1527–98) [Key]. In an attempt to exercise greater control over the prosperous Low Countries Philip introduced the Inquisition. Only during the course of revolt did Calvinism find its champions in the defenders of the Netherlanders' privileges.

In 1555 the Peace of Augsburg had brought a temporary lull to conflict in Germany. For the rest of the century the empire remained in uneasy peace as the Hapsburgs tried to re-establish their authority and the Calvinists (not recognized in the peace) gained ground. The Catholic princes of Germany, acting independently of the emperor, formed their own political league under Maximilian of Bavaria (1573–1651).

Thirty Years War
The signal for conflict was given when the Bohemian subjects resisted the decisions of the Emperor Matthias (1557–1619) to make the staunchly Catholic Ferdinand of Styria

(1578–1637) king of the Bohemians. The Bohemians wanted neither a Hapsburg nor a Catholic. In 1618 they cast two imperial councillors from the window of the council chamber at Prague and called upon the head of the German Protestants, Frederick the Elector Palatine (1596–1632), to defend their cause [8]. Thirty years of war ensued.

The imperial forces, assisted by Maximilian, crushed Frederick and Hapsburg power was restored. By 1629 Ferdinand, now emperor, was strong enough to impose an edict restoring to the Catholic Church all lands secularized since 1552, so undermining the territorial strength of the Protestants. But the strengthening of the Hapsburgs and the identification of Spanish and imperial Hapsburgs with Catholicism brought in turn all their enemies against them. The Swedes [7], then ironically the French, under the Catholic Cardinal Richelieu [9], fought with the Protestant cause against the Hapsburgs. After 30 years of turmoil the weary Emperor Ferdinand III (1608–57) signed the Peace of Westphalia, 1648, in which he recognized the Calvinist faith within Germany.

Philip II of Spain dominated European politics in the late 16th century. Philip was looked to as the champion of the Counter-Reformation Catholic Church. In practice, as more than one pope complained, he defended first and foremost the interests of the Hapsburgs. His identification of the Hapsburgs with the Catholic interest, using the Inquisition in Spain, complicated the religious and political life of Europe and he was involved constantly in war – against the Turks, the rebellious Dutch, the French Huguenots and the English. Philip made Spain a great power, but in the process he aligned against her almost all the states of Europe.

after more than 20 years of civil war in France the political and religious situation was reversed. Now the Catholics feared repression and the Guise feared subjection. Catholics, formerly apologists for divine monarchy, penned tracts justifying rebellion. But Henry of Navarre, who became king in 1589, finally brought peace to France. Renouncing his former Protestantism (saying "Paris is worth a Mass") he thus separated Protestantism from the House of Bourbon. By tolerating the Huguenots in the Edict of Nantes (1598) he appeased his former allies and ended the French civil wars.

6 The Protestant Henry of Navarre (1553–1610) became heir to the French throne in 1584 after the death of the Duke of Anjou. From that date and

7 A

7 Gustavus Adolphus [A] came to the throne of Sweden in 1611 and as a young man defeated Denmark,

Poland and also Russia. He built a fleet and a formidable army which transformed Sweden

into a great power. The threat to Sweden was the Hapsburg Catholic control of northern Germany [B]. The Polish war persuaded Gustavus to postpone a defensive war against

B

Magdeburg 1631 × Breitenfeld 1631
Lützen 1632 ⊠ Steinau 1633
Nuremberg 1632 ⊠
Nördlingen 1634 × ⊠ Donauworth 1632

☐ The Empire 1630–34
▨ Main areas of war
⇨ Sweden enters war
× Imperial-Allied victory
⊠ Protestant-Allied victory

the Hapsburgs in the Baltic, but growing Hapsburg strength after 1629 threatened Swedish security as well as Swedish Protestantism. The Swedes won resounding victories at Breitenfeld in 1631 and Lützen in 1632 but Gustavus Adlophus was killed in the latter battle and without his leadership the Swedes were routed at Nördlingen, 1634.

8 The Defenestration of Prague occurred in 1618 when the Bohemian Protestants hurled two imperial regents from a council window. In 1609 they had obtained from Emperor Rudolph II (1552–1612) a guarantee of religious equality and freedom of worship. Rudolph was succeeded by the old and sick Matthias, who was urged to name as his successor the king of Bohemia. Although they disliked Ferdinand of Styria, the Bohemians consented to his election. The crisis occurred after Matthias had visited Prague securing the election. In his absence the ten regents he appointed denounced the decree of religious freedom and demolished Lutheran churches.

9 The policies of Cardinal Richelieu (1585–1642), minister to Louis XIII, reflected the French conflict of interests at home and abroad. The French saw the Catholic Hapsburgs as their greatest threat. If Philip II and III had established strong controls in both Spain and the Low Countries, France would have been sandwiched between two branches of a hostile power. It was the dilemma of the French monarchy that its interests required Catholic orthodoxy at home but support of the Protestant cause abroad. From 1634 Richelieu brought France openly into war with the emperor (until 1648) and with Spain.

Politics in the age of Elizabeth

Elizabeth I (reigned 1558–1603) of England [Key, 1] believed that as a queen she belonged to a unique species and governed by divine ordinance. She told the House of Commons that matters were revealed to her "princely understanding" that could not be comprehended by "a knot of harebrains", and the head of the body politic was not to be ruled by the foot. But her theory of government was so intuitive and practical that she seldom needed to discuss it. Where James I would philosophize about the respective roles of king and Parliament, she met opposition with curses, demotions and sharp spells of imprisonment.

The organization of government

The heart of the Tudor system was the Privy Council. There, and in its offshoots such as the Star Chamber and High Commission, policy was decided in matters parochial as well as national. Effective government depended on the industry and loyalty of the principal secretary, and in this Elizabeth was scrupulously served by Lord Burghley [5], who was her secretary from 1558 to 1572, and Francis Walsingham (1530–90).

But the decisions of the Council could not be put into effect without the co-operation of the Commons and the unpaid magistracy of the justices of the peace (JPs) who had to enforce a growing corpus of social legislation. Most JPs belonged to the class of rising gentry [4] who were buying land and acquiring seats in Parliament. Thus their enthusiasm for legislation in Parliament reflected their willingness to enforce it.

Elizabeth and Parliament

Elizabeth reserved to the royal prerogative all decisions on national religion, foreign policy, her marriage and, later, the appointment of her successor. Religion was central to all these and the aggressively Protestant Commons wanted to amend the moderate religious settlement of 1559 and demanded a foreign policy hostile to the Catholic powers. Elizabeth would not allow the Commons to debate these issues, but they asserted that they had privileges of freedom of speech and freedom from arrest. Elizabeth, however, declared that the right of free speech meant merely the Commons' right to discuss what was set before them by the royal ministers, not to initiate legislation of their own. Ministers controlled debates through the Speaker, who was at that period a royal nominee and not a servant of the House; dissident members were occasionally imprisoned in the Tower of London.

But it was a losing battle. As inflation devalued the Crown's hereditary revenues from land and feudal dues, Elizabeth remained solvent only by selling lands and offices; and when after 1585 she had to meet the expense of a long war with Spain and rebellion in Ireland she became increasingly dependent on Parliament's support. Refusing assent to taxation was the Commons' ultimate weapon, and although the religious issue was quiescent in the 1590s, Elizabeth had to make concessions over monopolies. Measures for the regulation of trade belonged indisputably to the royal prerogative, and in theory a monopoly for the sale or manufacture of a product was considered a legitimate reward for enterprise. But Elizabeth had been selling monopolies as a

1 **Elizabeth**, seen here in her Armada portrait of 1588, was anxious to build up an image of majesty. But in practical politics she was down-to-earth, and let none of her subjects endanger her power.

2 **Sheep-farming** often replaced tillage as a result of enclosures. The ensuing drop in agricultural employment led Thomas More (c. 1478–1535) to say in his *Utopia*: "Sheep have become devourers of men".

3 The population of England rose steadily throughout the 16th century as plague became less virulent. Contemporaries thought that the country was overcrowded and suggested colonization as a means of disposing of the surplus. But the real cause of the unemployment and distress was inflation, a Europe-wide phenomenon of the period, and the failure of wages to keep pace with prices. Even though inflation steadied after the 1560s, when Elizabeth restored the currency after Henry VIII's debasement of the coinage in the 1540s, by 1600 prices were still more than five times greater than a century earlier. Despite the social problems caused by this inflation, it acted as a stimulus to a burst of industrial activity, including coal and iron mining.

4 Warminster Hall, on the border of Oxfordshire and Warwickshire was bought in 1572 by Richard Cooper, a successful yeoman. He rebuilt the house using local brown ironstone. The rural middle classes that developed in the 15th century became socially dominant because they bought much of the land available after the dissolution of the monasteries. Many were energetic and eager to succeed, and people complained that they disregarded the traditional obligations of landowners to the peasantry. This was the class that came to challenge the royal authority in Parliament, and in the 17th century was to take over much authority in the countryside from the traditional nobility.

financial expedient, and her surrender on this issue in 1601 was ominous for future attacks on royal authority.

Outside the area of political conflict the Tudor state exercised wide paternal powers. Imports might be prohibited to protect the home producer and exports banned in times of shortage. The Statute of Apprentices of 1563 was an attempt to settle wages and working conditions at a time of economic uncertainty. To assist the fishing industry, Protestant England still prohibited the eating of meat in Lent, and the sumptuary laws upheld the gradations of society by controlling the dress and diet of the lower classes. Despite Puritan opposition to Sunday activity, sports were officially encouraged on Sundays because they contributed to a healthy yeomanry which was considered important to national defence.

Social problems

Economic regulation failed to solve the growing problem of enclosures. The open-field system of strip farming was inefficient, and agriculture benefited from the consolida-tion of more compact fields. But the increasingly common enclosure merely for sheep-farming [2], which employed less labour and might deprive the peasantry of the common land on which they kept their beasts, was a source of hardship for many peasants. By the end of the century Elizabeth's governments had some success in maintaining tillage and checking enclosure to ensure social stability; but the merchants and gentry who bought land were anxious to exploit its full economic value, heedless of traditional rights that villagers might have on it.

The government was more successful in meeting its obligations to the dispossessed. Poverty and unemployment became a serious problem, aggravated by chronic inflation [3], and gangs of masterless men and beggars [6] in the towns caused widespread alarm. A series of measures culminated in the Poor Law of 1601, which distinguished between the sick and incapacitated poor, who were to be assisted from the parish rate; the able-bodied, who were to be provided with materials and work; and the wilfully idle, who were to be branded and whipped.

Elizabeth toured her country in triumphal processions almost every year of her reign. As well as creating a close rapport between the queen and her subjects, on which she prided herself, these progresses enabled her to gauge the dominant mood of the country. The cost of the tours was met by local noblemen whose hospitality she sought; they were thus an important way of reducing the huge expenses of her splendid court. There were many distant parts of England which she rarely visited.

5 William Cecil, Lord Burghley (1520–98), acted as adviser to Elizabeth for 40 years, including his times as lord treasurer after 1572. He too was cautious and recognized the limitations on action that were imposed by Elizabeth's unique position as an unmarried queen without obvious heirs. He personally sympathized with the Calvinists, but supported the Anglican settlement publicly, and was careful to maintain strict control over the House of Commons to prevent a decline in royal authority. But he sometimes complained that Elizabeth's policies were vacillating. His son Robert, Earl of Salisbury (1563–1612) inherited his authority in 1598 and served under James I.

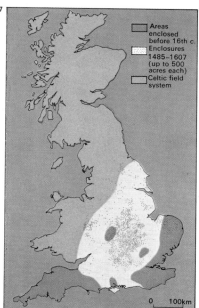

6 Beggars were increasingly numerous in Elizabeth's reign. Genuine unemployment was aggravated by the ancient system that left each parish responsible for its poor, so that a man travelling in search of work would be moved on quickly before he became a public liability. Despite much protest against heartless landlords, there were also thought to be many professional beggars who had ingenious methods of feigning dereliction and disease. Such beggars roamed the country in hordes. The Poor Law of 1601 tried to distinguish between these and genuinely needy people. It set a fixed rate from parishioners for workhouses. Begging was banned and "sturdy beggars" whipped.

7 The enclosures of the Tudor period were mostly confined to the old "champion" (midland) country and even there only 6% of the land was affected. Elsewhere enclosed fields had been used for centuries. Enclosure became less common after 1560.

Areas enclosed before 16th c.
Enclosures 1485–1607 (up to 500 acres each)
Celtic field system

0   100km

8 Education in Tudor England was reorganized after the monasteries, once the main source of lay education, were destroyed. But many grammar schools, for the children of the middle classes, were set up, as literacy became necessary.

9 Robert Devereux, Earl of Essex (1566–1601) was a glamorous nobleman who fascinated Elizabeth in her last years. After falling out of favour, he returned to London from Ireland in 1601 to organize a coup to regain power. On its defeat he was executed.

Elizabeth and the Armada

When Elizabeth I came to the English throne in 1558, France, not Spain was the immediate enemy. Philip II of Spain (reigned 1555–98), at that time a widower, had been married to Mary, Elizabeth's sister, and he proposed marriage to Elizabeth. Although Elizabeth declined the offer, as she was to decline all other offers of marriage, she and Philip had a common fear that the ambition of the Guise rulers of France would threaten their lands.

The threat of Mary, Queen of Scots

The danger to England was from the "auld alliance" of France and Scotland, where Mary of Guise (1515–60) was acting as regent for her daughter, Mary Stewart, Queen of Scots [5], who was married to the French king. Elizabeth connived at and unofficially assisted a Protestant revolution that overthrew French influence in Scotland in 1560, but Mary Stewart's husband died and in 1561 she returned to rule her native country, a beautiful young widow and the most attractive match in Europe. As a Catholic and the great-granddaughter of Henry VII, she was considered to be the

rightful queen of England for those of the Catholic faith who regarded Elizabeth as a Protestant bastard.

After a rash and unsuccessful campaign to recover Calais (1562–3), Elizabeth realized the weakness of her position and settled into the waiting game that thereafter characterized all her foreign policy. By procrastination, apparent indecision and calculated twists of policy she weathered crises that brought her advisers to despair. She was evasive on Mary's requests to be acknowledged as her successor; and when in 1568 Mary fled from Scotland, she kept her in protective custody for 19 years. So long as Mary lived, France or Spain would hesitate to make an all-out attack on England: it was not until war with Spain had begun that Elizabeth consented to Mary's execution.

War with Spain

In 1572, the French Huguenots (Protestants) were broken in the Massacre of St Bartholomew, and the revived civil war weakened France's activity as a protagonist in international politics. Since the excommunication of

Elizabeth by the pope in 1570, Philip, as secular leader of the Counter-Reformation, had a justification for war against her, and a subdued England would have meant improved routes for him to The Netherlands.

There were provocations on both sides, particularly the English privateers' attack on Spanish shipping and posts in the West Indies, and Catholic plots to murder Elizabeth. Yet the drift to war with Spain was gradual. English commercial ties with Spanish-controlled Flanders were too strong to be easily broken, and Philip gave little support to his envoys when they plotted against Elizabeth's life [2]. Elizabeth likewise was equivocal towards the seamen who captured Spanish treasureships and disrupted the trade routes. She willingly shared the plunder, but she disavowed responsibility. She was equivocal too, at first, towards the Dutch revolt against Spanish rule, because she disapproved in general of rebellious subjects and she had not yet the resources for a head-on conflict with Spain.

The assassination in 1584 of William the Silent, the leader of the Dutch rebels,

3 The English militia of levies organized county by county was the only means of defence on land, since there was no standing army. After 1557 every man between 16 and 60 was liable for training in the use of pike and musket. But to be efficient the militia needed energetic lord-lieutenants; much of its equipment was old and out of date. It was used to defend the Scottish and Welsh borders, and could not be used overseas.

1 The fall of Calais, the last English foothold in Europe, in 1558 after 200 years of English control, resulted from a war between France and Spain; England was allied to Spain by the marriage of Mary.

2 Edmund Campion (1540–81) was leader of the Jesuit mission in England that was associated with plots to kill Elizabeth to restore a Catholic monarch. He was betrayed and executed for treason.

4 Elizabeth sent the Earl of Leicester in command of an official royal army to The Netherlands to stiffen Dutch resistance to the Spaniards after the Catholics had captured Antwerp. He set out in December 1585, but returned a year later in disgrace. The expedition was costly, Leicester had quarrelled with his allies and angered Elizabeth by assuming the title of Governor of the United Provinces. This conflicted with the queen's wish to respect Spanish rights of sovereignty. Leicester returned to The Netherlands in June 1587, but was again recalled in November. He is shown here in command of the troops assembled at Tilbury in 1588. English troops fought in The Netherlands until the peace of 1608.

5 Mary, Queen of Scots (1542–87), the daughter of James V, was sent to France in 1548 to foil plans to marry her to Edward VI of England. She married the future French king Francis II (1544–60) instead. She later married Lord Darnley, her cousin. This was dangerous to Elizabeth because each had a claim to the English throne. Mary was driven from Scotland after the death of Darnley. Elizabeth imprisoned her in Sheffield Castle, to prevent her causing further trouble to England. But Mary was involved in several plots to assassinate the English queen and in 1587 Elizabeth at last consented to Mary's execution. She had been unwilling to do this because they were cousins and because it implied an attack on the royal authority.

prompted Elizabeth to send the Earl of Leicester on an official expedition to help the rebels in 1585 [4], thus provoking the long-feared war with Philip. John Hawkins (1532–95) had by that time reorganized the English navy. To meet the needs of the age he replaced the high-built carracks, fit to operate only in home waters, with fast ocean-going galleons longer in proportion to the beam [Key]. He armed them with long-range cannons, and used smaller crews, which meant less congestion and disease.

Francis Drake (c. 1540–96) employed some of these new ships in raids to harass the Spaniards at little cost. In 1585–6, after a landing in Vigo Bay, he plundered the Cape Verde Islands and attacked San Domingo and Cartagena in the Spanish Main. In 1587 he sailed to Cadiz and destroyed 30 ships of the Armada that Philip was preparing to carry an army to England.

The Spanish Armada
Such tactics delayed and weakened the Spanish invasion plan. Nevertheless it was a formidable force of 130 ships and 30,000 men, commanded by the Duke of Medina Sidonia (1550–1615), that rode into the Channel in July 1588 with the object of joining a Spanish army of 17,000 assembled in the Flemish ports [6]. England's land preparations were so dilatory and inadequate [3] that there could have been little effective resistance if this combined force had landed, or even if Medina Sidonia had interrupted his progress to seize a Channel port.

The English victory over the Armada [8] was a combination of good fortune and skilful tactics, and it was immediately apparent that the Spaniards would never again be able to threaten the existence of a Protestant England. But the war continued until 1604, with few dramatic successes on either side. English involvement in The Netherlands continued, and Elizabeth intervened against Spanish ambitions in France in the 1590s; the Spaniards, on the other hand, tried to intervene in a rebellion against English rule in Ireland [7] in 1596–7. The war became an increasing drain on the resources of both countries and the peace made by James I was well received at home.

The *Ark Raleigh*, later named the *Ark Royal* after being presented to the queen, was one of the finest of the new ships of Elizabeth's reign. It was a four-masted galleon, the main masts bearing topgallants. Walter Raleigh (c. 1552–1618) built it for his own privateers. During the Armada, it was the flagship of Lord Howard of Effingham, the English commander. Later it was named the *Anne Royal* after the queen of James I.

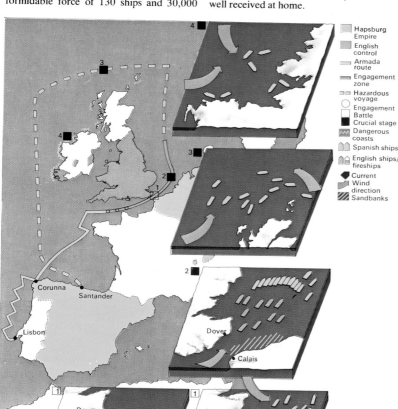

Hapsburg Empire
English control
Armada route
Engagement zone
Hazardous voyage
Engagement
Battle
Crucial stage
Dangerous coasts
Spanish ships
English ships/ fireships
Current
Wind direction
Sandbanks

Corunna
Santander
Lisbon
Dover
Calais
Dover
Calais
Dover
Calais

6 The Spanish Armada put out from Lisbon at the end of May 1588, but was immediately beset by storms which forced it into Corunna. It was not until July that it could sail for England. Lord Howard (1536–1624), the English admiral, sailed out of Plymouth when the Armada arrived eventually in the Channel, to secure the windward position, but the Spaniards continued eastwards. There were some serious encounters off Portland and the Isle of Wight, but when the Armada reached Calais, the army it intended to convoy to England was not ready. Fireships sent in by the English caused many Spanish ships to drift out to sea. Medina Sidonia, the Spanish admiral, tried to re-form off Gravelines, but the fleet was destroyed by the superior gunnery of the English at close quarters. As the wind veered south-west, the Spanish had to turn north, and were pursued as far as the Firth of Forth. Bad weather then wrecked the fleet, which limped home with more than half its ships missing. The English did not lose a fighting ship.

7 Rebellion in Ireland was dangerous towards the end of Elizabeth's reign. Turlough O'Neill (d. 1595), here seen acknowledging the authority of the English commander Henry Sidney in 1567, kept relative peace, but after his death a countrywide revolt occurred that the Spaniards tried to support in 1596–7. The Irish rebellion was more expensive to Elizabeth than the whole war with Spain.

8 Of the 130 ships of the Armada, only about 50 were designed as fighting craft. The rest carried equipment for the invasion. The galleons themselves were taller and less manoeuvrable than the English ships.

Exploration and trade in Tudor England

The sea-captains of the age of Elizabeth (1533–1603) – Drake, Raleigh and Hawkins – provide probably the most enduring myths of Tudor England, but their exploits were built on a history of exploration that went back nearly a century. The motives to explore were various; but the most important were the economic pressure of a rising population, inflation, and declining traditional markets. So too were sheer inquisitiveness and a thirst for adventure, and a national pride that would not permit England's enemies to enjoy the wealth of the New World unmolested.

The changing pattern of trade

Trade in wool and unfinished woollen cloth was England's vital export in the early sixteenth century. But in the 1540s the English currency was devalued by debasement and the exchange rate turned against the English exporter. The ancient "wool-staple", or official distribution centre of English wool in Europe, disappeared with the loss of Calais in 1558, and Antwerp itself declined during The Netherlands' long struggle for independence from Spain. The wool industry therefore was seriously affected and capital was diverted to the enclosure of land for improved agriculture rather than pasture, and to the development of coal and other mineral deposits. Traders therefore began to look overseas: merchants hoped to sell cloth to the heathen in undiscovered countries. Exploration was therefore central to economic expansion.

The start of English exploration

The first "English" explorer was an Italian by birth, John Cabot (c. 1450–c. 1500) [1], who had settled in Bristol. He set out in 1497, commissioned by Henry VII to discover lands not previously known to Christians, to circumvent the papal bull of 1493 that divided between Spain and Portugal all new territories. All subsequent Tudor voyages only acknowledged territory "effectively occupied" in 1493 as related to this bull. Cabot found Nova Scotia and the important cod fisheries of Newfoundland, and suggested a northwestern route to the Spice Islands and the fabled riches of the East that were to delude many future explorers.

Cabot's son Sebastian (1474–1557) unsuccessfully projected a northeastern passage to India, but the expedition of Hugh Willoughby (died 1554) and Richard Chancellor (died 1556) in 1553 took them to Moscow; the Muscovy Company that they set up traded with Russia and established an overland route to Persia. Earlier, in 1509, Sebastian had crossed the Davis Strait, northwest of Newfoundland, and come to Hudson Bay, which he thought to be the Pacific. In three voyages (1576–8) Martin Frobisher (c. 1535–94) [2] reached the same waters; while John Davis (c. 1550–1605) established that Greenland was separate from the mainland and sailed into Baffin's Bay. The Northwest Passage was never free of ice, and so it could not be, as was hoped, complementary to the Cape Horn and Cape of Good Hope routes. The belief in a great southern continent stretching from the south of Cape Horn to the East Indies was also doomed to disappointment. But in search of these far-ranging and often fictional objectives the Elizabethans sailed the known Earth and sometimes beyond it [4].

1 John Cabot and his son Sebastian discovered Nova Scotia off the North American coast in 1497, but believed that they had found Asia. John died on another voyage a few years later. In 1509 Sebastian sailed past southern Greenland into the Davis Strait, and may have gone as far as Hudson Bay. He later went to Spain and spent many years as pilot-major to the mercantile marine, making explorations of the South American coast. He returned to England in 1547, and helped to form a joint-stock company in 1553 to search for a northeastern passage. This company sponsored the expedition of Willoughby and Chancellor to Moscow.

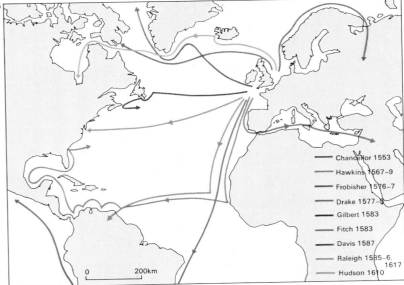

2 Martin Frobisher claimed to have found the elusive Northwest Passage in 1576, and a Cathay Company was set up to develop trade with China. The following year he thought he had discovered gold, and found backers for a new voyage to Greenland, almost unknown since the Viking voyages of the 10th century. He took some Eskimos, with whom he is shown fighting here, back to London, but the English climate killed them.

3 Navigational aids were vital for the success of ocean-going voyages and many navigation schools, such as Gresham College, were set up in Elizabeth's reign to teach sailors the new skills. These were as important to fishermen as to explorers. As well as the astrolabe, developed by the Portuguese from Arab designs, but which was much easier to use on land than on a ship pitching at sea, revolutionary innovations included Mercator's world map of 1569. Its projection allowed a compass course to be plotted as a straight line on a map. Another was John Davis's backstaff, which modified the cross-staff.

4 English exploration did not get fully under way until after 1550. Thereafter the main purpose of the voyages was to find a route to the lucrative markets of China and South-East Asia without infringing on the Spanish and Portuguese possessions that had been apportioned by papal bull in 1493. Thus the merchant companies such as the Muscovy Company sought to find new markets for English cloth in exchange for timber and other goods useful for the shipping industry rather than to establish colonies. Few other expeditions proved as profitable to English industry although raw materials brought home included tobacco, potatoes and many minerals. But the hope for the Cathay market was never realized and home industry did not profit from the American colonies until the mid-17th century. It was not until 1583 that Humphrey Gilbert attempted to set up the first permanent colony in the New World, in Newfoundland, but his expedition was destroyed by storms.

Chancellor 1553
Hawkins 1567–9
Frobisher 1576–7
Drake 1577–9
Gilbert 1583
Fitch 1583
Davis 1587
Raleigh 1585–6, 1617
Hudson 1610

0   200km

Voyages to the Caribbean brought the English into contact with the Spanish Empire. In 1562 John Hawkins (1532–95) [5] took Negro slaves at Sierra Leone and sold them to the Spaniards at San Domingo. A second, similar, voyage in 1564 seemed to have established a profitable commerce, but in 1567–8 a Spanish fleet caught him at San Juan d'Ulloa and only two of his ships reached England safely. Hawkins then left the sea to reorganize the navy, but Francis Drake (c. 1540–96), who had been with him at San Juan, returned to the Caribbean and in 1572 sacked Nombre de Dios and captured a treasure-train there.

Later voyages

In 1577 Drake set out in the *Golden Hind* sponsored by the queen, to find a south-western route into the Pacific. Sailing through the Magellan Straits, Drake plundered unguarded coastal towns, claimed possession of California, turned west for the Moluccas, and returned home in 1580 via the Cape of Good Hope [6]. In 1586–8 Thomas Cavendish (1560–92) became the second

Englishman to sail round the world.

Anthony Jenkinson (died 1611) traded in central Asia, Ralph Fitch (1583–1611) traded with Akbar, the great Mogul ruler of India, and James Lancaster (c. 1550–1618) returned with booty from the East Indies. From such individual enterprises companies were formed to exploit the commercial gains, notably the East India Company (1600). But colonization to establish permanent settlements, rather than trading posts, grew more slowly. Its motives were commercial, to obtain markets and raw materials; political, to weaken Spain; religious, to spread God's word; and social, to re-settle the unemployed. Humphrey Gilbert (c. 1539–83) died on a voyage to colonize Newfoundland, and his half-brother Walter Raleigh (c. 1552–1618) was imprisoned in the Tower of London before his settlement of Virginia took root [9]. Seduced by legends of fantastic wealth, Raleigh sailed to the Spanish Main (northern South America) in 1595, and the failure of a second expedition in 1617–18 to find Eldorado, the mythical golden city of the Incas, resulted in his execution.

Francis Drake captured a Spanish treasure ship in March 1579 in the Pacific Ocean, in the course of his voyage round the world, and thereby ensured the financial success of his trip. Sixteenth-century Spain depended on large shipments of gold and silver from the New World and their capture could be seen as a contribution to the defence of England and the Protestant faith. The expeditions also took home raw materials.

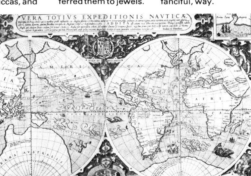

5 John Hawkins was the pioneer of the slave triangle that became the basis of English prosperity in the 18th century. He later became treasurer of the navy and introduced many innovations in the design and manning of the ships.

6 During his circum-navigation of the globe Drake crossed the Pacific from California to the Spice Islands. There he received a great welcome from the natives of the Moluccas, and took aboard three tonnes of cloves. He also made an agreement with the Sultan of Ternate for future supplies of spices which were so highly valued that Drake preferred them to jewels.

7 The *Mariner's Mirror* (1587) was one of several works in English on geography and navigation dealing with its subject in a scientific, rather than the familiar fanciful, way.

8 Richard Hakluyt the younger (c. 1552–1616) was a clergyman who, taunted in France that the English had achieved little in exploration, wrote *Principall Navigations* (1589–1600) to vindicate his countrymen by writing a detailed record of their voyages. Earlier he had written pamphlets to encourage settlers to go to Virginia. But it was hard to persuade investors to become involved in the new colonies, in which profits took far longer to accrue than in privateering expeditions. The joint-stock companies that were the usual form of investment meant that traders sought profits from single trips, not long-term advantages. But by 1600 permanent control of the East Indian trade seemed desirable.

9 The first settlements of Virginia were failures. Conceived of as a means to tap the vast resources of North America for England and to provide a safe market for English cloth, they were organized by Walter Raleigh (who never went to Virginia himself) and were commanded by Richard Grenville (c. 1542–91). Two expeditions – in 1585 and 1587 – were sent and established on Roanoke Island, but, as a result of the climatic extremes and Indian hostility, the first returned home with Drake after one winter and the second disappeared without trace. Each comprised more than 100 men and women, but Francis Bacon called them "the scum of people and wicked condemned men".

Elizabethan art and architecture

England in the later sixteenth century was artistically isolated. The primary reason for this was Henry VIII's break with the Church of Rome in the 1530s, which made cultural contact difficult with the leading artistic countries – Italy, France and Spain – all of which were Catholic. Thus English art developed its own strong and peculiar character, even though many leading artists and craftsmen were foreigners, often religious refugees. The main art forms were portraiture and country house architecture.

Holbein and English portraiture
In the 1530s the German painter Hans Holbein (1497–1543), one of the outstanding Northern European artists of his generation, settled in London. There he produced an unforgettable image of the king, was widely employed to paint portraits of courtiers and even commissioned by his royal master to record the features of prospective wives [1]. Holbein combined boldly economical design and virtuosity in depicting rich costumes with an astonishing power to recreate the physical presence of his sitters.

So popular was his style that he set his stamp on English portraiture throughout Elizabeth's reign. Elizabeth herself, acutely aware of the power of the visual arts, allowed only certain images to be used in her pictures. Many portraits of the queen made for the nobles' collections included symbolical elements emphasizing her power and the effects of her favour [Key].

Only in miniature painting did Holbein find worthy successors. Nicholas Hilliard and Isaac Oliver (died 1617) raised this art form to European pre-eminence. The conceits and metaphors of the Elizabethan poets have their parallel in the miniature, with its symbolic language of love [2].

Tudor architecture
A similar strain of fantasy on an incomparably greater scale than in the miniature can be found in the Elizabethan "prodigy" houses. Henry VIII's palace of Nonsuch, Surrey [3], introduced a novel flamboyance to domestic architecture. The decoration of its external walls, stucco figures in relief set between panels of slate, was the work of

French craftsmen who had previously been employed by Francis I at Fontainebleau.

Briefly, during the 1550s, the Duke of Somerset, Protector to Edward VI, and several of his political associates, introduced a more serious classicism: first in Somerset House, London, (1547–51) and then at its climax in John Thynne's great house at Longleat, Wiltshire, which gained its present form in the 1570s. Later Elizabethan architecture, however, developed towards an unclassical preoccupation with striking silhouettes, dramatic projections and recessions of mass, enormous areas of glass, and strapwork ornament. Wollaton [5] and Hardwick (1590–6) halls, both designed by Robert Smythson (c. 1536–1614), as well as such Jacobean mansions as Bramshill, Hatfield and Audley End all have this bizarre and inventive character. Some, such as Longford Castle [4], have symbolic forms.

Many magnates who built prodigy houses were impelled to such extravagance by the prospect of entertaining the queen, whose annual progresses took her far and wide through her kingdom. Richly decorated

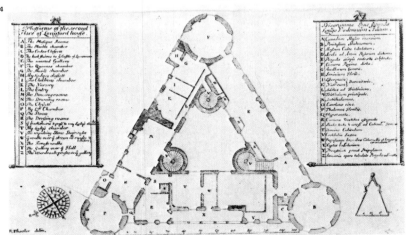

3 Nonsuch Palace, Surrey (built 1538–47), was the most extraordinary palace that Henry VIII built or enlarged. The lifesized stucco figures of gods and heroes set the fashion for Elizabethan plasterwork. Nonsuch was demolished in 1682–3.

1 Hans Holbein, the German portrait painter, was in England 1526–8 and 1532–43. His art was vividly realistic. This work, of 1538, shows Princess Christina of Denmark when she was 17, and being considered by Henry VIII as a prospective bride. The full-faced view gives the clearest and least flattering impression of her looks. Holbein's style was popular until 1600.

2 Nicholas Hilliard (c. 1547–1619) was the finest Elizabethan miniaturist. Portraits in small, often richly jewelled cases were given and worn as love tokens. This "Unknown Man", his shirt in loose disarray and anguish in his eyes, is tormented by the fires of love. Hilliard believed that miniature portraits were too small to allow the painter to use painted shadows and modelling.

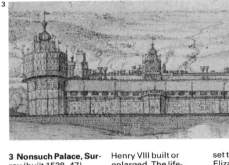

4 The plan of Longford Castle, Wiltshire (1580), in the form of a triangle with a round tower at each corner, symbolizes the Trinity, God the Father, Son and Holy Ghost. Such houses were designed rather for ceremonial forms of life than for comfort.

5 Wollaton Hall, Nottingham, built in 1580, was designed by Robert Smythson for Francis Willoughby, a great landowner of the Midlands and an early coal magnate. Set on a hilltop and flaunting towers of bizarre outline and a great central belvedere, Wollaton Hall epitomizes the self-confidence of the last years of the reign of Elizabeth. Although it has prominent classical pilasters, there is no classical spirit of balance and proportion. Few 16th-century architects' names are known today.

rooms were required, and plasterwork – white or picked out in heraldic colours – polished marbles, alabaster and stones, and even stained glass, all contributed to the effect of dazzling sumptuousness. The chimneypiece [7] usually carried the most elaborate concentration of carved ornament. By the late sixteenth century the great hall was beginning to lose the importance it had possessed in medieval times and more private rooms for the family, in particular the great chamber and the long gallery, had more attention lavished on them [6].

Sculpture and the applied arts

Another opportunity for an elaborate display of sculpture was the funerary monument [8]. Many imposing Elizabethan monuments survive, brilliant with heraldic colours and guilding. Artistically the best of these sculptures were fashioned by immigrants from the Low Countries – men such as Cornelius Cure (died 1607) and Maximilian Colt (flourished 1600–45) whose best known works are the monuments of Mary, Queen of Scots and Elizabeth I respectively in Westminster

Abbey. In general, however, the standard of Elizabethan figure sculpture was low. This mattered little because contemporary taste was universally naïve, delighting in rich textures, intricate detail and garish colouring.

The Tudor period was the heyday of the jeweller and goldsmith. Foreign artefacts were readily obtainable in England, and there is no clearly definable English style. Only miniature painting, considered a form of jewellery (Hilliard was himself a goldsmith by training and probably designed the settings for many of his miniatures), is an exception. Among full-scale goldsmith's plate, the most typically English was the standing salt [9]. Even Holbein designed such salts as well as other kinds of plate and jewels.

During Elizabeth's reign the arts in general had a hesitant start that was affected by the political and religious uncertainty of the 1560s, but developed greater and greater confidence. They did not reach their fullest flowering until the early years of the seventeenth century, when a monarch prepared to lavish patronage on the arts had taken the place of a parsimonious one.

The **Ermine Portrait of Elizabeth I** (r. 1558–1603) was painted by either Nicholas Hilliard or William Segar in about 1585. Portraits of the queen were commonly produced for the houses of the nobility, and they often included literary or poetic symbolism. In this work, the ermine is intended to denote purity, virginity and chastity, and the rubies on her necklace come from the Crown Jewels. Such portraits contributed to the assumption by Elizabeth of mythical status during her lifetime. Many other symbols were used in other paintings: a rainbow in one work was accompanied by the words "No rainbow without the sun", and there were eyes and ears in her clothes.

6 The long gallery, such as this one at Hatfield House, Hertfordshire, was a typical feature of the houses of Elizabethan nobility. Many were built in the roof-space of the house and had good light and views. Long galleries were used for exercise in bad weather and to display collections of portraits in a sumptuous environment.

7 The ballroom of Knole House, Kent, has a rich chimneypiece of variously coloured marble and alabaster (shown here) and a marble frieze in high relief of grotesque monsters, a typical effect of brilliant and barbaric splendour. The house was remodelled internally for the Earl of Dorset in 1605–8.

8 Funerary monuments, such as this one of Thomas Gresham (d. 1579), a merchant and founder of Gresham's Exchange, London, were objects of great expenditure for many Elizabethans. This one, built before Gresham's death, cost him £800. Its free-standing tomb chest of traditional form is remarkable for the purity and restraint of its classical decoration, qualities rare in Elizabethan art. Often funerary monuments included portrait sculptures of the dead man kneeling in prayer with his wife and children. Such portraits were stereotyped, but their setting was luxuriously decorated.

9 The standing salt, such as this example made in 1549, was a decorative object for the dinner table with its utilitarian purpose disguised. Precious materials, laborious workmanship and bizarre forms all held a fascination for Elizabethans. Jewellery and objects made of gold were easily portable and made good gifts; as a result, European style was more uniform in this medium than in any other in the 16th century. Nevertheless the standing salt was peculiar to England. Goldsmiths were legally bound to stamp each piece with the date, the place where it was made and the maker's initials, but unfortunately the key to Elizabethan marks has been lost.

English Renaissance literature

The new learning of Italy in the fifteenth century rapidly spread northwards and was promoted in England by the so-called "Oxford reformers", the humanist scholars John Colet (c. 1467–1519), William Grocyn (c. 1446–1519) and Thomas Linacre (c. 1460–1524). Traditional learning had undergone a change of purpose: it was no longer primarily aimed at elucidating the truths of the scriptures but, through the study of classical philosophy and history, at training future leaders in the arts of government. This reorientation was first summarized in *The Book Named the Governor* (1531) by Thomas Elyot (c. 1499–1546).

Translations from the classics
The new learning also promoted a great output of original works in Latin – from *Utopia* (1516) by Thomas More (c. 1478–1535), a tract on ideal politics and government, to *Novum Organum* (1620) on the ideal of education by Francis Bacon (1561–1626). Scholarly writing in Latin reached only a small audience and therefore translations from the classics, such as Thomas

North's version of Plutarch's *Lives* (1579) [2], became common.

Imitation, although not in the sense of slavish debasement, was another important aspect of the literature of the century. The brilliance of the Italians Francesco Petrarch (1304–74), Giovanni Boccaccio (1313–75), Ludovico Ariosto (1474–1533) and, later, Torquato Tasso (1544–95), was quickly taken as a model; Thomas Wyatt (c. 1503–42) and the Earl of Surrey (c. 1517–47) both distinguished themselves in their sonnets based on the pattern devised by Petrarch. It was the beginning of a great age of English poetry that continued through Spenser, Sidney and Shakespeare to the seventeenth-century poets Robert Herrick (1591–1674), Richard Lovelace (1618–57), Andrew Marvell (1621–78), John Donne and John Milton.

All kinds of poetry were produced during this extraordinary period – from the long romances such as Spenser's *The Faerie Queen* (1589–96) and the great epic *Paradise Lost* (1667) by Milton, to brief, gem-like lyrics from French and Italian models that were

intended to be sung [1]. The greatest poets of Elizabeth's reign were Edmund Spenser and Philip Sidney (1554–86), whose collection of sonnets *Astrophel and Stella* shows great mastery of the Petrarchan form. Spenser's main work, *The Faerie Queen*, was a long romance with allegorical overtones.

During this period there were also many anthologies such as *Tottel's Miscellany*, *The Paradise of Dainty Devices* and song books compiled by eminent musicians such as William Byrd (1543–1623) and Thomas Morley (c. 1557–1603).

In drama the revival of Greek and Latin learning introduced new classical elements. The old morality play had culminated in the plays of Christopher Marlowe (1564–93), whose three great tragedies, *Tamburlaine the Great*, *Doctor Faustus* and *The Jew of Malta*, are studies of heroic power and the vices that such power brings with it.

The Senecan influence
With the new learning, the Latin playwright Seneca (c. 4 BC–AD 65) became a favourite model for tragedy. Here were introduced the

1 The lute, a stringed instrument played by plucking with the fingers, characterized music of Elizabethan times, and English lutes were prized throughout Europe. Because singing was such a favourite entertainment at court, a normal gentlemanly accomplishment was the ability both to write his own songs – including lyrics – and to perform them to the accompaniment of his own lute.

2 Plutarch's *Lives* was translated into English in 1579 by Thomas North (the frontispiece is shown here). It provided one of the main sources for plot and character in the Roman plays of Shakespeare.

3 The authorized King James version of the Bible, based on earlier work by William Tyndale and Miles Coverdale, was produced in 1611. It is one of the great monuments of English literature in itself.

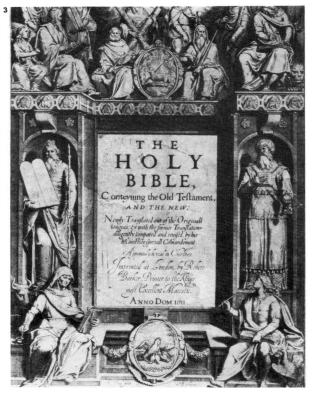

4 This chronological diagram shows the development of various forms of writing in the golden age of English literature that embraced the Tudor and Stuart periods. The interpreters of the new learning of the Italian Renaissance predated the dramatists and poets but all three reached a peak in the late Elizabethan period, which was sustained throughout the 1600s. Drama's achievements were most concentrated in time – between 1560 and the Civil War.

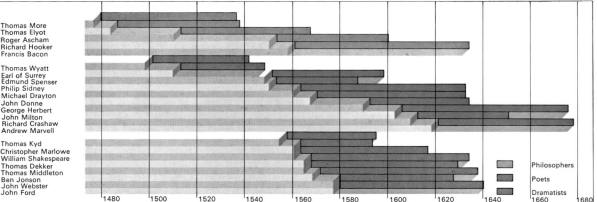

Thomas More
Thomas Elyot
Roger Ascham
Richard Hooker
Francis Bacon

Thomas Wyatt
Earl of Surrey
Edmund Spenser
Philip Sidney
Michael Drayton
John Donne
George Herbert
John Milton
Richard Crashaw
Andrew Marvell

Thomas Kyd
Christopher Marlowe
William Shakespeare
Thomas Dekker
Thomas Middleton
Ben Jonson
John Webster
John Ford

1480 1500 1520 1540 1560 1580 1600 1620 1640 1660 1680

Philosophers
Poets
Dramatists

themes of vengeance and retribution, with accompanying horror and violence; also stage mechanics to provide surprises.

The earliest English tragedy in the manner of Seneca was *Gorboduc* (1561) by Thomas Norton (1532–84) and Thomas Sackville (1536–1608). It started a tradition that culminated in *The Spanish Tragedy* (1592) by Thomas Kyd (1558–94), in which the Senecan elements are enlivened by a modern setting, stage devices and highly individualistic characters. In comedy the Latin influence of Plautus and Terence was of the greatest importance, as can be seen in two of the best early plays, *Ralph Roister Doister* (*c.* 1553) by Nicholas Udall (*c.* 1505–56) and *Gammer Gurton's Needle* (1575) by William Stevenson (died 1575).

English Renaissance literature is dominated by William Shakespeare. His earlier plays are adaptations of models in vogue; later comedies show a greater complexity in character and plot, as do the cycles dealing with kingship. Then come the great tragedies, followed by the very complex *Cymbeline, The Tempest* and *The Winter's Tale*.

Shakespeare's great successor was Ben Jonson (*c.* 1572–1637), who developed a series of characters of outstanding force, especially in *Volpone* (1607). He also exploited the masque [6], which incorporated mythology, courtly life, folklore and satire. In purely literary terms the masque found its highest expression in Milton's *Comus*.

The Metaphysical poets
The group of poets who flourished in the first half of the seventeenth century are known as the Metaphysical poets. Their language was complex and double-edged; their subject-matter contained the pious, the amorous and the reflective. They included Herrick, Lovelace, Marvell and Donne [7].

John Milton [Key] was the greatest poet of the seventeenth century and his masterpiece was the epic *Paradise Lost* [8]. Divinity, the moral and theological concern of the Church, was thus once again in the forefront of English writing. But from now on it was to be tempered by the study of man's character in his own right not merely as the creation of an almighty God.

KEY
A B C

The greatest figures of English Renaissance literature were William Shakespeare (1564–1616) [A], John Milton (1608–74) [B] and Edmund Spenser (1552–99) [C]. Spenser wrote mainly lyric poetry in a sensuous style; Shakespeare is the greatest verse dramatist in English history; and Milton stands as the unrivalled master of English epic poetry.

5 The Globe theatre was probably the largest in London in the late sixteenth and early seventeenth centuries. It was the golden age of English popular drama; the theatres, in total, attracted an estimated weekly audience of 21,000. The audiences consisted of ordinary people in the "pit" while the wealthy sat in the "orchestra". A protruding stage brought the audience into intimate contact with the action. Scenery was scant, though, and effects relied on imagination and context.

6 The *Masque of the Augurs* was designed and the sets built in 1622 by Inigo Jones (1573–1652). The masque shows the intimacy between architecture, theatre and music. It was an aristocratic, exclusive and purely courtly art, often containing disguised political references. The masque was of Italian origin and was an ancestor of modern opera. Even though pageants were not written dramas they have been recorded in designs, drawings and music.

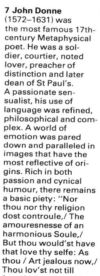

7 John Donne (1572–1631) was the most famous 17th-century Metaphysical poet. He was a soldier, courtier, noted lover, preacher of distinction and later dean of St Paul's. A passionate sensualist, his use of language was refined, philosophical and complex. A world of emotion was pared down and paralleled in images that have the most reflective origins. Rich in both passion and cynical humour, there remains a basic piety: "Nor thou nor thy religion dost controule,/ The amouresnesse of an harmonious Soule,/ But thou would'st have that love thy selfe: As thou / Art jealous now,/ Thou lov'st not till from loving more, thou free / My soule . . .".

8 "The Battle of the Angels", from an edition of Milton's *Paradise Lost*, reflects his interpretation of life as a struggle between the forces of good and evil that goes back to the first catastrophic fall of the rebel angels. Milton's interpretation was probably fostered by his experience as a supporter of Cromwell in the Civil War. His struggle was permanent and intense for a man as aware of the attractions of life and nature as was Milton, and he evoked the conflict in language that is at the same time exalted, heroic and yet direct: "A dungeon horrible, on all sides round / As one great Furnace flam'd, yet from those flames / No light, but rather darkness visible / Served only to discover signs of woe . . .".

Shakespeare and the Elizabethan theatre

William Shakespeare (1564–1616) [1] was born in the prosperous market town of Stratford-on-Avon, Warwickshire, where his father, a glove-maker, served on the local corporation and rose to become bailiff, or mayor. Little is known about William's early life, apart from the fact that in 1582 he married Anne Hathaway (1555–1623) from the nearby village of Shottery [Key]. In 1583 they had a daughter, with twins following two years later. The next record of him is in 1592 when Robert Greene, a rival dramatist, complained of him as an "upstart crow", a common actor who had the impertinence to be writing plays.

Actors and theatres

English drama developed from the miracle and morality plays, mostly performed by the craft guilds in the larger towns, although groups of professional actors had begun to play in market-places and inn yards. In 1576, James Burbage, a carpenter, built England's first permanent theatre. It was in Shoreditch, London, for the exclusive performance of plays and leased to the acting company of which his two sons [7], and later Shakespeare, were to be members.

The life of an early player was far from easy but a permanent theatre was a real advantage for the actors. Unless he had a patron, he could be prosecuted as a masterless man and a vagabond. Shakespeare's company was known as the Chamberlain's Men and, after 1603, as the King's Men. Public theatres were disliked by the Puritan preachers of the time, and also by the magistrates, who saw them as breeding grounds of riots, disease and prostitution. Frequently they were shut down, and strict regulations were imposed on their activities. Many theatres were also used for bear-baiting and prize-fighting as well as the production of dramas.

When the lease of the Shoreditch theatre expired in 1597, the company moved across the Thames to Surrey, where the authorities were less strict, and there built the famous Globe Theatre [2]. They were followed by their great rival, Philip Henslowe (died 1616), the business manager of the Admiral's Men, who built the Fortune Theatre to the same specifications. In Edward Alleyn [5] he had a fine tragedian – his style is burlesqued by the Player King in *Hamlet* – and, although there were other companies and other theatres, the competition between the Chamberlain's Men and the Admiral's Men stimulated artistic achievement. Henslowe's company had a success with a play, now lost, about Robin Hood; this was answered by Shakespeare's *As You Like It*. Shakespeare's constant working relationship with his company makes it highly unlikely that his plays were written by Bacon or by anybody else, as has been suggested.

Shakespeare's plays

Nobody really knows in which order Shakespeare wrote his plays. There seems to have been an early period of experiment, with varied results; and this was followed by the historical period, in which Shakespeare bade Englishmen look into their past as a guide to their present. After this came the golden comedies of his maturity. With the new century came the tragic period, during which even the comedy has an acrid taste and air of

1 Shakespeare is renowned for his non-dramatic poetry and, more especially, for his sonnets, which were probably written in the 1590s. The identities of the people to whom these poems were addressed ("W H" and the "dark lady") are a mystery. A common theme is death transcended by love and art.

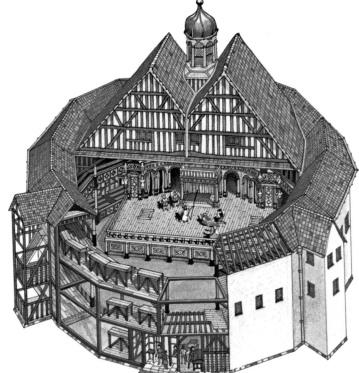

2 The Globe Theatre stood in this form between 1613 and 1644. It was a tiled and brick-built three-storey structure, surrounding an uncovered yard, where the poorer patrons could stand for a penny. It was a small building, with a diameter of only 25m (83ft), but with its galleries it could hold 2,000.

3 Henry V, as portrayed by Laurence Olivier (1907–   ) in the 1944 film, was an attempt to show life and conditions in the Elizabethan theatre, in addition to presenting this most stirring historical play. It reproduced the open stage, the quick reactions of the audience (some sitting on the stage), the boy actors and the frantic congestion backstage. Later scenes depicted the war in France.

4 Stratford Grammar School was founded in 1427 to provide a free education for the children of local guild members. It was closed in 1547 but it reopened six years later with a charter from Edward VI. The school provided the usual education in the classics, and it is likely that Shakespeare attended lessons here, even though Ben Jonson once jibed that Shakespeare knew "little Latin and less Greek". Otherwise he may have been a page in the house of a local nobleman.

disillusionment. Finally, there are the romances, with their promise of regeneration. These plays were more intimate and less rhetorical, partly because in 1608 the company obtained a new theatre at Blackfriars. This was an indoor, winter theatre, with a much smaller audience capacity and higher prices than at the Globe. By this time, Shakespeare was spending most of his time in Stratford. In 1597 he bought New Place, the handsomest property in Stratford. He earned nothing directly from his plays because they belonged to the company, but his share of the theatre profits made him a relatively wealthy and respected man.

Marlowe and Jonson

The early death of Christopher Marlowe (1564–93) removed the only rival whose genius might have shone as brightly as Shakespeare's. Marlowe, who worked for the Admiral's Men, was stabbed to death in a tavern brawl at Deptford, leaving behind a rich legacy of drama. The development of blank verse and the use of tragic themes in these plays prepared the theatre for Shakes-

peare's achievement.

Ben Jonson (1572–1637) [8], who created the English comedy of humours, was perhaps a little too pedantic and intellectual for the common taste, but in his youth he was as fiery as Marlowe. In 1598 he narrowly escaped the gallows for killing Gabriel Spenser, a fellow actor of Henslowe's company, in a duel.

Among Shakespeare's other rivals, George Chapman (c. 1559–1634), translator of Homer, aspired to philosophical tragedies that are somewhat over-rhetorical. The citizen comedies of Thomas Dekker (c. 1572–c. 1632), Thomas Middleton (1570–1627) and Thomas Heywood (c. 1574–1641), who claimed to have had a hand in 220 plays, were popular with London audiences, drawn from all classes. In the first decade of James I's reign, John Webster (c. 1580–1625) produced his powerful tragedies. With the invariably happy endings, the romantic tragi-comedies of Francis Beaumont (c. 1584-1616) and John Fletcher (1579–1625) marked a decline into triviality that lasted until the Restoration (1660).

Hewlands, or Anne Hathaway's Cottage, the place where Shakespeare's wife was born, is situated in Shottery, 1.5km (1 mile) from Stratford and is a highly popular tourist attraction. Shakespeare is probably the world's best known playwright and his works have been translated into most languages.

5 Edward Alleyn (1566–1626) was a great and popular tragedian, equal to Richard Burbage. He joined the Earl of Worcester's players when he was 19, but his best period came when he joined the Admiral's Men, under Philip Henslowe, later his father-in-law. Together they were proprietors of several playhouses. Alleyn retired a wealthy man in 1604, and founded Dulwich College, in London.

6 Richard Tarleton, who died penniless in 1588, was the principal comedian of the Queen's Men. He was a great favourite with Elizabeth and he was depicted as Yorick in *Hamlet.*

7 Richard Burbage (c. 1576–1619) was the son of the man who built the first playhouse. All Shakespeare's great tragic roles were written for him, the finest actor of his day.

8 *Volpone,* one of Ben Jonson's finest plays, is shown here in a recent production at the National Theatre, London. Many of his plays were satirical, although based on classical models. As well as being attached to Henslowe's company, Jonson was a success at the court of James I, and wrote many masques. He admired Shakespeare, although complaining that he wrote by nature rather than by art, and he wrote an affectionate tribute to him in the First Folio edition of Shakespeare's works (1623).

9 First performances of the plays of Shakespeare, Marlowe and Jonson have likely dates ascribed to them in this diagram. Together, these three men elevated the status of the playwright from one of mere scribbler or hack to that of man of letters.

9

Date of first performance	
1588–92	Henry VI Parts I, II, & III · Dr Faustus · The Rich Jew of Malta · Tamburlaine
1592–3	Richard III · The Comedy of Errors
1593–4	Titus Andronicus · The Taming of the Shrew
1594–5	Two Gentlemen of Verona · Love's Labour's Lost · Romeo & Juliet · Edward II
1595–6	Richard II · A Midsummer Night's Dream
1596–7	King John · The Merchant of Venice
1597–8	Henry IV Parts I & II
1598–9	Much Ado About Nothing · Henry V
1599–1600	Julius Caesar · As You Like It
1600–01	Hamlet · The Merry Wives of Windsor
1601–02	Twelfth Night · Troilus and Cressida
1602–03	All's Well that Ends Well · Sejanus
1604–05	Measure for Measure · Othello
1605–06	King Lear · Macbeth
1606–07	Antony & Cleopatra · Volpone
1607–08	Coriolanus · Timon of Athens
1608–09	Pericles
1609–10	Cymbeline · Epicoene
1610–11	Winter's Tale · The Alchemist
1611–12	The Tempest · Shakespeare
1612–13	Henry VIII · The Two Noble Kinsmen · Marlowe
1614	Bartholomew Fair · Jonson

The Stuarts and Parliament 1603–42

In 1603 the peaceful transfer of the English crown to James VI of Scotland (1566–1625), who thereby became James I of England [1], was greeted with joy and relief. The collapse of the new Stuart dynasty was not inevitable. There were, however, severe stresses in seventeenth-century England.

Deep-rooted problems
Royal service and its ensuing favours were important to the landed classes and they formed a staple issue in politics. Elizabeth I had played off the resulting factions against one another to preserve her freedom of action. The rebellion (1601) of the Earl of Essex (1566–1601) destroyed this system and left Robert Cecil (1563–1612), Burghley's son, as the Crown's dominant adviser. James's predilection for favourites eventually brought this monopoly to the Duke of Buckingham [4]. This monopoly encouraged corruption in the search for and exploitation of patronage and office.

The amount of administration required of government in defence and foreign policy and in management of the economy, social problems and reform in the Church had been steadily growing during the Tudor period. By 1600 the available machinery and money were inadequate. Central government was wasteful, and in the country the Crown depended on voluntary work by the gentry, who were deeply suspicious of central authority, hard to discipline and capable of wilful inertia. In Parliament the representatives of that same gentry were unwilling to vote the Crown an adequate share in the increasing national wealth. In 1610 an attempt at basic reform of royal finances (the Great Contract) proposed to exchange the Crown's profit from the wardship of minors and other feudal impositions for a regular peacetime tax. Its failure led the king to exploit for cash his rights of monopolies, customs, wardships and grants of honour.

Parliament's unwillingness to pay for what was needed meant that it had no effective control over the actions of the government. The king could defeat attempts to put pressure on him by dissolving Parliament, and there were real fears about its permanent survival. Before the 1640s it had little thought of any constitutional limitation of the king's power; agitation about parliamentary privileges arose directly from concern about the king's policies [5]. To have any influence, Lords and Commons had to work together, and in the 1620s much parliamentary protest was engineered by nobles who wished to oust Buckingham.

James's revenue and foreign policy
James I was not the man to make the necessary major reforms and in two areas he made matters worse. First, the royal finances had been affected by inflation, and Elizabeth left a debt of £100,000 net, 30 per cent of her annual peacetime revenue. By 1608, however, James's wanton extravagance had increased the royal debt to some £500,000. Second, in the field of foreign and religious policy, James ended Elizabeth's Spanish War in 1604 in favour of a sensible policy of avoiding European conflicts. As pressure grew against Protestants in Europe with the start of the Thirty Years War in 1618, the king's determined policy of peace and his attempts to negotiate with Madrid

1 James I told Parliament in 1610 "The state of the monarchy is the supremest thing on earth". On several occasions James, a genuine scholar, spelled out the implications of the theory of the divine right of kings. Both he and his audience took it for granted that the monarch, as the earthly representative of God, was the basis of all political authority. But as the Stuarts persisted in policies that seemed evidently wrong, the assumption eventually came into question. In its stead, Parliament began to assert the traditional authority of the Common Law, which was propounded in an extreme form by Edward Coke (1552–1634), who helped to draft the Petition of Right (1628).

2 The Gunpowder Plot (1605), was an attempt on the king's life by a few religious fanatics desperate at their failure to secure a relaxation of the laws against Catholicism on James's succession. Its discovery raised fear of Catholicism to fever pitch, and this hysteria was repeated during the crisis of 1640–42.

3 Charles and Buckingham went to Madrid in 1623, hoping to secure Charles's marriage to a Spanish princess. The plan was the culmination of James's policy of peace with Spain. But the policy was unpopular at home; Charles's return unmarried (shown here) was greeted with great celebrations. He and Buckingham then allied with the critics of James's foreign policy and helped first to bring down Lionel Cranfield, the lord treasurer, and his regime of strict economy, and then forced James into a war with Spain that he could not afford. Enthusiasm for the war soon declined.

4 George Villiers, Duke of Buckingham (1592–1628), was a younger son of a minor Leicestershire squire. In 1615 he captivated James by his physical charm and athletic prowess and he "jumped higher than ever Englishman did in so short a time, from private gentleman to a dukedom". The interest of Villiers and his family in corruption prevented any possibility of reform, and his moderate abilities did not justify the military and diplomatic authority he was given. He gave James and Charles uniformly bad advice and encouraged them in short-term decisions that always led to honour and profit for Buckingham. He was assassinated in 1628 after he had involved England in war with France as well as Spain, and inspired disastrous and much criticized expeditions against Cadiz (1625) and to La Rochelle three years later.

looked like appeasement of Catholicism [6] Although James was personally a convinced Calvinist, he flirted with the growing High-Church or Arminian faction in the Church, which stressed ceremony, tradition and episcopal authority. To most Englishmen, Arminians seemed little better than Catholics, and the king's toleration of them looked extremely dangerous [Key, 8].

Charles and the rise of opposition

Charles I (reigned 1625–49) [7], unlike his father, asserted his authority without regard for his dependence upon the co-operation of its subjects. In the Book of Orders, 1631, he attacked the local autonomy of the JPs, insisting on a uniform performance of their duties. Autocratic rule by the Earl of Strafford [9] in Yorkshire and Ireland seemed ominous for the country at large. Charles's archbishop, William Laud (1573–1645), imposed Arminianism on the Church in many areas, including Scotland in 1637. Immoderate opposition in the House of Commons in 1629 impelled Charles to rule without Parliament. Ship Money, an occasional local rate,

became a nationwide annual tax in the 1630s, extending taxation to yeomen and freeholders, hitherto largely exempt.

Although Charles increased his income to a million pounds a year, this was insufficient to oppose a rebellion against Laud in Scotland in 1639. Appeals for money were turned down, JPs refused to act, and the forces raised by Charles mutinied. The Scots occupied northern England, and to pay their wages off Charles was forced to call Parliament for the first time in 11 years.

The Long Parliament (1640–60), called after the Short Parliament (1640) failed to support Charles, was at first overwhelmingly traditionalist and quickly destroyed the alleged novelties of Charles I's "personal rule". A rebellion in Ulster in October 1641 forced Parliament to choose between trusting Charles with an army or insisting on revolutionary constitutional limitations before finding the money to quell the revolt. Polarization on the need for guarantees or the dangers of constitutional change, which would imply social change, split the political nation and led to the outbreak of war.

KEY

Preaching was the principal means of influencing public opinion or spreading information in the 17th century. Charles I said "People are governed by the pulpit more than the sword in times of peace". The regulation of sermons and religious lectures was an area of dispute between the Puritans and the Church authorities, especially the Arminians. St Paul's Cross, seen here, was the main preaching place in London and was carefully controlled by the government. Social and political implications, as much as theological ones, made religion a sensitive issue. James was reluctant to impose any aggressive religious policies, and often engaged in theological disputes.

5 John Eliot (1592–1632), once a protégé of Buckingham, led the outspoken opposition to Charles's policies in the Commons in 1629. He criticized government negatively and believed in the divine right of kings.

6 Henrietta Maria (1609–69), a French princess, married Charles in 1625. The match was unpopular, especially because he tolerated her Catholicism. She led a court faction against Laud and Wentworth.

7 Charles wore this elaborate costume in January 1640 in the masque *Salmacida Spolia* at a time when Scotland was in rebellion. It was light blue cloth embroidered with silver, white stockings, a silver cap and white feathers. The isolated culture of the court, symbolized by these Inigo Jones masques and Charles's patronage of such artists as Rubens and Van Dyck, presented the image of Charles, the embodiment of virtue, ruling an idyllic, contented and peaceful nation. But in this, his last masque performance, he played the Platonic king responding to adverse times with endurance and patience, a new image of himself that he retained until his death in 1649.

Matth.15.13. *Every plant which mine heavenly Father hath not planted should be rooted up.*

Of God, Of Man, Of the Divell.

Loe, here are three men, standing in degree, The least of these, the greatest ought to be.

The other two, of men and of the Devill. Ought to be rooted out for ever as vile.

8 This religious cartoon sets the orthodox Anglican cleric holding the humanly inspired Book of Common Prayer alongside the papist priest with his Devil-inspired superstitions and against the Puritan minister with the divinely inspired Bible. Thanks to Charles's intransigent support for Arminianism, the minority opinion that "root and branch" reform of the Church was necessary grew in strength in the 1630s. So, as well as questioning the king's right in law to command an army for Ireland and Long Parliament also challenged the validity of episcopal authority in general.

9 Thomas Wentworth, Earl of Strafford (1593–1641), was a critic of Buckingham in the 1620s, but entered Charles's service in 1628 fearing that the Commons might undermine royal authority. He was made the scapegoat for the "11 Years Tyranny". He was attainted by Act of Parliament (1641) and Charles could not protect him, under pressure from the London mob, which, as an organized pressure group increasingly threatened the king. Wentworth's execution, shown here, was a vital victory over Charles.

287

The English revolution

The English revolution (1640–60) was the first of the modern revolutions and it compares closely in many ways with the later French and Russian revolutions. Each political revolution follows its individual and complex pattern of development, but some common stages can be identified.

The revolutionary pattern

These stages include an initial period of political and intellectual excitement, with the collapse of the old regime; the emergence of a radical leadership as the political crisis becomes protracted; an eventual halt to the leftwards movement of the revolution; the consolidation of power by a strong-arm and often military-based party and a return towards political stability; and sometimes a formal restoration and a collapse of the revolutionary movement. These stages can be seen in the English, French and Russian revolutions, with the manifest exception that in Russia there has been no restoration of the Romanov dynasty.

The first stage of the English revolution [2] was the constitutional or "bloodless" revolution of 1640–41, which weakened the old absolutist regime. It began in November 1640, when Charles I (1600–49) summoned the Long Parliament to help him out of a financial crisis. Charles, isolated and deeply unpopular, was forced to agree to a wide-ranging set of constitutional reforms that gave Parliament a much more prominent place in the constitution. These changes were accompanied by much political excitement and debate, especially after the ending of press censorship [4] in July 1641.

The escalation of the crisis

The second stage saw the revolution move leftwards, as the political crisis escalated into civil war. The Long Parliament split into rival parties in the autumn of 1641, the Royalists (Cavaliers) forming a King's party in opposition to the Parliamentarians (Roundheads), who were prepared for further political and religious reforms. Charles attempted unsuccessfully to arrest his main opponents in January 1642 and then fled from London. He thus entered the first civil war (August 1642–April 1646) from a weakened position, because the Parliamentarians held London and the machinery of central government. The Parliamentarians were not, however, able to clinch their dominance until the reorganization of their forces into the disciplined and zealous New Model Army in January 1645. This successfully harnessed militant Puritanism to the Parliamentarian cause. The army, ably directed by Sir Thomas Fairfax (1612–71), won a decisive victory over the Royalists at Naseby in June 1645.

The third stage marked the apogee of the leftwards movement of the revolution. The victorious Parliamentarians themselves split at the time of their success in 1646 and 1647. The more conservative group were the Presbyterians, who, together with the Scots, then allied themselves with Charles. The more radical party, the Independents, backed by Oliver Cromwell (1599–1658) [8] and the army, attempted a settlement, failed and remained hostile to Charles. The second civil war (February–August 1648) was a brief but bitterly fought contest, which ended with Cromwell's overwhelming defeat of the Scottish army at Preston in August 1648.

1 Parliament was a well-established institution before 1640 with its own traditions, procedures and privileges. In the House of Commons' chamber there were a Speaker (centre), two committee clerks, the serjeant-at-arms (carrying the mace) and MPs.

2 In the first civil war the Royalists were strongest in areas remote from London in the north and west, while the Parliamentarians kept the capital and the populous southeast. Fighting followed for control of divided areas.

Parliamentarian counties
X Sites of major battles
Neutral counties
Royalist counties

OXFORD
LONDON
0   160km

3 The elegant portrait of Charles I by Anthony van Dyck concealed the king's small stature but emphasized his regal dignity and a certain melancholic determination. Van Dyck was appointed official court painter in 1632 by Charles, who was a famous art patron.

Ould Extreame   Goulden Meane   New Extreame

4 The end of press censorship in 1641 brought a flood of pamphleteering and debate. All traditional ideas were opened to discussion and argument. Before long, moderates on the Parliamentary side called for caution in reform, stressing the Aristotelian "golden mean". They warned of the "new extreme" among radical political and religious groups as much as the "old extreme" of the enemy.

It was followed by the purge of Parliament (Pride's Purge) in December 1648, the trial and execution of Charles on 30 January 1649, and the abolition of the monarchy and House of Lords in February of that year – a sequence of events that was unprecedented in European history.

At the same time Cromwell and the army leadership broke with the lower-class radical political party, known as the Levellers [6]. They pressed for fundamental constitutional changes, implicitly challenging the political and social power of landowning society. In the autumn of 1647, the Levellers briefly challenged Cromwell for control of the army itself. But Cromwell decided to crush the movement. A Leveller army mutiny was defeated at Burford (Oxfordshire) in May 1649. Deprived of its army base, support for the Levellers rapidly waned. But the difficulty of Cromwell's position – sandwiched between left and right – was accentuated: he was a regicide, as well as the man who had broken the Levellers.

The fourth stage represented the gradual return to political stability. The New Model Army was, in the 1650s, a force of more than 60,000 men. The importance of its power was shown in the brutal suppression of the Irish rebellion (1649–50) and the routing of a further Royalist uprising (1650–51).

Cromwell and collapse of army rule

Cromwell proved the most successful ruler (1653–8) because he was both a skilled politician and a brilliant general. His regime had some foreign policy successes, such as peace with the Dutch in 1654 and victory over Spain (1656–8), but this system did not last long after his death in 1658.

The fifth stage in the revolution therefore saw an initial period of political confusion with the collapse of army rule, followed in due course by the restoration of Charles II (1630–85) in May 1660 at the invitation of the Convention Parliament. The restored monarchy [9] was much weaker than it had been previously, and it was the landowning gentry as represented in Parliament who were the ultimate victors of the English revolution, ensuring the permanent end to absolutist monarchy.

The civil wars marked a breakdown in ordered political and social life and animosities heightened during the course of the fighting. There were not only political and religious divisions between the Royalists and Parliamentarians in 1642, but also social ones. The Royalists were satirized as courtiers and rakes; the Parliamentarians as lowborn moralists. The Royalists drew support from some of the gentry and most of the peerage, who often wore lavish costume and long hair, denoting them as men of social standing and wealth. The Parliamentarians also had considerable backing among landowners, but they derived additional support from urban merchants, tradesmen and artisans, who were often motivated by Puritan zeal and wore shorter hair and sober clothing.

5 Edgehill (23 Oct 1642) was the first major battle of the civil war and the majority of officers and soldiers were in action for the first time. The two armies were fairly evenly matched and in military terms the outcome was indecisive. By the end of the war, the king's forces were only a little more than half those of Parliament. The king marched south to set up his capital in Oxford, as London eluded him.

Infantry Cavalry Dragoons
1,000 720
2,150
2,800

Royalist
Parliamentarian

Guns
20
37
10,500
12,000

7 Officers wore civilian garb, whereas troopers on both sides wore some armour.

1 Beaver hat
2 Gorget
3 Partisan
4 Burgonet with face bar
5 Wheel-lock pistol
6 Pistol holster

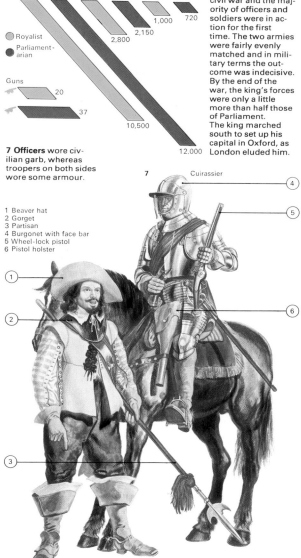

Cuirassier

Captain of infantry

6 A key development in the 1640s was the growth of the first organized, mainly lower-class political party in English history. This group, known as the Levellers, played an important political role in the years 1647–9 before eventually being defeated by Cromwell. The Levellers did not have one party leader, but a dominant part was played by "Freeborn John" – the spirited propagandist, John Lilburne (1614–57), shown here. The Levellers called for a considerable extension of the franchise and were suppressed by both king and Parliament.

8 Oliver Cromwell was a military genius and a pragmatic politician. He ruled as Protector and tried to reconcile discordant factions, staunchly upholding the principle of religious tolerance.

9 The restored monarchy in 1660 was much weakened and power was shared with Parliament. The "Cavalier Parliament" was opened amid much state pageantry on 22 April 1661 with Charles II riding in procession from the Tower of London. The landowning gentry had learned to fear a powerful monarchy and standing armies, and they soon became critical of Charles and his ministers.

The Commonwealth and Protectorate

The execution of Charles I on 30 January 1649 [1] inaugurated an era of constitutional experiment for England. Without a monarchy or a House of Lords, a republican government was established by the House of Commons – now reduced to a "Rump" composed chiefly of the most radical "Independent" members who had survived "Pride's Purge", the exclusion of members opposed to the king's execution. The Scottish Presbyterians, who refused to recognize the Commonwealth, were defeated at Dunbar in 1650, and the young Charles Stuart at Worcester in 1651; Oliver Cromwell [Key] also subdued Ireland in a brutally efficient campaign (1649–52). The Rump passed the Navigation Act (1651) providing for sea trade to be carried in British ships, and enforced it by a successful war at sea against the Dutch (1652–4) [3].

The Cromwellian Protectorate

Despite these military successes, Cromwell became impatient at the Rump's slowness to promote domestic reforms, and forcibly dismissed Parliament in April 1653 [4]. There followed a brief and ineffectual experiment (June-December 1653) with the Nominated, or "Barebones", Parliament, which was made up of religious zealots and was intended to lead the country towards truth.

Cromwell thus became the foremost political as well as military leader of the English Revolution. In December 1653, he was appointed Lord Protector, an office he held until his death. Having come to power by unorthodox means, Cromwell was faced with many difficulties, but he was powerful and successful, although not popular.

In the 1650s the "army of the saints", although politically radical, developed into a highly disciplined force, whose support Cromwell was always careful to retain. The survival of his regime was thus ultimately dependent upon the existence of a loyal standing army of 60,000 men. Cromwell was not, however, simply a ruthless seeker after power for its own sake. He was undoubtedly ambitious, but he was also motivated by a deep commitment to the principle of religious toleration. His domestic policies were not as revolutionary as his career in the 1640s had suggested; he opposed monarchial absolutism and believed in parliamentary rule. He was a pragmatist concerned with ensuring the survival of his experiment, and (as he himself said) not "wedded and glued to forms of government."

Foreign and domestic policies

One solution to a legacy of domestic problems after a period of revolutionary turmoil is to turn attention outwards with an aggressive foreign policy. Cromwell, like Napoleon, did this. He made peace with the Protestant Dutch in 1654 but then waged war against the Catholic power of imperial Spain (1655–8). This policy pleased some English commercial interests; but primarily it allowed Cromwell to invoke glorified memories of earlier wars against Spain under Elizabeth I. Cromwell achieved legal, administrative and educational reforms despite continuing financial problems posed by the war and standing army [7]. No dramatic social changes were attempted. In religious matters, Cromwell extended freedom of worship to all Puritan groups,

CONNECTIONS

See also
288 The English revolution
292 The restoration of the Stuarts
300 Europe 1500–1700
302 Science and technology 1500–1700
282 English Renaissance literature

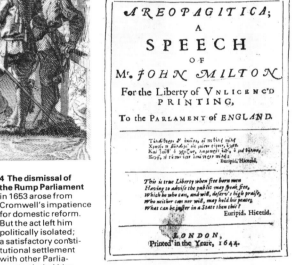

1 **The execution of Charles I** in 1649, after a trial, as a tyrant and traitor shocked many people at home and abroad. But it established the principle that a ruler was accountable to Parliament.

2 *Areopagitica* (1644) by the poet John Milton (1608–74) was a passionate plea for freedom of speech and liberty of conscience. Milton later acted as an apologist for the Commonwealth and Protectorate.

	Trade value	
	Dutch	English
1648	£254,191	£203,054
1650	£214,853	£276,066
1652	£138,561	£266,070
1654	£106,431	£364,486
1656	£176,838	£338,486
1658	£151,389	£352,704
1660	£186,205	£285,000

3 **The first Anglo-Dutch War** (1652–4), between the two main Protestant republics, sprang from longstanding commercial and colonial rivalries that had been exacerbated by the mercantilist-inspired Navigation Act of 1651, which permitted only English ships to serve English or colonial ports. The war resulted in disruption of the Dutch carrying trade and consequent English gains. The English navy under Robert Blake (1599–1657) defeated the Dutch at Portland (1652) and Texel (1653). These successes marked England's emergence as a major maritime power, and henceforth English foreign policy was marked by much greater concern for the significance of commercial and colonial interests.

4 **The dismissal of the Rump Parliament** in 1653 arose from Cromwell's impatience for domestic reform. But the act left him politically isolated; a satisfactory constitutional settlement with other Parliaments eluded him.

including the most radical ones. Members of the now-eclipsed Church of England [5] had to worship secretly, but actual persecution was directed only against Catholics, whose creed was thought to be too intolerant to be allowed to continue unchecked.

The return of conservatism
Although in the 1650s Cromwell had many critics, pressure for abolition of social distinctions and redistribution of wealth, which had appeared in the later 1640s among lower-classes and in the army, the Levellers and Diggers, had largely disappeared – partly as a result of a run of good harvests (1653–6) that reduced food prices and allayed economic discontents. Gentry landowners, finding their authority largely intact, were quiescent. An attempted Royalist uprising in March 1655, led by John Penruddock, attracted little support, but the government panicked and introduced the rule of the major-generals in June 1655. An experiment in direct military control in local affairs, this brought Cromwell's popularity with the gentry to its lowest point.

Cromwell, however, sought constantly to achieve a constitutional agreement with landowner representatives in Parliament [6] that would give his rule a solid basis. He called two parliaments (1654–5, 1656–8) and wrangled with both. But eventually he realized the extent of parliamentary opposition to the major-generals, and dismissed John Lambert, the army "strong man", in July 1658, after accepting a new constitutional settlement, the Humble Petition and Advice, in May 1657. It strengthened Parliament, restored the Upper House, which had been abolished in 1649, and sought to consolidate Cromwell's authority by making him king – an offer he refused under pressure from the army.

When death toppled Cromwell from power, his regime, in the hands of his son Richard, "Tumbledown Dick" (1626–1712) and the now-divided army leaders, did not long survive him. Stuart monarchy was restored in May 1660 [9] and the Church of England was revived. This ended an important era of upheaval and experiment in English history.

Oliver Cromwell (1599–1658) came from a minor landed family and entered Parliament in 1628. His military skill as a commander in the civil wars of the 1640s made him a natural leader of the parliamentary side. Determined to achieve constitutional reform, he supplied the strength of purpose needed to unify the nation without a king.

5 Puritans destroyed religious images in many churches and tried to curb Sunday recreations for all the people.

6 Public reporting of Parliament was stimulated by some expansion of press freedom in the period 1640–60. Cromwell's search for a political consensus on which to base the Commonwealth, and his difficulties with his two parliaments, were widely reported. The members of Cromwell's parliaments were drawn from the ranks of the respectable landed gentry society, but political discussion was common among all classes.

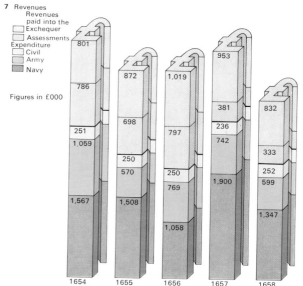

7 Revenues
Revenues paid into the
☐ Exchequer
☐ Assessments
Expenditure
☐ Civil
☐ Army
■ Navy

Figures in £000

	1654	1655	1656	1657	1658
Exchequer	801	872	1,019	953	832
Assessments	786	698		381	
			797	236	
Civil	251			742	333
Army	1,059	250	250		252
		570	769	1,900	599
Navy	1,567	1,508	1,058		1,347

7 Cromwell's finances, like those of the Stuarts, suffered from the chronic problem of rapidly increasing costs. This was aggravated by the need to support and find use for the army and navy. But Cromwell never had to face bankruptcy. This diagram shows the high proportion of expenditure on the armed services during the Protectorate. It excludes Scottish and Irish revenues which helped to redress the deficit balance.

8 The *Eikon Basilike*, first published in 1649, created the legend of Charles the Martyr. Written probably by John Gauden (1605–62) but supposed to be by Charles himself, it emphasized his dignity in adversity. Charles's posthumous reputation greatly assisted the cause of monarchy during the Commonwealth, but it is unlikely that he would have approved of the form of monarchy that his son accepted.

9 Events leading to the Restoration in May 1660 were complex. Richard Cromwell succeeded his father peacefully but faced difficulties from Parliament in 1659 and from the army, and resigned in May 1659. Power reverted to the divided army leadership which argued with the Rump, recalled for lack of any other source of authority, in the summer of 1659. An attempted Royalist rising in September 1659 received little support for fear of renewed civil war. Eventually General Monck (1608–70), army commander in Scotland, marched on London and summoned the Convention Parliament, which recalled Charles. The king's return was greeted with popular enthusiasm, but this followed rather than caused his restoration.

The Restoration of the Stuarts

The powers and position of the English monarchy, restored in 1660, were defined by the Convention Parliament (1660) and in the early sessions of the Cavalier Parliament (1661–79). Both these assemblies, like Charles II (1630–85) himself, wished to avoid the extremes of recent years. Parliament insisted on maintaining the limitations imposed on the Crown by the "constitutional revolution" of 1640–41. The king was not granted too large a revenue, so that he would be forced to call Parliament frequently to ask for more. Thus Charles had to summon Parliament almost every year until the very end of his reign. The gentry who sat in Charles II's parliaments were quite prepared to uphold the royal authority, provided the king used it in ways they approved.

The religious settlement

Charles's success as king therefore depended on his ability to keep the trust and goodwill of the House of Commons, above all by respecting the religious prejudices of the country. In the 1640s the old Church of England had been dismantled. In the resulting vacuum of religious authority new movements such as the Baptists and Quakers had grown and multiplied. The House of Commons of 1661, however, re-established the Church of England with severe laws against Protestant Nonconformists. Charles was not happy with this extremism. Apart from his personal Catholic tendencies, he was tolerant by nature. In 1660, before returning to England, he issued a declaration at Breda promising liberty to tender consciences, subject to the approval of Parliament. Parliament would not let him keep his promise, but Charles tried in 1662 and 1672 to procure greater liberty for Nonconformists.

More explosive was the problem of Catholicism, which became linked in the 1670s with questions of foreign policy. In 1670 Charles allied himself with Louis XIV (reigned 1643–1715) of France [Key]. In 1672 together they attacked the Protestant Dutch Republic, which was almost overrun by Louis's army. Louis was already feared as the archetype of absolutism and militant Catholicism, and Charles's alliance with him seemed doubly sinister when it became clear in the 1670s that Charles's brother James, Duke of York (1633–1701) [7], later James II, had become a Catholic. As Charles had no legitimate children, James was the heir apparent, and the prospect of the first Catholic monarch in England since Mary (reigned 1553–8) overshadowed the politics of the last half of Charles's reign.

Opposition to Catholicism

English Protestants had long identified Catholic rulers with both absolutism and cruel persecution [4]. In 1678 Titus Oates (1649–1705) [5], a disreputable adventurer, came forward with allegations of a "Popish Plot" to assassinate Charles and make James king. Three times between 1679 and 1681 the Commons passed bills to exclude the Duke of York from the succession, but neither the king nor the Lords would agree to them. Led by Anthony Ashley Cooper, Earl of Shaftesbury (1621–83) [6], the "Exclusionists" (who formed the nucleus of the embryonic Whig Party) mounted a strident campaign of petitions, propaganda and demonstrations. Echoes of the Civil War

1 The Treaty of Breda (1667) ended the 2nd Anglo-Dutch War. England thus renounced claims to the Dutch East Indies, but in return gained control of New York and New Jersey. Earlier, in 1660, Charles issued a declaration at Breda, in The Netherlands, promising a general pardon for activities during the revolution, as well as religious toleration, should he be invited back to the throne. He thus removed any obstacles to his restoration.

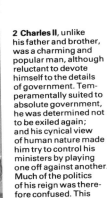

2 Charles II, unlike his father and brother, was a charming and popular man, although reluctant to devote himself to the details of government. Temperamentally suited to absolute government, he was determined not to be exiled again; and his cynical view of human nature made him try to control his ministers by playing one off against another. Much of the politics of his reign was therefore confused. This bust was made by Honoré Pellé.

3 The Royal Society was incorporated by Charles II in 1662. At that time interest in science was fashionable. The Society's early proceedings ranged from demography to botany, reflecting the polymathic interests of the members. The early Fellows included courtiers and dilettantes as well as men of outstanding talent such as Wren, Boyle and Newton. Above all, perhaps, the Society spread an understanding of "natural philosophy" to the ruling élite of England and thus contributed to the acceptance of the social thought of the Enlightenment and indirectly inspired the Industrial Revolution.

CAROLVS II SOCIETATIS REGALIS AVTHOR & PATRON

Rome's Scarlet whore doth here in Tryumph Ride,
And Spurns off Soveraign Crowns in height of Pride
Poor Christians and brave Cittes too shee Burns
And Stabbs and Poisons daily serve her Turns.

Behold our Church (like Esther here doth tend)
Her Supplication to the Faiths Defender
In vain Rome Plots, whilst Charles's Scepter Sway'd
May bled and Gibbet end all Traitors Days.

4 Fear of Catholicism during the Restoration was based on the Protestant feeling that not only was it erroneous and idolatrous theologically, but that it was a malign political force as well. It was thought that Catholic rulers, however gentle by nature, would be forced by the pope to persecute Protestants. The aggressive rule of Louis XIV in France seemed to show a clear link between Catholicism and absolutism, attacking Protestants both in his own kingdom and in the Dutch Republic. In particular the Revocation of the Edict of Nantes by Louis XIV in 1685 ended all toleration of Protestants in France. Many went into exile in England and their reports of Catholic intolerance added weight to the Protestants' fears of James II. Thus when James imprisoned for sedition seven bishops who refused to read his Declaration of Indulgence in May 1688 that promised toleration to the Catholics, as well as Dissenters, his attack on the Anglican Church, was seen as an attack on the rule of law.

agitations worried the more conservative opponents of Exclusion (who became known as Tories). They rallied to the king and helped Charles survive the exclusion crisis, permitting the peaceful accession of James II in 1685.

At first all went smoothly. James easily crushed a rebellion by James, Duke of Monmouth (1649–85) [8], an illegitimate son of Charles II and James's Protestant rival for the throne in the Exclusion crisis. But the Tories' co-operation with James soon ended. James hoped to give the Catholics religious liberty and political equality by repealing all the laws that prohibited Catholic worship and excluded Catholics from public office. The Tory Parliament of 1685 was strongly royalist but also strongly anti-Catholic. Annoyed by the Tories' hostility, James tried to persuade the Protestant dissenters to support a general toleration that would include Catholics. In 1686, despite opposition from Archbishop Sancroft (1617–93), he set up an ecclesiastic commission that was seen as an attempt to subject the Church to his religious policies. Meanwhile he stretched his powers as Charles I had done in order to admit Catholics to places in the army, the administration and the universities.

The "Glorious Revolution"

James's nephew, William III of Orange (1650–1702), was closely interested in developments in England. Already the dominant political figure in the Dutch Republic, William had a claim to the English throne through his wife Mary, James's elder daughter, and he was eager to use England's fleet and wealth in his lifelong struggle against France. By 1688 William feared that he might be cheated of the succession, because late in 1687 it was announced that James's queen was pregnant. By the time James's son was born, in June 1688, William had decided to invade England in response to an invitation from leading political figures. He landed with his army on 5 November [9]. James fled to France and in February 1689 Parliament offered the crown jointly to William and Mary, and the rule of law, enshrining the supremacy of the landed interest over the monarch, was assured.

Charles II usually followed Louis XIV, not vice versa as in this cartoon. Except for the period 1678–81, Charles was allied with Louis (whose power he envied) or was benevolently neutral. Louis gave him several subsidies to help him to stay independent at home, and to resist Parliament's demands for a war against Louis, whom most Englishmen saw as the main enemy. Initially Charles sought the French alliance to avenge England's failure to win the 2nd Anglo-Dutch War of 1665–7.

5 Titus Oates, the main witness in the Popish Plot, had spent a year as a Jesuit novice and the stories that he told were based on the information he collected at various seminaries. Despite inconsistencies, he spoke so compellingly that many believed him. The murder of the magistrate to whom Oates first recounted the plot appeared to confirm all that he said. In 1685, however, he was convicted of perjury and flogged. The crisis allowed Shaftesbury, a former supporter of Charles who now feared his conversion to Roman Catholicism, to drum up anti-Catholic agitation and turn it against James, the Catholic heir to the throne. He died in voluntary exile in Holland.

6 The Earl of Shaftesbury had been a royal minister in the 1660s and early 1670s. His organization of the Exclusionists into a coherent parliamentary group was important for the development of political parties.

7 James II served as an able commander of the fleet during the 2nd Anglo-Dutch War, but as a Catholic he lacked Charles's popularity, and quickly alienated all the support that he had at his accession.

VII

The late D of M beheaded on Tower Hill 15 July 1685

8 The Duke of Monmouth landed in Dorset in 1685 to assert his claim to the throne. He won little gentry support and was easily beaten at Sedgemoor. His followers were treated with severity by Judge Jeffreys (c. 1648–89) in the so-called Bloody Assizes, at which 200 were hanged and 800 transported. Monmouth himself was executed.

9 William of Orange landed at Brixham late in 1688 with an army of 15,000 men. But his invasion succeeded without bloodshed (hence its name, the "Glorious Revolution"), as James's army commander John Churchill (1650–1722), later Duke of Marlborough, went over to William's side at the last moment, removing James's last source of support in the country.

Pepys's London

The London of Samuel Pepys (1633–1703) [Key] was small (Marylebone and Mile End were in the country) and consisted of two cities: London, the commercial and financial centre; and Westminster, centre of government, politics and aristocratic society. The two were so different, culturally and politically, that there was often tension between them. Pepys, however, straddled both worlds. A Londoner by birth and a civil servant by profession, he moved easily between his office and lodgings near the Tower and the court at Whitehall. A man of quick intelligence, who paid careful attention to detail, he rose from a comparatively menial post to become James II's chief naval administrator.

Diary of a somebody
There was nothing very unusual about Pepys's rise to wealth and prominence. The Tudor and Stuart civil service offered a career open to talents. What was unusual about Pepys was his diary, which gives an unequalled panorama of life in London between 1660 and 1669: the politics of court and Parliament; meetings with naval contrac-

tors and mistresses; music and the theatre; food and drink; and tensions at the office and quarrels with the neighbours.

Insatiably curious and endlessly sociable, Pepys wanted to know everything that was going on and jotted it down, in shorthand, when he got home. On 25 March 1661 he told how, returning home, he "took up a boy that had a lanthorn, that was picking up rags, and got him to light me home, and had great discourse with him, how he could get sometimes three or four bushells of rags in a day, and got 3d a bushell for them, and many other discourses, what and how many ways there are for poor children to get their livings honestly". Gossip about Charles II's sexual cavortings is juxtaposed with details of admiralty business or the latest experiment performed before the Royal Society. National humiliations such as the Medway disaster [6] appear alongside details of a meal.

Congestion, crime and natural disasters
At the Restoration, London [1] was the largest city in England. Its population – not far short of 500,000 – was more than ten

times that of Bristol, its nearest rival. Little more than a century earlier, its population had been only 50,000 and it now displayed all the symptoms of unplanned urban growth. Within the old medieval walls rising population and property values encouraged landlords to build upwards and to overcrowd their property [3] despite municipal attempts to control them.

The fastest growth was to the east and north – Stepney, Whitechapel, Shoreditch, Clerkenwell – and south of the river in Southwark. There the speculative builder and slum landlord operated almost unchecked. Jerrybuilt tenements and insanitary courts housed artisans and porters, watermen and sailors, Huguenots and Irish.

As the population had far outgrown the flimsy machinery of law and order, crime flourished. Violence was never far below the surface in Restoration London. Drunken quarrels ended in murder. Courtiers and soldiers showed a casual disregard for human life. On public holidays law students from the Inns of Court and gangs of apprentices roamed the streets looking for trouble.

CONNECTIONS

See also
292 The Restoration of the Stuarts
296 Art and architecture in the 17th century
302 Science and technology 1500–1700
312 The age of Marlborough

In other volumes
90 History and Culture 2

1 London before the Great Fire was a chaotic jumble of houses. Old St Paul's dominated the skyline and London Bridge, lined with homes and shops, was the only bridge across the Thames. The city's streets, filthy with rubbish, were dangerous at night because of criminals. The River Thames served as the main highway. It was quicker and cleaner than the narrow and congested cobblestoned streets.

2 Charles II and his mistress Nell Gwynn converse in the presence of diarist John Evelyn (1620–1706). Such a scene, wrote Evelyn, made him "heartily sorry". He was shocked by the king's promiscuity.

3 Houses huddled against each other in London. Most dwellings were built of wood, (despite regulations against its use) and they had narrow frontages, steep roofs and projecting upper floors.

4 The plague of 1665–6 was carried by the rats that infested London's teeming tenements, and deaths exceeded 15 per cent of the population. By far the worst year was 1665. Carts trundled through the streets at night to collect corpses for communal burial. The court moved out of London as did many people. Houses touched by pestilence were vacated and the door marked with a red cross and with the words "Lord have mercy upon us".

Such a crowded, ill-housed population was vulnerable to the ravages of disease and fire, both of which struck terribly in the 1660s. The last great outbreak of bubonic plague (1665–6) [4] claimed more than 70,000 victims in London. Pepys wrote on 30 July 1665: "It was a sad noise to hear our bell to toll and ring so often to-day, either for deaths or burials; I think five or six times." The Great Fire of 1666 [5] destroyed most of the City of London. On 2 September Pepys watched the progress of the fire, "a most horrid malicious bloody flame".

Rebuilding and expansion

Out of this catastrophe came a chance to rebuild according to a single coherent plan [7] a great swathe of central London from the river up to London Wall and from Fenchurch Street to Fleet Street. From the ashes emerged a new city with broader, cleaner streets, brick-built houses, a much purer water supply and a whole series of churches [8] designed by the brilliant young architect Christopher Wren (1632–1723).

London had long been the greatest port in the kingdom, and under Charles II new trading companies were established, notably the Royal African and Hudson's Bay companies. But the main source of the impressive commercial growth of the 1670s and 1680s lay elsewhere. First, by protective legislation and naval power England began to challenge the Dutch dominance of the international carrying trade. This would at length enable London to replace Amsterdam as "the world's entrepôt". Second, these years saw a massive increase in imports and re-exports of sugar and tobacco from England's Caribbean and North American colonies. In addition, London was the hub of the growing internal trade within England.

Farther west, between the City of London and the City of Westminster, lay an area that had largely escaped the fire. Here were the palaces of the nobility, situated within convenient distance of Parliament and the royal palaces of Whitehall and St James's. The Strand had been developed much earlier and the main growth areas in Pepys's day were in Pall Mall, Piccadilly and St Giles's Fields – around the West End of the future.

Samuel Pepys held an important post in the Navy Office, and was an MP and president of the Royal Society. His diary shows him to have been cultivated, fastidious, ambitious and lecherous.

5 The Great Fire (1666) began in a baker's house in Pudding Lane and raged for three days. It destroyed 13,000 houses, St Paul's, 87 parish churches and many public buildings. Deaths were few but half the population was rendered homeless.

6 A Dutch fleet sailed up the Medway in June 1667, found defences almost non-existent, and burned, sank or captured several of the king's greatest ships. The flagship *Royal Charles* was among those captured. Meanwhile, wrote Pepys bitterly, the king was chasing a moth with his mistress. The Dutch Wars flared intermittently from 1652 to 1674.

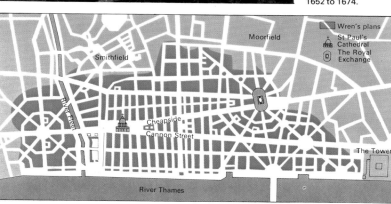

Moorfield

Wren's plans

St Paul's Cathedral

The Royal Exchange

Smithfield

Fleet Street

Cheapside

Cannon Street

The Tower

River Thames

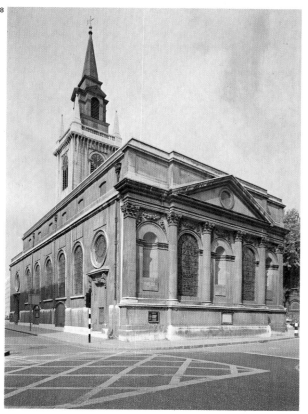

7 Several plans for rebuilding London were advanced after the Great Fire. Christopher Wren proposed that broad, straight streets should be laid down in a geometric pattern. A new St Paul's and Royal Exchange were the main focal points of his scheme, which featured several piazzas. But a shortage of money and conflicting interests meant that no scheme was adopted in full. Instead, the shape of the city was decided by a special committee.

8 St Lawrence, Jewry, was one of the 53 churches that Christopher Wren built in London following the Great Fire. He was appointed Surveyor-General in 1669 with the task of rebuilding the city. These churches have distinctive spires but Wren lavished most attention on their interiors in which he made an interplay of rhythms and light spaces. No authority was able to buy enough land to organize the rebuilding in a centralized manner; but stringent regulations limited private builders.

Art and architecture in the 17th century

Art in 17th century England falls into two sharply defined phases, separated by the unsettled years of the Civil Wars and Commonwealth from 1642 to 1660. In the earlier period, in particular, the vitality of the visual arts depended on the enthusiasm and cosmopolitan taste of successive monarchs, James I and Charles I, their families and courtiers. Whereas in the later sixteenth century only inferior European artists and craftsmen had been attracted to England, the Stuarts managed to employ painters of high calibre such as Rubens and Van Dyck, commissioned the great Italian sculptor Bernini and attracted to court good foreign sculptors. In the architect and designer Inigo Jones they had an Englishman who fully understood the lessons of the Renaissance.

The court of Charles I
The development of court taste was fostered by collecting. In the years around 1620 two leading political figures, the Earl of Arundel and the Duke of Buckingham (the favourite of Charles I), were collecting antique sculpture and European paintings, especially the work of the great Venetians, Titian and Veronese. English ambassadors and agents scoured Europe and Asia Minor on their behalf. Charles I's elder brother, Prince Henry, showed himself a discriminating patron and collector by 1610–12, and Charles himself built up an astonishingly rich collection during his reign. The Commonwealth sales of the 1650s dispersed most of this collection.

The major artistic figure at the early Stuart court was Inigo Jones (1573–1652). He acted as connoisseur, as designer of scenery and costumes for court masques, and above all as an architect whose influence reached out to town planning and garden design. Appointed Surveyor of the Royal Works (architect to the king) in 1615, he executed during the following 25 years a series of buildings that formed the basis of classical architecture in Britain for two centuries. In the Whitehall Banqueting House (1619–22) [1] he used classical columns and cornices to give scale and correct proportions to a façade; the Queen's House, Greenwich, (begun in 1616) was a version of a Palladian villa, with the richly fanciful interiors of the 1630s which are now largely destroyed. His chapel at St James's Palace (Marlborough House Chapel, 1623–5), and his church of St Paul's, Covent Garden (begun in 1631), both used the temples of the ancient Romans as models for modern ecclesiastical architecture. Covent Garden piazza, a formal square enclosed by terraces of houses set on open arcades, was the forerunner of the unified squares of Georgian London.

Sculpture and painting under the Stuarts
The influence of antiquity was also felt in sculpture, particularly in the later work of Nicholas Stone [2]. Hubert le Sueur (c. 1595–c. 1650) designed exceedingly elaborate monuments such as those of Henry VII's chapel in Westminster Abbey, and depended rather on recent French prototypes. Bernini executed a bust of Charles I, using a triple portrait by Van Dyck as a guide; his influence can be seen in post-Restoration sculpture, particularly the monuments of John Bushnell (died 1701). In general, figures on monuments became less stylized and were made to

1 In the Banqueting House, Whitehall, Inigo Jones first brought to England a fully understood Renaissance classicism. It has a single enormous room in which great ceiling paintings by Rubens were installed in 1635.

2 John Donne (1572–1631) requested that his monument should show him in his shroud rising to heaven at the Last Trump. He posed for the sculptor Nicholas Stone (c. 1587–1647). The monument shows in the soft drapery the study that Stone had made of antique sculptures brought to England by Arundel and Buckingham. Funerary monuments erected by Stone survive in many churches throughout Britain.

3 Anthony Van Dyck was a pupil of Rubens but developed a distinctive touch and elegant presentation flattering to his courtier patrons. This portrait of the "4th Duke of Lennox" (1634) has a pose reminiscent of Titian.

4 Cornelius Johnson was much influenced by Van Dyck but his figures are stiffer and less natural. Yet his ambitious family portrait of "Arthur Capel with his wife and children", of about 1639, is touchingly intimate. Capel was made Earl of Essex by Charles I, and was executed by the Parliamentarians in 1649 after serving as an ardent Royalist in the Civil War. In the background of the picture there is a remarkable formal garden, with classical statues and balustraded walks. Capel laid out this garden at his house, Hadham Hall, Hertfordshire. The classicism of Jones widely influenced country house and garden design even before the Civil War, although Jones himself probably designed few such country houses.

5 Ramsbury Manor, Wiltshire (dated 1683), is a perfect country house of the Restoration. The simple, regular design depends on good proportions for its effect. The architect was Robert Hooke (1635–1703), secretary of the Royal Society.

express emotion frankly. But Grinling Gibbons (1648–1721), the most famous sculptor of the century, specialized in virtuoso decorative wood-carving of fruit and flowers in the Dutch manner [8].

For painting, the presence of the Dutchman Anthony Van Dyck (1599–1641) in England from 1632 until his death was crucial. He transformed portraiture from the style of Holbein by his easy fluid technique, and by his sense of colour and form learnt from the Venetians as well as from his master, Rubens. He was the ideal court painter, and was kept fully occupied executing portraits [3].

Van Dyck's style influenced the work of his contemporaries, such as Cornelius Johnson (1593–1661) [4], as well as his successors, such as William Dobson (1610–46). The miniaturist Samuel Cooper (1609–72) [7], who worked from the late 1630s, surpassed Van Dyck in depth of character and created an astonishingly wide variety of poses for the tiny scale of his works. Restoration court portraiture was dominated first by Peter Lely [6] and then by Godfrey

Kneller (c. 1646–1723), both of whom were German by birth. The leading decorative painter of this period was the Italian, Antonio Verrio (c. 1639–1707) [8].

Wren and Restoration architecture

In architecture after the Restoration the classicism of Inigo Jones triumphed. One of Jones's successors as Surveyor of the Works (from 1669) was Christopher Wren (1632–1723), who brought to architecture the experimental spirit of the recently founded Royal Society. He rebuilt more than 50 London churches after the Great Fire of 1666, and fulfilled royal commissions for three palaces [8] and two magnificent hospitals, at Chelsea and Greenwich. Wren was therefore given vast scope to express his talents. His masterpiece was St Paul's Cathedral (1675–1710) [Key], which grew more ambitious as the building progressed. Wren's successors, Nicholas Hawksmoor (1661–1736) and John Vanbrugh (1664–1726), emphasized Baroque massiveness in their buildings, notably Blenheim Palace (1705–c. 1725) [9].

St Paul's Cathedral, London, was built by Christopher Wren. Its great dome is 111.2m (365ft) high and the building displays a restrained English version of the ornate Baroque style.

6 Peter Lely (1618–80) painted the "1st Earl of Sunderland" in 1643. He later became the main portrait artist of the Restoration.

7 Samuel Cooper revived the miniature, as in this portrait of the "Duke of Monmouth". He also painted Cromwell in old age.

8 Hampton Court Palace was partly rebuilt by Wren for William and Mary. Successive Stuart monarchs tried to make palaces as splendid as those of the kings of France and Spain. The formal gardens of Hampton Court, and the series of state rooms with illusionistic painting by Verrio and virtuoso woodcarving by Grinling Gibbons, are the finest surviving monument to these royal ambitions. The King's Great Staircase (shown here), leading to the King's State Suite on the first floor, has an iron balustrade by the Frenchman Jean Tijou, and Verrio's wall-paintings represent the triumph of the Protestant William III over the Catholic kings of Europe.

9 Blenheim Palace, Oxfordshire, was built after 1705 as a gift from a grateful nation to the Duke of Marlborough to reward him for a decisive victory over Louis XIV of France. It has a Baroque magnificence that English royal palaces never were able to achieve. The architect, John Vanbrugh, with the assistance of Nicholas Hawksmoor, carried the flexible and inventive style of Wren to new heights of grandeur and monumentality. The locally quarried stone used in the building is highly appropriate for such architecture, which aimed to survive for eternity. The grounds were laid out so that each clump of trees represented the position of one battalion at the battle of Blenheim.

Scotland from 1560 to the Act of Union

The religious upheavals in Scotland during the first half of the sixteenth century, initiated by the Reformation, found wide support for political as well as spiritual reasons. The new religious ideas were welcomed by those wishing to loosen the alliance with Catholic France in favour of England, and the nobles, who hoped to gain much from a weakening of the Church's power.

The Reformation and the Crown

With the abolition of papal authority in 1560, Calvinism quickly became the official belief in Scotland, but the ruler, Mary, Queen of Scots (reigned 1542–67), was a Catholic and a member of the French royal family [1]. On the death of her husband Francis II of France (reigned 1559–60), Mary returned to Scotland to try to establish her authority as queen.

She proved to be a clever politician but her power and public appeal were always limited by the tremendous influence of John Knox (1505–72), the leading preacher of the new church [3]. Eventually, in 1567, she was forced to abdicate by the nobility, who were angered at her turbulent marriages [2]. In 1568 she had to take refuge in England, and the nobles put her infant son James VI (reigned 1567–1625) on the Scottish throne with the expectation that a king so created would be their puppet.

On this point the nobles failed. James grew up to become the most effective ruler of any of the Stewarts. He did this without funds or force, but purely by negotiation and hard work, particularly helped by the fact that he was expected to inherit the English throne. He and his cousin, Elizabeth I of England (reigned 1558–1603), both wanted peace on the Border. In 1603, Elizabeth on her deathbed acknowledged James as her heir. James, known thereafter as James I, moved to England but continued to control Scotland as he said "by my pen" [Key].

James's son Charles I (reigned 1625–49) could not do this because he did not have the intimate knowledge of the country. His policy of personal rule made the Scottish nobility uneasy for their rights and powers while his religious policy was seen, rightly or wrongly, as contrary to Presbyterian principles. When Charles produced a prayer book for Scotland in 1637 [4] it provoked the Scots to formulate a claim for their traditional liberties – the National Covenant. Charles's attempts to bring an English army to suppress the Covenanters led to the Long Parliament, and thus contributed to the start of the English Civil War in 1642.

The English Parliament persuaded the Scots to join in the first Civil War (1642–6) by accepting the terms laid down by the Scots in the Solemn League and Covenant, of 1643. But after the war, power lay in the hands of the Parliamentarians' army, which wanted an independent church system with liberty for the individual congregation. Efforts by the Scots to insist on Presbyterianism led to the second Civil War (1648), the execution of Charles (1649) and occupation of Scotland by Oliver Cromwell (1599–1658) [6].

Stuart restoration and dethronement

When the English Parliament called Charles II (reigned 1660–85) back to the throne [7], it was taken for granted that he would rule in Scotland too. However, some of the extreme

1 On the death of Henri II of France in 1559 Mary, Queen of Scots' husband, Francis, ascended the French throne. Mary's association with Catholic France weakened support for her in Scotland.

2 Mary was executed in England in 1587. She had been driven out of Scotland in 1568 because of her association with the murder of her cousin and husband Lord Darnley following the death of her favourite Rizzio.

4 *The Arch-Prelate of S.t Andrewes in Scotland reading the new Service-booke in his pontificalibus assaulted by men & women, with Crickets stooles Stickes and Stones.*

3 John Knox compiled, with other Calvinist ministers, *The First Booke of Discipline* (1560) which was the blueprint for the creed and constitution of the Protestant Church in Scotland. It decreed a hierarchy of Church courts, substantial stipends for ministers of the new Church, and proposed a generous programme of public education. This last had only limited success because the nobility refused to surrender enough of the old Church's wealth to finance it.

4 The revised prayer book was introduced by Charles I in 1637 and caused widespread disturbances. Much of this was against its imposition on Scottish congregations, but many Scots also regarded the book as a source of dangerous innovations. Popular unrest led to the Covenant of 1638 – a signed agreement to defend the reformed religion. The rebellion of the Covenanters culminated in the First Bishops' War in 1639 which was peacefully settled in the same year. However, the Scottish assembly, organized to resolve the conflict, and the Scottish Parliament openly defied the king. Charles was refused funds for his army by the Short Parliament, and his forces were defeated in the Second Bishops' War in 1640.

Presbyterian Whiggamore Party refused to accept any government not chosen by them; this and Charles's policy of re-establishing episcopacy led to disturbances. In this period the Scottish economy became dependent on trade with England. Scotland had to take part in English wars, even when the opponents were The Netherlands (1665–7, 1672–4), its main trading partner, and economic nationalism in France and Sweden further deprived Scotland of overseas markets.

The Revolution of 1688–9 against the Catholicizing policy of James II (VII) (reigned 1685–8) was, like the Restoration, made in England and accepted in Scotland. But there was a larger and more effective resistance group this time, particularly in the Highlands, where many preferred James to the new King William III (reigned 1689–1702) because of William's reliance in Scotland on the unpopular house of Argyll. These Highland Jacobites, or supporters of the Stuart (as the Stewarts became known after 1603) line, defeated William's army in 1689 at the Battle of Killiecrankie in the southern Highlands, but were prevented by a regiment of Covenanters defending Dunkeld from breaking through into the Lowlands. Later, the remaining Jacobite clans were brought to temporary submission by the Glencoe massacre in February 1692 [8].

Events leading to the Union

With the failure of William III and the next ruler, Anne (reigned 1702–14), to produce an heir, there was a possibility that the Scottish Parliament might bring back James rather than accept the Hanoverian successor chosen by England. This reasoning seems to have persuaded the English to accept the idea of uniting the parliaments of the two countries. The increasing dependence on trade with England meant that economic sanctions could be used to compel the Scots to accept a union. There was a period of increased interest in the Jacobite cause, and hostility to England in the early 1700s, but in 1707 economic generosity by England and political generosity by Scotland brought about the union of the parliaments in Westminster [9], although leaving the laws and church systems of the countries distinct.

The English and Scottish Crowns were united in 1603 when James VI was crowned James I of England. But James was not able to effect a parliamentary union of the two kingdoms.

5 James Graham, Earl of Montrose (1612–50), was a leader of the Covenanters in the Bishops' Wars, although later, in the Civil War, he campaigned brilliantly for the king in Scotland.

6 General Monck (1608–70) commanded the English army in Scotland under the Protectorate. He took his army to London and made possible the restoration of the monarchy in 1660, with Charles II as king.

7 Charles II was crowned by the Marquess of Argyll (1606–61) [centre, right] in 1651, after compromising with the extreme Presbyterian Party. This ended in defeat by Cromwell in 1651.

8 The Massacre of Glencoe in 1692 took place after the Jacobite rising at Killiecrankie, three years earlier. William III demanded an oath of allegiance from the Highland clans, and the Macdonalds of Glencoe did not take it. As an example to other clans, 38 of the clan were slaughtered by government troops.

9 The Act of Union, in 1707, was made by commissioners from both countries and then passed into law by the separate parliaments. This process was relatively easy in England, but the Scottish Parliament needed a great deal of persuading by the Duke of Queensberry (1662–1711), shown attending Queen Anne's signing of the treaty.

Europe 1500–1700

In the first half of the sixteenth century the economy of Europe was dominated by a steady price rise; in the second, by unprecedented inflation. Price increases were felt first in Spain and resulted from the greatly increased imports of gold and silver from new mines in Mexico and Peru. The effects of this massive importation of bullion [5] were later offset by its export to other parts of Europe to liquidate Spain's unfavourable trade balances, to supply the needs of her armies in The Netherlands and by way of smuggling across the French frontier. Until these movements of bullion took effect general prices in the rest of Europe had risen more slowly than in the Iberian Peninsula.

Stimulus to international trade

This different rate of inflation in the European countries imparted a further stimulus to the expansion of international trade which had followed the discovery of the New World. The Baltic trade in corn with southern Europe, soon to be controlled and financed by the Dutch, was also promoted by the tendency of food prices to exceed those of manufactured goods. This was because an impressive growth in population in Europe [1, 2] – from 50–60 million in 1450 to nearly 100 million in 1600 – increased the demand for foodstuffs when supply was restricted by poor yields in corn-producing areas of eastern Europe.

Although industry still catered mainly for the provision of luxuries to the wealthy, commercial expansion led to a greater use of credit facilities and an extension in public banking, especially in Italy where Genoese financiers arranged the transfer of Spanish remittances to The Netherlands.

The Dutch "economic miracle"

In the seventeenth century many of these trends were reversed. Inflation was checked in the second decade by decreasing imports of bullion from the New World. International trade, buttressed by the defence requirements of the Thirty Years War (1618–48), later experienced a downturn and stagnated. Competition between the powers for a preponderant share of world trade led to the adoption by most governments, except the Dutch, of mercantilism, a system of protectionist trade policies.

The shift of the centre of commercial activity from the Mediterranean to the Atlantic seaboard, however, continued. The chief beneficiary, apart from England, was the Dutch Republic, which won virtual great-power status by its pre-eminence in the carrying trade, exploration and finance [8]. This "economic miracle", occurring in a period of general contraction, rested on Dutch control of the North Sea herring and the Newfoundland cod fisheries; on technical expertise in shipbuilding and insurance; on the elimination of the Portuguese from the Far Eastern spice trade and the Spanish from their monopoly of commerce with South America; on the international, exchange and credit facilities provided by the Bank of Amsterdam (founded in 1609) [11]; and on the policy of toleration which induced the prosperous victims of religious persecution in Spain and France – the Sephardic Jews and Huguenots – to settle in Amsterdam [Key].

Economic and naval warfare between the Dutch Republic (the United Provinces of the

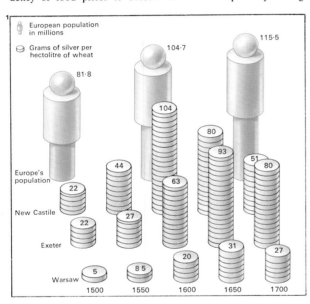

1 **The rapid rate of inflation** during this period (in which all food prices rose steeply) has no conclusive explanation. But population growth seems to have been a major factor.

2 **The growth of population** during the sixteenth century was considerable. The increase seems to have begun around 1450 and become more rapid after 1500. The recurrence of plague and the long duration of European wars had, by the mid-seventeenth century, reduced the rate of growth once more. The diagram illustrates the changing population figures of the major European nations between 1500 and 1700.

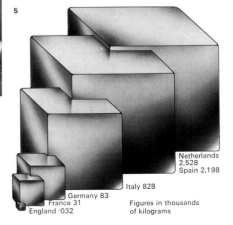

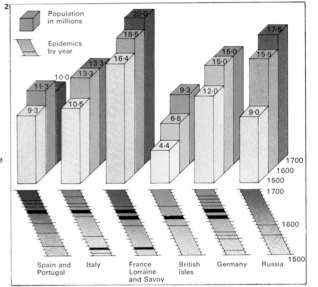

3 **Death by plague** had been common in Europe for almost two centuries. Periodic epidemics were a major check on population growth, but it is difficult to assess accurately the effects of plague in an age when routine diseases and accidents also made a significant contribution to the high mortality rate. Between a sixth and a third of the population died in each epidemic, although this figure could rise to as much as two-thirds, as happened in Germany in the 1630s. Plague also had some long-term effects; unstable social and economic conditions pushed up the age of marriage and thus lowered the birth-rate. Worst hit by the plague were the crowded, badly housed urban poor.

4 **In early modern Europe** life was hard for the ordinary man, as depicted by Bruegel the Younger. Beggars were a common sight and for the labourer the holy days of the Church were the only respite from the burdens of his day-to-day existence. But the Church also ordained that periods of abstention from any work or indulgence must be strictly observed. Such decrees were often not followed; in Charles I's reign, Archbishop Laud (1573–1645) encouraged sports on the sabbath.

5 **Silver mined in the New World** and imported by Spain had a considerable influence on political and economic life in Europe. How Spain distributed this bullion between 1580 and 1626 is shown here. Spending in Spain and The Netherlands reflects the huge sums used for defence. The rest of Europe, learning from Spain, looked to trade to provide the wealth necessary for military power.

Netherlands 2,528
Spain 2,198
Italy 828
Germany 83
France 31
England ·032

Figures in thousands of kilograms

northern Netherlands) and their main rivals, England and France, impaired Dutch commercial supremacy [6] in the late 1600s; it was finally undermined during the War of the Spanish Succession (1701–14).

European society, 1500–1700

The society of early modern Europe was principally agrarian with 90 per cent of the population deriving a living from the land. Farming was carried out under the jurisdiction of the manorial lord, although the farmer was also governed by various local customs. Inflation, however, was a social solvent that loosened the characteristic rigidity of sixteenth-century society where status was determined by law and not by wealth. The medieval concepts of a "just price" and a controlled economy also ceased to be valid. Until the end of the sixteenth century the pressure of population on the means of subsistence involved a fall in living standards and a decline in real wages.

The more volatile situation in the seventeenth century enabled the emergent middle classes to consolidate their wealth and to

improve their social status. Generally the people who profited most were those who could charge higher prices without in turn having to pay them. These were the farmers whose tenure was secure and those noblemen who could evict their tenants and exploit the land. For those who were landless or whose title to their land was insecure, the real effect of the price rises was eviction, vagrancy and perhaps death by starvation. Because a smaller work force was needed for pasture farming many were condemned to a life of wandering as beggars [4]. From the land-owners' point of view labour was relatively cheap and this helped to stimulate improvements in agricultural and manufacturing processes.

The seventeenth century saw the growth of mining, finishing industries (such as dyeing), tobacco-growing and even market gardening. But the place of "manufacture" in this period was the cottage and not the factory; cloth was distributed piecemeal to be spun, dressed or dyed by rural labourers or farmers, whose living standards improved as prices tended to fall.

FORUM AMSTELODAMENSE DEN DAM VOCANT.

Amsterdam, more than any other city, illustrates the revolution in the economic life of early modern Europe. During the 17th century the axis of economic activity shifted north – from a concentration on trade in the Mediterranean to a prevailing emphasis on Atlantic trade. As the centre of the Dutch carrying trade, Amsterdam became a mart, a world bank and a centre of insurance for traders.

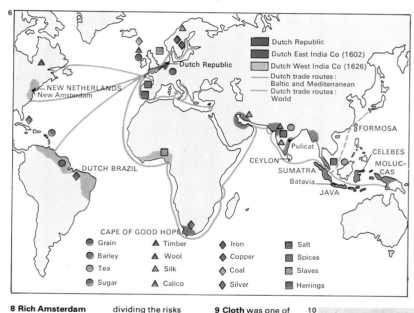

6
Dutch Republic
Dutch East India Co (1602)
Dutch West India Co (1626)
Dutch trade routes: Baltic and Mediterranean
Dutch trade routes: World

Dutch Republic
NEW NETHERLANDS
New Amsterdam
FORMOSA
Pulicat
CELEBES
CEYLON
MOLUC-CAS
DUTCH BRAZIL
SUMATRA
Batavia
JAVA
CAPE OF GOOD HOPE

○ Grain ▲ Timber ◆ Iron ■ Salt
○ Barley ▲ Wool ◆ Copper ■ Spices
○ Tea ▲ Silk ◆ Coal ■ Slaves
○ Sugar ▲ Calico ◆ Silver ■ Herrings

7

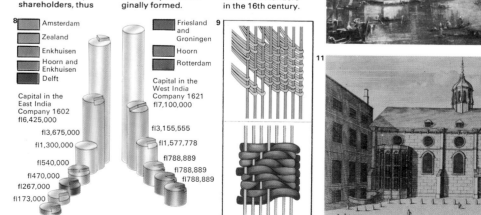

6 The Dutch held economic sway in Europe principally by carrying the products of other countries. The central areas of the carrying trade were the Mediterranean, the English Channel and the Baltic. The East Indies were the source of spices and luxury goods for re-sale in Europe. The West India Company had the more political aim of reducing Spanish trade by privateering.

7 Wealth and culture went hand in hand during the golden age of Dutch trading. Dutch art of the 17th century reflects the alliance, as in Rembrandt's celebrated portraits of wealthy Dutch merchants.

8 Rich Amsterdam merchants provided most of the backing for the Dutch East India and West India Companies. Both were joint-stock companies in which there were many shareholders, thus dividing the risks as well as the profits involved in colonial trade. The diagram illustrates the sources of the capital invested in the two companies when they were originally formed.

9 Cloth was one of the most important 17th-century manufactures. Italian silk damask-weaving techniques of the 15th century (shown here) were taken to England from Flanders in the 16th century.

10

11

10 Venice was one of Europe's most important trade centres at the start of the sixteenth century. However, the wars in the early part of the century and the colonization of the New World shifted the focus of trade to Atlantic ports such as Amsterdam and Bristol.

11 The growth of joint-stock trade over long distances gave rise to the need for more flexible instruments of credit and exchange. Significantly, the Dutch first broke from the tradition of raising money from private families such as the Fuggers, and the Bank of Amsterdam (shown here) was set up in 1609. The Bank of England, opened in 1694, was modelled on it.

8
Amsterdam
Zealand
Enkhuisen
Hoorn and Enkhuisen
Delft

Friesland and Groningen
Hoorn
Rotterdam

Capital in the East India Company 1602
fl6,425,000
fl3,675,000
fl1,300,000
fl540,000
fl470,000
fl267,000
fl173,000

Capital in the West India Company 1621
fl7,100,000
fl3,155,555
fl1,577,778
fl788,889
fl788,889
fl788,889

9

Science and technology 1500–1700

By the dawn of the sixteenth century the European Renaissance was well under way, and during the following two centuries the broad basis of modern science was laid. Knowledge of Greek science was widespread and an inquiring spirit that led to a critical examination of ancient ideas prevailed. As a result, the ancient practices of astrology and alchemy were being discarded; some scientists and philosophers realized that nature had to be investigated rationally.

The revolution in astronomy

The first fruits of this new approach came in astronomy. In 1543 the theory that the Sun, not the Earth, was the centre of the universe appeared in *De Revolutionibus* by Nicolas Copernicus (1473–1543). This theory profoundly affected man's view of himself and his place in nature. In astronomy the Copernican theory stimulated a spate of precise observations – notably by Tycho Brahe (1546–1601) – laying the basis for seventeenth-century discoveries with the telescope. Johannes Kepler (1571–1630) used Brahe's observations in his reinterpretation of the planets' motions in terms of ellipses instead of the complex Copernican system of circles. In England, Francis Bacon (1561–1626) advocated the empirical method but denigrated the use of mathematics for the interpretation of results, unlike René Descartes (1596–1650), whose work contains many of the philosophical ideas that were at the root of the new spirit of enquiry. Descartes wanted to lay a rational foundation for religion and science to give them a mathematical validity, proof against scepticism and superstition. He formulated an entire philosophical system that postulated a wholly mechanical universe in space completely filled with fluid matter [1].

The Italian astronomer, mathematician and physicist Galileo Galilei (1564–1642) became the first person to use the telescope to study the heavens. Galileo's conclusions about mechanics laid the foundations for later work, particularly Newton's [6]. But his belief in a universe governed by mathematically regular laws led to great hostility, not least from the Church. The genius of this period was Isaac Newton (1642–1727), whose *Philosophiae Naturalis Principia Mathemetica* is one of the most important works of modern science. In it he defined his laws of motion [Key], developing Kepler's and Galileo's work, and he first formulated the law of universal gravitation. Newton also made important contributions to mathematics, including the invention of calculus, although it had been formulated quite independently by Gottfried Leibniz (1646–1716), the German mathematician.

Discoveries in optics

Science gained new impetus in the seventeenth century from the invention of the telescope and the microscope. Galileo, using the telescope, was able to observe mountains on the Moon, spots on the Sun, the phases of Venus and the four larger satellites of Jupiter. Improved instruments followed, Newton perfecting the first reflecting telescope in about 1668. This led to further discoveries, as well as the use of the telescope as a celestial measuring instrument of great precision, and the establishment of national observatories in Paris and at Greenwich [2].

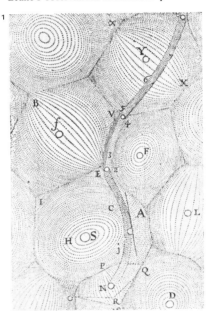

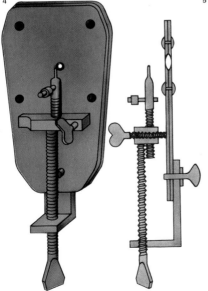

1 Descartes' idea of the universe rejected the theory of the existence of a vacuum and held that matter filling the universe was perpetually moving in vortices with stars at their centres. Some stars became planets with orbits in the vortex of another star. Comets wound their way between and across vortices, as in this engraving.

2 The Royal Observatory at Greenwich was founded in 1675 to compile a new star catalogue for navigational use. Designed by Christopher Wren, it became an important centre for accurate astronomical observations. The meridian of zero longitude still runs through Greenwich, but actual observation is now done in Sussex.

3 Vesalius, the great 16th-century anatomist, is shown holding a partly dissected human arm. The portrait is taken from his book *De Humani Corporis Fabrica*.

4 The powerful, single-lens microscope was designed in the 17th century by van Leeuwenhoek.

5 The discovery of the circulation of the blood was published by Harvey in his book *De Motu Cordis*, in 1628. Harvey is shown in Hannah's painting demonstrating the principle to Charles I.

The seventeenth century saw much fundamental work on optics. Willibrord Snell (1591–1626) determined the law of refraction of light. Christiaan Huygens (1629–95) used this knowledge to improve telescopes, also working with information from Kepler's study of lenses. Above all, Newton showed that white light could be split by a prism into components of every colour.

Developments in medicine

In 1538 Andreas Vesalius (1514–64) [3] produced his vast study of the human body – the first to go against the teachings of Galen (AD c. 130– c. 200). Vesalius' successor at Padua University, Bartolommeo Eustachio (1520–74), discovered the Eustachian tubes of the ears. In the next century Hieronymus Fabricius (1537–1619) laid the foundations of embryology and discovered valves in the veins, a finding that was used by William Harvey (1578–1657) [5] who, in 1628, announced his discovery of the circulation of the blood. Marcello Malpighi (1628–94) discovered the capillaries connecting veins and arteries and, like Jan Swammerdam

(1637–80) after him, used it also in embryological studies. The microscope greatly helped to advance medical and biological knowledge during a period when they became increasingly based on physiological experiment. In Holland, Anton van Leeuwenhoek (1632–1723) devised his own microscopes to study blood and spermatozoa as well as microscopic life forms [4].

In the sixteenth century botanical encyclopedias became common. In the next century Nehemiah Grew (1641–1712) used the microscope to study the sex organs of plants, and Robert Hooke (1635–1703), John Ray (1627–1705) and others began to reclassify the plant and animal kingdoms. In chemistry, Robert Boyle (1627–91) experimented on the physical properties of air and formulated his law on the relationship between the pressure and volume of a gas.

Rationalism and observation now replaced superstition and dogma as scientific guidelines. With this new spirit scientific societies could flourish; the Royal Society was founded in 1660 in London and the Académie des Sciences in Paris in 1699.

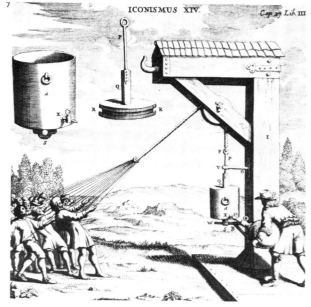

The third of Isaac Newton's three laws of motion states that to every action there is an equal and opposite reaction. An experiment designed to prove the validity of the theory is shown in this early 18th century book on Newton's laws. A metal globe emits a jet of steam in one direction and causes the "engine" to react by moving in the opposite direction.

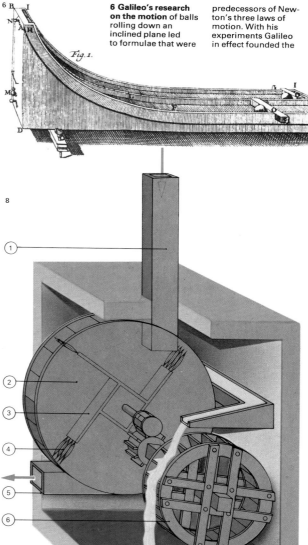

6 Galileo's research on the motion of balls rolling down an inclined plane led to formulae that were predecessors of Newton's three laws of motion. With his experiments Galileo in effect founded the science of dynamics. The illustration is of an 18th-century experiment in the same tradition.

7 The effects of a vacuum were demonstrated by Otto von Guericke (1602–86), the burgomaster of Magdeburg, at Ratisbon in 1654. A group of more than 50 men are trying to pull a plunger out of a cylinder from which the air has been exhausted. This is taken from von Guericke's book *Experimenta Nova*.

8 A 16th-century ventilator for a mine worked as follows. A water wheel [6] drove a fan through step-up wooden gearing. The blades [3] of the fan were tipped with feathers [4] and ran inside a drum [2]. Air was sucked down the ventilation shaft [1] by the fan and distributed by a duct [5] to the mine workers.

9 Printing from movable type radically improved the dissemination of knowledge about scientific discoveries in Europe by the 16th century. This woodcut of 1568 by Jost Amman shows a printing works with compositors setting up type in the background, while in the foreground the press is being operated.

Baroque art: style and content

Baroque was the most dynamic style of art and architecture produced by Western civilization in the seventeenth and eighteenth centuries. Flamboyant and emotional, it arose in Italy shortly before 1600, and by the late seventeenth century had reached most other parts of Europe, taking root especially in Germany and central Europe and spreading to Latin America. All these were Catholic regions with autocratic political regimes, and it is no accident that Baroque flourished there more than in the Protestant and relatively liberal states of northern Europe; its style made it a suitable vehicle for expressing the colourful, doctrines of Catholicism and the political principles of absolute monarchy.

The influence of the Renaissance

Two of the greatest Baroque artists, Peter Paul Rubens (1577–1640) and Gianlorenzo Bernini (1598–1680), were devout Catholics and supporters of absolutism. But it would be wrong to identify Baroque too closely with one particular ideology and social order. It also occurs in modified form, for example, in Britain and The Netherlands. Conversely, in

Catholic France under one of the most autocratic governments in Europe there was some resistance to it on both nationalistic and aesthetic grounds.

To understand Baroque as a style it must be seen against the background of the Renaissance, for the Baroque drew heavily on that movement and its successor, Mannerism. On the whole, classical forms were used throughout all three periods, although to a diminishing extent in the final stages of Baroque. ("Classical forms" are primarily the conventions for representing the idealized human body, seen in Greek sculpture, and features of classical architecture such as columns, pediments and friezes.) Renaissance artists employed classicism with restraint and with the dual aim of achieving clarity and realism. Mannerists abandoned these principles and pursued, instead, extremes of decorativeness, complexity and artifice. Baroque kept the complexity and some of the decorativeness but returned to the realistic style of the Renaissance.

Realism, space and movement were expressed in new ways: in painting, by

placing at least the main figures in the foreground; in sculpture, by stressing the roundness of forms and the details of their surface modelling; and in architecture, by using massive columns, overlapping pilasters and elaborate, deeply cut ornament. These features were underlined in all three arts by dramatic use of light and shade. Everywhere – in contrast to both Renaissance and Mannerist practice – there was a tendency for the barriers between the arts and between them and the real world to be broken down.

A rich and dynamic style

Because of its use of swelling forms, plenty of ornament and rich and glowing materials – much marble (often coloured), gilt and bronze – the Baroque is a heavy style. But it is also dynamic. Angels fly and saints soar up to heaven; men on the ground gesture and struggle; draperies flutter as if they had a life of their own. Baroque painters loved to depict a crowd; it is not surprising that their supreme large-scale achievement was ceiling decoration [5]. In architecture, vigorous, undulating façades, oval ground plans at

1 Bernini, foremost of the Baroque sculptors, first asserted his hold over his patrons and fellow artists by the brilliance of his technique. In his life-size marble of "Apollo and Daphne" (1622–5), for Scipione Borghese, Daphne is changed before our eyes into a laurel tree as the pursuing god is about to catch up with her.

2 Baroque realism was first expressed in a brutal and dramatic way by Michelangelo da Caravaggio, a north Italian painter who took Rome by storm in the 1590s. Although largely untutored he was a man of genius with a fiery temperament, well exemplified in his "David with the Head of Goliath" (c. 1606). His work had great influence on contemporary artists.

3 The full flowering of Baroque style in architecture as a field for structural even more than decorative ingenuity occurred in Germany in the 18th century. The church of the Vierzehnheiligen (Fourteen Saints) (1743–72) in northern Bavaria was designed by Balthasar Neumann on a plan of intersecting ovals, so that there is a constant sense of interpenetrating spaces, a sense enhanced in the vertical plane by endless vaults and arches. The interior appears to be all arches and projecting piers, with jutting curves, and no walls. The decoration of the altar is Rococo.

Vierzehnheiligen [3], designed by Balthasar Neumann (1687–1753), and broken pediments tend to replace the straight façades, square or circular ground plans and simple triangular or segmental pediments that were normal during the Renaissance.

Although Baroque artists invented few new forms, they displayed the utmost ingenuity in devising new types of decoration and in twisting traditional forms into unusual shapes. From its reverence for classical art the Baroque derived its erudition and sense of grandeur; from the Renaissance it took its understanding of form and its feeling for colour and light and shade; from Mannerism it inherited its love of complexity and decoration. But the mixture was new.

Development and spread of Baroque

The first signs of the new style were a return to realism [2] and a powerful use of light and shade, exemplified by the Italian painter Michelangelo da Caravaggio (1571–1610) and Annibale Carracci (1560–1609). Dynamism and a mastery of glowing colour emerged with Rubens [6], who brought the

Baroque to northern Europe after visiting Italy from 1600 to 1609. Then followed the "high Baroque" in Italy, dominated by a trio of artists active in Rome in the second and third quarters of the seventeenth century: the sculptor and architect Bernini [1], the architect Francesco Borromini (1599–1667), and the painter and architect Pietro da Cortona (1596–1669). After this the initiative passed northwards through the work of Guarino Guarini (1624–83) in Turin [Key] to central Europe and southern and central Germany, where Baroque architecture reached the ultimate in flamboyance and fantasy. Meanwhile, Baroque painting enjoyed a last, golden revival in eighteenth-century Venice with the ornate work of Giovanni Battista Tiepolo (1696–1770).

In Spain and Portugal, Baroque arrived first in a restrained form in painting and sculpture, achieving both a popular version in religious art and a courtly version in the portraits of Diego de Silva y Velázquez (1599–1660). Spanish and Portuguese Baroque architecture shows a cascade of surface ornament applied to a basic form [4].

Ingenuity and boldness in the handling of traditional forms, together with a sense of movement, are the hallmarks of Baroque architecture as seen in Guarino Guarini's Dome surmounting the Holy Shroud Chapel, Turin (1667–90).

4 Spanish Baroque architecture commonly has separate units or "blocks" placed on top of each other, the vertical effect so created being emphasized by pepperpot domes, turrets and slender grouped pilasters. The Reloj Tower (1676–80) of Santiago da Compostella Cathedral, exemplifying this, was designed by Domingo de Andrade. Its encrusted decoration is also typical.

5 In late Baroque, the time-honoured Italian ambition to paint a many-figured fresco ceiling decoration was carried out with greatest virtuosity. In the dining-room of the Würzburg Residenz, Germany, the 18th-century Venetian painter, Giovanni Battista Tiepolo not only worked on an oval field but dispensed with an internal painted architectural framework, so that the whole vast design occupies a single unit of space and spills out on to the white and gilt stucco surround. He received the commission from the Prince-Bishop in 1751, and the subject is "Apollo conducting Beatrice of Burgundy to Frederick Barbarossa".

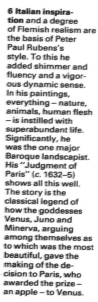

6 Italian inspiration and a degree of Flemish realism are the basis of Peter Paul Rubens's style. To this he added shimmer and fluency and a vigorous dynamic sense. In his paintings, everything – nature, animals, human flesh – is instilled with superabundant life. Significantly, he was the one major Baroque landscapist. His "Judgment of Paris" (c. 1632–5) shows all this well. The story is the classical legend of how the goddesses Venus, Juno and Minerva, arguing among themselves as to which was the most beautiful, gave the making of the decision to Paris, who awarded the prize – an apple – to Venus.

Baroque art: patronage and development

The chief patrons of Baroque were kings, popes, cardinals, the higher aristocracy, diocesan councils and representatives of religious orders. The two latter groups, although neither necessarily cultivated nor wealthy, were more active in their patronage than they had been since the fifteenth century. Baroque thus became the first style since the Middle Ages to develop both a popular and a sophisticated form.

Religious background to Baroque

The popular form is naturally most evident in outlying districts. While each region produced its own variations the local monastery or parish church [2] with slim bell-tower, simple Baroque doorway and Baroque stucco-work in the interior is a familiar sight throughout the Catholic world from Poland to Peru. Popular Baroque, unless mixed with non-European influences is, however, essentially a coarsened and simplified version of its sophisticated forms and it is these that must be examined to understand the social background and ideological content of the style.

Baroque arose under the stimulus of the Counter-Reformation. This, a movement of reform within the Catholic Church to reaffirm Catholic doctrine in the face of Protestant hostility, began in the 1530s long before the emergence of Baroque art in the normal meaning of the term. The Counter-Reformation was accompanied by the founding of several new, militant religious orders, most notably the Society of Jesus (Jesuits), founded by Ignatius de Loyola (1491–1556).

St Ignatius (he was later canonized and is frequently represented [5] in Baroque art) was, like many of the principal Catholic figures of the period, a Spaniard. In their writings the Spanish religious teachers evolved meditative techniques to turn the mind's eye upon the details of Christ's Passion and so train the soul for its ultimate reward – union with Christ. This reward was achieved through a good death (at its highest, through martyrdom) or through experiencing visions of Christ, the Virgin and the saints. The realization of these experiences depended on cultivating a heightened emotional state, and both the visions themselves and the state of mind required to produce them were well suited to representation by the means of Baroque art [1].

Art and the emotions

It was many years before Baroque emerged, however. The Counter-Reformation in its early stages was a puritanical movement and the religious orders were also poor and indifferent to the arts; official policy was confined to ensuring that religious pictures and statues were kept free of indecency and heresy. But in the late sixteenth century a few artistically minded prelates, such as Cardinal Federico Borromeo, began seeing the possibilities of using art not only to demonstrate Catholic doctrine but also as an aid to worship. They believed that vivid and attractive representations of the sacred stories and the visions of the saints would help ordinary Christians to believe, and share the visions.

Art, in other words, was to be used to reach the mind through the emotions – the fundamental technique of propaganda as it has been revived in modern times (the word derives from an organization founded in this

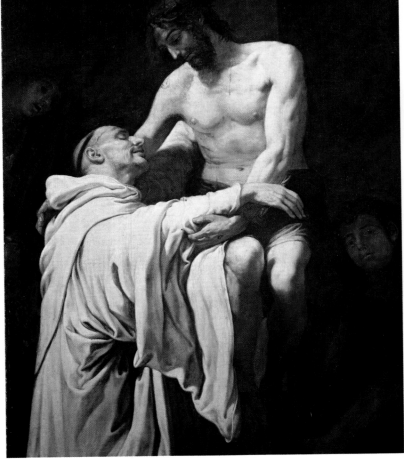

2 **S. Francisco at Ouro Preto** (1766–94), in the then Portuguese colony of Brazil gives some idea of the type of small church that sprang up in the farthest corners of the Catholic world in the Baroque period. More sophisticated than some, with its exquisite decoration and round towers set back from the façade, it is by a mulatto architect, Lisboa, called Aleijadinho ("little cripple").

1 **The spirit of devotion** in the time of the Counter-Reformation is seen at its most extreme in Francisco Ribalta's "Vision of St Bernard" (c. 1625). This is typical both of Spanish religion and Spanish painting. The saint draws so close in spirit to the suffering Christ that He appears to bend down from the cross to embrace him. Such vivid imagining reflects the ideals of Ignatius de Loyola's *Spiritual Excercises,* and the painter uses all his art to convey the experience.

3 **Diego de Velázquez,** like Bernini and Van Dyck, conceived it his task when portraying royalty – here, Philip IV of Spain (1644) – to ennoble his sitter, but he used more subtle, less flamboyant means. The face is stubborn and guarded rather than beautiful and, although the colours glow, the pose is restrained. Besides Velázquez, who painted many portraits of him and his family, Philip patronized Francisco de Zurbarán (1598–1664) and Rubens.

period, the Council for the Propagation of the Faith, the *Propaganda fide*). It was here that the Baroque style, with its realism and dynamism, proved to be so effective an instrument. It could translate a religion, stirring in its appeal and richly provided with incident and heroic characters, into visual terms and do this with energy and panache. A worshipper entering a church and gazing at the solemn pictures on the altars, the glittering statues against the piers and the vision of heaven on the ceiling was made to feel almost literally in the presence of God.

Power and glory of the Catholic Church

From the 1620s onwards, art was increasingly called upon not only to proclaim the faith but also to demonstrate the power and glory of the Catholic Church. Supported now by art-loving cardinals and some popes, Baroque painters, sculptors and architects became busily engaged in embellishing and building churches. The greatest single achievement of this kind was the transformation inside [4] and outside of St Peter's, Rome into its present state – the principal task of the sculptor, painter and architect Gianlorenzo Bernini (1598–1680).

In the same way, Baroque was well suited to the purposes of secular rulers, who saw themselves as divinely ordained. In the mid-seventeenth century a group of monarchs – Philip IV of Spain [3] (1605-65), Charles I of England [6] (1600–49), the Archduke Leopold-Wilhelm, Regent of Flanders, Pope Urban VIII (1568-1644) (who acted in many ways like a secular monarch) – were all active and knowledgeable patrons of the arts and of artists such as Velázquez (1599–1660) and Van Dyck (1599–1641), and all more or less favoured the Baroque style.

They were succeeded in the late seventeenth century by the most powerful monarch of all, Louis XIV of France (1638–1715), who turned French classicism towards the Baroque [Key], and in the eighteenth century by the princes of many German states. Where kings and princes led, dukes and lesser nobles followed, dotting the countryside of Europe with magnificent Baroque palaces [7] and houses which were eloquent symbols of aristocratic power.

Louis XIV summoned Gianlorenzo Bernini to Paris in 1665, among other things to sculpt his portrait bust – a meeting of luminaries conspiring to produce a quintessentially Baroque work. Chantelou, a Frenchman, noted how Bernini altered nature to bring out the effect of majesty in the king.

5 Loyola founded the Society of Jesus in 1534. The Gesù, mother church of the order in Rome, was begun to a stark design by Vignola (1507–73) in 1568. The magnificent chapel to Loyola in the Gesù was executed in 1695–9 by Andrea Pozzo (1642-1709), with a silver statue of the saint by Pierre Legros.

In 1797 the statue was melted down to pay Napoleon after his Italian campaign. It is now replaced by a partly silver copy, shown here. The long passage of Counter-Reformation Catholicism from heroic début through ostentatious triumph to later humiliation is faithfully reflected in the chequered history of the chapel.

6 Charles I, King of England, joined his fellow European monarchs to become one of the keenest patrons of the age. No other English king has rivalled him in this respect. Here he is portrayed (*c.* 1635) in the hunting field as an elegant gentleman – yet every inch a king – by his court painter, Sir Anthony Van Dyck.

7 There is a hint of Versailles in the palace of the Upper Belvedere, Vienna, built by Lukas von Hildebrandt for Prince Eugène (1721–24). (No ambitious patron of the period, let alone one who had helped to defeat Louis XIV, could ignore the Sun King's palace.) But the design is more playful and festive.

4 The visual climax of St Peter's, Rome, and hence one of the most sacred monuments of Catholicism, is the *Cathedra Petri* at the west end (St Peter's, unusually, is orientated towards the west). Bernini designed this in 1657- 66 as the final stage of his transformation of the Basilica's interior. The *Cathedra* takes its name from the ancient wooden chair of St Peter allegedly encased in the bronze throne that forms the centrepiece. Below are four Fathers of the Church, also in bronze, and above are gilt angels on white stucco clouds surrounding the Dove of the Holy Ghost set in a coloured window. Gilt metal rays extending from this window complete the scheme.

French art in the 17th century

Seventeenth-century French art, like most aspects of French life and thought, was dominated by one special quality: a profound belief in order and reason. There was much talk of "rules" and academies were founded for the first time with the object of setting standards (the Royal Academy of Painting and Sculpture was started in 1648 although it became effective only from the 1660s). Typical of this approach is the painter Nicolas Poussin's (1594–1665) statement: "Things having the quality of perfection ought not to be viewed hastily but with care, discrimination and intelligence."

Classical sources

Ancient Greek and Roman forms in architecture and sculpture were regarded above all as models of harmonious organization. They were pre-eminently creations of a rational mind, fitting together to form wholes with the logic of a mathematical proposition. Harmony in the relationship of parts is the essence of classicism. In France the idea was pursued with vigour. The French were also unusually interested in theory and insisted that the rules of proportion in architecture laid down by the Roman writer Vitruvius should be adhered to.

When French artists used forms drawn from Greek and Roman art they saw them as possessing even more clarity and order than they contained already [Key]. Their works consequently lack the grace of contemporary Italian art. The French classicists were equally opposed to the confusion of Mannerism and the licence of the Baroque.

Development of architecture

The first major artist of the period was an architect, Salomon de Brosse (1571–1626), and architecture alone has a continuous history in and around Paris throughout the century. Most French architects did not even visit Italy as they could gain all they needed from books and from the work of their sixteenth-century predecessors, some of whom were Italian-trained. De Brosse introduced symmetry on all four sides of a building and was the first for some time to design in terms of the mass of the building rather than decoration of its sur-

face. These qualities were inherited by his greatest successor, François Mansart (1598–1666), who, in a building such as Maisons [1], expressed most purely and completely the principal characteristics of the French classical spirit in this period: clarity combined with subtlety; restraint with grandeur; obedience to a strict code of rules combined with flexibility within them; and the elimination of inessentials.

Up to this point (the 1650s) the most important buildings had been executed for the rich bourgeois who had amassed fortunes in the new France of Cardinal Richelieu (1585–1642); and Parisian town houses are as significant in the architectural history of this period as country châteaux. The patrons of the French landscape painter, Nicolas Poussin, were also drawn from this class.

After 1660, however, when Louis XIV came of age, developments became increasingly centred on the crown. The first major undertaking to reflect this was the erection of the east front of the Louvre (1667–70) [6]. This was symbolic because the most celebrated artist in Europe, Gianlorenzo Bernini

2 In Nicolas Poussin's "Arcadian Shepherds" (c. 1639), gestures are limited to lines parallel to the picture plane and space is plotted with absolute precision.

3 The serene art of Claude Lorrain conjured up a mood of harmony and repose. In his "Hagar and the Angel" (1646) the angel comforts the servant Hagar.

1 In the 17th century the typical French château consisted of a solid block with frontispiece, short wings and high-pitched roof. It appears in its most massive and classical form at Maisons, near Paris, built 1642–6 by François Mansart for the Treasury official, René de Longueil. The walls are divided into two nearly equal storeys by a heavy entablature running all round the building, and each storey is articulated by repeated groups of pilasters. The roof line is varied by chimney-stack and dormer windows.

(1598–1680), had been summoned to Paris to design it but his proposals were eventually rejected in favour of those submitted by a French architect, Louis Le Vau (1612–70). Thus national pride and classical taste overcame the desire for Baroque splendour.

Development of painting

French painting meanwhile had found its most interesting and significant expression in two centres remote from Paris: Lorraine and Rome. In Lorraine, Georges de la Tour (1593–1652) painted candle-light scenes of religious subjects [4] in the dark manner of Michelangelo Caravaggio (1571–1610) (whose works in the original he may never have known), but treated them with a solemnity, precision and restraint that were typically French. In Rome, Poussin [2] and Claude Lorrain (1600–82) [3], who resided there for most of their lives, developed a classicizing style in intimate contact with ancient Roman and Italian Renaissance sources, although Poussin, at least, worked mainly for French patrons and decisively influenced French painting. His saturation in the moral attitudes and visible remains of the ancient world gave a mood of austere gravity to his work. He took great care to arrange his figures in such a way that every gesture and expression would tell.

Claude Lorrain was a more genial artist. His speciality was the ideal landscape of which he was, although not the inventor, the greatest exponent. The principle of ideal landscape is the production of an image of nature more beautiful and more harmonious than nature. To this Claude added an enchanting poetic light.

The last quarter of the century is dominated by Louis XIV's creation of the palace of Versailles [5]. It was designed by Le Vau and decorated by a team of painters, sculptors (whose art became important for the first time in the period) and designers under the supervision of the head of the Academy, Charles Lebrun (1619–90). Although the general lines of the buildings and their ornamentation are classical, the need for pomp and splendour at last proved irresistible and turned French art, temporarily, in the direction of the Baroque.

This plate from a French architectural textbook of 1650 underlines the link between French classicism and its ancient sources.

4 Biblical characters were given contemporary reality by the Italian Caravaggio, who painted them dressed in clothes worn by the poor of his time. Georges de la Tour borrowed this idea, perhaps indirectly, for his "Penitent Magdalene" (c. 1650), adding a note of French refinement.

5 The Salon de Vénus by Charles Lebrun (c. 1671) is part of the king's state apartments at Versailles – the first section of the palace to be built. With its heavy gilt stucco mouldings and architectural wall treatment, it shows how French classicism became part-Baroque.

6 Louis Le Vau's east front of the Louvre, Paris, 1667–70, which completed the palace apart from 19th-century additions, is almost a manifesto of the French classicism of its date – the period between Maisons and Versailles. It is clear and regular, yet magnificent.

The golden age of Dutch painting

The flowering of painting in The Netherlands during the seventeenth century is one of the most remarkable in the history of art. In many ways the Dutch school seems uniquely "modern" in the Europe of its time: realistic, unencumbered by rules, bourgeois and based on the operation of a free market rather than the system of ecclesiastical and aristocratic patronage prevailing in other countries. But its development, like that of the society it mirrored, was the result of a reshaping of past traditions, as well as of some entirely new forces.

The new republic

Seventeenth-century Dutch realism was rooted in an attitude that was already evident in fifteenth-century Flemish painting. A fondness for minute detail and for narrative, together with a certain popular appeal and an interest in depicting everyday life remained even in sixteenth century art, despite the Italianate forms of the Renaissance.

The whole Netherlands, including Flanders, was under the suzerainty of Spain in the sixteenth century, but the Dutch, many of whom were Protestants, fought a long war against the Spanish army of occupation and by the early seventeenth century had established an independent republic – the "United Provinces". The new republic rapidly became what was from many points of view the most progressive nation in Europe, with a flourishing economy, an advanced technology, a powerful merchant class and a habit of political and even, to some extent, religious toleration. At the same time it was conservative enough to maintain the ancient rights of the individual provinces and towns and also of the various guilds.

The guild of painters in each town was able to regulate entry to the profession and decide how and where artists could sell their works – normally through auctions at fairs. More broadly, the guild system favoured a traditional, craft-orientated approach to art, in contrast to the more intellectual approach fostered in aristocratic societies under Renaissance influence. The association of art with craft in turn encouraged the distribution of pictures among a wide cross-section of the community and prices were generally low.

Society also had some negative effects on art. The establishment of Protestantism as the official faith meant the exclusion of religious paintings [Key] and sculptures from churches, and their popularity with private buyers steadily declined. The absence of a royal family also largely deprived Dutch painters of the opportunity for large-scale baroque decorative schemes. Nor were painters, in contrast to architects, much interested in the classical forms which pervaded art in most of the rest of Europe.

Realism in portraiture and genre scenes

All this left Dutch artists free to concentrate on depicting, with increasing realism, themselves, their possessions and their surroundings. As with other mercantile societies, the seventeenth century in The Netherlands was a great age of portraiture. The highest category was the group portrait, usually of the officers of some militia company [1] or professional guild; both Rembrandt van Rijn (1606–69) and the Dutch portrait painter next to him in stature, Frans Hals (c. 1581–1666), made memorable contributions

1 Militia companies, formed to defend Dutch towns against Spanish troops in the 16th century, became social clubs after the Dutch Republic was established early in the 17th century. Frans Hals was one of their best portraitists in, for example, "The Banquet of the Officers of the Company of St George" (1627).

3 Low life, a genre inherited from the 16th-century Flemish artist Bruegel, was practised in Flanders and The Netherlands by Brouwer; his "Boor Asleep" was painted in Flanders c. 1635, but is also typical of Dutch painting. Its earthy realism, spiced with wit, is a feature of the first half of the period.

2 When he and they were old, Rembrandt saw the rich Dutch merchants and their wives as Old Testament patriarchs – proud, stiff and severe, yet with a sense of the pathos that affects all mankind. His portrait of Margareta Trip (c. 1661), an arms dealer's wife, shows his mastery of light, shade and brushwork.

4 After 1650, Dutch paintings of scenes from daily life became more refined and a new pictorial type was invented: the bourgeois domestic interior, the main subject of which were women. "Lady and Gentleman at the Virginals" (c. 1660), by Jan Vermeer, a master-painter of daylight, is a subtle example.

in this field, as well as that of the single portrait [2] – canvas after canvas of serious-faced, strongly characterized individuals, the sombreness of their clothes relieved only by white collars and cuffs.

The Dutch also loved pictures of themselves at home and at parties or in taverns; these were normally not portraits but genre scenes from daily life ranging from refined interiors showing women alone or conversing with gentlemen or making music [4], by artists such as Jan Vermeer (1632–75) and Gerard ter Borch (1617–81), through the rowdy feasting and drinking scenes of Jan Steen (1626–79), to the tavern brawls of Adriaen Brouwer (c. 1606–38) [3] and Adriaan van Ostade (1610–85). One aspect of daily life conspicuous by its absence is work, except light domestic employment or the traditionally paintable rural activities of minding herds and fishing.

Material possessions are often prominent in genre scenes but are presented even more blatantly in still-life painting: the glass and silver goblets, pottery bowls and jugs, imported Turkish carpets, and varieties of

flowers, of which the Dutch were so fond, together with all kinds of food. Still life also gave an unrivalled opportunity for that cool, painstaking realism in the rendering of detail which Dutch painters brought to such an extraordinary pitch [8].

Landscapes and seascapes

Finally, in landscape and marine painting, the Dutch invented a method of their own. Gone were the contrived if beautiful conventions of previous ages and other schools. It is true that they introduced other conventions, but these were mere fashions in composition and subject-matter rather than systems for creating imaginary visions more varied and more seductive than nature itself. Seventeenth-century Dutch landscapes and seascapes, by artists such as Jan van Goyen (1596–1656) and Jacob van Ruisdael (c. 1628–82), are windows on an actual world [7], a world of dunes, estuaries and canals, of flat coastal plains and dark inland woods, with grey water and high, grey cloud-filled skies – a world that can be recognized in the countryside of The Netherlands today.

The most typical and most exceptional of Dutch painters, Rembrandt, shows the human side of his religious faith in a detail of his etching "Christ Healing the Sick" (1643–7).

5 Realism in Dutch landscape painting was first achieved by simplification, with concentration on tone and light at the expense of form and colour, as in Van Goyen's "View of Emmerich" (1645).

6 Architecture was one field of Dutch art where classical rules were followed. The Dam Palace of Amsterdam, originally the town hall, was designed by Jan van Campen in the style of Palladio.

Begun in 1648, the year in which the republic formally became independent, it is covered with symbols in sculpture of the maritime and commercial supremacy achieved by the Dutch at this time.

6

7 The most powerful and versatile of Dutch landscapists was Jacob van Ruisdael. After an era of greyness and austerity, he brought back colour and a sense of form to the treatment of dunes, forests and windmills. His "Wooded Landscape" (c. 1660) suggests a contrast of life and death.

8 The Dutch love of material things is shown in this sumptuous "Still Life" by Willem Kalf (c. 1660).

7

8

The age of Marlborough

In January 1689 William III controlled the government and armed forces of England. He alone could maintain order and prevent confusion and was therefore in a strong enough position to prevent Parliament from imposing new restrictions on his power. The Declaration of Rights, read to William and Mary before they were offered the Crown, condemned James II's abuses of power and otherwise was so vague as to be unenforceable. William's position seemed as strong as that of Charles II in 1660, but with one crucial difference – Charles was granted a revenue for life; William was not, and as a result was to rely on Parliament for money.

England at war
From 1689 to 1697 and again from 1702 to 1713 England was involved in European wars of unprecedented scale. England declared war on France in 1689 because James II had fled there and hoped to recover his throne with the help of Louis XIV. But once involved, the English found themselves caught in the meshes of European power-politics. The English navy was already

formidable but William III also turned the English army into the great fighting force that John Churchill, Duke of Marlborough (1650–1722) [3] was to use so brilliantly in a series of battles [4, 8]. Under Charles II England had been neither a major European power nor a military power, but by 1713 it was both. The War of the Spanish Succession concerned the future of the ailing Spanish Empire, and the treaty of Utrecht that ended the war in 1713 established Britain as the major colonial power.

The huge sums needed to sustain the war intensified the Crown's dependence on Parliament. Never before had the Commons voted so many taxes. The Triennial Act (1694) meant that new parliaments were called every three years, and were thus less easily managed by the government. Members of Parliament naturally wanted to know how their taxes had been spent, and demanded more and more detailed accounts. The Commons began to take a much more constructive attitude towards finance, voting taxes, not to spend on the war, but to pay interest on long-term loans from the public.

Parliamentary taxes constituted excellent security and, as a result, the government's credit was good. It could therefore borrow vast sums much more cheaply than could the other European monarchies. But the king's credit now depended so much on Parliament that its management became more important than ever. Those politicians who could control the Commons were in a position to make the king heed their advice.

The war also led to a great expansion of both the armed forces and the revenue administration. Ministers thus had many more rewards at their disposal than in the past; they used offices and pensions to buy support in and control of the Commons.

Whigs and Tories
Because the government had to borrow more money, those institutions that could lend on a large scale became increasingly powerful. These included the great trading companies and the Bank of England, established in 1694 [5]. The rise of this "moneyed interest" broke the landowners' traditional monopoly of political power. Country gentlemen saw

1 **William III** (r. 1689–1702) and Mary (r. 1689–94) were invited jointly to the throne in order to save the country from James II's Catholicism; William, who had been fighting Louis XIV for many years in The Netherlands, saw English naval strength as vital for him to defeat France permanently. This print shows him holding a globe on which Belgium, England and Scotland are marked as "free", and France and Spain are "to be freed".

2 **The Quakers**, whose Synod is shown in session, were one of several Protestant sects that had sprung up in England by 1660 and asked only for toleration. Before 1640, although there had been disputes about the form of the Church of England, only the most extreme questioned the need for a single national Church. Anglican magistrates and clergy regarded the new sects as socially and politically subversive – Quakers refused to doff their hats to their social superiors – and so Dissenters were persecuted sporadically under Charles II. The persecution ended with the 1689 Toleration Act, although Dissenters were still excluded from public office.

3 **John Churchill, Duke of Marlborough,** owed his rise first to the favour of James II and then to the great influence of his own wife Sarah Jennings (1660–1744) over her close friend Anne, James's younger daughter; Anne became queen when William died (1702). It was Marlborough's great credit with Anne, more than his as yet unproven military genius, which led William to groom Marlborough from 1700 to 1702 to succeed him as leader of the great coalition against France.

4 **The War of the Spanish Succession** (1702–13) saw England, Holland and Austria joined against France. At Blenheim (1704) after a quick march across Germany, Marlborough eliminated Bavaria, and prevented Louis from knocking out the Austrians by attacking Vienna. From 1705 Louis' efforts were concentrated in the southern Netherlands where, after some striking allied successes, at Ramillies (1706) and Oudenarde (1708) he was bogged down in a war of attrition until 1713.

the wars as a conspiracy to divert their hard-earned money into the pockets of bankers, contractors and civil servants. As the landowners were mostly Tories and the moneyed men mostly Whigs, this feeling added to the new bitterness of party politics. High taxation eventually made the wars unpopular and Marlborough himself was removed from his command and retired to Blenheim Palace [Key] as the result of a political vendetta, soon after a sweeping Tory election victory in 1710.

The other great political issue of the period 1689–1714 was religion. The Toleration Act of 1689 allowed most Protestant Dissenters [2] to worship freely but not to hold public office. The Church of England lost its monopoly of religious worship and of education. In 1695 the clergy also lost the last vestiges of their control over the press. Even so, the universities were still closed to Dissenters. The rigid Anglicans, concentrated in the Tory Party, bitterly resented this erosion of the Church's authority. They attacked Dissenting schools and occasional conformity (whereby a Dissenter took the sacrament in an Anglican church in order to qualify for office). The Tories' views might be reactionary but they had a great deal of popular support. In 1710 Dr Henry Sacheverell was impeached by the Commons after an intemperate sermon. The sermon sold a hundred thousand copies, Sacheverell became a popular hero and his Whig prosecutors were routed in the general election of 1710 [6].

The whirl of party politics

That particular period experienced the most vigorous electoral politics seen in England before 1832. The electorate was volatile and independent – and predominantly Tory. Other than the support of the moneyed interest the Whigs had one great electoral asset, fear of Jacobitism – of a return of Catholic rule if James II's son could seize power. The Whigs exploited this fear after the failure of the Jacobite rising of 1715. They had already won over the new king, the Hanoverian George I (1660–1727) [7]. But after the political excitement of Anne's reign came the relative political stagnation of the age of Walpole and the Whig supremacy.

Blenheim Palace, which stands in a beautiful park in Oxfordshire, was designed for Marlborough by John Vanbrugh (1664–1726). It testified to the gratitude of Parliament to a great general and to the profitability of high office.

5 An early bank-note issued by the Bank of England in 1699 is illustrative of the bank's steadily growing capital. The Bank of England was one of the first commercial banking companies to be established in England. It was originally incorporated to lend the government £1,200,000, and empowered to issue paper money. It established the National Debt as a means of financing the war, leaving the debt for later generations to pay.

The Mytre in one hand and league in t'oth Shew that the Tubster is a fickle Brother.

6 Henry Sacheverell (c. 1674–1724) became famous in 1709 for his sermon "The perils of false brethren", in which he stated that subjects should offer no resistance to their governments, and he criticized the Glorious Revolution of 1688–9 as an act of resistance to the divinely sanctioned monarch. In particular he violently opposed the toleration of Dissenters, arguing for a strong episcopacy, and by implication supported the agreement ("league") by which Louis promised to help James II to regain his throne. The Whigs impeached Sacheverell for sedition but the London mob rioted on his behalf and sympathy for him contributed to the Tory electoral victory in 1710.

7 The ceiling of the Painted Hall now in the Royal Naval College at Greenwich depicts the two foreign rulers who symbolized England's deliverance from Catholic rule: William III and George I (r. 1714–27). The latter was named heir to the throne after it became clear that Anne would leave no heir. The ceiling was painted by James Thornhill (1675–1734) and the dining hall is one of the most magnificent frescoed rooms in Britain.

8 The allied victory at Blenheim was made possible by two principal factors. The first of these was Marlborough's bold tactics. The French centre was weak and relied on marshy ground between the rival armies to impede the progress of the English and Austrian forces. But Marlborough's cavalry picked and floundered its way across with the aid of planks and brushwood and then routed the French infantry. The second factor was guile – even deceit – which Marlborough used to hoodwink the English and Dutch governments into allowing their troops to give battle so far from home.

The age of Louis XIV

In 1660 France was internally divided, rent by faction. The rebellion of the "officers" and nobles in the Fronde – years of chronic civil unrest from 1648 to 1653 – had presented a threat of civil war and driven Cardinal Mazarin (1602–61), the first minister during Louis XIV's minority, from Paris. The work of Richelieu in re-establishing the authority of the monarchy had collapsed. France needed a strong adult king. In March 1661 Mazarin died and the young Louis, then 23 years old, decided to dispense with a first minister and to rule as well as reign. By that decision he restored to the crown the charisma surrounding it and the obedience owed to a divinely appointed king. He was determined to restore authority and majesty to the Bourbon dynasty in France and in Europe and, further, to end the disorder that affronted a dynasty ordained by God.

The theme of order is important in understanding Louis XIV's policies at home and abroad. Louis' obsession with order is depicted best at Versailles [3]: the architecture of the palace and the plan of the gardens and fountains follow the rules of symmetry;

elaborate ceremonial accompanied the king's actions throughout the day. Louis reduced the powers of nobles, *parlements* and the various provincial and national interest groups which he felt had swelled beyond their true station. Centralization extended court authority into the provinces. Agents of the central government (*intendants*) supervised regional affairs while the nobles were kept busy with entertainment at Versailles. Order demanded too the eradication of heresy. The French Protestants (Huguenots) lost the right to practise their faith and were forced to conform or go into exile by the revocation of the Edict of Nantes in 1685.

Foreign policy

The same belief in order and justice lay behind Louis' foreign policy. He aimed to restore to France all territories to which it once had a claim and to extend the nation to its "natural frontiers". What Louis saw as rights, the rest of Europe regarded as naked aggression. War characterized the reign of Louis XIV and campaigns were almost continuous from 1667 onwards. The War of

Devolution, concerned with Louis' claim to the Spanish Netherlands, was fought during 1667 and 1668; the war against Holland from 1672 to 1678; the War of the League of Augsburg from 1688 to 1697; and the War of the Spanish Succession from 1702 to 1713. The needs of war directed all departments of government to require new taxes and increases in the traditional *taille* (land tax) and *gabelle* (salt tax) to further administrative and technological developments.

Economic policy

The economic policies of Jean Baptiste Colbert were a response to the needs of war. France needed to be self-sufficient if she was not to depend on her enemies, especially the Dutch. Colbert attempted to stimulate the growth of native French industries [5] – iron and textiles – and especially the production of luxury goods, such as silk and lace. When these fashionable goods were acquired from the Mediterranean and the East Indies they were often brought to France by the Dutch and had to be purchased with bullion. As contemporary economic attitudes associated

1 A majestic image was a central facet of the absolutism of Louis XIV. The pre-eminence of the king was seen as part of divine and natural order, the establishment of his power a duty to God. Louis described that pre-eminence in his memoirs: "All eyes are fixed on him [the King] alone; it is to him that all wishes are addressed; he alone receives all respect; he alone is the object of all hopes . . . no one can raise himself but by gradually coming close to the royal person or estimation . . . one can even say that the splendour emanating from him in his own territories spreads as by communication into foreign provinces."

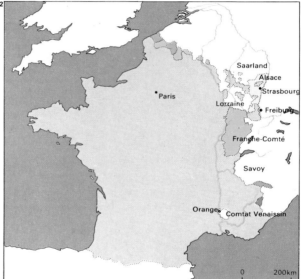

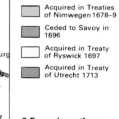

France in 1661

Acquired in Treaty of Aix-la-Chapelle 1668

Acquired in Treaties of Nimwegen 1678–9

Ceded to Savoy in 1696

Acquired in Treaty of Ryswick 1697

Acquired in Treaty of Utrecht 1713

2 France's northern and southern defences were strengthened by the acquisition of various frontier towns by treaties. Louis XIV's foreign policy was directed to annexing to France areas that would consolidate the frontiers. The map shows the territories gained over the course of 50 years.

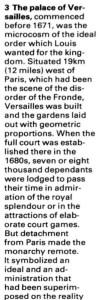

3 The palace of Versailles, commenced before 1671, was the microcosm of the ideal order which Louis wanted for the kingdom. Situated 19km (12 miles) west of Paris, which had been the scene of the disorder of the Fronde, Versailles was built and the gardens laid out with geometric proportions. When the full court was established there in the 1680s, seven or eight thousand dependants were lodged to pass their time in admiration of the royal splendour or in the attractions of elaborate court games. But detachment from Paris made the monarchy remote. It symbolized an ideal and an administration that had been superimposed on the reality of political life.

power with the possession of bullion these imports could not be tolerated. More trade and industry meant more power and wealth. Mercantilist economic theory meant economic war and because of this Colbert embarked upon a tariff policy [6] to hamper the Dutch. The French East India Company was formed in an attempt to wrest trade from the Dutch in the colonies which remained an important source of bullion.

For much of his reign Louis appeared to have fulfilled his desires. Order was established at home, territories were conquered and recognized as French possessions by treaty, military victories were won. Colbert's new industries enjoyed some success and France emerged as a rival to the legendary commercial supremacy of the Dutch and the English. The glory of the Bourbon dynasty was recognized: Versailles and French absolutism became the model for other monarchs to emulate.

But the achievements were won at the cost of great strain at home and the enmity of almost every power in Europe. As the cost of war reached unprecedented levels royal

finances began to fail. Louis never felt strong enough to tax the nobility and there was no developed system of credit in France. When in 1688 William of Orange, the Stadholder of Holland, became King of England the might of two naval powers was joined and the financial resources of the two most advanced commercial nations in the world were placed at the disposal of Louis' bitterest enemy.

Reaction and rebellion
As war continued and defeats piled up there was mounting reaction in France. The peasants, who bore the brunt of taxation but who were the least able to pay it, rebelled. Towards the end of the reign, shortages of grain became acute. Serious crises in 1693–4 and 1709–10 forced up grain prices and led to many deaths in country areas. The nobility reacted against Bourbon ambition and ostentation. Many trumpeted the virtues of the humbler life of pastoral simplicity and rural retreat. Louis' absolutism, centred on Versailles, became more and more remote from the nation, a symbol of the strength and the limitations of autocratic rule.

The glory of Louis XIV (1638–1715), the Sun King, is symbolized in his emblem.

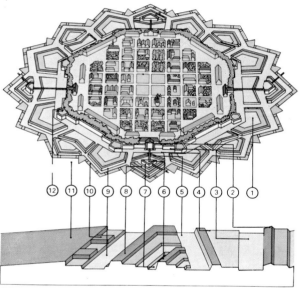

4 In the reign of Louis XIV fortification methods were improved by Sebastian de Vauban (1633–1707), as demonstrated by Neuf Brisach fortress at Alsace. Situated vulnerably on flat ground, its defences extended in depth round the whole circuit by detaching bastions [1] from the inner line, filling the gaps with *demilunes*, surrounding *ravelins*. First came the curtain [2], then a moat [3], *tenaille* [4] and second moat [5]. Next a *ravelin* [6], moat [7], *demilune* [8], outer moat [9], covered way [10] and lastly the *glacis* [11], sloping to outer ground level. The inner rampart was commanded by angled "bastion towers" [12].

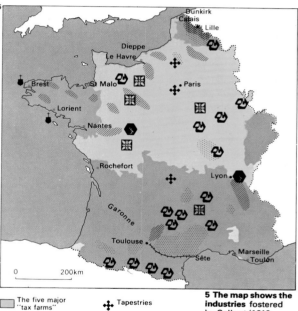

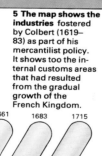

6 Colbert sought to bring order to the multiplicity of internal and external tariffs which impeded French trade. He also intended to employ a protective tariff to defend newly developed French industries and to hit at Dutch and English competition. The first tariff (1664) was primarily a product of the first aim of consolidation. The second, of 1667, strongly protected the French textile industry. The tariff brought retaliatory duties against French wine, and economic rivalry brought war with the Dutch in 1672. At the Peace of Nimwegen (1678), economic questions were settled in favour of the Dutch and the customs tariffs of 1664 and 1667 were lifted.

6 A Fine English and Dutch woollens, piece of 25 ells
B Fine Spanish woollens, piece of 30 ells
C Flemish tapestries per hundredweight
D Lace and embroidered linen per pound
E Tanned ox leather per dozen hides
F Tin plate per barrel

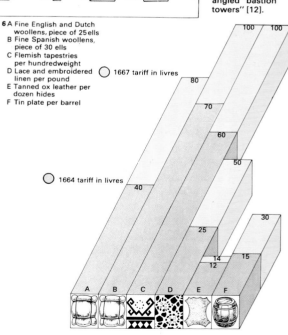

○ 1667 tariff in livres
○ 1664 tariff in livres

The five major "tax farms"
"Foreign" provinces
Provinces not under French control
● Ports
┈ Canal du Midi
░ Linen
░ Cloth

✛ Tapestries
✚ Lace
↓ Shipbuilding
♺ Iron works
⬡ Silk

5 The map shows the industries fostered by Colbert (1619–83) as part of his mercantilist policy. It shows too the internal customs areas that had resulted from the gradual growth of the French Kingdom.

7 One of the major financial problems of the monarchy stemmed from the difficulty of collecting taxes and ensuring the moneys due actually arrived at the Treasury. Under Colbert the process of tax collection was improved, although taxes increased. Following Colbert's death in 1683 and with growing reaction against the costly wars of Louis, the situation deteriorated. By 1715 more than half the income due failed to arrive.

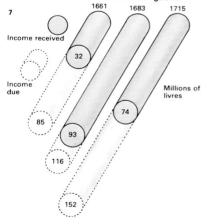

Income received
Income due
1661   1683   1715
32   74
93
85
116
152
Millions of livres

French classical literature

All the French writers of the sixteenth century attacked in one way or another the culture of the Middle Ages. The emphasis was on self-awareness, the development of individualism and innovation. This culminated in the critical scepticism of Michel de Montaigne (1533–92) [Key]. The same qualities are also seen in Clément Marot (1496–1544) [1], whose irreverent humour and obstinate insistence on his right to think for himself are often reminiscent of François Rabelais (c. 1494–c. 1553) [2]. Jean Calvin (1509–64), the Protestant reformer, was equally concise and ironical: his polemical writings showed that the French language could be as effective for argument as for narrative.

The Renaissance in literature

The spirit of Renaissance humanism had begun to permeate French literature. This at first led to the establishment of Greek and Latin writers as the supreme models and to many translations, and remained dominant until the "Quarrel between the Ancients and Moderns" in the latter half of the 1600s.

Apart from that of Rabelais, Calvin and Montaigne, little distinguished prose was produced in the sixteenth century. But in poetry supremely important innovations were made, first by Marot, then by the Lyons poets. Their chief representative was Maurice Scève (c. 1510–64), whose complex masterpiece *Délie* (1544) depended to a great extent on Petrarch's example. Scève was regarded as a pioneer by the Pléiade, a group of poets (Rémy Belleau, Jean Antoine de Baïf, Etienne Jodelle, Pontus de Tyard and Jacques Peletier as well as Pierre de Ronsard and Joachim du Bellay [3]) who took their name from a constellation of seven stars, which was also the name given to the seven most eminent Greek poets of the reign of Ptolemy II in Alexandria.

The Pléiade intended to rehabilitate French poetry and put the French language on an equal footing with Latin, Greek and Italian as a vehicle for it: the group's inspiration was obviously classical. Later the court poet François de Malherbe (1555–1628) turned sharply against the aims of the Pléiade because of its excessive complexity and erudition; but it had made its mark on him and thus influenced the course of poetry.

French classical theatre did not reach its apogee until the seventeenth century, but Robert Garnier (1534–90) wrote humanist (if over-literary) tragedies such as *Les Juives* (1583) which were performed by itinerant actors – paving the way for the dramatists who succeeded him – and he exercised a strong influence in England.

"Ancients" versus "Moderns"

The seventeenth century was above all a period of classicism, although individualism was by no means entirely suppressed. However, generalizations about human beings were drawn from careful examination of individuals and eccentricities were not encouraged. The quest for a perfect language continued, resulting in the foundation of the Académie Française [7].

The battle lines of the Ancients and Moderns were drawn with Nicolas Boileau (1636–1711) [9] as the chief representative of the Ancients and the poets Jean Desmarets (1596–1676) and Charles Perrault (1628–1703), a poet and critic most famous

CONNECTIONS

See also
314 The age of Louis XIV
308 French art in the 17th century
282 English Renaissance literature

In other volumes
34 History and Culture 2

1 Clément Marot was a court poet. He wrote, at first, in the fashionable late 15th-century style but soon abandoned this to become an important innovator. He began to fashion his poems on Latin and Greek models and introduced almost all of the new styles that the Pléiade announced as theirs. In particular, he introduced the sonnet into the language. He knew Calvin and joined him in Geneva when he was attacked again in France (after already having been imprisoned for eating meat in Lent). He was a witty, graceful, epigrammatic writer – one of the best court poets and an important transitional figure.

2 This illustration by Gustave Doré of Rabelais' *Gargantua and Pantagruel* (c. 1532–52) shows Gargantua poised above Notre-Dame Cathedral. In 1530 Rabelais abandoned the monastic life and became a doctor. His contemporaries saw him mainly as an eminent physician and humanist. His work changes as it proceeds. The first two books tell a fantastic tale of a family of giants, and also satirize the clergy and educational methods of the times. The third book is more topical and the fourth the most complex. Rabelais was a realist who believed in the equal importance of all human beings. His great work, uneven though it is, anticipated many of the problems of 20th-century novelists. He has incomparable value as a supremely imaginative writer and as a lucid and commonsensical portrayer of his times.

3 The two most gifted members of the Pléiade group were Pierre de Ronsard [A] and Joachim du Bellay [B]. Du Bellay wrote the manifesto of this group in his *Défense et Illustration de la Langue Française* (1549). His advice to authors was this: take Greek, Latin and Italian models to make French "illustrious"; create new words to enrich the language; use ancient and Italian forms and reject those of the Middle Ages; above all, seek formal perfection. Ronsard's greatest achievement was in his lyrical poetry and his mastery of the ode and sonnet. Du Bellay wrote the first sonnet sequence in French, *L'Olive* (1549), which was influenced by Petrarch's poetry.

4 Mme de La Fayette [left], the novelist, Mme de Sévigné, the observer, and La Rochefoucauld the moralist represent three great strands of French writing in the 17th century. It should also be noted that all three were aristocrats, for *belles lettres* were almost an aristocratic preserve. Another dominant fact of French literary life was the near equality between the sexes and the markedly social nature of the circumstances of much literary production.

for his enchanting fairy tales (1697), as champions of the Moderns. The virtues of fairly rigorous imitation of classical models were set against the merits of exploring modern and progressive ideas. Finally (1700), Boileau acknowledged to Perrault that contemporary French writers could – if they recognized the classical virtues – be as good as any ancient authors.

Drama and prose

Poetry flourished in the tragic drama (although the great tragedians also wrote comedies), first in the rhetorical Pierre Corneille (1606–84) and then with Jean Racine (1639–99) [6]. The comedy was developed to perfection by Molière (1622–73) [5], and the fable was mastered by La Fontaine (1621–95) [8].

Fiction, a product of the expansion of literacy and the desire (especially among those who frequented the *salons*) for entertainment, properly begins (even though Rabelais is, with Cervantes, the real father of the novel) with *L'Astrée* (1607–27) by Honoré d'Urfé (1567–1625). It is a long pastoral story, but with some psychological realism and evident moral intention. One of its admirers, Marie de La Fayette (1634–93) [4], wrote the first great French novel, *La Princesse de Clèves* (1678). This tale of passion is remarkably modern in that it is quietly realistic, psychologically acute and obliquely critical of society: its structure and style were supreme, according to Stendhal and others.

The picture of the age given by the major writers is completed by the letter-writers, moralists, aphorists and writers on religion. Among these are Marie de Sévigné (1626–96) [4] whose feminine and intelligent letters give an incomparably lively and accurate picture of Paris, the centre of culture; François de La Rochefoucauld (1613–80) [4], whose *Maxims* are far more than merely cynical in that they expose the real nature of motives; and Blaise Pascal (1623–62), the mathematical genius whose posthumous *Pensées* (written some years before his death but not published until later) defend the influential brand of critical Catholicism called Jansenism and anticipate the "open" tone of contemporary religious discussion.

KEY

The *Essays* (1580–95) of Michel de Montaigne contain the purest expression of the spirit of 16th-century France. Montaigne – an aristocrat by birth – invented the essay. Although a sceptic he would not offend religious orthodoxy and took the conservative side in the religious wars. The curiosity he displayed was a typical feature of those men of his time who were fighting against the stultifying effects of the Middle Ages. In his essays he touches on almost every known subject in a direct and lucid style: on anthropology, morality, judicious behaviour, education and above all on wise and sound judgment.

5 **Molière** was the stage name of Jean Baptiste Poquelin who is on the left in this anonymous painting (1670). Molière, an actor-manager, transformed the art not merely of French but of European comedy. He is possibly the supreme comedian of manners, always ironically tolerant of faults and follies.

6 **A scene from Racine's tragedy *Phèdre*** is shown here. Racine achieved classical dramatic perfection (observation of Aristotle's three unities of time, place and action) and yet he sacrificed no psychological power. His plays are deeply moving and his eloquent verse rhythms heighten the emotion.

7 **The Académie Française** was established by Cardinal Richelieu after some authors took to meeting secretly (about 1630) to discuss literary questions. Its purpose was to define and thereby to perfect the French language.

8 **Jean de La Fontaine** (1621-95), an adherent of the "Ancients" in the famous "battle", drew from many sources, including Aesop, for his *Fables*, an illustration from which is shown here. By attributing human responses and motivations to animals he created situations that he could use to expose and satirize human failings both large and small. He was not a true moralist, but rather an amused, detached and subtle observer of human nature and social behaviour.

9 **The influential critic and poet** Nicolas Boileau, although not a great poet, was a legendary figure in his own lifetime – and the chief advocate of the "Ancients" in their quarrel with the "Moderns". His many associates included Molière and Racine. His *L'Art poétique* (1674) became famous throughout Europe and his rules were taken as law. The poem preaches, in a neat epigrammatic manner, all the classical virtues of poetry.

The origins of opera

Opera developed from three distinct sources and became established in Italy about 1600. *Sacre Rappresentazioni*, dramas with music, usually presented in the vernacular, were influenced by the splendid court pageants mounted to celebrate feast days and royal birthdays during the fifteenth century. Simultaneously, part-songs and madrigals enjoyed great popularity and were often linked in cycles to tell a story.

Opera takes shape

A sung story, spectacle and drama were forged into opera by the *Camerata*, a group of artists who were concerned with the role of music in Greek classical drama. The *Camerata* invented the recitative, a declamatory style of singing in which the notes and rhythms were allied to verbal accentuation. This was initially accompanied only by the harpsichord and was designed to enhance the meaning of the words.

The great master of the madrigal, Claudio Monteverdi [Key], adopted the formal recitative but combined it with the madrigal style to provide the basis of arias or set songs

which punctuated the recitatives. In addition Monteverdi employed a small orchestra and decorated the sung melody with runs and other musical devices to increase the expressiveness of the form. His *Orfeo* (1607) is accepted as the first operatic masterpiece with a known score.

By 1637 opera was established as a secular entertainment in Florence; and in Venice the first commercial opera house was opened. Early opera houses were equipped with complicated machinery for lavish stage effects [2, 3, 4, 5]. Orchestras became larger and, with the development of the violin, they became increasingly important in opera.

The spread of opera

Royal marriages between Italy and France took opera to the latter country, which then became the centre of a controversy about the pre-eminence of words or music. Jean-Baptiste Lully (1632–87), an Italian, adapted the recitative to the unique cadences of the French language and indulged the court's love of dance and spectacle in his operas [6]. Jean-Philippe Rameau (1683–1764), who

succeeded Lully at the French court, used the chorus and orchestra to give opera a new grandeur and dignity [7]. The French approach to opera was essentially literary, drama taking precedence over the music.

In Italy an increase in musical resources and the advance of vocal technique to a stage where it rivalled the violin in agility concentrated public interest in opera. The works of Alessandro Scarlatti (1660–1725) established the form of the *da capo* aria (in three sections, the last one being a repetition of the first) and the overture.

English opera likewise derived from court entertainments but was slow to adopt the new musical forms. Henry Purcell (1659–95) set English to the new "recitative musick" with great success, and his *Dido and Aeneas* is regarded as the first true English opera [8].

This germ of a tradition was not further developed. Instead, English opera was dominated by one of the first truly international composers, Georg Friedrich Handel (1685–1759). Born in Germany, Handel worked briefly in Italy before coming to England. He absorbed all the major musical

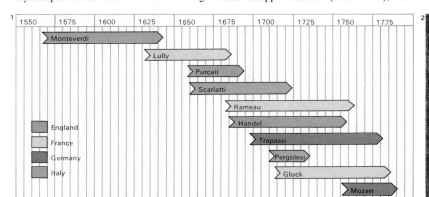

1 Some of the major composers who established or extended opera are listed here. It can be seen that from its origins in Italy, opera's centre of development moved first to France and later seemed to settle in Germany, although Gluck did his most important reformist work in Paris. In the 18th century there was a concentration of activity that produced a kind of operatic heyday.

2 One of the first opera houses was that of Cardinal Barberini, which was built in his palace in Rome in 1632. The first opera performed there, Landi's *Il S. Alessio*, made several departures from tradition. It dealt with a 15th-century legend rather than a classical subject and also contained comic scenes. The first commercial opera house opened its doors in Venice in 1637 and in the following 60 years at least 16 more opera houses were established in Venice alone. These early houses were extremely lavish and equipped with all manner of ingenious stage machinery.

3 "Cloud machines" were used by Giacomo Torelli (1608–78), the leading stage technician of his day. This one has been lowered to introduce dancers, while the upper one contains an orchestra.

4 A recent revival of Cavalli's *La Calisto* at Glyndebourne, Sussex, recreated in modern terms some of the stage effects used at the time of its first performance. For example, gods and goddesses were always required to float above the stage, to descend from the flies or heaven, while devils logically entered through trapdoors from below stage. Adherence to these rules in modern productions has helped to revive interest in some long-neglected works.

5 Some of the early stage effects were achieved with great simplicity and cunning. In 1638 Nicola Sabbatini, whose inventive devices were widely used, wrote an illustrated treatise on stage mechanics. Despite technical advances, it is doubtful whether modern producers have been able to rival the spectacle of some of those early presentations. Revival of interest in 17th-century music has recently led to a new concern with authenticity in staging the operas.

forms and traditions and brought *opera seria* (opera sung throughout) to its peak.

Vienna was the true home of *opera seria*. It used sparsely accompanied recitatives to tell the usually ennobling plot which was much interrupted by ornately embellished, often repeated, arias. The major influence in *opera seria* was Pietro Trapassi (1698–1782), known as Metastasio, the poet whose libretti perfected this artificial form. This, too, was the age of the singer – for whom arias were conceived primarily as vehicles for vocal brilliance – and especially the *castrato* (a singer castrated in boyhood to preserve the soprano or contralto range of his voice).

From Gluck to Mozart

Christoph Wilibald Gluck (1714–87), a Bohemian working in Vienna and Paris, was the major reformer after *opera seria* fell into disfavour. He strove for simplicity and directness in his operas, qualities that are embodied in his *Orfeo ed Euridice*.

Comedy had so far played little part in opera, although the practice of inserting comic interludes had long been popular in

Italy. *Opera buffa* – full comic opera – emerged in 1733 when Giovanni Battista Pergolesi (1710–36) presented *La Serva Padrona*. *Opera buffa* required a lighter musical touch using a speedier recitative, simple melodies and sparkling orchestration.

All these forms were familiar to and employed by Wolfgang Amadeus Mozart (1756–91). His great operas, *The Marriage of Figaro, Cosi Fan Tutte* and *Don Giovanni*, are a dazzling amalgam of styles, welded together into supreme musical-dramatic entities. His command of a large orchestra, his perfection of vocal ensemble, his gift of melody and his sure sense of psychology lifted opera to new eminence. He transformed the popular *singspiel* – a combination of spoken dialogue and sung material – into high art in *The Magic Flute*, and at the end of his life wrote magnificently in the older *opera seria* form in *La Clemenza di Tito*.

Mozart's operas provided the basis for late eighteenth and nineteenth century development, during which time the stylistic division between German and Italian opera became absolute [11].

KEY

The father of opera, **Claudio Monteverdi** (1567–1643), was born in Cremona, where he served as a choirboy. He wrote more than 250 madrigals that are masterpieces of the form and which employed new, expressive harmonies. His first opera, *Orfeo*, at once indicated his ability to marry the vocal and instrumental traditions of his time with the new recitative. He became a priest in 1632 and wrote much church music, including the celebrated *Vespers*. A few examples of his enormous operatic output have survived; they include *Orfeo*, *Il Combattimento di Tancredi e Clorinda*, *The Return of Ulysses* and *The Coronation of Poppea*. These works have provided musical scholars with many problems, but they are now widely known, albeit in differing versions. He spent the last 30 years of his life in Venice where, in addition to his religious duties, he was deeply involved with commercial opera.

6 Jean-Baptiste Lully was a French musician who was born in Italy. Early in his career he collaborated with the playwright Molière in the production of opera-ballets. His first true opera was performed in 1672, after which he wrote 20 more, including *Armide* (1686), whose title page is shown here. He used accompanied recitatives where the words meant more than the music, and very formal arias.

7 Jean-Philippe Rameau was Lully's successor, and he continued the formal tradition of French opera. He was a controversial composer who played a leading part in the enduring quarrel about the supremacy of words or music. He was a great musical theorist and his ideas brought a new flexibility to the recitative.

8 Henry Purcell was organist at Westminster Abbey and regularly provided music for plays and masques. He wrote much fine church music and many songs. His only true opera, *Dido and Aeneas*, the first masterpiece of the form produced in England, was written for performance by a Chelsea girls' school. Several 20th-century composers, including Tippett and Britten, have been influenced by his music.

9 Georg Friedrich Handel is universally known for his great English oratorio, *Messiah*. He was, however, a prodigious composer of operas such as *Serse* and *Alcina*, and other music.

10 Carlo Farinelli (1705–82) was the most celebrated of the Italian "male sopranos" or *castrati*. Women in Vienna and London are said to have fainted with excitement at his singing.

11 Wolfgang Amadeus Mozart was a child prodigy who was attracted to the theatre early in life. He had already written music for the stage when, at the age of 12 years, he wrote his first *opera buffa, La finta semplice*, followed in the same summer of 1768 by the operetta *Bastien und Bastienne*. Much later, his friendship with the celebrated actor, singer and theatre manager, Emmanuel Schikeneder (1751–1812), led to their collaboration in the *singspiel The Magic Flute*, first performed in 1791.

Medieval Russia 900-1600

The first Russian state emerged out of the Slavic settlements northeast of European Byzantium between the sixth and ninth centuries AD. The Russian Slavs occupied a large belt of territory bounded by the Carpathian Mountains, the Baltic Sea and the headwaters of the Volga, Don, Dnieper and Dniester rivers. In the 800s, fierce Viking merchant-warriors, called Varangians, conquered the area and established a federation of city states whose centre was Kiev and which was ruled by the grand prince of Kiev. The existing Slav ruling class assimilated the Varangian princes and established a lucrative trade with Constantinople [1]. In contrast to the barter economy of the medieval West, money and credit systems were used and a prosperous commercial civilization grew up.

Byzantine influence in Kievan Russia

By 1100, bolstered by commercial wealth and contact with the Byzantine Empire, Kievan Russia had become a powerful state and the centre of a flowering culture. From her chief trading partner Russia accepted the cultural heritage of Byzantine Orthodox Christianity. Missionaries introduced Orthodox liturgy by means of a language that, using the Cyrillic alphabet, became the basis of modern Russian. Grand Prince Vladimir I (c. 956–1015) adopted Orthodoxy as the official religion and a metropolitan bishop arrived from Constantinople and set up an ecclesiastical organization. Byzantine styles of building [2] and icon painting flourished [3].

The Russian political and social structures contrasted sharply with their Western counterparts. Three governmental elements managed to co-exist – the ruling Riurik dynasty had to share power state-wide with a council of noblemen (*duma*) and with the town meetings (*veche*) at local level. The princes and their relations stood at the top of the social order, followed by the merchant-soldier-landowner class of boyars [9], then landowning peasants, tenant farmers and slaves. Kievan peasants were as a rule free to buy, sell and inherit land in their own right and owed no vassal-like allegiance to the great landholders. Kievan law recognized few class differences and most non-slave citizens were equal in the eyes of the law.

During the twelfth and thirteenth centuries Russia was profoundly affected by events beyond her frontiers. The growth of competing Venetian trade routes, raids by Asiatic tribes, the decline of Byzantium, and the rise of Poland-Lithuania and the Mongol Empire all contributed to Kiev's disintegration; there was also internal political and dynastic strife. Russian links with Constantinople loosened and the economic strength of the Kievan state began to decline. The great city of Novgorod with its mercantile democracy and rich hinterland was able to establish its independence [8], but western Russia fell under Polish-Lithuanian (and hence Roman Catholic) control and eastern Russia was overrun by the Mongol Empire, which was expanding rapidly under Genghis Khan (1167–1227) and his successors.

The emergence of Moscow

It was under Mongol rule that the principality of Moscow (Muscovy) first became important and then asserted political control over most of what is now European Russia. The

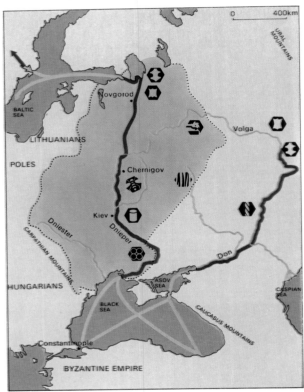

Eastern end

1 Trade was essential to the Kievan economy. Kiev owed its importance to its location on the international trade route that connected the Baltic Sea with the Mediterranean – the water road from the Varangians to Byzantium. Great annual trading convoys floated down the River Dnieper and, via the Black Sea, to Constantinople. At the Byzantine capital the products of the Russian forests – chiefly furs, honey and beeswax – were exchanged for spices, wines, perfumes and weapons. The river route was protected against nomads by soldiers.

2 Construction of Sophia Cathedral, Kiev, began in 1037, during the reign of Grand Prince Yaroslav ("the Wise", c. 1036–54). One of the first major Byzantine-inspired churches in Russia, it set an example for innumerable Orthodox churches built during the next nine centuries.

A brick, cross-domed basilica, its square plan was based on that of the Hagia Sophia in Constantinople. A Russian innovation was the arrangement of 13 cupolas – 12 smaller domes surrounding a large central dome, while the eastern end of the church ended in five semi-circular apses.

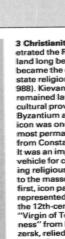

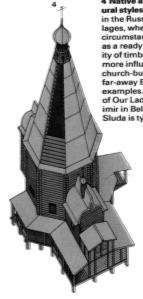

4 Native architectural styles emerged in the Russian villages, where local circumstances such as a ready availability of timber had more influence on church-building than far-away Byzantine examples. The Church of Our Lady of Vladimir in Belaya Sluda is typical.

3 Christianity penetrated the Russian land long before it became the official state religion (c. AD 988). Kievan Russia remained largely a cultural province of Byzantium and the icon was one of the most permanent gifts from Constantinople. It was an important vehicle for conveying religious truths to the masses. At first, icon painting, represented here by the 12th-century "Virgin of Tenderness" from Byelozersk, relied heavily on the Byzantine style.

5 Muscovite icon painting reached its peak in the 14th and 15th centuries. The most famous masterpiece of this period, the "Old Testament Trinity", was painted in about 1411 by Andrei Rublev (c. 1370–1430), Russia's greatest iconographer.

grand dukes of Moscow, descended from a branch of the Riurik family, were originally the tax collectors of Sarai (the Mongols' western headquarters on the lower Volga). By treaty, conquest, purchase and the strategic marriage of their offspring, the grand dukes emerged with the largest territory [6] and as the head of a coalition that eventually drove out the Mongols. The grand dukes' nicknames reveal the variety of their methods – Iuri ("Long Arm", reigned *c.* 1149–57), Vsevolod ("Big Nest", referring to his fertility, reigned *c.* 1176–1212), Ivan I ("Money Bags", reigned *c.* 1328–40).

Development of national monarchy

Between 1460 and 1600 Moscow vastly increased in size and strength and its rulers took on the trappings of a national monarchy. Ivan III ("the Great", 1440–1505), with a series of brilliant diplomatic and military campaigns, dealt the final blow to the disintegrating Mongol Empire in 1481 by forming an alliance with various Mongol confederations. He began to see his newly independent realm as a successor to the

recently vanquished Byzantine Empire. He married Sophia, niece of the last emperor, and adopted the Byzantine double-headed eagle as a royal symbol. Ivan IV ("the Terrible", 1530–84) conquered fresh Lithuanian and Mongol territory and crowned himself "Tsar (Caesar) of all the Russias" [7].

The fall of Constantinople to the Ottoman Turks in 1453 aided the establishment of a separate Russian Orthodox Church. The seat of the Church had moved from Kiev to Moscow in the fourteenth century and a distinct Russian style in architecture [4] and painting [5] had emerged there by 1589, when the first Russian Orthodox patriarch was sworn in.

Modern Russia owes its political and social character to Moscow rather than to Kiev. The Russian state, consolidated as a mercantile economy, gave way to an agrarian one and while feudalism was on the wane in the West the roots of serfdom were being sunk in Muscovy. To compound the problem, no class of townsmen or independent farmers emerged that was capable of setting limits on the power of the crown.

KEY

The crown of Vladimir Monomach (1053–1125) was a gift from the Byzantine emperor. The notion that it symboliclly bestowed imperial succession rights on the Russian dynasty gained credence during the national awakening of the 15th century. This bolstered Moscow's claim to be the secular and spiritual successor to fallen Constantinople. Vladimir was one Kievan ruler whose folk-hero status extended into Muscovite times. He married Gyda, the daughter of Harold of England, who had fled soon after the Battle of Hastings (1066).

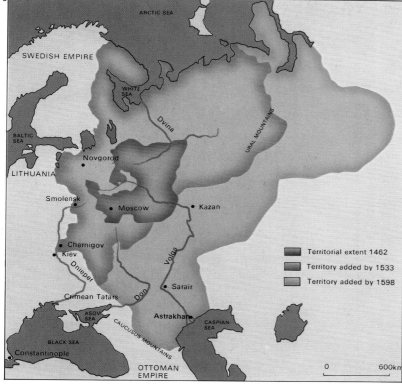

6 The expansion of the Muscovite principality (founded mid-12th century) was helped by the absence of mountain barriers within European Russia. By 1452 Muscovy controlled headwaters of the major rivers leading to the White, Black, Azov and Caspian seas. Ivan III and Vassily III conquered Novgorod (1478) and Smolensk (1514). Ivan IV established control over areas on the Don and Volga and extended Russian territory eastwards across the Urals.

Territorial extent 1462
Territory added by 1533
Territory added by 1598

0   600km

7 Ivan "the Terrible" earned his name. His childhood coincided with a period of court intrigue and open conflict among boyars. The young Ivan's exposure to these confrontations – often culminating in stranglings and dramatic chases through the palace – stunted his moral growth. At 13 years of age, four years before taking the throne, he ordered the murder of the high-ranking boyar Andrei Shuisky. Near the end of his reign he killed his own son in a quarrel. Whether because of childhood experience, madness, spinal disease or calculated attempts to destroy the power of the boyars, Ivan ruled in an arbitrary fashion. He was subject to alternating bouts of sadism and religious melancholia and his tactics against internal "enemies" included executions, property confiscations and mass deportations. This portrait is by V. Vasnetsov.

8 The city of Novgorod and its province became an independent republic early in the 12th century. Novgorod's economic strength was based on a flourishing handicrafts industry and trade in forest products. Its ruling institution, as in other Russian towns, was the *veche*, made up of all free citizens, which was responsible for foreign policy and the election of civil and military authorities. Novgorod held off the Lithuanians and Mongols for over three centuries, but finally succumbed to the superior force of Muscovy in 1478.

9 The Russian boyar class developed in Kievan times out of the intermarriages between the Varangians and the native Slavic aristocracy. Under Moscow they were given large tracts of land as rewards for military service. Leading boyar families had hereditary access to privileged positions in the state administration. This woodcut by Michael Peterle (1576) shows a procession of boyars and merchants at the Austrian court of Maximilian II.

321

Early modern Russia

The seventeenth and eighteenth centuries saw the transformation of Russia from a medieval kingdom into a powerful, modern state. Her population more than doubled, from 15 million to 35 million; she acquired territory from Sweden, Poland, Lithuania and Turkey until, by 1700, Russia was the largest European state [8]. She also developed a modern army and navy and a centralized civil service. The two rulers chiefly responsible for this metamorphosis were Peter the Great (1672–1725) [Key] and Catherine the Great (1729–96) [6].

Military and economic expansion

Russia's territorial expansion was achieved with a new army and navy based on compulsory service, equipped with new weapons and led by trained officers. By the 1670s Russia had the largest army in Europe. Military growth and the increasing cost of administering and equipping the armed forces – particularly with artillery – strained both the economy and the administrative structure inherited from the Riurik dynasty. New sources of revenue had to be tapped and repeated attempts were made to improve the tax-collecting system [2].

To increase taxable wealth and satisfy military and naval needs, Russia's rulers actively encouraged the growth of trade and industry. A nationwide market was formed as local and regional trade barriers were eliminated [7]. Foreign trade – especially with Britain and Holland – increased during Catherine II's reign from 21 million roubles to 71.3 million, thanks to the acquisition of ice-free ports on the Baltic and Black seas. Russian iron and hemp found large British markets, and hemp was also sought after by the fledgling United States. High import tariffs and borrowed French mercantilist doctrines helped Russia's balance of trade and fewer goods were imported. With European money and technical knowledge, Russia became more self-sufficient.

The new aristocracy

The increasing volume and complexity of state affairs demanded the creation of a modern civil service. Theodore III (1656–82) abolished the Muscovite system of choosing military officers and civil officials according to the positions occupied by their fathers. Peter I ("the Great") accelerated the decline of the upper nobility (boyars) by creating a civil service staffed by career officials recruited from the lesser gentry. Promotion through the military and administrative ranks was achieved according to merit. Anyone advancing halfway up a scale of 14 ranks automatically became a nobleman. Thus, in theory, nobility became a mark of impersonal service to the state.

Peter I set up separate administrative "colleges", based on the Swedish system. The most powerful of these dealt with the army, the navy, foreign affairs and finance, and reflected his priorities. Alexander I (1777–1825) brought this arrangement into line by introducing ministries. By the beginning of the 1800s there were more than 18,000 civil servants, whose upkeep absorbed 10 per cent of the state budget.

The new service nobility soon grew into an increasingly privileged class (the *dvorianstvo*) whose sons found it easier than those who were not nobles to reach the top rungs of

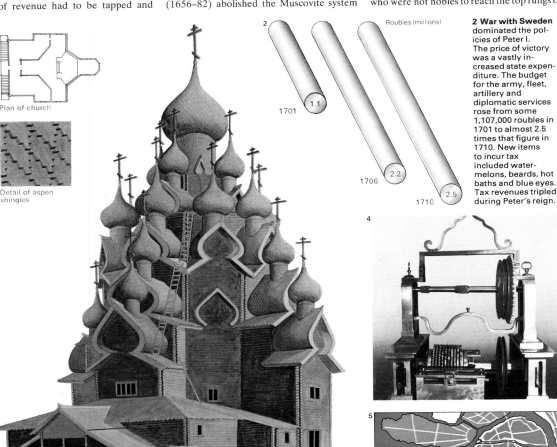

1 The cultural gap between the educated élite and the masses is a theme that has pervaded the history of modern Russia. No more striking contrast could be offered than that between the baroque and rococo buildings of St Petersburg and local, wooden church architecture. The latter could be quite splendid, as in the case of the Church of the Transfiguration on Kizhy Island in north Russia (1714), with its 22 aspen domes and its sculptural unity of composition. Village life could go on for decades and remain largely unaffected by the artistic and literary currents that reached St Petersburg from the West. While Catherine II and her court spoke French and lived in Italianate palaces, the villager spoke Russian and lived in a hut. While culture among the ruling classes became increasingly secular, the Church provided the only example of civilized culture experienced by the masses.

Plan of church

Detail of aspen shingles

2 Roubles (millions)
1701 1.1
1706 2.2
1710 2.5

2 War with Sweden dominated the policies of Peter I. The price of victory was a vastly increased state expenditure. The budget for the army, fleet, artillery and diplomatic services rose from some 1,107,000 roubles in 1701 to almost 2.5 times that figure in 1710. New items to incur tax included watermelons, beards, hot baths and blue eyes. Tax revenues tripled during Peter's reign.

3 Peter, portrayed as a cat in this derisive cartoon, inspired much hostility by breaking with tradition and shaving his beard.

4 Peter's lathe was one of his many Western acquisitions – his interest in European craftsmanship is legendary. A restless, vigorous person, with great manual dexterity, he became master of dozens of crafts, including shipbuilding, and was as much at home working on the wharves as he was conducting affairs of state. While travelling incognito in Europe he disguised himself as a carpenter.

Central radial avenues
Canals
Gardens
1 Peter and Paul Fortress
2 Winter Palace
3 Admiralty
4 Nevsky Prospect

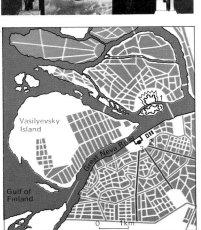

Vasilyevsky Island

Great Neva River

Gulf of Finland

5 St Petersburg was founded in 1703 on Baltic marshlands. Peter mercilessly requisitioned over 100,000 labourers each year to build the city. With its canal system and Western architecture, St Petersburg became known as the "Venice of the North".

the bureaucratic ladder. Catherine II ("the Great") confirmed their right to own serf-populated estates and their exemption from taxation, corporal punishment and even the obligation of service. It was left to the peasants, who formed 90 per cent of the population, to bear the brunt of the tax burden. The institution of serfdom, which made them chattels of the landowners, was formerly recognized by the law code of 1649. Throughout the eighteenth century landowners gained increasing powers, including the right to punish serfs by military conscription or exile to Siberia. Laws against runaway serfs were tightened and the state's role in enforcing them became paramount.

There was nothing in the tradition or self-interest of the *dvorianstvo* to prompt them to seek limitations on the power of the crown. Indeed, they depended on its strength to guarantee their position as officers and administrators and their wealth as serf-owners. Nor was any other social class able to challenge the tsar. The power of the old aristocracy had lessened, the urban middle class remained small and dependent on royal

favour, and the clergy became further reduced in power as Russia became more Westernized and secular.

The Church's declining power

The Russian Orthodox patriarchate enjoyed a brief revival during the reign of Michael (1596–1645) when Patriarch Philaret, Michael's father, ruled Russia as the "second lord" at the side of his weak son. But when Patriarch Nikon tried to assert a measure of independence from Tsar Alexis (1629–76), he was demoted by the Church Council of 1666. Peter I abolished the patriarchate in 1721 because it was largely opposed to his reforms and replaced it with a Holy Synod of bishops under a state-appointed layman.

The political humiliation of the Church did not eliminate its cultural influence, particularly in rural areas, but the nobility became more receptive to secular, European artistic and literary styles. The eighteenth century saw the emergence of a Westernized élite who spoke better French than Russian and designed new buildings in the contemporary styles of France and Italy [9, 10].

Peter I, more than any other person, was responsible for the conversion of the medieval tsardom into the modern Russian state. He, with Catherine II, stood out from the rest of the Romanov dynasty because of keen intelligence and determination. Peter's contributions to Russia's modernization were many: he rebuilt the army on a permanent basis, created a navy, reformed the tax system, expanded mining and manufacturing and remodelled the civil service. His statue, commissioned by Catherine II and erected in St Petersburg's Senate Square, symbolizes Peter's enormous strength and, significantly, faces due west.

6 Catherine II was the philosopher and educator of modern Russia. A student of the works of Montesquieu and Blackstone and a friend of Voltaire, she considered herself an "enlightened despot", much like Austria's Joseph II and Prussia's Frederick II. Most of Catherine's good intentions were corrupted by the realities of power, but she did lay the groundwork for the Russian state school system. By the end of her reign (1796) there were 22,210 pupils and 760 teachers in 288 primary and secondary schools. One result of her work was the development of an intellectual class in the 19th century.

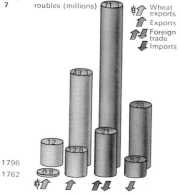

7   roubles (millions)

- Wheat exports
- Exports
- Foreign trade
- Imports

71.3
43.3
28
21
1.2
2.8
8.2
0.8

1796
1762

7 Catherine's economic policies furthered the expansion of Russian trade and investment. During her reign most internal trade barriers were abolished and she entered into trade agreements with Britain, Poland, Denmark, Turkey, Austria, Naples, Portugal and France. In the same period foreign trade grew enormously: the percentage growth in exports was 230, in imports 250.

8   Territorial extent by 1689
  Territory added by 1725
  Territory added by 1796

SIBERIA
St Petersburg
Moscow
Kiev
OTTOMAN EMPIRE
AFGHANISTAN
MONGOLIA
CHINA
0   1,500km

8 The Romanovs' urge to acquire large new territories was part and parcel of an insatiable desire for new maritime outlets. By 1700 Russia was the largest of the European states. Peter won a length of the Baltic coast after two decades of war with Sweden. Catherine II annexed areas of Poland and Lithuania, giving Russia Austrian and Prussian borders for the first time. War with Turkey yielded a Black Sea coast and rights of commercial passage into the Mediterranean.

9 An ornate rococo style was favoured for domestic architecture by aristocrats of 18th-century Russia. Illustrated here is the lavish Knight's Dining Room, now restored to its original design, in the Great Palace at Pushkin.

10 The stateliness of Renaissance and baroque public buildings complemented the domestic architecture of St Petersburg (now Leningrad). The city's Western feel is the result of its canals as much as of its unity of style.

Enlightened despotism

The wave of new ideas that swept Europe in the eighteenth century, and has come to be known as the Age of Enlightenment, had its origins in the scientific and rationalistic movements of the seventeenth century. The spirit of rational enquiry that typified the writers and thinkers of the eighteenth century also had important political repercussions. Radical criticism of existing institutions, values and practice were the characteristic features. In Europe these ideas were more or less accepted by powerful monarchs, creating "enlightened despotism".

The influence of writers

The most common feature of the ideas of the *philosophes*, as this group of thinkers was called, was their faith in reason and the critical spotlight they cast upon the accepted institutions and practices of the age. Among the most influential writers and thinkers were Voltaire [Key], Charles-Louis Montesquieu (1689–1755), Denis Diderot (1713–84) and Jean Jacques Rousseau (1712–78), whose ideas on politics and society attracted a large following among the educated classes of Europe. Although the political theories of such thinkers would in ideal circumstances have led many of them to favour constitutional government [4] and consultative institutions, they were often prepared to act as advisers to powerful absolute rulers such as Frederick the Great of Prussia (1712–86) and Catherine the Great of Russia (1729–96). In this capacity and in their general writings they advocated a number of specific reforms, such as the introduction of equality before the law, the abolition of serfdom, religious toleration and the reduction of noble and clerical privilege.

The "enlightened despots", however, did not form a consistent and coherent group. Many European rulers adopted the ideals of the *philosophes* because they were useful in their own domestic political arrangements. Thus the application of enlightened legislation was varied and conditioned by individual circumstance. For many monarchs the need to increase revenues was central to their aims, which made them favour intellectual attacks upon noble and clerical privileges such as exemption from taxation. Similarly, European rulers had a vested interest in the efficient economic exploitation of the lands under their control. Hence many of the policies of the "enlightened despots" can be explained in terms of the traditional doctrine of *raison d'état* (for the good of the country).

Implementation of Enlightenment ideas

Many European monarchs practised to some degree the policies advocated by the *philosophes*. Among the most sincere was Joseph II (1741–90), Holy Roman Emperor, in whose 10-year reign as sole ruler a large number of reforms were initiated [8].

Frederick the Great of Prussia, who succeeded to the throne in 1740, was keenly interested in the ideas of the *philosophes* and presented himself as an exponent of their ideals. In practice he was a firm and authoritarian ruler who placed the interests of the Prussian state and of his own power before the ideals of the Enlightenment. He emancipated the serfs on his own estates, although for military rather than humanitarian reasons, but he failed to eliminate serfdom elsewhere in Prussia because it

1 State revenues 1715 and 1815 in £ millions

Great Britain — Prussia
France — Dutch Republic
Russia — Spain
Austria

Total increase in state revenues
1715 21·9
1815 138·5

1 The rapid growth of agriculture and industry in 18th-century Europe helped state revenues to rise rapidly. Greater power wielded by the European states also helped them to secure greater taxation from their subjects. Because Great Britain's economic growth outstripped that of other countries, she was able to expand government revenue faster and to higher levels than many larger countries.

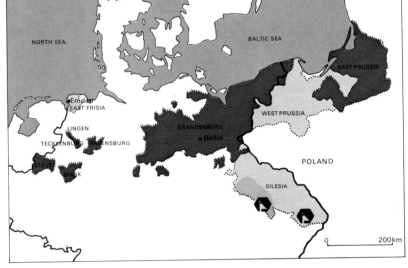

Brandenburg-Prussia 1740
Territories gained 1740–86
Coalfields
Textiles
Boundary of Holy Roman Empire

2 Frederick the Great's acquisition of Silesia in 1740 gave him an economically valuable area, to which he added East Frisia in 1744 and West Prussia in 1772, to expand Prussia's borders.

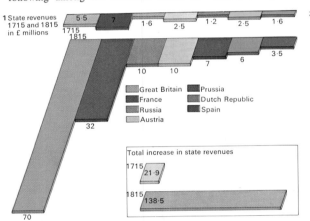

3 Monarchs still played an important part in leading their nations in war. George II (1683–1760), shown here, was the last British monarch to lead his troops in person, during the Battle of Dettingen in Germany in 1743.

4 The "Tobacco Parliament" was an informal gathering of political advisers with Frederick I of Prussia. The idea of such assemblies sprang from the Physiocratic concept of "legal" despotism in which elected landowners would guide the monarch in his deliberations. In spite of support from writers and intellectuals, the tendency towards strong government militated against such elected bodies. The centralizing reforms of Frederick and Joseph II, for example, led to the extinction of provincial administration and institutions. Even where constitutional bodies existed, however, they tended to represent the interest of the propertied classes (especially landowners and merchants) only.

would alienate the nobility. He did, however, reform the legal system and establish more humane punishments. His economic policies were strictly mercantilist [2] and drew little from the progressive economic ideas of the Physiocrats (a school of political economists). He used state monopolies and protectionist tariffs to raise extra revenue and to foster established industries.

Similarly, Catherine the Great of Russia was an admirer of the ideas of the Enlightenment, maintaining correspondence with Voltaire and entertaining Diderot at court. Her most idealistic proposal was a reform of the Russian law code, issued as the "Instruction" of 1767; to discuss it she called together a Legislative Commission representing the whole nation. The reform, however, broke down because of differences of opinion among the delegates and the outbreak of war with Turkey. In spite of her superficial desire to adopt enlightened policies such as the emancipation of the serfs, she was forced to compromise with existing vested interests. Her "Charter of the Nobility" in 1785 placed the serfs even more firmly under the control of

the landowners and established the nobility in their status, forging an alliance between them and her dynasty that was to last almost to the Russian Revolution.

Other influences in Europe

Elsewhere in Europe the ideals of the Enlightenment were adopted rather unevenly [6]. In Portugal, Sebastião Pombal (1699–1782), as chief minister from 1751–77, applied himself to strengthening the state and its economy by expelling the Jesuits and attacking noble privilege. He standardized administration, adopted free trade policies and granted civil rights to the Jews. On the other hand he also maintained a rigorous police system and threw hundreds of people into prison.

"Enlightened despotism" took many forms and therefore is not a precise description of the great variety of motives and policies adopted by European rulers in the eighteenth century. However, its legacy of humane and rational legislation, in theory if not in practice, laid foundations for liberal governments of the post-revolutionary era.

Voltaire [left] was the pen-name of François Marie Arouet (1694–1778). One of the great French *philosophes*, he was on occasion a guest at the "enlightened" courts of Europe. The degree of respect that he was accorded by Frederick the Great of Prussia is clear in this painting of the two. Voltaire was the greatest playwright of his time, writing more than 50 plays. He was exiled to England in 1726 after an argument with a powerful nobleman. From 1734 to 1749 he lived with Mme du Châtelet, one of the most educated women of the day. After her death he lived in Berlin and Switzerland, returning to France just before his death.

5 Maria Theresa's long reign over Austria (1740–80) laid the foundations for the rule of her son Joseph II. She helped to transform the diverse Hapsburg dominions into a centralized nation state and initiated many progressive reforms in the spheres of education, law and the Church. Her son completed her work by emancipating the serfs in 1781, imposing administrative uniformity upon the state and stimulating rapid economic development.

6 Enlightened principles also influenced lesser monarchs such as Charles III, seen here (centre) entering Madrid, king of the Two Sicilies (1735–59) and of Spain (1759–88). In Naples he sought to bring solvency and order to a poverty-stricken state, while in Spain he provided an enlightened government with the aid of able ministers. Poverty was tackled through workhouses while schools, roads and canals were constructed and education secularized.

Kingdom of Poland
Land acquired by Russia
Land acquired by Brandenburg-Prussia
Land acquired by Austria
= 1 million people

7 Poland was repeatedly dismembered and partitioned in the 18th century. Her elective monarchy proved a considerable weakness and led to the involvement of her powerful foreign neighbours in her internal struggles. With a backward economy, a small army and little revenue she was in no position to defend her frontiers. All her monarchs in the 18th century were the nominees of foreign powers. The first partition occurred in 1772, when Prussia, Russia and Austria took a total of a third of her former land area and half her population. In 1793 more was seized and in 1795 the remainder was divided between the three neighbouring powers.

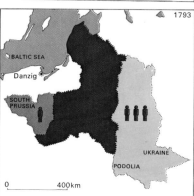

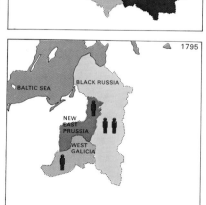

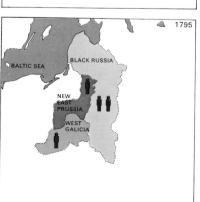

Proposed tax reforms, 1789

8 Joseph II planned to commute the feudal labour obligations of the peasants to their landlords, the state and the Church into a new tax based on a fixed percentage of their gross yearly income. This would form a new land tax which was to apply equally to all subjects in the empire. This diagram shows the percentage distribution of a peasant's income before and after the proposed scheme. However, the reform was abolished in 1790, on Joseph's death, before it became effective.

325

Time Chart

It is the true office of history to represent the events themselves, together with the counsels, and to leave the observations and conclusions thereupon to the liberty and faculty
of every man's judgment. **Francis Bacon,** *Advancement of
Learning* (1605)

What a day may bring a day may take away. **Thomas Fuller, MD,**
Gnomologia (1732)

Time flies over us, but leaves its shadow behind. **Nathaniel
Hawthorne,** *The Marble Faun* (1860)

There is no history of mankind, there are only many histories
of all kinds of aspects of human life. And one of these is
the history of political power. This is elevated into the
history of the world. **Karl Popper,** *The Open Society and Its
Enemies* (1950)

4000-2000 BC The first civilizations

Principal events

Assisted by the invention of writing and the wheel, the world's first urban civilizations grew up and flourished in Mesopotamia, and later in Egypt in the fourth millennium BC. Mesopotamian city states emerged from ancient agricultural and religious settlements, encouraged by the immigration of the Sumerian people into the area, and grew rich from agriculture and long-distance trade. With growth, however, came conflicts between cities, though none achieved permanent supremacy. In Egypt early unification and centralization led to the Pyramid Age with its celebration of the pharaoh's authority. In both areas hereditary monarchies were set up c.3000 BC, with a bureaucracy that placed emphasis on public works, especially canal building. Towards 2000 BC, Sumeria was threatened by barbarian invasions, while Egypt declined from internal stresses, as Minoan civilization emerged.

4000-3800 BC

Farming settlements found in the lower Mesopotamian plain since 5000 BC probably included the sites of the future royal cities of Eridu, Uruk, Nippur and Girsu by 4000 BC.
The need for irrigation led to a more concentrated population and complex social systems.
The site of Babylon was settled by Sumerians c.4000.
The Nile cultures were based on farming villages c.4000.

3800-3600 BC

The Creation, in Jewish tradition, dates from 3761 BC.

3600-3400 BC

Uruk (modern Warka), the greatest Sumerian city, already possessed many features of the city state by 3500 BC. At least twelve autonomous cities, including Ur, Lagash, Umma and Kish, developed over the next millennium.
The pastoral Sumerians moved into the Mesopotamian plains and encouraged the growth of this civilization, c.3500, building a network of canals for irrigation.

3400-3200 BC

The Nile Valley provinces (nomes) had been merged into two separate kingdoms – Upper and Lower Egypt – by 3300 BC.
City states began to develop in Syria and Palestine c.3300.
The Proto-literate period, when writing was first used in Sumer c.3500-3000 BC, coincides with the semi-legendary rule of the First Dynasty of Kish.
Mesopotamian influence is thought to have stimulated Egyptian cultural development c.3400 BC.

National events

Neolithic society emerged from Mesolithic hunting communities in Britain after 4000 BC, producing a megalithic (large stone) culture similar to those emerging, apparently independently, throughout Europe. An advanced civilization using bronze developed in Wessex and built Stonehenge c.2000 BC

Mesolithic man was a hunter-gatherer and had lived in Britain since 10000 BC, using flints and boats for hunting and fishing.

The first traces of farming found in Britain date from c.3700 BC.

Neolithic man domesticated his animals and cultivated crops, building causeway camps and barrows in Wessex on the Salisbury Plain c.3500 BC

Megalithic tombs and standing stones were built in Maeshowe and Nympsfield c.3300 BC. Similar monuments were also built in Denmark and Brittany.

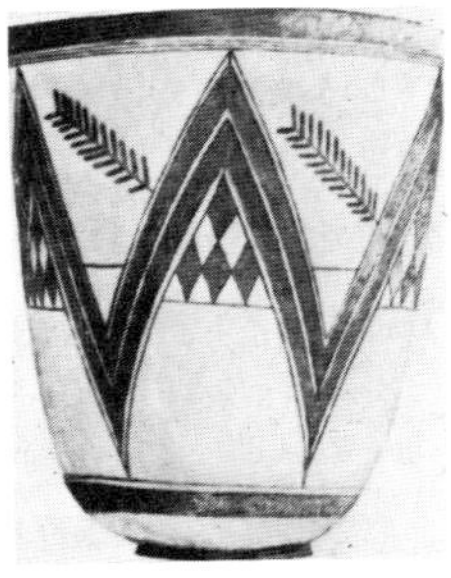

Susa ware, 4thc. BC

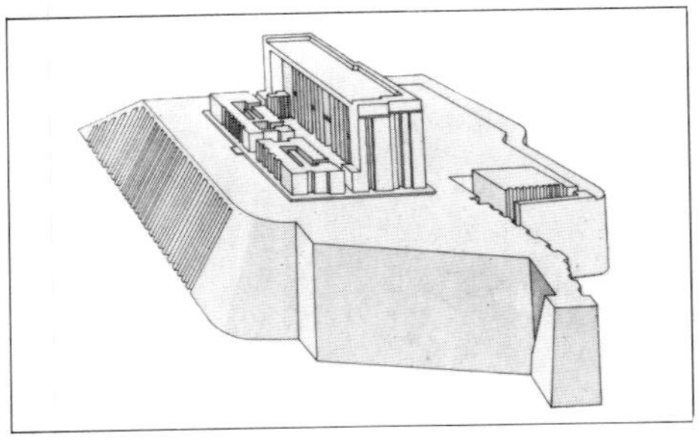

White Temple, Uruk, Mesopotamia 4thc. BC

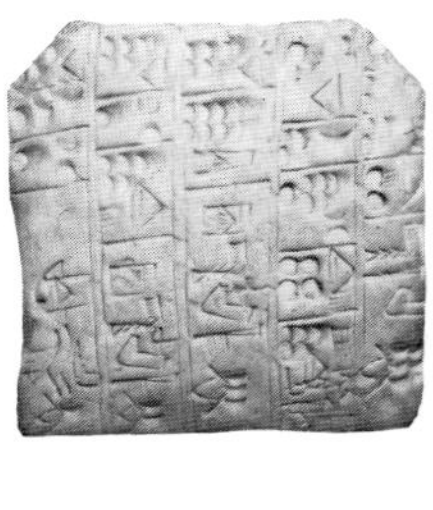

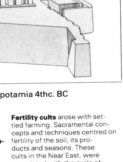

Sumerian cuneiform, 4thc.

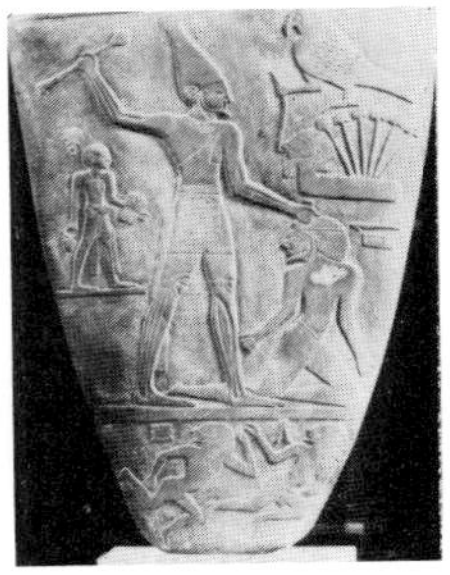

Palette of Narmer

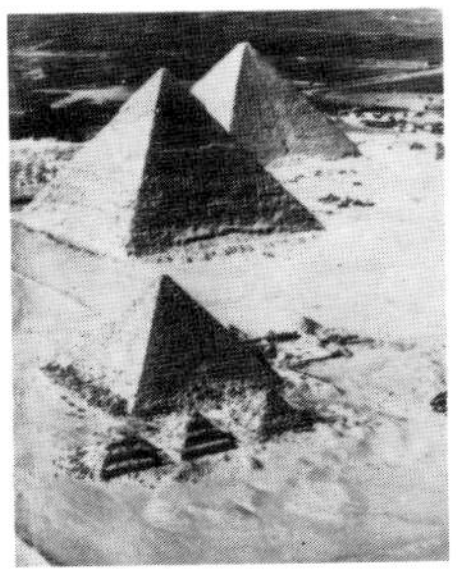

Pyramids of Giza

Religion

The development of religious ideas in early history was closely related to the rise of settled agriculture and the emergence of the first states and empires.

The change from a hunting economy to one based on arable agriculture was reflected, first of all, in the rise of fertility cults in which the central figure was a mother or earth goddess.

With the growth of urban civilizations in the fertile valleys of the Nile and the Euphrates, a priest-dominated society grew up with a system of gods, each related to a particular city or region. As the authority of the state and the priesthood became more centralized this was reflected in the changing importance of particular gods, and in Egypt in the rise of the doctrine of divine kingship and the construction of increasingly elaborate temples and royal tombs culminating in the Age of the Pyramids.

Fertility cults arose with settled farming. Sacramental concepts and techniques centred on fertility of the soil, its products and seasons. These cults in the Near East, were associated with the cycle of death and rebirth and took as their chief divinity a sexual mother goddess or a non-sexual creator – the earth goddess, known as Ninna in Mesopotamia.

Burial cults had existed since early prehistory. A specific site was often marked by a mound and sacrifice and ritual eating of the dead were frequently involved. The placing of artefacts in the graves indicated a desire to ensure the continuity of life. Cave paintings at Lascaux and Trois Frères, France, dating from c.20000 BC, show witch doctor-like figures among animals hunted by Cro-Magnon man.

Before the third millennium BC peoples of Mesopotamia worshipped nature gods in human form, each god being associated with a city temple and the temples themselves occupying a central place in city life. The gods were organized as a democratic council which reflected the political relations between the various city states.

The religion of Egypt before the foundation of the dynasties was based on totemism, the idea that there is a relation between kinship groups and specific animals and plants. Independent principalities each had their own totem: **Horus** the falcon was that of Bedhet in the north, while the god **Seth,** represented by a he-goat, protected Naqadah in the south. Above these local gods was the sun god, Re, the source of all life.

Literature

Writing developed in Mesopotamia as the Sumerians tried to simplify and regularize earlier picture writing and ideograms; a system for depicting sounds rather than ideas was probably invented by temple clerks in response to the need to record tribute payments and wages in the Mesopotamian city temples. With the development of the regular cuneiform style of script, writing became a skill for every aspiring man to acquire.

Literature had its origins in oral chronicles such as the *Gilgamesh* epic in Sumeria and in written prayers in Egypt, where a hieroglyphic script developed after 3400 BC and poetry emerged during the Pyramid Age. By 2000 BC, China was creating independently an elaborate system of word signs.

The Sumerian language was in use by 5000 BC and a pictographic script developed by 4000 BC. Although this communicated ideas by the use of pictures, it gradually began to take on a more formal appearance with agreed symbols standing for ideas. This simplified the task of the Sumerian picture writer.

The first use of writing is attributed to the Sumerian city of Uruk. Simplification of the characters in earlier pictographic script led them to the idea of using conventional symbols to represent the sound of a word rather than the idea it conveyed. Motifs on painted pottery indicate that a script incorporating phonetic elements was in use in Uruk by 3700 BC.

Temple clerks recording wages, tribute and stores had developed after 3500 BC some 2,000 signs, which were engraved on clay tablets. The linking of these signs with sounds made it easier to write names and abstract ideas as well as lists of objects. As Sumerian words were largely of one syllable, the system is called a *syllabary*.

An Egyptian hieroglyphic script developed after 3400 BC, possibly influenced by trading contacts with Sumeria. A hieroglyph could represent either a sound, an idea, or an identifying mark attached to another sign. The syllabic signs did not indicate differences in vowel sounds, as did the Sumerian script.

Art and architecture

In Egypt and Mesopotamia, the development of pottery and small domestic articles, cosmetic implements and jewellery occurred in the Neolithic period, but with the growth of states and technological advance sophisticated metal crafts and stone sculpture developed. In the absence of stone, builders in Mesopotamia used as their basic medium the baked mud brick which they later decorated with ceramics and copper reliefs, while the finest sculpture and much beautiful jewellery was produced in metal. In Egypt, stone was used for a series of monuments, culminating in the pyramids, as well as for a highly sophisticated tradition of sculpture with its own rules of proportion that persisted for over 1,000 years.

Throughout this period in China and after c.3000 BC in the Aegean, the manufacture of fine decorated ceramic ware anticipated the artistic achievements of subsequent centuries.

The appearance of painted pottery coincided with the late Neolithic/early Chalcolithic period – a transition period between the late Stone Age and the early uses of copper. Richly decorated pottery dating from c.4000 BC has been found in Anatolia at Hacilar and in Assyria at Arpachiyah.

In Egypt black-topped polished bowls (Badarian ware) and terracotta figurines were produced. Ivory combs and cosmetic articles also date from this period.

The Sumerians in Susa, capital of Elam (SW Iran), became associated with a variety of remarkably fine pottery vessels, on which sharp geometric devices were brilliantly interwoven with stylized figures of birds, animals and men.
Egyptian Amratian culture, c.3600 BC, showed technical advances on the Badarian period. Decorated ivory and bone combs were found and figures of animals such as hippopotamuses appear on pottery.

Undecorated stone vases from Egypt's Gerzean period superceded vessels of the Amratian culture. Spherical and cylindrical jars were light and skilfully hewn out of solid blocks of hard stone by means of flint borers. Votive objects, tomb paintings and palettes depict battles, ships, animals and vase bearers.

Music

Music probably originated in man's desire to express himself more richly and formally than he could in speech alone; ritual chants rapidly developed into musical forms with special meanings. Widely separated cultures produced similar kinds of instruments, adapting natural objects to musical uses: bone flutes and whistles found in Hungary and Russia date from c.25,000 BC

The harp, in prehistory, probably developed from the archer's bow, played over a covered pit to add resonance. In Mesopotamia musicians played flutes, as well as drums and rattles.

Drum and reed pipe music bloomed in Mesopotamia. Called bull and reed music, it symbolized strength and weakness with the use of drums for a vigorous beat and pipes for the melody.

Religious music was performed by musicians chanting and playing on reed pipes, flutes, drums and tambourines as part of the liturgy of temple worship in the Sumerian city states.

Science and technology

The elaborate civilizations of Mesopotamia and Egypt depended for their development on the settled agriculture practised by Neolithic peoples in these regions since c.8000. Without it they could not have sustained either the increased population or the specialization of urban life. Once secure these societies spawned a remarkable series of technological advances. The 4th millennium BC saw the invention of the plough, the wheel, the sailing boat and methods of writing. Stone tools gave way to those of copper and bronze, a hard alloy that came into use c.2500 BC.

Scientific method as we know it – the systematic testing of theories about the material world – did not develop until much later but the technical knowledge of these early societies was very sophisticated. The pyramids remain one of the finest engineering feats of all time and in west Asia, as later in China and Mesoamerica, the mathematics used by priests provided the basis for the development of other sciences.

Neolithic, New Stone Age, settlements prospered in Egypt, Mesopotamia between the rivers Tigris and Euphrates, and in other parts of the East, between 8000 and 3500 BC. Stone tools included polished stone axes and a type of flint sickle mounted in an animal's lower jawbone. The flints were mined. Buildings were reed and wattle huts or made of hand-moulded clay bricks dried in the sun.
Clay seals were used c.4000 in the Middle East, to place the owner's name on pots.

Land transport vehicles in Sumeria included sledges.
The wheel was invented in Mesopotamia during the period of the establishment of city states. It took two forms: a stone potters' wheel, and a cartwheel, made from a single, solid piece of wood.

Copper, fashioned into beads as early as 6000 BC in northern Europe, was smelted from ores or melted as the native metal over wood fires from 4000 BC in Sumeria.
Kilns were introduced c.3400 in Sumeria. Many pots were fired at once, and raised above the fire, thus protecting the painted designs from wood ash.
Shadow clocks originated in Sumeria c.3500 BC

Ploughs take their first form, that of a forked branch, in Mesopotamia and Egypt, the forks being held by the ploughman and the sharpened end, or share, being drawn through the soil by oxen. Horses were not used. Such ploughs were shown c.3000 BC in Egyptian picture writing, although used earlier.
The Egyptian Copper Age began in the Upper and Lower Kingdoms c.3200, and lasted until 2000 BC, after which iron and bronze artefacts were made.

3200-3000 BC

The Delta Kingdom of Lower Egypt was conquered by Menes (Narmer) *c.* 3100, who came from the south and unified Egypt into a single monarchy. He is attributed with the founding of the First Dynasty which he ruled from his new capital at Memphis.
The Phoenicians, a Semitic-speaking people, began to settle the coast of Syria *c.* 3000 BC.
Copper was widely used throughout the Near East *c.* 3000 BC.

3000-2800 BC

Cretan Neolithic culture gave way to bronze-based culture *c.* 3000 BC.
Sumerian cities came to be ruled by hereditary kings from 2900 onwards. The Archaic Tablets of Ur came from this Early Dynastic or Classical Sumerian age.
Public works in Egypt, especially canal construction, led to the growth of the Egyptian bureaucracy in the Early Dynastic Period, when a national government first developed.

2800-2600 BC

Gilgamesh, the legendary king of Uruk, *r. c.* 2750 BC. **Records of Sumerian kings** began with Mebaragesi of Kish *c.* 2700.
The Old Kingdom Of Egypt, a 500-year period of stability and cultural splendour, began *c.* 2700 with the reign of Zoser. Egypt also expanded towards Nubia, *c.* 2600.
Akkadians came to dominate the northern Mesopotamian plain in the Early Dynastic II period.

2600-2400 BC

Conflict between Sumerian cities such as Ur, Kish and Lagash reached a climax *c.* 2500.
A prosperous culture emerged at Yang-shao in China *c.* 2500.
Royal power reached its zenith in the Egyptian Old Kingdom *c.* 2500 under the pharaohs Khufu and Khaphre.
A sea-going Minoan civilization developed in Crete *c.* 2500.

2400-2200 BC

Sargon the Great built Akkad in northern Mesopotamia, conquered Sumer and created an empire stretching from the Persian Gulf to the Mediterranean *c.* 2350. His soldiers settled at Ashur, the future Assyria.
Urukagina, King of Lagash, introduced reforms but was ousted by Lugalzaggisi of Umma.
The Indus Valley civilization around the cities of Harappa and Mohenjo-daro emerged *c.* 2300.
The Gutians destroyed the Akkadian Empire *c.* 2230.

2200-2000 BC

Gudea of Lagash restored disrupted Sumerian commercial prosperity in southern Mesopotamia.
Ur-Nammu of Ur drove out the Gutians and established a brief Sumerian renaissance.
After Egypt had expanded into Nubia and west Asia on a large scale, its Old Kingdom ended in anarchy with the collapse of the central government, 2181.
King Mentuhotep of Thebes re-united Egypt *c.* 2060.
The first Minoan palaces were probably built on Crete *c.* 2000.

Stone temples were built at Ggantija and Haġar Qim on Malta *c.* 3000 BC by a Neolithic culture.

Silbury Hill, a man-made earth mound near Avebury, Wiltshire, was built *c.* 2500 BC.
Chieftains probably emerged in Wessex at this time.

Stonehenge was built *c.* 2000, testifying to the astronomical knowledge and social organization of Wessex. Gold and bronze objects were placed in tombs.

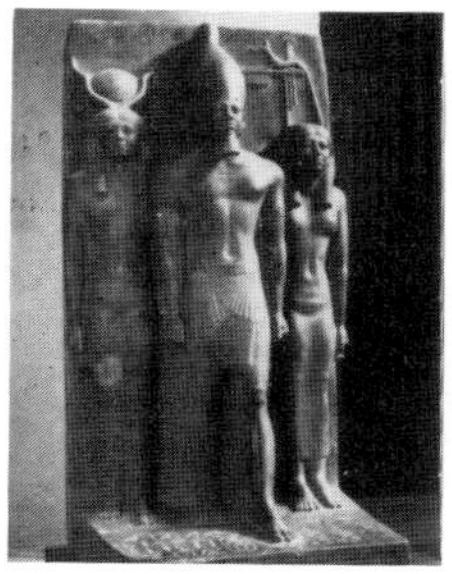
Egyptian sculpture, *c.* 2500

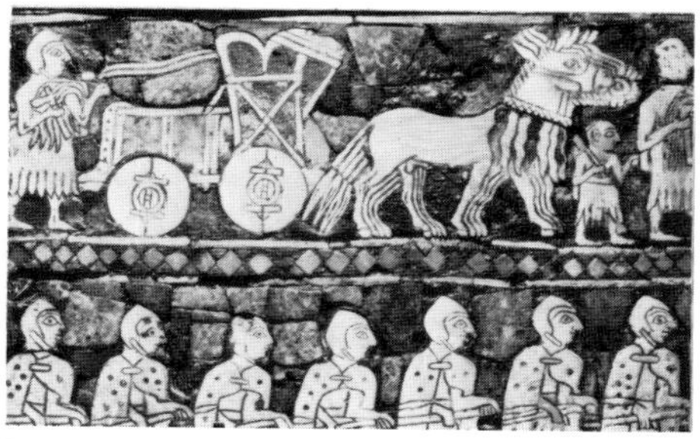
Royal standard of Ur, Mesopotamia

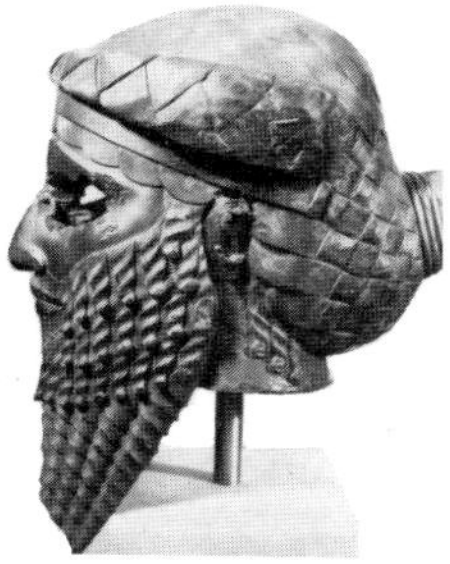
Bronze bust of Sargon 1

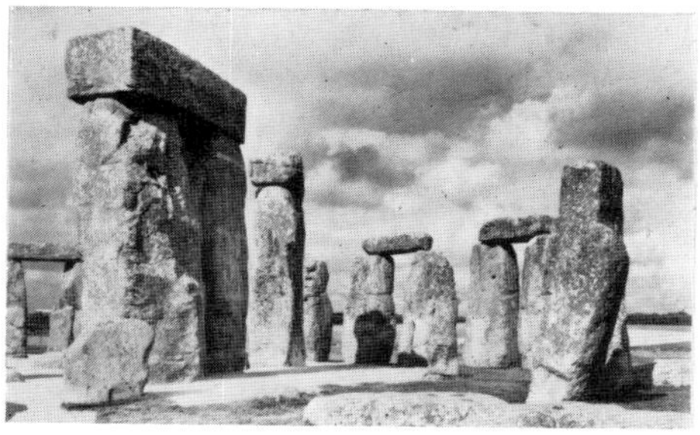
Stonehenge, Wessex, *c.* 2000 BC

The religion of Mesopotamia reached its classical form with the rise of more centralized political units in the early dynastic period. There were four main gods: **Anu**, god of heaven; **Enlil**, god of the winds; **Ninhursag**, goddess of birth; and **Enki** (Ea), god of water. Hierarchical relationships between the gods reflected the growing separation between the strata of Mesopotamian society. Divination of dreams and interpretation of entrails were practised.

The priests at Memphis in Lower Egypt established the Memphite theology after the unification of Upper and Lower Egypt. Their god **Ptah** was believed to have created the world and was known as the patron of craftsmen. The creation myth associated with him is more abstract than those of the pre-dynastic period, and testifies to the sophistication of the Memphite priesthood.

In the Egyptian early dynastic period the king became associated with **Horus** the falcon, deity of Hierakonpolis in Bedhet.
Classical Egyptian religion described an optimistic vision of an ordered cosmos, itself an expression of the predictability of life in Egypt governed by the regular flooding of the River Nile.
The first pyramid tombs were built in Egypt *c.* 2700 BC.

The concept of divine kingship was well established by 2500, as was the existence of a specialized priesthood. Both contributed to the force of royal authority. The king became identified with the god **Horus**, who by this time was associated with the whole land of Egypt.

With the decline of the Old Kingdom, the idea of survival after death was extended to include people other than royalty for the first time. This may have been a reflection of the growing power of the nobility.

A cuneiform script was in use in Sumeria by 3200 BC. It consisted of vertical, horizontal and oblique strokes made with a sharpened wooden stylus on a wet, hand-sized clay tablet. The name comes from the Latin *cuneus* (wedge) and refers to the wedge-shaped strokes of the stylus.

As cuneiform spread, writing began to serve a wide range of social needs, although there is no indication that it was used for anything but practical purposes. The Babylonians and Assyrians kept lists and inventories for business and legal purposes. There is an Egyptian record of farming procedures.

Literature had yet to emerge but writing was becoming an important tool of social advancement and literary form was evolving in the oral tradition of the Sumerian-Babylonian epic of *Gilgamesh*, mankind's first great poem.

Another script was evolving on the Indian continent, not yet settled by Aryans – the Indus (or proto-Indian) script, found on seals dating back to *c.* 2500 BC in which each sign seems to have had a single phonetic value.
In Sumeria, the Akkadians produced a simplified script of only 550 symbols, seen in the legal code of Urukagina *c.* 2400.

The first literature dates from *c.* 2300 BC in the prayers of Egyptian pyramid texts. Also preserved in papyri is the "Pessimistic Literature", which includes the *Prophecy of Neferty* and *Admonitions of an Egyptian sage*, the *Tale of an Eloquent Peasant* and *A Dialogue of a Desperate Man with his Soul.*

A Chinese script emerged though it was not standardized and no examples survive. A "concept script" in which each idea had a corresponding sign, it replaced a system of knotted cords and was used to record commands and perhaps chronicles and poetry.
In Babylon, the first known library, composed of clay tablets, existed by 2000 BC.

Mesopotamian cylinder seals dating from the **Protoliterate period** were used in the business of temple administration and bear the miniature prototypes of the relief friezes that were to become important in Sumerian art, and reached a high degree of craftsmanship by the Akkadian period.
The Palette of Narmer, *c.* 3100 BC, a carved slate tablet from Hierakonpolis in Egypt, shows the king wearing the crowns of both kingdoms.

Complex tombs for Egyptian notables were constructed. These *mastaba* consisted of underground funerary chambers with stone or brick structures above.

In Mesopotamia, a typical temple of the Ubaid period, 2900–2800 BC, had a façade decorated with niches dedicated to the cult of the god Enki. Sculpture of the period consisted of terracotta statuettes of both men and women.

The outstanding advance of Egyptian Old Kingdom architecture was the building, under the direction of royal architect Imhotep, of Zoser's step pyramid at Sakkara *c.* 2700. The Great Pyramid at Giza and the Great Sphinx of Kafre were built *c.* 2500 BC.

Egyptian royal sculpture concentrated on idealized figures with an emphasis on set proportions.

Early Minoan art was characterized by marble statuettes of goddesses (Cycladic idols) *c.* 2500 BC and vases made from Cretan and imported stone.
Mesopotamian decorative arts – in particular the use of gold and copper, lapis lazuli and other fine inlays – achieved a high degree of craftsmanship.
In China the painted ceramics of the late Neolithic Yang-shoa culture, *c.* 2500, have geometric patterns painted in black and red pigments.

Narrative reliefs and stelae proclaimed the achievements of Mesopotamian culture in the Akkadian period, while **King Sargon's bronze bust** is one of the greatest examples of ancient portrait sculpture.
The earliest Indus Valley cities of Harappa and Mohenjo-daro were constructed of fire-baked bricks and utilized such features as corbelled arches. Among the few known vestiges of Indian art are seals with animal motifs and figurines.

A group of diorite statues of Gudea, the famous ruler of Lagash, *c.* 2130, are the finest works of Sumerian artistic revival.
The temple of Ur – a ziggurat dedicated to the moon goddess Nanna – was built by the Sumerian King Ur-Nammu.
Minoan pottery *c.* 2200-2000 BC is represented by ceramics with a creamy white glaze over a dark ground.

Egyptian pottery depicts instruments like those used in Mesopotamia, including harps, drums, sistra (metal rattles on a U-shaped frame) and reed pipes of various lengths.

Vertical (end-blown) flutes, sistra and tambourines were played in processional music, suggesting the possible use of music in courtly ritual in Mesopotamia and Egypt *c.* 3000 BC.

A harp from ancient Egypt has been unearthed, dating *c.* 2500 BC. It has a lower sound chest to improve its resonance. About the same time, doubled reed flutes were played, probably in unison.

Two kinds of harp, dating from *c.* 2400 BC, were uncovered during the excavation of the royal tombs at Ur, in Mesopotamia. One had a lower sound chest and the other an upper sound chest.

Antiphonal forms, in which two choirs or a priest and choir chant responses, appeared in the ritual music of Sumerian temples under the Akkadian ruler Naram-Sin *c.* 2200 BC.

Boats, as depicted on Egyptian pre-dynastic pottery, had square sails and many oars. Boats of bundles of reeds navigated the Nile *c.* 3000 BC.
Metal-moulding was practised in Sumeria by 3200 to make copper and bronze axes with moulded sockets for holding the shafts. Previous models had weaker sockets of folded metal.

Horse-drawn chariots were recorded in Mesopotamia *c.* 3000 BC.

Bronze alloys were widely made in Mesopotamia by 2800 BC; they were a mixture of copper and tin ores and were fashioned into ornaments, tools and weapons. Copper ores were plentiful and widespread in Syria; the chief ore, tin, being found as an alluvial deposit in rivers and lakes. Some Sumerian bronzes are very hard as they accidently contain silicon.
Cotton was grown in India *c.* 3000 BC.

Populations in Mesopotamia and Egypt had grown by 2800 BC owing to improved agricultural methods. Despite its primitive appearance and action the fork-branch plough brought greatly increased crop yields.
The first calendar of 365 days was invented by the Egyptians.
The first pyramid, of King Zoser, was built *c.* 2700. Its construction involved a practical knowledge of geometry which was not formulated in theory for many centuries.

Mesopotamian metallurgy advanced significantly by 2600, for example in the development of soldering techniques, used to make the ornaments found in the royal tombs at Ur.
The Great Pyramid of Khufu was finished *c.* 2500. It is 146.6m (481ft) high and covers 5 hectares and yet is accurate to within a fraction of an inch.
Egyptian wooden boats, are shown on tomb walls from 2500.

Weaving with looms was practised well before 3000 BC in west Asia and Egypt. By 2300, horizontal looms, with the warp thread pegged on the ground, were usual in the Near East.
Weight standards and accurate scales were used in Egypt from 2200 BC. For example, the dried fish eaten by miners in Egypt was measured by their masters using stone weights.
Sewage and drainage systems were built in Harappan cities.

Ziggurats became most refined *c.* 2000 BC. These Mesopotamian buildings served both as storehouses for grain and as platforms for astrological and astronomical observations.
The Bronze Age reached the Neolithic settlements and nomadic cultures of western Europe *c.* 3000-2000 BC.

2000-1200 BC Hittites and Assyrians

Principal events

An influx of Aryan tribes, at the beginning of the second millennium BC, disrupted the civilizations of Sumeria, the Indus, and to a lesser extent Egypt, while adapting well to the existing cultures, especially in Babylon. After 1600 the Egyptians, the Hittites in Anatolia and the Assyrians all developed large-scale military organizations to sustain their growing imperial ambitions. In the eastern Mediterranean new civilizations began to emerge in this period. The Cretan Minoans created a sea-based empire and a flourishing, peaceful civilization based on Knossos, and Mycenae began to establish itself as a power in southern Greece where olive and vine farming formed the basis of future economic development.

After the fall of the Minoan civilization Mycenae took over much of its maritime power and culture, but with further invasions from the north *c*. 1200 BC she too declined.

2000-1920 BC

The brief Sumerian renaissance centred on Ur continued until *c*. 1950 BC when Semitic Amorites overran much of Sumeria. This was the beginning of a long period of instability in Mesopotamia.
In Egypt the Middle Kingdom reached its height with the 12th dynasty, 1991-1785, after Amenemhat I had subdued the nobility and restored prosperity. Building, art and international commerce flourished.

1920-1840 BC

Senusret III, 1887-1849 BC, further consolidated royal authority in Egypt by suppressing provincial rulers (nomarchs) and assisted the rise of a bureaucratic and trading middle class.
Unrest in Sumeria centred on conflict between the cities of Isin and Larsa, during which the area broke down once more into independent city states.
The Semitic language of the Amorites gradually superseded Sumerian in Mesopotamia, between 2000 and 1700 BC.

1840-1760 BC

Hammurabi the Great of Babylon, *c*. 1792-1750 BC, an Amorite, subdued the other cities of Mesopotamia and built an empire from the north Euphrates to the Persian Gulf, ruling with a code of laws based on principles absorbed from Sumerian culture.
The Middle Kingdom of Egypt ended *c*. 1786 BC, weakened by an influx of the "Hyksos", a Semitic people from Syria.
The Indus civilization, already in decline, was destroyed by invading Aryans *c*. 1760 BC.

1760-1680 BC

The Hyksos became firmly established in the Delta region of Egypt and adopted Egyptian culture. By 1700 BC a dynasty of Hyksos pharaohs was established.
The Babylonian Empire slowly crumbled under Hammurabi's son *c*. 1700 while culture and religion flourished.
A natural disaster on Crete caused the Minoan palaces to be rebuilt *c*. 1700 BC.
The Hittites, an Aryan people, grew powerful in Anatolia.

National events

The Bronze Age in Europe, which stretched from 3000 BC to 750 BC in three main phases, became more widespread in western Europe in the third millennium. Copper and bronze working entered Spain *c*. 2000 and came to flourish in southern England, apparently as a result of trading contacts with Europe.

The so-called "Bell-Beaker" folk, originating in the middle Rhine area, made contact with the Wessex culture *c*. 2000 and introduced metallurgical skills.

During the Early Bronze Age, *c*. 2000-1700, Wessex flourished because of its control of the trade routes taking British gold and copper to the rest of Europe.

At this stage most of the bronze used in England was imported, already worked as cups and other useful objects.

In the Middle Bronze Age, *c*. 1700-1300, a highly skilled indigenous metal industry emerged but the objects made were mainly luxury articles.

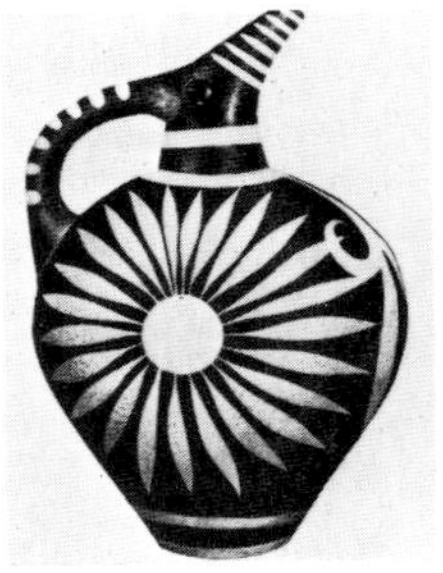

Minoan jug, *c*. 1800 BC

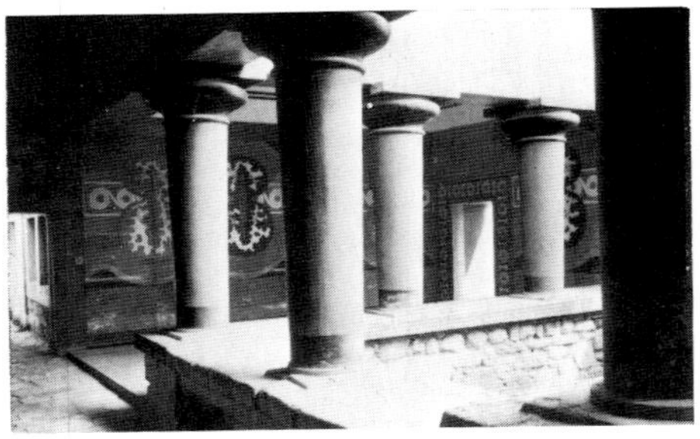

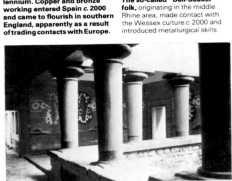

State apartments of the Palace of Knossos, Crete

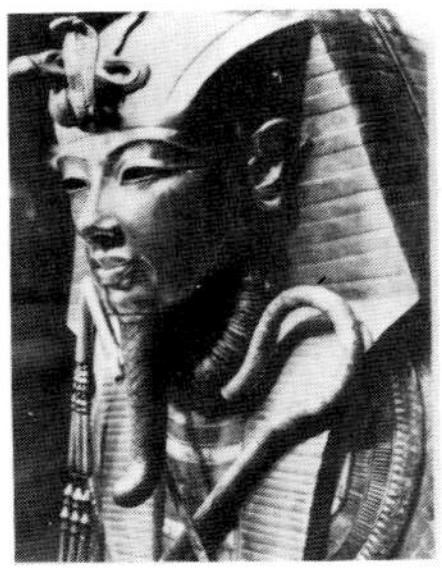

Mask of Tutankhamen

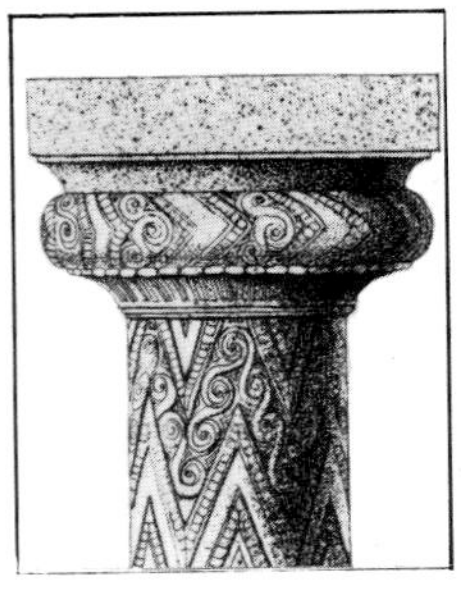

Mycenaean capital, 14thc.

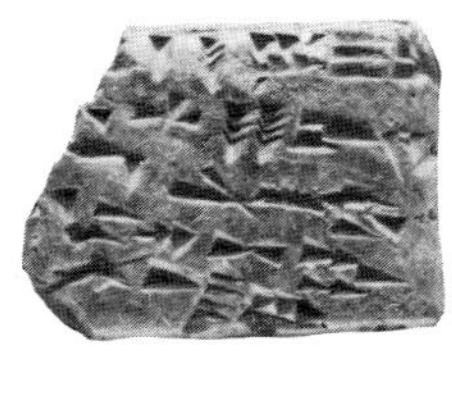

Early Ugaritic script, 14thc.

Religion and philosophy

The incursions of Aryans and other invading peoples into the main centres of civilization in the Near East disrupted the established religious traditions, dispersed some of their elements and introduced new ones. The gods of the newcomers reflected their warlike nature, and the worship of their gods evolved into ecstatic sacrificial cults. Their impact upon Egypt, however, was transitory, and the traditional religious system continued and developed, interrupted by a brief but interesting monotheistic interlude under Pharaoh Akhenaton in the early 1300s BC. Once the more turbulent areas of Mesopotamia had settled to orderly lives of commerce and agriculture in city-centred communities, the first codes of law and concepts of citizenship were devised. In this period the basis of the Judaic tradition, with its emphasis on ethical monotheism, was laid among the Israelite tribes.

Amenemhat I, the founder of the Egyptian 12th Dynasty, claimed descent from **Amun,** a local god of his native Thebes. From this time Amun, a father of the gods, and the Heliopolitan god Re were identified as **Amun-Re,** emphasizing the change in the royal family and confirming the divine right of the king to rule all of Egypt.

The Canaanite religion emerged in Palestine with **El** as supreme god, and **Baal,** god of rain, vegetation and fertility sharing the central position. The Canaanite religion was an important influence, both negatively and positively, on Israelite culture. It is possible that **Yahweh,** the sole god of Israel, was an Israelite name for the Canaanite El.

The migrations of Aryans and other peoples from the Black Sea area helped to disperse religious ideas and practices. The Hurrians who invaded Upper Mesopotamia at this time transmitted elements of Sumerian beliefs northwards to Hittite areas.

The complex law code devised by Hammurabi the Great of Babylon *c*. 1792- *c*. 1750, was created in response to the needs of increasing trade, usury and commerce. It sought to end blood feud and personal retribution and replace these with a secular state code based on the idea of citizenship. For example, one of the articles of the code stated that if a man's home fell down and someone was injured, then the owner was to be held responsible.

Literature

Cuneiform became more sophisticated after 2000 BC but a more important development was the emergence of the first consonantal (BCD) script, far simpler to master than earlier syllabic systems of writing, none of which have remained in use. The Syrian Ugaritic script, however, which developed in the mid 2nd millennium BC also had three vowel signs although the five-vowel alphabet would not be elaborated until after 1000 BC by the Greeks who would draw on a variety of Semitic scripts.

Literature of the period ranged from narrative and love poetry in Egypt to historical narrative among the Hittites, the religious *Vedas* in India and the ethical and divinatory *I Ching* and philosophy in China.

The Egyptian Coffin Texts 2040-1786, found on coffins and papyri, include spells, ritual texts and mythological stories. Their purpose was to give the dead person power in the afterlife, and after 1570 they would evolve into a more unified text, the Egyptian ***Book of the Dead.***

The Babylonian ritual poem, the *Epic of Creation*, first written about 2000 BC, had reached a classic form as part of the ceremonies associated with the new year. It told how the god Marduk slew the sea monster Tiamat, and created men as servants of the gods. Babylonian literature of the period is infused with a sense of metaphysical pessimism.

The ancient Greek script, Linear B, was deciphered only in AD 1953 by Michael Ventris (1922-56). It flourished at Mycenae in the 12th century BC but dates back earlier than this and may derive from Linear A, an undeciphered script used by the Minoan civilization of Crete. Linear A dates from 1700 BC and is a syllabic script.

Art and architecture

The brilliant civilization which emerged in Crete reached its peak of cultural achievement between 1900 and 1500 BC with palaces of a highly functional design and decorative arts whose grace and vitality reflected a long period of peaceful development. Minoan fresco painting, sculpture and painted pottery were characterized by a humane outlook and a love of nature and movement. The influence of Minoan art extended to Mycenae.

Temple architecture revived in Egypt with the Middle Kingdom and enjoyed its golden age in the 14th and 13th centuries. The New Kingdom ruler, Akhenaton, who made sun worship the sole cult during his reign, built some of the finest of these and introduced a revolutionary naturalism into royal portrait sculpture.

Chinese bronze workmanship was the most advanced in the world and calligraphic art was beginning to develop.

In Egypt, the establishment of the Middle Kingdom in 1991 BC was marked by an economic and artistic revival. The Great Temple of Karnak, built during the 12th dynasty, *c*. 1991-*c*. 1785, showed a high level of craftsmanship in tomb reliefs, gold ornamentation and paintings. Middle Kingdom sculpture adheres rigidly to rules dictating proportion and posture devised in the Old Kingdom, despite a new element of naturalism in royal portraiture.

The Bell-Beaker culture in central and western Europe made good-quality red ceramic beakers decorated with horizontal bands of geometric patterns. **In China,** wheel-turned **Lung-shan** black pottery (named after a site in Shantung) replaced the Yang-shao type at the end of the Neolithic period. With thin walls and a metallic, burnished finish, it marked a great technical advance and was commonly used for ritual purposes and funerary ware.

The Minoan palaces at Knossos, Phaestos and other Cretan sites were rebuilt on a grander scale in the Middle Minoan period, 1900-1600, with more varied architectural features such as light shafts and efficient sanitation. By about 1760 BC, Minoan potters were producing fine Kamares ware pottery in graceful and varied shapes with a profusion of floral and geometric motifs. Craftsmen specialized in small works such as faience figurines.

A major revival of Mesopotamian art marked the rule of **Hammurabi,** *c*. 1792- *c*. 1750. Old palaces and buildings were strengthened and new ziggurats constructed. To the north, the city of Mari, partly built under the ruler **Zimrilim,** *c*. 1779- *c*. 1761 BC, is remarkable for its size. Its 200 rooms cover four hectares (10 acres) and the fine painted decorations include narrative pictures such as the Investiture of Zimrilim.

Music

The development of a metal-working technology in ancient civilization enabled craftsmen to make metal instruments based on older instruments made from organic materials and stone. Bells of bronze replaced stone chimes, and bronze, copper or silver trumpets replaced hollowed horns. A metal tube with a more cylindrical bore than animal horn gave a brilliant tone.

The yellow bell, or huang chung, was the name given to an absolute (fixed) pitch produced by a bamboo pipe of set length. It is attributed to a mythical Chinese emperor of *c*. 2000 BC.

Science and technology

Trade and warfare were the main stimuli of technological advance in the 2nd millennium BC. Larger sailing ships were used to bring tin from the Mediterranean countries to Mesopotamia and to carry away bronze objects made using the tin; radical improvements took place in chariot design, but the greatest technological event was the mastery of iron by Hittite smiths. Although they lacked heat enough to melt the metal, the Hittites made iron implements by hammering them out of the heated ore. The resultant metal, albeit flawed by slag, could be tougher and harder than bronze. Weapons, sword blades in particular, benefited while metalwork for decorative purposes in iron, bronze, gold or silver became highly refined, as the objects found in the tomb of Tutankhamen show. Bronze vessels of superlative craftsmanship were made in China too under the Shang dynasty.

The shaduf, a device for raising water from one level to another with a bucket, appears on Mesopotamian seals *c*. 2000. This is an Egyptian invention still used in the Nile region.
Early Chinese technology is suggested by the finding of jade plaques, which could only have been worked effectively with metal tools.
Iron weapons and ornaments, dating from 2000 BC, have been found in the Near and Middle East.

Early iron technology involved repeated hammering of the ore until most of the slag was beaten out. Wood or charcoal fires are not hot enough to melt iron. This can only be accomplished in some kind of blast furnace, which was not developed until the Middle Ages.

Cosmetics, in Egypt, already used in the 4th millennium included perfumery oils extracted from fruits by pressing them through a cylindrical cloth bag, held upright with sticks. Filter pressing methods of this kind are still used in the food and chemical industries.

Bronze-casting in Mesopotamia followed the *cire perdue* method, a one-off process necessarily reserved for valuable items. Objects were modelled or sculptured in wax, covered with clay, and the wax melted out to make a mould for the molten metal. Hollow objects were made by moulding the wax around a clay centre.

1680-1600 BC

The Minoans established a sea-based empire in the eastern Mediterranean under their semi-legendary King Minos c. 1650 BC, creating the Minoan golden age centred on Knossos.
The Babylonian Empire was increasingly threatened by the influx of Aryan Kassites from the north c. 1600 BC.
The Shang dynasty, which introduced writing to China, developed an urban civilization c. 1600 BC.

1600-1520 BC

The Hittites plundered Babylon in c. 1550 under King Mursilis I. In their wake the Kassites ruled there for 400 years.
The Hyksos were driven from Egypt in c. 1570 BC by the Theban kings Kamose and Amosis, who established the New Kingdom, sparking off a growth of nationalist feeling.
Mycenaean civilization was growing on mainland Greece, and has left rich "shaft-graves".
Minoan civilization reached its height c. 1550 BC.

1520-1440 BC

Mitanni, a kingdom of Aryan Hurrians in northern Mesopotamia, the **Hittites** and **Assyria** all grew as military powers c. 1500 BC.
Thutmose I of Egypt established an empire in the Near East between 1520 and 1510, and began the construction of the valley of the tombs of the kings at Thebes.
Minoan civilization was destroyed c. 1450, probably by an earthquake at Thera.

1440-1360 BC

After the volcanic eruption at Thera, c. 1450, Cretan civilization revived and continued to spread to mainland Greece.
Mitanni conquered Assyria c. 1440 BC to become a military power equal to Egypt.
Amenhotep III, 1417-1379 BC, extended the Egyptian Empire and brought peace and prosperity at home, which was threatened by the attempted religious reforms of his successor, Akhenaton, and by internecine murders.

1360-1280 BC

Under Ashur-uballit I, Assyria again became a military power.
Tutankhamen ruled as pharaoh in Egypt, c. 1348-1340 BC.
Hittites destroyed the empires of Mitanni c. 1360 BC, conquered north Syria and Aleppo and built a major empire in the Near East.
Mycenaean civilization reached its height c. 1320 BC.
Ramesses I re-established the Egyptian Empire in the Near East c. 1319 BC. Rameses II fought the Hittites at the battle of Kadesh c. 1299 BC.

1280-1200 BC

A truce was agreed between Egypt and the Hittites c. 1270 as both came under pressure from migrations of "Sea Peoples".
The Trojan War reflected stresses in Mycenaean culture from Dorian invasions, c. 1200.
Shalmaneser I of Assyria, r. 1274-1245, took Babylon from the Cassites and defeated the Hittites and the Hurrians.
Moses led the Jews from Egypt c. 1250 BC.
Olec civilization emerged in Mexico c. 1200 BC.

Amber-bead spacers originating in Wessex have been found in Mycenaean shaft-graves of this period, suggesting considerable long-distance trading contacts.

The number of bronze objects found in British tombs suggest an increase in British wealth and trade.

The Late Bronze Age, c. 1300-700, saw an increase in the use of metal for everyday objects.

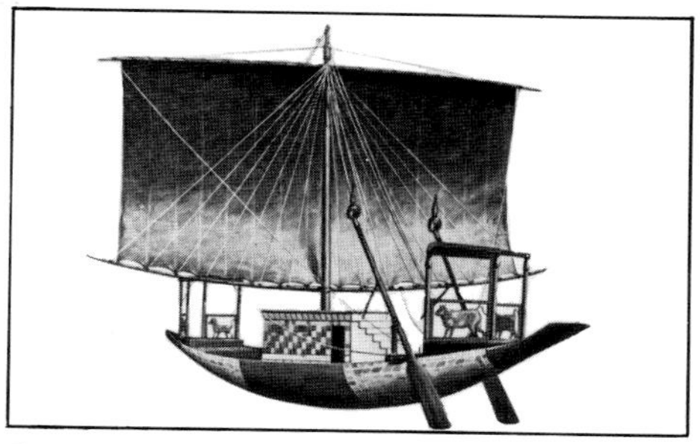
Egyptian ship, c. 1300 BC

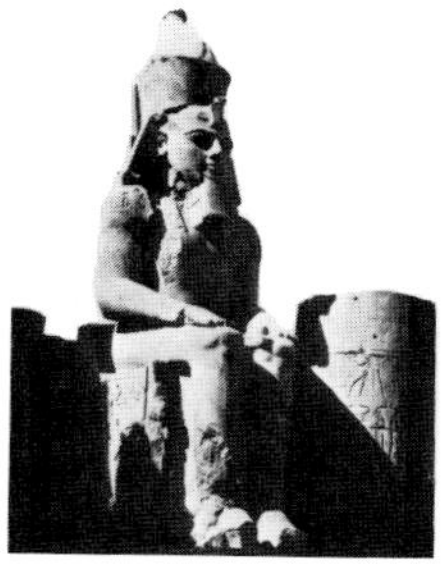
Rameses II at Luxor

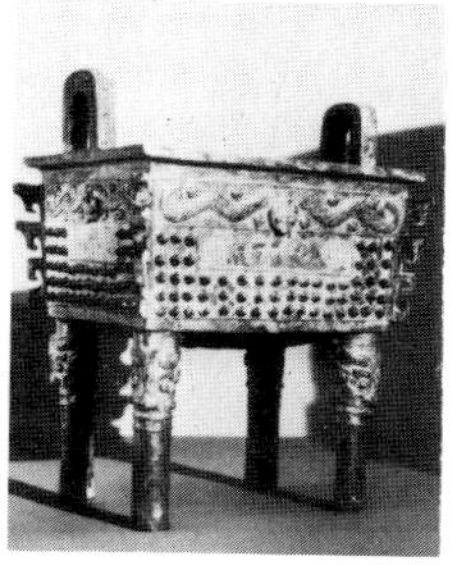
Shang bronze ritual vessel

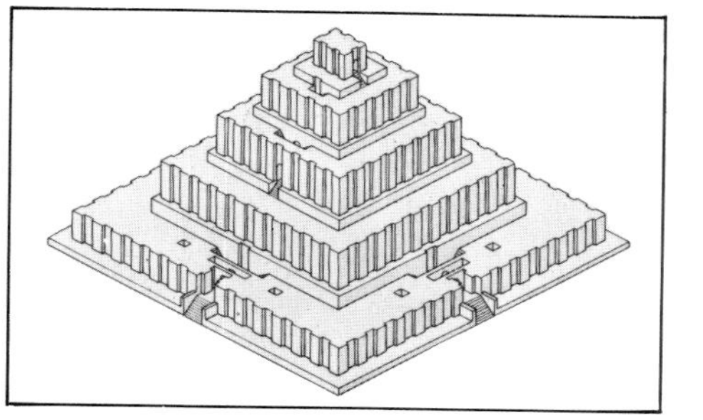
Ziggurat at Elam, 13thc. BC

The royal palace at Knossos was a centre for the worship of nature-gods with human and animal characteristics. These deities included a fertility goddess associated with snake worship and possibly a bull-god. The extant palace is decorated with a bull's horn motif and is thought to have provided the model for the Labyrinth in the Greek myth of the Minotaur.

The religion of the Hittites, who overthrew Babylon in the mid-1500s, derived from many sources. Several of their gods were attributed characteristics that varied locally, and they indiscriminately absorbed the best gods of other tribes. However, the mother goddess and the weather god were always retained, the dominance of the latter reflecting the importance of rain to fertility.

Egyptian religion went through a short-lived phase in which only one god was worshipped in the reign of **Akhenaton**, c. 1379-1362. He suppressed the older gods including Amun-Re and instituted **Aton**, represented by the solar disc, as the only god. During his reign the only other god acknowledged was King Akhenaton himself, who was thought to be eternally revitalized by Aton's rays. After Akhenaton's death, however, Egypt reverted to its traditional gods.

The Aryan invaders of India brought with them a religion which came to be embodied in the ***Vedas***, a set of sacred hymns codified by the end of the 2nd millennium BC. The Vedic pantheon of nature gods included **Indra**, the storm god, **Agni**, the fire god, and **Soma**, the intoxicating ritual juice; these and many other gods were involved in a complex mythology and an elaborate system of ritual.

The Israelites fled Egypt c. 1250 BC under the leadership of Moses, who instituted worship of a single god, **Yahweh**, to whom the tribes were bound by a covenant promising them possession of Canaan (Palestine). The Israelites quickly conquered much of Canaan and their monotheistic religion became common to the "twelve tribes" of Israel.

The first known Phoenician inscriptions, found at the city of Byblos, date from c. 1600 BC. From Byblos, a main trading centre for papyrus, the Greeks took their word for books, *biblia*. The Phoenicians developed the simplest of all the consonantal scripts, reducing the number of symbols used to represent sounds to 22.

Egyptian love songs had evolved into a sophisticated literary form by 1200 BC. Although the surviving songs date from the New Kingdom (c. 1570–1085) many are clearly from older sources. They are vigorous, direct and lyrical in their appeal to the senses: "When the wind comes it desires the sycamore tree; When you come near to me, you will desire me".

Hittite literature would flourish between 1600 and 1200 BC. It was written in cuneiform or, for private communication, in an older pictographic script. There are royal decrees, treaties, a law code, religious instructions and some Sumerian and Babylonian tales. The literary style is distinguished by laconic vigour and lack of verbosity.

Cuneiform had become entirely syllabic with fewer and more simplified characters, each having a phonetic value. This development culminated in the **Syrian Ugaritic script**, one of the first scripts with vowel signs. The earliest examples of writing in this Semitic language date from 1400 BC and describe Canaanite mythology.

The origin of the Greek alphabet is attributed in mythology to the Phoenician Cadmus (son of Agenor, king of Tyre), who is said to have brought 16 letters to Boeotia c. 1313 BC. Evidence from Mycenae indicates that this legend may have some basis in fact. The Phoenician consonant signs certainly provided a model for the Greeks, who later added vowel signs.

The *Vedas*, verse hymns dealing with sacrificial and magical formulae, were written in Sanskrit between 1500 and 1200 BC in India. The foremost is the *Rigveda*. The hymns include incantations and spells for good health and long life. Indian literature of this period was primarily religious in inspiration.

Minoan culture was approaching its golden age. Wheel-thrown pottery decorated with figures, a wide range of gold jewellery and fine quality seals were made, together with miniature sculpture in bronze, terracotta figurines and ivory carvings. Carved vases of stone and marble appeared c. 1600 BC with relief decoration, some in the shape of bulls' heads, reflecting the Minoan passion for bull sports and the religious significance of this animal.

Egyptian art experienced its classical flowering under the New Kingdom, c. 1570-1085. A standard temple plan, often on a monumental scale, was established, with floral motifs as characteristic decoration on the columns. A spirit of freedom produced lighter and more elegant sculpture, coupled with precise rendering of detail, but formal rules were preserved. Fine quality work in precious materials reflected the influence of new trade links.

The peak of Minoan culture was reached, its influence extending to Greece, particularly in Mycenae. The Minoan's love of depicting nature in their art is exemplified in the lively frescoes of the palaces, in works such as the unique painted limestone coffin, c. 1450, from Hagia Triada, Crete, and in the richly decorated pottery which flourished c. 1500-1450, depicting marine creatures of many varieties, as in the Octopus Jar from Heracleon.

A more naturalistic style of Egyptian royal portraiture was encouraged by **Akhenaton**, r. c. 1379-1362, who built a new palace and temple to the sun god at Tell el Amarna. A head of his consort **Nefertiti** is one of the most beautiful works of this short-lived Amarna style. The temple of Amun and Colossi of Memnon were built under **Amenhotep III**, 1417-1379. Examples from **Ras Shamra** (Ugarit) show a high standard of Phoenician decorative art.

The tomb of Tutankhamen with its rich furnishings included a sarcophagus with a gold and lapis lazuli funerary mask.
Under China's Shang dynasty at An-yang, mastery of bronze casting produced distinctive vessels, drums and bells, some with calligraphic ornamentation.
Tholoi (beehive tombs), including the Treasury of Atreus, with great vaulted ceilings, were built at Mycenae, in the 14th century.

Ramesses II, c. 1304-1237 completed the colonnaded hypostyle hall at Abydos, with fine funerary reliefs. His rock temple at Abu Simbel was one of the most grandiose achievements of Egypt's New Kingdom. Hewn from a pink sandstone cliff with an entrance 32m (105ft) high, it extended 61m (200ft) into the mountain and was flanked by four massive statues of the king. He also added to the Temple of Luxor and erected a colossus of himself in the forecourt.

A lute-like instrument with fretted fingerboard appears in a wall painting in an Ancient Egyptian tomb dating from about 1520 BC. Earlier types are found in Mesopotamian pottery c. 2000.

The oldest known Chinese instruments are suspended stone chimes and globular bone flutes, dating from about 1500 BC. They were played in the early part of the Shang dynasty.

Bamboo culture in South-East Asia produced an unusual music, mixing the sounds of blown pipes, such as flutes, and struck pipes in the form of bamboo xylophones.

Copper and silver trumpets found in the tomb of Tutankhamen date from about 1320 BC. The brilliance of their tone would have contrasted with other instruments in use at the time.

Vedic chant, a sung form of ancient Hindu scriptures, was established by 1200 BC and is the world's oldest continuous musical tradition. It was based on a three-note scale system.

Ships underwent improvement in the second millennium BC. A major impetus for sea trade came from the Mesopotamians, who had probably become cut off from their major sources of tin in Syria and so imported the metal to make bronze. Many vessels sailing the eastern Mediterranean were built from planks and could be made up to 12m (40ft) long. Minoan ship design was particularly influential.
Sea battles took place towards the end of this millennium.

Ploughs were improved c. 1600 in Mesopotamia, by the invention of a share and sole that dug deeper furrows.
Cementation steel was made by the Chalybes, a subject people of the Hittites. This, the earliest form of steel, is made by repeatedly hammering red-hot iron together with charcoal until carbon enters the iron.
Fine metalwork in iron, copper bronze, gold and silver, with filigree and inlay work reached a new peak in Egypt.

Chinese bronze urns and vases appeared suddenly c. 1500 BC, under the Shang dynasty with no previous evidence of a metal technology (except for jade carving). Shang bronzes were moulded in sections to extremely complex designs.
Glass bottles appeared in Egypt c. 1500 BC. Glazed beads and glass imitations of precious stones have been found dating from a thousand years earlier, but this is the first evidence of work with molten glass.

An Egyptian water clock, c. 1400 BC, had bucket-shaped vessels from which water drained by way of small holes in the bases. Hours were marked inside the vessels.
Mathematics may have developed from linear measurement used in the division of land in Egypt and Mesopotamia. Measurements were often made in units based on parts of the body, such as the Egyptian cubit, from the elbow to the fingertip.

Currency, in Egypt, took the form of copper ingots in the shape of a stretched ox-hide. These ox-hide ingots were often transported by ship, as we know from Egyptian wrecks.
Egyptian chariots were improved by increasing the number of wheel spokes from four to six, and the movement of the axle rearwards so that the rider's weight was more evenly distributed. This prevented see-sawing movements over rough ground, giving a smoother ride.

Cavalry soon challenged the charioteer in war. Saddles and reins were developed in south Turkey, but stirrups were not used for 1,000 years and did not reach Europe until 8th c. AD.
Hittite ironsmiths scattered with the destruction of the Hittite Empire, c. 1200 BC, with far-reaching consequences. The smiths had kept their techniques secret for hundreds of years, but knowledge of them now began to spread, reaching eastern Europe by 1000.

1200-700 BC Iron swords and the alphabet

Principal events

Barbarian invasions continued to strike the Near East, obliterating the power of the Hittites and Mycenaeans and limiting Egyptian and Assyrian military ambition. In the same period, however, the smaller trading societies, particularly the Jews and Phoenicians, flourished. The Phoenicians built colonies throughout the Mediterranean and the Jews established their distinctive identity and claim to the region west of the river Jordan. After 900 the military power of Egypt and Assyria recovered, financed by tribute from their subject peoples, but the focus for cultural development moved to Greece where the adaptation of the Phoenician alphabet marked the end of the Dorian-imposed dark ages.

In India, the Aryans overran the Ganges area and established a caste system based on the Vedic religion. In China the Shang dynasty fell to their former subjects, the Chou, but this had little cultural effect.

1200-1150 BC

The Sea Peoples invaded the eastern Mediterranean from the Caspian Sea area and destroyed the Hittites *c.* 1200 BC. Some settled on the Canaanite coast to become the Philistines. **Ramesses III** *c.* 1198-1166, repelled their invasion of Egyptian soil *c.* 1190 BC, after which Egypt withdrew into cultural and political isolation. **The Canaanites**, a Semitic race, settled in Syria and developed a flourishing culture based on the production of purple cloth.

1150-1100 BC

Nebuchadrezzar I, *r.* 1124-1103, of Babylon restored stability in Mesopotamia, facilitating the recovery of Assyrian trade disturbed by the Hyksos. **The Egyptian monarchy** fell under the growing influence of the priesthood of the sun god Re, the Amun, causing political and economic stagnation *c.* 1100. **The Shang dynasty in China** consisted of 30 kings in fraternal succession but declined through internal unrest *c.* 1100 BC.

1100-1050 BC

Tiglath-Pileser I of Assyria conquered Mesopotamia and the eastern Mediterranean, defeating Babylon and exacting tribute from the Phoenician city states. After his death in *c.* 1077, **the Aramaeans** took Babylon and destroyed Assyrian power, driving the Canaanites south. **The Philistines**, a trading people, conquered **the Jews** who had settled in Palestine after leaving Egypt and were at this time a loose confederation of tribes ruled by the Judges.

1050-1000 BC

Aramaean rule in Assyria produced little military or cultural activity. **In Greece**, monarchical city states including Athens, Thebes and Sparta, developed *c.* 1000 based on wealth derived from trade and agriculture. **Saul** became the first king of the Jews *c.* 1020 BC with powers limited by religious tradition. **Aryan rule** was established in north India by 1000 BC. **The Chou dynasty** was set up in China in 1027 BC.

National events

The Bronze Age was a period of relatively peaceful evolution. Bronze tools became commonplace in Britain from the 12th century onwards and villages were organized with divisions of land for arable farming purposes. The introduction of the plough in the 12th century led to an increase in population.

The expansion of the **"Urnfield" people** westwards from central Europe *c.* 1200 onwards led to close contacts between England and Europe.

Cremation in urnfields was normal in Britain by 1100 BC. **Iron metallurgy** spread through Europe after the fall of the Hittite kingdom *c.* 1200.

A fairly simple **subsistence economy** developed in southern England based on barley and livestock agriculture using oblong field patterns.

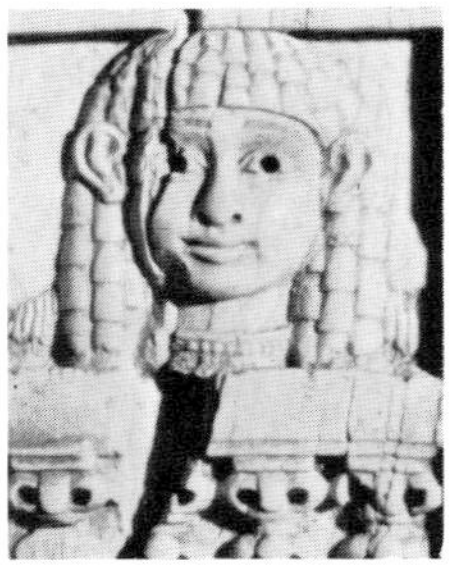

Phoenician ivory, Nimrud

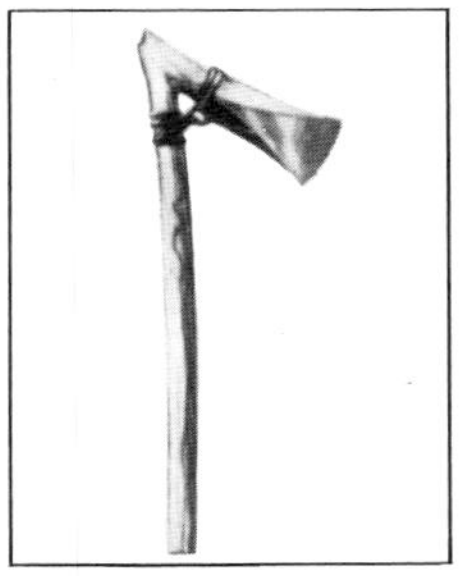

Bronze axe, *c.* 1000 BC

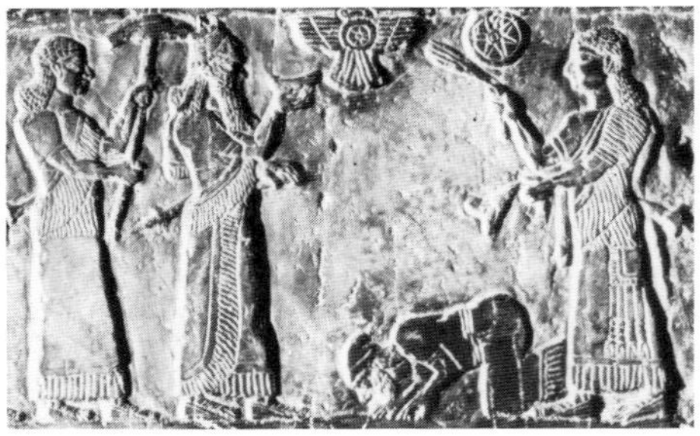

Black obelisk of Shalmaneser III

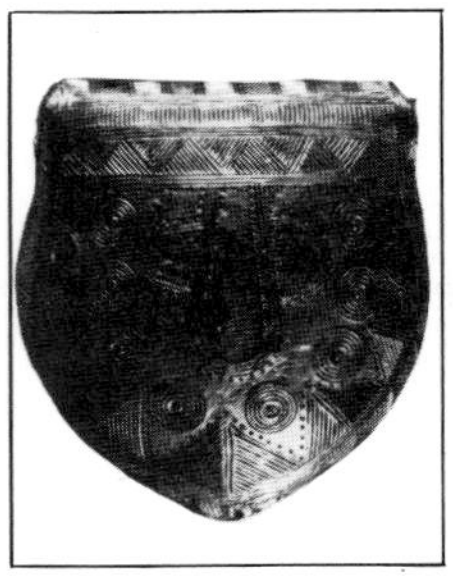

Hallstatt art: pendant

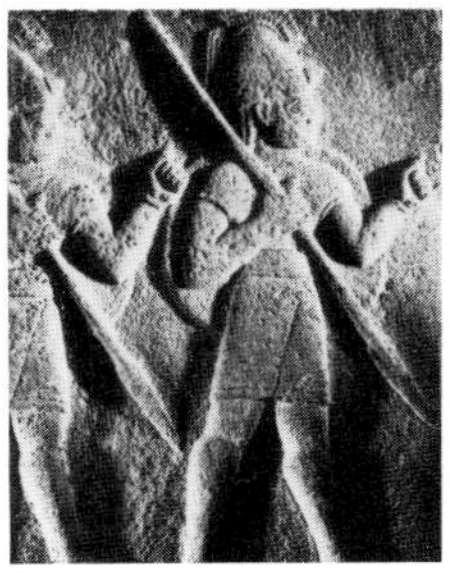

Hittite relief, *c.* 800 BC

Religion and philosophy

Beginning with the period of the Judges, Israelite history shows a continual effort by certain individuals and nomadic groups to defend the purity of the religion of Yahweh against its dilution by the pagan affiliations of her central rulers. This unceasing resistance to the addition of other gods of their faith reinforced the distinctive features of the Judaic tradition.

In China the emergence of a secular philosophy foreshadowed the development of later religious sytems, with their characteristic lack of emphasis on the supernatural.

The invasion of Greece by a succession of northern tribes led to a joining of new Olympian gods with older deities. This varied religious atmosphere, contrasting sharply with the rigid, priest-dominated society of Egypt, would play an important part in the emergence of the brilliant culture of 5th-century Greece.

Successive invasions of Greece culminating in that of the Dorians, *c.* 1200, brought Aryan gods such as Zeus, Apollo and Hermes, who largely replaced the more nature-oriented gods of the Minoan-influenced Mycenaeans. These gods, who in Greek mythology became associated with Mount Olympus, bore a far more arbitrary relation to human affairs than the original Mycenaean deities.

Chinese philosophy emerged during the Chou dynasty, 1027-221 BC, as increasing control over nature and the growth of social stability led to a demystification of thought. Irrigation replaced prayers for rain, and heaven (*T'ien*) was seen as rewarding virtue, thus giving man the power to control his own destiny through being virtuous.

The Israelites began to assimilate Canaanite ideas during the period of the Judges, which ended *c.* 1050 BC. The Yahweh of Moses and his nomadic followers absorbed features of the Canaanite deities as Israelite society became more settled and structured. Religious purists such as **Rechabites** and **Nazarites** opposed this degeneration of the monotheistic ideal.

Religion in Iran before Zoroaster (6th century BC) bore similarities to the early Vedic religion of India. Many of the Iranian pantheon of gods coincided with Vedic ones, including **Mithras**, the cult of fire, and **Haoma**, the sacred liquor.

Literature

The Greek alphabet, in which letters were used to represent vowels for the first time, had developed from Phoenician forms by the 8th century BC. It provided the most flexible and economical way of writing yet devised. At about the same time the ballads of the Trojan wars, which had emerged in oral form in the 10th century, were compiled and written down by Homer in the *Iliad* and the *Odyssey*. It is not entirely clear whether Homer was a single man or the name for a group of poets. But the later Greeks, who drew strongly on Homeric traditions, regarded him as an individual, the father of Greek poetry. Similarly, in Mesopotamia, the *Gilgamesh* epic neared its final form, and a Chinese poetic tradition grew up.

Ten thousand Hittite cuneiform tablets constituting the state archives at Boğazköy, the capital, survived the destruction of the empire *c.* 1200 BC. These represent the main source of information on Hittite history and culture.

The collection of myths and folklore that coalesced into the *Gilgamesh* epic in Mesopotamia was now approaching its final form. It combined religious elements with story themes that were to become widely popular throughout the Middle East, including references to a flood, the quest for immortality and the friendship of two great warriors.

Traces of the *Gilgamesh* epic can be found in the Trojan ballads, which culminated in the poetry of Homer, as well as in Hebrew and other classical literature. The adventures of Gilgamesh, king of Uruk, and his friendship with Enkidu, a wild man sent to destroy him, were the centre of a pessimistic poetic cycle that combined realism with myth.

The Greek alphabet is thought to have begun evolving after 1050 BC as the Greeks modified symbols they had borrowed from the Phoenicians to suit the sounds of their own language. The name alphabet comes from the first two symbols, *alpha* and *beta*. By using signs for vowels as well as for consonants the Greeks made a crucial advance, enabling any word to be written.

Art and architecture

The trading societies that emerged during this period helped to spread artistic styles and techniques in the eastern Mediterranean. In Greece foreign influences imported especially by the Phoenicians led to the adoption of complex figurative images in pottery where previously only geometric patterns had been used. The influence of the Phoenicians was similarly felt in the Hebrew kingdom of King Solomon, who employed their craftsmen to build his Great Temple at Jerusalem.

Monumental architecture ceased to be built in Egypt but in Assyria the political resurgence of the 9th and 8th centuries led to the restoration of Nimrud and the building of Sargon's palace at Khorsabad.

In Central America the isolated Olmec civilization produced colossal sculpture without the aid of metal tools.

With the decline of the Egyptian New Kingdom major architectural programmes ended and the sarcophagus of **Ramesses III**, *d.* 1166, is one of the last major works in the classical New Kingdom style. **The earliest form of Greek art** was the Proto-Geometric style of pottery decoration painted with zigzags and wavy lines. This pottery probably originated in Athens, the leading city at the end of the Bronze Age.

San Lorenzo, the earliest Olmec site, was established *c.* 1150. The chief Olmec art forms were large stone monuments,including colossal heads, some weighing over 40 tonnes.

In China the artistic traditions of the Shang dynasty were perpetuated by the Chou, *c.* 1122-221. Jade and ritual bronze vessels became increasingly elaborate, palace architecture developed and roof-tiling and bricks were introduced. Wall painting probably began during this period.

Egypt's Late Dynastic period was dominated by the High Priests of Amun, and primarily religious artefacts were made. Metal was increasingly used to make figurines and larger statuary was often made in the harder stones such as schist and basalt.

Music

Noticing that there was a mathematical relationship between the length of a pipe or a string and the pitch it produced, musicians in the ancient civilizations linked this relationship with the underlying order of the universe and phenomena of the natural world. This aspect of musical theory would later be expressed in Pythagoras' concept of the harmony of the spheres.

Secular music was established as an important part of the life of the Assyrian court *c.* 1200. Minstrels were highly regarded and music held a recognized place in court entertainment.

Pentatonic scales became prevalent in the East. The five notes of these scales, still characteristic of Eastern music, were often related to north, south, east, west and centre.

The reed mouth organ or shêng developed in China. It has several bamboo pipes rising from a wind chest into which air is blown through a mouthpiece.

Science and technology

The major technological advance of the 2nd millennium BC – a radical improvement in the quality of wrought iron – was a major factor in the expansion of the later Assyrian Empire. Assyrian ironsmiths were able to make a sharp edge using a process of tempering which involved repeated hammering and quenching in water. For the first time, effective iron swords and axes could be made; these weapons, together with siege towers and the use of cavalry, greatly contributed to the image of Assyrian indomitability. Sharpened iron was first used effectively in agriculture with the introduction of iron ploughshares in Mesopotamia. Iron blades withstood wear far better than bronze-shared ploughs and could cut deeper furrows, which led in turn to greater crop yields.

In South America the Chavin produced beautiful objects in hammered gold, and Chinese bronze and ceramic technology was further refined.

Phoenician sea trade, by 1200 BC, supplanted that of Minoan Crete and would later focus on the docks of Tyre and Sidon. The Phoenicians are thought to have developed the bireme *c.* 1100. **Early food technology** included the preservation of fish by drying, smoking and salting, thus allowing it to be stored. Such methods were used widely in the Bronze Age, but in particular by the Greeks and Phoenicians who were great eaters of fish.

Vitreous enamelling was an achievement of the later Mycenaeans. This process involves fusing glass materials on to a metallic base and first appeared in Cyprus in the form of glass decorated gold rings.

Iron ploughshares, developed in Mesopotamia *c.* 1100 BC, constituted a further leap forward in agricultural technology. The new wrought iron was sufficiently hard to take a sharp cutting edge.

The mass production of iron tools was a major feature of the Assyrian Empire. The Assyrians were not great innovators, but they used techniques developed by Hittite and other subject artisans. **Early South American farming** settlements appeared on the coasts of Ecuador and Peru *c.* 1000 BC. The simple technology of these Neolithic peoples included building in mud, brick and stone, cultivation of the potato and maize which was first grown there *c.* 3000 BC.

1000-950 BC

David, king of the Jews, 1000-961 BC, defeated the Philistines. His successor, Solomon, 961-922, built the temple at Jerusalem and a trading fleet in the Indian Ocean.
Hiram I, king of Byblos, 969-36, consolidated the Sidonian states and assisted Phoenician trade by building a harbour at Tyre, his new capital.
Damascus and Geshu were founded by the Aramaeans.

950-900 BC

Assurdan II, *r.* 935-913, briefly restored Assyrian military authority *c.* 935 BC. But by 912 Assyria was at its smallest size.
Egypt re-emerged as a military power reconquering Palestine in 918, after Shoshenk, *c.* 935-914, had reunited Egypt by making his son high priest.
The Jewish kingdom was divided on Solomon's death into the kingdom of the Israelites in the north and Judah in the south, following religious and political opposition to his rule in the north.

900-850 BC

The Phoenician city of Byblos grew up *c.* 900 BC and became the centre of the cult of Baal.
Egypt and Assyria fought in Syria-Palestine 900-830 BC.
At the **battle of Qarqar**, 854, the Aramaeans and Israelites, inspired by Elijah, defeated Shalmaneser III of Assyria.
In Greece there was a gradual shift from monarchies to oligarchies in most of the city states with the exception of Sparta.
The Chavin culture flourished in Peru *c.* 900- *c.* 200 BC.

850-800 BC

Shalmaneser III of Assyria, *r.* 858-824, defeated the forces of Damascus and Israel and exacted tribute from the Phoenician cities.
Damascus came to dominate the Aramaean states and subdued the Israelites *c.* 820 BC.
The Medes, an eastern people noted for their horse rearing, were first mentioned in Babylonian records *c.* 835 BC.
The Phoenicians colonized the eastern Mediterranean, and established Carthage *c.* 814 BC.

800-750 BC

Assyrian military power declined under a succession of weak monarchs *c.* 800 BC and with it Assyrian wealth.
A Greek renaissance occurred under the stimulus of trading contact with the Phoenicians.
Judah played an important role in a military alliance against the Assyrians *c.* 769 BC.
Jeroboam, 780-740, brought prosperity to the Israelites.
The caste system was now firmly established in India.
Rome was founded in 753 BC.

The increase in **luxury bronze goods** used in England and found in graves marks a rise in trading contacts with the Continent of Europe.

750-700 BC

Tiglath-Pileser III, *r.* 744-727, Shalmaneser V, *r.* 726-722, and Sargon II, *r.* 721-705, restored Assyrian military power, founding a standing army and often moving subject peoples. Babylon was conquered and Damascus paid tribute to Assyria.
Greek cities founded colonies in Sicily and southern Italy *c.* 750.
The Kushite kingdom in Nubia was founded *c.* 800 BC with its capital at Napata. In 725 BC, it overran Egypt and a Kushite dynasty, 725-656, was founded.

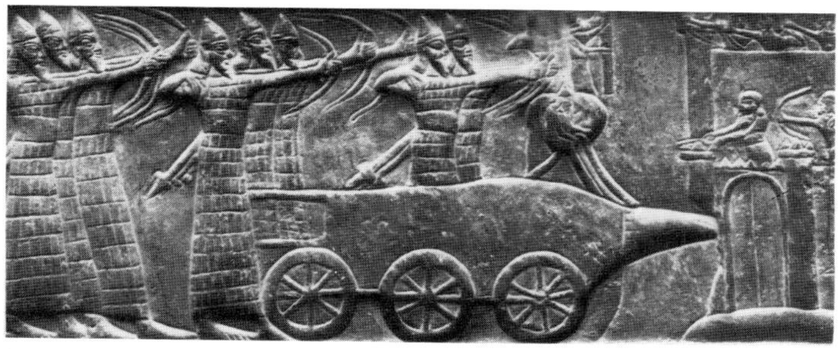
Gates of Shalmaneser III: Assyrian war machines

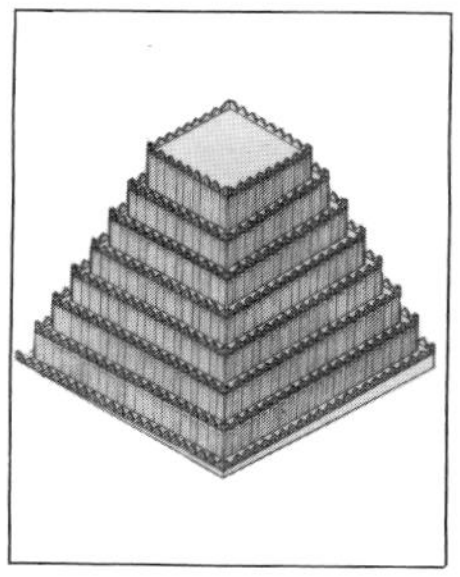
The Ziggurat at Khorsabad

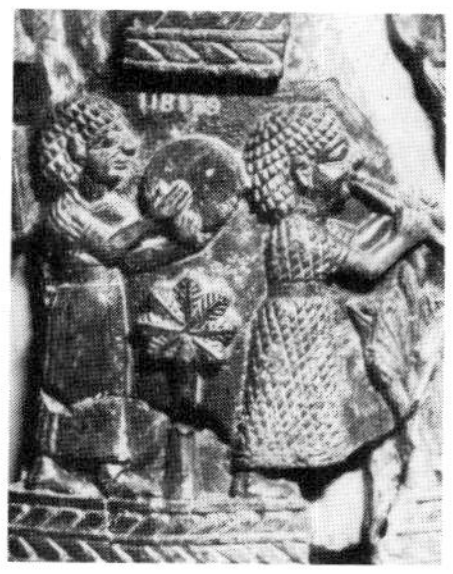
Phoenician musicians, 8thc.

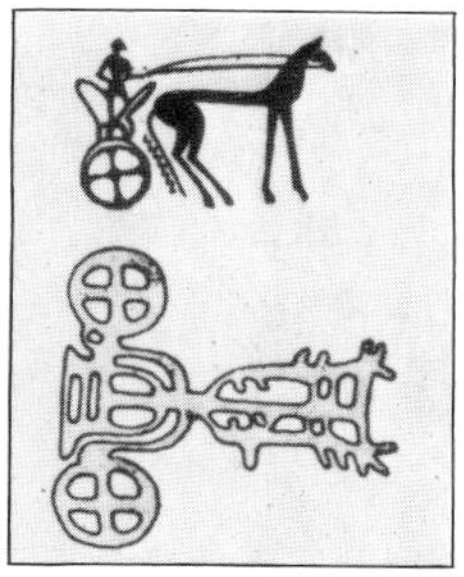
Chariot from Greek amphora

The foremost Vedic writings were the *Rigveda*, a collection of hymns and sacred formulae of "mantras" that formed the liturgical basis for a priesthood. Cremation of the dead came to replace burial, and there was a differentiation of priestly functions into those relating to actual sacrificial procedures and those relating to the ritual chanting of the sacred hymns.

Primitive Japanese religion, Shinto, was based on a love of nature. The powers of nature, **Kami**, were seen as beneficent rather than awesome, and pollution was "biological" rather than moral. Pollution from contact with death or menstruation had to be removed by ritual cleansing. The art of divination by burning bones was practised.

Jezebel, wife of the King of Israel Ahab, *r.* 874-853 BC, built a temple to the Canaanite god Baal. This aroused opposition among the zealous followers of Yahweh led by Elijah, *fl. c.* 875. Elijah's disciple **Elisha** inspired the slaying of Jezebel and the complete overthrow of the royal family.

The orgiastic cult of the nature divinity **Dionysus, or Bacchus**, reached Greece from Thrace and Phrygia. His followers, mostly women known as **Maenads** (mad ones), would take to the hills in ecstasy under the god's inspiration and wander about in *thiasori*, or revel bands. Dionysus was god of fruitfulness and vegetation, and was especially known as the god of wine.

The concept of caste that had emerged in India was elaborated by the highest priestly caste, or **Brahmins**. There were four main castes, covering occupations of priests, nobles, merchants and labourers. Brahmins further developed the earlier Vedic traditions in the *Brahmanas*, prose commentaries on the Vedas. The *Aranyakas* (Books of the Forest) foreshadowed later trends towards a mystical ascetic religious life.

The 8th-century prophets of the Old Testament, **Amos, Hosea and Isaiah**, castigated the moral turpitude of the Israelite rulers and their syncretist tendencies. The prophets rejected contemporary and foreign standards and urged other worldly ideals, prophesying that the dilution of the old religion would lead to the fall of the kingdom of Israel. This prediction was fulfilled when the Assyrians conquered Israel in 721 BC.

Hebrew literature flourished in the tenth century with the composition of the mystical *Song of Songs*, a poetic drama full of lyrical beauty celebrating nature and love. It is attributed in the Bible to Solomon, king of Israel, but its origins may be even older.

The Trojan cycle of ballads which Homer would immortalize in his *Iliad* and *Odyssey*, had probably begun to evolve in the 10th century although they may well not have been written down. The cycle told of the twelfth-century war between Greece and Troy, the wrath of Achilles and the wanderings of another Greek hero, Odysseus.

The Moabite Stone, or Mesha Stele, was erected at Dibon *c.* 850 BC by King Mesha, who composed the inscription on it to commemorate his successful revolt against Israel in the ancient land of the Moabites. It approximates Hebrew but the script is the Phoenician one from which the Greeks derived the alphabet.

The *Upanishads* in India summed up much of the wisdom of earlier Hindu scriptural writing and expressed it in the form of a dialogue between teacher and pupil which would provide the basis for the major philosophical branches of Hinduism. The dialogues bring out the essential unity of Brahman (god) and Atman (soul).

The Iliad and ***Odyssey***, Greek poems of 24 books each, belong in style to the 8th century BC but little is known of their authorship. They combine ancient legends with a vivid evocation of scene and event and masterly delineation of character. Their literary magic is generally attributed to a single poet, Homer, who may have lived in Asia Minor.

The first surviving Greek inscription, the Dipylon vase from Athens, is dated *c.* 710: "Who now of all dancers performs most gracefully, he shall receive this".
Hesiod, *c.* 750-700 BC, a didactic poet from Boeotia, wrote the *Theogony* and *Works and Days*, providing indispensable information on Greek myths, religion and agriculture.

The Scandinavian Iron Age developed a high level of metal craftsmanship in grave goods. The most outstanding example is the Sun Chariot from Trundholm, Denmark, *c.* 1000. It shows a horse and six-wheeled chariot, with a bronze-gilt solar disc.

The first Temple of Jerusalem was completed *c.* 950. Built by Phoenician craftsmen under the direction of King Solomon, it was based on Canaanite and Phoenician models. The main building, decorated with massive carvings in ivory, wood and gold, was flanked by three-storey chambers.
The 'Megaron B' temple at Thermon, one of the earliest major examples of Greek architecture, has the characteristic form of later Greek temples.

Nimrud in Mesopotamia was restored and enlarged by Ashurnasirpal II, 883-859, who built at least two temples and four palaces, decorated with winged human-headed lions in carved stone and reliefs showing the king himself. Ivories from the northwest palace show the widespread assimilation of Phoenician craftsmanship.

The Black Obelisk, known as "Jehu's stele", describes the campaigns of Shalmaneser III, *r.* 858-824. The frieze shows King Jehu of Israel making obeisance to him – the earliest representation of a Semite in traditional costume.

Pottery in Greece flourished between *c.* 900 and 750 BC, decorated with a wide range of human and animal figures, which prefigured the classical narrative style. Bronze work, and the small terracotta figurines made for the new sanctuaries at Delphi and Olympia, herald the emergence of Greek sculpture. New motifs such as floral designs were included in geometric pottery decoration as a result of increased trade with Cyprus and the Near East.

The art of the late Assyrian Empire flowered under **Tiglath-Pileser III**, *r.* 744-727, and **Sargon II**, *r.* 721-705. Tiglath-Pileser's palace at **Nimrud** was decorated with reliefs in the epic tradition, but freer in style. Sargon's palace at **Khorsabad**, was a sophisticated structure covering 9 hectares with its own drainage system. Man-headed winged bulls guarded the entrance and 2m (7ft) reliefs depicted members of the court.

Bronze trumpets or lurs date from about 1000 BC. Found in Danish bogs, lurs were made with conical bores ending in flat discs. They are usually found in pairs.

Psalms were the central feature of the music of the first temple of the Jews. Responses between priests and congregation established the pattern of many later forms of Christian music.

Lyres and harps were used to accompany Jewish temple songs. Trumpets and cymbals were played to signal special moments or interludes in the liturgy.

The lyre was popular in Greece. A large form with from three to twelve strings called the kithara was plucked with a plectrum while the player's other hand dampened the strings' vibrations.

Professional bards in Greece recited or sang epic poems to their own lyre accompaniment, while shepherds played pan-pipes made of reeds of various lengths bound together side by side.

Assyrian military technology, 1000-700 BC, was stimulated by constant warfare. The Assyrian Ashurnasirpal II, was the first to use cavalry units to any extent in addition to infantry and war chariots. The Assyrians also developed siege weapons for attacking the mud walls of enemy cities. Battering rams that rocked to and fro were not very successful, but iron-shod beams, which were raised and allowed to fall, were extremely effective.

Chinese chariots had wheels with many more spokes than the chariots of the Middle East, but otherwise differed very little from them. The Chinese, however, who had acquired chariot design from nomads who lived to their west, still placed the axle centrally beneath the platform.
Peruvian gold ornaments of the Chavin culture date from *c.* 900 BC. Goldsmiths used stone hammers to beat gold into stone moulds, probably without attempting to melt it.

Iron mines in Italy were worked from *c.* 900 BC, and would later be taken over by the Etruscans.

Leather manufacture originated with the animal skins taken by men of the Old Stone Age. Leather technology, however, developed slowly and few of the dates are certain. The Egyptians treated hides with fat, to increase durability, before 1500 BC. Tanning, or soaking the hides in a solution containing vegetable material to make leather, probably evolved much later, as did a method for hardening leather with alum, but all these techniques were in use *c.* 800 BC.

Trade in glassware became widespread, extending as far as the Atlantic coasts. Simple glass technology, such as the manufacture of glass beads, followed this trade and was practised in Britain *c.* 800 BC.
Crops grown in the Middle East included wheat, barley, flax and, latterly, cotton.

Siege towers were a later development of Assyrian military inventiveness. These wooden towers on wheels, often armour-plated and fitted with battering-rams, were used to attack the walls of besieged cities. Although clumsy, they were undoubtedly effective; defenders could retaliate only with spears and arrows, grappling at the battering-rams with iron hooks.
Biremes, ships with double banks of oars, are pictured in Assyrian reliefs of 700 BC.

700-300 BC The birth of philosophy

Principal events

Middle Eastern civilization entered a phase of turmoil between the 6th and 4th centuries BC with a series of short-lived empires established by the Babylonians, Persians and Greeks, while the Magadha Empire grew up in India.

The most striking feature of the period, however, was cultural – a massive shift towards the systematization of thought which took place in literature and the sciences and was manifested above all in the founding of many of the Eastern religions and the principal schools of Western philosophy.

Athens, where culture and democracy flourished in the context of a prosperous city state, stood at the centre of the first brilliant flowering of European urban culture, and although her power even in Greece was limited by Spartan and Macedonian militarism, the conquests of Alexander would carry this culture throughout the Middle East.

700-660 BC

Sennacherib II, 704-681, of Assyria made Nineveh his capital after destroying the rebellious Babylon. Esarhaddon, his successor, *r.* 681-669, rebuilt Babylon and attacked Egypt, captured Memphis and drove the pharaohs back to Kush in 671.
Tyrants (non-hereditary rulers, mostly of the merchant classes) appeared in many Greek cities and in Athens in 683 BC, assuring growth and prosperity.
Twelve Etruscan cities flourished in Central Italy *c.* 675.

660-620 BC

Sparta became dominant in the Peloponnese after subduing the Messenians in 630 BC.
After the reign of **Ashurbanipal**, 668-*c.* 627 BC, a time of military activity in Egypt and artistic splendour, Assyrian fortunes declined suddenly. The **Chaldean Nabopolassar** led a successful revolt in Babylon.
Josiah, 640-609, inspired a successful political and religious uprising in Judah, and **Phoenicia** won her independence from Assyria in 627 BC.

620-580 BC

Nabopolassar of Babylon destroyed Assyria with Median help in 612 and took Nineveh.
Sparta introduced barrack life and military education *c.* 610.
Nebuchadrezzar II, *r.* 604-562, built a Babylonian empire in Syria-Palestine, taking the Jews to Babylon as prisoners in 586 BC, and ending Pharaoh Neko II's imperial aspirations.
Solon, who became the archon in Athens in 594, smoothed the growing tensions between the aristocracy and the merchants.

580-540 BC

Babylonian power waned after Nebuchadrezzar died in 562.
The Peloponnesian League was founded on Spartan military strength in 560 BC.
Pisistratus, *c.* 600-527, the tyrant, secured Athenian authority in eastern Greece.
Cyrus the Great, *d.* 529, founded the Persian Empire, defeating Media in 549 and Ionia in 547.
The Kushite kingdom reached southwards to Khartoum *c.* 550.
The Magadha Empire was established in Bihar *c.* 542.

National events

The Iron Age reached Britain and, as elsewhere in northern and eastern Europe, stimulated the development of a tribal Celtic culture, made homogeneous by the supremacy of the druids. A rising population and pressure on the land ended the tranquillity of the Bronze Age.

700-660 BC: **The so-called Hallstatt culture**, developing in Bavaria *c.* 720, reached Britain *c.* 675 introducing iron weapons, mainly through trading contacts.

660-620 BC: **Urnfields** were gradually replaced by barrow tombs, of which many rich examples from this period have been found in southern England.

620-580 BC: **The use of iron** made land-clearance easier and thus facilitated population growth.

580-540 BC: **Trading contact** with Europe apparently declined. The earliest hill-forts date from this period, for instance Beacon Hill in Buckinghamshire.

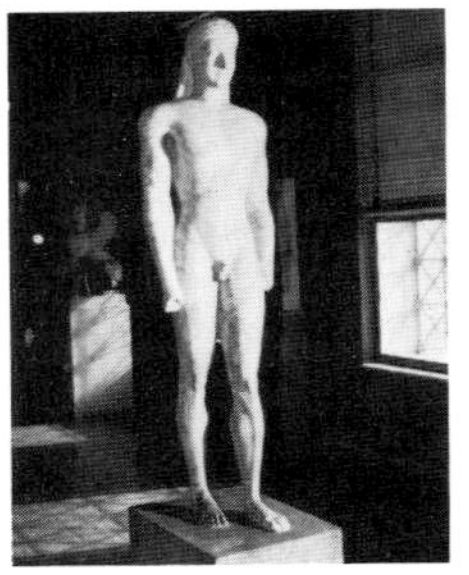

Greek *Kouros* figures, 6thc.

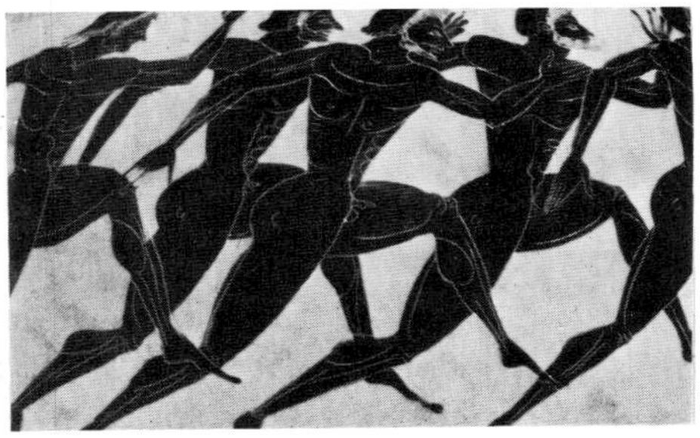

Greek black-figure vase decoration

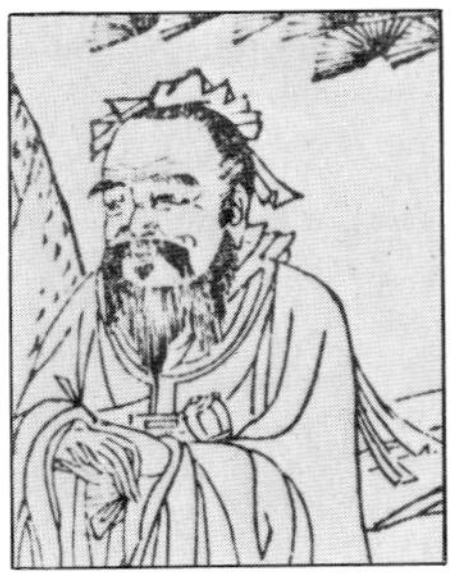

Ming portrait of Confucius

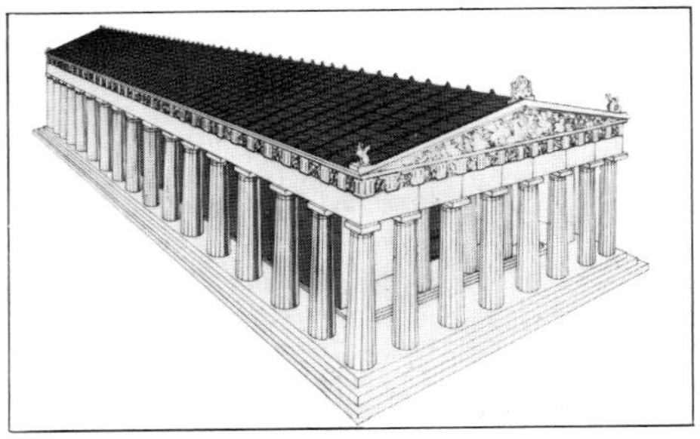

The Parthenon on the Acropolis, Athens

Religion and philosophy

Within a remarkably short time, a number of diverse and highly sophisticated religions of world significance arose. Buddhism and Jainism in India marked a break with Brahmanic ritualism. Zoroastrianism in Persia established religious themes which were to spread westwards with far-reaching consequences. In China, both Taoism and Confucianism established the doctrine of harmony as their central idea.

The flowering of Greek philosophy is one of the most extraordinary episodes in the history of thought. A questioning of accepted ideas, including the tenets of religion, and an emphasis on rational argument resulted in Greek thinkers from Thales to Aristotle raising most of the issues which have occupied Western philosophy up to the present time. Central concerns were the nature of reality and the basis of virtuous conduct.

660-620 BC: **Thales of Miletus**, *c.* 640-*c.* 546, believed that the essence of all matter was water. His attempt to find simple material causes was seminal to Greek thought.
Zoroastrianism, proclaimed in Iran by **Zoroaster**, *c.* 600, proposed a dualistic cosmology of the spirit of good, Ohrmazd, and of evil, Ahriman, between whom man is free to choose. The core of Zoroastrian scriptures, the *Avesta*, was the *Gatha* hymns.

620-580 BC: **Taoism** is thought to stem from the work of the Chinese philosopher Lao Tze, *c.* 604-*c.* 531, to whom the *Tao-te Ching* anthology is attributed. The Tao is an imperceptible state of void and undifferentiated being exemplified in the childlike innocence in man. Taoism led to quietism and a retreat from worldly affairs for both rulers and subjects. This was in accordance with the notion of Tao as the all-embracing void, the oneness of all things.

580-540 BC: **Jainism** was developed in India by **Mahavira**, *c.* 560-*c.* 468, a member of the Kshatriya noble caste. It emphasized self-denial and non-violence, and rejected **Vedic** authority in reaction to the dominance of Brahmin ritual. Within the mainstream of Vedism **the first *Upanishads*** were written epitomizing the doctrines of samsara (rebirth) and karma (inescapable consequences). Unity with the cosmos through contemplation provided the only escape from suffering.

Literature

From religious mysteries and epic poems, literature began to develop secular forms of many kinds. The most varied and accomplished writing emerged in Athens where theatre became a medium into which most poetic energy flowed, whether tragic, epic, comic or lyrical. Poetry flourished and prose writing developed in the fields of history and philosophy, although Plato's use of the Socratic dialogue indicated the all-pervading influence of the dramatic form.

Systematization of previous literary developments occurred in all the major cultures of the world. For the first time, the *Gilgamesh* epic was collected, as were the Indian epics, while in China the teaching of philosophers was written down.

700-660 BC: **Etruscan inscriptions**, mostly liturgical or funerary, dating from *c.* 700 BC, have been found at Magliano, Italy, and elsewhere – indicating the existence of a literature that has been lost. The language is incomprehensible and of unknown origin but the alphabet is Greek-based and led to the development of a Latin alphabet.

660-620 BC: **A library** of 20,000 tablets was established at Nineveh by **King Ashurbanipal**, *r.* 668-*c.* 627, who collected Assyrian, Babylonian and Sumerian writings, among them the *Gilgamesh* epic and religious and scientific works.

620-580 BC: **In the *Bhagavad Gita***, or Song of the Beloved, Hindu sacred literature took the form of poetic dialogues on the soul. In Greece, the period saw the rise of **burlesque plays** with religious themes in which the chorus consisted of satyrs. The **earliest existing Latin inscriptions** – on the Black Stone of the Roman forum and the Manios clasp – date *c.* 600.

580-540 BC: The poetic style of **Sappho** was echoed in the simpler lyrics of **Anacreon of Teos**, *c.* 570-485, but he had no successors in Greece where literature served a public function linked to religion.
Aesop, a slave from Thrace who died *c.* 564, wrote popular animal fables to illustrate moral points, some possibly derived from Oriental sources.

Art and architecture

Although the Assyrians and Babylonians produced much striking monumental art in the 7th and 6th centuries BC the most significant artistic development of the period was the evolution of Classical Greek art in which a new realism superseded the stylization common to the ancient Near Eastern art.

A narrative style grew up in pottery decoration drawn with a fluid hand in black- and red-figure work. In sculpture the traditional *kouros* figure gradually took on a more relaxed pose and a representational style developed – the single most vital step in the emergence of European art.

Greek architects, especially in Athens, designed simple but subtle buildings using austerely ornamented lintels and columns as their basic units and counteracting the effects of foreshortening with the help of mathematics.

700-660 BC: **Assyrian wall reliefs** of the 7th century were incised on stone instead of moulded, giving finer detail to scenes of savage conquest of the expanding Assyrian Empire.
Saite artists of the early 7th century attempted to revive the brilliance of the Old Kingdom. Sculpture and bas-relief were elegant, and the new use of hard stone made for a studied and severe style.

660-620 BC: **Olmec sculptures** at La Venta between 800 and 400 BC produced basalt monuments carved in elaborate relief, depicting scenes of historical and contemporary events.
A gold scabard from Litoi, *c.* 650 showed typical Scythian designs of ceremonial weaponry characterized by the designs combining different animals to make a mythical beast.

620-580 BC: **Babylon** was rebuilt by Nebuchadrezzar between 612 and 538. The decorated glazed bricks of the **Ishtar Gate**, and the famous **Hanging Gardens** were intended to outshine the brilliance of Assyrian palaces.
Etruscan tomb frescoes began in the mid 6th century. Those at Tarquinii depicted scenes from the life of the dead man in a realistic style.

580-540 BC: **Attic black-figure pottery** achieved technical excellence by the mid-6th century and came to predominate in Greek vase painting. Mythological scenes were depicted in black glaze against a red background.

Music

Music theories developed in the East and the West, and especially in China, as complex scales were devised, but little is known of the style of music produced at this period.

The seven-note scale, later to become the basis of most European music, entered Europe from the Near East through Greece *c.* 550 BC. In India, the basis for the râga was evolving.

700-660 BC: **Terpander**, the Greek composer, *fl.* *c.*675 BC, was a founder of classical Greek music. He is sometimes credited with having completed the octave.

660-620 BC: **Scale theory** was developed in Babylonia in the Chaldean period (626-538 BC). Mathematical division of strings produced a four-note scale, which was associated with the four seasons.

620-580 BC: **The kettledrum** appears as a bowl-shaped drum beaten with sticks in a relief dating from *c.* 600 BC found in Persia.

580-540 BC: **Pythagoras** introduced Chaldean scale theory into Greece *c.* 550 BC. He based a system of tuning on the fact that a string stopped at two-thirds its length sounds a fifth higher than its full length.

Science and technology

By the 5th century BC the Athenian Greeks had established a rich and complex pattern of manufactures, particularly in pottery and textiles with trading contacts from the Black Sea to the Rhine. They did not introduce many advances in technology, which they preferred to regard as the preserve of slaves. Their theoretical writings, however, have provided the basis for much Western European science even where, as in the case of Aristotle, this consisted in the systematization of earlier thinking or the elaboration of untested hypotheses.

Greek mathematics began in the 6th century with Pythagoras but by the 4th century the focus of the science had moved to Alexandria where scientific study flourished.

In China, where military technology may have received a spur from constant attacks by nomadic invaders, the crossbow was invented and a way was found to melt and cast iron.

700-660 BC: **Greek silver coins**, stamped with an owl design, came into usage in 700 BC. Early coins from Lydia, in Asia Minor, were made of electrum, a gold and silver alloy. Coinage was developed, because the barter system was inadequate to deal with the growing trade between the countries of the Middle East and the Mediterranean.
Greek silver mines at Laurion were heavily worked by the Athenians, using prisoner or slave labour.

660-620 BC: **Central European technology** thrived at Hallstatt, Austria, with the mining, manufacture and export of iron and salt

620-580 BC: **Anaximander**, *fl.* 6th c. who believed the Earth to be cylindrical in shape, is thought to have produced the first map of the known world.
The **potter's kick wheel** may have been invented at this time, although pottery designs still showed potters using the old two-man turntable method.
Indian mathematical texts of the 6th-3rd centuries BC deal with simple geometric forms, and calculations involving large numbers.

580-540 BC: **Thales**, *fl.* 580 BC, "Father of Greek philosophy", made detailed observations on methods of triangulation navigation.
Pythagoras, *c.* 580-500 BC, and his school, studied medicine, astronomy and musical scales and mathematics, particularly the theory of numbers.
It is debatable whether Pythagoras invented the theorem that bears his name.

540-500 BC

Cyrus took Babylon in 538 BC, thereby assisting the Jews' return to Jerusalem, and Egypt in 525 BC. Darius I, 548-486, benevolently ruled a centralized empire from the Indus to the Mediterranean, divided into regions (satrapies) for administrative purposes. **Rome expelled her Etruscan kings** in 509 BC and became an independent republic. **Cleisthenes** reorganized Athenian local government, and laid the basis for democracy in 508.

500-460 BC

Athens checked Darius' invasion of Greece at Marathon, 490. A second Persian invasion by **Xerxes**, *c.* 519-465, in 480 was stopped by the Spartans at Thermopylae and the Athenians at Salamis. **The Delian League** was founded in 478 BC, reflecting Athenian ascendancy in eastern Greece. **Celtic culture** spread in Europe *c.* 500 BC. **The Greeks defeated Carthage**, 480, and the Etruscans, 474, and thus won control of the sea.

460-420 BC

Athenian power and culture was at its height under Pericles, *c.* 490-429, who assisted Egypt in an abortive revolt against the declining Persian Empire, 456-454. **Rome expanded** into central Italy and the plebeians won new constitutional rights 445 BC. **Buddhism** became popular in India, especially among the merchant classes. **The Peloponnesian war** between Athens and Sparta began in 431 resulting from political and cultural tensions between them.

420-380 BC

The Peloponnesian war ended in naval defeat for Athens at Aegospotami in 405 BC. **The Romans** captured Veii, an Etruscan city, in 396, but Rome herself was sacked by marauding Celts in 390 BC, who also hastened the Etruscan decline. **Socrates**, *c.* 469-399, was put to death in Athens in 399 BC. **The Chou dynasty** in China declined in the long "Warring States period", 475-221.

380-340 BC

Athens defeated Sparta in 371 in alliance with Thebes, which became leader of the opposition to Sparta. **Philip of Macedon**, *r.* 359-336, built up his military strength in northern Greece. Many of the Persian satrapies had become semi-autonomous. **The Persian Artaxerxes III**, *r.* 359-338, restored royal authority and re-established Persian rule in Egypt in 343 BC, thus ending the last native pharaoh dynasty.

340-300 BC

Alexander of Macedon, having assured his authority in Greece, defeated Darius, of Persia *d.* 330, at Issus in 333, and crossed the Indus in 327. He died at Babylon in 323 after turning back to consolidate his authority. His empire had fallen apart by 306. **Chandragupta**, *r. c.* 321-*c.* 297 created the Maurya dynasty at Magadha in India. **Rome** had effectively destroyed Etruscan power by 300 BC.

Celtic La Tène culture, which emerged in Switzerland *c.* 500 and spread quickly throughout northern Europe, had close ties with Etruscan civilization.

Wheat became an important crop in England *c.* 500. **Salt**, useful both for preserving meat and as a medicine, and **wool** were traded in Britain *c.* 500.

Increasing population led to new pressure on available land *c.* 450, and increased warfare between rival tribes in Britain.

Many hill-forts such as Badbury Rings were built and fortified, and a warrior class emerged. Swords became more common than daggers.

The druidic religion, based on fertility cults, became firmly established *c.* 350 BC.

The druids acted as priest-kings, providing an overall cohesion and cultural unity to the tribes of Celtic society.

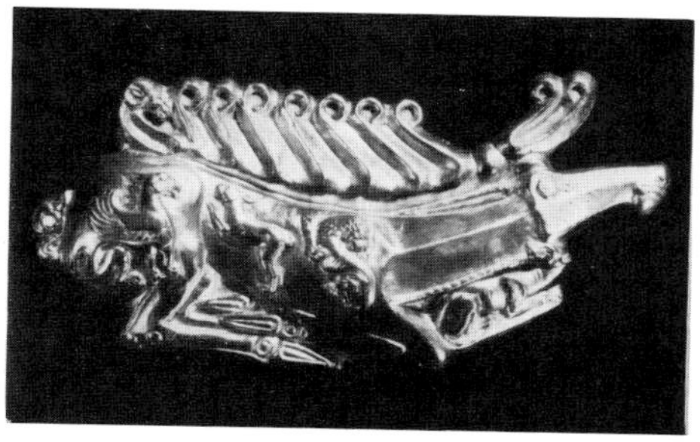

Scythian gold plaque, 5thc.

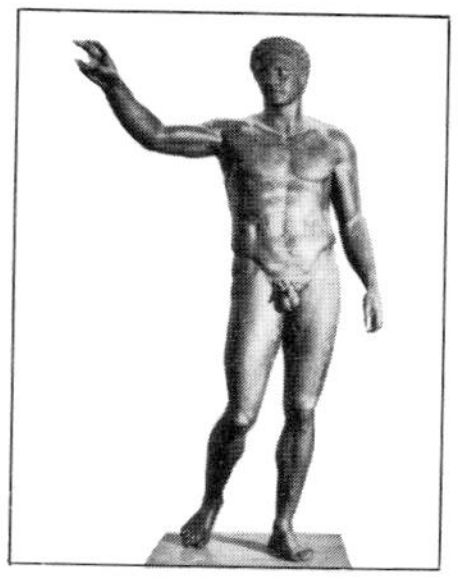

Classical Greek sculpture

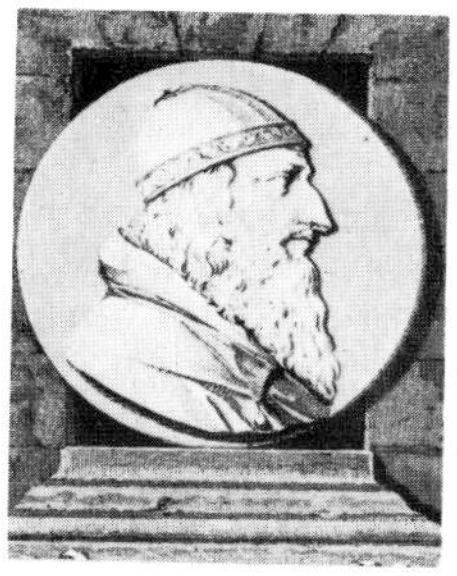

18thc. Portrait of Aristotle

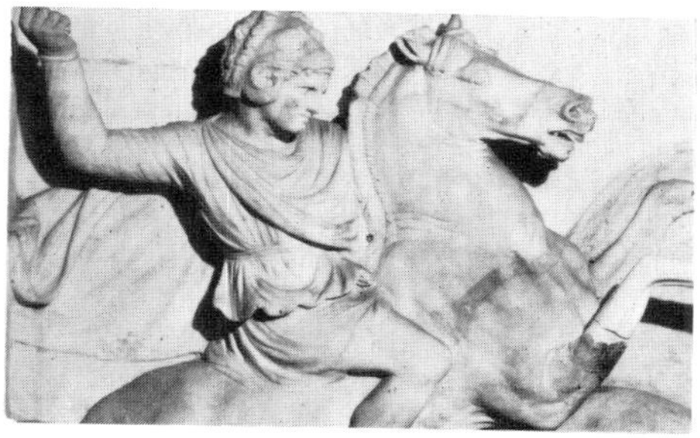

Alexander the Great in battle

Confucius, or K'ung Fu-tzu, 551-479, taught social ethics in China. His doctrine was taken up by the rulers and governed the Chinese way of life for over 2,000 years. It embraced elements of traditional Chinese religion and emphasized aristocratic social virtues and conduct harmonious with the heavenly order. It stressed awareness of fate and the decrees of heaven. In Greece **Pythagoras' theory of numbers and music** quickly developed into a mystical cult.

Buddhism was founded in India when **Siddhartha Gautama**, *c.* 563-*c.* 483 BC, began propagating the insights he achieved through long periods of contemplation. He taught that suffering can be avoided only by following an eight-fold path of moral conduct, non-violence and meditation, leading to a state of perfect enlightenment, nirvana. **Zeno of Elea**, *c.* 495-*c.* 430, a Greek philosopher, originated the dialectic and supported his argument with paradoxes.

The Greek Sophists, led by **Protagoras**, *c.* 485-*c.* 410, were agnostic towards the gods. **Socrates**, *c.* 469-399, argued that no one can possibly do that which he knows to be wrong. He followed this principle to the point of political dissent for which he was tried and condemned to death. In China, **Mo Ti**, *c.* 470-391, taught pacifism and universal love; he also established a dialectical method of argument.

Plato, *fl.* 4th century BC, founded the Academy in Athens, where philosophy was taught to young members of the Athenian aristocracy, in, *c.* 387 BC. He advocated subordination of the individual to the all-powerful republic, and also maintained that phenomena perceived by the senses are merely impure copies of the perfect reality of eternal Ideas.

The Cynics in Greece, such as **Diogenes**, *c.* 412-323, believed that happiness needed the repudiation of human values and the adoption of a simple animal-like existence. **Aristotle**, 384-322, rejected Plato's idealism, urging a more detailed empirical examination of natural and social phenomena and the doctrine that good consists in individuals achieving the states appropriate to their natures.

Zeno, 334-262, of Citium, founded Stoicism, claiming virtue to be the only good and wealth, illness and death of no human concern. **Epicurus**, 342-270, of Samos, believed pleasure to be the essence of a happy life. **Mencius**, 372-289 BC, a Confucian philosopher, saw man as inherently good and urged filial piety.

Classical Greek tragedy began with **Aeschylus**, *c.* 525-456. He developed drama from choral cult songs by introducing dialogue between the actors when tragedy became a regular feature of the spring Dionysiac festivals. His greatest work is the *Oresteia* trilogy. **The Boeotian Pindar**, *c.* 522-*c.* 440, wrote patriotic poems often celebrating athletic prowess.

Sophocles, *c.* 496-406 BC, and Aeschylus continued the tradition of classical tragedy. In his plays like *Oedipus Rex* Sophocles retained a functional chorus but shifted the centre of interest to the actors. **During the time of Confucius**, 551-479, the five classic *Ching* books reached their final form.

Euripides, *c.* 480-406 BC, last of the great writers of tragedy, dealt with social issues as well as myths, reflecting a growing humanism in Greek drama. **Herodotus**, *c.* 485-425, emerged as the first major historian with a lively account of the Persian Wars, while **Thucydides**, 460-400 BC, took a more rigorous approach to the history of the Peloponnesian War.

Aristophanes, *c.* 448-*c.* 388, was the best comic dramatist and topical satirist at the Athenian drama contests. His *Lysistrata*, 411 BC, deals with a strike by women aiming to end war. **The philosopher Socrates**, 469-399, invented the cross-questioning (dialogue) teaching method but his sceptical approach to religion brought him a death sentence.

Plato, *c.* 427-347, continued the Socratic method in a masterly series of prose dialogues, including the *Republic*, which achieve the quality of a drama of ideas. His pupil **Aristotle**, 384-322 BC, made important contributions to literary criticism, although his books are mainly lecture notes, in *Rhetoric* and *Poetics*, which analysed classical drama.

The New Comedy flourished in Athens from *c.* 330. This was a comedy of manners, using stock characterization and avoiding touchy subjects. **Menander**, *c.* 342-292 BC, was its best exponent but was less popular than **Philemon**, *c.* 365-265 BC. Greek prose style was meanwhile brought to its zenith by the Athenian orator Demosthenes, *c.* 383-322 BC.

Painted grey ware of the urban Ganges cultures, *c.* 500, was a hard wheel-turned pottery decorated with linear and dotted patterns. **Bronze and ceramic vessels** and ornamental and ritual jade carvings were the primary Chinese art forms under the Chou dynasty, 1000-200 BC. **Cyrus' funerary monument** at Parsagadae, 529 BC, anticipated Achaemenid success with its artistic traditions of Greece and Mesopotamia.

Early Celtic bronze wine flagons from the Moselle region, *c.* 460 BC, show how classical and eastern elements were assimilated to produce a new and purely Celtic art form. The human mask was a typical motif. **The Palace of Darius**, 522-486, records Persia's victory over the Median kings in the bas-relief friezes on the gigantic columns.

Greek art became increasingly independent of foreign influences and more humanistic in style, reaching its High Classical period between 450 and 400. The *Cretan boy*, *c.* 460, from the Acropolis showed a relaxation of the 6th-century *kouros* pose. Pottery after 530 was dominated by the red-figure technique. Doric architecture became less severe in style. The **Parthenon**, 447-432, built entirely of marble, was the least conventional of this style.

The Greek Late Classical period was between 400 and 323. The more complex Ionic capitals superseded the Doric style, with the richly ornate designs of the **Erechtheum**, *c.* 430. The monumental gate building of the Acropolis, the **Propylae**, was begun *c.* 430. A new naturalism dominated sculpture, as seen with the transformation of the archaic kore in the figure of *Victory* from Olympia *c.* 420.

Idealized grace and beauty was characteristic of Late Classical Greek sculpture. The sensual possibilities of carved marble were explored by **Praxiteles** in *Hermes with young Dionysus*, 350, at Olympia. The nude female form was introduced, and Praxiteles' *Aphrodite from Cnidus*, 350, initiated a feminine ideal of narrow shoulders and broad hips.

Corinthian columns were first used on the exterior of Greek architecture with the monument at Athens, built *c.* 335 to celebrate a victory. **La Tène art** from a 4th-century grave at Waldagesheim shows how classical motifs decorating neck torques and bracelets had superseded earlier Celtic styles. **The Appian Way** between Rome and Capua was begun *c.* 312.

The ancient Greeks developed modes (patterns of sounds in descending order, a basis for tunes) with distinct moods, and named them after Greek tribes, such as the Lydian and Phrygian.

The rise of drama in Greece linked dance, music and poetry. The chorus performed in an area called the orchestra in front of the stage, after which the modern orchestra is named.

A Chinese text showed how the chromatic scale (of 12 half tones) can be derived from the cycle of fifths but in practice they used it only to transpose pentatonic scales into various "keys".

The *Ramayana*, a book of Hindu myths, recorded nine basic moods associated with scales in Indian music c. 400 BC. Similar to Greek modes, this music anticipated the râga.

Aristoxenus, a pupil of Aristotle, tabulated Greek modes *c.* 320 BC, after the fall of Athens had resulted in a decline in the status of music and the use of modes in practice.

Alexander the Great's invasion of India brought with it new instruments and a developed theory. The introduction of the lute *c.* 300 BC and new theories affected Indian music deeply.

Anacharsis the Scythian and **Theodorus of Samos**, *fl.* 6th century, are thought to have developed the key, a metal anchor, with grappling flukes, a lathe and an improved bellows. **Iron welding**, by hammering the red-hot metal, is associated with the name of Glaukos of Chios. Before his time iron sections were joined by elaborate lappings and flanges. **Heraclitus**, *c.* 540-*c.* 480, held that fire is the fundamental principle of the universe.

Athenian culture, reflected in the volume and variety of their trade and industry, was well advanced; producing large quantities of metals, oil and cloth. Pottery, too, was of the highest quality in design and manufacture and was in great demand abroad. **Anaxagoras**, *c.* 500-*c.* 428, who came to Athens from Ionia in *c.* 480, gave the first scientific explanations of celestial events, especially eclipses. He influenced much of Aristotle's scientific works.

Democritus, *c.* 460-370, held the theory that matter was composed of atoms. **Quarries on Mount Pentelicon** provided the Athenians with fine milky-white marble, with which they built the Parthenon. **Hippocrates** of Cos, *c.* 460–*c.* 377, called the Father of Medicine, explained mind and body conditions in terms of "humours", glandular secretions whose imbalance caused disease. He emphasized dietary and hygienic factors in the maintenance of good health.

Chinese cast iron appeared *c.* 400 BC. The Chinese, unlike early Western ironsmiths, were able to melt iron to cast it, helped by the high phosphorous content of their iron, which lowered the melting-point; however, it also made the iron brittle. Two centuries passed before the Chinese could produce a satisfactory cast iron.

Aristotle systematized knowledge in the realm of science, logic, politics and ethics. His scientific thinking, although often merely speculative, was enormously influential. For example, his belief that heavenly bodies move in perfect circles governed Western thinking until the 17th century AD. **Eudoxus**, *c.* 408-355 BC, studied mathematical proportions and developed a method of successive addition to determine irregular areas; this theory was the forerunner of calculus.

The elements of Aristotle, **earth, air, water and fire**, in fact proposed by earlier Greeks, represent an early attempt to systematize nature. These elements could be used to produce each other; for example, smoke (air) and ash (earth) could be made by burning wood (fire). **Epicurus**, 342-270, advanced an atomic theory, similar to that of Democritus.

300BC-AD100 Rome conquers the West

Principal events

Despite the constant political unrest that followed Alexander's fall, a cosmopolitan Hellenistic culture spread throughout the Middle East. This was absorbed by Rome, which now emerged as a great power, creating eight provinces by 146 BC, including Macedonia and Spain. Conquest, however, brought with it chronic social conflict in Italy which finally helped to destroy the Roman Republic and led to the establishment of the empire in 28-27 BC, a constitutional solution which left the problems of expansion untouched. By AD 100 the empire stretched from Egypt to Britain, bringing to Rome new manpower, art forms and the many religious cults she incorporated, among them Christianity.

In the East Buddhism and Confucianism grew influential; the former spread by Ashoka in southern India, the latter becoming an integral part of society in China where it would remain so for 2,000 years.

300-260 BC

Alexander's empire had broken into four parts by 297 BC: the Hellenistic kingdoms of Macedonia and Thrace, the Seleucid dynasty in Syria and the Ptolemaic in Egypt (whose invasion of Palestine in 301 BC revived old tensions with Syria).
Rome gained control of southern Italy and with the defeat of Tarentum in 272 BC came into conflict with Carthage in Sicily.
Ashoka, *c.* 274-*c.* 236 BC, expanded the Mauryan Empire and promulgated Buddhist principles.

260-220 BC

In the first Punic War, 264-241, Rome built her first fleet, took Sicily and the Lipari Islands and defeated Carthage, which expanded into Spain.
Conflict continued between the flourishing **Hellenistic kingdoms**.
Ptolemy II, *r.* 285-246, extended Alexandria in Egypt.
Bactria left the Seleucid kingdom, and developed a combination of Greek and Buddhist social philosophy, *c.* 250.
Olmec civilization in Peru began to decline *c.* 250 BC.

220-180 BC

In the **second Punic War, Hannibal**, 247-183, invaded Italy across the Alps in 218 but retreated after Roman aggression in Spain *c.* 206. His alliance with Macedon, 215, involved Rome in eastern Mediterranean politics.
The Roman nobility took control of the wealth from the new provinces while smallholders suffered from military service in the new standing army.
Huang-Ti, *r.* 221-210, completed the Great Wall of China in 214.

180-140 BC

Rome invaded and defeated Macedon, and after further unrest annexed Greece in 147.
Carthage was totally destroyed by a Roman army in the third Punic War, 149-146.
The Arsacid dynasty ruled in Parthia, stretching from the Euphrates to the Indus *c.* 150.
Under **Menander**, *r.* 155-130, the Indo-Greek kingdoms reached over much of northern India.
The Han dynasty in China consolidated imperial authority over the provinces.

National events

The Roman Empire came closer to Britain, which was torn by conflicts between Celtic tribes. Contact with Europe, and especially Gaul, increased. After Julius Caesar's abortive expedition in 55-54 BC, the divided Britons were able to put up only minimal resistance to the Roman conquest.

La Tène culture flourished in England, producing highly decorated military equipment.

Italian silver coins have been found in Cornwall dating from this period, when Britain was exporting hides, gold, tin and slaves to Europe.

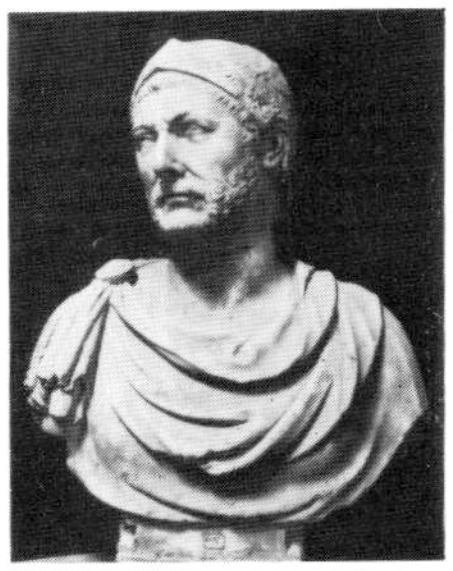
Roman bust of Hannibal

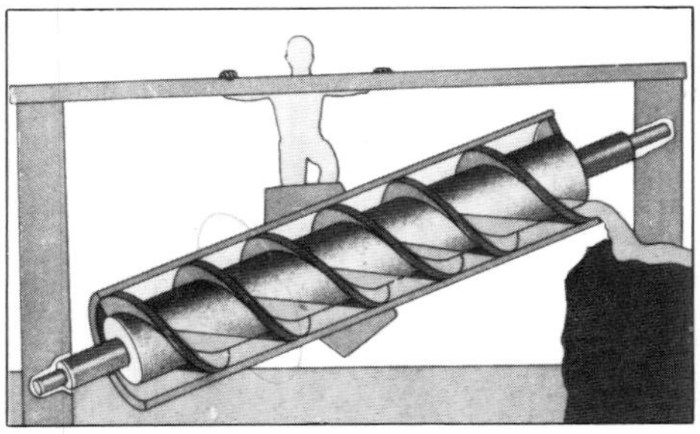
Archimedean screw

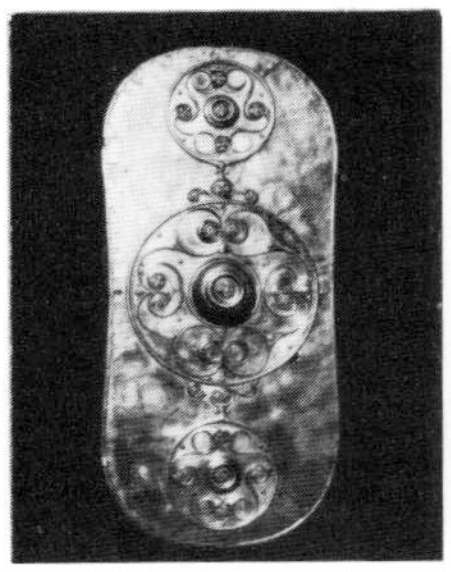
La Tène art: shield

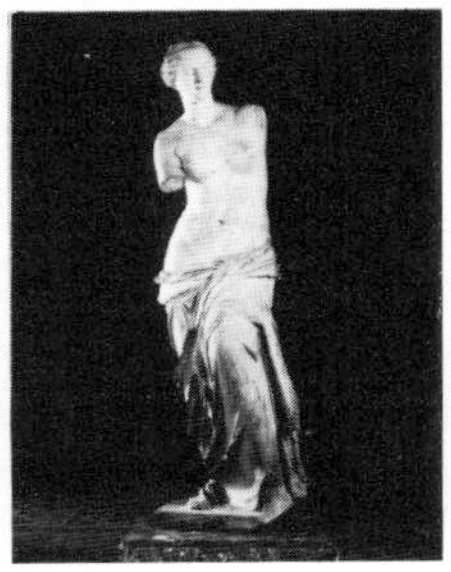
The Aphrodite of Melos

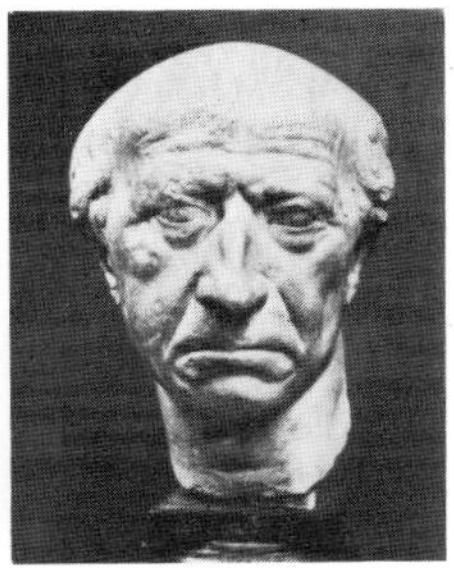
Roman bust: old Republican

Religion and philosophy

Christianity grew from being one of several Jewish nationalist sects into a more universally significant religion, as Paul spread it throughout the Roman Empire and introduced Hellenic ideas. The Christian emphasis on the individual conscience and on love brought persecution in Rome where religion was primarily a public or political concern.

In India a proliferation of Hindu sects provided more popular forms of worship than the traditional Brahminical cult and important additions to the sacred literature expressed further compromises with the everyday needs of worshippers.

New schools of thought grew up in China, while Confucianism became adopted as the official state religion under the Han dynasty *c.* 140 BC.

The Yin-Yang school of Chinese philosophy and cosmology, the leading exponent of which was **Tsou Yen**, 340-*c.* 260 BC, considered the universe to comprise five elements, metal, wood, water, fire and earth. The Yin-Yang school thought the universe was governed by the two complementary forces of yin, female and passive, and yang, male and active.

Ashoka, *r. c.* 274-*c.* 236 BC, established Buddhism as the state religion in the Mauryan Empire.
The developing Shinto religion in Japan had a hierarchy of deities presided over by the sun goddess **Amaterasu** and her descendants, the imperial family. Past heroes became mythological figures and each clan venerated its own deity.
In China Hsun Tzu, 313-238 BC, taught that human nature was fundamentally evil. Goodness required training.

Mystery religions and Eastern astrology took hold in Rome in response to the stress of the second Punic War, 218-201 BC. Novel gods, such as the Great Mother of Asia Minor, **Cybele**, and **Dionysus** superseded the traditional deities of Greek origin.
The introduction of the Stoic philosophy to Rome at this time with its emphasis on Fate, encouraged the development of mystical and astrological thought.

Within the Vedic tradition yogic thought was codified by **Patanjali**, *c.* 150 BC, into the four volumes of the *yoga sutras*, rules dealing with transcendental trance states and mystical liberation. **Bhakti** (popular devotional cults) emerged as well as cults centred on the gods Vishnu, Shiva and Krishna. The *Bhagavad Gita* was added to the earlier epic poem of the *Mahabharata*, and in it a new emphasis on salvation through the performance of duty emerged.

Literature

Although Greek literature itself declined its influence spread first in the wake of Alexander and then through Rome, which built its culture to a large extent on that of Greece.

A Latin literature arose in the third century BC based on Greek models in poetry, drama and history although under the late republic a number of writers, among them Caesar, Livy and Virgil, set out to glorify specifically Roman culture and history. The New Testament was written in Greek, still at the end of the first century the lingua franca of the eastern Mediterranean.

The Chinese classical literature flourished while in India the epics and Buddhist scripture were finalized.

The last classical writer, Chü Yüan, 343-277, is traditionally described as the author of the celebrated *Ch'u Elegies*, the most famous of which is the *Heavenly Questions*.
Bucolic and pastoral poetry by **Theocritus**, *c.* 300-260 was imitated by his fellow Greeks **Moschus**, *fl.* 150, and **Bion**

A Latin literature emerged, based largely on Greek poetry, drama and history.
Naevius, *c.* 270-*c.* 201, was the first Latin epic poet. A freed Greek slave, **Livius Andronicus**, *c.* 284-*c.* 204, wrote plays and translated the *Odyssey*. **Ennius**, 239-169, historian and playwright, was important for his efforts to adapt to Latin the methods of writers such as Euripides.

The Chinese Classical period came to a dramatic end with the reign of Ts'in, 221-210. He burnt many Confucian texts.
Latin drama emerged with **Plautus**, *c.* 255-184, who used Greek metres and plots from the New Comedy. *Miles Gloriosus*, *c.* 205, was his best play.
Terence, *c.* 190-159, was less popular but more influential with work such as *Adelphi*.

The Indian epic *Ramayana*, with between 20,000 and 40,000 couplets in its various versions, is attributed to the 2nd century BC. Although incorporating much religious material, it is primarily the story of the Aryan victory over the indigenous Indians. It was the forerunner of court poetry and had much influence on literary developments throughout southern Asia.

Art and architecture

Throughout the Hellenistic world private patronage helped artistic production and collectors' demands were so great that copying and pastiche developed, although many artists explored new ideas. Styles like that of the Pergamon school, marked by a new and masterful handling of emotion, were still for the most part associated with places rather than individuals. The Roman Empire absorbed the art of both the Hellenistic world and Italy. Etruscan iconography was copied, and the basic design for Roman architecture came from Etruscan models. Roman municipal architecture reflected both a strong civic pride and the varied leisure pursuits of her urban elite. Baths, theatres and basilicae, constructed with careful attention to practical requirements, also served to create a dramatic framework for public life. Monuments were erected to commemorate the victories of the late republic and early empire.

The Colossus of Rhodes, one of the Seven Wonders of the ancient world, was a huge bronze representation of the sun god Helios which was built astride the harbour at Rhodes between 292 and 280 BC.

The Temple of Horus at Edfu in Egypt, begun by **Ptolemy III** in 237, was planned with one main axis, typical of Egyptian temple architecture.
The Hellenistic period in Greek art, *c.* 323-1st century BC, developed the sensuous possibilities of marble sculpture. The naturalistic poses of the bronze "Eros" from Tunisia and the "Sleeping Hermaphrodite" are typical of three-dimensional realism of 3rd-century Greek sculpture.

Town planning played an important role in Greek architecture. The chaotic market places of Ephesus and Miletus were replaced by public squares.

The "Venus de Milo", *c.* 150, was a Pergamene pastiche of preceding sculptural styles. The classical features coldly echoed 5th-century tradition, while the slight twist of the torso accommodated the new taste for multiple-view figures. This Hellenistic Aphrodite became the classic source for the Roman models.
The Great Altar of Zeus at Pergamon was begun *c.* 170. The exterior frieze depicts the battle of gods and giants.

Music

Music stagnated in the West under the Roman Empire, but the Jews maintained their vocal tradition from which the Christian Church borrowed heavily later. In the Arab world, music was still a lively art. The progress of music in the East had little influence on Western musicians, but Western incursions affected Eastern music to some degree.

The Indian vina, from which the sitar later developed, was a lute-like plucked instrument thought to have evolved from the instruments carried by Alexander's invading soldiers.

The first keyboard instrument was the hydraulis, a water-powered organ. It was built by Ctesibius, who was working in Alexandria from 246-221 BC.

The Imperial Court of the Han dynasty in China employed more than 800 musicians to impart a rich panoply of sound to the rituals of state occasions.

A Greek hymn to Apollo composed at Delphi survives from this date.

Science and technology

In the Hellenistic world Alexandria became the focus of scientific work particularly in mapping, astronomy and mathematics. At the same time Greek technology found its greatest mathematician and experimental physicist in Archimedes.

As Rome expanded she developed the use of concrete and the arch in the building of bridges, roads and aqueducts, creating a series of civil and military engineering projects that surpassed in scale any since those of the Assyrians and Egyptians. Nevertheless, whether because of the widespread use of slave labour or the stifling effects of a powerful bureaucracy the Romans, like the Athenians, failed to exploit many inventions of their time.

In China, iron metallurgy improved still further, surpassing any in the West. Chinese astronomy was also active, and the preparation of many useful drugs, from plants, became a speciality of Chinese medicine.

Euclid wrote his *Elements of Geometry* at Alexandria *c.* 300. Alexandria became a centre of learning in the century that followed Alexander's death, acquiring a great library and museum. Alexandrian inventions were based on the known principles of the siphon, gear-wheel, spring, screw, lever, pulley, cam and valve. The lighthouse of Pharos, a Greek achievement, stood 76m (250ft) high, with a 56km (35-mile) beam.

Aristarchus, 310-230 BC, was the first to maintain that the Earth moved around the Sun.

Archimedes of Syracuse, *c.* 287-212 BC, discovered fundamental laws of floating bodies, made advances in mathematics and was the greatest of Greek inventors: the "Archimedean" screw is still used in Egypt to raise water.
The crossbow was invented in the 3rd century BC in China. This weapon had a cocking and trigger mechanism similar to that of children's toys today.

A map of the world produced by Eratosthenes, *c.* 276-*c.* 194, a librarian at Alexandria in the 3rd century BC, was a great improvement on its predecessors. He calculated the diameter of the Earth to within a few hundred kilometres, and did work in number theory.
Glass-blowing spread to Alexandria from Syria via the Romans and for two centuries was the most active of the technologies. Larger vessels and dishes were now made, by blowing the molten glass at the end of an iron blowpipe.

A piston bellows with a double action was among Chinese inventions of the 2nd century BC. It provided a regular, steady air draught for the production of higher quality cast iron.

140-100 BC
Han emperor, Wu Ti, *r.* 140-87, established Confucianism as the basis of Chinese civil administration, 136 BC.
The Gracchus brothers proposed radical land reform to relieve Roman unemployment and poverty 133 BC. Gaius Gracchus, 157-86, introduced state control of grain imports and allowed landless men into the army in 112.
The Senate feared the destruction of both the constitution and the nobility's power, and so opposed these measures.

100-60 BC
After Marius' death in 86, **Sulla**, 138-78, rescinded his reforms.
Cicero, 106-43, prosecuted Verres, *d.* 43, for corruption while he was governor of Sicily in 70, and denounced Cataline's attempt to gain a consulship by force in 63 BC, defending the constitutional basis of the republic which depended upon no one man gaining supreme power.
Pompey, 106-48, and **Julius Caesar**, 100-44, rose to power after Pompey's victories had led to the annexation of Syria.

60-20 BC
The Han dynasty in China, 206 BC-AD 220, expanded into central Asia *c.* 60 BC.
Julius Caesar subdued Gaul, 58-51 BC, and after defeating Pompey became "dictator" in 45. Civil war followed his assassination in 44 by a pro-senatorial conspiracy.
Octavian (Augustus), 63 BC-AD 14, took Imperial authority, 27 BC, to centralize power and prevent the unrest recurring.
Egypt became a Roman province in 30 BC.

20 BC-AD 20
Augustus strengthened the Roman Empire in the north and east and brought peace to Rome.
Tiberius, *r.* AD 14-37, did the same although imperial power became increasingly dependent on the approval of the Praetorian Guard (the emperor's bodyguard).
Judea became a Roman province in AD 6.
The Kushite kingdom in Nubia was in decline *c.* AD 10.
Mexican lake houses were built.

AD 20-60
Tiberius and Claudius, *r.* 41-54 AD, expanded the empire, instituted social reforms and consolidated imperial power, although the danger of palace revolutions increased.
Jesus of Nazareth was crucified in Jerusalem *c.* AD 30.
The Christian cult was taken to Asia Minor, Greece and Rome by Paul, *fl.* 1st century AD.
Buddhism was accepted as the official religion of China by **Emperor Ming**, *r.* AD 58-75.

AD 60-100
Nero, *r.* AD 54-68, rebuilt Rome after a fire in AD 64, which was blamed on the Jews and Christians, who were unpopular because they refused to recognize the emperor's divinity.
Peter and Paul were executed in the ensuing persecution *c.* 64.
Vespasian, *r.* AD 69-79, became emperor after a civil war which followed Nero's death.
Jewish religious revolt, 66-70, was defeated by the Romans.
Mongol invaders brought iron and rice to Japan by 100 AD.

Chariots were introduced to Britain *c.* 100 BC.

Celtic Britain was divided between many warring tribes.
Close trading contact was established with Gaul followed by population movements.

Julius Caesar invaded Britain in 55-54 BC, but retreated.
Druids organized the opposition to the Roman invasion.

Large-scale tribal kingdoms developed with conflicts between the Catavellauni to the north and Atresates south of the Thames.

Emperor Claudius established a Roman province in the prosperous southeast of Britain.
The Fosse Way made a frontier with the Celtic world AD 45.

After the defeat of **Boadicea**, *d.* AD 62, Roman power reached Scotland. By AD 90, several legions were withdrawn and a town-based administration set up.

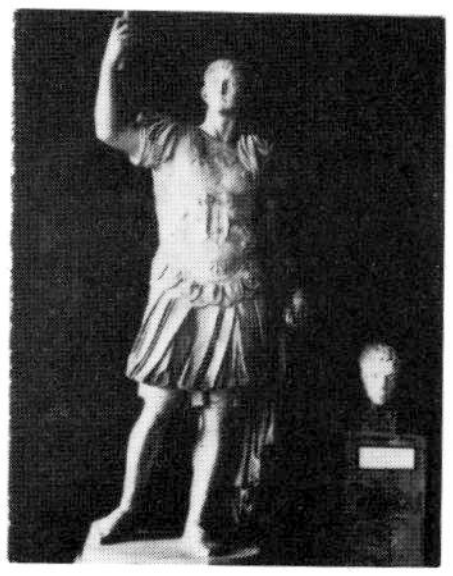

Julius Caesar

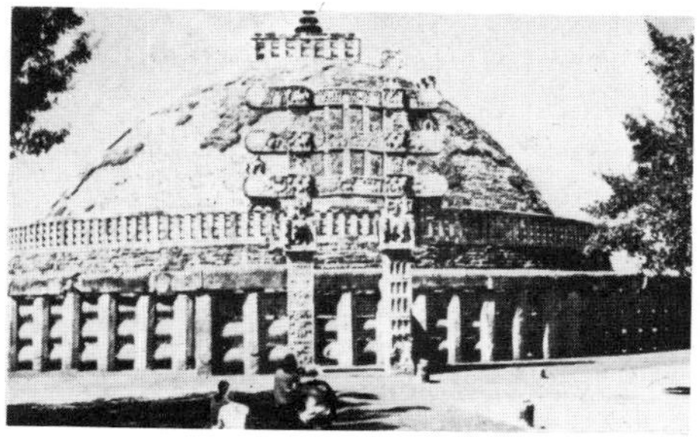

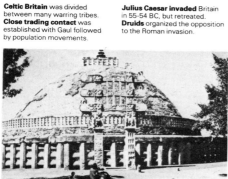

The Great Stupa of Sanchi

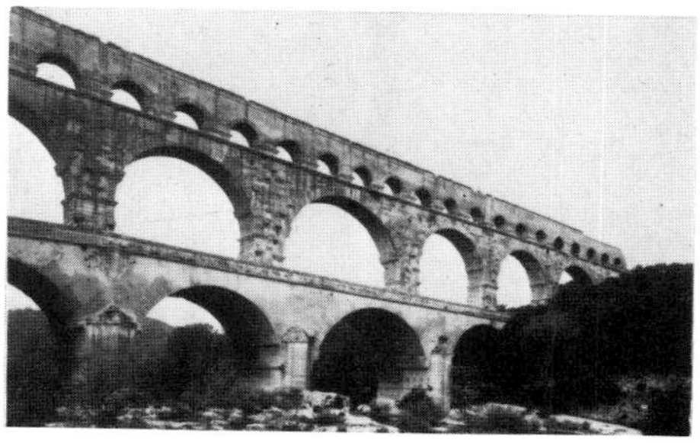

Roman civil engineering: the Pont du Gard, Nîmes

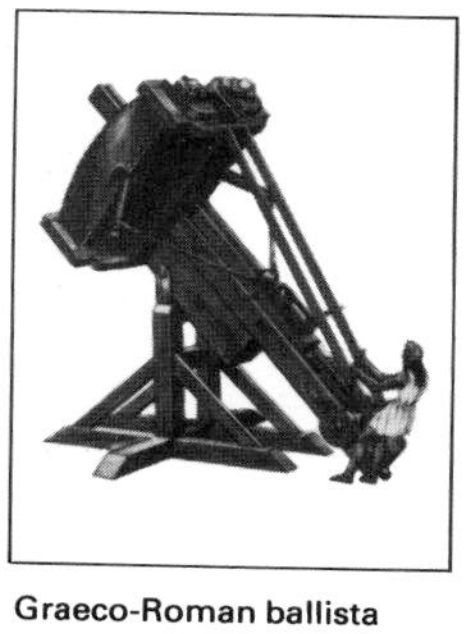

Graeco-Roman ballista

Tung Chung Su established Confucianism in China as the state cult *c.* 140 BC, combining elements of the Yin-Yang school and Confucianism. He taught that Heaven, Earth and Man formed a triad that the emperor, ruling by decree of Heaven, must maintain in harmony. This idea was important throughout Chinese imperial history, being associated with the stable order of Chinese society which lasted for 2,000 years.

The Pharisees in Israel, opposed the adoption of Hellenistic culture by their conservative Sadducean rulers and were accused of arid formalistic legalism because of their emphasis on the regulation of all aspects of life in accordance with Jewish law. However, they enjoyed the allegiance of much of the population because of their personal austerity and asceticism.

In Rome Marcus Cicero, 106-43 BC, and **Lucius Seneca**, *c.* 55 BC-*c.* AD 40, developed an **Eclectic philosophy**, drawing on Platonist, Stoic, Epicurean and Aristotelian sources.
Titus Lucretius, *c.* 99-55 BC, composed a long poem, *De rerum natura* (On the nature of things), in which he elaborated the Epicurean theory of physical atomism – a doctrine derived from Democritus that the world is composed entirely of microscopic particles.

Jesus Christ was born in Bethlehem in Judea in 4 BC and crucified in Jerusalem in AD 30.

After the Crucifixion in AD 30, the early Jewish/Christian sect developed a unique emphasis on the resurrection of a Messiah and the imminent transformation of the world at the dawning of a millennium of universal love. **Paul of Tarsus**, *fl.* 1st century AD, who saw the death of Jesus as a universal sign reflecting cosmic forces, spread Christian ideals through the Roman Empire. Contact with Hellenic thought turned Christianity into a world religion.

After the death of Paul, AD 64, and the destruction of Jerusalem in AD 70, the **Pauline version of Christianity**, with its more transcendental significance, became completely dominant. A few Christians who still upheld Jewish law became a small sect without links with either synagogue or Gentile Church. The oral tradition of early Christianity was gradually replaced by composed narratives, the first of which was the Gospel of St Mark.

Fu poetry, influenced by Chü Yüan, was brought to perfection *c.* 100 BC. Used mainly for description, it combined elements of prose and poetry in a form that was freer than that of the more personal **sao poetry** which continued to develop.
Ssu-ma Chien, *c.* 145-86, wrote the first dynastic history of China. His *Historical Record* is notable for its objectivity.

Cicero, a politician and philosopher, brought Latin oratory to a peak. His letters are a model of literary style.
Latin poetry flourished with **Lucretius**, *c.* 99-55, who gave emotional body to Epicurean philosophy in *De rerum natura*, and **Catullus**, 84-54 BC, who showed a mastery of technique and lyrical intensity.

Julius Caesar, *c.* 100-44 BC, wrote his history *Commentaries on the Gallic War* in a style of exemplary clarity.
Roman history was idealized in the *Aeneid*, 70-19 BC, by **Virgil**, second only to Homer as a model for Western poetry. He also wrote fine pastoral poetry, the *Eclogues* and *Georgics*.
Horace, 65-8 BC, combined elegance with humanity in his *Odes*.

Ovid, 43 BC-AD 17, the supreme Roman poet of love, and one of the most influential of classical writers, developed erotic verse into a major form in his long poem *Metamorphoses*. It was completed in AD 8, the year of Ovid's banishment, partly for his witty but irreverent *Art of Love*. His later poetry is sceptical, often elegaic.

Seneca the Younger, 4 BC-AD 65, made nine melodramatic adaptations of such Greek tragedies as *Oedipus*. They influenced Spanish literature and the revenge tragedies of Jacobean England. A noted orator and philosopher, Seneca exercised power in Rome in AD 54-62 but was finally ordered to commit suicide.

Plutarch, *c.* AD 46-120, the Greek biographer, wrote *Parallel Lives of Illustrious Greeks and Romans*, a work that approached history from the viewpoint of the characters of the men and women who made it. Shakespeare and many other European writers drew from its vivid portraits of life in Rome and Greece.

Roman sculpture at first reproduced and imitated the styles of the past. The "Dying Gaul", *c.* 100, was an accurate copy of the Pergamene original.
The steam baths of Stebiae, *c.* 120, were an early example of domed Roman architecture.

Roman arches were used for tenement houses and theatres in the highly populated cities of the 1st century BC.
Silverware brought to Rome following the sack of Corinth began a new taste for luxury articles in Roman circles.
"The Old Republican", a portrait bust *c.* 75 BC, captures Roman realism in a grim projection of asceticism and authority.

Wall paintings at the Pompeian **Villa of Mysteries**, *c.* 40 BC, portrayed Dionysiac themes. Roman villas of the 1st century BC increasingly introduced walled gardens and Greek peristyles.
The "Augustus" from **Prima Porta**, *c.* 20 BC, displays a naturalistic classicism characteristic of Augustan portrait sculpture.

Roman temples derived their typical high podia and deep porches from Etruscan architecture, while the **Maison Carrée** at Nîmes, *c.* 16 BC, has colonnades built in imitation of a Greek façade.
Roman bas-reliefs like that on the **Ara Pacis**, 9 BC, often used a combination of real and mythological figures to evoke contemporary Roman history.
Thermae, magnificent municipal buildings, were developed at Rome *c.* 20 AD.

Roman columns acquired a new function in the support of arches and when used as a free form. The Tuscan and Composite orders developed as more ornate variants of the Doric and Ionic.
Roman villas of the 1st century AD were often decorated with wall paintings which, like that at the **Villa Albani**, made use of idyllic scenes to evoke the peace of the countryside.

The Arch of Titus, a triumphal arch of the kind developed in the 2nd century BC, portrayed Roman victories in the Judean War, AD 70.
Stupas, typical Buddhist edifices, derived their dome shape from Vedic tombs. Stupa I at Sanchi, *c.* AD 100, was embellished with a square base, balconies and ornamental gateways.

Music notation had been devised in China, according to Symaa Chian, 163-85 BC, a chronicler who tells of a music master who wrote down a zither tune.

An Imperial Office of Music was founded in China *c.* 100 BC. The office supervised such activities as standardizing pitch and the building and administration of music archives.

Buddhist monks arrived in China *c.* AD 50 from India, bringing with them chant and decorative melodic features that were incorporated into Chinese music of later periods.

Destruction of the Second Temple, AD 70, led to the dispersal of the Jews. To keep their identity, secular music was discouraged and only singing permitted.

Peruvian technology during the last three centuries BC developed the moulding of elaborate pottery and metal casting.
Hipparchus, 190-120 BC, was the greatest Greek astronomer of his time. He estimated the Sun to be millions of miles away, instead of hundreds as previously thought. He also made a catalogue of stars.

Only two furnaces both using charcoal and a draught could produce iron in a malleable (workable) form in which it could be forged with carbon to produce steel. These were the **Catalan iron furnace** with two bellows and the **two-storey shaft furnace**. Neither, however, produced sufficient heat to melt the iron completely.
Lucretius, *c.* 95-55 BC, a Roman poet, wrote *De rerum natura*, a scientific treatise praising Epicurus and his ideas.

The Julian calendar of 45 BC, introduced under Julius Caesar, took a base year of 365.25 days. It was designed by the astronomer Sosigenes of Alexandria, and was inaccurate by a mere 11 minutes per year. This calendar was not supplanted until the 16th century AD, by which time it was inaccurate by ten days.
Water mills were a feature of Roman technology, as early as the 1st century BC, but were only fully described a century later, by Vitruvius.

Strabo, *b.c.* 64 BC, a Greek historian and geographer, wrote on the uses of materials.
A roller bearing for a cart wheel is another example of sophisticated engineering of the 1st century BC; made entirely from wood it was found in Denmark but probably made in Germany or France.

Metallurgical developments of the Romans include the manufacture of brass and the amalgamation of mercury with gold in the extraction of gold from its ores.
Roman civil engineering left an impressive record, including a 5.6km (3.5 miles) mountain tunnel, many aqueducts and 85,000km (53,000m) of roads.
The city of imperial Rome received a million cubic metres of water each day through lead piping, which in turn went to cisterns and centrally heated baths.

Chinese science was very active under the Han dynasty. Astronomers recorded eclipses and observed planetary motions. Mathematicians constructed "magic squares" of numbers that add up to the same answer in any direction and influenced arithmetic and algebra.
Paper was invented in China *c.* 100 AD.
Chinese inventions of this time include a camera obscura, and convex and concave mirrors.

100-400 Early Christianity

Principal events

The Roman Empire reached its greatest extent under Trajan but further expansion became impossible due largely to pressure from barbarian migrations in the north and east, which brought increased economic and social instability to Rome. Various defensive measures were adopted, such as allying with the barbarians, but Rome remained weak because of the dependence of most emperors on the support of the army. The reign of Theodosius saw the beginning of the close identification of the interests of the Christian Church and the empire in the east.

Cultured and prosperous civilizations arose in India, where the Guptas set up a northern empire and the strong southern Chola kingdom traded with Rome. In Central America the Mayas entered their classic period, while China suffered from instability and lack of central authority.

100-30

Emperor Trajan, *r.* 98-117, expanded into Dacia. His heir **Hadrian**, *r.* 117-38, pursued an essentially defensive policy, suing for peace in order to limit the eastern boundaries of the empire. He established his personal authority in Rome and travelled widely, but his eastern policy was unpopular because it caused a drop in the import of slaves.
The Western Satrap dynasty in Malwa, India, made Ujjain a centre of Sanskrit learning.

130-60

Emperor Antoninus Pius, *r.* 138-61, continued Hadrian's peaceful policies and quelled opposition from the Roman Senate.
The dispersion of the Jews followed the ruthless suppression of a Zealot revolt in Jerusalem in 135.
Migrating Goths settled on the northern Black Sea coast.
Taoism became popular in China, stimulated by military and social instability during the decline of the **Han dynasty** and the introduction of **Buddhism**.

160-90

Plague brought back by troops returning from the wars with Pathia depopulated the Roman Empire in 166-7.
Marcus Aurelius' reign, 161-80, marked the high point of **Stoicism** as the dominant philosophy, with its emphasis on the empire as a "common weal".
Marcomanni from Bohemia crossed the Danube, 167, and were settled by the Romans in areas depopulated by the plague.
The persecution of Christians in Rome increased *c.* 170.

190-220

Praetorian scheming prevented the establishment of a strong emperor, 192-3, until **Severus**, *r.* 193-211, reformed the army and reinforced provincial administration.
Caracalla, *r.* 211-17, expelled the Goths and Alamanni and in 212 bestowed citizenship on most free inhabitants of the empire, a token of Rome's reliance on provincial talent.
The Han dynasty in China fell in 220 and was replaced by three separate kingdoms.

National events

Under Roman rule Britain was prosperous. Celtic civilization assimilated Roman culture and urban life flourished. Hadrian's Wall established a permanent frontier in the north and British produce, particularly woollens, brought trading contacts with the rest of the empire in Europe.

Hadrian, *r.* 117-38, consolidated the conquest of Britain, building a wall from Solway to Tyne, 122-7. His large garrisons brought prosperity to England.

Urbicus, the governor, built the Antonine Wall from Clyde to Forth, 140-42. **The St Albans theatre** was built, 140, as a place of recreation.

After Caledonians took the Antonine Wall, 180, the Romans concentrated on defending Hadrian's Wall.

The Roman governor Albinus rebelled against **Severus** in 196, who then divided Britain into two provinces with York as the northern capital, *c.* 208.

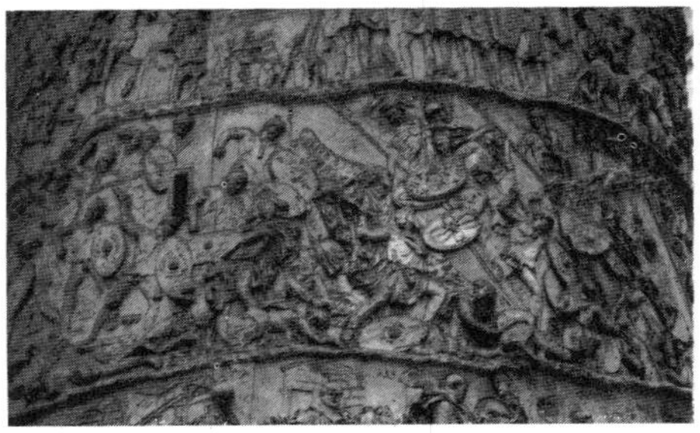

Trajan's Column: Romans fighting the Dacians

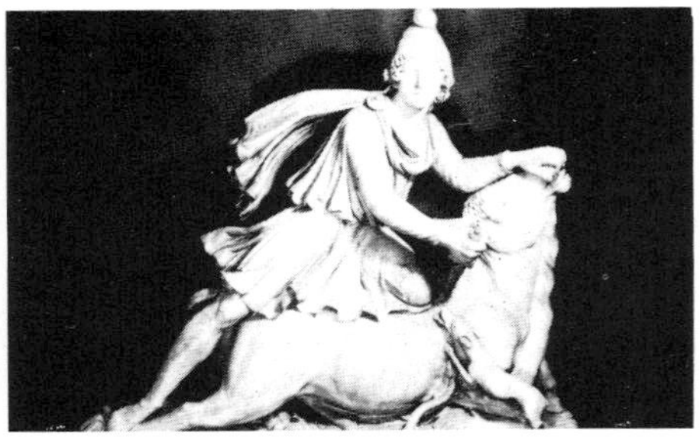

Late Roman sculpture: "Mithras slaying the Bull"

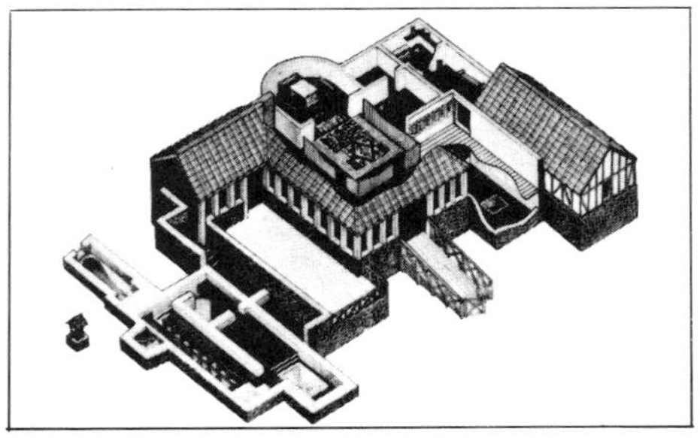

Roman villa at Lullingstone, 1stc. – 4thc.

Religion and philosophy

The early Church clarified and developed a sophisticated theology in response to attacks from religions whose origins lay outside the Judaeo-Christian tradition, including Gnosticism, Manicheanism, Montanism and Mithraism. By 400, Theodosius the Great had finally established Christianity as the religion of the empire but with the barbarian invasions began a long struggle against paganism. Missions were sent out to the Germanic tribes, and those who were converted largely adopted the Arian form of the Christian faith. Under Pope Damasus the Roman see claimed primacy over the five patriarchates of Rome, Constantinople, Antioch, Jerusalem and Alexandria, basing its claim on the "Petrine" text.

Buddhism began to influence the development of Hinduism in India and similarly affected religion and society in China.

Gnosticism, a diffuse movement based on a variety of religions, some earlier than Christianity, absorbed Christian ideas, giving the Gospels equal weight with Greek and Oriental texts. Many Gnostic sects were proclaimed heretical by the early Church and Gnostic interpretations of the scriptures forced the Church to establish authoritative versions of the Gospels and to consolidate the basis of a universal Church.

The Mahayana school of Buddhism was founded by **Nagarjuna**, *c.* 150-*c.* 250, in India. This school, often known as "The Great Vehicle", departed from the traditional Hinayana or Theravada, "Little Vehicle", doctrines in holding that laymen as well as monks could achieve nirvana through the intervention of saints. This development resulted from the impact of the Brahmin religion on Buddhism.

Montanus, who appeared in Phrygia *c.* 172, preached that prophecy and revelation had not ended with the death of Jesus. This belief and the expectation of the Second Coming of Christ threatened the stability of the Church, and many Montanists were excommunicated *c.* 177.
Buddhism reached China in the 1st century AD and began to exert an influence there in the 2nd century, when it received official patronage.

Origen, 185-254, became head of the Christian Catechetical school at Alexandria *c.* 212. This was the most famous of the Christian schools and offered a wide curriculum, including Greek, philosophy and science. Origen, a Platonist, furthered the synthesis of the Christian Gospels and Greek philosophy, emphasized the study of the Bible as essential to a proper understanding of Christianity.

Literature

With the diffusion of Christianity the scriptures began to spread in translation, notably in Latin, although fragments exist of translations into Gothic (mainly from the New Testament) made by Bishop Wulfila. The influence of the Gospels was literary as well as religious, setting the stage for the Christian allegorical treatment of pagan literature that was to dominate European literature until the 17th century. Homer, Virgil and Ovid were all interpreted in this way. Classical Latin literature petered out after AD 200.

In China, paper (chih) was invented but took nearly a millennium to reach Europe, where papyrus gradually gave way to stitched parchment or vellum.

The Latin satirical tradition had begun with **Lucillius**, *c.* 180-102 BC, but culminated with **Juvenal**, *c.* AD 55-*c.* 140, whose verse *Satires* on folly and Roman corruption profoundly influenced Western satirists. Tacitus, *c.* 55-*c.* 120, like Juvenal, vividly recounted the cruelties of the period up to the death of Domitian in 96, in his *Histories*, 104-9, and *Annals*, 117.

Asvaghosa, *c.* 80-150, was the first known poet and dramatist to write in classical Sanskrit. He was a Brahmin convert to Buddhism who wrote two epic poems, *Saundaranada* and *Buddhacarita* (The Life of Buddha), a philosophical work that became the source for later studies of Buddha's life.

The prose romance became a popular literary form in both Latin and Greek. *Satyricon*, a romance of Nero's Rome written by **Petronius** *c.* 60, gained popularity during this time.
Elements of science fiction were introduced into a parody of traveller's tales by a Greek writer, **Lucian**, *c.* 115-200, living in Syria. His *True History* describes a trip to the moon.

Apuleius, *b. c.* 125, wrote the only Latin prose romance that survives in full. His *Metamorphoses*, now known as *The Golden Ass*, relates the hilarious adventures of a man magically transformed into a donkey.

Art and architecture

The Imperial and Hellenistic styles of Rome gradually lost ground, to be replaced in western Europe by the more mysterious and magic art of the Christian period. A recognizable Christian style in art and architecture had developed by the 4th century, by which time the empire was divided into east and west, presaging the lasting division between the Byzantine and Western Christian traditions in art. Into this world of changing imagery the barbarian invaders brought new decorative abilities and tastes which were also assimilated into the art of Rome.

The Persian culture of the Sassanids and the Indian culture of the Ghandara region were influenced in part by Rome and this influence penetrated China during the late Han period, producing the basis for a recognizable style. Indian art flourished under the Guptas, during whose dynasty the Ajanta cave paintings were done.

The ascending spiral bas-relief narrative on **Trajan's column**, 113, relates and glorifies the emperor's military victories during the Dacian campaign.
The Pantheon, *c.* 118, is an architectural realization of the climax of imperial grandeur.
Monumental stone tomb sculpture appeared in China during the Han dynasty, probably due to foreign influence. The tomb of **Ho Ch'uping**, *c.* 117, includes a figure of a horse trampling a barbarian.

The Temple of Mithras, 2nd century, in London, is a typical architectural design of the period with its small size, basilican plan and central apse. Temples were common throughout the periphery of the empire.

One example of **Roman imperial sculpture** is the bronze equestrian portrait of Emperor Marcus Aurelius, *c.* 173, which still dominates the Campidoglio in Rome.
Sculpture from Gandhara in northwest India of the 2nd to 4th centuries exemplifies the meeting point of Graeco-Roman and Buddhist canons of beauty. Delicate reliefs depict the life of the Buddha – the first time he is represented figuratively.

Early Christian painting remained stylistically in the tradition of Roman decoration as can be seen in the fresco "The Celestial Refreshment", *c.* 200, in the catacomb of St Calixtus in Rome.
The Synagogue of Dura Europus in northern Mesopotamia is decorated with symbolic frescoes on the subject of Ezekiel in the valley of the bones, typical of the art resulting from the development of mystical religions in the Middle East.

Music

Plainsong, a form of religious chant, developed in Europe. St Augustine, 354-430, warned of the "peril of pleasure" in this music, whose austere unaccompanied line would be the basis of later European developments in polyphony (two or more related melodies played together) and harmony (chord progressions). Eastern music, with its sensuous sonorities, reached its peak.

The Chinese zither, adopted by Buddhist monks from *c.* AD 100, brought more instrumental colour to their music. Zither players produced sliding runs and delicate harmonic overtones.

Greek modes lived on in the **plainsong chants** of the Christian Church in an adapted form, ascending rather than descending as the Greek modes did.

Buddhism became a vital force in China from *c.* AD 200. Its chants were accompanied by the music of elaborate percussion orchestras of bells, gongs, triangles, drums and cymbals.

Science and technology

In the Graeco-Roman world, there was a decline in science and technology, although a brief revival started in the reign of Constantine the Great. The "occult sciences", astrology and alchemy, were held in great esteem, forming the basis for much technological innovation.

Most Western scientists of the 4th century were engaged in translating, collecting and commentating on the works of earlier thinkers, rather than making observations or doing experiments of their own.

In China, however, in spite of unsettled times, scientific thought progressed as advances were made in mathematics, astronomy and medicine, while materials technology remained active and productive.

In Central America Mayan culture began its classic period. This would produce remarkable advances in mathematics and astronomy and massive stone buildings constructed without the aid of metalworking.

Menelaus, a Greek mathematician, *fl.* 100, wrote the first work on non-Euclidean geometry.
Hadrian's Wall was built in Britain, 122-7. It was 118.3km (73.5 miles) long, with many forts.
Surgical instruments were well developed in Rome, as described by Celsus in the 1st century; but no pain-killing drugs were available to sufferers.
There is evidence that the **wheelbarrow** was invented in China at the beginning of the 2nd century.

Ptolemy, the Hellenic astronomer and geographer, wrote the *Almagest c.* 150. This became the "bible" of astronomers for the next 1,400 years, although it contained few new discoveries. Like Hipparchus and other Greek astronomers, Ptolemy accounted for the erratic paths of planets by suggesting that they moved in epicycles (small circles centred on the rim of a planet's orbit). Ptolemy's *Geography*, which included Africa and Asia, had great subsequent influence.

Galen, *c.* 130-*c.* 200, a surgeon and philosopher of Alexandria, wrote over 500 works on medical subjects. His experiments on animals led to the science of physiology. Galen's knowledge of the body was influenced by the works of Aristotle and Hippocrates, who believed in vital substances or essences at work in the body. Despite practical knowledge of the circulatory system, he postulated that blood vessels carried the blood to the skin where it was transformed into flesh.

Alchemists of the first two centuries include **Dioscorides of Cilicia** *fl. c.* 60, who described the processes of crystallization, sublimation and distillation of substances. He also described the use of minerals for medical purposes.
Alchemy, a pseudo-science of obscure origins, sought a philosopher's stone thought capable of changing base metals into gold, and the elixir of life which would preserve youth indefinitely.
The abacus is recorded in use in China *c.* 190.

220-50

The murder of **Alexander Severus**, *r.* 222-35, instigated a period of military control over the Roman emperor and factional warfare among the troops.
Official persecution of Christians began under **Decius**, *r.* 249-51, as the worship of living rulers became proof of loyalty.
The new Sassanid Empire in Persia, founded by **Ardashir I**, *r.* 224-41, took Armenia from Rome in 232.
The first written records in Japan date from *c.* 230.

250-80

The Goths took Dacia in 257. A series of capable emperors from Illyria began with **Claudius II "Gothicus"**, *r.* 268-70, who defeated the Gothic invasion of the Balkans and settled the Goths in the Danubian provinces. In 271, **Aurelian**, *r.* 270-75, drove out the Alemanni who had invaded Italy but abandoned the Roman province of Dacia.
China was nominally reunited under the **Western Ch'in dynasty** in 265.

280-310

Diocletian, *r.* 284-305, divided the empire into eastern and western spheres in 285 with two equal emperors. In 285 the western capital moved to Milan to defend the northern frontiers more easily.
Rome recaptured Armenia and Mesopotamia in 297.
The Mediterranean economy continued to collapse under heavy Roman taxation, *c.* 300.
The Franks, Alemanni and Burgundians crossed the Rhine.

310-40

Constantine, *c.* 285-337, became interested in Christianity and granted religious freedom to all religions in 313. He founded Constantinople as his eastern capital in 330 and gave Christian bishops a major administrative role in the empire.
The Gupta dynasty united much of northern India under **Chandragupta**, *c.* 320, and introduced a classic period of urban culture in north India.

340-70

Julian, 331-63, tried unsuccessfully to organize a pagan Church, and campaigned against the Franks.
The Persians recaptured Mesopotamia in 364.
Samudra Gupta, the Indian emperor, conquered Bengal and Nepal and broke the power of the tribal republics in northwest India. This marked a victory for caste over tribe.
The Pallava dynasty was set up in southern India *c.* 350.
Japan conquered Korea *c.* 360.

370-400

Roman absorption of the Germanic tribes was reaching its limits.
The Visigoths crossed the Danube in 376 and were settled as military allies by **Theodosius the Great**, *r.* 379-95.
Stilicho, *c.* 368-408, a Vandal Roman commander, defeated a Visigoth invasion under Alaric.
Christianity received official support from the emperor **Theodosius**.
Persian power was at its zenith under **Shapur II**, *r.* 309-79.

Many towns whose public architecture reflected their prosperity and importance as administrative centres were fortified *c.* 250, including Colchester.

Roman villa-based estates provided the foundation for a more efficient agricultural system than the former Celtic villages.

Carausius' control of channel shipping enabled him to declare Britain independent in 285, but **Constantius** restored imperial authority in 297.

British woollen cloth was popular throughout the empire.

Theodosius expelled an invasion of Picts, Irish and Saxons, 369.
Luxury products found in Britain attest to the wealth of this period.

The Romans began to withdraw from Britain in 383 after repelling a second barbarian invasion. Christianity was popular among the upper classes.

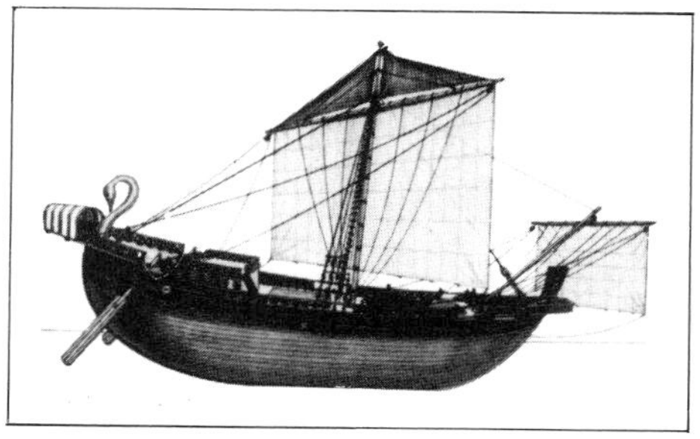

Roman grain ship

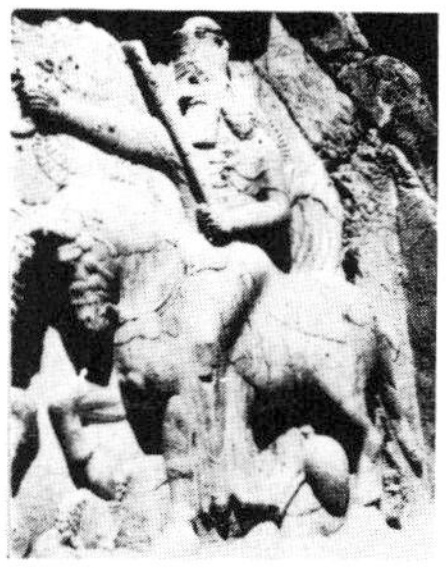

Relief at Naqsh-e-Rustan

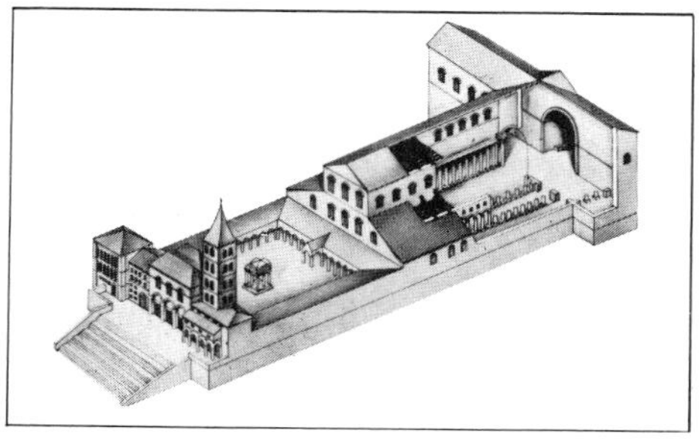

Early Christian churches: St. Peter's, Rome

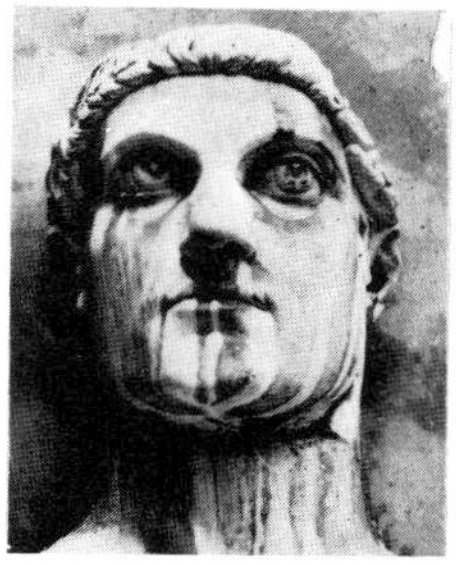

Constantine the Great

Neo-Platonism was founded by **Plotinus**, 205-70; his belief in the superiority of ideas over mundane reality fostered the Christian conception of heaven, widely influencing Christian and Islamic thought.
Mithraism, a cult based on the recognition of the two powers of good and evil, became popular with Roman legionaries and received official patronage.

Manicheanism, founded by **Mani**, *c.* 216-76, a dualistic religion combining the teachings of Zoroaster, Buddha, Jesus and Gnosticism became widespread from Europe to China. Mani held that knowledge of oneself and God guaranteed salvation (light) and liberation from one's present fallen condition (darkness). The soul had to be kept pure and in communion with God, both of which could be achieved by an abstemious life.

The Desert Fathers in Egypt formed the earliest Christian monastic orders. They included **St Anthony**, 250-355, who organized a group of hermits in 305, and **St Pachomius**, *c.* 290-346, who founded many communal monasteries, in which monks lived and worked together.
Neo-Taoism was created in China by **Kuo Hsing**, *d.* 312, and **Wang Pi**, 226-49, who believed in controlling the emotions and in an ultimate all-uniting principle of non-being.

At the Council of Nicaea, in 325, called by Constantine the Great, a group of bishops from all over the Christian world issued a creed stating that God and Christ are of one identical substance. **The Arian-ist heresy**, stating that only God was divine and that Christ was created as other men, was condemned by the Council. Constantine continued to support Arianism in spite of this.

The Latin Fathers of the Church, **Jerome**, *c.* 347-420, and **Ambrose**, 340-97, began their life's work of theological writing and furtherance of monasticism in the West. Jerome was baptized in 366, after he had studied Latin literature. His Latin translation of the Bible, the Vulgate, is still important today.

The Nestorian Church continued to develop, but separately, in Asia Minor.
The spread of Buddhism in China was greatly speeded by **Kumarajiva**, 344-413, who translated Mahayana Buddhist texts from Indian into Chinese.
Theodosius the Great, *r.* 379-95, extirpated Arianism and linked the Christian Church with the Roman state.

Early Tamil literature, associated with the kingdoms of southern India, dealt with the themes of courtly love and kingship. Its earliest works are the *Eight Anthologies* and possibly the *Ten Songs*, both written *c.* 100-500 by the third of the legendary Sangam literary academies which are said to have lasted for thousands of years.

Valluvar, *c.* 200-*c.* 300, was the author of the classic Tamil poem the *Tirukkural* (sacred couplets). The work is a collection of aphorisms dealing with government, society, virtue and love, and has proved almost impossible to translate. The outlook of the poet is so varied that several religious groups in the Tamil region have claimed him as their own.

A more uninhibited and individualistic writing sytle evolved in China with Taoist and Buddhist thought. The poet **Lu Chi**, *fl.* *c.* 300 was the first to express this movement towards original creativity. Simple language styles and folk-songs were used by the poet **Ts'ao Chih**, *c.* 300, and later developed in the poetry of **Tao Ch'ien**, 365-427.

Runes were the early Germanic script, used for magical charms and riddles. One of the earliest surviving examples, dating from the 4th century, is the *Mojbro Stone* from Upland, which says that a man was slain on his horse.

The golden age of classical Sanskrit began with the rule of the Guptas, 320-535. The poet and dramatist **Kalidasa**, 388-455, excelled in the epic genre of the *kavya* school.
The childhood and licentious youth of **Saint Augustine**, 354-430, before the time when he became a Christian convert in 387, is described in his *Confessions*, 397-401.

Realistic portraiture flourished throughout the Roman Empire. Paintings recalled Egyptian mummy portraits in both style and technique. Most exquisite were the delicate miniatures on gold glass medallions, among them the "Family of Vennerius Keramis", *c.* 250.
Sassanid Persia reached its cultural zenith during the reign of **Shapur I**. The rock carvings at Naqsh-i Rustam, 242-73, record the humiliation of the Roman emperor Valerian.

Mohican art flourished on the coast of Peru, *c.* 200-500, and was notable for its naturalistic ceramics, particularly of warrior figures.
Mayan culture grew in Central America from about the first century AD and lasted for the next 900 years. The architectural monuments of this civilization and the cities of Palnque, Copan, Uaxactan and Yaxchilan were built from the 3rd century onwards.

Roman architecture was at its most massive in the early 4th century with the Palace of Diocletian at Split and the Baths of Diocletian and Basilica of Maxentius in Rome.
The fixed hieratic expression on the colossal head of Constantine in the Forum marked a break from Hellenistic realism and heralded the formalized style of Byzantine art.

The old church of Saint Peter in Rome was built in 330, but destroyed during the Renaissance. The first religious building designed specifically for the needs of Christian worship, its basilican shape determined the layout of the majority of Western churches. **Sta Costanza**, 323-37, also in Rome, is an early example of the alternative centrally planned style of Christian architecture.

Hsieh Ho's *Six Canons of Painting* is the earliest work on the theory of art, written in the mid-4th century. The Taoist **Ku K'ai Chih**, *c.* 344-406, produced masterly landscapes and genre paintings, conforming to Ho's artistic definitions.
Gupta art flowered with some of the greatest paintings at the Ajanta caves in the north Deccan. Massive Buddhist stupas were built, with a marked stylistic influence from central Asia and China.

Roman art became increasingly stiff and formalized, as with the ivory diptych of **Stilicho** (a Vandal leader in the Roman army) and his wife Serena, *c.* 396. The Jonah Sarcophagus in the Lateran in Rome shows the merging of late Roman classical style with Christian motifs.

Heroic poems were sung among the German tribes by bards who accompanied themselves on harps. The songs were narrative lays of couplets set to music.

The harp, Europe's main musical instrument, was regarded as a precious possession. Later versions of the instrument became the national emblems of Ireland and Wales.

Psalms used by the Christian Church in its liturgy were among the first Christian chants. They were sung as responses by two choirs, or a priest and congregation sang alternate verses.

Persia under the Sassanid dynasty, 224-642, was rich in musicians and well developed instruments. Azádá's songs were celebrated in poems. Trumpet, lute and mouth organ flourished.

Chinese music was further enriched by foreign influences. After the conquest of Kutcha in Turkestan, 384, drums, cymbals and Persian harps with upper sound chests were imported.

In China the use of paper became widespread during the period of the Three Kingdoms, 220-64.
Diophantus of Alexandria, *fl.* 250, wrote the *Arithmetica*, of which six volumes survive in Greek manuscripts. He was the first to introduce symbols into Greek algebra. His numerical equations, together with the Hindu system of numbering, influenced the development of Arabian algebra.

In China Huang Fu wrote a treatise on acupuncture, in use since 2500 BC.
Chinese mathematical books describe the Pythagorean theorem; solve problems involving square roots; and give the value of π as 3.1547.
Shafts on chariots and carts first appeared in Europe in the 3rd century although they had been used in China for many hundreds of years.

Hippology – the science of breeding and managing horses – flourished under the Romans.
Clinker-built boats, made from overlapping wooden planks fastened with iron rivets, were developed in northern Europe in the 3rd century.
By order of Diocletian, Roman emperor from 284 to 305, all books on the working of gold, silver and copper were burnt to prevent counterfeiting. The effect was to increase interest in alchemy and magic as a method of turning base metal into gold.

Mathematics – developed by the Central American civilization of the Mayas, between the 4th and 10th centuries – was the first to make use of a symbol for zero. Mayan arithmetic was based on the number 20 and is notable for calculations involving very large numbers. One reason for this may have been the smallness and cheapness of the Mayan unit of money, the cocoa bean.
Yu Hsi studied the equinoxes *c.* 330 and was one of the first astronomers to describe the precession of the equinoxes

Mayan calendars, superior to those of early Christianity, were developed in order to calculate the year more accurately for religious purposes.
Mayan astronomy was in some ways very advanced, owing to the Mayan concern with time. Thus the Mayas calculated the length of a year on Venus and used it partly to work out the dates of religious festivals.
Pappus of Alexandria, *c.* 300-50, rewrote and commented on the works of earlier mathematicians such as Euclid and Diodorus.

Chinese astronomers of the 4th century believed, fairly correctly, that the blue of the sky was an illusion and that the Sun, Moon and stars float freely in space.

400-700 The new barbarian kingdoms

Principal events

After the fall of Rome in 476, the Western Roman Empire divided into a galaxy of unstable "barbarian" kingdoms which adapted Roman institutions, while the Byzantine Empire became cut off from the west despite Justinian's brief expansion c. 550. The growing independence of the papacy and the new monastic movement made Christianity a powerful political weapon among the barbarian kingdoms, so that national conversion and the suppression of heresy had a more than religious significance.

The teachings of Mohammed brought a new unity and aggression to the Arabs, who threatened Constantinople and expanded towards India.

The T'ang dynasty completed the development of the Chinese imperial system, on which Japan modelled its own, while India split into smaller kingdoms with the fall of the Guptas, although the classical era they initiated outlasted them.

400-30

The western capital of the Roman Empire retreated to Ravenna in 402.
The Visigoths under **Alaric** sacked Rome, 410, and invaded Spain in 415 under **Ataulf**, displacing the Vandals, who then moved to Africa.
The Franks and Burgundians, who created the first barbarian kingdom inside the empire, occupied Flanders and the Rhineland in 406.

430-60

The Huns' attack on Gaul led by **Attila**, *c.* 406-53, was defeated by a Roman/Visigothic alliance in 452, and their invasion of Italy stopped on Attila's death.
The Vandals attacked Rome in 455 from North Africa and annexed the Mediterranean islands.
After St Patrick's conversion of Ireland the Irish monasteries developed into centres of Christian learning.

460-90

The Western Roman Empire ended in 476 when **Odoacer**, *d.* 493, set up a barbarian kingdom in Italy. But **Theodoric the Ostrogoth**, *r.* 489-526, invaded Italy in his turn in 488.
The Frankish king Clovis I, *r.* 481-511, defeated the Roman governor in Gaul in 486 and set up the Merovingian dynasty.
In China political fragmentation prevented the development of Chinese culture, while Buddhism won many converts.

490-520

Odoacer surrendered in 493 to Theodoric, who set up an Ostrogothic kingdom that was initially recognized by Byzantium, 497. He built his capital at Ravenna.
Clovis was baptized in 497, becoming the first non-heretical barbarian king and thus winning the support of the papacy and the emperor against the heretical Germanic tribes. In 507 he drove the Visigoths into Spain.

National events

Anglo-Saxon invaders replaced Roman culture with their own and set up seven predominantly agricultural kingdoms. Christianity was reintroduced from both Ireland and Rome, and was popular particularly with the warrior aristocracy, acting as a unifying force on the disparate kingdoms.

Roman troops were evacuated in 410, and Angles, Jutes and Saxons brought to England by the Romano-British civilians to repel the Picts *c.* 429.

Widespread Angle and Saxon invasions from the Rhineland destroyed Roman town life and Christianity. The invaders settled along the eastern rivers.

The Anglo-Saxon invasion continued westwards. The invaders stayed mainly in lowland areas.

The kingdom of Wessex was founded *c.* 495 and **King Arthur** is said to have organized the Celtic defence against the Anglo-Saxon advance *c.* 520.

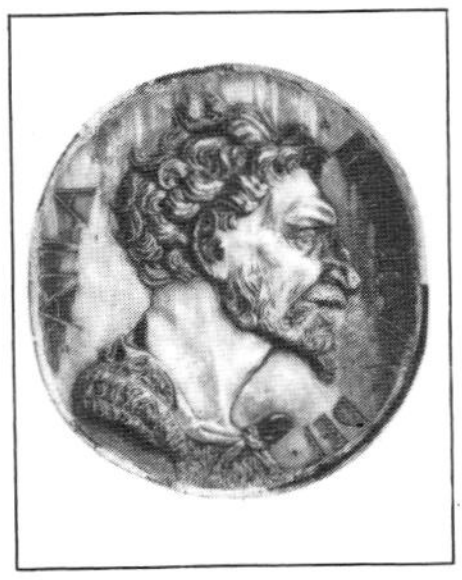

Attila the Hun

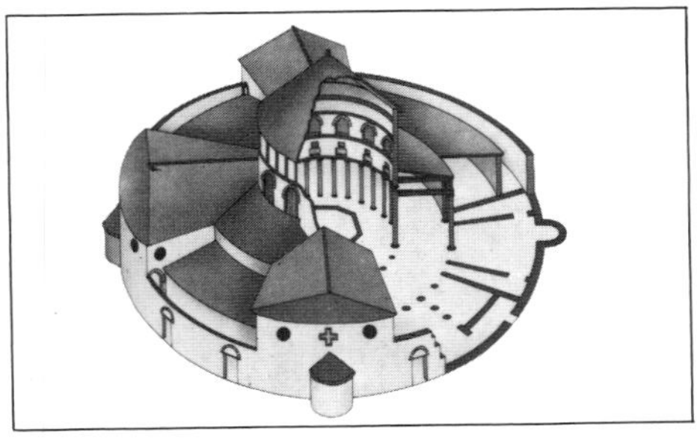

Early Christian churches: S. Stefano Rotundo, Rome

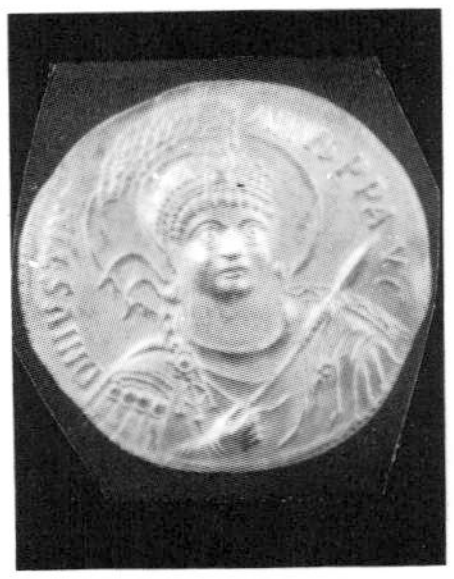

Coin of Justinian the Great

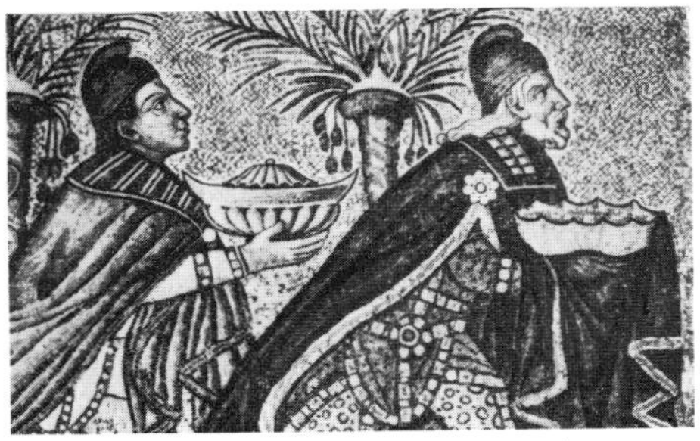

Mosaics from S. Apollinare, Ravenna, 6thc.–7thc.

Religion and philosophy

With its consolidation and the removal of the threat of alien ideas, the Christian Church turned in upon itself and became engaged in a series of fierce internal doctrinal disputes centring on the many interpretations of the nature of Christ. At the same time Western monasticism emerged with the founding of the Benedictine rule.

Islam arose as a small sect in the early 7th century and quickly became a powerful cohesive movement with an aggressive evangelical mission. By 700 it had spread throughout the Middle East. After the death of the prophet Mohammed, however, it became subject to internal schismatic tendencies deriving from the conflicts between the temporal and spiritual aspects of the Islamic religion.

Buddhism advanced beyond the borders of the Indian subcontinent; by 700 it had become firmly established in both China and Japan.

Nestorius, consecrated Bishop of Constantinople in 428, maintained that Christ was both divine and human.
The Council of Ephesus, 431, asserted the unity of Christ and declared this view heretical.
St Augustine of Hippo, the greatest theologian of Christian antiquity, combined the New Testament with Platonism. In *The City of God*, *c.* 410, he put forward the doctrine of predestination.

The Council of Chalcedon, 451, in response to the claim of **Pope Leo I** to universal supremacy, declared the Patriarch of Constantinople to be of equal authority. This council also emphasized that Christ had both a human and a divine nature, countering the doctrine of **Monophysitism**, which stated the essential unity of Christ.
St Patrick, entrusted by Pope Celestine I to convert the Irish people, landed in County Wicklow, Ireland, in 432.

The Shakta and Tantra cults became important in India, emphasizing mystical speculations on divine fertility and energy. These doctrines were regarded as unorthodox by religious teachers. Tantrism was also an important trend in the Buddhist tradition.
Under the Guptas, Vaishnaism flourished as a separate cult distinct from **Shivaism**.
The **Yogacara** school of meditative techniques flourished within Buddhism.

Buddhism grew in China at the expense of the more elite cult of **Confucianism**. There were two schools of Buddhism, the **T'ien-T'ai sect**, rationalists who sought to integrate Hinayana and Mahayana Buddhism, and the **Mahayana Amitabha sect**, who believed that salvation required reflection on the **Amitabha Buddha** as well as general meditation.

Literature

The classical tradition of literature largely disappeared with the fall of Rome in 476 but survived in Byzantium and in Christian monasteries where a few late Latin works were influential. Western European literature centred around the heroic myths of the Germanic invaders and the Celts, sung by bardic poets whose verse forms Christian writers later adapted to religious poetry. With the founding of Islam in the Arab world the Koran was collected but the re-emergence of Arabic poetry would await the prosperous dynasties of the 8th century.

In India and in China under the T'ang dynasty lyric poetry flourished, both religious and secular, while a Japanese writing and literature emerged.

Buddhist sacred literature, in its earliest complete forms, appeared in the 5th century Pali texts that collected together the *Jataka Tales* (birth stories). These 547 tales consisted of prose and verse fables about the former births of the Buddha, often in animal forms. Similar tales are found in Aesop and in non-Buddhist Indian literature.

Japan assimilated Chinese civilization in the first four centuries AD and evolved a writing system of extreme complexity by adapting the script of the monosyllabic Chinese language to convey the phonetics of Japanese. The earliest texts are the 8th-century histories *Kojiki* and *Nihon Shoki*, but include songs and myths probably from the 5th century.

The Jewish Haggadah texts in Palestine and Babylonia used legends, stories and anecdotes to illustrate ethical and theological matters dealt with in the Talmud. This material, with its lively embellishment of such Old Testament stories as that of Noah, influenced the similar treatment of biblical tales in the miracle plays of medieval Western Europe.

An oral literature of heroic verse known as **Heldenlieder** developed among the tribes of western Germany. From these songs and from pagan hymns and laments emerged later epic narratives, notably the story of Siegfried and Brunhild which was incorporated into Germanic epics like the 13th-century *Song of the Nibelungs* and into the heroic lays of Iceland.

Art and architecture

The fall of the Western Empire in 476 enabled new art forms combining Celtic, Scandinavian, German and Roman styles to develop in northern Europe, reaching their high point in the exquisite illuminated manuscripts produced by the Irish and Northumbrian monasteries.

Byzantine art, a sacred and stylized offspring of late Roman art, spread from Greece to Italy, blossoming in the 6th century with the building of Hagia Sophia in Constantinople and S. Vitale at Ravenna – the main cultural centre in Italy after the fall of Rome.

Middle-Eastern culture, divided until the 7th century between Byzantine and Persian influences, later collapsed before the onslaught of Islam, which absorbed certain elements of church design but forbade the use of representational imagery.

Japanese art during the Asuka period developed a style of its own distinct from that of China and Korea.

The Mausoleum of Galla Placidia in Ravenna, *c.* 425, shows a Byzantine influence in its plan and its decorations. The mosaics were made over a period of one hundred years and illustrate the shift from the light, decorative qualities of Rome to the sombre and awe-inspiring images of a wholly Byzantine style.

The hieratic and stylized form of Roman art can be seen in the ivory carvings "Scenes from the Passion", dating from the early 5th century. Classical ideals of proportion and anatomy were no longer considered important.
Christian architecture was a blend of Roman and indigenous styles, AD 400-600. In Egypt monasteries with frescoes were built at Bawit and Sakkara and the basilica of St Mena was constructed near Alexandria.

The church of St. Stefano Rotondo in Rome, 468-83, is exceptional for its entirely circular plan, although the centrally planned style continued a tradition which reached back to the Pantheon.
Chinese art during the Six Dynasties period, 220-589, developed the Han tradition of monumental stone sculpture. In 460 a series of rock-cut shrines were begun in the caves of Yunkang, which contain a 13.7m (45ft) figure of a standing Buddha.

Manuscript illumination was an important art form of the early Christian period. Only four religious texts survive, including the *Vienna Genesis*, a luxurious work on purple ground, and the *Rossano Gospels*, the earliest illustrated version of the New Testament. Both texts date from the early 6th century.

Music

In India, râgas were well established by the fifth century, having evolved from traditional melodies and scale theory that utilized many seven-note scales and complex rhythms to evoke various moods. In the West, by 600, Christian monks had developed plainsong to a level of accomplishment, codified by Pope Gregory I, that placed it lastingly in the liturgy.

The marimba, played today by the Bantu in Africa, developed from a xylophone introduced to Africa by Indonesian immigrants in the 5th century.

Japan adopted music and dance that were to die in their countries of origin. Supple Indian and Chinese forms were considered female, and Korean and Manchurian forms male in character.

Irish song, carried through Europe by minstrels and monks, revitalized musical composition. Unlike classical verse, which might be sung to a melody repeated as often as the poem required, Irish poetry – with its lines of irregular length but equal accents – demanded specific settings. This inspired musicians to greater feats of ingenuity and expression.

Science and technology

Chinese science and technology was by far the most active and inventive of this time. Under the T'ang dynasty the sciences and arts were encouraged and science was no longer hampered in any way by religious dogmas or prohibitions. Chinese attitudes to medicine were particularly enlightened: even before the 5th century medical treatment was regarded as a public service and was administered by the state. In astronomy, practical chemistry and mathematical calculation China also led the world.

By contrast Western science had dwindled to commentaries, and even these often met with discouragement of an extreme kind. Boethius, one of the last major Western commentators on science and philosophy, was executed in 524 by Theodoric the Ostrogoth for advocating a return to political and intellectual liberty. Although overshadowed by China, Indian mathematical, astronomical and medical sciences also advanced.

The university of Constantinople was founded in 425 by **Theodosius II**, 401-50, the Roman emperor of the East, who later (438) produced the **Theodosian Code**, a systematization and simplification of the Roman legal code.

Chinese scientific instruments of the 5th century included water-driven armillary spheres, which revolved in phase with the stars, and a compass, originating in the 2nd century, whose pointer was a metal spoon balanced on its bowl. From the 9th century these spoons were replaced by magnetic needles.

Boethius, 480-524, wrote on the four advanced "arts", geometry, arithmetic, music and astronomy. Two of these manuscripts survive, *De Institutione Musica*, and *De Institutione Arithmetica*.

Indian astronomical literature shows an upsurge which lasted a century, beginning with the publication in the late 5th century of the work of the astronomer **Aryabhata**, 476-550. This mentions rotation of the Earth and the epicyclic movements of planets. He also obtained an extremely accurate calculation for π.
Metal stirrups, invented in 5th-century Korea and used by the Avars on their incursions from Asia, were first seen in southern Europe *c.* 500.

520-50

The Byzantine emperor Justinian the Great, *r.* 527-65, temporarily reconquered North Africa in 534 and Italy in 554, and codified Roman law. His alliance with the papacy led to the suppression of heresy in the empire.
Chosroes I, *r.* 531-79, brought Persia to its greatest strength in a protracted war with Byzantium.
The Gupta dynasty in north India fell in 535.

550-80

Byzantium had reconquered most of the Mediterranean seaboard by 560, but by 571 the Lombards had taken Italy and settled in the North.
The Frankish kingdom stayed divided because of the Merovingian custom of equal division of inheritance between the king's sons.
Persia took southern Arabia from the Abyssinians in 576.
The introduction of Buddhism to Japan in 552 marked the start of a period of Chinese influence.

580-610

Pope Gregory I, *r.* 590-604, assisted papal authority by defending Rome against the Lombards.
Persia and Byzantium were at war in Syria-Palestine, 602-28.
The Sui dynasty reunified China in 589 by conquering the southern Chen dynasty.
In Japan, the Soga clan rose to power in 587, introducing a paternalist, Chinese-style constitution.
Irish missionaries worked in Scotland and Germany *c.* 600.

610-40

The Muslim era began with the flight of Mohammed, *c.* 570-632, from Mecca to Medina in 622. His ideas brought a new unity, sense of responsibility and aggression to the diverse Arab traders and tribesmen. After Mohammed's death **Caliph Omar**, 581-644 (the head of Islam), expanded the Islamic realm in the Near East.
The T'ang dynasty was founded in China in 618, ruling with a large and powerful imperial bureaucracy.

640-70

Disputes about the authority and succession of the caliphate under **Othman**, *d.* 656, and **Ali**, *d.* 661, led to civil war, which destroyed the unity of the Ummah and led to the establishment of the **Umayyad dynasty** at Jerusalem in 638. The Muslims then took Iran and Egypt, 642, Armenia, 653, and Afghanistan, 664, ruling as an autocratic but tolerant minority.
Japan entered a period of reform in 646, imitating Chinese society.

670-700

The Islamic world was divided by disputes which led to the emergence of the Sunni, Shi'ite and Khawarij sects, reflecting the problems of succession and the growing discontent at the prosperity of the Meccans, which was increasing at the expense of other Muslims.
A 30-year truce was concluded between the Byzantine and Muslim empires after the failure of the Muslim blockade of Constantinople in 673-8.

The kingdom of Bernicia was founded from Anglo-Saxon settlements in the north *c.* 547. It ultimately reached the Firth of Forth.

St Columba, 521-97, set up an Irish monastery at Iona, 563. The seven Anglo-Saxon kingdoms (**the heptarchy**) were established by 550.

St Augustine arrived to preach Roman Christianity, 597, in Kent, which was the chief of the seven kingdoms *c.* 600.

St Aidan, *d.* 651, introduced Irish Christianity to Northumbria, 635. Its king, **Edwin**, had won the overlordship of the heptarchy except Kent in 626.

The Synod of Whitby, 664, secured the victory of Roman Christianity in Northumbria.
The Sutton Hoo ship burial took place in East Anglia *c.* 650.

Anglo-Saxon society comprised king, thanes, freemen and slaves. The Northumbrian monastic culture produced the **Lindisfarne Gospels**.

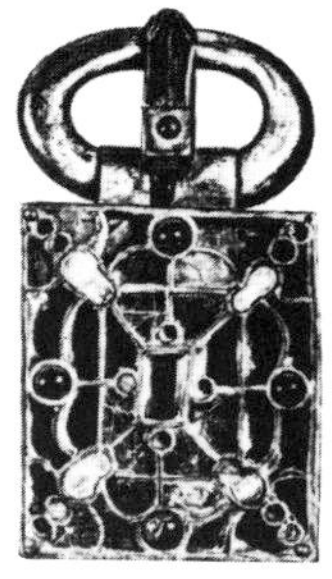

Merovingian buckle, 6thc.

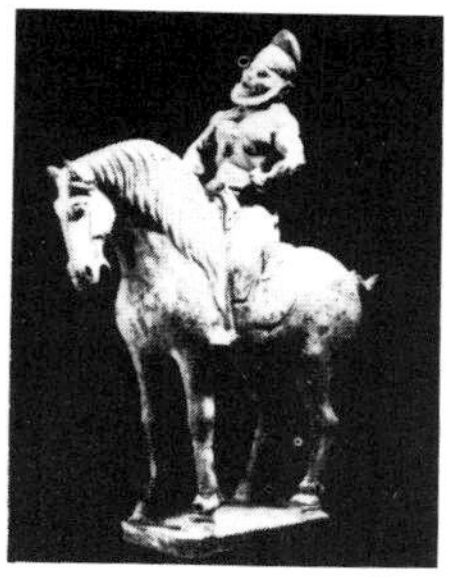

T'ang pottery figure

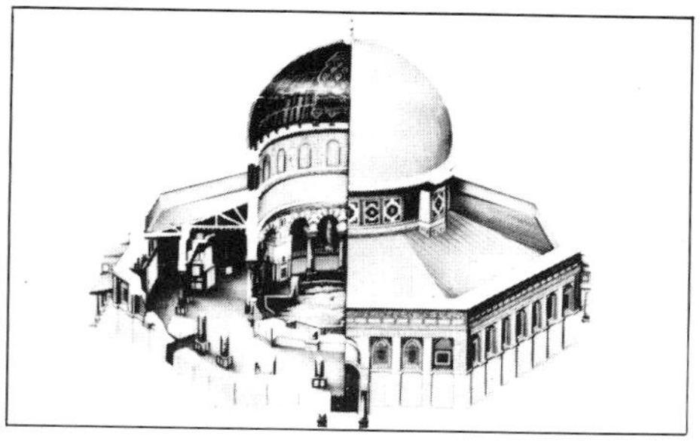

Early Islamic architecture: the Dome of the Rock

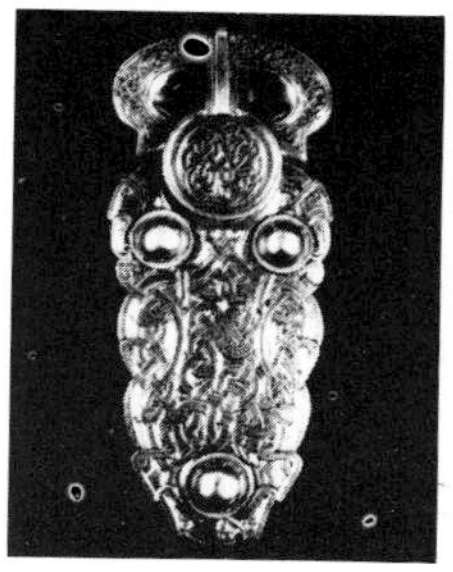

Gold buckle, Sutton Hoo

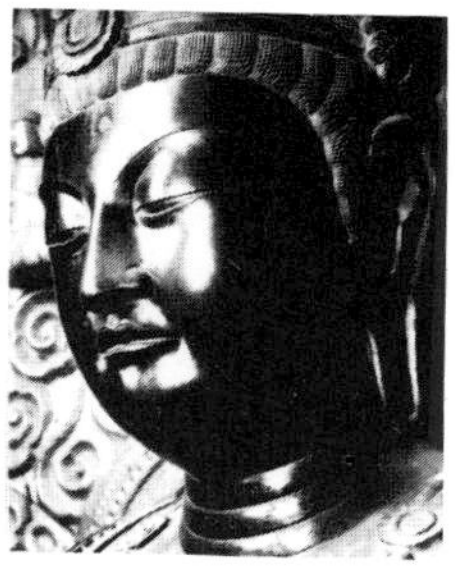

Japanese Buddha, 7thc.

St Benedict, *c.* 480-547, founded the first Benedictine monastery *c.* 529. He laid down a complete set of rules for monastic life, including a period of probation before full membership of the monastic community, prohibition of ownership and, most important of all, rules for obedience, celibacy and humility. His monastic ideal was of a self-contained and self-sufficient community.

Buddhism, supported by the **Soga clan**, was officially introduced into Japan in 552.
Mazdakism, founded by **Mazdak**, *fl.* 560, in Iran was an offshoot of **Manicheanism**. Mazdak held that good (light) acted by free-will and evil (darkness) by chance. Light could only be released into the world by asceticism, vegetarianism and non-violence.
The prophet Mohammed, was born in *c.* 570 in Mecca.

Mohammed, founder of Islam, received his first prophetic call in 610. Thereafter he began to proclaim his message publicly. His revelation was of a majestic being, the one God, Allah, whose command was that Mohammed was to be his prophet. This and subsequent revelations form the content of the *Koran* which emphasizes generosity, the goodness and power of God and retribution on the Day of Judgment.

In 622 Mohammed and a small following emigrated to Medina after opposition and harassment in their native Mecca. The **Ummah**, or Muslim community, claimed supremacy over tribal or familial loyalties, regarding all Muslims as brothers. In so doing it helped to make Arab society more cohesive. The crucial concept of **jihad** (holy war) was instituted at this time by Mohammed and led to the conquest of Mecca in 630. Mohammed died in 632.

Divisions appeared in the Ummah over the succession after the death of Mohammed. The supporters of his son-in-law **Ali** were the forerunners of the major **Shi'ite** division of Islam. The puritanical **Kharijites** who opposed Ali withdrew from the main body of the Ummah. The Shi'ites stressed leadership, the Kharijites community and permanent religious aggression.

The Monothelites were condemned as heretics at the Ecumenical council at Constantinople in 681. The heresy concerned the divinity and humanity of Christ. The Monothelites, following the decision reached at Chalcedon, claimed that although Christ had two natures He had a single will. The Council insisted on Christ's duality by asserting that both a divine and a human will were in Christ's person.

Aristotelean logic was translated into Latin by **Boethius**, 480-524, the last great Roman writer. A Christian who served as a minister under Theodoric the Ostrogoth, he was condemned to death and in prison wrote *De consolatione philosophiae*, 523, a treatise in verse and prose on free will, good and evil which helped spread Greek thought in the Germanic world.

Alliterative bardic verse romanticizing the heroism of Celtic warriors had become an established literary form by the middle of the 6th century. **Nennius**, *c.* 800, attributed to a Welsh bard, **Taliesen** (possibly a mythical figure), odes and lays in praise of **King Urien**, *fl.* 547–59. These were later collected with others in the 14th-century *Book of Taliesen*.

An Irish bard, Dallan Forgaill, *d.* 597, is credited with the *Eulogy of St Columba*. Its vigorous alliterative style is also found in Irish sagas about the hero **Cú Chulainn**, possibly 7th century, known as the Ulaid cycle. Ireland had a class of professional poets, the filiad.
A Welsh poet, Aneirin, *c.* 600, celebrated northern British heroes in *Y Gododdin*.

In India, classical Sanskrit literature, which had thrived under the rule of the Guptas, 320-535, reached its late flowering in the poetry of **Bhartrhari**, *c.* 570-651, a philosopher who wrote three collections of verses, the *Sātaka*, on the sensual pleasures of love, the nature of justice and the means of liberation from earthly existence.

The *Koran* reached its final form, 651-2. Written partly in rhymed prose reflecting the mood of Mohammed during his life as a solitary visionary preacher, it was regarded by Muslims as the perfect word of God. Its style and thought permeated the literature of Islam, an expansionist force that took many Persian stories to Europe.

The first named English poet, Caedmon, *fl.* 670, used the metre and diction of Old English pagan verse to compose poems on biblical and religious themes at the monastery of Whitby. A nine-line *Hymn on the Creation* is the most generally accepted of several works attributed to him in a 10th-century MS. He was an untutored herdsman, according to the Venerable Bede.

The age of Justinian saw the flowering of Byzantine architecture. The architecture and mosaics of **S. Vitale**, Ravenna, 526-48, were the splendour of Italy. The great cathedral of **Hagia Sophia** in Constantinople, 532-7, was an architectural and engineering triumph.
King Theodoric's mausoleum in Ravenna, *c.* 530, is surmounted by a colossal domed monolith, a fitting tribute to the half-barbarian, half-civilized king.

The distinctive Japanese art style of the Asuka period, 552-645, culminated with the temple complex of Hōryūji, Nara. The courtyard with its Pagoda, Konda (Golden Hall) and Kodo (for meetings) was based on the traditional Chinese and Korean layout.
The solid ivory throne of Maximian, archbishop of Ravenna, was carved in Constantinople and was a gift from Emperor Justinian *c.* 550.

The Basilica at Turmanin in Syria is a typical eastern variation of Roman Christian architecture, which was common throughout Syria and Palestine before the rise of Islam.
The art of Sassanid Persia of the 5th and 6th centuries shows a combination of Byzantine and Irano-Buddhist styles. Metalwork was highly developed and decorated with complex motifs and intricate filigree work.

The Great Chalice of Antioch is typical of Christian metalwork from the Roman provinces. It probably dates from the early 7th century, when there was an enormous output of silverware and fine gold jewellery.
The Ashburnham *Pentateuch* is a masterpiece of vivid narrative illumination dating from the late 6th or early 7th century. It is not known where it was made, nor whether by a Jewish or Christian illuminator.

The Sutton Hoo treasure comes from the grave of an East Anglian king who died in 654. It includes superb examples of Anglo-Saxon decorative metalwork.
Christian scholars and artists who took refuge in Ireland during the period of the Anglo-Saxon invasions produced an abstract and extremely ornate style of illumination, the 7th-century *Book of Durrow* being one of the best surviving examples.

The Dome of the Rock in Jerusalem is the first great Islamic architectural monument. Construction began in 688 in the reign of Abd al-Malik, but the design was a creative adaptation of Christian church buildings.
Book illumination reached great heights in Northumberland. The *Lindisfarne Gospels*, *c.* 700, combine the Roman narrative tradition with the decorative skill of the Celts.

Confucian ceremonies in China closely integrated music, dance and poetry. Chinese court music and dance expressed the form and calligraphy of poems around which they were created.

Harps of six to twelve strings were played by European musicians. The instrument became a symbol of their calling. A six-string example, found at Sutton Hoo, dates from about 640.

Pope Gregory I supervised the compiling and codifying of plainsong *c.* 600, giving his name to Gregorian chant, an unaccompanied and unharmonized style that has persisted to the present day.

Under the T'ang dynasty, 618-907, orchestral suites and programmatic works, some describing battles, were composed in China. Music-dramas incorporating folk-song developed.

Classical Arab music evolved richly under the Umayyad caliphs, 661-750, in Damascus. **Ibn Misjah**, died *c.* 715, codified its theory, embracing eight modes for lute music. Arab modes were nearly identical to Greek modes, but were performed with rich embellishments characteristic of the sinuous ornamentation of much Arab visual art and architecture.

Justinian, Byzantine emperor, *r.* 527-65, closed the Athenian university because the teachers were not Christians.
John Philoponus, *fl. c.* 530, speculated that a projectile would gain momentum from the mechanism which fired it, thus arriving at a crude idea of inertia.
Palaeontology was furthered in China in 527 with a book by **Li Tao Yuan** in which he described animal fossils.
Indian decimal notation began in the 6th century; on later inscriptions a dot signified zero.

The Ma'daba mosaic, the oldest-known map of the Holy Land, shows the area from ancient Byblos to Thebes and has a street plan of Jerusalem. It was made in Palestine *c.* 550.
Silk production was attempted at Byzantium in the 6th century after silkworm eggs had been smuggled out of China and taken there, reputedly by Nestorian monks.
Abacus calculators are described in a mathematical work thought to have been written by **Chen Luan** *c.* 570.

The diagnosis of disease in China in the 7th century was documented by **Chao Yuan Fang**, *c.* 610, who wrote a treatise listing 1,720 diseases classified into 67 groups.

Chinese surgical treatment in the 7th century included the removal of cataracts.
Windmills, probably invented in Persia in the 7th century, may have had their origin in wind-driven prayer wheels. Another theory, unproven, is that they were inspired by ships' sails. The axis of a wheel, driven by some 6 to 12 sails, was mounted on the first storey of a Persian windmill. Stone wheels used for grinding corn were located on the storey above.

Greek fire, used in the defence of the Byzantine Empire in the 7th century, was a highly inflammable substance of uncertain composition. Probably a mixture of pitch, naphtha and potash, it could be discharged from tubes in the prows of ships.
Fine metalwork including cloisonné, enamel and lathe-turned jewellery was found at the Sutton Hoo ship burial dating *c.* 650, showing that metallurgy in the Dark Ages was not only used to make swords.

Swords were the most advanced product of Burgundian and Frankish metallurgy in the 7th and 8th centuries. Their blades were expertly forged, with strips of decorative metal running along the whole length. Handles and scabbards were inset with jewels and welded decorations.
In northern England the tides and moon were studied by the **Venerable Bede**, *c.* 673-735, who also wrote a treatise on finger reckoning.

700-1000 Islam reaches India and Spain

Principal events

Invasions from the Muslims in Spain, Vikings in the north and Magyars in the east destroyed much of Europe's culture and economic strength, though Charlemagne's conquests east of the Rhine brought Germany within the European orbit.

The Muslim world reached from Spain to Afghanistan by 736, and the papacy, although relatively isolated by Muslim control of the Mediterranean, used its new states for political ends, reviving the Roman ideal by crowning its main supporters Holy Roman Emperor. Royal authority in Europe at this time was often precarious, based only on the personal allegiance of a provincial nobility whose power was strengthened by the need to defend the kingdom's frontiers.

In China constant warfare weakened the T'ang armies and the Sung dynasty gained control, while in Japan the Heian period marked a moment of transition to a society run on feudal lines.

700-30

Pope Gregory II, *r.* 715-31, appointed St Boniface, *c.* 680-754, to convert Germany.
The Umayyad Arabs took Spain in 715.
Leo III, *r.* 717-41, defeated the second Arab siege of Constantinople, 717-18, and began the iconoclastic controversy in 726 asserting the religious authority of the emperor and limiting the spread of monasticism.
The Nara period in Japan began with the establishment of a capital at Heijō in 710.

730-60

Charles Martel, *c.* 688-741, stopped the Muslim invasion of Europe at Poitiers, 732, and assisted Boniface in Germany.
His son **Pepin**, *r.* 747-68, campaigning in Italy, established papal temporal power by a donation of land to the papacy, 756.
Al Mansur, *r.* 754-75, founded the **Abbasid Caliphate**, defeating the Umayyads in North Africa and the Near East, 750.
The Gurjara-Pratihara dynasty defended India against the Muslims after 740.

760-90

Charlemagne, *r.* 771-814, united France and conquered Italy in 774, northern Spain in 777, Saxony in 785 and Bavaria in 788.
Baghdad became the Abbasid capital in 762.
An Umayyad dynasty emerged at Cordoba in Spain, 756, tolerating Jews and Christians.
Scandinavian trade with Byzantium began *c.* 770.
Turkish and Tibetan tribesmen threatened western China *c.* 763.

790-820

Charlemagne was crowned Holy Roman Emperor, 800, reviving the idea of a Western Roman Empire. Byzantium recognized the title in 812.
The Bulgar kingdom reached its peak under **Krum**, *r.* 808-14.
Ghana was an important trading kingdom, bringing gold from southern Africa to the Sahara.
Emperor Kammu, *r.* 781-806, instituted the Heian period in Japan, 794-1185, in which indigenous feudalism superseded the Chinese-based social order.

National events

A single king emerged and ruled England until Viking invasions took over much of eastern England, where a Danish society developed. Although monastic culture declined, secular learning flourished under Alfred, 849-99, and town life with a money economy recovered in spite of the Vikings.

Ethelbald of Mercia, *r.* 716-57, was overlord of all England, except Northumbria.

Offa, *r.* 757-96, brought Mercian power to its zenith and made the rulers of sub-kingdoms renounce their kingship.

Offa built a dyke between Mercia and Wales, and created a unified currency, bearing his picture, which was coined in many separate towns.

The Vikings sacked Lindisfarne, 793, and overran Ireland, 802-25. **Egbert of Wessex**, *r.* 802-39, inherited the Mercian supremacy.

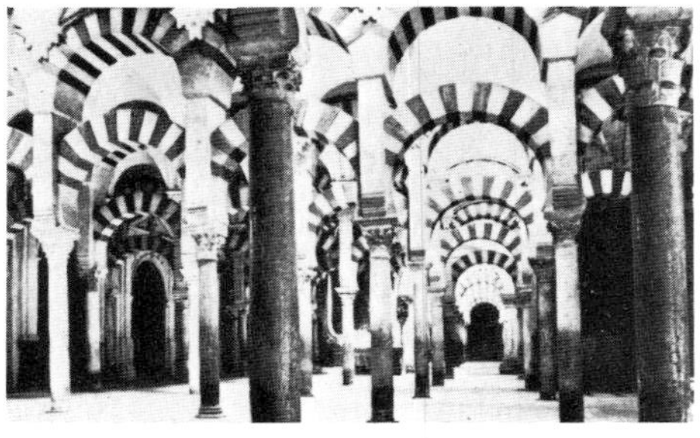

Islamic architecture: Mosque at Cordoba, 8thc. interior

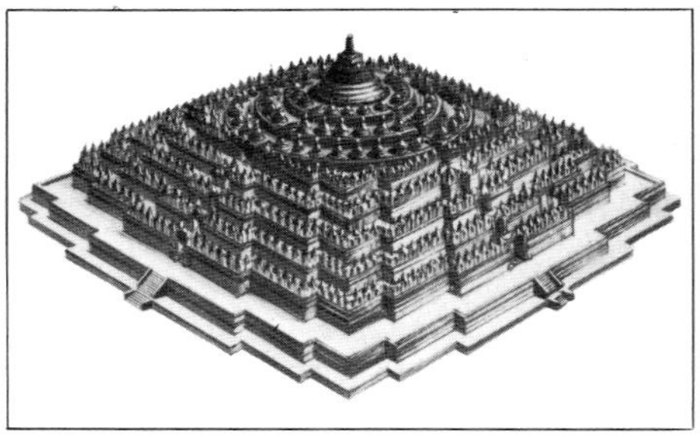

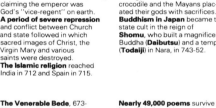

The stupa of Borobudur, Java, 8thc.–9thc.

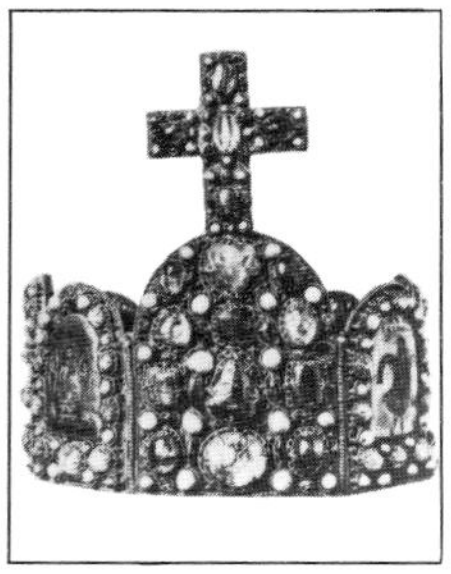

Crown of Charlemagne

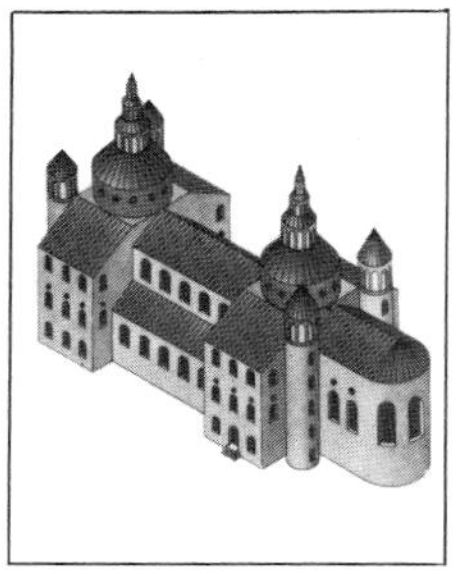

Carolingian church, *c.* 800

Religion and philosophy

The Christian Church continued the struggle to assert its authority over the secular powers of the Holy Roman and Byzantine empires while the assertion of the primacy of the Roman popes over the Eastern Church led to an increasing separation of Eastern and Western forms of Christianity. In the West, papal sanction of Charlemagne's empire brought the Church additional prestige. The practices of the clergy, however, were becoming increasingly lax, and would eventually prompt the Cluniac reform movement.

In the Islamic world, the Sufi movement was founded and grew, emphasizing an austere mysticism in response to the rational ideal and the reason of orthodox Islam.

The spread of Buddhism within Japan continued and won official support.

Mayan religion reached its most elaborate hierarchical form at the height of the empire's power in Central America.

Iconoclasm as a movement began, 726, when the Byzantine emperor **Leo III** prohibited the use of icons as idolatrous, claiming the emperor was God's "vice-regent" on earth.
A period of severe repression and conflict between Church and state followed in which sacred images of Christ, the Virgin Mary and various saints were destroyed.
The Islamic religion reached India in 712 and Spain in 715.

The Classical period in Mayan culture in Central America reached its height. Mayan cosmology saw the earth as a crocodile and the Mayans placated their gods with sacrifices.
Buddhism in Japan became the state cult in the reign of **Shomu**, who built a magnificent Buddha (**Daibutsu**) and a temple (**Todaiji**) in Nara, in 743-52.

The new Anglo-Saxon humanism was introduced in France by the Northumbrian monk **Alcuin**, *c.* 732-804, who met **Emperor Charlemagne**, 781, and became an important figure in the **Carolingian Renaissance**. Alcuin encouraged the study of the liberal arts. His revision of the liturgy of the Frankish Church was carried throughout Charlemagne's empire and he created a new edition of the Vulgate.

The Tendai and Shingon sects were founded in Japan *c.* 805 by Buddhist monks returning from a visit to China.
Sankara, 780-820, the most important member of the new **Vedanta school of philosophy** in India, affirmed the one true reality (**Brahma**) as the source of all things. He also wrote commentaries on the *Upanishads* and *Brahma Sutra*.

Literature

Chinese literature of the T'ang dynasty reached its finest form in the evocative poetry of Li Po, Tu Fu and Wang Wei in the 8th century. With the later decline of the dynasty, social criticism and an elegiac mood appeared. Chinese influence on Japanese literature gave way to new vernacular forms of Japanese verse and prose.

The spread of Islam led to more sophisticated themes in Arabic poetry and to an extension of Arabic influence into Persian literature.

The epic saga took shape in Norway and Iceland. In England scholastic Latin developed and the growing power of Anglo-Saxon vernacular literature showed itself in the saga of *Beowulf*, in religious poetry and in the Anglo-Saxon Chronicle.

The Venerable Bede, 673-735, wrote his *History of the English Church and People*, a major source of information on England between 597 and 731. He drew on wide sources in creating a work of literary and historical value. **In India, the Sanskrit dramatist Bhvabhutti**, *fl.* 730, wrote three outstanding plays, two of which tell the story of **Rama**.

Nearly 49,000 poems survive from China's golden age of poetry, the **T'ang dynasty**.
Tu Fu, 712-70, showed his mastery of imagery in such lines as "Blue is the smoke of war, white the bones of men". Equally famous is **Li Po**, 701-62, who wrote of wine and companionship. **Wang Wei**, 699-759, was a painter and poet of nature. The 8-line *shih* predominated.

Beowulf, the greatest surviving Anglo-Saxon epic poem, dates from the period between 700 and 1000. A vivid narrative of a warrior's struggles against dragons and monsters of the sea, it is based on north European heroic legend, with elements of moral and religious significance probably added by Christian writers after the Angles brought it to England.

A rebirth of European learning took place under **Charlemagne**, *r.* 771-814, who encouraged the copying of old manuscripts. His biography was written by the German monk **Einhard**, 770-840, in personal and political terms. Charlemagne's court at Aachen attracted scholars such as **Alcuin**, *c.* 732-804, an Anglo-Latin writer and cleric with a humanistic outlook.

Art and architecture

After the period of confusion that followed the decline of the Roman Empire, European art again flourished. A Germanic decorative style subordinating realistic representation to stylized patterns is found in jewellery, Viking carving and Celtic manuscripts. In architecture, elements of the Romanesque developed, based on a combination of Roman, Byzantine and Carolingian art, replacing the utilitarian basilicas of the early northern churches with more complex structures using a system of bays, often with vaulted roofs.

Islamic art entered its classical age in the 9th century, the religious ban on figurative art producing a wealth of geometric designs in architectural detail, while Islamic and Christian styles mingled in Spain. Buddhist art flourished in the East, contributing to a mingling of cultural styles as Chinese influence reached Japan, while China itself felt the impact of Indian ideas.

Byzantine icons have survived from Sinai, Constantinople and Rome. The early beginnings of defined painting schools can be seen in the life-size "Enthroned Virgin and Child", *c.* 705, commissioned by Pope John VII.
Chinese Buddhist sculpture combined the traditional linear delicacy with the Indian sense of form, resulting in such superb statues as the seated stone Buddha, 711.

The Iconoclastic age lasted in the Byzantine Empire from 726-843. In order to stop the cult of images and discourage monasticism, all figurative representations, except of the Cross, were either defaced or destroyed.
The earliest Orissan-style temples were built at Bhuvanesvar in east India, 700-800. A hollow terraced pyramid supported a conical beehive-shaped spire.

The Great Mosque at Cordoba was built by Spain's Arab conquerors, 785-990. The naves use elegant star vaulting and the whole was intricately decorated with coloured marbles and precious stones.
The *Book of Kells* was produced in Ireland at the end of the 8th century. It is the finest and most elaborate of early Western illuminated manuscripts.

A Viking earth barrow, *c.* 800, contained the Oseberg ship, as well as a cart, several sledges and numerous small decorated objects. The delicate interwoven wood carvings of figures and abstract motifs are typical of northern art.
Charlemagne's Palace Chapel at Aachen in Germany was consecrated in 805. Local Roman remains and the church of S. Vitale in Ravenna were used as models in an assertion of the continuity of the empire.

Music

The establishment of the Divine Office and Mass by the 9th century encouraged the development of chants more complex than Gregorian chant. At the same time, the Muslim invasions of Western Europe brought schools of singing, lute playing and musical theory which would have a lasting influence on European music over the next five hundred years.

The first compositions by known European composers took the form of tropes, melodic passages added to the liturgy either as new music or as variations on the preceding plainsong melody.

The Arabs in conquering Spain brought with them **the lute** (the first fretted instrument to arrive in Europe), **the rebec** (an ancestor of the violin) and the violin type of bow.

"Ut Queant Laxis" – written *c.* 770 – was an early medieval hymn tune in the then unusual form of six separate phrases, each starting a step or half step higher than the previous one.

Arab music entered its golden age under **Harun ar-Rashid**, *c.* 764-809, whose musical tastes are revealed in *The Arabian Nights*. A style of romantic song flourished in the period.

Science and technology

The rise of Islam transformed the course of European science and philosophy. The Arabs were heirs to the Hellenic Greeks and acknowledged their role as custodians of that culture. Following the Athenian tradition they founded schools for wide-ranging, unprejudiced and objective study, most important of which was the Academy of Science at Baghdad. A great respect for Greek learning, and particularly for Aristotle, may have held them back from even greater discoveries, but some Arab scientists rejected Aristotle, arguing for a more experimental approach to science. With the spread of the Arab Empire, Arabic became the language of science outside the Far East, absorbing elements of Indian astronomy while benefiting to a lesser extent from achievements in China. Many Arabic texts on astronomy, chemistry and mathematics retained their influence until modern science began in Europe with the work of Galileo and Newton.

Mayan science, with its detailed astronomical observations and advanced use of mathematics, reached its peak.

Jabir, or Gebir, *c.* 721-815, the "father of Arabic chemistry" left evidence of a systematic approach to this science, relatively uncluttered by alchemical superstitions. For example, Jabir described the manufacture of nitric acid and how it may be used in extracting silver and gold from their ores or salts.
Gunpowder, probably invented in China in the 8th century, was used initially to make fireworks and only much later in weaponry.

Printing with blocks from which the letters stand out in relief was invented in Japan in or prior to the 8th century.
Bells and organ pipes, made at this time from bronze, indicate an advance in European metalworking.

Arab paper was made in Baghdad for the first time, 793, following the capture of Chinese papermakers during the battle for the city of Samarkand in 751.
Viking ships of the 9th century were clinker-built (using overlapping planks) with square sails, a single steering oar aft and many rowing oars. Their narrow hull shape made them faster than Mediterranean ships.
The Baghdad Academy of Science replaced Jundishapur, Persia, as the centre of scientific learning *c.* 800.

820-50

The Carolingian Empire was divided into three at the Treaty of Verdun in 843.
Scandinavians, having founded Kiev and Novgorod, absorbed Byzantine culture and religion through trading contacts, *c.* 850.
Al-Mamun the Great, *r.* 813-33, set up a House of Knowledge in Baghdad and encouraged the most glorious epoch of the Abbasid dynasty.
The Abbasid capital moved to Samarra in 836.

850-80

Frequent invasions and the weakness of the monarchy gave new power to the provincial nobility in the Carolingian states and in Italy caused a decline in papal authority.
Roman and Byzantine Christianity officially split in 867.
Basil I of Byzantium, *r.* 867-86, attacked the Muslims in Mesopotamia and stimulated a revival of Byzantine civilization.
The Bulgarians were converted to Christianity in 865.

880-910

Urban development in northern Europe, stimulated by long-distance overland trade, was disrupted by Norse raiders *c.* 900.
The Bulgarians warred constantly with Byzantium under **Symeon I**, *r.* 893-927.
The Chola dynasty displaced the Pallavas in India in 888.
The T'ang dynasty in China fell in 907 and was followed by a period of weak imperial authority and constant barbarian invasions.

910-40

Rollo, *c.* 860-932, founded an independent dukedom of Normandy in 911 and was baptized in 912.
Henry I, *r.* 919-36, became the first Saxon king to rule a unified Germany, whereas the French monarchy was weak.
Umayyad culture reached its zenith in Spain under **Abd ar-Rahman III**, *r.* 912-61.
The rise of a military class in Japan resulted in civil strife in the provinces, 935-41.

940-70

Otto I, *r.* 936-73, ended the recurrent Magyar invasions at the battle of Lechfeld in 955 and became the first Saxon Holy Roman Emperor in 962.
The Northern Sung dynasty, founded in 960, brought a more modern humanism to Chinese government, social organization and thought.
A Muslim **Ghaznavid** dynasty grew up in Afghanistan in 962.

970-1000

Hugo Capet, *r.* 987-96, became king of France and reasserted royal authority over the nobility, pope and emperor.
Venice was given trading privileges in the Byzantine Empire in 992.
Viking invasions of Europe reached their peak c. 1000, threatening southern France and Italy.
Basil II of Byzantium, *r.* 976-1025, took Greece from the Bulgarians in 996.

Egbert became king of all England in 828. The Danes sacked London, a small market and trading town, in 836.

Alfred of Wessex organized English opposition to the Danes and won peace by ceding much of eastern England (the Danelaw) to the invaders.

Alfred stimulated the growth of learning and vernacular literature. His son **Edward**, *r.*899-925, reconquered much of the Danelaw.

Athelstan, *r.* 925-39, completed the reconquest, but permitted the survival of Danish customs.
Local government was organized in shires and hundreds.

A monastic decline had set in with the Viking raids but c. 940 a clerical revival occurred, providing a useful network for spreading royal authority.

Danish attacks overran much of England again in the reign of Ethelred II, 979-1016, but the monastic movement brought an artistic revival to Winchester.

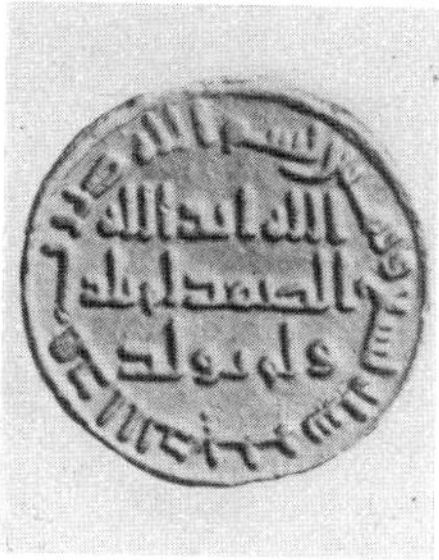
Arab gold dinar, 9th.–10thc.

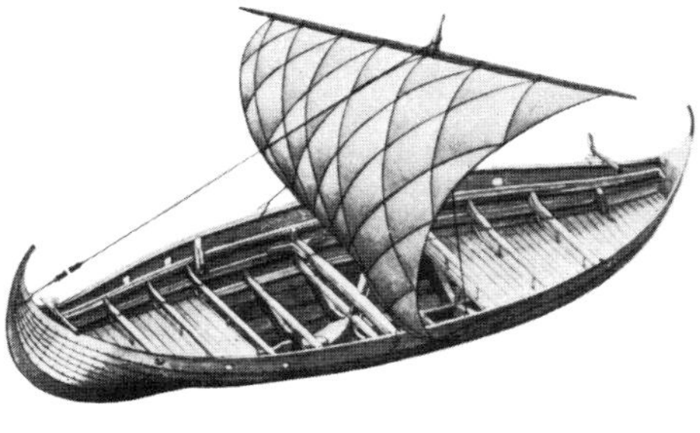
Viking ship

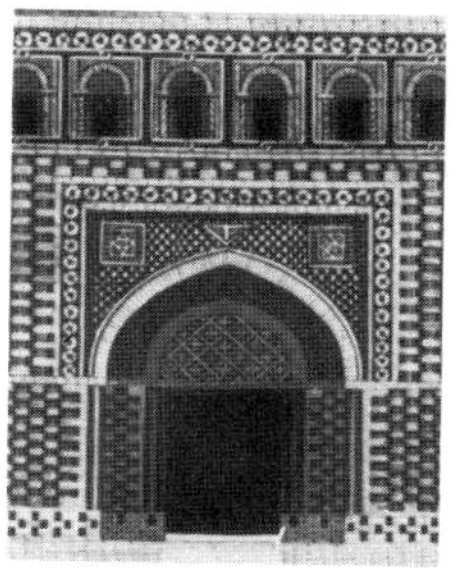
Islamic tomb façade, *c.* 900

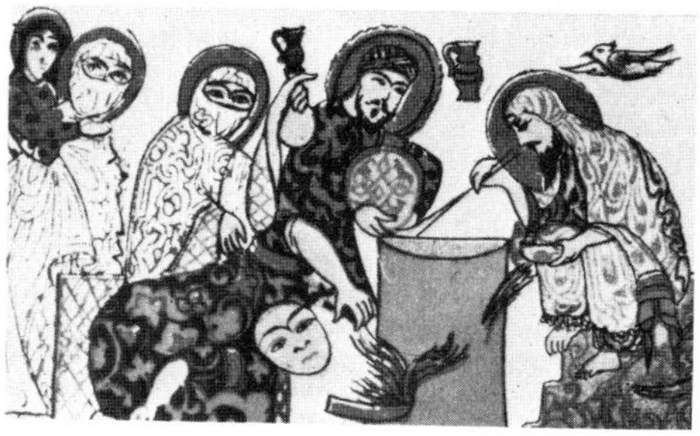
Arab manuscript showing preparation of perfumes

Ahmad Ibn Hanbal, 780-855, within the **Sunni branch of Islam**, founded the most orthodox of the four schools of Islamic law, which holds that the *Koran* as interpreted by the Islamic community contains the answers to all moral questions. In 833 **Hanbal** was imprisoned for refusing to accept **Mutazili** rationalist doctrines. **The Ch'an school**, the precursor of Japanese **Zen Buddhism** developed in China.

The Fourth Council of Constantinople was called in 869-70 **by Basil I**. It deposed Photius, patriarch of Constantinople, who had challenged the Pope's authority in the East, and reinstated Ignatius, thus ending the schism with Rome.

The Photian schism ended in 880 and the Greek Church made peace with the pope.
The Abbey of Cluny in France was founded, 910, marking a revival of the monastic movement. It was here that the **Cluniac reform movement** began, which introduced the notion that the Church hierarchy has a responsibility for clerical discipline and formed the basis of a widespread attack on abuses and corruption in the Church.

Sufism, a mystical literary and philosophical movement within Islam, stressed divine love through the immediate personal union of the soul with God. It developed a reaction against more orthodox interpretations of the *Koran*, and **Al Hallāj** who was crucified in 922 for his teachings became revered as a Sufi martyr.

Sa'adia ben Joseph, head of the **Jewish academy in Babylon**, is known as the father of Jewish philosophy. He defended orthodox Judaism by reaffirming a belief in one God against gnostic dualism. He also repudiated the earlier **Koraite** rejection of the Talmud (the oral tradition of law) in favour of the Torah (the original scriptures that were given to Moses).

The Vikings, whose incursions into the Christian world reached a peak *c.* 1000, worshipped gods similar to those of the Germans. There were two tribes of gods, one of them (**the Aesir**) led by **Odin**, who lived in castle Valhalla where he was joined by heroes killed in battle and assisted by them in a perpetual fight against wolves.

Arabic literature had a strong tradition of lyrical desert poetry, which re-emerged at the peak of the **Abbasid Empire**, 786-861. The lyrics of **Abu Nuwas**, *c.* 762-815, reflected the town life of the caliphates, while Islam influenced the religious poetry of **Abu al-Atahiya**, 748-826. Another poet, **Abu Tamman**, *c.* 807-50, edited the fine Hamasu anthology.

Vernacular literature in both prose and verse was created in Germany and Britain, best shown in the plain narrative style of the *Anglo-Saxon Chronicle*, a history begun during the reign of **Alfred the Great** *c.* 871-99. The Welsh monk **Asser**, *d.* 909, wrote a biography of Alfred. The heroic **Edda** lays began to develop in Iceland after 860.

Classical Japanese literature emerged in the **Heian period**, 794-1192. The *Kokinshu*, 905, was an anthology of short poems with themes of love and nostalgia, showing the flexibility made possible by the phonetic *kana* script. Ladies of the Heian court wrote witty prose, notably the *Pillow Book of Sei Shonagon c.* 1000.

Lyric and elegiac Anglo-Saxon poetry survives in a manuscript known as the *Exeter Book*. This includes individualistic poems such as *The Seafarer* as well as work by an earlier poet, **Cynewulf**, *fl.* 850. *The Dream of the Rood* was a notable poem on the Crucifixion.

A critical note had entered Chinese poetry of the 9th century in the didactic verse of **Po-Chü-i**, 772-846. With the continuing decline of the T'ang dynasty and the unrest of the 10th century, nostalgia suffused the ***tzu* poetry** of **Li Yu**, 937-78. The *tzu* poets adapted the irregular structure and colloquial language of Chinese folk verse, usually sung to a tune.

A revival of Persian poetry using the Arabic alphabet produced the national epic *Shah-Nama* (Book of Kings) by **Firdausi**, 935-1020, who used legend and history in verse that became a model for Arab epics.
An Anglo-Saxon historical poem with a central theme of feudal loyalty was *The Battle of Maldon c.* 995.

The overwhelming size and grandeur of **Mayan religious architecture** can be seen in the **Pyramid of the Sun** at San Juan Teotihuacan, which rose from a base of over 213m (700ft) in diameter to the height of 66m (216ft).
The constructional and geometric skills of Islam are seen in the spiral-ramped minaret of the **Malwiyya Mosque**, begun at Samarra in 848.

The Middle Byzantine age, 867-1025, saw a second flowering of Byzantine art with the energetic redecoration of pre-Iconoclast churches. The mosaic of the "Madonna and Child" in the church of Hagia Sophia dates from 867. Figurative representation became increasingly stylized with the characteristic Byzantine distortion of a face – a small mouth, a long nose and huge, wide-open eyes.
The early German abbey of Corvei was begun in 873.

Phnom Bakheng became the new administrative and religious centre of Cambodia during the Angkor period, 889-1434. The "mountain temple" design has a single base supporting six tower-like structures.

During the Chola period, 907-1053, in India, improved metal-casting techniques enabled notable achievements in figurative images, especially in portraying the complex and balanced poses of the dancing Shiva.

An Imperial Academy of painting was founded in western China during the Ten Kingdoms period, 907-80. **Ching Hao**, 900-60, wrote an essay on landscape painting which stressed the metaphysical implications of the art.

Romanesque architecture after 950 possessed a grandiose quality which derived partly from the use of stone vaults below the roofs and partly from a more unified concept of the church which developedin responseto theneeds of the clergy, monks and pilgrims who used them. Two main plans were adopted – that of an ambulatory with radiating chapels as at St Martin at Tours, 918, and the chapels on either side of the main apse at Cluny Abbey, 981.

Plainsong notation, which originated in Europe in the 9th century, first consisted of marks like accents over syllables to denote a rise and fall in pitch; they did not indicate by how much.

Organum, the practice of singing extra lines of music at intervals of a fourth or a fifth above or below plainsong, appeared in the 9th century. This was primitive polyphony.

Pitch notation was required to communicate to singers the relationship of two parts in an organum. A Flemish monk called **Hucbald**, 840-930, first used letters to denote pitch.

Organs were installed in abbeys and cathedrals of Europe by the 10th century. They were played to support parts of the organum sung by the choir, and followed the sung lines.

The tambura, a 4-stringed lute-shaped instrument, developed in India in the 10th century as a drone accompaniment to melodic instruments. The tambura gave a repeated chord below the melody.

Chinese temple music under the Northern Sung dynasty, 960-1279, involved huge choruses with orchestras of zithers and mouth organs in an organum style of complex sonority.

Spanish metal mines were taken over *c.* 850 by the Moors, who also prepared pure copper by reacting its salts with iron – a primitive forerunner of modern electroplating methods.
Al-Farghani, or Alfraganus, *d.* 850, wrote the *Elements*, a summary of Ptolemaic astronomy studied in Europe until 1600.
Algebra, as a word, first figures as al-jabr, meaning transposition, in a treatise by the Arab mathematician, **Al-Khwarizmi**, *d.c.* 850. The Arabs based their algebra on both Greek and Indian maths.

Al-Rhazi, a physician and encyclopedist, *d.* 920, and **Al-Khindi**, a scientist and philosopher, *d.* 873, were exceptional in objecting to alchemical and Aristotelian dogmas. They sought new concepts of the nature of motion and heat and encouraged the use of experiments to solve scientific problems.
Bardas reorganized the University of Constantinople in 863 for the teaching of science. Soon afterwards the teaching was again suppressed by **Basil II**, *r.* 976-1025.

Cotton and silk manufacture was introduced into Spain and Sicily by the Moors in the 9th and 10th centuries.
Lateen sails, triangular fore-and-aft sails which may have appeared in the eastern Mediterranean in the 2nd century, were brought to the West in the 9th century by the Arabs.

Cordoba, in Spain, reached its height as a centre of Islamic science in the 10th century under **Abd ar-Rahman III**, *r.* 912-61. **Optical lenses** of four kinds were described by **Than Chhiao** in China *c.* 940.

The alembic, an apparatus for distilling chemicals and perfumes, was illustrated in Arabic books of this time. The alembic played an important part in Arab chemistry and strongly influenced its development.

The windmill reached Muslim Spain from Persia in the 10th century.
Mining in Christian Europe centred on the Harz Mountains in the 10th century, where the Saxons mined copper and iron.
Gerbert, a French mathematician, 940-1003, who became Pope Sylvester II, is thought to have introduced the astrolabe and Arabic (Indian) numerals into Europe from Cordoba. He has also been credited with the invention of a mechanical clock in *c.* 996.

1000-1250 The Crusades

Principal events

Europe now began to take the offensive, expanding geographically and economically, her population rising. A new spirit of confidence, epitomized by the cosmopolitanism of Norman culture, brought a series of attacks on the Muslims in Spain and in Syria, where the Crusades provided an aggressive outlet for the military nobility of the flourishing feudal system. The papacy reached the height of its power during the reign of Innocent III, 1198-1216, in spite of continuing opposition to the gradual concentration of its power both from within the Church and from secular rulers.

In the 13th century Genghis Khan set up a Mongol Empire in China, swept across Asia and threatened Europe and North Africa, creating the largest empire ever known and bringing a new peace and unity to Asia in his wake. He did not, however, conquer India, where the various Muslim rulers built up their authority in the north.

1000-25
Basil II, r. 976-1025, briefly restored Byzantine authority in Syria, Crete and south Italy and destroyed the Bulgarian army.
Canute, 994-1035, built a unified Danish Empire comprising England, Norway and Denmark.
Mahmud, the brilliant Muslim ruler of the Ghaznavid Empire in Afghanistan, 997-1030, plundered and annexed the Punjab.
The Chola dynasty of Tamil kings unified southern India and took Ceylon and Bengal, 1001-24.

1025-50
William I, a vassal of the French king, became Duke of Normandy in 1035, organizing Normandy on full feudal and military lines.
The Umayyad dynasty in Spain fell as a result of racial and religious pressures in 1031. The support of **Pope Leo IX**, r. 1049-54, for monastic reform stimulated the concept of papal supremacy over secular rulers.
Yaroslav, r. 1019-54, brought Kievan Russia to its peak (promoting education and building.)

1050-75
Ferdinand of Castile, r. 1035-65, recovered Portugal from the Muslims in 1055.
William of Normandy conquered England in 1066, while another Norman kingdom was established in southern Italy in 1068, finally ousting the Byzantines who also lost Georgia and Armenia to the Seljuk Turks, 1063-72.
The Berber dynasty of Almoravids built a kingdom in Algeria and Morocco, 1054.
The Ottoman Empire began with the capture of Anatolia in 1071.

1075-1100
Pope Gregory VII, r. 1073-85, and Emperor Henry IV, r. 1056-1106, clashed on the investiture issue, over the respective rights of the Holy Roman Emperor and the papacy in appointing bishops.
The Almoravids annexed Moorish Spain, 1086, but **Alfonso VI**, r. 1072-1109, retook Toledo. The First Crusade, 1096-9, captured Jerusalem and set up Frankish kingdoms in the Near East.
Alexius I, Byzantine emperor, r. 1081-1118, recovered some territory in Asia Minor.

National events

Attracted by English wealth the Danes and Normans invaded, turning Anglo-Saxon society into a more flexible, feudal, system. Opposition to the king's authority grew among the barons and found expression in Magna Carta, while industry and town life prospered, offsetting the decline of Anglo-Saxon culture.

The English paid regular tribute to the Danes until **Canute** was generally accepted as king by the English in 1016 and appointed his leading supporters earls.

Edward the Confessor, r. 1042-66, ruled England supported by the Danish earls and brought the Normans to the English court.

William I of Normandy conquered England in 1066, bringing Cluniac reform with him and introducing a fully feudal society with grants of land for military service.

The Domesday Book, 1086, recorded land use and tenure in full for taxation purposes. **William I**, r. 1066-87, built stone castles and fostered the growth of towns.

Bayeux tapestry: William the Conqueror and companions

The Great Church at Cluny

Crusading knights: Hospitaller, Teutonic Knight, Templar

Early Gothic: Laon Cathedral

Religion and philosophy

In the emerging struggle for power between the Church and the rulers of the new European states, the papacy succeeded in asserting its right to judge the morality of secular political actions at the same time as it took the lead in the reform movements within the Church.

In both the Muslim and Christian world, there was a revival of philosophy and a return to the Greeks, especially Aristotle. This was essential to the rise of scholasticism, an important philosophical movement within the Catholic tradition, based on the notion that dialectical reasoning as well as faith and revelation could illuminate the mysteries of Christian belief.

The Mahayana form of Buddhism, which allowed lay salvation, spread from China to Japan. There it evolved into a popular devotional cult centred on ritual chanting, in sharp contrast to the elitist monasticism of Zen, which was also emerging within Japanese Buddhism at this time.

Saint Symeon (Simon), c. 949-1022, "The New Theologian", developed the orthodoxy within the Greek Church on meditation and revelation in a mystical direction.
Pope Benedict VIII, r. 1012-24, promulgated a decree against clerical marriage and concubinage at the Council of Pavia in 1018.

Avicenna, 980-1037, also known as Ibn Sina, was an eclectic Muslim thinker and physician. He wrote The Book of Healing, a monumental encyclopedia elaborating mainly Aristotelian theories of philosophy and medicine.
Buddhism became firmly established in Tibet in 1038.
Pope Leo IX, r. 1049-54, issued stern decrees against simony (the purchase of ecclesiastical office), thereby identifying the papacy with Cluniac reform.

The schism between the papacy and the Greek Christian Church was fixed in 1054, when **Pope Leo IX** closed Greek churches in southern Italy for unorthodox practices, such as the use of leavened bread in the Mass.
Berengar of Tours, 999-1088, argued that reason could justify the contravention of authority. He denied the doctrine later known as transubstantiation, but was finally forced to recant, 1059.

The Dictatus Papae of 1075 by **Pope Gregory VII** (Hildebrand), r. 1073-85, decreed that popes were able to depose emperors.
Roscelinus, c. 1050-1120, was an early proponent of the scholastic tenet of nominalism, holding that the qualities we ascribe to objects, like colour, do not exist in reality but are just the product of thought or language. This led him to deny the unity of the Trinity, a position he was forced to recant at Rheims in 1092.

Literature

Of the European literatures, French was the most influential in the development of new literary forms in the 11th and 12th centuries, producing the chanson de geste in written form, the Arthurian romance tradition and the lyrical vernacular poetry of the troubadours, all of which soon became international. The common heritage of warfare against the Muslims in Spain was the subject of the French Chanson de Roland and also of the great Spanish epic El Cid.

In the Near East the solitary genius of Omar Khayyám flourished, while in Japan the late Heian period saw the emergence of underivative Japanese styles including the literary diary of which The Tale of Genji, written by a lady at court, is the best-known example.

The greatest of all Japanese novels, The Tale of Genji, was written by the court lady **Murasaki Shikibu**, 978-c. 1031; it is an elaborate, realistic tale of court life.

The Sung period, 960-1279, in China was mostly an age of prose. Its great writers were **Ou-Yang Hsiu**, 1007-72, and **Sung Chi**, 998-1061, who collaborated on a Confucian history, **Su Shih**, 1037-1101, widened the subject matter of tzu (song form) poetry and introduced vernacular words, thus contributing to Yan "drama", which resembled opera.

In Persia the scientist, mathematician and poet **Omar Khayyám**, c. 1048-1122, wrote The Rubaiyat (quatrains), which expresses a rational, pessimistic and hedonistic philosophy – ideas then unacceptable to orthodox Islam. It is not certain how many of the almost 500 quatrains were written by him.

The chansons de geste, epic poems consisting of a series of stanzas using a single rhyme and celebrating the history of the Age of Charlemagne, were sung by travelling musicians. The earliest written example, The Song of Roland, dates from c. 1100.

Art and architecture

The transition from Romanesque to Gothic architecture involved a structural and visual change in the aisled church, beginning in France and England. Separate inventions – stronger pointed arches at Cluny and rib vaults at Lessay – were then combined as in the vaults of Durham, which were supported by buttresses beneath the gallery roofs. External flying buttresses, first used at Notre Dame, allowed Gothic architecture to develop. With these the building became an independent frame in which larger windows were inserted.

Bar tracery produced the lovely patterns of French 13th-century architecture which spread throughout Europe, reaching Cologne in 1248 and England with the additions to Westminster Abbey in 1245.

Castles developed from the primitive motte-and-bailey to the sophisticated designs of Crusaders' permanent garrisons, such as the Krak des Chevaliers (first fortified 1110) in Syria.

Ottonian architecture in Germany took its cue from **St. Michael's Hildesheim** (designed apparently by Bishop Bernward) with its unvaulted double choirs and arcades of square piers alternating with round, short columns.
Dravidian architecture reached a peak of sophistication under the Chola period in India. The great **Temple of Shiva** at Tanjore with its pyramid and dome-shaped finial profoundly influenced South-East Asian architecture.

The Muslims raided west India between 1000 and 1026, defacing many of the temples. This led to the building of the most important **Gujarat temples** with characteristic colonnaded halls and "pyramids" of massed cupolas.
Wulfic's Rotunda of St. Augustine's Abbey, Canterbury, 1049, marked the end of English architectural isolation, both this and Edward the Confessor's original **Westminster Abbey**, 1055-65, used Continental models.

The Byodo-in Temple, 1053, in Japan has the brilliance and delicacy of ornament typical of the **Fujiwara** culture. The Phoenix Hall houses a wooden Amida Buddha by the contemporary sculptor **Yocho**.

Durham Cathedral, unlike all previous church architecture, was vaulted throughout. It used a new and more stable combination of round and pointed arches and had buttressed arches beneath the gallery roof.
The Bayeux tapestry, c. 1080, whose continuous narrative describes the Norman victory over the English and the events preceding it, was sewn to adorn Bayeux Cathedral, though it was probably made in Canterbury.

Music

Polyphony developed further in both the religious and secular music of Europe in the Middle Ages, having long existed in folk music, particularly in Britain. At the same time set musical forms, like the ballade, virelai and rondeau, evolved from songs and dances. Both developments reflected the medieval delight in uniting contrasting elements in a consistent whole.

A cantus firmus was used as a fixed melody about which a line of embellishment could be worked. In this could be seen the origins of counterpoint (two or more related tunes played together).

Guido d'Arezzo, c. 997-c. 1050, advocated the use of the staff (a grid of horizontal lines) in notation and made simple rules for defining relative pitch of notes, later revived as the tonic solfa.

Troubadours appeared in Provence late in the 11th century, singing to their own harp accompaniment. They set stanzas of poems to music, producing complete compositions in new forms.

Science and technology

Arabic science and philosophy reached its height in the 11th century with the work of such major figures as Avicenna, al-Biruni and Alhazen in the Middle East and Averrhões in Spain, but soon afterwards it declined. It was at this time, in the early 12th century, that the influence of Arab science began to show in Europe with the introduction of Arabic numerals. These were used in the already powerful business world of Italy which, unlike China and the Arab lands, was to develop an economy based on money. Other signs of the power Europe was to achieve were rapid growth in the silk and glass industries in the south and the use of coal and the beginnings of cast iron manufacture in the north. This technology owed a heavy debt to Chinese expertise, brought to Europe at this stage via the Arab world but later derived directly from China, which would trade extensively with Italy after the visits of **Marco Polo**.

Avicenna (Ibn Sina), 980-1037, and al-Biruni, 943-1048, two of the greatest Arab encyclopedists of science, both lived in Persia in the early 11th century. **Avicenna** wrote on astronomy, physics and medicine, which he also practised, and his theory and methods were taught in Europe for the next 700 years. **Al-Biruni** wrote on mathematics, astronomy and astrology, geography and history, and was the first botanist to analyse the structure of flowers by methods important to plant classification.

Illustrated botanical texts were published in China in the 11th century. These had medical as well as botanical importance since the pharmacology of drugs obtained from plants was a highly advanced science in China.
Alhazen, or Ibn al-Haitam, c. 965-1038, wrote Optical Thesaurus, the first important work on dioptrics (the optics of the eye), which influenced the work of **Roger Bacon**, the 13th-century English scholar.

Mould boards, curved boards on ploughs, which overturn the ploughed earth and thereby improve soil structure and aeration, came to be used in Europe from the 11th century onwards, although they had been known in China for 1,000 years.
Omar Khayyám, c. 1048-c. 1122, a Persian poet and mathematician, solved cubic equations by geometric methods c. 1075, and worked at the sultan's court in Merv reforming the Muslim calendar.

Chinese medical texts written in the 11th century include one of a qualifying examination for doctors and enlarged editions of medical pharmacopoeias.
Alcohol was probably first distilled from wine at Salerno in the 12th century. Although fully able to do so, the Arabs had not made alcohol because it was prohibited by the Koran.
Indian commentators on science in the 11th and 12th centuries described the medical uses of yoga meditational techniques.

1100-25

The Seljuk Empire gradually split into separate regencies, 1100-25. **The Concordat of Worms**, 1122, brought a compromise to the investiture controversy. **Louis VI of France**, *r.* 1108-37, granted urban charters to many French towns, which like towns throughout Europe were growing. In **Manchuria** the Jurchen tribes overthrew the Khitai with Chinese assistance, 1116, and destroyed the Chinese Sung dynasty, 1136. **The Khmer Empire** in Cambodia reached its peak, *c.* 1100

1125-50

Alfonso VII, *r.* 1126-57, resumed the conquest of Spain while the Muslim dynasties of Spain and North Africa fought each other. After the fall of the **Frankish kingdom of Edessa** to the Seljuk Turks, 1144, the disorganized Second Crusade failed to halt the Turkish advance, 1147-9. **The communal movement** of north European towns claiming independence from royal authority reached Rouen, 1145. **Kiev** declined after the death of **Vladimir Monomach**, 1125.

1150-75

Henry II of England, *r.* 1154-89, added Aquitaine and Gascony to the Angevin Empire in France, and heightened the conflict of secular and papal authority by having **Becket**, Archbishop of Canterbury, murdered in 1170. **Saladin**, *r.* 1169-93, united the disparate Muslim tribes in Egypt and Syria under the Egyptian **Ayyubid dynasty**. **Civil war in Japan** between the local clans, 1156-81, accelerated the decline of imperial authority over the feudal magnates.

1175-1200

The Seljuks took Anatolia, 1176, and Saladin took Jerusalem, 1187, causing the Third Crusade, 1189-91, which rewon the city. **Muhammad of Ghur**, *r.* 1176-1206, took Delhi and Bihar in India. A Muslim kingdom was set up at Delhi on his death. **Yoritomo's** defeat of the Taira clan, 1185, in Japan inaugurated the Kamabura period. **Emperor Frederick I** (Barbarossa), *r.* 1152-90, was defeated by the league of Lombard towns in his invasion of Italy, 1176.

1200-25

Venice, which sought control of the eastern Mediterranean, persuaded the **Fourth Crusade**, 1202-4, to take Constantinople, after which the Latin Empire of the East was set up, 1204-61. **King John of England**, *r.* 1199-1216, lost his French lands, 1204. **Alfonso VIII**, *r.* 1170-1214, defeated the Almohads, 1212, who then declined in Africa and Spain. **The Mongols**, under **Genghis Khan**, *r.* 1206-27, had invaded China, Persia and southern Russia by 1225.

1225-50

Assimilation of native ideas by the ruling minority created a fusion of Muslim and Hindu cultures in northern India by 1230. **The Mongols** annexed the Chin Empire in China, 1234, overran eastern Europe and set up the Tatar state of the Golden Horde on the lower Volga in 1242. **Alexander Nevsky**, *r.* 1236-63, prince of Novgorod, defeated the Teutonic Knights, 1242. **Jerusalem** was finally lost to the Turks in 1244 and the Seventh Crusade, 1248-50, achieved little.

Henry I, *r.* 1100-35, stimulated economic development and created the first royal administration. Saxon and Norman integration was complete by 1125.

The civil war for the succession between **Stephen**, *r.* 1135-54, and **Matilda** reduced England to anarchy as the barons fought among themselves.

Henry II, *r.* 1154-89, reaffirmed royal authority over the barons by appointing lesser knights as sheriffs, 1170, but failed to assert his authority over the clergy.

Henry II reformed the common law, introducing the jury system, but **Richard I**, *r.* 1189-99, failed to develop Henry's policies.

John, *r.* 1199-1216, flagrantly ignored baronial interests and was forced to sign Magna Carta in 1215, subjecting the monarchy to the rule of law.

National sentiment developed among the nobility and townsmen in opposition to the pro-papal attitudes and expensive foreign policies of **Henry III**, *r.* 1216-72.

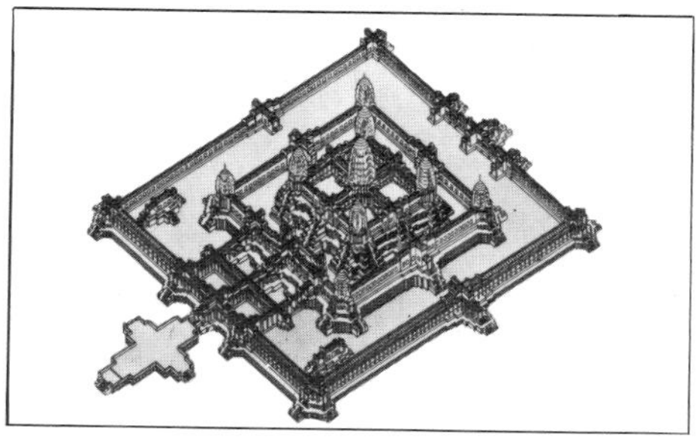

Classical Khmer architecture: temple at Angkor Wat

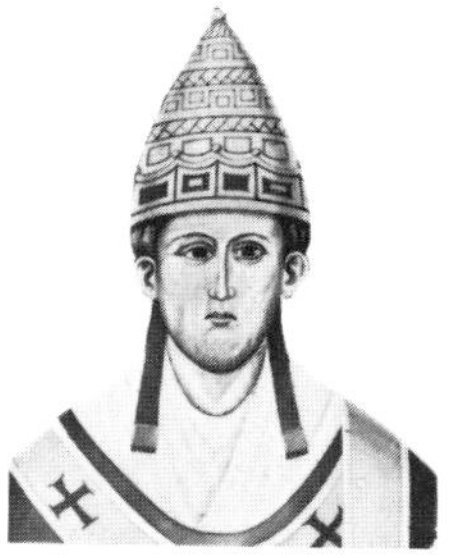

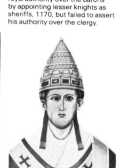

Pope Innocent III

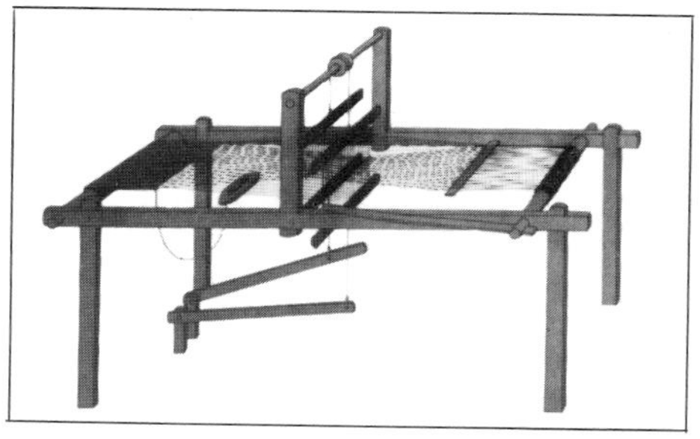

European treadle loom, 13thc.

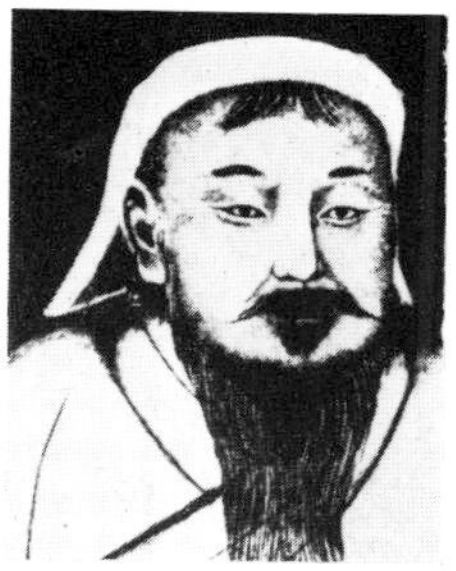

Genghis Khan

St. Anselm, Archbishop of Canterbury 1093-1109, one of the first scholastic philosophers, sought to establish the existence of God by reason, arguing that God must necessarily exist since He is perfect and it is more perfect to exist than not. **Peter Abelard**, 1079-1142, French theologian and philosopher, advocated reason as a source of truth. His nominalist ideas led to his condemnation at the Synod of Soissons in 1121 for his views on the Trinity.

St. Bernard, 1090-1153, Cistercian Abbot of Clairvaux in France, strongly encouraged mysticism and contemplation in opposition to the scholastic rationalism prevalent in Western Christendom. **Honen**, 1133-1212, founder of the Pure Land Sect in Japan, joined the **Tendai Sect** in 1148. He later proclaimed that the only way to salvation was **Nembutsu** (calling the name of the Amida Buddha). **Gratian**, a Benedictine monk, compiled the *Decretum Gratiani*, a collection of canon law *c.* 1140.

Averrhöes, 1126-98, the Islamic scholar, began writing his influential commentaries on Aristotle in 1169. He also argued that reason could serve to establish religious truths. **The Waldenses**, founded by **Peter Waldo** in 1170 in southern France, rejected the licence of the official Church and adopted a simple way of life, electing their own priests.

Zen Buddhism was introduced into Japan in 1191 by the monk **Eisai**, 1141-1215. Zen stressed personal instruction by a master, rather than the study of scriptures, as the way to enlightenment. His techniques included sudden physical shocks and meditation on paradoxical statements. **Neo-Confucianism** emerged in the 12th century in China. **Chu-Hsi**, 1130-1200, one of its most influential exponents, completed his commentaries, *The Four Books*, in 1189.

Islam became firmly entrenched in India with the establishment of the Delhi kingdom in 1206. **The True Pure Land Sect** (Jodo Shin) was founded in Japan in 1224 by **Shinran**, 1173-1262. For him, salvation came only through faith and the Buddha's grace. Because it rejected monasticism and ascetic practices, this became, and still is, the largest Buddhist sect in Japan.

The Franciscan and Dominican orders of friars, devoted to the care of the poor and the sick, spread quickly, 1225-30. **Nicherin**, 1222-82, a Japanese Buddhist monk, added a highly nationalist element to Japanese Buddhism. By 1250 he had proclaimed the *Lotus Sutra*, the central writing of the Mahayana tradition, as the supreme Buddhist scripture. He desired to end Buddhist sectarianism in order to regenerate and unify Japan against the threat of Chinese invasion.

A miracle play was performed at Dunstable *c.* 1100. In such Latin plays, performed in churches and drawing on both the scriptures and the lives of saints, lie the roots of the medieval drama as later practised.

The first bardic period of Hindi literature began in India. Among the important early epics is the *Prithvi Rah Raso*.

The lyrical poetry of the troubadours grew up in 12th century France. Written in a Provençal dialect and sung to music, it lauded a concept of love as a knightly duty then fashionable in the southern French courts. ***Mystère d'Adam*** *c.* 1175, marks a major development towards popular drama; it is in French, not Latin, uses the vernacular and was later played outdoors.

The long Middle High German epic *Nibelungenlied*, which has survived in thirteenth-century manuscripts, was written by an unknown Austrian; its hero is Siegfried, it has connections with Scandinavian legends and has influenced many writers and composers, notably Wagner. **Chrétièn de Troyes**, *fl.* 1165-80, developed the prose romance in *Cliges and Lancelot*.

The German minnesinger tradition, parallel to the Provençal courtly poetry, is exemplified in the songs of **Walther von der Vogelweide**, *c.* 1170-1230, a wanderer and a beggar. He discarded the older strict form, as did his contemporary, **Wolfram von Eschenbach**, *c.* 1170–*c.* 1220, author of the great German romance, the grail-story *Parzival*, *c.* 1210.

The Icelandic Classical period culminated in the work of **Snorri Sturluson**, 1179-1241, who wrote the *Edda*, a handbook which set out the Icelandic myths and the types of poetic diction used in old Norse poetry. **Literature in Japan** declined with the Kamakura period, 1192-1333, but war tales, especially the *Heike monogatari*, *c.* 1215-50, became an established form.

Chinese landscape painting reached its zenith under the patronage of **Hui Tsung**, *r.* 1101-25. Great care was lavished on tiny details in an attempt to reveal the inner life of the objects shown. Li Chieh's treatise on **Sung architecture** is a blend of learning and practical instruction on survey geometry, uses of building materials and decorations, and includes recipes for coloured glazes for floor and roof tiles.

At Autun Cathedral in France Gislebertus sculpted all the nave capitals, *c.* 1125-35, the west door tympanum depicting the Apocalypse. **Abbot Suger** rebuilt the choir and westwork of **St. Denis**, near Paris, *c.* 1140-44. The first example of mature Gothic, its slender pillars and pointed arches allowed big lancet windows with stained glass in the apse chapels; statues adorned the porch.

A change in the design of Cistercian monasteries followed the death of **St Bernard of Clairvaux** in 1153. After the harsh simplicity of **Fontenay**, built in 1139, **Clairvaux III**, 1153, and **Pontigny** apse, *c.* 1185, are richer and more imposing. **External flying buttresses**, first used at Notre Dame, Paris, *c.* 1163, enabled clerestory as well as ground-floor windows to be treated as a frame, with a thin web of stone and glass between.

In the second Angkor period the Cambodian capital of **Angkor Thom** was rebuilt, 1181-95, followed by temples in **Angkor Wat**. The ashlar façades were deeply carved to resemble gigantic faces. **High Gothic architects** used the new construction techniques to varied aesthetic ends and made structure itself ornamental. Most important were **Chartres** and **Bourges**, 1195, **Canterbury** 1174, and **Lincoln**, 1192.

The Chimu subdued the Peruvian Tiahuanaco culture *c.* 1200. Adobe buildings at **Chan Chan**, the capital, have trapezoidal doors and intricate geometrical surface designs. Pottery played an important part in decoration. **An international style** known as **Rayonnant Gothic** was born in 1220 at Amiens Cathedral. **The massive "Black Pagoda"**, a Jain temple of the sun, was begun at Kanarak in Orissa *c.* 1200. Only the base, carved with erotic reliefs, survives.

The Sainte Chapelle, Paris, was built, 1240-8, as **St Louis'** palace chapel and to house the Crown of Thorns relic. The walls are like continuous sheets of glass and made the design a symbol of prestige.

Three- and four-part polyphony was composed round a *cantus firmus*, but early in the 12th century two parts were more usual. The harmony often used sounds dissonant to modern ears.

Trouvères in northern France developed on similar lines to the troubadours, producing *formes fixes* (set structures of contrasting phrases), among them the ballade, virelai and rondeau.

The *conductus* developed as processional music in a chordal style late in the 12th century. Composed for voices or instruments, it was based on original themes rather than plainsong.

The minnesinger created a tradition of German song inspired by the art of the troubadours. Notre Dame choir school in Paris flourished under the great masters Léonin and Pérotin.

Muslim rule in northern India after 1206, strengthened secular music and featured the use of the **sitar**. Southern Indian music remained restrained and classical, favouring the **vina**.

"Sumer is icumen in", an English song of astonishing form, written *c.* 1240, was the first recorded canon. It is a four-part round over a two-part repeated pattern in the bass parts.

Silk manufacture bagan in south Italy in the 12th century as a result of Arab influence and by the 13th century water-powered silk mills were in operation. **Stained-glass windows** of the early 12th century demonstrate the high-level glass technology found in Europe. Glass was coloured by the addition of particular metal salts: those of copper for green; copper or gold chloride for red; iron or silver for yellow and cobalt for blue.

Coal was used at Liège for iron-smelting after about 1150.

Averrhöes, or Ibn Rushd, 1126-98, the leader of Arabic science and the major encyclopedist of his day, worked in Cordoba. His scientific writings maintained the authority of Aristotle. **Maimonides**, 1135-1204, the greatest medieval Jewish thinker, worked at the court of Saladin and wrote on medicine, theology and philosophy. He described diseases and cures in a way which we now recognize as that of psychosomatic medicine.

Stückofen, the precursors of blast furnaces, operated in Styria, central Europe, as early as the 13th century. These furnaces, 3 or more metres high, burned charcoal, which produced a reducing atmosphere suitable for iron- and steel-making. **Cast iron**, made by melting and moulding the metal, was first produced in Europe in the 13th century. It was made possible by higher furnace temperatures. **Old London Bridge** and the Avignon bridge were built *c.* 1175.

Leonardo Fibonacci, the greatest medieval mathematician, wrote the first Western textbook on algebra in about 1200. **Universities** founded in Europe in the early 13th century included those of Paris and Oxford. **Roger Bacon**, *c.* 1214-92, was one of the few important experimenters in medieval English science. He had an extensive knowledge of astronomy and medicine and employed lenses to correct defective vision.

Frederick II, Holy Roman emperor, *r.* 1212-50, a serious student of natural science, wrote a treatise on falconry that is a model of natural history for its combined learning and observation. **Stern-mounted rudders** were first fitted to European ships at this time, although the Chinese had invented them centuries earlier. **Navigational charts** came to be first used by Western sailors in the 13th century.

1250-1400 The Mongols unify Asia

Principal events

In Europe the crusading mentality gave way to a more flexible, commercial society, epitomized by the rise of the Italian city states, the Hanseatic League of trading towns in the north and merchant and trade guilds. Kings came to depend more closely on popular support and called parliaments in which they consulted a wider section of the population, including townsmen. After 1300, the growth of population and prosperity gave way to famine and plague which reduced many towns and introduced a period of retrenchment. The population decline, however, increased the bargaining power of the labourers, later enabling the peasantry to escape serfdom.

The Mongols brought prosperity and trade to much of Asia, but their empire was fragmented by religious conflict. Their benevolent rule in China was replaced by the native Ming dynasty in 1368.

1250-65

Gold currencies were introduced in Florence and Genoa, 1252, and bankers, such as the Bardi in Florence, flourished.
Chinese silk became available in Europe in 1257 along the silk route opened up as a result of Mongol expansion.
Kublai Khan, *r.* 1260-94, set up the Yüan dynasty in China.
In England, de Montfort's Parliaments, 1264-5, reflected the improved status of townsmen and lesser knights.

1265-80

Louis IX of France, *r.* 1226-70, the most powerful and respected monarch in Europe, died in Tunis on the ninth and last Crusade.
Mongol peaceful rule in Asia inspired a Venetian trader, **Marco Polo**, 1254-1324, to visit China in 1271-95.
Rudolf of Hapsburg, from an old Swabian family, was elected King of Germany in 1273 and thus became founder of the Hapsburg dynasty.

1280-95

The defeat of the Mongol invasions of Japan in 1274 and 1281 strengthened the Japanese military clans.
The Danish Magna Carta, 1282, united royal power.
Tripoli and Acre fell to the Mamelukes, 1289-91.
The Yüan dynasty in China restored canals and built roads.
Osman, 1259-1326, founded the Turkish Ottoman principality in Bithynia in 1290.
The western Mongols rejected the Khan's authority in 1295.

1295-1310

Venice was governed by a narrow oligarchy of merchants who consolidated their power by crushing the popular and patrician revolts of 1300 and 1310.
A conflict over papal authority led **Philip IV**, *r.* 1285-1314, of France to call one of the first Estates-General in 1302 to appeal for national support.
Military anarchy in Italy drove the papacy to Avignon in 1309.
The African Empire of Mali, based in Sudan, flourished.

National events

Although Parliament was first called to voice responsible opinion in support of Simon de Montfort, *c.* 1208-65, its powers gradually grew as the monarchy needed extra taxation for war. In spite of the ravages of the plague England became more prosperous as a result of war with France.

Opposition to Henry III led to civil war, 1264-5, in which de Montfort, the rebel leader, called a broadly based Parliament, including townsmen.

Edward I, *r.* 1272-1307, reformed the common law in the Statute of Westminster, 1275, and encouraged the development of the woollen trade in London.

Edward conquered Wales by 1284 and campaigned in Scotland without lasting success. He expelled the Jews in 1290.

Edward summoned the "Model Parliament", which provided a convenient means of carrying royal decisions to the country.

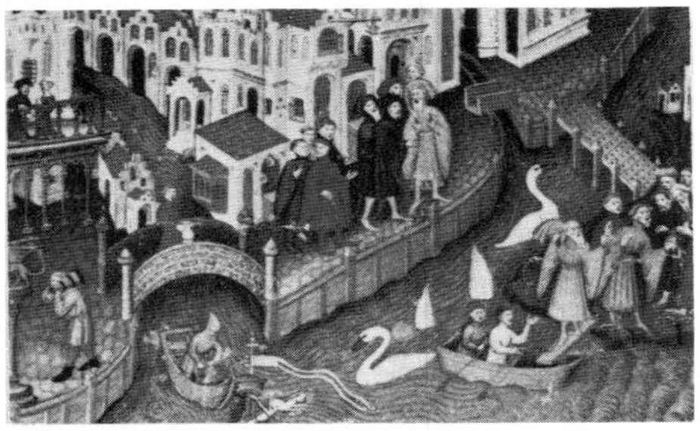

The harbour of Venice, late 12thc.

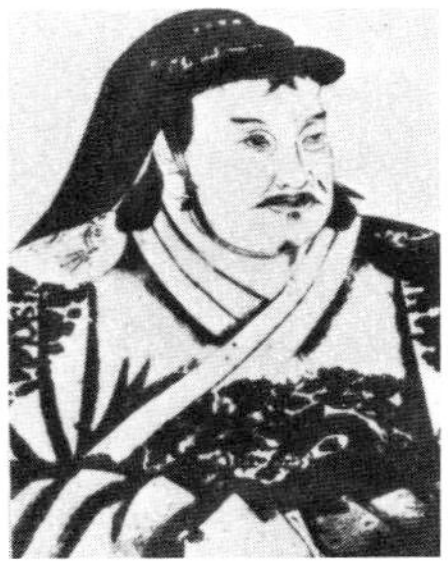

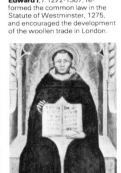

Kublai Khan

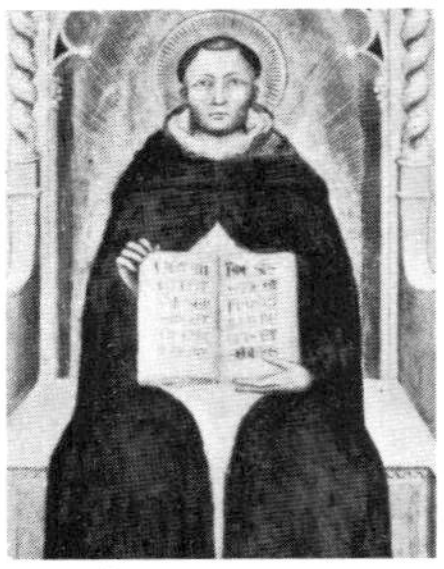

St Thomas Aquinas

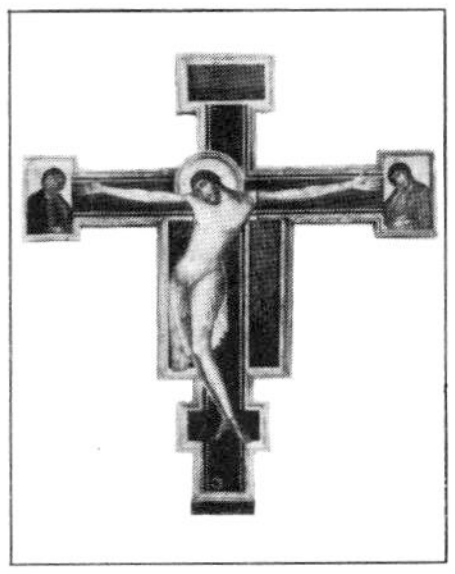

"S. Croce Crucifix", 1283

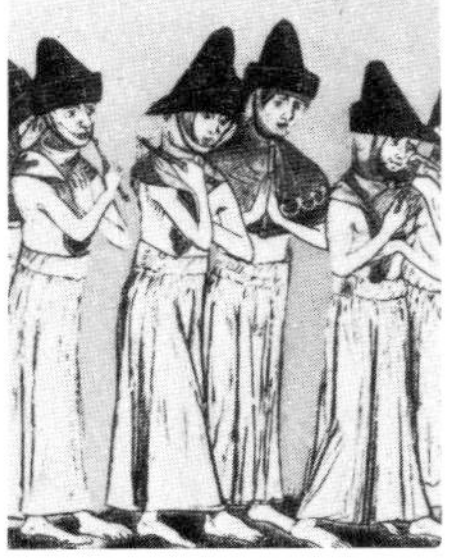

The Black Death: flagellants

Religion and philosophy

The scholastic tradition in European philosophy, which sought to strengthen religious faith with the help of reason, culminated in the work of Thomas Aquinas, Duns Scotus and William of Ockham, who all looked to the work of Plato and Aristotle. Ockham, however, was also an empiricist, disputing the self-evidence of the principles of Aristotelian logic, like the final cause, and of Christian teachings, like the existence of God.

The reign of Pope Boniface VIII marked the summit of papal power. Following the move to Avignon, 1305, the power of the papacy declined.

The vernacular writings of Langland and Wycliffe, both Englishmen, foreshadowed the Reformation in condemning priestly corruption while advocating spiritual as well as social equality.

The Islamic world produced its most original thinker, the scholar Ibn Khaldun.

St Thomas Aquinas, *c.* 1225-74, the greatest scholastic philosopher, stated his belief in the power of reason in *Summa contra Gentiles*, 1264, in which he presented arguments designed to convince the non-believer of the power and truth displayed in Christianity. This work, together with his *Summa Theologica*, 1266-73, was influential in giving a strong Aristotelian basis to Catholic philosophy.

Roger Bacon, *c.* 1214-92, a Franciscan philosopher interested in science, magic and mathematics, in 1272 wrote *Compendium Studii Philosophiae*, attacking clerical influence. He was unusual in valuing experiment as a worthwhile and useful source of knowledge.
Madhava, 1197-1276, an Indian thinker whose life was remarkably similar to that of Christ, denied the Sankara doctrine of the illusory nature of the world.

Duns Scotus, *c.* 1265-1308, an English scholastic philosopher and a Fransciscan monk, drew on the work of Plato. He was a realist, denying the nominalist view that the qualities we perceive, such as the colour green, are merely products of thought and do not exist in the real world. He rejected the idea of predestination and inclined to the Pelagian view that man can alter his fate by his conduct.

Pope Boniface VIII instituted the Jubilee year of 1300, when plenary indulgence was granted to those visiting Rome.
The papacy moved to Avignon in 1309, under Clement V, where it remained for nearly 70 years. The French influence over the papacy in this period marked the beginning of the decline in its temporal power.

Literature 1250-1400

Italian writers of the early Renaissance, particularly Dante, Petrarch and Boccaccio, drew on the passionate and poetical faith of St Francis, the philosophical theology of Aquinas and the new lyricism of the French troubadours to forge a brilliant literature. Honoured by the princes of the Italian states, they explored in allegories, love poems and philosophical writings the contradictions between classical humanism and Christian ideals. Italian vernacular poetry and prose was the latest to emerge among the Romance languages, but the work of writers such as Boccaccio opened the way to the vivid portrayal of contemporary life that marked the work of Chaucer in England and became characteristic of Renaissance literature.

Laudi (praises to God) became a common form of religious song in Italy during the period following the death of **St Francis**, 1226. The Franciscan friar **Jacapone da Todi**, *c.* 1230-1306, was the greatest poet of this style. Written in an Umbrian dialect, his ardent mystical laudi counterposed a love of God with a harsh awareness of the secular world.

The ghazal – a 7th-century form of Arabic love poetry celebrating mystical and worldly love in mono-rhymed verses without logical sequence – was developed by Persian Sufi mystics, notably **Rumi**, 1207-73, in his *Divan*.
Roman de la Rose, a French poem of 22,000 lines in 8-syllable couplets, completed by 1280, included an elaborate allegory on the psychology of love.

One of the major figures in Catalan literature, Raymon Lully, 1232-1315, was a poet, mystic, philosopher and theologian who produced 243 works. *Blanquerna*, 1289, is notable as a philosophical study of Utopia and the forerunner of the novel.
The exploits of the 13th-century Tannhäuser, a knight and poet of the **Minnesinger** school, were described in legend and ballad.

Marco Polo recorded in a Genoese prison, *c.* 1298, the story of his travels in China and Asia. His account was the basis of Western knowledge of China.
Heinrich von Meissen, *c.* 1250-1318, was a representative of the school of middle-class poets who succeeded the knightly Minnesingers, adapting Minnesinger traditions to poems dealing with theology and philosophy.

Art and architecture

Art and architecture in Europe between 1250 and 1400 show their initial indebtedness to France, a nation that had achieved success both politically and artistically. German patrons sent for architects to build "in the French manner", while Italian masons grafted details from Rayonnant Gothic architecture onto buildings that were essentially the piled masses of Italian Romanesque. Everywhere much time, skill and money were lavished to make buildings bigger and more ornate and objects more intricate and naturalistic, both in religious and secular spheres. The period saw the "birth" of Italian painting in the works of Cimabue, Giotto and Duccio and the beginnings of modern sculpture with Nicola Pisano and his son, Giovanni.

Cultures in South-East Asia became more distinct, yet borrowed freely from each other. Comparatively little survives from India at this time.

The French Rayonnant Gothic style, characterized by circular windows with wheel tracery, was exemplified on the western façade of **Rheims Cathedral**, begun in 1255, and also in Spain at **Leon Cathedral**, 1255-1303. The choir of **Old St Paul's**, London, begun in 1256, also incorporated French features.
Nicola Pisano, *c.* 1225-*c.* 84, the greatest sculptor of his generation, completed a font for the baptistery at Pisa, 1260.

The influence of the Four Great Masters on landscape painting of the Mongol **Yüan dynasty** in China, 1264-1368, brought a greater robustness and broader colour spectrum to this important art form.
The Benin (Nigeria) bronze-casters' "lost wax" technique was developed in the late 13th century, introduced by tradition from Ife.

Cimabue, *c.* 1240-1302, one of the first great Italian painters, worked towards the realistic depiction of physical form and human emotion, as in his "Sta Croce crucifix", 1283.
In England one of the earliest lierne vaults, characterized by small ribs running from one major rib to another, was built at **Pershore Abbey**, Worcs., 1288. Lierne vaults became a purely decorative, typically English device.

The frescoes by Giotto, 1266-1337, in the Arena (Scrovegni) chapel in Padua, painted *c.* 1305, show solidity, naturalistic detail and perspective, and represent the turning point in Italian painting. By contrast **Duccio di Buoninsegna**, *c.* 1260-1315, the first great Sienese painter, summed up the mastery of the Byzantine tradition in his "Maesta" for Siena Cathedral, commissioned in 1308.

Music

Late medieval European music was increasingly complex and brilliant, requiring more exact systems of notation. The beginning of the Renaissance produced an easing of the Church's nearly exclusive hold on serious composition and, as patrons began to sponsor secular music, compositions began to show signs of greater individuality and independence.

The motet, a polyphonic form with different words sung simultaneously in the various parts, in which the *cantus firmus* was reduced to a repeated rhythmic phrase, developed after *c.* 1250.

Notation was developed for shorter time values as intricate parts were introduced to overlie lines of long notes. The new notation helped to clarify the time relationship between the parts.

In China, during the Yüan dynasty, 1264-1368, music was encouraged. Opera developed in the theatre with recitative (musically declaimed words), arias and melodies for set moods.

The madrigal emerged in Italy. Usually set for two or more voices, it was in a strict poetic structure corresponding to the *formes fixes* of France. It was unrelated to the 16th-century madrigal.

Science and technology

European trade and industry, although violently arrested in the mid-14th century by the Black Death, expanded rapidly in the first century of this period. Italian galleys carried cargoes of glass, silk and finished metal goods to northern Europe and elsewhere, returning laden with textiles from the Hanseatic cloth towns and metals from the mines of central Europe.

Intellectual life, including that of the newly founded universities of Oxford and Paris, was also vigorous. Although in science Scholasticism was still the rule, signs of a breakthrough to a more experimental approach began with the works of Oresme and Buridan in Paris and William of Ockham in England, whose ideas conflicted with those of the scholastic's ultimate scientific authority, Aristotle.

By the end of this period Arab science was limited to the teachings of a few wandering scholars, and China, the home of accurate scientific reasoning and technology, had declined due to an unwieldy bureaucracy.

Gold florins were first struck in Florence in 1252.
The first cannons, employed by the Moors perhaps as early as 1250, were simply iron buckets charged with gunpowder and filled with stones. They were ignited by means of a touch-hole near the bottom of the bucket.
Vincent of Beauvais, *d.* 1264 was a major encyclopedist. His *Speculum Majus*, unequalled in length until the 18th century, summarized the scientific and philosophical views of the major scholastic writers.

Commercial fishing, encouraged by the many meatless fastdays of the Christian calendar, grew rapidly during the Middle Ages in Europe. The Hanseatic League fishery, in the Baltic, reached its peak, 1275-1350, with catches of 13,000 tonnes of herring a year.
The spinning wheel may have been invented by 1280 but was not commonly introduced into Europe until the 14th century, when it replaced the distaff and loose spindle. It is pictured in the Luttrell Psalter of 1338.

Spectacles, with convex lenses, were first recorded by an Englishman, Roger Bacon, 1286. By the early 14th century they were factory-made in Venice.
Albertus Magnus, *c.* 1200-80, a German encyclopedist, classified plants by their structures.

Stanches, or navigation weirs, which maintain a depth of water for ships, were built in European rivers and canals and include one built on the Thames in 1306.
Linen clothes were widely worn in the 14th century for the first time. This led to an improvement in personal cleanliness and an associated decline in diseases such as leprosy.
Gunpowder for artillery appeared in Europe *c.* 1300.
Watermarks were first used in papermaking in Italy in the late 13th century.

1310-25

Bad harvests in 1315 brought famine to much of Europe, slowing population growth.
In north India, where Muslim dynasties ruled from Delhi since 1206, a Turkish Tughluk dynasty was founded in 1320.
In Mexico the Aztecs founded the capital Tenochtitlan in 1325 and began to colonize Central America.
Uzbeg, *r.* 1312-41, converted the Mongol Golden Horde to Islam and brought Mongol prosperity to its height.

1325-40

The Ottoman Empire expanded into Thrace, 1326-61, threatening Constantinople.
The Hundred Years War of England and France broke out in 1337 as a result of the rival claims to the French throne of **Edward III**, *r.* 1327-77, and **Philip of Valois**, *r.* 1328-50.
The Hanseatic League grew politically powerful *c.* 1340.
Victory at Rio Salado in 1340 by Alfonso XI of Castile, *r.* 1312-50, ended the African threat to Spain.

1340-55

Italian economic decline followed the fall of the Bardi bankers after the English monarchy repudiated its debts.
Cola de Rienzo, 1313-54, was murdered after his attempt to set up a Roman republic independent of the papacy.
The Black Death destroyed up to half Europe's population between 1348 and 1352, totally disrupting commerce.
During the Hundred Years War England profited from pillage and the ransom of captives.

1355-70

The Holy Roman Empire was changed by papal decree, the **Golden Bull**, from a monarchy to an aristocratic federation.
The ransom of John II of France by England provoked the **Jacquerie**, a peasant revolt against war taxes which was violently suppressed in 1358.
The Ming dynasty in China was created after a popular revolt against the Mongols, 1368.

1370-85

The Ottomans took Adrianople.
Rival popes were created in Rome and Avignon, 1378, after the breakdown of negotiations over plans to reform the papacy.
Popular revolts in Florence in 1378 and England in 1381 were suppressed.
Constitutional reform in Florence marked the beginning of Florentine power in 1382.
Moscow emerged as the focus of Russian opposition to the Mongols after the defeat of the Tatars at Kulikovo in 1380.

1385-1400

Portugal assured its independence by defeating Castile at Aljubarota in 1385.
Tamerlane, *r.* 1369-1405, conquered Central Asia and defeated the Golden Horde, 1391, destroying the kingdom of Delhi in 1398 and delaying the Ottoman advance westwards into Europe, despite a victory over the Serbian alliance in 1398.
The Ming dynasty began to develop a naval empire *c.* 1400.
Japanese prosperity was primarily based on piracy *c.* 1400.

Robert Bruce, *r.* 1306-29, assured Scottish independence at Bannockburn in 1314 by defeating **Edward II**, *r.* 1307-27.

The nobility deposed and killed Edward II, 1327. **The need to defend the interests of the woollen trade** in Flanders helped to bring war with France, 1337.

The Black Death quickened the decline of serfdom and **Edward III**, *r.* 1327-77, attempted unsuccessfully to control wage rates and labour problems.

English pillaging in France brought new wealth to all classes and professional soldiers emerged in the army alongside the feudal nobility.

The Peasants' Revolt, 1381, failed in its attempt to end villeinage immediately.
Parliament claimed the right to impeach royal ministers, 1376.

Richard II's absolutist reign, 1377-99, was ended by noble opposition led by Henry of Bolingbroke, 1367-1413, who became Henry IV, 1399.

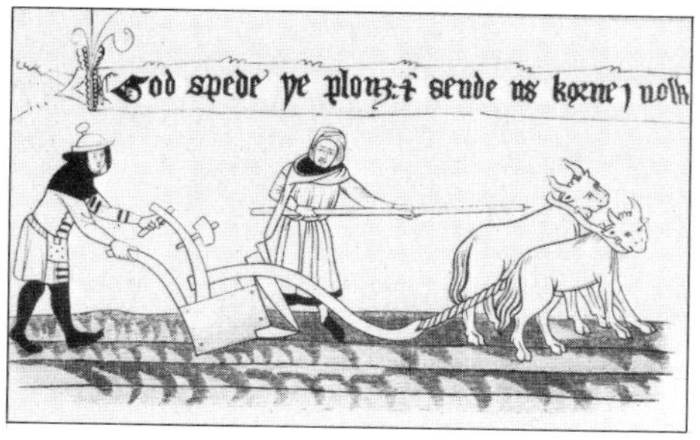

Piers Plowman frontispiece

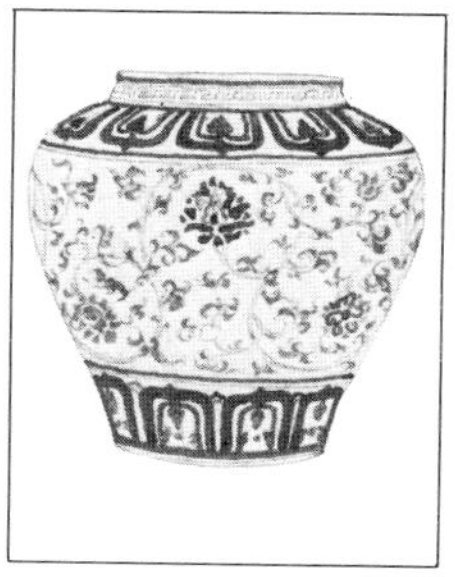

Ming vase

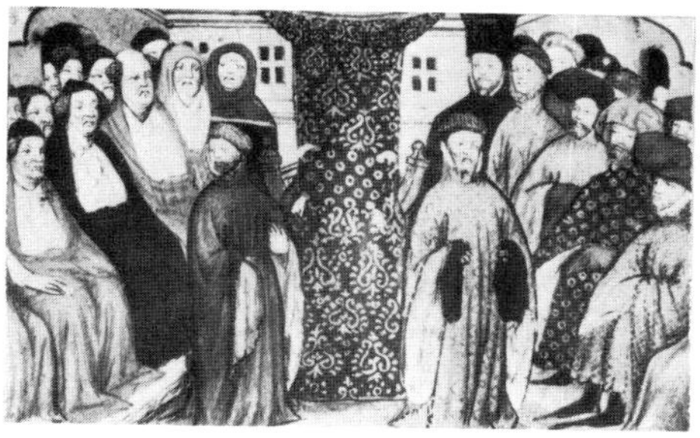

The English Parliament deposing Richard II

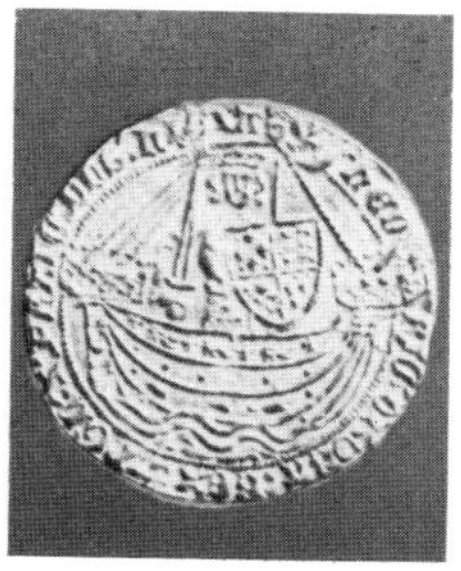

English gold coin, *c.* 1400

Giovanni Monte Corvino, 1247-1328, established the first Christian missions in China and baptized **Khaistan Kuluk**, *r.* 1307-11, the third great Khan of the Chinese Yüan dynasty.
Marsiglio of Padua, *c.* 1275-1342, wrote *Defensor Pacis*, 1324, a famous treatise espousing the supremacy of lay power over the Church and claimed that all power derives from the people, for whom the ruler is a delegate.

William of Ockham, *c.* 1300-*c.* 47, an Englishman, was the last of the great scholastic Franciscan philosophers. He broke with the Aristotelian realism of **Aquinas** and took a nominalist position. His importance lies in his development of a sophisticated logic and epistemology of more than a purely theological significance, which was to have a great effect on later secular philosophy.

The Flagellant movement arose in response to fear of the Black Death, but was condemned by **Pope Clement IV**, 1349. The Flagellants sought to avoid divine wrath by whipping themselves thrice daily. They began in Italy and spread to Germany and the Low Countries, where they toured the countryside proclaiming flagellation as the way to salvation.

The Sufi branch of Islam spread into India, Malaya and Africa south of the Sahara.
Piers Plowman, a vernacular poem probably by the English parson **Langland** *c.* 1362, attacked corruption in the state and Church. The poem is an appeal on behalf of the poor and a plea for spiritual equality.

John Wycliffe, *c.* 1328-84, and his followers the Lollards, a religious group with noble supporters in England, spread ideas that were unacceptable until the Reformation. In *On Civil Dominion*, 1376, Wycliffe proposed a propertyless Church and argued for direct access to God for individuals.
In the Netherlands disciples of Gerhard Grote, 1340-84, who espoused a non-ritualized, humane Christianity, formed the Brethren of the Common Life.

Ibn Khaldun, 1322-1406, the Islamic scholar, was unique in the medieval era. The greatest social thinker until modern times, he based a theory of society on social cohesion and a cyclical pattern of growth and decay. In his masterwork, the *Muquaddimah*, he outlined a philosophy of history and laid the foundations for what he called "a science of culture".

The theme of spiritual love developed by such Tuscan poets as **Guido Calvacanti**, 1260-1300, was given expression by **Dante Alighieri**, 1265-1321. In his *Divina Commedia*, begun *c.* 1307, he describes his journey through *Inferno*, *Purgatorio* and *Paradiso*, giving insight on medieval views and religious beliefs. It made Tuscan Italy's literary medium.

Petrarch (Francesco Petrarca), 1304-74, gave passionate form to Italian love poetry in his "Canzoniere", sonnets and madrigals inspired by his unrequited love for Laura.
An admirer of Roman and Greek ideals, his humanistic outlook influenced other writers.
The Persian mystic poet Hafiz, 1320-88, used complex lyrical imagery in ghazal form.

The Italian novella was developed by **Giovanni Boccaccio**, 1313-75, in *The Decameron*, 1353, a collection of 100 witty and bawdy tales set in the time of the Black Death. Their humanism and breadth of social and psychological observation had an enormous influence on Renaissance literature everywhere. Boccaccio was influenced by Graeco-Roman styles.

A great Christian allegorical poem, *Piers Plowman*, attributed to **William Langland**, *c.* 1332-1400, brought to Middle English the alliterative tradition of Anglo-Saxon verse in a series of 11 dream visions.
A more mysterious allegory was the anonymous *Gawain and the Green Knight*, *c.* 1370, an Arthurian romance. *Pearl* was found in the same manuscript.

The No play emerged in Japan in a classic form established by **Kanami Motokiyo**, 1333-84, and his son **Zeami Motokiyo**, 1363-1443, who wrote most of the 100 plays that survive from this period. No drama is formal in style, incorporates music and dancing and is performed without scenery by males who wear masks to portray women, old men or supernatural beings.

The first truly native English poetry was created by **Geoffrey Chaucer**, *c.* 1345-1400, influenced by French and Italian styles. His best works include *Troilus and Criseyde*, *c.* 1385, and *Canterbury Tales*, *c.* 1395.
Confessio Amantis, *c.* 1390, by English poet **John Gower**, 1325-1408, told moralistic stories of courtly love.

The Paris school of manuscript illumination flowed in the work of **Master Honoré**, *d.* 1318.
Second generation "decorated" Gothic architecture in England developed with the building of Ely Cathedral Lady Chapel, 1321-48, whose undulating blind arcading and curvilinear tracery derived from geometrical forms.
Giovanni Pisano, *c.* 1245-1314, completed the pulpit in Pisa Cathedral, 1310, synthesizing Gothic and classical elements.

In the Muromachi (Ashikaga) period in Japan, 1338-1573, painters followed previous traditions such as continuous narrative scrolls. Renewal of contact with China and Korea introduced new techniques such as the art of painting.
The Perpendicular style of architecture in England first appeared in Gloucester Cathedral cloister, where the ribs of the vault spread out into fan-vaulting.

Italian painting followed the Sienese tradition in the works of **Simone Martini**, *c.* 1284-1344, and of **Pietro and Ambrogio Lorenzetti**, both active in the first half of the 14th century. Papal patronage in the palace of Avignon brought numerous Italian artists to France. Giotto's earlier detailed studies of nature influenced the decoration of the **Tour des Anges**, *c.* 1340, by **Matteo Giovanetti**.

Potters of the Ming dynasty in China, 1368-1644, discovered underglaze painting using imported Persian cobalt. Their harmonious blue designs on white ground balanced their favourite opulent shapes and sinuous line.
Italian architecture's continuity with Romanesque was shown in the new design for the east end of Florence Cathedral by **Francesco Talenti**. Only external details such as windows are borrowed Gothic.

English Perpendicular architecture matured. Canterbury Cathedral nave had smoothly shafted columns rising to the lierne vaults.
The Hindu Vijanagar dynasty of the Deccan, India, 1336-*c.* 1614, favoured an almost baroque style. Groups of small buildings were characteristic and columns were often sculptured with groups of figures and animals.

The rebuilding of Milan Cathedral, 1387, in the northern Late Gothic style showed the influence of and enthusiasm for French ideas. Building continued throughout the Renaissance.
Tamerlane's mausolea at Samarkand, built in the decade after Baghdad's capture, 1393, inspired Timurid architects, 1405-1500, by their tall domes on high drums and colourful glazed relief-tile decoration.

Ars nova was a term coined by **Phillipe de Vitry**, 1291-1361, to describe new and freer forms of music. The earlier forms became known as *ars antiqua*. Religious music still favoured triple time as symbolic of the Trinity, but growing acceptance of **duple** (beats in groups of two) time advocated in *ars nova* implied an acknowledgment of the equality of secular music.

Guillaume de Machaut, *c.* 1300-77, was the chief figure of *ars nova*. The complex forms he used involved modulation (changing key), intricate cross-rhythms and great independence of line.

Meistersinger in Germany took over the lyric art of the aristocratic Minnesinger. Meistersinger were traders and craftsmen who founded guilds to set and keep up standards for their art.

Dissonances and great embellishment in music were part of the general concern with richness and diversity seen in European art of the time. Paris produced the best examples.

Drainage mills, windmills which operate drainage scoops, were invented in the 14th century. (In the 16th century the Dutch used such mills for recovering land for agriculture.)
Chaulmoogra oil, for the treatment of leprosy, was first seen in 14th-century China. It was the only effective treatment for leprosy until the 20th century.

Salt-glazing of pottery was practised in the Rhineland from the 14th century. The potter threw salt over wares in the final stages of firing in the kiln; this produced a fine glaze which sealed off the wares.
The cross staff, a primitive form of sextant, was popularized for use in navigation by **Levi ben Gerson** of Provence, 1288-1344.

The Black Death of 1348-52 caused a severe decline in European trade and power, badly affecting labour-intensive industries such as agriculture, mining and fishing, which did not recover for over a century.
Double-entry book-keeping methods are known to have been used by the Massari family of Genoa *c.* 1340, although they were probably used before that by the Hanseatic League, the Medicis and the Fuggers.
Iron cannons were used by the Germans from 1350 onwards.

Oresme, *c.* 1325-82, and **Buridan**, 1300-58, in Paris, criticized Aristotle's doctrine of motion. They were influenced by the idea of *impetus*, conceived by **Philoponus**, *c.* 530, and later developed into a theory of motion by Galileo.
Mechanically wound steel crossbows were developed.

Lock gates on Dutch canals date from at least 1373. By 1400, locks were an integral part of navigation and drainage systems of Italy and Germany.
Geoffrey Chaucer, an English writer, *c.* 1345-1400, described what may be the first scientific work in English, *The Equatorial Planetarie*, which deals with a device for predicting the paths and positions of planets.
Forged iron guns weighing 272kg (600lb) were used by Richard II, *r.* 1377-99, to defend the Tower of London.

Weight-driven clocks, often employing elaborate striking mechanisms, appear in Europe at this time. The earliest surviving clock in England is in Salisbury Cathedral, installed in 1386.
Arab observatories were among the last achievements of Arab science and include that of **Ulugh-Beg**, 1394-1449, the grandson of Tamerlane, in Samarkand. His astronomical tables were used by Western astronomers until the Renaissance.

1400-1500 Printing and discovery

Principal events

In spite of a generally static economic climate the move towards national sovereignty increased at the expense of papal authority. The process of the consolidation of the European states continued and the power of the monarchs over the nobility grew gradually with the help of ostentatious artistic patronage and ambitious foreign wars. In Spain, united under Ferdinand and Isabella, the Moors were finally expelled and Ivan I established the power of Moscow by bargaining with the Tatars.

Byzantium fell to the Ottoman Turks in 1453, closing the eastern Mediterranean to Christian traffic, but European expansion began to the west as the Spanish and Portuguese thrones sponsored the exploration of alternative routes to India round the coast of Africa.

In China, the Ming dynasty made contact by sea with India and Africa and fought to protect its weak northern frontiers.

1400-10

Tamerlane's victory at Angora in 1402 brought temporary disorder to the Ottoman Empire.
Chinese naval expeditions to India and Africa for commercial and military prestige began in 1403.
Burgundian ambitions led to a French civil war with the Armagnacs in 1404.
Venice seized Vicenza, Padua and Verona, to become the dominant power in northern Italy.
Florence won access to the sea by buying Pisa in 1405.

1410-20

Mehmet I, *r.* 1413-21, reunited the Ottoman Empire and consolidated power in the Balkans.
Henry V of England, *r.* 1413-22, captured Normandy, 1417-19, after his victory at Agincourt in 1415.
The papal schism was ended at the Council of Constance, 1414-17, where **Huss**, 1369-1415, the Bohemian religious reformer, was burnt for heresy.
Henry the Navigator, 1394-1460, began his systematic exploration of the African coast.

1420-30

The Bohemian Hussites under **John Ziska**, *c.* 1370-1424, were defeated in a series of imperial crusades, 1420-33.
Henry V of England was recognized as heir to the French throne, 1420. **Joan of Arc**, *c.* 1412-31, then inspired a new French national unity in support of **Charles VII**, *r.* 1422-61.
Peking became the Ming capital in 1421.
Murad II, *r.* 1421-51, led an Ottoman attack on Constantinople in 1422.

1430-40

Alfonso V of Aragon, *r.* 1416-58, campaigned in Italy and took Naples in 1435.
The banking and wool merchant Medici family controlled Florence, 1434-94.
Hapsburg control of the Holy Roman Empire became virtually hereditary with **Albert II**, *r.* 1438-9.
John VIII, *r.* 1425-48, of Byzantium, inspired serious opposition by accepting the primacy of the pope in 1439.

National events

Unstable royal authority resulted in civil war, the Wars of the Roses. Traditional feudal ties were breaking down and a strong independent yeomanry emerged. Although Henry Tudor introduced few innovations, his reign prepared the way for the later assertion of royal authority.

Henry IV, *r.* 1399-1413, whose seizure of power had relied on Parliament, suffered from parliamentary demands to supervise his expenditure.

Henry V, *r.* 1413-22, united Parliament and the country behind his nationalist appeal for a successful invasion of France.

The weakness of **Henry VI**, *r.* 1422-61 intermittently, resulted in a rise in the activity of noble factions.

The rise of **"bastard feudalism"** reflected the short-term political interest of the rising knightly classes.

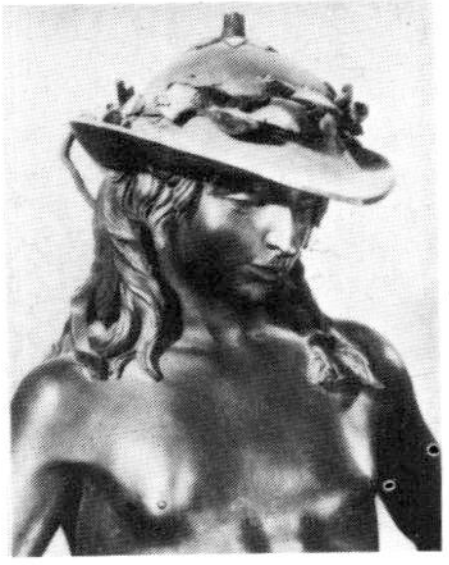

"David" by Donatello

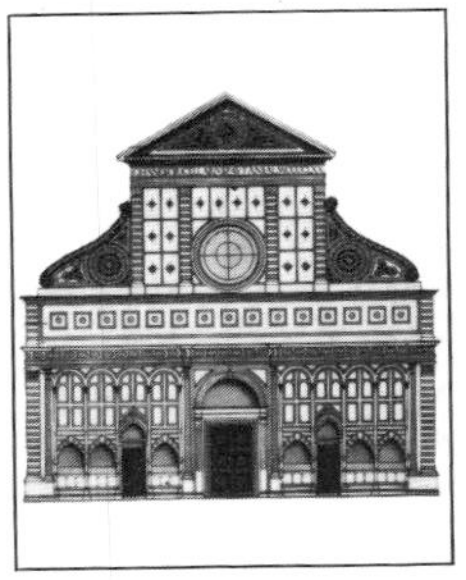

Façade of S. Maria Novella

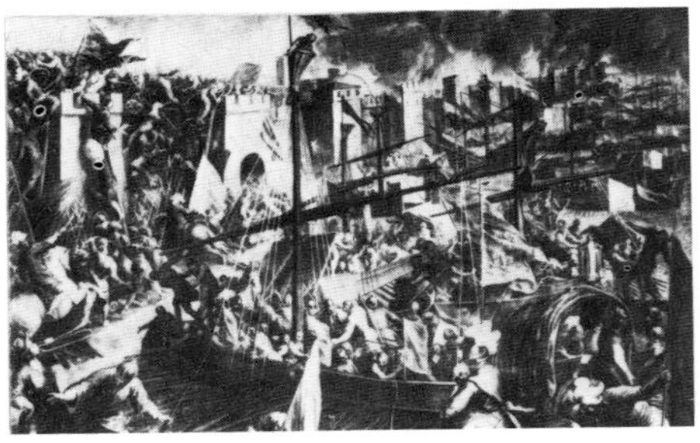

The conquest of Constantinople by the Turks

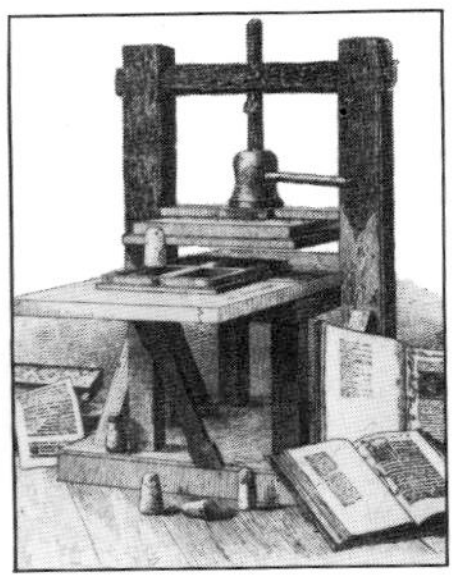

Gutenberg's printing press

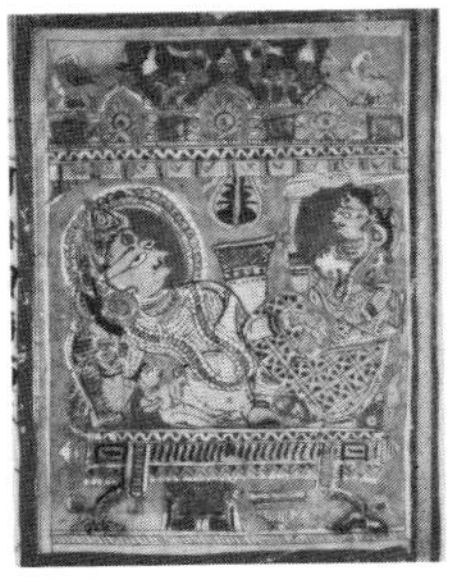

Jain manuscript, 14th.–15thc.

Religion and philosophy

The relationship of Church and state was a major subject of controversy in 15th-century Europe, while the corrupt practices and moral laxity of the established religious orders came under attack. Reformers and critics of religious authority spelt out many of the themes that would be elaborated in the Protestant Reformation of the next century.

Savonarola, an Italian monk, denounced corruption in Florence and the abuse of political power, calling for a regeneration of spiritual values and a steadfast devotion to asceticism.

In Bohemia the Hussites identified religious reforms with Bohemian nationalism while humanist writers in Italy, England and Holland argued for the separation of religious and secular law and the freedom of conscience of the individual. In another sense the power of the universal Church was challenged in Spain where the crown set up the Spanish Inquisition.

The Chinese emperor Ch'eng-Jsu, *r.* 1403-24, sponsored the publication of an 11,095-volume encyclopedia in 1403.
The Council of Pisa, 1409, attempted to resolve the Great Schism in the papacy. This had arisen in 1378, with Urban VI in Rome and Clement VII in Avignon as rival claimants, backed by the empire and France. **John XXII**, however, the compromise candidate, satisfied no one and the schism continued until 1417.

Pope Martin V, *r.* 1417-31, whose election ended the Great Schism, moved the papacy permanently to Rome in 1420 and consolidated Church unity.
John Huss, *c.* 1369-1415, a Bohemian follower of Wycliffe, criticized the papacy for the sale of indulgences (absolutions from sin) and urged a literal interpretation of the Bible. He denied the infallibility of an immoral pope, and claimed the supremacy of the state over the Church.

The Hussites were Bohemian followers of John Huss. They believed that the laity should receive both the wine and bread in communion instead of bread alone. **Thomas à Kempis**, *c.* 1379-1471, wrote the *Imitation of Christ* *c.* 1425. This simple book, emphasizing the need for a moderate asceticism, was considered at the time to be the most influential Christian work since the Bible.

Nicholas of Cusa, 1401-64, in his *De Concordantia Catholica*, 1433, argued for the General Council's authority over the pope. However, the council's lack of power led him to reverse his position by 1437. Cusa also contributed to the sciences and philosophy. He wrote *Of Learned Ignorance*, in 1440, arguing against the possibility of ever attaining eternal truths.

Literature

Compared with the vitality and initiative of Boccaccio and Chaucer, European writers of the early 15th century produced less distinctive work. Learning rather than literature held sway and the revival of interest in classical studies led by the Humanist scholars in Italy had its main impact only after 1454, when the development of printing by Gutenberg in Germany produced a rapidly increasing flow of books. Two outstanding writers who drew on medieval traditions were Villon, France's first great lyrical poet, and Malory, who dominated English prose with an adaptation of Arthurian legend. Lively vernacular poetry emerged in Scotland with Henryson and Dunbar and also in Florence and Naples late in the century.

The Mabinogion collection of Celtic tales and heroic legends was preserved in the Welsh *Red Book of Hergest*, *c.* 1375-1425. These anonymous stories contained a wealth of ancient mythology. They fused narrative with dialogue, conveying the vitality of the oral tradition from which these tales emerged, probably during the 11th century.

Miracle plays based on biblical themes or the lives of saints were enacted in popular style in England and Europe.
John Lydgate, *c.* 1370-1451, English imitator of Chaucer and Boccaccio, wrote the *Troy Book* and *Siege of Thebes*, *c.* 1420.
Perez de Guzman, *c.* 1376-*c.* 1460, Spanish historian and poet, examined the theory of history and role of the historian.

Alain Chartier, *c.* 1390-1440, wrote the allegorical poem *La Belle Dame Sans Merci* in 1424. An attack on courtly love, it reflected political unrest in France after the defeat at Agincourt. In *Le Quadrilogue invectif*, 1422, a political pamphlet, he called for French solidarity to combat the turmoil of the Hundred Years War, using prose form to convey his plea.

The Italian Leon Battista Alberti, 1404-72, a brilliant Renaissance figure who was an architect, sculptor and musician, wrote *Della Famiglia*, 1434, containing a theoretical treatise within a discussion of household affairs. Styled on Latin models, it displayed a pessimistic view of contemporary life. He also published works on ethics, jurisprudence and architecture.

Art and architecture

Fifteenth-century European art was profoundly affected by the artistic Renaissance that emerged in Italy – a stylistic revolution characterized by a revival of interest in Greek and Roman antiquities that brought with it a new interest in the anatomy of the human form, in proportion and in perspective, combined with a new sense of human dignity and confidence. Beginning in Florence, and fostered by widespread court patronage both ecclesiastical and secular, it spread rapidly to other parts of Italy and culminated at the end of the century in the masterpieces of Mantegna, Botticelli, Bellini, Leonardo, Michelangelo and Bramante.

The Gothic style in architecture still flourished even in Italy and took new forms with the Perpendicular style in England, while International Gothic brought a new realism to European painting and sculpture.

The International Gothic style introduced a new realism into the painting of landscape, costume and animals, exemplified in the wings of the altarpiece at Dijon, 1399, by **Melchior Broederlam**, *d. c.* 1410.
Gothic and Renaissance styles were linked in the bronze doors of the Florence Baptistery sculptured by **Lorenzo Ghiberti**, 1378-1455, from 1403 to 1452.
Burgundian sculpture was characterized by the "Well of Moses" 1401, by **Claus Sluter**, *d.* 1406.

The Duc de Berry commissioned the *Très Riches Heures*, *c.* 1415, from the Limbourg brothers. His extensive patronage included the less-known *Très Belles Heures* and he built twelve elegant castles.
The design for Innocenti (Foundling) Hospital in Florence, 1419, by **Filippo Brunelleschi**, 1377-1446, began the architectural Renaissance in Italy and established Brunelleschi's reputation as one of the finest Renaissance architects.

Masaccio, *c.* 1401-28, the first of the great *quattrocento* painters, used simplicity, naturalism and light in a new way in his Brancacci Chapel frescoes, Florence, 1425-8.
One of the masterpieces of the International Gothic style in Italy was the "Adoration of the Magi", 1423, by **Gentile da Fabriano**, *c.* 1370-1427.

Donatello, 1386-1466, an Italian and one of the greatest figures of 15th-century art, [illegible] bronze [illegible] 1435. [illegible]tion- [illegible]

Music

European music was dominated by the brilliance of the Franco-Flemish composers, the first great musical school. The Church favoured an international style of music and would admit no other styles, but national composers successfully challenged the Franco-Flemish school in the quality of their work, especially in the field of polyphonic songs.

Under the Ming dynasty, 1368-1644, music declined in China. Long pieces, interspersing new material with a refrain, were played on the zither and tunes modulated for special effect.

Composers set parts to imitate each other. Polyphony related melodies to the *cantus firmus*, and **counterpoint** used rhythmically related tunes to combat the turgid rhythm of much polyphonic music.

Choral polyphony using four independent parts now grew up. The voices (parts) were finely blended and the harmony euphonious, avoiding the dissonances common in earlier music.

Science

Important changes occurred in the economic and industrial organization of Europe. The Hundred Years War ended and with the Renaissance feudal methods of exchange gave way to more dynamic systems of trade. Technological change – in agriculture, mining, textiles and glassmaking – continued, bringing a steady expansion of industry. The breakthrough of the century was the creation of Gutenberg's book-printing industry at Nuremberg.

Ships and navigational instruments had been undergoing steady improvement and by the end of the century provided explorers with the means to sail to all parts of the globe. Maritime successes stimulated the founding of schools of navigation, which produced men trained in mathematics and curious about science but relatively untrammelled by the religious ideas that had ruled the minds of educated people of earlier medieval times.

Technical treatises on military engineering and ballistics abounded in the early 15th century, especially in Germany and Italy. Among the most famous was the *Bellifortis* of 1405 by **Conrad Kyeser**, a German.
Archimedean screws, used for lifting water in Dutch polder dams, are known from 1408.
Perspective, used first in painting but later in scientific and architectural drawings, was discovered in the early 15th century by **Filippo Brunelleschi**, 1377-1446.

Drift nets up to 110m (360ft) long, towed behind fishing boats, were introduced by the Dutch fishing industry in 1416. These nets greatly improved the size of herring catches. The fish were preserved by a salting process improved by the Dutch.
Navigation was studied by experts from many nations at the court of **Henry the Navigator**, 1394-1460.

Nicholas of Cusa, 1401-64, wrote that the Earth, and not the heavens, revolved daily, a refutation of the accepted Ptolemaic astronomical system. Nicholas's idea was based upon philosophic notions and not on observable scientific data.
Hollow-post mills, invented *c.* 1430 in Holland, were an improved form of windmill in which the size of the rotating sail arms was reduced and a shaft was passed from them through a hollow post to drive machinery in a building below.

1440-50

After losing Serbia, the Ottoman Turks defeated a Hungarian crusade against them in 1444.
Charles VII created a French standing army free from feudal obligations.
Wars in Italy caused a rise in diplomatic activity, and artistic patronage became a major factor in a ruler's prestige *c.* 1440.
Japan, under Ashikaga rule since 1336, underwent a period of cultural refinement *c.* 1440.

The flourishing woollen industry moved away from the towns, where economic decline had made the guilds increasingly restrictive.

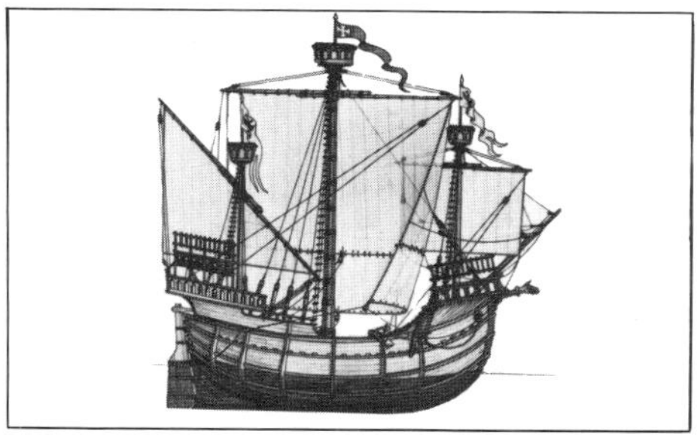

Carrack, 15thc.

In 1440, **Lorenzo Valla**, 1407-57, attacked papal political claims by asserting that the *Donations of Constantine*, an anonymous document that supposedly granted universal temporal power to the papacy, was a forgery. As a humanist of the Italian Renaissance, Valla accused the medieval philosophers of deliberately misunderstanding and poorly interpreting the works of Plato and Aristotle.

Bengali literature, which had existed in India since the 10th century, was enriched by the rhymed version of *Ramayana*, made *c.* 1440 by Kirttivasa, *b.* 1385, and by the lyrical *Song of Krishna* by **Chandidas**, 1417-77.
In Spain, the Marquis of Santillana, 1398-1458, wrote Italian-style sonnets that enriched the poetic tradition.

In Italy Domenico Veneziano, *d.* 1461, represented the most advanced stage of mid-century Florentine painting in his "St Lucy" altarpiece, 1445.
Rogier van der Weyden, 1400-64, a major mid-15th century Flemish artist, produced a more emotional style than van Eyck. His great "Deposition" was painted *c.* 1435.
Fra Angelico, *c.* 1387-1455, and **Fra Filippo Lippi**, *c.* 1406-69, both linked Gothic and Renaissance styles.

The use of the interval of a third (long established in England) standardized harmony in polyphony but the increasing preoccupation with harmony itself led to dull and static rhythms.

Johann Gutenberg, *c.* 1400-68, of Strasbourg, began printing with movable metal type, a process invented in Korea in the 15th century. Gutenberg's books were the first to be printed in this way in the West, yet no printed work bearing his name exists. Letters were cast in type metal, composed into sentences on a type stick and set up as pages of type before being inked for the press. It is possible that Gutenberg designed his press along the lines of wine and linen presses.

1450-60

The alliance of Florence, Naples and Milan, 1450, inspired by Medici diplomacy, ensured the balance of power among the Italian states.
The fall of Constantinople in 1453 to the Ottoman ruler **Mehmet II**, *r.* 1451-81, ended 1,000 years of Byzantine rule.
George Podiebrad, 1420-71, ended the Bohemian religious wars with conciliatory policies.
The Wars of the Roses between the houses of Lancaster and York began in England, 1455.

The widespread enclosure of common land to promote sheep farming began in 1453. A strong independent smallholder class developed.

The Indian mystic Kabir, 1440-1518, attempted to merge some aspects of the Hindu creed with Sufist Muslim ideas. Kabir, originally a weaver from Benares, rejected Hindu beliefs in idols and castes but accepted the institutions of reincarnation and eventual release. His followers were known as **Kabirpanthis**. This movement was a forerunner of the movement of Sikhism.
The first printed Bible was produced in Mainz in 1456.

Medieval French verse forms were infused with vigour and blunt realism in the lyrical poetry of **François Villon**, 1430-*c.* 63, in which he recalls his wasted life. He was awaiting execution when he wrote *Ballad of a Hanged Man*.
Diego de San Pedro, *fl.* 1450, was best known for his sentimental novels that influenced the evolution of the Spanish novel.

The study of perspective absorbed **Paolo Uccello**, *c.* 1397-1475, whose famous "Battles", 1454-7, also possess an eerie, dream-like atmosphere.
The frescoes of S. Francesco Arrezzo, 1452, by **Piero della Francesca**, *c.* 1410-92, were outside the mainstream of Italian painting and closer to the diffused naturalism of Flemish art with their mathematical precision and use of light and shade.

National forms evolved in polyphonic song, with the *frottola* in Italy and the *lied* in Germany matching the richness of the established *chanson* in France.

Instrument-making in Europe became centred on Nuremberg *c.* 1450, and Augsburg *c.* 1475.
Calendar reform was undertaken *c.* 1450 under the direction of the astronomer **Puerbach**, 1423-61. The Julian calendar, commissioned by Julius Caesar and accurate to 1 day in 128 years, was wrong by 10 days in 1450. However, revision was not finished until 1582.
Quadrants, for determining latitude at sea, were used by European seafarers *c.* 1456.

1460-70

Venice fought the Turks for control of the Mediterranean, 1463-79.
Louis XI, *r.* 1461-83, aided French unification by ending provincial and urban privileges.
The Onin War, 1467-77, resulted from a succession dispute among the Ashikaga in Japan. This was a prelude to a century of war.
The kingdom of Songhay, based on the Middle Niger region, reached its zenith under **Sonni Ali**, *r.* 1464-92.

Edward IV, *r.* 1461-83, a Yorkist, won the throne from Henry VI in the Wars of the Roses, relying on popular support in London for his authority.

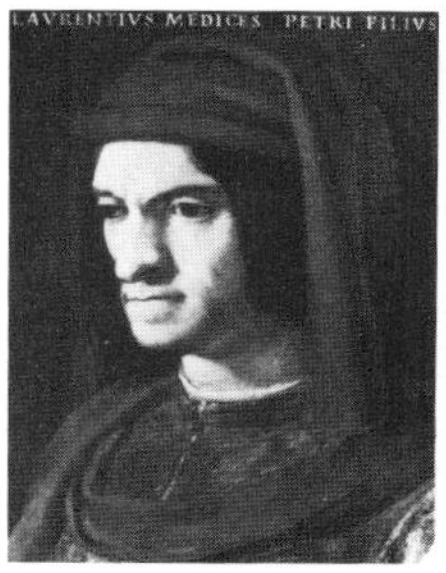

Lorenzo de Medici

The Unitas Fratrum (Bohemian Brethren), founded by Peter of Chelchich, *d.* 1460, broke with the Utraquists in 1467. They were a militantly democratic sect who, like the Taborites, rejected subordination to Rome.

Scottish poetry flourished with *The Testament of Cresseid* by **Robert Henryson**, *c.* 1425-1508, a tragic and powerful sequel to Chaucer's poem. **William Dunbar**, *c.* 1460-1520, was less earnest but more versatile. His *Dance of the Seven Deadly Synnis* is similar to Villon in its macabre vigour. The Scottish poets combined romance and satire with idiomatic language.

Leon Battista Alberti, 1404-72, writer, musician, painter and architect, crystallized Renaissance ideas on architectural proportions and harmonious design. His use of classical elements for the church of San Sebastiano, Mantua, 1470, deprived it of an "ecclesiastical" flavour.
Hans Memling, *c.* 1440-94, a German painter settled in Bruges, was, after 1465, a successful Flemish painter and influenced later Italian art.

The Franco-Flemish school included Guillaume Dufay, *c.* 1400-74, Jean d'Ockeghem, *c.* 1430-95, and Josquin des Prés, *c.* 1450-1521, who often used popular tunes in his work.

Carracks, the earliest form of modern sailing ships, are illustrated on a French seal of 1466. These ships had three or four masts, raised decks fore and aft, and a stern rudder and tiller for steering; by 1500 they weighed as much as 600 tonnes. They supplanted the trading galleys for ocean voyages, and a military version followed – the galleon.

1470-80

Ivan the Great, *r.* 1462-1505, adopted the title of tsar in 1472 and subjected Novgorod to Muscovite rule in 1478.
Burgundy was reunited with France, 1477.
In spite of the Pazzi plot to assassinate him, **Lorenzo de' Medici**, 1449-92, ruled in Florence and exhausted the stagnant economy with his flamboyant foreign policy.
The marriage of Ferdinand and Isabella united Aragon and Castile in 1479.

After Henry's brief restoration, 1470-1, **Edward IV** continued to rule parsimoniously in order to assert royal independence from nobility and Parliament.

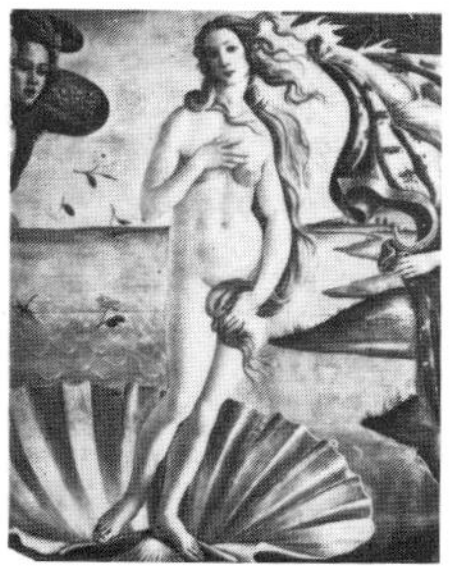

"Venus" by Botticelli

Sir John Fortescue, *c.* 1385-*c.* 1479, in *De Laudibus Legum Angliae*; *c.* 1470, praised English over Roman law and introduced the principle of "innocent until proven guilty".
Set up in 1478 by Ferdinand and Isabella with the reluctant permission of the pope, who regarded it as a breach of Church privilege, **the Spanish Inquisition** persecuted converted Jews and Muslims as well as Catholic intellectuals, among them Ignatius de Loyola.

Arthurian legend was unified in the epic prose romance *Morte d'Arthur*, 1469-70, by **Sir Thomas Malory**, *d.* 1471. Its admirably plain style and its creative adaptation of medievalism to modern thought deeply influenced later writers. It was published in 1485 by **William Caxton**, a key figure in the development of English printing, which he began in 1476.

The equestrian monument to Bartolomeo Colleone, in Venice, commissioned in 1479 and executed by **Andrea del Verrocchio**, *c.* 1434-88, showed a masterly rendering of movement and a use of light and shade that anticipates Michelangelo.
In Mantua Andrea Mantegna, 1431-1506, painted his fresco "Camera degli Sposi" 1474, and in Florence **Sandro Botticelli**, 1445-1510, produced his "Spring", *c.* 1478, with abstract colours and unreal light.

The mass attained a great variety of structure, although the use of the *cantus firmus* throughout, in many ingenious modifications, brought unity to the form.

Rifles were first made *c.* 1475, according to armoury records in Turin and Nuremberg. These were muzzle loaders, in which the lead bullet was made slightly larger than the bore so that it had to be forced into the barrel, giving a tight fit.
Tables for navigation were revised by the German astronomer **Johann Müller**, 1436-76.

1480-90

Ivan the Great ended the Tatar threat to Moscow, 1480, and annexed Tver in 1485.
The Spanish Church and the **Inquisition** came under royal control after a concordat with the pope in 1482.
The Portuguese Bartolomew Diaz, 1450-1500, rounded the Cape of Good Hope, 1478-8.
The Wars of the Roses ended with the dominance of **Henry VII** (Tudor), *r.* 1485-1509 who established royal independence from baronial support.

After a revival of the wars, **Henry VII**, *r.* 1485-1509, the first Tudor king, brought permanent peace by uniting the rival families in his marriage.

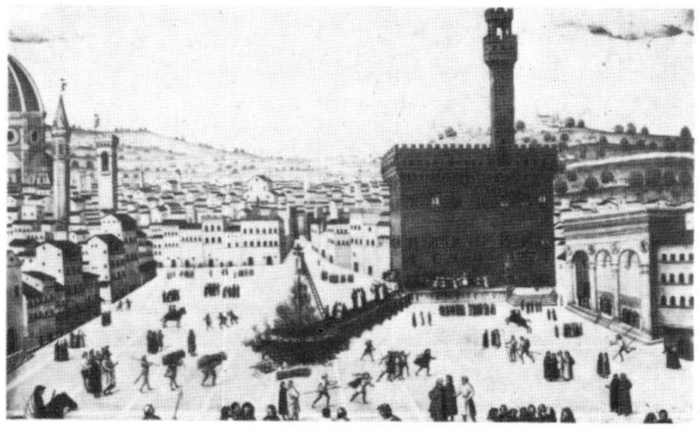

The burning of Savonarola in Florence

Rodolphus Agricola, *c.* 1443-85, was an early Dutch humanist who influenced Erasmus. In his lectures at Heidelberg, given from 1484, he expounded a philosophy emphasizing the freedom of the individual and the intellectual and physical development of the self.
The existence of witchcraft was admitted by the Church in 1484 and its practices condemned. *Malleus Maleficarum*, 1487, described witchcraft and encouraged its suppression.

Humanist poetry emerged in Italy where **Luigi Pulci**, 1432-84, treated the heroic Charlemagne theme irreverently in *Morgante*, *c.* 1480. Another Florentine poet, **Angelo Poliziano**, 1454-94, wrote the first secular play *Orfeo*, 1480. **Matteo Maria Boiardo**, *c.* 1441-94, wrote Latin eulogies and lyric love poems including his epic *Orlando Inammorato*.

English Perpendicular Gothic style with its extremely intricate vaulting is seen in the **Divinity School** at Oxford University, completed 1480.
The rebirth of Venetian painting began with **Giovanni Bellini**, *c.* 1430-1516, whose "Madonna and Saints", 1488, has resonant colours and novel lighting.
Medici patronage in Florence produced the first great Renaissance villa, **Poggio a Caiano**, begun *c.* 1482 by **Giuliano da Sangallo**, 1445-1516.

Keyboard instruments improved in Europe. A Flemish painting of 1484 depicts an organ with a chromatic keyboard. The clavichord (a forerunner of the piano) had a range of up to four octaves.

The voyages of Diaz, Vasco da Gama, Columbus and Magellan, in the late 14th and 15th centuries, encouraged the founding of navigation schools in Portugal and Spain. These schools produced a new group of expertly trained mathematical and nautical technicians, which greatly influenced the standing of science in Europe.

1490-1500

Spain captured Granada, the last Moorish outpost, and expelled 200,000 Jews, 1492.
Columbus, with Spanish support, discovered the Bahamas and Cuba, 1492, on his search for a western route to India.
Vasco da Gama reached India around Africa, 1497-8.
Charles VIII of France, *r.* 1483-98, invaded Italy, 1495, but was expelled by an alliance including the empire, the papacy and Venice formed to protect Italy from foreign domination.

Henry VII ensured his military control over England and asserted royal authority over the nobility in the new Star Chamber court.

The French statesman **Phillippe de Comines**, 1445-1509, argued that taxes needed sanction of the Estates-General, the representative body of nobles, gentry and clergy.
With the defeat of the Medicis in Florence, **Savonarola**, 1452-98, established city rule free from corruption and along democratic lines. His sermons criticizing aristocratic and papal corruption led to his death at the stake in 1498.

German satirical writing reached the common man in the popular and influential *Ship of Fools*, 1494, by **Sebastian Brant**, 1458-1521, which mocked vice in rhyming couplets.
In Persia, the death of the poet and mystic **Jami**, 1414-92, ended the classical period of Persian Sufi poetry. He was notable for such romantic verse as "Salaman u Absal".

The two giants of the Italian **Renaissance** emerged. **Leonardo da Vinci**, 1452-1519, painted his "Last Supper" in the refectory of Sta Maria delle Grazie, Milan, in 1495-8. His rival **Michelangelo**, 1475-1564, sculpted the St Peter's "Pietà" in 1499 when 24 years old.
Donato Bramante, 1444-1514, one of the greatest architects of the High Renaissance, designed the spacious gallery of **Sta Maria delle Grazie**, Milan, dating from 1492.

Music printing began in Germany but was developed fully in Venice by **Ottaviano dei Petrucci**, who patented his process in 1498. His technique led to the birth of music publishing.

Leonardo da Vinci, 1452-1519, painter, sculptor, architect, engineer and scientist, began service as a military engineer with Cesare Borgia, 1476-1507, in 1499. Working mainly in Milan and Florence, his scientific drawings included animal, human and plant anatomy, rocks and optical systems. He also conceived and drew a helicopter, a mobile canal cutter and several kinds of pumps. However, most of his designs were never built, as mechanics lagged behind his inventiveness.

1500-1600 The Reformation

Principal events

The Reformation brought a new dimension to Europe's dynastic wars and social conflicts. As the Italian states declined, Spain, invigorated by wealth from the New World, led the Catholic offensive against England, the Netherlands and the Protestant German princes. Royal authority increased with the decline of papal authority, but the religious and political debates, and the new wealth from confiscation of church lands, enabled an eloquent and powerful middle class to challenge royal power.

European expansion continued. Much of the American coastline and the Far East was reached by all the major powers although only the American civilizations succumbed to the explorers. Japan experienced vigorous expansion and the Moguls brought a stable and flourishing culture to India with the establishment of an extensive empire under Babur and Akbar.

1500-10

The Italian wars provided an opportunity for conflict between the Hapsburgs and the Valois (French kings) until 1559. This caused a decline in the prosperity and autonomy of the Italian cities.
The Portuguese claimed Brazil and established regular trade with India, 1500. The Spanish introduced African slaves to the West Indies, 1501.
Ashikaga prestige in Japan was in decline.

1510-20

Russia took Smolensk from Poland in 1514.
Charles V, 1500-58, created the Hapsburg Empire, inheriting the Spanish crown, 1516, and being proclaimed Holy Roman emperor in 1519.
Ferdinand Magellan, *c.* 1480-1521, sailed through the Pacific Ocean, 1519-21, and **Hernán Cortés**, 1485-1547, conquered the Aztecs in Mexico, 1519-21. **Portugal** controlled the import of spices from the East Indies *c.* 1520.

1520-30

Portuguese traders reached China, 1520-1.
Frederick III of Saxony, *r.* 1486-1525, led the princely support for Luther, 1483-1546.
Peasant revolts in Swabia, inspired by Luther's example and by discontent with feudal obligations, were ruthlessly suppressed in 1525.
Babur, 1483-1530, founded the brilliant Mogul Empire in north India in 1526.
The Medici were driven out of Florence in 1527.

1530-40

Protestantism spread throughout northern Europe. **Henry VIII** of England, *r.* 1509-47, dissolved the monasteries, 1536-9.
Francisco Pizarro, *c.* 1471-1541, a Spaniard, conquered Peru for booty, 1531-6.
The Afghan Sher Khan, *r.* 1539-45, expelled the Mogul emperor Humayan and reformed the administration.
Suleiman the Magnificent, *r.* 1520-66, brought Ottoman power to its zenith.

National events

The power of Parliament increased steadily as a series of changes in religious policy and threats from abroad caused the monarchy to rely increasingly on the support of the newly educated gentry class. In the same period, the political power of the nobility began to decline.

1500-10: **Henry VII's thrift** made him virtually independent of Parliament, although his methods of taxation such as "Morton's Fork" were unpopular.

1510-20: **Henry VIII**, *r.* 1509-47, ruled as a magnificent Renaissance prince, engaging in costly dynastic wars with France until 1520.

1520-30: **Henry's desire for a divorce** from Catherine of Aragon was thwarted by political opposition from her nephew the emperor and from the pope, 1525-9.

1530-40: **Henry defied the pope** and remarried, 1533. His reliance on Parliament to claim supremacy over the Church greatly increased its prestige.

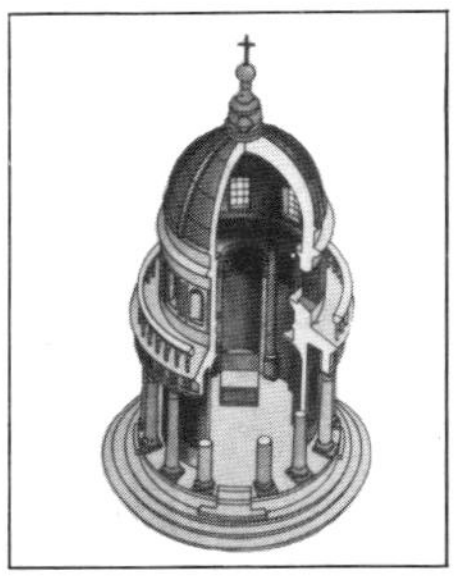

Tempietto of S. Pietro

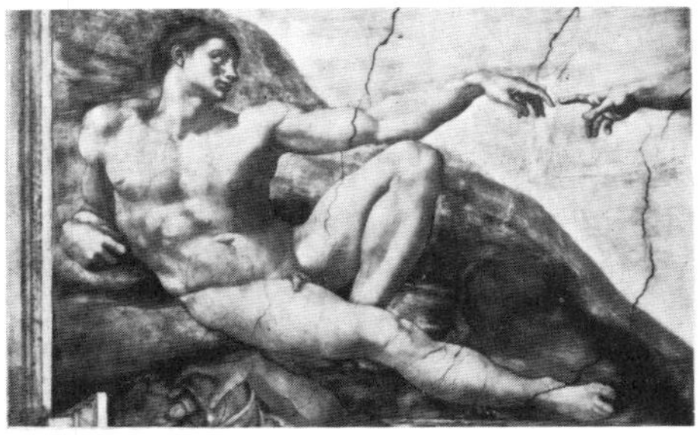

Sistine chapel frescoes: the "Birth of Adam"

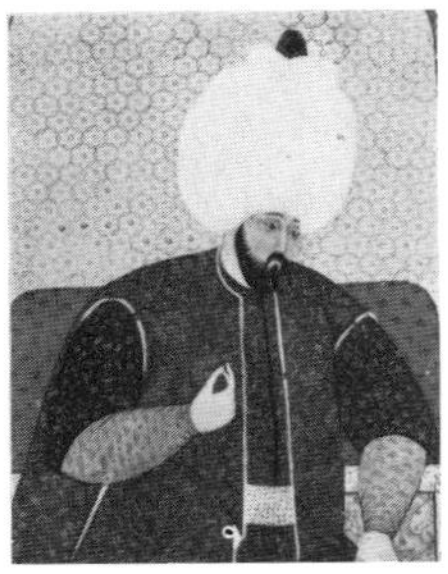

Ottoman Emperor Selim I

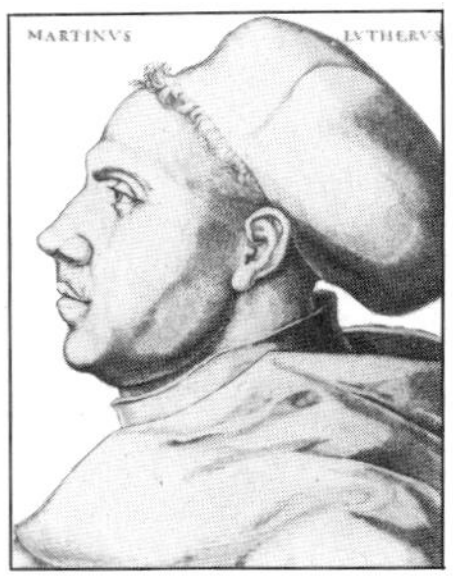

Martin Luther

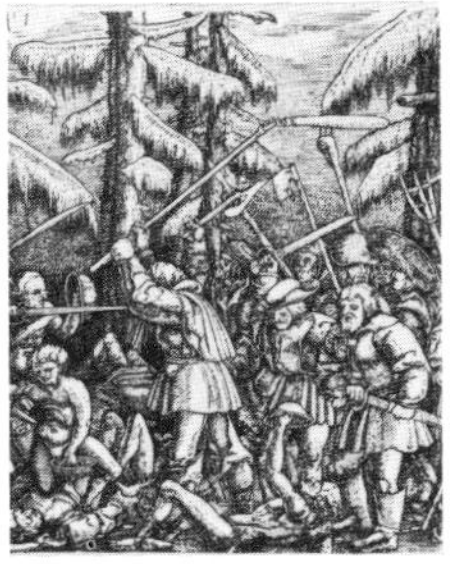

The German Peasants' war

Religion and philosophy

The Protestant Reformation took place in western Europe, arising from objections to many of the doctrines and practices of the medieval Church. Reformers attacked the worldliness of clergy, the stifling of intellectual progress and the inability of the Church to provide spiritual leadership. Luther stated that faith alone was the basis of salvation, believing that no intermediary between man and God could alter his salvation. Calvin in Geneva also rejected the power of his Church to alter who was saved and who damned by God. The general questioning of religious authority gave a new dimension to the already critical question of the relation of Church and state, leading for example in England to rapid changes in official religion. By 1600 the Reformation had spread to almost all of northwest Europe, and there were also large numbers of Protestants in France, Poland and Hungary.

1500-10: **Erasmus**, *c.* 1466-1536, a humanist scholar, wrote *In Praise of Folly*, 1509, which satired church corruption and scholastic philosophy.

1510-20: The ***Utopia***, 1516, of **Thomas More**, *c.* 1478-1535, depicted an imaginary island lacking the evils of Europe.
Machiavelli, 1469-1527, wrote *The Prince*, 1513, the first ruthlessly pragmatic analysis of politics.
Martin Luther, 1483-1546, affixed his *95 Theses* to the door of Wittenberg Castle Church in 1517.
Sikhism, a combination of Hinduism and Islam, was founded *c.* 1519 by **Nanak**, 1469-1533.

1520-30: **Luther** was excommunicated in 1520. At the Diet of Worms, 1521, he argued for "justification by faith alone", the doctrine that no intermediary priest can aid salvation. Luther's attacks on the Catholic Church led to a rejection of papal authority and marked the start of the Protestant Reformation. This doctrine was adopted by many princes for its political implications. Luther translated the Bible into the vernacular *c.* 1525.

1530-40: **The Anabaptists**, who prophesied the imminent end of the world, gained control of Munster in Germany in 1534.
The Memonites, a Dutch sect, shared the Anabaptists' belief in pacifism and pastoralism.
In 1534 Henry VIII of England assumed full authority over the English Church.
Ignatius de Loyola, 1491-1556, formed the Catholic Jesuit order in 1540.

Literature

The spread of the Renaissance in the sixteenth century brought moments of great brilliance to national literatures, particularly those of England, France, Italy, Spain and Portugal. In England, the work of Wyatt and the Elizabethan poets and dramatists culminated in the genius of Shakespeare, who created a body of lyric poetry and drama of unmatched scope and power. Rabelais and Montaigne dominated French writing and poets such as Ronsard began the move towards classical themes in French literature. Pastoral idealism found expression in the Iberian Peninsula and Italy, and epic poetry flourished.

In China the novel form emerged and, in both India and Turkey, Islamic influence revitalized literary traditions.

1500-10: **Commedia dell'arte** developed from earlier peasant traditions in Italy. Actors improvised farce from a set scenario using stock characters such as Pedrillo, who became the French Pierrot, and the stupid but agile Harlequin. This boisterous form of theatre had little literary merit but influenced later drama, especially the comedies of **Molière** in France.

1510-20: The Portuguese dramatist **Gil Vicente**, 1470-1536, wrote naturalistic plays, full of intrigue and psychological insight. The innovate English poet and humanist **John Skelton**, 1460-1529, wrote scathing attacks on the court and clergy.
His German contemporary **Ulrich von Hutten**, 1488-1523, used dialogues to champion the cause of the Reformation.

1520-30: **Italian court** life and etiquette were vividly portrayed in the *Libro del Cortegiano*, 1528, by **Baldassare Castiglione**, 1478-1529.
Portuguese poetry reached its peak with the epic *The Lusiads*, 1572, by **Luis Camoes**, *c.* 1524-80.

1530-40: **Ludovico Ariosto**, 1474-1533, poet, dramatist and satirist, published *Orlando Furioso*, 1532, the greatest Italian epic of romantic chivalry.
Meistersang, a form of poetic song based on minstrel tradition, was popular in Germany. It was enlivened by the work of the devout Lutheran poet **Hans Sachs**, 1494-1576, a cobbler with a talent for comic verse.

Art and architecture

With the work of Michelangelo, Leonardo, Raphael and Bramante in the early 16th century, Italian Renaissance art reached a climax in the development of perspective, the analysis of the human form and the celebration of classical models. By 1520, however, this peak was past. Mannerism followed with its lack of harmony, distorted forms and search for novelty. Later in the century, the naturalistic experiments of Caravaggio, and Carracci's reassertion of classical canons in new dramatic compositions, pointed towards a new style – Baroque. Italy remained the official arbiter of taste in Europe, but the styles were more readily absorbed by the still solidly Catholic France and Spain. In the Protestant north the Reformation replaced Church patronage with that of merchants, encouraging the growth of secular art forms. Exploration in the New World carried European art abroad.

1500-10: **The Renaissance** reached its height in Florence and Rome *c.* 1500-20. **Leonardo da Vinci**, 1452-1519, painted the "Mona Lisa", 1503-6, achieving a more naturalistic effect by leaving outlines blurred. **Michelangelo**, 1475-1564, also in Florence at this time, completed the statue of "David", 1504. **Donato Bramante**, 1444-1514, built the Tempietto of S. Pietro, Rome, 1502, and was invited by Pope Julius II to design the new St. Peter's, 1506.

1510-20: **Michelangelo** in Rome completed the Sistine Chapel frescoes, 1512, and **Raphael**, 1483-1520, the Stanza frescoes in 1514, with their dazzling use of perspective.
Leonardo left for Amboise, France, 1516.
English Gothic art in its final stage was seen in Henry VII's Westminster Chapel, 1503-19. His tomb in the chapel by **Torrigiano**, 1472-1528, was the first use of Italian Renaissance motifs in England.

1520-30: **German painting** showed two trends: **Grunewald**, 1480-1528, painted the Isenheim altarpiece, *c.* 1512-16, in the late Gothic style but **Dürer**, 1471-1528, who had visited Italy, made use of Rennaissance ideals in his "Four Apostles", 1526.
Venetian painting broke from the Renaissance emphasis on drawing and perfect form. The Pesaro altarpiece, 1519-26, by **Titian**, *c.* 1487-1576, used a dramatic juxtaposition of contrasting colours and diagonals.

1530-40: **The Reformation** had interrupted patronage in Basel and forced **Hans Holbein**, 1497-1543, to seek work in England, where he arrived in 1526 with a letter of introduction from Erasmus to Sir Thomas More. After 1532 he settled there and painted court portraits.
The Wu school in China, including **Wen Cheng-ming**, 1470-1559, and **T'ang Yin**, 1470-1523, worked away from the Imperial Academy, painting ink landscapes with genre scenes.

Music

The growing Protestant Church in Europe redefined the liturgy for its own use and sacred music began to be performed by lay people in church and in their homes, widening the basis of religious music. City councils and individual patrons established their own groups of musicians, raising instrumental music to the same status as choral music.

1500-10: **The single or solo line** drew the interest of composers of the early 1500s, reacting to the increasing complexity of polyphony. Their interest is seen in their airs and lute songs.

1510-20: **The fantasia**, toccata and variations and the ricercar (forerunner of the fugue) were new instrumental forms devised to exploit the individual qualities of musical instruments.

1520-30: **German hymns or chorales** were composed in the 1520s and were firmly established by the end of the century. Set in four or five parts, they were often written to existing popular tunes.

1530-40: **Consorts of instruments** (viols or recorders) were cultural perquisites found in many wealthy homes. Families of instruments were usually played separately to give euphonious sonorities.

Science

A scientific revolution began in Europe in the 16th century and with it a long-held conception of the nature of the universe died. The century began with the later work of Leonardo da Vinci – a series of brilliant inventions which came to little because the scientific principles needed to realize them were hardly known – and culminated in Kepler's exact scientific calculations, based on Copernicus' idea of a Sun-centred universe and the precise astronomical observations of Tycho Brahe. This work finally destroyed the Aristotelian picture of the universe as a group of perfect crystal spheres centring on, and revolving about, the Earth, and opened the way for Galileo.

Advances were also made in medical science, particularly in anatomy, chemistry, larger-scale iron production and mining technology. Despite opposition from the Church, by 1600 science was firmly based on the experimental method and had turned its back on theology.

1500-10: **T.B. von Hohenheim Paracelsus**, *c.* 1493-1541, professor of medicine at Basel, made advances in chemistry although his system of iatrochemistry (chemical doctoring) was a mixture of observed fact and superstition.
The coach was invented in Hungary, probably in the early 16th century, and appeared in England in the 1580s.
Dissection of the human cadaver has been practised for a century in Europe but systematic dissections in the schools of Padua, date from the early 16th century.

1510-20: **Coins** containing copper mixed with gold or silver came into use in Europe in the early 16th century as a result of the great increase in prices caused in part by large imports of Spanish silver from Peru. **Henry VIII of England**, *r.* 1509-47, in particular, debased the currency in this way. Although later recoinage partly improved the real value of money, alloying became the rule.
A mass production technique for casting small brass objects was practised in Italy at this time.

1520-30: **Blast furnaces**, able to produce large quantities of cast iron, gradually evolved from earlier Stückofen. Cast iron so made was mostly used in weaponry, an industry in which England led Europe, selling to any customer who could pay the price, whether friend or enemy.
Coal became a major fuel in mid-16th century industrial Europe as the price of wood soared and forests disappeared. Coal mines opened in Liège and Newcastle.

1530-40: **Telesio**, an Italian who lived 1509-88, proposed the first system of physics to rival Aristotle's. His theory argued that heat and cold were the motive powers of the universe, an idea which influenced the work of the English philosopher **Francis Bacon**, 1561-1626.
Andreas Vesalius, 1514-64, advanced knowledge of internal anatomy in his book *De Humani Corporis Fabrica*, 1543. He accepted the chair of anatomy in Padua, 1537, and shocked the Church by dissecting corpses.

1540-50

Jean Calvin, 1509-64, established a puritan theocracy at Geneva in 1541.
The Catholic Counter-Reformation inspired Charles V to conduct the Schmalkaldic War, 1546-7, against the Protestant princes.
Brittany was united with France in 1547.
The Portuguese were the first Europeans in Japan, 1542, where the Jesuit **Francis Xavier**, 1506-52, founded a mission in 1549.

The sale of the monastic lands encouraged the rise of the middle gentry class. Inflation was accelerated by debased coinage and hit the poorest classes.

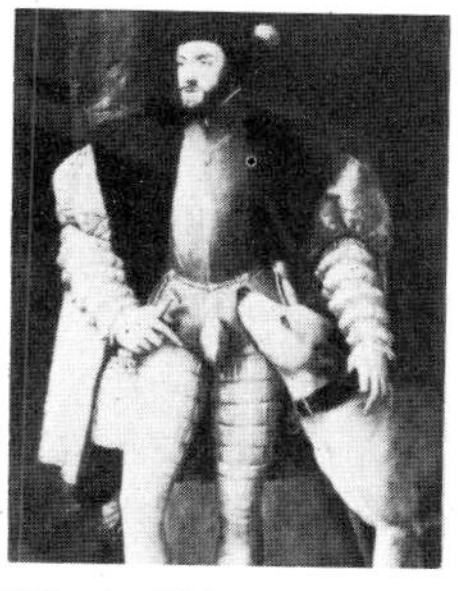

"Charles V" by Titian

Jean Calvin 1509-64, promoted the Reformation in Geneva, 1541. He espoused the doctrine of predestination – that God had already elected those to be saved but it was believed that exemplary conduct signified election.
Decrees issued by the Council of Trent on Church reform in 1545, initiated the Catholic, or Counter, Reformation.
Thomas Cranmer, 1489-1556, issued the Church of England's *Book of Common Prayer*, 1549.

La Pléiade, a group of seven French poets, of whom the greatest was **Pierre de Ronsard**, 1524-85, established the Alexandrine metre of a 12-syllable line and emphasized the dignity of the French language while turning to classical themes. Their manifesto was written by **Joachim du Bellay**, 1522-60, in *Défense et Illustration de la Langue Française*, 1549.

French Renaissance art copied Italian models as the painters **Rosso**, 1494-1540, and **Primaticcio**, 1504-*c.* 70, and the architects **Vignola**, 1507-73, and **Serlio**, 1475-1564, came to France to work on the Palace of Fontainebleau, 1528-60.
Mannerist painting in Italy, like the "Madonna with the Long Neck", 1534-6, by **Girolamo Parmigianino**, 1503-40, shows an elongation of figures, lack of harmony and a search for the new and unusual.

The lute became popular as an accompanying instrument. It could be used to accompany the new contrapuntal madrigal style that grew up in Italy after 1530. Later, the English adopted it.

A Sun-centred universe was proposed in the book *De Revolutionibus Orbium Caelestium* by **Nicolas Copernicus**, 1473-1543, published in the year of his death. In this revolutionary work, the Earth, Moon and planets, and outside them the stars, orbited around the Sun in circles. This theory is the basis of modern cosmology.
Zoological and botanical works were published in the mid-16th century by the French biologists **Gesner**, 1516-65, **Belon**, 1517-64, and **Rondelet**, 1507-66.

1550-60

The Peace of Augsburg, 1555, permitted each German prince to decide the religion of his subjects
After Charles V's abdication in 1556, **Philip II of Spain**, *r.* 1556-98, took over the Catholic offensive, while **Elizabeth I**, *r.* 1558-1603, confirmed England's Protestantism.
The influx of American silver to Spain accelerated inflation and caused hardship to the poor, but encouraged the rise of a European middle class.

Edward VI, *r.* 1547-53, faced peasant risings, 1549-50, and introduced a Protestant prayer book. **Mary**, *r.* 1553-8, restored Catholicism, burning heretics.

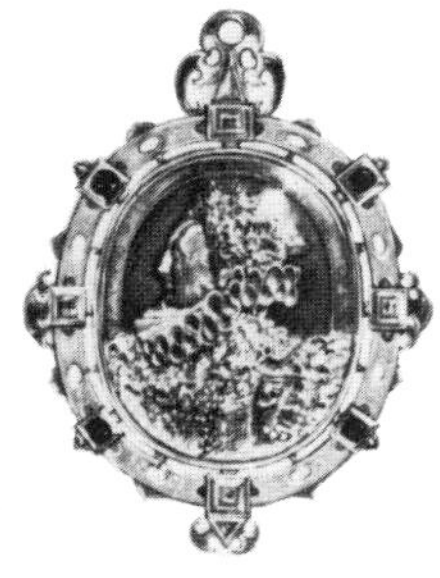

Elizabeth I: Armada jewel

Many English Protestant bishops, including Cranmer, were burned at the stake in the reign of Queen Mary. Elizabeth, her successor, re-established the Protestant Church but continued to burn heretics.
The Holy Roman Empire acknowledged Lutheranism in the Peace of Augsburg, 1555.
Protestantism in Scotland was united by the Calvinist **John Knox**, 1513-72, and became the national faith by Act of Parliament in 1560.

One of the great comic prose works of world literature, *Gargantua and Pantagruel* was completed in 1552 by **Francois Rabelais**, *c.* 1494-*c.* 1553. This bawdy, satirical tale of two grotesque giants was an erudite allegory vigorously attacking established institutions and conventional wisdom, mocking superstitious fears and defending free will.

The historian and painter **Vasari**, 1511-74, published the *Lives of the most excellent Painters, Sculptors and Architects* in 1550. **Palladio**, 1508-80, designed the Villa Rotunda, Vicenza, *c.* 1550, beginning work *c.* 1566. With its four porticoes and symmetrical plan, it is an example of his search for classical and harmonious proportions. **Benin bronze figures** of West Africa adopted freer poses as a result of contact with Portuguese culture.

Sacred polyphony declined in influence after the Council of Trent, 1545-63, regularized the musical forms suitable for the mass of the Roman Catholic Church.

Georg Bauer (Agricola), 1494-1555, a German doctor, gave a full description of mining, smelting and chemistry in *De Re Metallica*, published at Basel in 1556. Agricola's book is still the major source on the state of technology in the later Middle Ages.
Discoveries of metals in the 16th century included that of mercury, *c.* 1550, in Peru. Zinc, bismuth, cobalt and nickel were other metals used in alloys or mixtures.

1560-70

The French wars of religion began, 1562, between the Catholics and the Protestant Huguenots (mostly nobles and townsmen in west and south France).
The Calvinist and predominantly mercantile Dutch provinces began a long war of independence from Spain, 1568.
Nobunaga, 1534-82, introduced a dynamic period of Japanese centralization and expansion.
Akbar, *r.* 1556-1605, expanded the Mogul Empire and created a tolerant cosmopolitan culture.

Under Elizabeth I, *r.* 1558-1603, Protestantism was reinstated. The Statute of Apprentices marked the end of the authority of trade guilds.

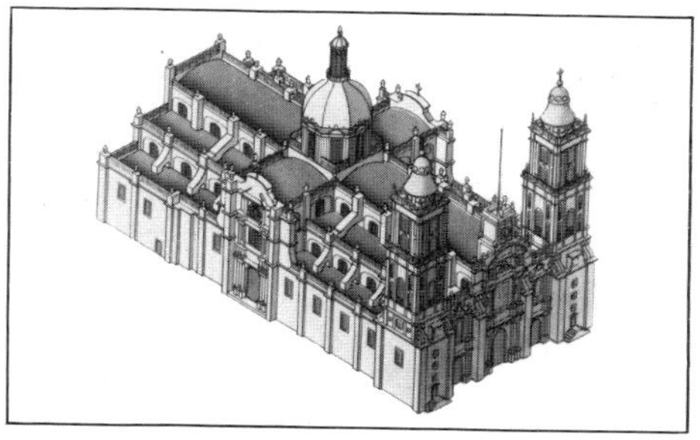

Renaissance Mexico city cathedral

The adoption of the 39 Articles in 1563, combining Protestant doctrine with Catholic church organization, finally established the Church of England. There were many dissenting groups, among them the Puritans, who opposed church ritual, the Separatists, who rejected Anglicanism entirely, the Presbyterians, who had synods instead of bishops, and the Brownists, a communistic sect. All but the Brownists and Catholics were tolerated.

The English poets **Thomas Wyatt**, 1503-42, and the **Earl of Surrey**, *c.* 1517-47, wrote in sonnets and blank verse – forms perfected by the Elizabethan poets **Shakespeare, Walter Raleigh**, 1552-1618, and **Edmund Spenser**, 1552-99.

Flemish painting saw the emergence of the individualist **Pieter Bruegel the Elder**, *c.* 1520-69, one of the greatest landscape painters and a remarkable satirist, whose series "The Months" dates from 1565.
Indian Mogul art assimilated the Persian tradition of miniature painting, which emphasized sumptuous decoration and lively colour patterns. This was combined with indigenous styles in the illustrations of Akbar's life in the *Akbar-nama*.

Japanese music began to win its individual character with the popularization of national forms of vertical bamboo pipe *(shakuhachi)*, three-stringed guitar *(samisen)* and zither *(koto)*.

Letter symbols for algebra and trigonometry were pioneered by **Vieta**, a French mathematician, 1540-1603. Words had previously been used for variables; the substitution of letters such as *x* and *y* greatly speeded up calculations and also removed many previous ambiguities.
Gerhard Kremer Mercator, 1512-94, published a map, 1568, using a projection of the world that has since borne his name.
The potato was introduced to Europe from South America by the Spaniards *c.* 1570.

1570-80

A European alliance defeated the Ottoman fleet at **Lepanto** in 1571, but Venice failed to use the opportunity to regain control of the eastern Mediterranean.
The Portuguese began their settlement of Angola, 1574.
The Dutch provinces, with increasing involvement in trade outside Europe, united in opposition to Spain, 1579.
Drake, *c.* 1540-96, an Englishman, circumnavigated the world, 1577-80.

Calvinists frequently attacked Elizabeth's religious policy in Parliament.
The Dutch wars upset English trade in the Low Countries.

Jean Bodin, 1530-96, a major French political theorist, published his *Six Books of the Commonwealth* in 1576, arguing that the basis of any society was the family. His most important contribution was an analysis of sovereignty. He argued that in any state sovereignty was necessary to prevent anarchy and that the exercise of monarchical power in conformity with the natural law was unquestionable, as it had divine authorization.

Michel de Montaigne, 1533-92, began his *Essays* in 1580.
In China the realistic, erotic novel the *The Golden Lotus* was published *c.* 1575.
John Lyly, *c.* 1554-1606, wrote *Euphues*, 1578-80, an early novel of manners.
Torquato Tasso, 1544-95, published *Jerusalem Liberated* in 1575 and the pastoral romance *Aminta* in 1573.

The brilliant Monoyama period in Japan, 1573-1615, is seen in the castle at Azuchi, built for **Nobunaga**, 1576-9, which contained large rooms decorated with murals. Screens painted with strong colours on gold ground came into fashion.
Spanish colonial architecture in the late 16th century, like Mexico Cathedral, 1563-1667, was based on contemporary Spanish mannerist styles but derived a pre-Columbian flavour from the native Indian labour.

Javanese fleeing the spread of Islam reached Bali and kept early traditions of Indonesian music in the works for the gamelan orchestra (mostly tuned percussion instruments).

Sir Thomas Gresham, 1519-79, established by will the first British institute for teaching science, which later housed the Royal Society.
Chinese pharmacology was summed up by **Li Shih-Chen** in his *Great Pharmacopoeia*, 1578. Chinese medicine was completely conservative and few new treatments were reported.

1580-90

Portugal and Spain were united on the death of Philip I of Portugal, 1580.
England assisted the Dutch revolt in 1585, executed the Catholic **Mary, Queen of Scots** in 1587 and defeated the **Spanish Armada** in 1588.
Pope Sixtus V, *r.* 1585-90, a supporter of the Counter-Reformation, began the internal reform of the papacy.
Hideyoshi, *r.* 1584-98, expelled the Portuguese missionaries in 1587.

War was declared with Spain in 1587. A newly constructed English fleet defeated an invasion by the Spanish Armada, 1588.

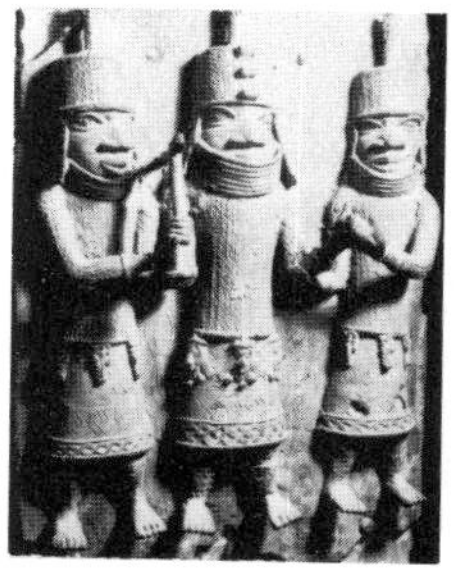

Benin bronzes, 16thc.

Akbar, the greatest Mogul emperor of India, *r.* 1556-1605, attempted to establish "**Din Illahi**" as a universal religion acceptable to his many Hindu subjects. Vegetarianism and other Hindu practices were supported by Akbar. Although the Din Illahi movement was influential for some time after Akbar's death it was discouraged by **Emperor Aurungzebe**, *r.* 1658-1707, and eventually collapsed under the 18th-century Muslim revival.

English drama entered its great period. **Christopher Marlowe**, 1564-93, in his *Dr Faustus*, *c.* 1588, perfected the blank verse of *The Spanish Tragedy* by **Thomas Kyd**, 1558-94.
Sir Philip Sidney, 1554-86, in his verse-prose *Arcadia*, 1590, drew on a tradition of pastoral romance established in Spain by **Jorge de Montemayor**, *c.* 1520-61, in his *Diana*.

English court portraiture and domestic architecture was given impetus by the Reformation.
Longleat, Wilts., 1567-80, was a house built in the rectangular style with large expanses of windows by **Smythson**, *c.* 1536-1614.
The Mannerist style emphasizing the bizarre and the tortuous spread through Europe; to Spain with **El Greco**, 1541-1614, to Germany in the works of artists like **Spranger**, 1546-1611, and to France with the second **Fontainebleau School.**

Equal temperament (based on equal half-tone divisions) was proposed by Prince Tsai-Yu in his *Handbook of Music*. It predated the West's recognition of its importance to harmony.

Decimals were introduced to mathematical calculations in physics by **Simon Stevin**, 1585.
Tycho Brahe, 1546-1601, and his assistant **Johannes Kepler**, 1571-1630, extended Copernican theory. Brahe made accurate observations of planetary movements. Using these results, Kepler calculated the actual orbits of the planets, which he found to be ellipses and not perfect circles. Kepler's results established astronomy as an observational science, free from any religious considerations.

1590-1600

Henry IV, *r.* 1589-1610, ended the French wars of religion and granted equal rights to Catholics and Huguenots in the **Edict of Nantes**, 1598.
The Dutch took over much of Portugal's former trade with the East Indies, 1595.
Japan invaded Korea, 1592-3 and 1597-8, but was expelled by the Chinese.
Spanish power gradually declined owing to the stagnation of her internal economy and the lack of a middle class.

Economic depression and the need for taxation to pay for the war with Spain served to make Elizabeth increasingly dependent on Parliament.

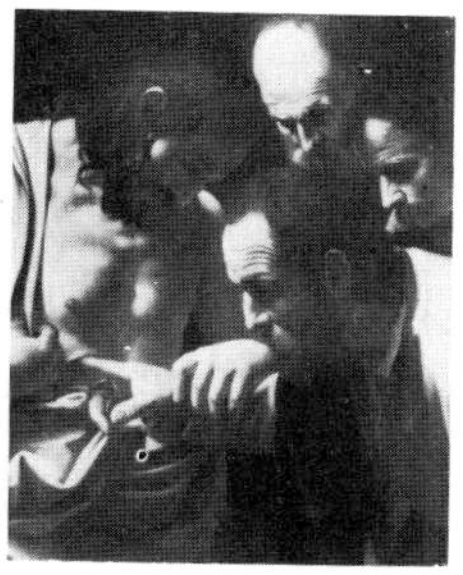

Work by Caravaggio

A Protestant movement, opposed to Calvinism, grew up in the United Provinces, denying the doctrine of predestination and arguing for religious tolerance. The movement came to be called **Arminianism**, after **Jacobus Arminius**, 1560-1609, who defended the Arminians in a controversy with his colleague **Gomarus.**
The Edict of Nantes, 1598, granted liberty of worship to the Huguenots, the French Protestant sect.

William Shakespeare, 1564-1616, consummate master of the English language, began *c.* 1591 to produce a stream of historical dramas and comedies revealing a remarkable range of human experience and thought. By 1600 he had written some 20 plays, including the comedies *As You Like It* and *A Midsummer Night's Dream* and the romantic tragedy *Romeo and Juliet*.

The precursors of Italian baroque painting were radically opposed to each other. "The Loves of the Gods", 1597, in the Palazzo Farnese by **Annibale Carracci**, 1560-1609, returned to the classical ideals, but with an emotional, anecdotal appeal and complex composition. "Doubting Thomas", *c.* 1600, by **Caravaggio**, 1571-1610, introduced a vivid realism and simplicity seen in the portrayal of Christ and the Apostles as ordinary men.

Sonata Pian'e Forte, 1597, by Giovanni Gabrielli, 1557-1612, was composed for two consorts, the first ensemble piece in which instrumentation was specified. It also used changes of volume.

Galileo, 1564-1642, wrote in 1597 stating his agreement with the Copernican system. Three years later, **Giordano Bruno** was burned by the Inquisition as a heretic for propagating the same idea.

1600-1660 Galileo and the new science

Principal events

The political and religious tensions generated by the Reformation in the previous century were brought to a head in the Thirty Years War, which involved most of the European powers and left the Holy Roman Empire in particular devastated from constant military activity. Although England remained out of the war, the same conflicts over religion and constitutional authority led to the execution of Charles I in 1649, but the establishment of the Commonwealth proved no solution.

Colonial trade expanded throughout the world bringing skirmishes and trade wars in India, America and Europe as the European powers jostled for supremacy, regarding control of trade as a tangible form of political power.

In China the Manchu dynasty brought strength and prosperity, while Japan withdrew into isolation after experiencing the disruptive impact of Christianity and European trade.

1600-6

Power struggles in Japan resulted in the Tokugawa (Edo) period, 1603, which advanced education and economic growth. **Charles IX**, *r.* 1604-11, a Protestant, succeeded to the Swedish throne after the deposition of his Catholic predecessor.
A period of anarchy in Russia resulting from rivalry between the boyars (nobility) began under **Boris Godunov**, *r.* 1598-1605, who was opposed by a pretender, the false **Dmitry.**

1606-12

In Japan the Tokugawa introduced **Confucianism** as the official religion, 1608, and Dutch traders arrived, 1609, rivalling the Spanish and Portuguese in the Far East.
The settlement of the French wars of religion was threatened by the murder of **Henry IV**, *r.* 1589-1610.

1612-18

The influence of the English East India Company extended to India, ousting the Portuguese as a rival to the Dutch.
Persecution of Christianity began in Japan, 1612, although trade with Europe increased.
The accession of Mikhail Romanov, *r.* 1613-45, in Russia established royal authority by ending local autonomy and strengthening serfdom.
A group of Tungus tribes in Manchuria grew powerful under **Nurhachi**, 1615-16.

1618-24

The Thirty Years War began in 1618 after a nationalist and Protestant revolt in Bohemia. By 1619 **Emperor Ferdinand**, *r.* 1619-37, had restored Catholicism in Bohemia.
Spanish troops invaded the Protestant Palatinate to ensure a route to the Netherlands.
The Pilgrim Fathers landed in North America in 1620.
Batavia was established by the Dutch as the centre of their Eastern spice trade, 1619.

National events

The Protestant gentry objected to the Stuart kings' claim to rule by divine right, and to the use of feudal levies to circumvent Parliamentary control over royal expenditure. Civil war followed and, after the execution of Charles I, Cromwell set up a shortlived Puritan Commonwealth.

The Gunpowder Plot, 1605, was a Catholic attempt to blow up the Houses of Parliament. **James I**, *r.* 1603-25, ended the war with Spain in 1604.

James I encouraged colonial development in North America. **A regular financial grant** to the king was discussed in Parliament but was never agreed.

Walter Raleigh, 1552-1618, was executed to placate Spain after the failure of his expedition in search of El Dorado.

Edward Coke, 1552-1634, supported the common law against absolute monarchy. **Francis Bacon** was impeached for corruption, 1621.

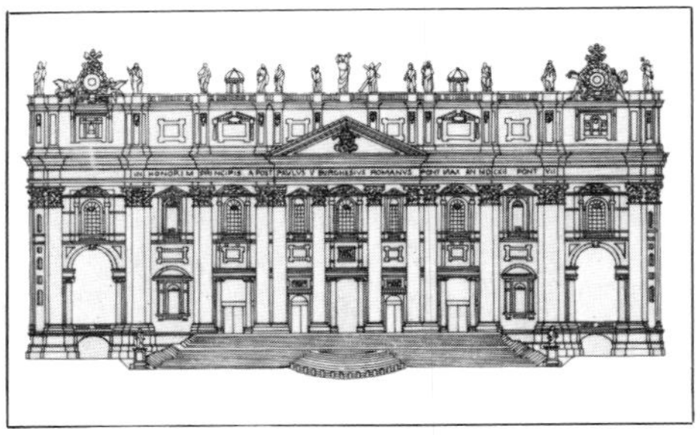

Baroque: façade of St. Peter's Rome by Maderna

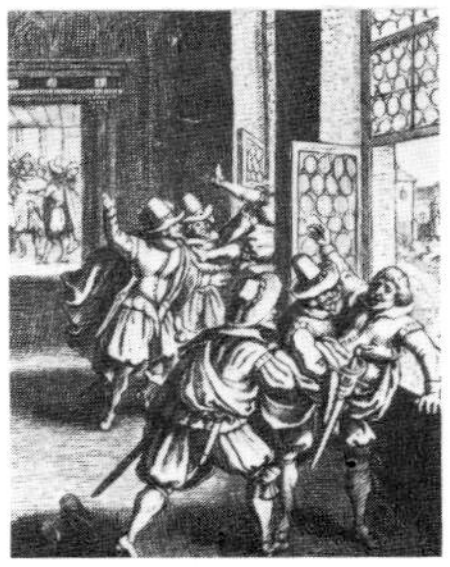

Defenestration of Prague

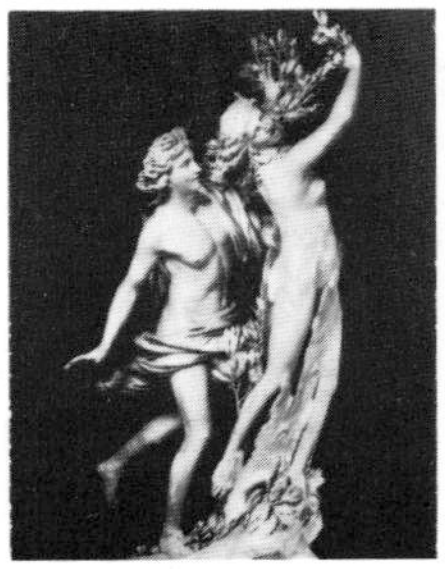

"Apollo and Daphne"

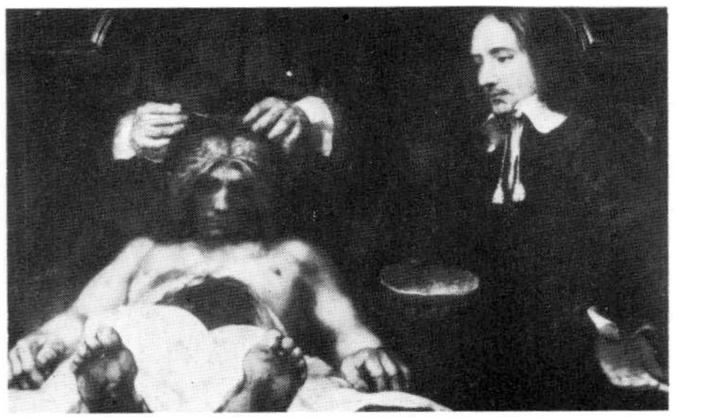

"The Anatomy Lesson" by Rembrandt

Religion and philosophy

As the basic assumptions and methodology of the natural sciences underwent a dramatic change with Galileo's suggestion that the workings of natural phenomena could be described exactly, philosophical and religious thought was also transformed. Theories of society based on the natural condition of man were common and resulted in the concept of the social contract, which could be renounced if the ruler rejected his duties to his subjects. Such ideas were used to justify widespread political revolts.

Descartes and Hobbes laid the foundations of modern philosophy by attempting to return to first principles, using only scientific or mathematical tools, and the same reliance on reason brought the beginnings of Deism, which would become popular in the 18th century.

In England in the Civil War, utopian ideas linking political and religious aims abounded.

Faustus Socinus, 1539-1604, in Poland, argued that Christ, though sinless, was not divine. He inspired the **Polish Unitarian movement**, which denied the existence of the Holy Trinity.
Johannes Althusius, 1557-1638, a Dutch Calvinist, said in 1603 that voluntary agreement should be the basis of political association. He advocated federalism and republican government.
The Tung-lin Academy, founded in China, 1604, revived Confucianism and attacked graft.

John Dee, 1527-1608, Elizabeth of England's astrologer, helped to revive interest in mathematics in England. As a magician and scientist he was a leading representative of the Hermetic tradition of alchemical study. This tradition, which sought to establish mystical connection among empirical phenomena with the help of experimentation, influenced the Cambridge Platonists and the development of Newtonian science.

Francisco Suarez, 1548-1617, a Spanish Jesuit, argued in *On Laws*, 1612, that a contract between ruler and subject was the basis of sovereignty. He hoped to refute James I of England's claim to rule by divine right.
The Dutch rejected Arminianism at the Synod of Dort, 1618. But after the publication of Arminius' works in 1629, they were granted freedom of worship in the United Provinces.

Francis Bacon, 1561-1626, elaborated a sophisticated method of establishing scientific truths, using observation and experiment to test hypotheses, in *Novum Organum*, 1620. He argued for the usefulness of scientific knowledge in giving man mastery over nature and conceived a scientific Utopia in *The New Atlantis*, 1627, a book that foreshadowed later developments in mid-17th century scientific thought.

Literature

The Elizabethan age in English literature culminated in the later work of William Shakespeare whose plays and sonnets epitomize the innovation and humanism of the Renaissance, while in Spain Cervantes wrote his picaresque *Don Quixote*.
As the century wore on the religious and political conflicts between Royalists and Puritans were reflected in English literature with the poetry of Marvell and the genius of Milton.

In France an effort to systematize the rules of language and literature was made by the newly founded Académie Française, which would stand until the 19th century, and an interest in classical models produced the tightly organized psychological dramas of Corneille and the accomplished verse of Malherbe.

Miguel de Cervantes, 1547-1616, blended and transcended the realistic and idealistic veins of Spanish prose writing in *Don Quixote*, published in two parts in 1605 and 1615. Its satirical theme of an amiable landowner who fancies himself an adventurous knight had a universality and a delicate juxtaposition of humour and sadness that influenced many later novelists.

Ben Jonson, *c.* 1572-1637, poet, critic and playwright, wrote *Volpone*, 1607, *The Alchemist*, 1610, and other comedies notable for their honesty. Other English dramatists were **John Webster**, 1580-1625, with *The White Devil*, 1608, **Thomas Middleton**, 1570-1627, **Thomas Dekker**, *c.* 1570-*c.* 1632, and the prolific **Francis Beaumont** and **John Fletcher**, *fl.* 1606-16.

Shakespeare's profound tragedies, including *Hamlet*, *Lear* and *Othello*, dealt with heroes trapped as much by the human condition as by their individual flaws of character. In 1608 he began writing his last, enigmatic plays, in which a spirit of reconciliation appears, among them *The Tempest*, *c.* 1612. His *Sonnets*, were printed in 1609.

The English metaphysical poets, who included **George Herbert**, 1593-1633, explored the unity of flesh and spirit in a style that influenced modern poetry. Erudition, wit, reason and passion were best combined in the devotional and love poems of **John Donne**, 1572-1631, (*Anniversaries*), who as Dean of St Paul's from 1621 preached a series of fine sermons.

Art and architecture

Baroque art emerged in Italy in the 1600s and reached its peak in the mid-17th century in the works of Bernini, Pietro da Cortona and Borromini. Its stylistic emphasis was on unity of composition, so that the parts were subordinate to the whole, an effect most expertly achieved in sculpture and architecture. Throughout the 17th century, Baroque spread from its basically Roman origins to Catholic Europe but had least influence in northern Protestant countries in spite of the achievement of Rubens and Van Dyck.

Bourgeois Dutch art flourished during the long war of independence from Spain, while native English painting was relatively unaffected by European developments, although Charles I patronised many continental artists. In France the tradition of rationalism produced the restrained classicism of Poussin and Claude.

Indian art flourished at the height of Mogul power.

Parisian town planning, like the Place Royale, 1605, with its smaller terraced houses, was the result of Henry IV's policy to support the new merchant classes and improve traffic circulation.
Painting in China within the traditional schools was dominated by literati and theorists, like **Tung Ch'i-ch'ang**, 1555-1636, who, in his "Dwelling in the Ch'ing-pien Mountains", emphasized the spiritual message of landscape.

Art and architecture in Mogul India reached its greatest achievement during the reigns of **Jahangir**, *r.* 1605-27, and **Shah Jehan**, *r.* 1628-58. Painting was characterized by the realism and vigour of Jahangir's picture albums, which were primarily portraits and depictions of the hunt. **The Taj Mahal**, 1632-43, the most renowned structure in India, was built by Shah Jehan as a mausoleum for his dead wife.

Italian baroque painting was dominated by the influence of the Carracci and Caravaggio. **Guido Reni**, 1575-1642, painted "Aurora", 1613, in the Carracci style. In Flanders, **Peter Paul Rubens**, 1577-1640, shows the influence of Caravaggio in his work "Descent from the Cross", 1611-14.
Palladian architecture was introduced to England by **Inigo Jones**, 1573-1652, whose Queen's House, Greenwich, 1616, is thoroughly classical.

The baroque style was epitomized in the magnificent sculptures of **Gianlorenzo Bernini**, 1598-1680, whose "Apollo and Daphne", 1622-5, established him as the greatest sculptor since Michelangelo.
Reality, allegory and myth are combined in one of **Rubens** masterpieces, the gigantic Medici cycle painted for the Luxembourg Palace, Paris, 1622-5.

Music

Many of the forms of music current today had their beginnings in 17th-century Europe.
The suite was developing to provide the basis of the later sonata, and opera and ballet were evolving from court entertainments. Italy was the centre of the stage, and interest in the solo line pressed forward the development of a new style of madrigal and fine singing styles.

Dances for lutes and consorts became popular, providing musical forms such as the *pavane*, *galliard*, *allemande* and *gavotte*, which were later gathered into composite pieces called suites.

La Favola d'Orfeo, 1607, by Claudio Monteverdi, 1567-1643, is the earliest European opera extant. The form arose from a search for a new way to express the ideals of classical drama.

The violin made its orchestral appearance in the *Vingt-quatre violons du roi*, set up by Louis XIII, *r.* 1610-43, as a court band. Later, bands of several consorts had up to 35 players.

Figured bass developed in Italian lute songs. Beneath the melody was written a base line with figures and signs to indicate the harmony of the inner parts without writing chords in full.

Science and technology

Religious dissent marked this period in Europe and from it rose the beginnings of modern science. The century was only a few weeks old when the Italian Giordano Bruno was burnt at the stake for heresy. He had conceived of the universe as infinite in time and space and filled with a multitude of suns each bearing planets, everything being in constant motion. His views were a major threat to orthodox theology at a time when the Catholic Church was threatened by the Reformation. Bruno's death probably persuaded Galileo, another Italian, to retract his belief that the Earth moves, and helped to shift the scene of progress towards the Protestant countries of Northern Europe.

The concept of scientific method was established and practical endeavour stimulated invention and enquiry. For example, the pumps required to clear water from mines prompted the investigation of air pressure and helped understanding of the heart's action.

De Magnete, a study of magnetism and electricity, was published in 1600 by an Englishman, **William Gilbert**, 1504-1603. He suggested that the Earth was a giant magnet with its own magnetic field.
Galileo Galilei, 1564-1642, studied the motions of falling bodies and discovered that they accelerated constantly towards the Earth. Galileo, the father of experimental science, drew conclusions from observation and experiment only without theological speculations.

The telescope was invented by the Dutchman Hans Lippershey, *c.* 1570-1619, in 1608.
Astronomia Nova, published in 1609 by Johannes Kepler, 1571-1630, argued that the planets moved around the Sun in ellipses and at varying speeds.
The moons of Jupiter and phases of Venus were discovered by Galileo in 1610.

The Art of Glass Making, 1612, by the Italian Antonio Neri, was one of many handbooks that helped the spread of technology.
John Napier, 1550-1617, introduced logarithms in 1614. Logarithmic tables prepared by **Henry Briggs**, 1561-1631, greatly facilitated their use.
Sanctorius, 1561-1636, founded the study of metabolism with his *De Medicina Statica*, 1614. He weighed himself over thirty years, recording changes in weight, pulse and temperature.

Harmonice Mundi, 1619, by Kepler, returned to the ancient concept of the harmony of the spheres in trying to find a relationship between music and astronomy. This work nevertheless contained a third law of planetary motion.
Francis Bacon published *Novum Organum*, 1620, in which theories are drawn from hypothesis and tested by observation and experiment.

1624-30
England, the United Provinces, Denmark and France allied against the Hapsburgs, 1625.
In France, Richelieu, 1585-1642, rebuilt royal power, attacking the Huguenots, 1628.
French settlements in the West Indies began in 1625, exporting sugar and tobacco, and emigration to Canada was encouraged among traders and fishermen.
The Tungus Manchus overran Korea, ousting the Ming dynasty from the Liao Basin, 1627.

1630-36
The Dutch East India Company seized part of Brazil for its sugar and silver, 1630.
Gustavus Adolphus, *r.* 1611-32, of Sweden invaded the Holy Roman Empire, 1630, to protect the Protestant cause against ruthless Catholic suppression.
Magdeburg was sacked in 1631 by the Catholic general **Tilly**, 1559-1632.
The war in Europe dislocated previous patterns of trade and industry and the search for colonial wealth increased.

1636-42
After the Shimabara revolt of the Christian peasantry, 1637, Japan cut her foreign trade and cultural contacts.
France first entered the Thirty Years War in 1639.
Spain was weakened by the establishment of Portuguese independence and a Catalan nationalist revolt in 1640.
England was close to civil war in 1641 after constitutional opposition to royal absolutism.

1642-8
The New England Confederacy was founded in 1643 for defence against the Indians.
The Manchus set up the **Ta Ch'ing** dynasty at Mukden 1644, replacing the Ming dynasty.
The English Civil War, 1642-6, resulted in military victory for Parliament and the puritans after the reorganization of their army, in 1645. Attempts to find a constitutional settlement failed.
France confirmed her new military superiority by defeating Spain at Rocroi, 1647.

1648-54
The Peace of Westphalia ended the Thirty Years War in 1648 with every participant exhausted.
The Fronde, a series of noble and peasant uprisings in France, tried to substitute government by law for royal power and voiced economic grievances but was crushed, 1648-53.
Charles I, *r.* 1625-49, of England was executed and a Commonwealth set up under **Oliver Cromwell**, 1599-1658. His Navigation Act, 1651, led to war with the Dutch, 1652.

1654-60
The rise of Brandenburg and Russia as military powers brought a new conflict in the Baltic and Poland, 1655-60.
The Venetians drove the Turks from the Dardanelles, 1656, following a period of anarchy among the Ottomans.
Anarchy after Cromwell's death led to the restoration of the English monarchy, 1660.
The war between France and Spain ended, 1659, emphasizing the Spanish decline and the rise of French power.

Charles I, *r.* 1625-49, wed the Catholic **Henrietta Maria**, 1625, and led an abortive expedition to assist the Huguenots, 1627.

Charles ruled without Parliament, 1629-40, relying in part on feudal dues for finances.
Thomas Strafford, 1593-1641, ruthlessly subdued Ireland.

Charles recalled Parliament, 1640, to finance an expedition to end Scottish religious revolt. He faced attacks on his economic and religious policies.

Charles raised the royal standard at Nottingham and sparked off the Civil War, 1642. The superiority of Cromwell's army forced him to surrender, 1646.

Charles was executed in 1649.
The Navigation Act, 1651, made the colonies economically dependent on the mother country.

Cromwell's dependence on the army was not popular and after his death anarchy followed. The monarchy was restored in 1660.

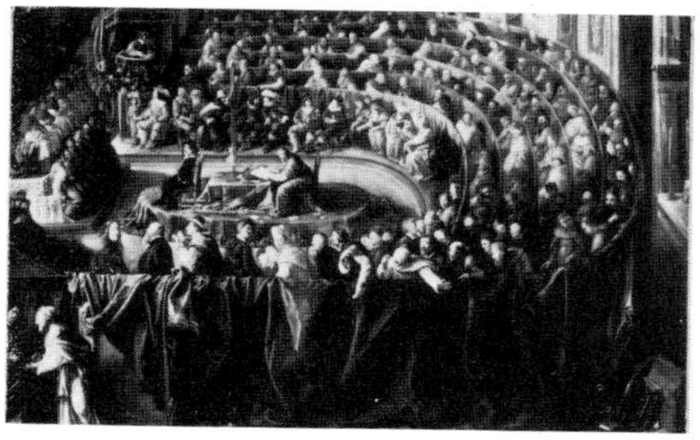

The trial of Galileo

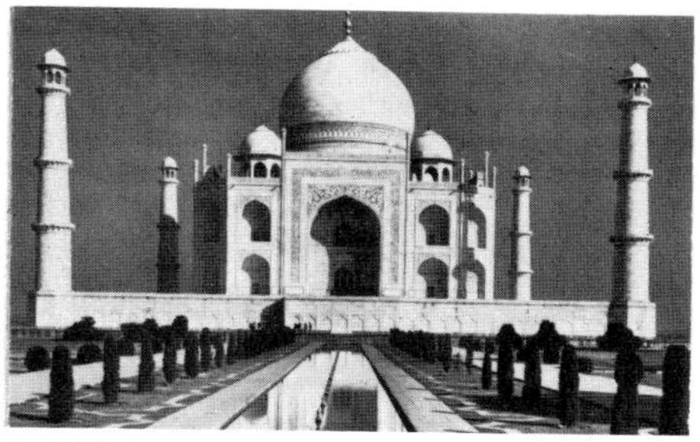

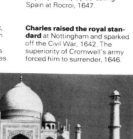

The Taj Mahal

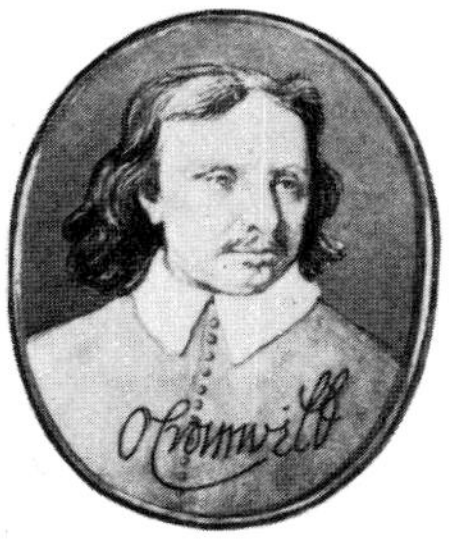

Oliver Cromwell

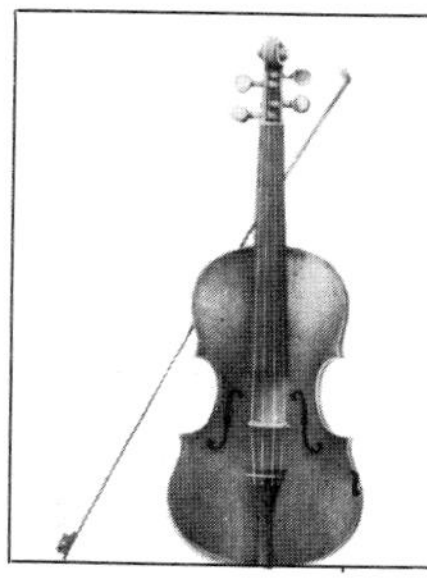

Late 17thc. violin

Herbert of Cherbury, 1583-1648, attempted to establish a belief in God based on rational enquiry rather than faith in *On Truth*, 1624. His belief that the basic tenets of religion were reasonable and universal was central to the growth of Deism.
Hugo Grotius, 1583-1645, a Dutchman, developed the theory of international law in *On Law*, 1625. He aimed to make war more humane, arguing that nations, like individuals, are bound by natural law.

Galileo Galilei, 1564-1642, an Italian, began modern science by uniting mathematics with physics. He distinguished real or "primary" qualities such as mass, from subjective "secondary" qualities such as colour. The religious opposition to his work highlighted the challenge of experimental science to the Aristotelian world-view, both philosophically and politically.

Cornelius Jansen, 1585-1638, a Frenchman, attacked the Jesuits and proclaimed strict predestinarianism, while staying within the Catholic Church, in the *Augustinus*, 1640. **Blaise Pascal**, 1623-62, supported the Jansenist movement in France, where it appealed to the nationalist opposition to papal power. The Jesuits rejected these views because they implied the denial both of free will and of the universality of redemption.

René Descartes, 1596-1650, who founded modern philosophy, attempted to establish a philosophical system from first principles alone, relying on mathematical logic and using systematic doubt as his method. He espoused a total dualism between mind and matter, arguing that the physical world was governed by deterministic laws, while the clarity and distinctness of ideas established their truth independently of any experience.

Utopian social and religious ideas flourished in England after the civil war. **George Fox**, 1624-91, founded the pacifist and egalitarian Friends, or **Quakers**, in 1652, while the **Diggers**, an agrarian communistic group, believed that religious ideas had diverted man from asserting his political rights in this world. **The Levellers**, another Puritan group, led by John Lilburne, 1614-57, demanded an egalitarian and republican society.

In *De Corpore*, 1655, Thomas Hobbes, 1588-1679, following Descartes' mathematical method, suggested that the universe comprises material particles moving in a void. This atomism also occurred in his political tract, *Leviathan*, 1651, in which he argued that in a state of nature men would fight because of their natural selfishness; they could only escape by means of a contract whereby they renounced their freedom to a supreme ruler.

Spanish drama was dominated by the popular and prolific **Lope de Vega**, 1562-1635, whose ingeniously plotted verse plays mixed comedy and tragedy.
Pedro Calderón, 1600-81, added deeper characterization in plays that reflected the richly ornate **culteranismo** style of the poet and satirist Luis de Gongora, 1561-1627.

The Passion play at Oberammergau, Bavaria, the most famous survival of its genre, was inaugurated in 1634. It has been performed every ten years except for three wartime interruptions.

French writers applied strict classical rules under the influence of **François de Malherbe**, 1555-1628. **Pierre Corneille**, 1606-84, successfully adapted these in a series of tragedies in Alexandrine couplets, starting with *El Cid*, 1637. His artificial but powerful plays based on Spanish and Roman heroes made drama the chief form of French classical literature.

English prose had acquired a new eloquence in *Anatomy of Melancholy* by **Robert Burton**, 1577-1640. This tradition was extended by **Sir Thomas Browne**, 1605-82, whose *Religio Medici*, 1643, was a reflective study of a doctor's spiritual life. An equally individualistic writer, **Izaak Walton**, 1593-1683, began to write *The Compleat Angler*.

English poetry reflected the political conflict of Puritans and Royalists. The Cavalier lyricists included **Sir John Suckling**, 1609-42, **Robert Herrick**, 1591-1674, and **Richard Lovelace**, 1618-57, whose best work was collected in *Lucasta*, 1649. On the Puritan side, **Andrew Marvell**, 1621-78, wrote poems on nature during the 1650s.

The greatest Dutch poet, Joost van den Vondel, 1587-1679, turned from satire to write his religious drama *Lucifer*, 1654.
In Germany, literature revived after the Thirty Years War, 1618-48. Poetry was much influenced by Lutheranism and the poet and mystic **Paul Gerhardt**, 1607-76, wrote outstanding hymns, including "O sacred head sore wounded".

Classicism in French painting was developed by **Nicolas Poussin**, *c.* 1594-1665, whose "Triumph of David", 1626, shows an abstraction and modelling based on antique ideals.
Realism and the skilful use of colour and light began to appear in Spanish painting in such works as the "Scenes from the life of St Bonaventura", 1629-30, by **José Ribera**, 1591-1652, and in the works of **Francisco de Zurbarán**, 1598-1664.

Roman high baroque painting was represented in the works of **Pietro da Cortona**, 1596-1669, whose masterpiece, the ceiling of the Gran Salone, Palazzo Barberini, painted in 1633-9, was a skilful illusion, its centre seemingly open to the sky.
Anthony van Dyck, 1599-1641, working at the English court, brought sophistication and elegance to English portraiture in the "Equestrian portrait of Charles I", 1633.

Dutch art found its greatest painter in Rembrandt van Rijn, 1606-69, whose psychological insight and technical virtuosity produced "The Night Watch", 1642. **Jan Vermeer**, 1632-75, painted domestic interiors and **Frans Hals**, *c.* 1581-1666, lively portraits.
The greatest of the baroque architects, Francesco Borromini, 1599-1667, produced his masterpiece of spatial ingenuity, **S. Carlo alle Quattro Fontane**, 1634-44.

French landscape painting as developed by **Claude Lorrain**, 1600-82, involved the formal arrangement of trees and a panoramic background as the setting for diminutive foreground figures, as in "Hagar and the Angel", 1646.
Individualist schools in China broke away from traditional painting. **Kung Hsien**, *c.* 1620-89, painted vast landscapes of great originality, such as his "A Thousand Peaks and a Myriad Rivers".

French classical architecture was initially developed by **Francois Mansart**, 1598-1666, whose Château of Maisons Laffitte, with its elegance, clarity and cool restraint, epitomized his subtly proportioned style.
Classical compositions and the use of indirect lighting are combined in the highly personal style of **Georges de la Tour**, 1593-1652, whose "St Sebastian", *c.* 1650, suggests the influence of Caravaggio.

Realism and a superlative handling of colour distinguish the works of the Spanish court painter **Diego de Velazquez**, 1599-1660. His "Las Meninas", 1656, an informal royal group, represents the culmination of his remarkable style.
Bernini's genius as an architect was affirmed in his Piazza of St Peter's, begun in 1656, which was both simple and original in design and reflected the dignity and grandeur of Mother Church.

Fugue developed, principally in Germany, as a contrapuntal treatment of one main theme. It remained the dominant form for solo organ until the 1700s but also had wider applications.

Bel canto, a lyrical and agile style of singing, developed in Italy. *Castrati*, men castrated before puberty, were renowned for their high, sweet, powerful voices, often used in opera.

Dynamic markings, such as *p* (piano) and *f* (forte), were used for the first time in 1638 by **Domenico Mazzochi**, 1592-1665, in Italy. He was quickly followed by other composers.

Ballet developed at the French court in the reign of Louis XIV, *r.* 1643-1715, who first danced it in 1651. Brought from Italy, ballet had been known at the French court since *c.* 1581.

The violin was perfected in Italy by the Amati, Stradivari and Guarneri families from 1650 to 1740. The great brilliance of violin tone soon overwhelmed the softer viols, which died out.

The koto became the national instrument of Japan. Its strings and movable bridge produced various five-note scales. Its solo music was often composed in the form of variations.

Johann Glauber, *c.* 1603-68, discovered many chemical compounds, including benzene, acetone and hydrochloric acid.
William Harvey, 1578-1657, discovered the circulation of the blood in 1628, but this was not confirmed until later improvements in the microscope took place. By studying valves Harvey realized that blood must flow in one direction only. His mechanistic view of man perfectly complemented Galileo's mechanistic universe.

In ***Dialogues concerning two World Systems***, published in 1632, Galileo presented the evidence for a heliocentric solar system in which the Earth moves. In 1633 Galileo was forced by Inquisition to retract his views.
Fen drainage in England since the 1620s had increased farm land. Fertilizer experiments also aided agriculture. English trade and industry prospered, especially coal production, iron mining and metallurgy.
The slide rule was invented in 1632 by Oughtred, 1575-1660.

Discours de la Méthode, 1637, by René Descartes, 1596-1650, established the deductive method, by which theories are deduced from observations and experimentally tested for validity. He also invented co-ordinate geometry, in which position can be described mathematically, an advance vital to the growth of engineering and the calculus.
Two New Sciences, published by Galileo, 1638, dealt with dynamics and established, more than any other work, experimental science.

Blaise Pascal, 1623-62, invented an adding machine in 1642. He also discovered the principles of hydraulics and investigated the theory of probability, showing that chance can be assessed mathematically.
Evangelista Torricelli, 1608-47, demonstrated in 1643 that air pressure is sufficient to hold up a column of mercury about 76 cm high, thus producing the first barometer. This discovery laid down the fundamental principles of hydromechanics.

The air pump, developed *c.* 1650 by Otto von Guericke, 1602-86, was used to show that sound cannot cross a vacuum. In 1654 Guericke conducted a famous experiment in which two teams of horses tried and failed to separate two evacuated hemispheres, thus demonstrating the power of air pressure, later to be harnessed in the first steam engines.

Christiaan Huygens, 1629-95, a Dutchman, invented the pendulum clock from 1656.
Academia del Cimento, the first scientific research institute, was founded in Florence, 1657.

1660-1720 The age of Louis XIV

Principal events

Louis XIV's schemes for the expansion of France brought him into conflict with the major European powers. The spectacle of his rule as an absolute monarch dominated 17th-century European politics, arousing the envy of lesser rulers including James II of England who was expelled in 1688 for trying to emulate him. This second English Revolution finally confirmed the victory of Protestantism and the rule of Parliament which would serve to inspire the Enlightenment thinkers of the following century, particularly in France itself.

Outside Europe the major powers fought for colonies – valued for their dual role as sources of raw materials and luxury goods like tobacco, sugar and spices and as markets for the produce of the home country.

The Mogul Empire in India declined after Aurungzebe had made the dynasty unpopular with his policy of intolerance towards Hinduism.

1660-66

Louis XIV began his personal rule in 1661 marked by a suppression of noble authority and the creation of a bureaucracy for local government. **K'ang Hsi**, *r.* 1662-1722, introduced a period of Chinese cultural splendour. **The English acquired Bombay** in 1661 and took New Amsterdam from the Dutch in 1664. **The Spanish colonies** became a prize sought after by the major naval powers in the reign of **Charles II**, 1665-1700.

1666-72

Louis XIV invaded the Spanish Netherlands but was opposed by the United Provinces, 1667-8. **The English and Dutch** fought an indecisive trade war, 1665-7. **Russia defeated Poland** for the Ukraine, 1654-67. **The Mogul Emperor Aurungzebe** revoked Hindu toleration in 1669, causing unrest in India. **The English** founded the Hudson Bay Company for the exploration of North America.

1672-8

The French again attacked the Dutch in 1672, backed by riches gained through the mercantilist economic policy of **Jean Baptiste Colbert**, 1619-83. They were opposed by Spain and the empire, who feared French strength in the north. **A two-party system** emerged in England in the 1670s.

1678-84

Brandenburg sent an expedition to West Africa in 1680. **Louis XIV** moved his court to Versailles to consolidate his independence from the nobility and the Parisians, 1682. **K'ang Hsi** took Formosa, 1683, which had been wrested from the Dutch by a Chinese pirate in 1661. **The Turks** besieged Vienna, 1683, but were defeated at **Mohacs**, 1687. **Robert de la Salle** explored the Mississippi for France.

National events

Attempts by Charles II to establish absolute power and by James II to restore Catholicism led to the Glorious Revolution of 1688. This brought the final victory of the constitutional idea and the supremacy of Parliament over the monarch. Protestantism was supreme and the rights of the gentry assured.

Charles II, *r.* 1660-85, restored peace, confirmed the supremacy of the Anglican Church in 1661 and promoted the growth of overseas trade.

London was rebuilt following the **Plague** and the **Great Fire**, 1665-6. Charles distrusted Parliament and made a secret alliance with Louis XIV.

Parliament forced through the **Test Act**, 1673, which aimed to prevent Catholics from attaining office, hoping thereby to exclude James from the throne.

After the scare of a Catholic coup (**The Popish Plot**), 1678-81, further attempts to exclude James failed. Charles revoked London's charter in 1683.

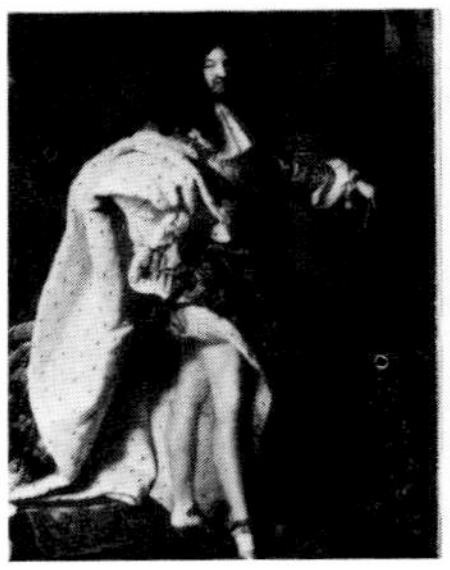
Louis XIV

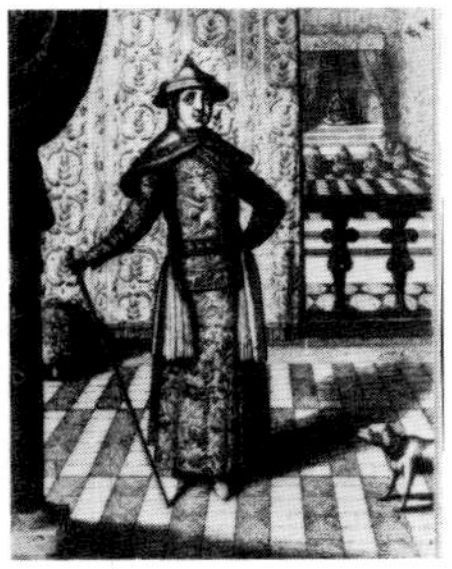
Chinese Emperor K'ang-hsi

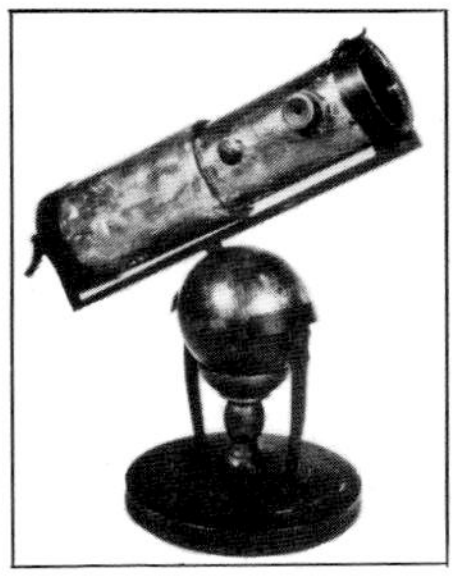
Newton's first telescope

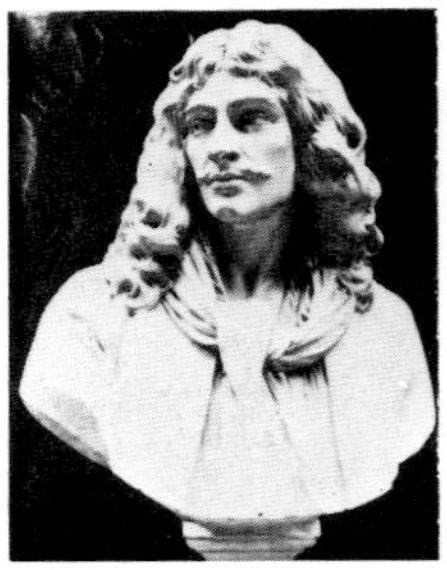
Molière

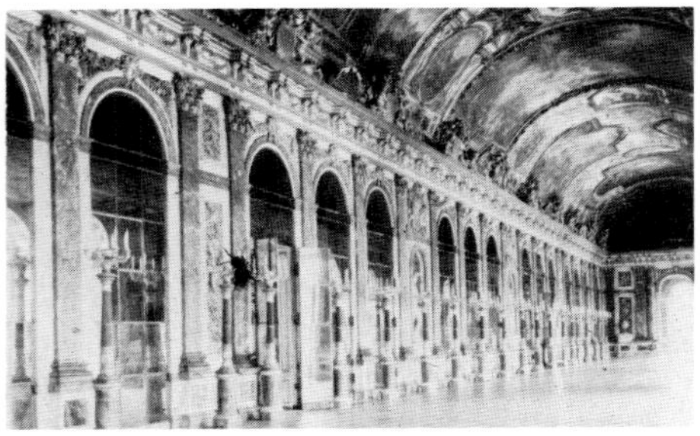
Versailles: the Hall of Mirrors

Religion and philosophy

European theories of knowledge and politics underwent important changes in the latter half of the 17th century, at a time when Newton's revolutionary ideas on the workings of the universe were transforming Western science. In Britain Newton himself, Locke and Berkeley took the empiricist position that knowledge was obtained by experience alone, in direct contrast to the rationalist views of thinkers like Spinoza and Leibniz who argued that knowledge of the world could be obtained by deductions from certain key principles like the nature of substance. Empiricism would dominate British philosophy hereafter and was to have a major influence on the thinkers of the French Enlightenment. Among political theorists, Locke and Pufendorf argued that political authority depended upon consent and took the form of a contract between the people and the king.

Mercantilists, such as **Thomas Mun**, 1571-1641, in England and **Jean Baptiste Colbert**, 1619-83, in France held that governmental regulation of the economy was necessary to increase the power of the state, since a nation's economic power depended on the bullion at its disposal. A key factor was the monopolization of colonial trade by the mother country.

The Swede Pufendorf, 1632-94, based his concept of natural law on "socialitas", the essentially social nature of man. He believed that agreement was the basis of political relationships and that human dignity implied the equality of all men. **The Old Believers** broke with the Russian Church in 1667 to counteract the reforms of the patriarch **Nikon**, 1605-81, who introduced Greek practices and reformed the parish clergy.

Spinoza, 1632-77, a Portuguese Jew, attempted to find a rational explanation of the universe and argued that since God cannot be other than He is then the world, His creation, cannot be other than it is. In his *Ethics*, 1675, he held that free will was an illusion which would be dispelled by man's recognition that the world was completely determined. He supported democracy as the most natural form of government, and rejected Descartes' dualism of mind and body.

Ralph Cudworth, 1617-88, an Englishman, published his *True Intellectual System*, 1678, admitting mental as well as material forces to science. He belonged to the **Cambridge Platonist** group of Christian humanists associated with the religiously tolerant "Latitudinarian" followers of Arminius. **Jacques Bossuet**, 1627-1704, upheld Louis XIV's absolute monarchy against Protestantism, arguing that any legally formed government is sacred.

Literature

Neo-classical drama, based on logic and Graeco-Roman stylistic rules, reached a peak in France in the tragedies of Racine and comedies of Molière. After the comic licence of early Restoration drama, English writers such as Dryden also turned to classical models, laying the ground for the Age of Reason. English journalism began with Addison and Defoe and satire developed with Pope.

The period also saw the publication of the chief work of the two greatest English Puritan writers, Milton and Bunyan, as well as developments in baroque poetry and picaresque prose in Germany and the emergence of new prose and verse forms in France with La Fontaine.

In Japan, Basho emerged as the supreme haiku poet.

German baroque literature was dominated by the influence of **Andreas Gryphius**, 1616-64, whose comedies and religious poems were collected in 1663. Another baroque writer was **Hans Grimmelshausen**, *c.* 1621-76, whose *Simplicissimus*, a graphic account of the experiences of the peasantry in the Thirty Years War, is regarded as the start of the German novel.

The Greek "unities" of action, time and space were given dramatic form in the French classical drama of **Jean Racine**, 1639-99, and **Molière**, 1622-73. Racine's *Andromaque*, 1667, blended poetic style with tragic passion. In comedies such as *Le Misanthrope* and *Tartuffe*, Molière exposed upper and middle class hypocrisies, mastering both plot and dialogue.

John Milton, 1608-74, an English poet who was politically prominent on the Puritan side, published in 1674 his final version of *Paradise Lost*, written, 1658-63, in strong blank verse, showing man as obsessed by sin. **In Mexico**, a Spanish nun, **Juana Inez de la Cruz**, published *A Nosegay of Poetic flowers*. Her works are among Latin America's best.

A forerunner of the English novel, *Pilgrim's Progress*, 1678, was an allegorical journey through life, told in plain prose with a wealth of narrative detail that overrode the narrow puritanism of its author, **John Bunyan**, 1628-88. **Madame de La Fayette**, 1634-93, wrote *La Princesse de Clèves*, the first French court romance of psychological depth.

Art and architecture

France replaced Italy as the centre of the arts in Europe. They were dominated by the royal patronage of Louis XIV, who rebuilt Versailles using the talents of Lebrun, Le Vau, Hardouin-Mansart and Le Nôtre. Baroque architecture was at its purest in Italy, at its most restrained in England and at its most extravagant in Spain and Portugal. Painting during the latter half of the 17th century produced few masterpieces though the works of Murillo in Spain, Pozzo in Italy, Claude in France and the landscapists in Holland were exceptional. In the American colonies architecture and painting adapted European styles to their own conditions.

The Rococo style emerged in France in the late 17th century, bringing to interior decoration the use of swirls, scrolls and conches in design, and finding a stylistic parallel in the elegant paintings of Watteau, dealing with life at court.

Spanish painting was represented by the works of **Bartolomé Estebán Murillo**, 1617-82, who founded the Seville Academy and became its first president, 1660. Eight of his 11 paintings for the almshouse of St Jorge, 1661-74, are regarded as his masterpieces. **The greatest exponent of Baroque, Gianlorenzo Bernini**, 1598-1680, went to Paris to redesign the Louvre, 1665. His plans were rejected, but he made a superb bust of Louis XIV.

Dutch landscape painting was exemplified in "The Avenue of Middelharnis", 1669, by **Meindert Hobbema**, 1638-1709, and in "Windmill at Wijk", *c.* 1670, by **Jacob van Ruisdael**, 1628?-82, a great Dutch landscapist. **The palace of Versailles** in France was first remodelled in 1669 by **Louis Le Vau**, 1612-70, France's leading baroque architect. The park and gardens at Versailles were designed by **André Le Nôtre**, 1613-1700, from 1662.

The classical landscape tradition of Poussin in France continued with **Claude Lorrain**, 1600-82, whose "Evening", 1672, expresses a questioning melancholy. **Jules Hardouin-Mansart**, 1646-1708, officially supervised building at Versailles after 1678. **Christopher Wren**, 1632-1723, the greatest English architect, began work on St Paul's Cathedral, 1675. It is a classical work with baroque overtones.

The Poussinistes/Rubensistes controversy was sparked off by the French Academy's publication of rules for painting, 1680. **André Félibien**, 1619-95, defended the orthodox view which valued drawing, idealism, formalized rules and the work of Poussin. **Roger de Piles**, 1635-1709, led the revolutionaries and argued the importance of colour, imagination and the works of Rubens. The Academy was officially associated with the ideas of Poussinistes.

Music

Baroque music grew up in Europe in the second half of the 17th century in the princely states of northern Italy and Germany. The freewheeling melodic lines and firm harmonic structure in the works of such composers as Diderik Buxtehude, 1637-1707, and Johann Sebastian Bach, 1685-1750, paralleled the ornamented but firm qualities of Baroque architecture.

The Restoration in England saw the introduction of the first public concerts in the modern sense. But music declined there for two centuries after the death of **Henry Purcell**, 1659-95.

The trio sonata was developed by Germans and Italians, using a quick first movement adapting *aabb* dance form with sections in contrasting moods and keys, and a slow second movement.

The chorale prelude, a free composition based on a hymn tune, exploited the varied capabilities of the organ. **Buxtehude's** chorale preludes influenced young composers such as **Bach**.

Continuo was played on a keyboard instrument – often a harpsichord – filling in the harmony between treble and bass lines, as in the cantatas of **Alessandro Scarlatti**, 1660-1725.

Science and technology

Isaac Newton's account of the workings of the universe surpassed Galileo's and provided a new framework for scientific thought. His exceptional insight into nature found definitions for concepts such as inertia and gravity that cannot easily be sensed. Newton's view of the universe, as one obeying set laws, accorded with the spirit of Protestant enquiry into the purpose of creation, in opposition to the Catholic world of personal salvation and divine intervention. Scientific advance in England and Holland was also stimulated by wealth from their growing trade.

Scientific communities grew up and provided scientists with the means to pool their researches, facilitating the spread of information and ideas internationally, while increasing the scientist's stature by granting him royal patronage. However, the growth of these communities contributed to the new division in men's minds between the impersonal sciences and the humanities.

The Royal Society was founded in London, 1660, and the French **Académie Royale des Sciences**, in 1666. **Marcello Malpighi**, 1628-94, used a microscope to discover capillary blood vessels in 1661, thereby confirming Harvey's theory of blood circulation. **Robert Boyle**, 1627-91, a British physicist, found that gas pressure varies inversely with volume (Boyle's law, 1662). His book *The Sceptical Chymist*, 1661, defined the concepts of element, alkali and acid.

Isaac Newton, 1642-1727, conceived of gravity, 1664-6, correctly concluding that it obeys an inverse square law. He discovered the spectrum 1666, and invented the reflecting telescope, 1671. **Francesco Redi**, *c.* 1626-97, disproved previous theories of the spontaneous generation of lower animals by showing in 1668 that flies are needed to produce the eggs of maggots. **A calculating machine** that could multiply and divide was made by Leibniz, 1646-1716, in 1671.

Greenwich Observatory, founded in 1675 principally to improve navigation, marks the standard meridian of longitude. **The speed of light** was calculated for the first time in 1675 by **Olavs Roemer**, 1644-1710, and shown to be finite. **A single-lens microscope** was made by **Anton van Leeuwenhoek**, 1632-1723, a Dutch biologist who discovered protozoa, 1677, and bacteria, 1683. **The calculus** was independently developed by Leibniz and Newton.

The pressure cooker was invented, 1679, by **Denis Papin**, 1647-1712. Papin also experimented on steam engines, using both the vacuum made by condensing cylinders and the power produced by the expansion of steam as water boils. **John Ray**, 1627-1705, laid the ground work for modern plant classification in his *Historia Generalis plantarum*.

1684-90

The Edict of Nantes, granting freedom of worship to Huguenots in France, was revoked by Louis XIV in 1685. Many Huguenots emigrated. **Russian eastward expansion** led to conflict with China, 1683-9. **James II of England**, *r.* 1685-8, was expelled for trying to restore Catholicism. **The Bill of Rights**, 1689, confirmed a constitutional monarchy. **In Japan**, the Genroku year period, 1688-1704, saw the rise of a merchant culture.

1690-96

William of Orange, *r.* 1689-1702, who reigned jointly with his wife, James II's daughter **Mary**, brought England into the war against France. **Peter the Great**, *r.* 1689-1725, began his policy of Russian expansion towards Azov for an outlet to the Black Sea and visited western Europe. **English trade in India** grew and a factory was set up in Calcutta in 1690. **European sugar traders** competed in the West Indies.

1696-1702

Charles II of Spain died in 1700 leaving **Philip, Duke of Anjou** and grandson of Louis XIV, as heir to his lands. This led to the **War of the Spanish Succession**, 1702-13. **Hungary** was recaptured from the Turks and by 1699 was restored to Austrian control. **Frederick II**, the Elector of Brandenberg, assumed the title King of Prussia with the consent of the emperor and became **Frederick I of Prussia**, *r.* 1701-13.

1702-8

The French won control of the *asiento* contract in 1702, which allowed them to transport Negro slaves to the Spanish colonies. **Portugal** joined the alliance against France, acting as a base for operations in Spain, 1703. **The Duke of Marlborough**, 1650-1722, defeated the French at **Blenheim**, 1704. **After Aurungzebe's death** in 1707, the Mogul Empire disintegrated as local princes asserted their autonomy, seeking assistance from European traders.

1708-14

The Sikhs became militant and made the Punjab virtually independent of Mogul rule, 1708. **A mass emigration** of Germans to America began in 1709. **War between the native Brazilians** and the Portuguese erupted after France attacked Rio de Janeiro in the course of the Spanish War of Succession. **The Treaty of Utrecht**, 1713, confirmed that France and Spain should not be united and left Britain in control of the *asiento* slave trade.

1714-20

The South Sea Company was set up in 1710 to increase British trade with South America. **The English East India Company** won trading concessions over rival companies from the Mogul emperor in 1717. **Frederick William I of Prussia**, *r.* 1713-40, laid the foundations of Prussian military power by setting up a standing army. **Louis XIV** died in 1715, with France's economy exhausted. **Manchu rule in Tibet** was assured by 1720.

James II attempted to bring a Catholic restoration and was expelled by an alliance of the landed gentry and merchant classes in 1688.

William III, *r.* 1689-1702, defeated James at the **Boyne** in Ireland, 1690. **The Bank of England** was founded, 1694, to finance William's wars.

The Stock Exchange was founded, 1698. The unpopularity of the war gave the Tories a majority in the Commons, 1701.

England and Scotland were formally united, 1707. England won Gibraltar in 1704, after coming to a trading agreement with Portugal in 1703.

The Tories formed a government, 1710, and reduced involvement in the war. They also tried to exclude merchants from Parliament.

George I, *r.* 1714-27, suppressed Jacobite risings in Scotland, 1715.

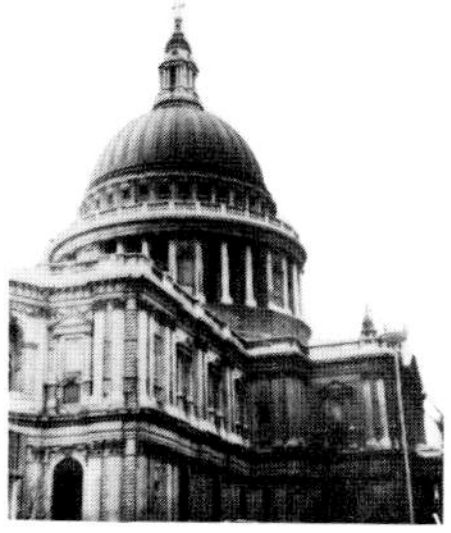

St. Paul's Cathedral

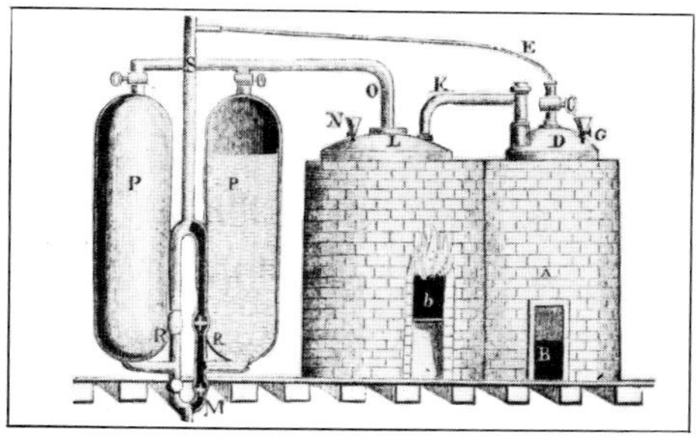

Thomas Savery's steam engine

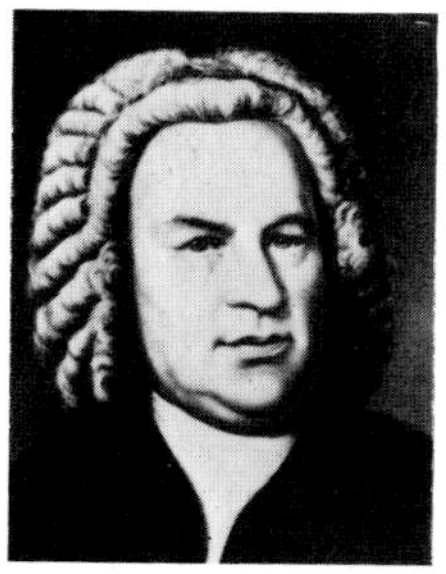

Johann Sebastian Bach

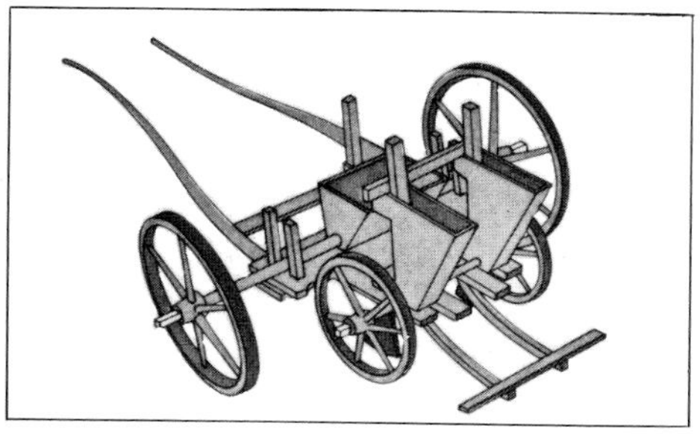

Jethro Tull's seed drill

Isaac Newton, 1642-1727, published the *Principia* in 1687. He defended the idea of a gravitational force by arguing that science should merely establish observed regularities, without speculating about underlying mechanisms. His view that the same set of laws, comprehensible with the aid of the physical sciences, apply throughout the universe was fundamental to the development of the mechanistic and optimistic philosophy of the 18th century.

John Locke, 1632-1704, produced the first thorough empiricist study in *An Essay Concerning Human Understanding*, 1690. He denied the existence of innate knowledge, arguing that the mind was a "tabula rasa" (a blank slate) that was only filled in by sensory experience. His *Two Treatises on Government*, 1690, which justified the English Revolution of 1688, claimed that rulers' legitimacy depended on their protecting the citizens' rights.

Govind Rai, 1666-1708, the tenth Guru of the Sikh religion, began a strategy of armed resistance to Mogul persecution and in 1699 gave the common surname Singh (meaning "lion") to the Sikhs. He also introduced the strict practices of the Sikhs, who were pledged to wear a turban, to carry a knife and never to cut their hair. The Sikhs eventually dominated the Punjab.

In *The Grumbling Hive*, 1705, **Bernard de Mandeville**, 1670-1733, argued that all individual actions are motivated by self-interest but the net effect of many such actions is the general good. This idea influenced later *laissez-faire* economists. **The Earl of Shaftesbury's** *Letter concerning Enthusiasm*, 1708, helped to popularize **Deism**, or Natural Religion. Deists criticized formal religions, intolerance and extremism.

In *The Principles of Human Knowledge*, 1710, **Bishop Berkeley**, 1685-1753, starting from the belief that all knowledge must come from perception, went beyond Locke and argued for an extreme idealism. He claimed that all we perceive is in the mind alone. As a result, to exist is merely to be perceived, and thus the continuing existence of the external world depends on God's external perception of it. Berkeley thus hoped to refute atheism definitively.

The metaphysical views of **Gottfried Leibniz**, 1646-1716, were summed up in the *Monadologie*, 1714. He saw the universe as comprising an infinity of "monads", dimensionless entities endowed with souls, in pre-established harmony with each other. Leibniz held that God had chosen this as the best of all possible worlds and that the evil in it was necessary. He also worked influentially in symbolic logic.

The French poet Jean de La Fontaine, 1621-95, read his *Discours en Vers* to the Academy in 1684. His verse *Fables*, begun in 1668, conveyed human insights through the old tradition of animal stories. **In Japan**, the succinct three-line poetic form called **haiku** reached a peak in the poetry of **Matsuo Basho**, 1644-94.

Restoration English drama, dominated since 1660 by the influence of the poet, critic, satirist and playwright **John Dryden**, 1631-1700, culminated in *Love for Love*, 1695, by **William Congreve**, 1670-1729. This comedy of manners improved on the comedies of **William Wycherley**, 1640-1716, whose bawdy play *The Country Wife*, 1675, rejected puritan morals.

John Dryden ended a fruitful career with *Fables Ancient and Modern*, 1699. He also wrote a significant political satire on Monmouth and Shaftesbury, *Absalom and Achitophel*. His clear, elegant verse and prose influenced many, including **Samuel Butler**, 1612-80, whose *Hudibras*, 1663-78, was a satire on Puritanism. *All for Love* was Dryden's best play.

English journalism arose to satisfy the new middle class market. **Daniel Defoe**, 1660-1731, journalist, novelist, merchant and spy, issued the *Review*, 1704; **Richard Steele** the *Tatler*, 1709; and **Joseph Addison** the *Spectator*, 1711. Addison and Steele were informed and sensible essayists on literary, political and social issues of the period.

English Classicism found its wittiest poet in **Alexander Pope**, 1688-1744, whose *Rape of The Lock* was published in its full form in 1714. The main defenders of classicism in France were **Nicolas Boileau**, 1636-1711, **Jean de la Bruyère**, 1645-96, and **Jacques Bossuet**, 1627-1704.

The tradition of the picaresque novel (recounting exploits of an adventurer), which derived from Spain, was used by **Alain Le Sage**, 1668-1747, in his *Gil Blas*, 1715. **The romantic Japanese dramatist Chikamatsu Monzaemon**, 1652-1725, wrote *Love Suicides*, the last of many successful plays both in **kabuki** (song-dance) and **jojuri** (puppet) forms.

The arts in France under Louis XIV were dominated by **Charles Lebrun**, 1619-90, who was director from 1663 of the French Academy and was also responsible for the Versailles Galerie des Glaces, completed 1684, and the Salons de la Guerre and de la Paix, 1686. **Venetian architecture** was represented by the Sta Maria della Salute, 1687, by **Baldassare Longhena**, 1598-1682. It was classical in conception but had baroque overtones.

Spanish Baroque style in architecture was derived from the works of **José Churriguera**, 1665-1725, whose east end of St Esteban, Salamanca, 1693, shows the extravagant surface decoration and richly guilded ornament called Churriguesque. **The leading exponent of the Baroque style of illusionist decoration** in Italy was **Andrea Pozzo**, 1642-1709, whose ceiling of S. Ignazio, Rome, 1691-4, was a masterpiece of perspective and trompe l'oeil.

Baroque architecture in England was exemplified in Castle Howard designed by **John Vanbrugh**, 1664-1726, from 1696. He worked with **Nicholas Hawksmoor**, 1661-1736, on this and other buildings. **The beginnings of the Rococo** style were seen in the "arabesques" and "grotesques" designed by **Jean Bérain**, 1640-1711, and **Claude Audran**, 1658-1734.

The grandeur and formal design of Versailles were emulated throughout Europe in the 18th century with the founding of St Petersburg, Russia, 1703, and in England with Blenheim Palace, 1705, built for the Duke of Marlborough by Vanbrugh.

European artists, like **Gustavus Hesselius**, 1682-1755, from Sweden, settled in Philadelphia in 1711 and executed realistic portraits and history paintings. **A triumph for Rubensistes** was evident in the vast ceiling of the Chapel at Versailles, 1708, by **Antoine Coypel**, 1661-1722, which is in the manner of Roman baroque illusionism.

The "fête galante", a new genre of painting characterized by exquisite scenes of pleasure and dalliance, was introduced by **Antoine Watteau**, 1684-1721, in the "Departure for the Island of Cythera", 1717. **The first phase of the Rococo in France**, 1700-20, largely in sculpture and interior design, was exemplified in the fountains of **Gilles Marie Oppenordt**, 1672-1742, designed about 1715 and showing twisting figures, shells and scrolls.

Baroque composers' awareness of modulation through a cycle of fifths brought more harmonic interest to their music, but tuning problems grew when harmony wandered far from the home key.

The concerto was developed by **Arcangelo Corelli**, 1653-1713, and others as a concerto grosso for a group of instruments and orchestra or for a virtuoso solo performer with orchestra.

Fugues in organ music were often paired with a free composition for contrast, giving the prelude and fugue or toccata and fugue found in many works by Buxtehude and Bach.

German suites by Bach and others mixed free forms such as prelude and toccata with dance forms such as *allemande, sarabande, minuet, gavotte* and *gigue*.

The pianoforte was invented. It is usually attributed to the Italian **Bartolommeo Cristofori**, 1655-1731, who in 1709 substituted hammer action for the harpsichord's plucking action.

Italian became the usual operatic language in Europe, though France still kept its own opera. **Georg Friedrich Handel**, 1685-1759, composed Italian opera in England after 1719.

Newton's *Principia*, probably the most important book in science, was published in 1687. The first section dealt with the behaviour of moving bodies and enunciates Newton's three laws of motion, as well as the principles of gravitation. The second deals with the motion of bodies in fluids, and also wave motion. The third utilizes the principles expounded in the earlier sections to explain the motion of bodies on the Earth and in the universe. It was a revolutionary conception.

Christiaan Huygens, 1629-95, a Dutch physicist, put forward a wave theory of light, 1690. Newton at this time proposed a particle theory. Later science would prove them both right.

The first practical steam engine was invented, 1696, by **Thomas Savery**, 1650-1715, a British engineer. **Thomas Newcomen**, 1663-1729, another British engineer, invented the atmospheric steam engine, used until 1934 to pump water from mines. Both engines had a great drawback: the cylinder had to be cooled at each stroke, wasting 99% of the heat from the fuel. **Agriculture** was improved by sowing seeds in rows with a drill invented in 1701 by **Jethro Tull**, 1674-1741, in England.

Opticks, published by Newton in 1704, encapsulated his work on light. His particle theory of light held great sway for a century before Huygen's wave theory was revived. **Edmond Halley**, 1656-1742, British Astronomer Royal, proposed the idea that comets orbit the Sun and, using Newton's principles, correctly predicted in 1705 the return of the comet that now bears his name.

Jesuit missionaries made an accurate map of China, 1708. **High quality iron** was produced in 1709 by **Abraham Darby**, 1677-1717, a British iron worker. The iron was smelted with coke and moulded in sand for cheap production, making the cast iron steam engine an economic proposition. **Francis Hawksee**, an Englishman, made the first accurate observations of capillary action in glass tubes in 1709. **Prussian blue**, a coloured dye, was produced from 1710.

The mercury thermometer was invented in 1714 by **Gabriel Fahrenheit**, a German physicist, 1686-1736. **Jethro Tull** brought the horse-hoe to England from France. **Thomas Lombe**, 1658-1739, an Englishman, patented a machine to make thrown silk in 1718.

INDEX

Anthony, St, 339R
Anthropology, 22
Antioch, 114, 176–7, *177*, 341A
Antonine emperors, 99
Antonine Wall, 100, 182, *182*, 338N
Antonius Pius, 99–100, 338P
Antony, Mark, 97–8, *97*
Antony and Cleopatra
   (Shakespeare), 285
Antwerp, 278
   banking, *238*
   High Renaissance, 260–1
Anu, 329R
An-Yang, 44–5, *45*
Aphaia, temple carving, *71*
"Aphrodite from Cnidus"
   (Praxiteles), 335A
Aphrodite of Melos, *336*
Apollo, 332R
"Apollo and Daphne" (Bernini),
   304, 352A
"Apollo conducting Beatrice of
   Bergundy to Frederick
   Barbarossa" (Tiepolo), *305*
Apollo Fountain, 261, *261*
Apollonius of Perga, 83
Apollonius Rhodius, 74
Apollo of Veii, *88*
Apostles, 110–11, *110–11*
Appian Way, 335A
Apprentices, Statute of, 275, 351N
Apprenticeship, medieval, 193
Apuleius, 113, 338L
Aquae Sulis. *See* Bath, England
Aquileia, *99*
Aquinas, St Thomas, 186, 188–9,
   188–9, 346R
*Arabian Nights' Entertainments,
   The*, *149*, 342M
Arabs
   Abbasid dynasty, 148, *148*, 342–3P
   and Carolingians, 152–3, *152*
   coinage, *149, 159*
   conquests, *134*, 137, 143–4, 148–9
   language, 145
   music, 341–2M
   science, 342–4S, 347S
   Umayyad. *See* Umayyad dynasty
Aragon, 171
Aramaeans, 28, 56, 332–3P
Aranyakas, 333R
Ara Pacis (Altar of Peace), *106*, 107,
   337A
Ararat, Mount, 29, *30*
Arbela. *See* Gaugamela
Arcadia (Sidney), 351L
Arcadians, 68
"Arcadian Shepherds" (Poussin),
   308
Arcadius, 136
Archaeology
   development, 22–3, *22–3*
   techniques, 20–1, *20–1*
"Archangel Raphael and Tobias,
   The" (Stoss), *253*
Archimedean screw, *82*, 83, 348S
Archimedes, *82*, 83, 336S
Architecture
   African, *229*
   Asian, *226–7*
   Athenian, 78–9, *78*
   Baroque, 304–5, *304–7*, 354–5A
   Carolingian, *152–3*
   Dravidian, 118, *118*, 344A
   Dutch (17th century), *311*
   Egyptian, 48–9, 329–31A
   Elizabethan, 280–1, *280*
   French (17th century), 308–9,
   308–9, 314, 353–4A

Architecture (continued)
   Gothic. *See under* Gothic art
   Greek, *40*, 76–7, *76, 78*, 79,
   333–5A
   Islamic, *145, 148*, 150, *150*,
   341A, 343A
   Japanese, *131–2*
   Mannerist, *264*
   Mesoamerican, *230–1*
   Minoan, 38
   Palladian, 258–9, *258*, 296,
   352A
   Renaissance, 253, *253*, 261, *261*
   Florentine, 250, *250–1*
   French, 262–3, *263*
   High, 256–9, *256–9*
   Italian, 254–5, *255*
   Restoration, 296–7, *297*
   Rococo, 355A
   Roman, *92*, 97–8, *106, 106–9*, 108,
   115, 337A, 339A
   Romanesque, 172–5, *172–5, 191*,
   343A
   Russian, *320, 322–3*
   South American, *232–3*
   Spanish, *170–1*, 355A
   Sumerian, *30*
   Turkish, *220–1*
   *See also individual architects and
   buildings by name*
Ardabil carpet, *150*
Ardagh Chalice, *141*
Ardashir I, 339P
Arena Chapel, Padua, *217*, 346A
Areopagitica (Milton), *290*
Arginusae, Battle of, 70
Argyll, Marquess of, *299*
Ariadne, *112*
Arianism, 122–3, 136, *137*, 339R
Arinna, 52
Ariosto, Ludovico, 282, 250L
Aristarchus, 83, 336S
Aristophanes, 75, 335L
Aristotle, 63, 68, 82, 335RLS, *335*
   influence of, 188–9, *189*
   as tutor, 80, *188*
Aristoxenus, 335M
Arithmetic, early, 116
Arithmetica (Diophantus), 339S
Ark Royal (ship), *277*
Armada, Spanish, 272, 277, *277*
Armada portrait of Elizabeth I, *274*
Armenia, 98–9, *98*, 224, 339P, 341P,
   344P
Armide (Lully), *319*
Arminianism, 287, 351–2R
Arminius, Jacobus, 351R
Arms and armour
   during English Revolution, *289*
   Frankish, *136*, 341S
   Greek, *68*
   Japanese, *131*
   Norman, *181, 204*
   Roman, *91, 97*
"Arnolfini Wedding, The"
   (van Eyck), *252*
Arnolfo di Cambio, 217, *250*
Arouet, François Marie. *See*
   Voltaire
Arras, Congress of, 191, 203
Ars nova, 347M
Art
   Anglo-Saxon, 164–5, *164–5*
   Athenian, 76, 78
   Baroque, 304–7, *304–7*
   Byzantine, 146–7, *146–7*
   Carolingian, *152–3*
   cave. *See* Cave art
   Celtic, 67, *67*, 86, *86–7*, 140–1

Art (continued)
   Chinese, 128–9, *128–9*, 332A,
   338A, 340A, 345A
   Classicism, 304
   Dutch (17th century), 301, 310–11,
   310–11, 352–4A
   Egyptian, *34–5*, 151, 328A,
   330–2A
   French (17th century), 308–9,
   308–9, 351–5A
   Gothic. *See* Gothic art
   Greek, *40–1*, 76–9, *76–9*
   Indian, 120–1, *120–1*
   Irish, *66–7*, 67, *141*
   Islamic, *144–5*, 150–1, *150–1*,
   342A
   Italian. *See under* Italy
   Japanese, 132–3, *132–3*
   Mannerism, 264–5, *264–5*, 304,
   350–1A
   Mesopotamian, 151, 329–30A
   Minoan, 38–9, *39*, 329–31A
   Mogul, 351–2A
   Phoenician, *54–5, 332*
   prehistoric, 24–5, *24–5*
   Renaissance, 304
   Flemish, 252–3, *252–3*
   Florentine, 250–1, *250–1*
   French, 262–3, *262–3*
   German, 260–1, *260–1*
   High, 256–9, *256–9*
   Italian, 254–5, *254–5*
   Netherlandish, 261, *261*
   Roman, 106–9, *106–9*, 336–9A
   Romanesque, 172–5, *172–5, 191*
   Turkish, 151, *221*
   See also Architecture, Metalwork,
   Painting, Pottery, Sculpture *and
   individual works and artists*
Artaxerxes I, *61*
Artaxerxes III, 335P
Arthashastra (Kautilya), 62
Arthur, King, *20–1*, 138, 340N
Arthur, Prince of Wales, *185*, 266
"Arthur Capel with his wife and
   children" (Johnson), *296*
Art of Glass Making, The (Neri),
   352S
Art poétique, L' (Boileau), 317
Arundel, Earl of, 296
Aryabhata, 340S
Aryans, 32, 60, 330PR, 332P
Ascham, Roger, 282
Ashburnham *Pentateuch*, 341A
Ashikaga dynasty, 132
Ashoka, 62–4, *62*, 336PR
Ashokan pillars, 63, *63*
Ashur, 57
Ashurbanipal, 50, 57, *57*, 334L
Ashurnasirpal II, 54, 56, *57*, 333A
Ashur-uballit I, 56, 331P
Asia
   Marco Polo in, 222–3, *222–3*
   Southeastern empires, 226–7,
   226–7
   western, early, 28–9, *28–9*
   See also individual civilizations
Asia Minor, 52, 344P
Asiento, 355P
Askia the Great, 229
As-Saheli, *228*
Assassins, *177*, 224
Asser, 343L
Assurdan II, 333P
Assyria, 330–1P, 333PA, 334A
   boat, *56*
   carvings, *56–7*
   conquests, 53, 56–7
   cuneiform, *28–9*

Assyria (continued)
   Empire, 50
   hunting, *56*
   painting, *57*
   music, 332M
   technology, 332-3S
   tools, 332S
Assyro-Babylonia, mythology, *32*
Astarte (Anat, Tanit), *55*
Astrée, L' (Urfé), *317*
Astrolabe, 235, *278*, 343S
Astrology, 338S
Astronomia Nova (Kepler), 352S
Astronomical "computer", 83, *83*
Astronomy
   Arabic, 347S
   Chinese, 337S, 339S
   English, early, 117
   European, 302, *302*
   Greek, 82-3, *82*
   Indian, 340S
   Mayan, 339S
"Astronomy" (Gianbologna), *265*
Astrophel and Stella (Sidney), 282
Asvaghosa, 338L
As You Like It (Shakespeare), 284,
   285, 351L
Atahualpa, 236, *236*
Ataulf, 340P
Athelstan, King of Wessex, *139*,
   343N
Athelstan (Danish commander). *See*
   Guthrum
Athena, Temple of, *71, 73*
Athens, 332P, 334–5PS
   Acropolis, 61, *73*
   architecture, 78–9, *78*
   art, 76, 78
   citizenship, 72
   city state, 61
   coin, *70*
   Darius' invasion, 69
   and Delian League, 70
   democracy, 69, 71–2, *72*
   economy, 72
   Empire, *70*
   military organization, 73, *73*
   navy, *70*
   painting, 77, 335A
   Parthenon, 78, *78–9*, 335A
   Pnyx, *72*
   slaves, 72, *72*
   theatre, 335L
   Tower of Winds, *83*
Aton, 48, *48*, 331R
Atreus, Treasury of, *41*, 331A
Attica, 68, 76
Attila the Hun, 137, *137*, 340P, *340*
Aubrey, John, 22
Aucassin et Nicolette, 207
Audran, Claude, 355A
Augsburg, 349S
   Peace of, 269, *269*
Augur, 93
Augustan emperors, 98, *99*
Augustine of Canterbury, St, *102*,
   139, 164
Augustine of Hippo, St, 113, 134,
   339L, 340R
Augustinus (Jansen), 353R
Augustulus, Emperor, 137
Augustus (Octavian), 98, *99, 106–7*,
   337P
"Augustus", Prima Porta, 337A
Auld Alliance, 276
Aulos, 75
Aurelian, 114, 339P
Aurignacian art, 25
Auroch (cave painting), *24*

357

Picture Credits

Every endeavour has been made to trace copyright holders of photographs appearing in *The Joy of Knowledge*. The publishers apologize to any photographers or agencies whose work has been used but has not been listed below.

Credits are listed in this manner: [1] page numbers appear first, in bold type; [2] illustration numbers appear next, in parentheses; [3] photographers' names appear next, followed where applicable by the names of the agencies representing them.

History and Culture
16 [1] Keystone Press Agency; [2] Novosti Press Agency. **17** Trustees of the National Gallery. **19** Bodleian Library, Oxford. **20–1** [Key] Michael Holford; [1] Professor Leslie Alcock; [2] Professor Leslie Alcock; [4] Aerofilms; [7] Aerofilms. **22–3** [Key] Geoff Goode/by permission of The Society of Antiquaries of London; [1] Geoff Goode; [2] Picturepoint; [3] Smithsonian Institution National Anthropological Archives; [4] Mansell Collection; [5] Radio Times Hulton Picture Library; [6] Dr M. H. Day/property of the Government of Tanzania; [7] Times Newspapers; [8] Marion Morrison. **24–5** [1] Colorphoto Hans Hinz, Basle; [2] Achille B. Weider, Zurich; [3] Photoresources; [4] Photoresources; [5] Picturepoint; [6] D. F. E. Russell/Robert Harding Associates; [7] published by permission of the Danish National Museum; [9] Federico Arborio Mella. **26–7** [3] Jon Gardey/Robert Harding Associates; [6] Picturepoint; [12] C. M. Dixon. **28–9** [Key] P. Hulin; [2] P. Hulin; [3] P. Hulin; [4] P. Hulin; [6] P. Hulin; [7] P. Hulin; [8] Staatliche Museen, Berlin. **30–1** [Key] Ronald Sheridan; [1] P. Hulin; [4] Ronald Sheridan; [5A] Michael Holford/British Museum; [5B] Michael Holford/British Museum; [6] Photoresources/British Museum; [8] Photoresources/Istanbul Archaeological Museum. **32–3** [Key] P. Hulin; [1B] P.

Hulin; [2] Source unknown; [3] Photoresources/British Museum; [4] P. Hulin; [6] Mansell Collection; [7] Giraudon/Louvre; [8] Lauros Giraudon/Louvre. **34–5** [Key] Michael Holford; [2] Werner Forman Archive; [3A] Museum of Fine Arts, Boston, Havard Boston Expedition Fund; [3B] Ronald Sheridan; [5] Ronald Sheridan; [6] Roger Wood; [7] Ronald Sheridan; [8] Ronald Sheridan. **36–7** [Key] All pictures Ann and Bury Peerless; [2] J. Allan Cash; [3] Ann and Bury Peerless; [6] J. Powell/Karachi Museum; [7] Source unknown; [8] Ann and Bury Peerless. **38–9** [Key] Ronald Sheridan; [5] Leonard Von Matt; [6] Ronald Sheridan; [7] Ronald Sheridan; [8] Ronald Sheridan; [9] Ronald Sheridan. **40–1** [Key] Ronald Sheridan; [3] Mrs Alan Wace; [4] Photoresources; [7] Photoresources; [8] Hirmer Fotoarchiv, Munich/National Museum, Athens. **42–3** [Key] Michael Holford; [3] Jarrolds; [4] Michael Holford/BM; [5] Michael Holford/University Museum of Archaeology & Ethnology, Cambridge; [6] Michael Holford/University Museum of Archaeology & Ethnology, Cambridge; [7] Michael Holford/University Museum of Archaeology & Ethnology, Cambridge; [8] Picturepoint. **44–5** [Key] William MacQuitty; [1] Paolo Koch; [2] William MacQuitty; [3] William MacQuitty; [4] William MacQuitty; [5] William MacQuitty; [7] William MacQuitty; [8] William MacQuitty/Shanghai Museum. **46–7** [Key] Instituto Nacional de Antropologia e Historia, Mexico/Norman Hammond; [1] Norman Hammond; [2] Norman Hammond; [3] Norman Hammond; [4] Tony Morrison; [5] G. Bushnell; [7] Tatiana Proskouriakoff/Peabody Museum; [8] Norman Hammond. **48–9** [Key] Barnaby's Picture Library; [1] Werner Forman Archive/Charles Edwin Wilbour Fund, Brooklyn Museum; [2A] Ronald Sheridan; [2B] Ronald Sheridan; [4] Ronald Sheridan; [5] Michael Holford; [6] Erich Lessing/Magnum; [7] Werner Forman Archive/Cairo Museum. **50–1** [Key A] Source unknown; [Key B] Source unknown; [2] Werner Forman Archive; [3] Werner Forman Archive; [4] Werner Forman Archive; [5] Roger Wood; [6] Alan Hutchinson; [7] Werner Forman Archive. **52–3** [Key] P.

Hulin; [1] Hirmer Fotoarchiv, Munich; [2] Richard Ashworth/Robert Harding Associates; [3A] Ronald Sheridan; [3B] Werner Forman Archive/Schimmel Collection, New York; [4] Ronald Sheridan; [5] Richard Ashworth/Robert Harding Associates; [6] P. Hulin; [7] Mansell Collection; [8] Photoresources. **54–5** [Key] Hirmer Fotoarchiv, Munich/Baghdad Museum; [1] Ronald Sheridan; [2] Trustees of the British Museum; [4] Ronald Sheridan; [5] Michael Holford/British Museum; [6] Photoresources; [7] Michael Holford; [8] Source unknown. **56–7** [Key] Staatliche Museen, Berlin; [2] Ronald Sheridan; [3] Michael Holford/British Museum; [4] Mansell Collection; [5] Ronald Sheridan; [6] P. Hulin; [7] P. Hulin; [8] Michael Holford/British Museum; [9] P. Hulin. **58–9** [Key] Ronald Sheridan; [2] Camera Press; [3] Ronald Sheridan; [4] C. M. Dixon; [6] Scala; [7] Photoresources/British Museum; [8] Source unknown. **60–1** [Key] William MacQuitty; [2] William MacQuitty; [3] William MacQuitty; [4] Ray Gardner/Trustees of the British Museum; [5] William MacQuitty; [8] William MacQuitty; [9] William MacQuitty. **62–3** [Key] India Office Library and Records/John F. Freeman; [2] Ann and Bury Peerless; [3A] Ann and Bury Peerless; [3B] Ann and Bury Peerless; [4] Ann and Bury Peerless; [5A] Trustees of the British Museum/Ray Gardner; [5B] Trustees of the British Museum/Ray Gardner; [7] Ann and Bury Peerless; [8] Ann and Bury Peerless. **64–5** [Key] Ann and Bury Peerless; [1] A. F. Kersting; [3] Ann and Bury Peerless; [4] Ann and Bury Peerless; [5] Source unknown; [6] Musée Louis Finot, Hanoi; [7] Bill and Claire Leimbach/Robert Harding Associates; [8] Museum of Fine Arts, Boston/Ross Collection. **66–7** [Key] Anne Ross; [2] Anne Ross; [3] Anne Ross; [4] Anne Ross; [5] National Museum of Ireland; [6] National Museum of Ireland; [7] Photoresources; [8] Anne Ross/Musée de Chatillon-sur-Seine; [9] Photoresources; [10] National Museum of Ireland; [11] Anne Ross. **68–9** [Key] Trustees of the British Museum; [3] Erich Lessing/Magnum; [5] Michael Holford; [6] Metropolitan Museum of Art, New York, Purchase 1947,

Joseph Pulitzer Bequest. **70–1** [Key] Metropolitan Museum of Art, New York/Rogers Fund 1906; [1] Photoresources/Acropolis Museum, Athens; [6] Photoresources/National Museum, Athens; [7] Ronald Sheridan; [8] Mauro Pucciarelli. **72–3** [Key] Ronald Sheridan; [1] Edwin Smith; [3] published by permission of the Danish National Museum; [4] Michael Holford; [5] Photoresources; [6] Wadsworth Atheneum, Hartford, Connecticut; [7] Michael Holford; [8] Metropolitan Museum of Art, New York, Rogers Fund 1914. **74–5** [Key] Trustees of the British Museum/Ray Gardner; [1] Michael Holford/British Museum; [2] Staatliche Museen, Berlin; [3] Dimitrios Harissiadis; [4] Dimitrios Harissiadis; [5] Hirmer Fotoarchiv, Munich; [7] Scala. **76–7** [Key] Ronald Sheridan; [2] Roger Wood; [3] Michael Holford; [4] Hirmer Fotoarchiv, Munich; [5] Trustees of the British Museum; [6] Photoresources; [7] Photoresources; [8] Photoresources/Delphi Museum. **78–9** [Key] Russell Ash; [2] Ronald Sheridan; [4] Joseph Ziolo/Olympia Museo; [5] Photoresources; [6] Photoresources/Vatican Museum; [7] Joseph Ziolo. **80–1** [Key] Trustees of the British Museum/Ray Gardner; [1] Michael Holford; [2] Photoresources/National Museum, Naples; [5] Photoresources/British Museum; [6] Bodleian Library, Oxford. **82–3** [Key] Ronan Picture Library/Royal Astronomical Society; [1] Ronan Picture Library; [3] Ronan Picture Library; [5] Michael Holford/Ann Mowlem; [7] Michael Holford/National Maritime Museum. **84–5** [Key] British Museum; [2] Aerofilms; [3] Eileen Tweedy/BM; [5] C. M. Dixon; [7] Aerofilms. **86–7** [Key] Anne Ross; [2] Anne Ross; [3] Anne Ross; [4A] C. M. Dixon; [4B] C. M. Dixon; [5] Anne Ross; [6A] Anne Ross; [6B] Anne Ross; [7] Anne Ross; [8] Anne Ross; [9] Anne Ross. **88–9** [Key] Photoresources; [2] Photoresources; [4] Michael Holford; [5] Michael Holford; [6] Leonard von Matt; [8] Leonard von Matt. **90–1** [Key] Photoresources; [2] Source unknown; [3] Vatican Library/Octopus Books; [4] Aquileia Museum Rome/Fototeca Unione; [5] K. E. Lowther; [6A] Trustees of the British Museum/Ray Gardner; [6B] Trustees of the British

Museum/Ray Gardner. **92–3** [Key] Mansell Collection; [2] K. E. Lowther; [4] Sonia Halliday/F. H. C. Birch; [5] Scala; [6] C. M. Dixon; [7] Mansell Collection. **94–5** [Key] Scala; [1] Scala; [2] Scala; [3] Scala; [4] Scala; [5] Scala/National Museum, Naples; [6] Scala; [7] C. M. Dixon; [8] Mansell Collection; [9] Photoresources. **96–7** [Key] Mansell Collection; [1] Trustees of the British Museum/Ray Gardner; [2] Scala; [3] Trustees of the British Museum/Ray Gardner; [4] Mansell Collection; [5] Scala; [8] Giraudon; [9] Roger-Viollet. **98–9** [Key] Photoresources; [2] Angelo Hornak; [3] Scala; [4] Angelo Hornak; [6] Photoresources; [7] Photoresources; [8] Michael Holford/Gerry Clyde; [9] Mansell Collection. **100–101** [Key] Woodmansterne/Colchester Museum; [1] "Heritage of Britain", Copyright 1975/Reader's Digest Association Ltd, London; [3] David Strickland; [4] C. M. Dixon; [5] Museum of London; [8] David Strickland. **102–3** [Key] Michael Holford/Corinium Museum [2] "Heritage of Britain" Copyright 1975, Reader's Digest Association Ltd, London; [4] "Heritage of Britain" Copyright 1975, Reader's Digest Association Ltd, London; [5] Aerofilms; [7] Trevor Wood/Robert Harding Associates; [8] British Museum; [9] Picturepoint. **104–5** [Key] Picturepoint; [2] Aerofilms; [3] Michael Holford; [4] "Treasures of Britain"; Copyright 1968, Drive Publications Ltd., London; [5a] Aerofilms; [5b] Picturepoint; [7] Museum of London; [8] Picturepoint. **106–7** [Key] Trustees of the British Museum; [1] Mansell Collection; [2] Scala/Vatican Museum; [3] Lauros Giraudon; [4] Russell Ash; [5] Scala; [7] Photoresources; [8] Photoresources. **108–9** [Key] C. M. Dixon; [2] Scala; [3] Picturepoint; [4] C. M. Dixon; [5] Picturepoint; [6] C.M. Dixon/British Museum; [7] Trustees of the British Museum/Ray Gardner; [8] Michael Holford. **110–11** [Key] Scala; [2] Ronald Sheridan; [3] C. M. Dixon; [4] Ronald Sheridan; [5] Scala; [6] Scala; [7] Scala. **112–13** [Key] Scala; [1] Bodleian Library, Oxford; [2] Photoresources; [3] Michael Holford/British Museum; [4] Sotheby's; [5] Snark International; [6] Scala/Bibliotheca Vaticana; [7] Photoresources; [8] Scala. **114–15** [Key] Scala; [1] Giraudon/Louvre; [2] Ronald Sheridan; [3A] Mansell Collection; [3B] C. M. Dixon/Louvre; [5] Trustees of the British Museum/Ray Gardner; [6] Photoresources; [7] Angelo Hornak. **116–17** [Key] Ronan Picture Library; [2] Mansell Collection; [3] Mansell Collection; [4] Ronan Picture Library; [7] Mansell Collection; [9] Ronan Picture Library. **118–19** [Key] Hamlyn Group Picture Library; [1] Ann and Bury Peerless; [2] Ann and Bury Peerless; [3] Ann and Bury Peerless; [4] Ann and Bury Peerless; [5] Trustees of the British Museum/Ray Gardner; [6] Peter Fraenkel; [8] Mansell Collection; [9] Oriental Art Archives, University of Michigan. **120–1** [Key] from *The Art of Indian Asia* by Heinrich Zimmer/Gunvor Moitessier; [1] From *The Art of Indian Asia* by Heinrich Zimmer; [3] Ann and Bury Peerless; [4] Trustees of the British Museum; [5] Michael Holford/Musée Guimet, Paris; [6] Victoria and Albert Museum/Carltograph; [7] Michael Holford/Victoria and Albert Museum; [8] Michael Holford/Victoria and Albert Museum; [9] Michael Holford/Victoria and Albert Museum; [10] Ann and Bury Peerless. **122–3** [Key] William MacQuitty; [1] William MacQuitty; [3] Mary Evans Picture Library; [4] William MacQuitty; [5] William MacQuitty; [6] William MacQuitty; [7] William MacQuitty; [8] Robert Harding Associates; [9] William MacQuitty; [10] William MacQuitty. **124–5** [Key] British Museum/Hamlyn Group Picture Library; [1] Royal Ontario Museum, Toronto/ Hamlyn Group Picture Library; [2] Howard Sochurek/T.L.P.A. © Time Inc 1976/Colorific; [3] Source unknown; [4] William MacQuitty; [5] Metropolitan Museum of Art, New York/Munsey Bequest, 1924; [6] Victoria and Albert Museum/Godfrey New Photographics; [7] Sally and Richard Greenhill; [8] Bildarchiv Foto Marburg. **126–7** [1] Trustees of the British Museum; [2] Victoria and Albert Museum; [3] Robert Harding Associates/Witty, Times Newspapers Ltd; [4] William MacQuitty; [6] William MacQuitty; [8] Source unknown; [9] William MacQuitty; [10] Mansell Collection. **128–9** [Key] National Palace Museum, Taiwan; [1] Trustees of the British Museum; [2] Trustees of the British Museum; [3] After "Ma-Wang-Tui I Hao Han Mu"; [4] Trustees of the British Museum; [5] Trustees of the British Museum; [6] National Palace Museum, Taiwan; [7] reproduced by permission of the Syndics of the Fitzwilliam Museum, Cambridge; [8] National Palace Museum, Taiwan; [9] Werner Forman Archive, Palace Museum, Peking. **130–1** [Key] Bradley Smith; [2] International Society for Educational Information; [3] Ministry of Foreign Affairs, Japan; [4] International Society for Educational Information; [5] Ministry of Foreign Affairs, Japan; [6] International Society of Educational Information; [7] Ministry of Foreign Affairs, Japan; [8] Bradley Smith. **132–3** [Key] Sakamoto/Joseph P. Ziolo; [1] K. Ogawa/Joseph P. Ziolo; [2] K. Ogawa/Joseph P. Ziolo; [3] Sakamoto/Joseph P. Ziolo; [4] Sakamoto/Joseph P. Ziolo; [5] Sakamoto/Joseph P. Ziolo; [6] Sakamoto/Joseph P. Ziolo; [7] Sakamoto/Joseph P. Ziolo; [8A] Zauhopress/Joseph P. Ziolo/Musée D'Atami; [8B] Zauhopress/Joseph P. Ziolo/Musée D'Atami; [9] Victoria and Albert Museum; [10] Michael Holford. **134–5** [Key] Bodleian Library, Oxford; [2] Kunsthistorisches Museum Vienna; [3] Michael Holford; [5] Bavaria Verlag; [6] Joseph P. Ziolo; [7] National Museum of Ireland; [8] Scala; [10] Burgerbibliothek, Bern. **136–7** [4A] Lauros Giraudon; [4B] Photoresources; [4C] Ashmolean Museum, Oxford; [5] C. M. Dixon; [6] Snark International; [7] Scala; [8] C. M. Dixon. **138–9** [Key] British Library; [1] Bodleian Library, Oxford; [4] Courtesy Miss Mercie K. Lack, A.R.P.S./Aldus Books Ltd., [6] Michael Holford; [7] The Antiquaries Journal; [8] Bodleian Library, Oxford. **140–1** [Key] Anne Ross; [2] Trinity College, Dublin/The Green Studio; [3] Trinity College Dublin/The Green Studio; [4A] National Museum of Ireland; [4B] National Museum of Ireland; [5A] National Museum of Ireland; [5B] National Museum of Ireland; [6] C. M. Dixon; [7] C. M. Dixon. **142–3** [5A] Trustees of the British Museum/Ray Gardner; [5B] Trustees of the British Museum/Ray Gardner; [6] C. M. Dixon; [10] with courtesy of the Trustees, National Gallery, London. **144–5** [Key] Ronald Sheridan; [1] Spencer Collection, The New York Public Library, Astor Lenox and Tilden Foundations/Hamlyn Group Picture Library; [2] Edinburgh University Library; [3] Ronald Sheridan; [4] Camera Press; [5] Ronald Sheridan; [6] Gerry Clyde/Michael Holford; [7] Photoresources. **146–7** [Key] Osvaldo Bohm/Biblioteca Marciana; [1] E. J. W. Hawkins; [2] C. M. Dixon; [3] Hirmer Fotoarchiv, Munich/Bibliothèque Nationale; [4] Sonia Halliday; [5] Mansell Collection; [6A] Scala; [6B] Scala; [7] Sonia Halliday; [8] Sonia Halliday; [9] Hirmer Fotoarchiv, Munich; [10] Hirmer Fotoarchiv, Munich/Victoria and Albert Museum. **148–9** [Key] Snark International; [1] Edinburgh University Library; [3] Dr Georg Gerster/John Hillelson Agency; [5] Madame Solange Ory; [6] C. M. Dixon; [7] Radio Times Hulton Picture Library; [8] Tunisian National Tourist office. **150–1** [Key] Roloff Beny, Rome; [2] Victoria and Albert Museum; [3] The Pierpont Morgan Library; [4] Victoria and Albert Museum; [5] Peter Fraenkel; [6] Courtesy of the Smithsonian Institution, Freer Gallery of Art, Washington, DC; [7] Ronald Sheridan; [8] Werner Forman Archive; [9] The Metropolitan Museum of Art, New York. **152–3** [Key] Mansell Collection; [1] Mansell Collection; [2] Giraudon/Louvre; [4] Snark International; [5] Trustees of the British Museum; [6] Snark International; [7] Snark International. **154–5** [Key] Bodleian Library, Oxford; [1] British Library; [2] "Heritage of Britain" Copyright 1975, Reader's Digest Association Ltd., London; [3] Picturepoint; [6] Weidenfeld & Nicolson Archives/D.O.E.; [8] Weidenfeld & Nicolson Archives/BM; [9] Studio 28, Reykjavik; [10] Bodleian Library, Oxford; [11] British Library. **156–7** [Key] Photoresources; [2A] Crown Copyright, reproduced with permission of the Controller, Her Majesty's Stationery Office; [4] National Museum of Iceland/Leifur Porsteinson; [7] Forhistorisk Museum, Denmark; [8] Spectrum Colour Library; [9] Werner Forman Archive. **158–9** [Key] Bodleian Library, Oxford; [2A] Trustees of the British Museum/Ray Gardner; [2B] Trustees of the British Museum/Ray Gardner; [2c] Trustees of the British Museum/Ray Gardner; [2D] Trustees of the British Museum/Ray Gardner; [3] Scala; [4] Ronald Sheridan. **160–1** [Key] Roger Viollet; [3] Photo Meyer, Vienna; [4] Magnum/Erich Lessing; [5] Hatle Werbung; [6] Vatican Library; [8] Corpus Christi College, Cambridge; [9] Burgerbibliothek Bern/Gehard Howald; [10] Bavaria Verlag. **162–3** [1] Bildarchiv der Österreichischen Nationalbibliothek; [2] Source unknown; [3] Bavaria Verlag; [4] Bildarchiv Foto Marburg; [5] Bavaria Verlag. **164–5** [Key] Ronald Sheridan; [1] Eileen Tweedy; [2] Fotomas Index; [3] Colour Centre Slides Ltd/British Museum; [4] David Strickland; [5] Bodleian Library, Oxford; [6] Bodleian Library, Oxford; [7] Weidenfeld & Nicolson Archives/National Monuments Record; [8] Weidenfeld & Nicolson Archives/BM; [9] Michael Holford. **166–7** [Key] Phaidon Press Ltd.; [1] Bodleian Library, Oxford; [3] Picturepoint; [4] British Museum; [5] Michael Holford; [7] Michael Holford; [8] Mansell Collection; [9] Weidenfeld & Nicolson Archives/Courtauld Institute of Art. **168–9** [Key] British Tourist Authority; [1] Michael Holford; [2] Public Record Office; [6] Aerofilms; [7] Trustees of the British Museum; [8] Perfecta Publications Ltd. **170–1** [Key] Michael Holford; [1] Ampliaciones Y Reproducciones Mas; [2] Ampliaciones Y Reproducciones Mas; [3] Ampliaciones Y Reproducciones Mas; [4] Ampliaciones Y Reproducciones Mas/Barcelona University; [5] Ampliaciones Y Reproducciones Mas; [7] Ampliaciones Y Reproducciones Mas; [8] Camera Press; [9] Ampliaciones Y Reproducciones Mas. **172–3** [1] Bildarchiv Foto Marburg; [2] Edistudio; [4] Roger Viollet; [5] Mansell Collection; [6A] Photo Zodiaque; [6B] Photo Zodiaque; [7] Giraudon; [8] Michael Holford. **174–5** [Key] Photo Zodiaque; [1] A. F. Kersting; [4] Trustees of the British Museum; [5] Corpus Christi College, Cambridge; [6] Photo Zodiaque; [7] Architectural Association; [8] Photo Zodiaque; [9] French Tourist Office. **176–7** [Key] Aerofilms; [1] Snark International; [7] A. F. Kersting; [8] Michael Holford; [9] Corpus Christi College, Cambridge; [10] Michael Holford. **178–9** [Key] Bodleian Library, Oxford; [1] Bodleian Library, Oxford; [2] Weidenfeld & Nicolson Archives/BM; [3] Picturepoint; [4] Picturepoint; [5] Weidenfeld & Nicolson Archives/BM; [7] Weidenfeld & Nicolson Archives/Mansell Collection; [8] British Museum; [9] Mary Evans Picture Library; [10] Picturepoint. **180–1** [Key] Irish Tourist Board; [1] National Museum of Ireland, Dublin; [3] Board of Trinity College, Dublin; [4] Dr. Peter Harbison; [5] Commissioners of Public Works in Ireland; [7] British Library; [8] Angelo Hornak. **182–3** [Key] Permission of the Duke of Roxburghe/George Rainbird Ltd./Robert Harding Associates; [1] Picturepoint; [2] Ian Yeomans/Susan Griggs Picture Agency; [3] British Tourist Authority; [4] National Museum of Antiquities, Scotland; [6] Public Records Office, London; [7] Popperfoto; [8] Master & Fellows of Corpus Christi College, Cambridge. **184–5** [1] Picturepoint; [2] National Museum of Wales, Cardiff; [3] British Museum; [4] Picturepoint; [5] Radio Times Hulton Picture Library; [7] Picturepoint; [8] Cardiff City Corporation; [9] National Portrait Gallery. **186–7** [2] Bodleian Library, Oxford; [3] Bodleian Library, Oxford; [5] Mansell Collection; [6] Giraudon; [8] Bildarchiv Foto Marburg; [9] Scala; [10] Ronald Sheridan. **188–9** [Key] Mansell Collection; [1] Scala; [2] Trustees of the British Museum; [3] Scala/Vatican; [4] Bodleian Library, Oxford; [5] Scala; [6] Scala; [7] Bodleian Library, Oxford; [8] Bodleian Library, Oxford; [9] Ronan Picture Library. **190–1** [2] Roger Viollet; [3] Mauro Pucciarelli, Rome; [5] Scala. **192–3** [Key] Photographie Giraudon; [1] British Library; [2] Weidenfeld & Nicolson Archives/Courtesy Corporation of London; [3] Bodleian Library, Oxford; [6] Mansell Collection; [7] A. F. Kersting; [9] A. F. Kersting. **194–5** [Key] Weidenfeld & Nicolson Archives/British Library; [1] Picturepoint; [2] By permission of the Syndics of the Fitzwilliam Museum, Cambridge; [3] Aerofilms; [4] David Strickland; [5] Bodleian Library, Oxford; [6] Bodleian Library, Oxford; [7] Edwin Smith. **196–7** [Key] Weidenfeld & Nicolson Archives/BM; [1] Bodleian Library, Oxford; [3] Weidenfeld & Nicolson Archives/British Library; [4] Weidenfeld & Nicolson Archives/British Library; [6] Bodleian Library, Oxford; [7] Aerofilms; [8] Picturepoint; [10] British Tourist Authority. **198–9** [1] C. Wilson; [2] C. Wilson; [3] C. Wilson; [5] C. Wilson; [6] C. Wilson; [8] C. Wilson; [9] C. Wilson; [10]

Artwork Credits